CAMBRIDGE STUDIES IN
ADVANCED MATHEMATICS 12

THE LOGARITHMIC INTEGRAL I

Already published

$$\int_{-\infty}^{\infty} \frac{\log M(t)}{1+t^2} \mathrm{d}t$$

THE LOGARITHMIC INTEGRAL I

PAUL KOOSIS
McGill University in Montreal
formerly at the University of California, Los Angeles

PUBLISHED BY THE PRESS SYNDICATE OF THE UNIVERSITY OF CAMBRIDGE
The Pitt Building, Trumpington Street, Cambridge CB2 1RP, United Kingdom

CAMBRIDGE UNIVERSITY PRESS
The Edinburgh Building, Cambridge CB2 2RU, UK
40 West 20th Street, New York, NY 10011–4211, USA
10 Stamford Road, Oakleigh, Melbourne 3166, Australia

www.cambridge.org
Information on this title: www.cambridge.org/9780521309066

First published 1988
First paperback edition (with corrections) 1998

A catalogue record for this book is available from the British Library

Library of Congress Cataloguing in Publication data

Koosis, Paul.
The logarithmic integral.
(Cambridge studies in advanced mathematics; 12)
1. Analytic functions. 2 Harmonic analysis.
3. Integrals, logarithmic. I. Title. II. Series,
QA331.K7393 1988 515.4 85-28018

ISBN-13 978-0-521-30906-6 hardback
ISBN-10 0-521-30906-9 hardback

ISBN-13 978-0-521-59672-5 paperback
ISBN-10 0-521-59672-6 paperback

Transferred to digital printing 2005

Pour le Canada

Notice

In this paperback edition of volume I a number of small errors – and some actual mathematical mistakes – present in the original hard-cover version have been corrected. Many were pointed out to me by Henrik Pedersen, my former student; it was he who observed in particular that the hint given for Problem 28 (b) was ineffective. I wish to express here my gratitude for the considerable service he has thus rendered.

Let me also call the reader's attention to two annoying oversights in volume II. In the statement of the important theorem on p. 65, the condition that the quantities a_k all be > 0 was inadvertently omitted. On p. 406 it would be better, in the last displayed formula, to replace the difference quotient now standing on the right by $\dfrac{\mu(x+\Delta x)-\mu(x-\Delta x)}{2\Delta x}$.

March 22, 1997
Outremont, Québec

Contents

Preface

The two volumes that follow make up what is meant primarily as a book for reading. One reason for writing them was to give a connected account of some of the ideas that have dominated my mathematical activity for many years. Another, which was to try to help beginning mathematicians interested in analysis learn how to work by showing how I work, seems now less important because my way is far from being the only one. I do hope, at any rate, to encourage younger analysts by the present book in their efforts to become and remain active.

I have loved $\int_{-\infty}^{\infty}(\log M(t)/(1+t^2))\,dt$ – the logarithmic integral—ever since I first read Szegő's discussion about the geometric mean of a function and the theorem named after him in his book on orthogonal polynomials, over 30 years ago. Far from being an isolated artifact, this object plays an important role in many diverse and seemingly unrelated investigations about functions of one real or complex variable, and a serious account of its appearances would involve a good deal of the analysis done since 1900. That will be plain to the reader of this book, where some of that subject's developments in which the integral figures are taken up.

No attempt is made here to treat anything like the full range of topics to which the logarithmic integral is relevant. The most serious omission is that of parts of probability theory, especially of what is called prediction theory. For these, an additional volume would have been needed, and we already have the book of Dym and McKean. Considerations involving H_p spaces have also been avoided as much as possible, and the related material from operator theory left untouched. Quite a few books about those matters are now in circulation.

Of this book, begun in 1983, all but Chapter X and part of Chapter IX was written while I was at McGill University; the remainder was done at UCLA. The first 6 chapters are based on a course (and seminar) given at McGill during the academic year 1982–83, and I am grateful to the mathematics department there for the support provided to me since then out of its rather modest resources. Chapters I–VI and most of the seventh were typed at that department's office.

Chapter VII and parts of Chapter VIII are developed from lectures I gave at the Mittag–Leffler Institute (Sweden) during part of the spring semesters of 1977 and 1983. I am fortunate in having been able to spend almost two years all told working there.

Partial support from the U.S. National Science Foundation was also given me during the first year or two of writing.

I thank first of all John Garnett for having over a long period of time encouraged me to write this book. Lennart Carleson encouraged and helped me with research that led eventually to some of the expositions set out below. I thank him for that and also for my two invitations to the Mittag–Leffler Institute. For the second of those I must also thank Peter Jones who, besides, helped me with at least one item in Chapter VII. The book's very title is from a letter to me by V.P. Havin, and I hope he does not mind my using it. I was unable to think of anything except the mathematical expression it represents!

It was mainly John Taylor who arranged for me to come to McGill in the fall of 1982 and give the course mentioned above. Since then, a good part of my salary at McGill has been paid out of research grants held by him, Jal Choksi, Sam Drury, or Carl Herz. Taylor also came to some of the lectures of my course as did Georg Schmidt. Robert Vermes attended all of them and frequently talked about their material with me. Dr Raymond Couture came part of the time. The students were Janet Henderson, Christian Houdré and Tuan Vu. These people all contributed to the course and helped me to feel that I was doing something of value by giving it. Vermes' constant presence and evident interest in the subject were especially heartening.

Most of the typing for volume I was done by Patricia Ferguson who typed Chapters I through VI and the major part of Chapter VII, and by Babette Dalton who did a very fine job with Chapter VIII. I am beholden to S. Gardiner and P. Jackson of the Press' staff and finally to Dr Tranah, the mathematics editor, for their patience and attention to my desires regarding graphic presentation. The beautiful typesetting was done in India.

August 13, 1987
Laurel, Comté Argenteuil, Québec.

Introduction

The present book has been written so as to necessitate as little consultation by the reader as reasonably possible of other published material. I have hoped to thereby make it accessible to people far from large research centres or any 'good library', and to those who have only their summer vacations to work on mathematics. It is for the same reason that references, where unavoidable, have been made to books rather than periodicals whenever that could be done.

In general, I consider the developments leading up to the various results in the book to be more important than the latter taken by themselves; that is why those developments are set out in more detail than is now customary. My aim has been to enable one to follow them by mostly just reading the text, without having to work on the side to fill in gaps. The reader's active participation is nevertheless solicited, and problems have been given. These are usually accompanied by hints (sometimes copious), so that one may be encouraged to work them out fully rather than feeling stymied by them. It is assumed that the reader's background includes, beyond ordinary undergraduate mathematics, the material which, in North America, is called graduate real and complex variable theory (with a bit of functional analysis). Practically everything needed of this is contained in Rudin's well-known manual. My own preference runs towards a more leisurely approach based on Titchmarsh's *Theory of Functions* and the beautiful *Leçons d'analyse fonctionnelle* of Riesz and Nagy (now available in English). Alongside these books, the use of some supplementary descriptive material on conformal mapping (from Nehari, for instance) is advisable, as is indeed the case with Rudin as well. The Krein–Milman theorem referred to in Chapters VI and X is now included in many books; in Naimark's, for example (on normed algebras or rings), and in Yosida's. In the very few places where more specialized material is called for, additional references will be given. (Exact descriptions of the works just mentioned together with those cited later on can be found in the bibliographies placed at the end of each volume.)

Although the different parts of this book are closely interrelated, they may to a large extent be read independently. Material from Chapter III is, however, called for repeatedly in the succeeding chapters. For finding one's way, the descriptions in the table of contents and the page headings should be helpful; indices to each volume are also provided. Throughout volume I, various arguments commonly looked on as elementary or well-known, but which I nonetheless thought it better to include, have been set in smaller type, and certain readers will miss nothing by passing over them.

The book's units of subdivision are, successively, the chapter, the § (plural §§) and the article. These are indicated respectively by roman numerals, capital letters and arabic numerals. A typical reference would be to '§B.2 of Chapter VI', or to 'Chapter VI, §B.2'. When referring to another article within the same §, that article's number alone is given (e.g., 'see article 3'), and, when it's to another § in the same chapter, just that §'s designation (e.g., 'the discussion in § B') or again, if a particular article in that § is meant, an indication like '§ B.2'. Theorems, definitions and so forth are not numbered, nor are formulas. But certain displayed formulas in a connected development may be labeled by signs like (∗), (†), &c, which are then used to refer to them within that development. The same signs are used over again in different arguments (to designate different formulas), and their order is not fixed. A pause in a discussion is signified by a horizontal space in the text.

About mistakes. There must inevitably be some, although I have tried as hard as I could to eliminate errors in the mathematics as well as misprints. Certain symbols (bars over letters, especially) have an unpleasant tendency to fall off between the typesetters' shop and the camera. I think (and hope) that all the mathematical arguments are clear and correct, at least in their grand lines, and have done my best to make sure of that by rereading everything several times. The reader who, in following a given development, should come upon a misprint or incorrect relation, will thus probably see what should stand in its place and be able to continue unhindered. If something really seems peculiar or devoid of sense, one should try suspending judgement and read ahead for a page or so – what at first appears bizarre may in fact be quite sound and become clear in a moment. Unexpected turnings are encountered as one becomes acquainted with this book's material.

It is beautiful material. May the reader learn to love it as I do.

I

Jensen's formula

On making the substitution $t = \tan(\vartheta/2)$ and then putting $M(t) = P(\vartheta)$, the expression

$$\frac{1}{\pi}\int_{-\infty}^{\infty} \frac{\log M(t)}{1+t^2}\,\mathrm{d}t$$

goes over into

$$\frac{1}{2\pi}\int_{-\pi}^{\pi} \log P(\vartheta)\,\mathrm{d}\vartheta.$$

We begin this book with a discussion of the second integral.

Suppose that $R > 1$ and we are given a function $F(z)$, analytic in $\{|z| < R\}$. If $F(z)$ *has no zeros* for $|z| \leqslant 1$ we can define a *single valued* function $\log F(z)$, analytic for $|z| \leqslant R'$, say, where $1 < R' < R$. By Cauchy's formula we will then have

$$\log F(0) = \frac{1}{2\pi}\int_{0}^{2\pi} \log F(\mathrm{e}^{\mathrm{i}\vartheta})\,\mathrm{d}\vartheta,$$

so, taking the real parts of both sides, we get

$$\log|F(0)| = \frac{1}{2\pi}\int_{-\pi}^{\pi} \log|F(\mathrm{e}^{\mathrm{i}\vartheta})|\,\mathrm{d}\vartheta.$$

What if $F(z)$ *has* zeros in $|z| \leqslant 1$? Assume to begin with that there are none *on* $|z| = 1$, and denote those that $F(z)$ does have inside the unit disk by $a_1, a_2, \ldots, a_n$. According to custom, a zero is *repeated according to its multiplicity* in such an enumeration. Put

$$\Phi(z) = \frac{F(z)}{(z-a_1)(z-a_2)\ldots(z-a_n)}.$$

Then $\Phi(z)$ has no zeros in $\{|z| \leqslant 1\}$, so, by the special case already treated,

$$\log|\Phi(0)| = \frac{1}{2\pi}\int_{-\pi}^{\pi} \log|\Phi(e^{i\vartheta})|\,d\vartheta$$

$$= \frac{1}{2\pi}\int_{-\pi}^{\pi} \log|F(e^{i\vartheta})|\,d\vartheta - \sum_{k=1}^{n} \frac{1}{2\pi}\int_{-\pi}^{\pi} \log|e^{i\vartheta} - a_k|\,d\vartheta.$$

Here we make a side calculation. For $|a_k| < 1$ we have

$$\frac{1}{2\pi}\int_{-\pi}^{\pi} \log|e^{i\vartheta} - a_k|\,d\vartheta = \frac{1}{2\pi}\int_{-\pi}^{\pi} \log|1 - \bar{a}_k e^{i\vartheta}|\,d\vartheta,$$

and this $= \log 1 = 0$ by the case already discussed ($F(z)$ without zeros in $|z| \leqslant 1$)! Combined with the previous relation this yields

$$\log|\Phi(0)| = \frac{1}{2\pi}\int_{-\pi}^{\pi} \log|F(e^{i\vartheta})|\,d\vartheta.$$

Especially, if $F(0) \neq 0$,

$$\log|F(0)| - \sum_{k=1}^{n} \log|a_k| = \frac{1}{2\pi}\int_{-\pi}^{\pi} \log|F(e^{i\vartheta})|\,d\vartheta.$$

The sum on the left can be written differently. *Call $n(r)$ the number of zeros of $F(z)$ in $|z| \leqslant r$ (counting multiplicities).* Then, if $F(0) \neq 0$,

$$-\sum_{k=1}^{n} \log|a_k| = \int_0^1 \frac{n(r)}{r}\,dr.$$

Indeed, since $n(r) = 0$ for $r > 0$ *close* to 0,

$$\int_0^1 \frac{n(r)}{r}\,dr = n(1)\log 1 - \int_0^1 \log r\,dn(r) = -\sum_{k=1}^{n} \log|a_k|.$$

We therefore have

$$\log|F(0)| + \int_0^1 \frac{n(r)}{r}\,dr = \frac{1}{2\pi}\int_{-\pi}^{\pi} \log|F(e^{i\vartheta})|\,d\vartheta.$$

In case $F(z)$ is regular in a disk including $\{|z| \leqslant R\}$ in its interior and $F(0) \neq 0$ we can (provided that $F(z) \neq 0$ for $|z| = R$) make a change of variable in the preceding relation and get

$$\log|F(0)| + \int_0^R \frac{n(r)}{r}\,dr = \frac{1}{2\pi}\int_{-\pi}^{\pi} \log|F(Re^{i\vartheta})|\,d\vartheta.$$

This is Jensen's formula.

The validity of Jensen's formula *subsists* even when $F(z)$ *has* zeros on the circle $|z| = R$. To see this, observe that then $F(z)$ *will not* have any zeros on the circles $|z| = R'$ with $R' < R$ and sufficiently close to R, for $F(z)$ is analytic in a disk $\{|z| < R + \eta\}$, $\eta > 0$, and not identically zero ($F(0) \neq 0$). So, for such R',

$$\log|F(0)| + \int_0^{R'} \frac{n(r)}{r} \mathrm{d}r = \frac{1}{2\pi} \int_{-\pi}^{\pi} \log|F(R'\mathrm{e}^{\mathrm{i}\vartheta})| \mathrm{d}\vartheta.$$

As $R' \to R$, the left side clearly tends to $\log|F(0)| + \int_0^R (n(r)/r) \mathrm{d}r$ – the integral on the left is a *continuous* function of its upper limit because $n(r)$ is *bounded*. We need therefore merely verify that

$$\int_{-\pi}^{\pi} \log|F(R'\mathrm{e}^{\mathrm{i}\vartheta})| \mathrm{d}\vartheta \longrightarrow \int_{-\pi}^{\pi} \log|F(Re^{\mathrm{i}\vartheta})| \mathrm{d}\vartheta$$

as $R' \to R$. The idea here is the same whether $F(z)$ has *several zeros* on $|z| = R$ or *only one*, and in order to simplify the writing we just treat the latter case. Suppose then that $F(\alpha) = 0$ where $|\alpha| = R$, and there are *no other zeros* in a ring of the form $\{R - \eta \leqslant |z| \leqslant R + \eta\}$, $\eta > 0$. On this ring we then have $|F(z)| \geqslant \text{const.}|z - \alpha|^m$, if m is the multiplicity of the zero at α, so, since $|F(z)|$ is also bounded above there,

$$|\log|F(R'\mathrm{e}^{\mathrm{i}\vartheta})|| \leqslant \text{const.} + m \log^+ \frac{1}{|R'\mathrm{e}^{\mathrm{i}\vartheta} - \alpha|}$$

for $R - \eta \leqslant R' \leqslant R$. (*Here, for* $p > 0$, $\log^+ p$ *denotes* $\log p$ *if* $p \geqslant 1$ *and* 0 *if* $p < 1$.) The expression on the right is, however, $\leqslant \text{const.} + m \log^+(1/|R\mathrm{e}^{\mathrm{i}\vartheta} - \alpha|)$, independently of R', when the latter quantity is close to R:

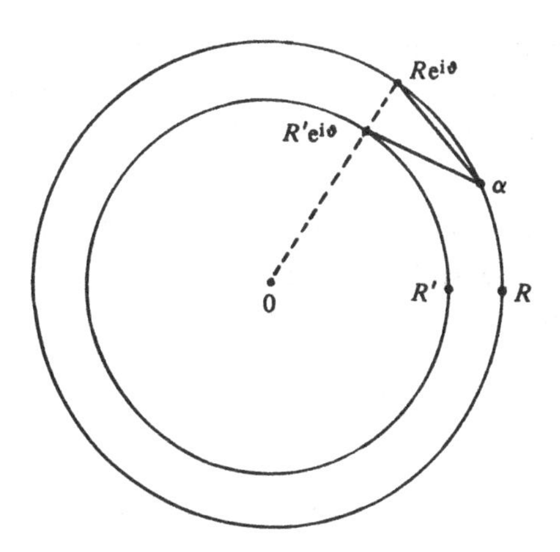

Figure 1

(The constants of course will be *different*; the relation between them need not concern us here.) In other words, for $R' \to R$ the expressions $|\log|F(R'e^{i\vartheta})||$ are bounded above by the *fixed* function const. + $m\log^+(1/|Re^{i\vartheta}-\alpha|)$ of ϑ, *which however, has a finite integral over* $[-\pi, \pi]$, *as we easily check directly.* Since also $\log|F(R'e^{i\vartheta})| \to \log|F(Re^{i\vartheta})|$ pointwise as $R' \to R$, we have

$$\int_{-\pi}^{\pi} \log|F(R'e^{i\vartheta})|\,d\vartheta \longrightarrow \int_{-\pi}^{\pi} \log|F(Re^{i\vartheta})|\,d\vartheta$$

by *Lebesgue's dominated convergence theorem.* This is what we needed to complete our derivation of Jensen's formula. (We see that the *same computation* which shows that

$$\int_{-\pi}^{\pi} \log|F(Re^{i\vartheta})|\,d\vartheta \;>\; -\infty$$

also establishes the *convergence* of $\int_{-\pi}^{\pi} \log|F(R'e^{i\vartheta})|\,d\vartheta$ to that quantity as $R' \to R$!)

Here is a first application of Jensen's formula.

Theorem. *Suppose that* $F(z)$ *is analytic and* $\not\equiv 0$ *for* $|z| < 1$, *and that the integrals*

$$\int_{-\pi}^{\pi} \log^+|F(re^{i\vartheta})|\,d\vartheta$$

are bounded for $0 \leqslant r < 1$. *Then for any* r_0, $0 < r_0 < 1$, *the integrals*

$$\int_{-\pi}^{\pi} \log^-|F(re^{i\vartheta})|\,d\vartheta$$

are bounded for $r_0 < r < 1$.

Notation. For $p \geqslant 0$, we write (as remarked above) $\log^+ p = \max(\log p, 0)$. *We also take* $\log^- p = -\min(\log p, 0)$, *so that* $\log^- p \geqslant 0$ *and* $\log p = \log^+ p - \log^- p$. (Everybody means the same thing by $\log^+ p$, but, regarding $\log^- p$, usage is not uniform.)

Proof of theorem. Without loss of generality (*henceforth abbreviated* 'wlog'), let $F(0) \neq 0$. (Otherwise work with $F(z)/z^k$ for a suitable k instead of $F(z)$.) By Jensen's formula,

$$-\infty < \log|F(0)| \leqslant \frac{1}{2\pi}\int_{-\pi}^{\pi} \log^+|F(re^{i\vartheta})|\,d\vartheta$$

$$-\frac{1}{2\pi}\int_{-\pi}^{\pi} \log^-|F(re^{i\vartheta})|\,d\vartheta, \quad 0 < r < 1.$$

By hypothesis, the right-hand side is

$$\leqslant \text{const.} - \frac{1}{2\pi}\int_{-\pi}^{\pi} \log^{-}|F(re^{i\vartheta})|\,d\vartheta.$$

The desired result follows by transposition.

Corollary. *Under the hypothesis of the theorem, suppose that*

$$F(e^{i\vartheta}) = \lim_{r\to 1} F(re^{i\vartheta})$$

exists a.e. Then

$$\int_{-\pi}^{\pi} \log^{-}|F(e^{i\vartheta})|\,d\vartheta \;<\; \infty.$$

Proof. Fatou's lemma.

Remark 1. Actually, the hypothesis of the theorem *forces* a.e. existence of

$$\lim_{r\to 1} F(re^{i\vartheta}).$$

This is a fairly deep result, and depends on Lebesgue's theorem on a.e. existence of derivatives of functions of bounded variation. In the situations we will mostly consider, the existence of this limit can be directly verified ('by inspection'), so the deeper result will not be needed. Therefore we do not prove it now. The interested reader can work up a proof by using the subharmonicity of $\log^{+}|F(z)|$ together with an argument from Chapter III, §F.1, so as to produce a positive measure ν on $[-\pi,\pi]$ for which

$$|F(z)| \;\leqslant\; \left|\exp\left\{\frac{1}{2\pi}\int_{-\pi}^{\pi}\frac{e^{i\vartheta}+z}{e^{i\vartheta}-z}\,d\nu(\vartheta)\right\}\right|, \quad |z|<1.$$

After this, one applies results from §F.2 of Chapter III to the analytic function $\Phi(z)$ within $|\quad|$ on the right, and then to the ratio $F(z)/\Phi(z)$.

Remark 2. The idea of the corollary is that if $|F(z)|$ is *not too big* in $\{|z|<1\}$ (especially if $|F(z)|$ is *bounded* there), then the boundary values $|F(e^{i\vartheta})|$ *cannot be too small* unless $F \equiv 0$.

Problem 1

(a) Let $F(z)$ be entire, $F(0) = 1$, and $|F(z)| \leqslant Ke^{A|z|}$ for all z, where A and K are constants. If $n(R)$ denotes the number of zeros of F having modulus $\leqslant R$, show that, for all R, $n(R) \leqslant \mathrm{e}AR + \text{const.}$ (Here, the constant depends on K.)

*(b) Show that in the relation established in (a) the coefficient $\mathrm{e}A$ of R cannot in general be diminished. (Hint. Fix $R = m/\mathrm{e}$ with m a large integer. Compute the maximum value of $(x/R)^{\mathrm{e}R}\mathrm{e}^{-x}$ for $x \geqslant 0$. Then look at a function which has m equally spaced zeros on the circle $|z| = R$ and no others.)

II

Szegő's theorem

A. The theorem

Szegő's theorem is a beautiful result in approximation theory, obtained with the help of Jensen's formula. Its proof also uses a limit property of integrals involving the Poisson kernel (for the unit disk) which is now taught in many courses on real variable theory. The reader who does not remember that result will find it in §B, together with its proof.

Theorem (Szegő). *Let* $w(\vartheta) \geqslant 0$ *belong to* $L_1(-\pi, \pi)$. *Then the infimum of*

$$\frac{1}{2\pi}\int_{-\pi}^{\pi}\left|1 - \sum_{n>0} a_n e^{in\vartheta}\right| w(\vartheta)\,d\vartheta,$$

taken with respect to all possible finite sums $\sum_{n>0} a_n e^{in\vartheta}$, *is equal to*

$$\exp\left(\frac{1}{2\pi}\int_{-\pi}^{\pi}\log w(\vartheta)\,d\vartheta\right).$$

Note: $\int_{-\pi}^{\pi}\log^+ w(\vartheta)\,d\vartheta$ *is finite* if $w \in L_1(-\pi, \pi)$. So $\int_{-\pi}^{\pi}\log w(\vartheta)\,d\vartheta$ either *converges*, or else *diverges* to $-\infty$.

Proof of theorem. By the inequality between arithmetic and geometric means,

$$\frac{1}{2\pi}\int_{-\pi}^{\pi}\left|1 - \sum_{n>0} a_n e^{in\vartheta}\right| w(\vartheta)\,d\vartheta$$

$$\geqslant \exp\left\{\frac{1}{2\pi}\int_{-\pi}^{\pi}\left(\log\left|1 - \sum_{n>0} a_n e^{in\vartheta}\right| + \log w(\vartheta)\right)d\vartheta\right\}.$$

Jensen's formula applied to $F(z) = 1 - \sum_{n>0} a_n z^n$ shows that this last expression is always

$$\geqslant \exp\left(\frac{1}{2\pi}\int_{-\pi}^{\pi}\log w(\vartheta)\,d\vartheta\right);$$

the desired infimum is thus $\geqslant$ the latter quantity. We must establish the reverse inequality.

Write $w_N(\vartheta) = \max(w(\vartheta), e^{-N})$. By Lebesgue's monotone convergence theorem and the finiteness of $\int_{-\pi}^{\pi} \log^+ w(\vartheta)\,d\vartheta$ we have

$$\frac{1}{2\pi}\int_{-\pi}^{\pi} \log w_N(\vartheta)\,d\vartheta \underset{N}{\longrightarrow} \frac{1}{2\pi}\int_{-\pi}^{\pi} \log w(\vartheta)\,d\vartheta.$$

It will therefore be enough to show that for any N and any $\delta > 0$ there exists some finite sum $1 - \sum_{k>0} A_k e^{ik\vartheta}$ such that

$$\frac{1}{2\pi}\int_{-\pi}^{\pi} \left|1 - \sum_{k>0} A_k e^{ik\vartheta}\right| w(\vartheta)\,d\vartheta \;<\; \exp\left(\frac{1}{2\pi}\int_{-\pi}^{\pi} \log w_N(\vartheta)\,d\vartheta\right) + \delta.$$

To this end, put first of all

$$(*) \qquad F_N(z) = \exp\left\{\frac{1}{2\pi}\int_{-\pi}^{\pi} \frac{e^{it}+z}{e^{it}-z} \log\left(\frac{1}{w_N(t)}\right) dt\right\}$$

for $|z| < 1$. We have

$$(\ddagger) \qquad F_N(0) = \exp\left(\frac{1}{2\pi}\int_{-\pi}^{\pi} \log\left(\frac{1}{w_N(t)}\right) dt\right).$$

Since $w_N(t) \geqslant e^{-N}$, $|F_N(z)| \leqslant e^N$ for $|z| < 1$. Indeed, taking real parts of the logarithms of both sides of $(*)$ gives us

$$\log|F_N(re^{i\vartheta})| = \frac{1}{2\pi}\int_{-\pi}^{\pi} \frac{1-r^2}{1-r^2-2r\cos(\vartheta - t)} \log\left(\frac{1}{w_N(t)}\right) dt.$$

On the right side we recognize the *Poisson kernel* (that's the *real reason* for using $(e^{it}+z)/(e^{it}-z)$ in $(*)$, aside from the fact that we *want* $F_N(z)$ to be *analytic* in $\{|z| < 1\}$). As one knows,

$$\frac{1}{2\pi}\int_{-\pi}^{\pi} \frac{1-r^2}{1+r^2-2r\cos(\vartheta - t)}\,dt = 1;$$

the integrand is obviously *positive*. We see that $\log|F_N(re^{i\vartheta})| \leqslant N$ by the previous formula.

Now we use another, much *finer* property of the Poisson kernel, established in §B below. According to the latter,

$$\frac{1}{2\pi}\int_{-\pi}^{\pi} \frac{1-r^2}{1+r^2-2r\cos(\vartheta - t)} \log\left(\frac{1}{w_N(t)}\right) dt \longrightarrow \log\left(\frac{1}{w_N(\vartheta)}\right)$$

for almost all ϑ as $r \to 1$. So $|F_N(re^{i\vartheta})| \to 1/w_N(\vartheta)$ a.e. for $r \to 1$. However, $|F_N(z)|$ is bounded above and $w(\vartheta) \in L_1(-\pi, \pi)$. Therefore, by *dominated* convergence,

$$\frac{1}{2\pi}\int_{-\pi}^{\pi}|F_N(re^{i\vartheta})|w(\vartheta)\,d\vartheta \longrightarrow \frac{1}{2\pi}\int_{-\pi}^{\pi}\frac{w(\vartheta)}{w_N(\vartheta)}\,d\vartheta$$

as $r \to 1$. The right-hand side is *clearly* $\leqslant 1$. Given $\varepsilon > 0$ we can therefore get an $r < 1$ with

$$(\dagger) \qquad \frac{1}{2\pi}\int_{-\pi}^{\pi}|F_N(re^{i\vartheta})|w(\vartheta)\,d\vartheta < 1 + \varepsilon.$$

Fix such an r.

By the very *form* of the right side of $(*)$, $F_N(z)$ is *analytic* in $\{|z| < 1\}$; it therefore has a Taylor expansion there. And, by $(\substack{*\\*})$, $F_N(0) \neq 0$. Letting $S(z)$ be any *partial sum* of the Taylor series for $F_N(z)$, we see that *for our fixed r*,

$$\frac{S(re^{i\vartheta})}{F_N(0)} \quad \text{is of the form } 1 - \sum_{k>0} A_k e^{ik\vartheta},$$

the sum on the right being finite. Since $F_N(z)$ is regular in $\{|z| < 1\}$ and $r < 1$, we see by $(\dagger)$ that we can choose the partial sum $S(z)$ so that

$$\frac{1}{2\pi}\int_{-\pi}^{\pi}|S(re^{i\vartheta})|w(\vartheta)\,d\vartheta < 1 + 2\varepsilon.$$

Hence

$$\frac{1}{2\pi}\int_{-\pi}^{\pi}\left|1 - \sum_{k>0} A_k e^{ik\vartheta}\right| w(\vartheta)\,d\vartheta$$

$$= \frac{1}{2\pi}\int_{-\pi}^{\pi}\left|\frac{S(re^{i\vartheta})}{F_N(0)}\right| w(\vartheta)\,d\vartheta \;\leqslant\; (1 + 2\varepsilon)\cdot\frac{1}{F_N(0)},$$

which equals

$$(1 + 2\varepsilon)\exp\left(\frac{1}{2\pi}\int_{-\pi}^{\pi}\log w_N(t)\,dt\right) \quad \text{by } (\substack{*\\*}).$$

This is enough, and we are done.

Remark. This most elegant result was extended by Kolmogorov, and then by Krein, who evaluated the infimum of

$$\frac{1}{2\pi}\int_{-\pi}^{\pi}\left|1 - \sum_{n>0} a_n e^{in\vartheta}\right| d\mu(\vartheta)$$

for all finite sums $\sum_{n>0} a_n e^{in\vartheta}$ when μ is *any finite positive measure*. It turns out that the singular part of μ (with respect to Lebesgue measure) *has no influence here*, that the infimum is simply equal to

$$\exp\left\{\frac{1}{2\pi}\int_{-\pi}^{\pi}\log\left(\frac{d\mu(\vartheta)}{d\vartheta}\right)d\vartheta\right\}.$$

I do not give the proof of this result. It depends on the construction of Fatou–Riesz functions which, while not very difficult, is not really part of the material being treated here. The interested reader may find a proof in many books; some of the older ones which have it are Hoffman's and Akhiezer's (on approximation theory). The newer books by Garnett (on bounded analytic functions), and by me (on H_p spaces) both contain proofs.

B. The pointwise approximate identity property of the Poisson kernel

Theorem. Let $P(\vartheta)\in L_1(-\pi,\pi)$, *and, for* $r<1$, *write*

$$U(re^{i\vartheta})=\frac{1}{2\pi}\int_{-\pi}^{\pi}\frac{1-r^2}{1+r^2-2r\cos(\vartheta-t)}P(t)\,dt.$$

For almost every ϑ, $U(z)$ *tends to* $P(\vartheta)$ *uniformly as* z *tends to* $e^{i\vartheta}$ *within any sector of the form*

$$|\arg(1-e^{-i\vartheta}z)| \leqslant \alpha < \frac{\pi}{2}.*$$

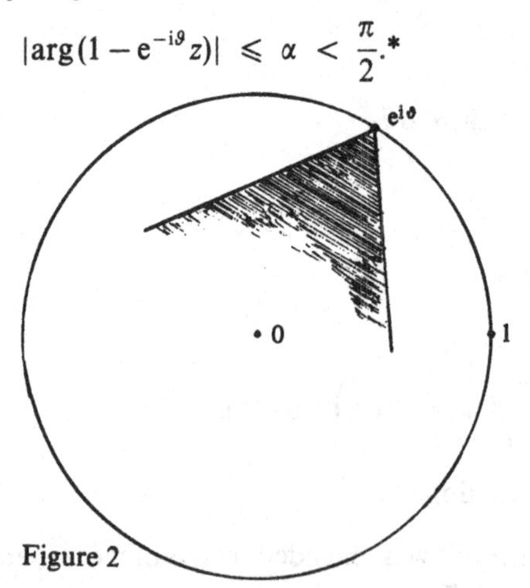

Figure 2

Remark. We write '$U(z)\to P(\vartheta)$ a.e. for $z \angle\!\!\to e^{i\vartheta}$.' Some people say that $U(z)\to P(\vartheta)$ a.e. for z tending *non-tangentially* to $e^{i\vartheta}$, others say that

* It is clear that for z of modulus $>\sin\alpha$ in such a sector we have $|\arg z-\vartheta|\leqslant K(1-|z|)$ with a constant K depending on α.

$U(z)\to P(\vartheta)$ uniformly within any Stoltz domain as $z\to e^{i\vartheta}$ (for almost all ϑ).

Of course, the theorem includes the result that $U(re^{i\vartheta})\to P(\vartheta)$ a.e. for $r\to 1$, used in proving Szegő's theorem.

Proof of theorem. We will show that $U(re^{i\vartheta_r})\to P(\vartheta)$ for $r\to 1$ if $|\vartheta_r-\vartheta|\leqslant K(1-r)$, *whenever* $(d/d\vartheta)\int_0^\vartheta P(t)\,dt$ *exists and equals* $P(\vartheta)$, *hence for almost every* ϑ, by Lebesgue's differentiation theorem. The rapidity of the convergence will be seen to depend only on the value of K measuring the opening of the sector with vertex at $e^{i\vartheta}$, and *not* on the particular choice of ϑ_r satisfying the above relation for each value of r.

Without loss of generality, take $\vartheta=0$, and assume that

$$\int_0^\vartheta P(t)\,dt=\vartheta P(0)+o(|\vartheta|)$$

for $\vartheta\to 0$ (from above *or* below!). Pick any small $\delta>0$ and write

$$\frac{1}{2\pi}\int_{-\pi}^{\pi}\frac{1-r^2}{1+r^2-2r\cos(\vartheta_r-t)}P(t)\,dt$$

as

$$\left(\frac{1}{2\pi}\int_{\delta\leqslant|t|\leqslant\pi}+\frac{1}{2\pi}\int_{-\delta}^{\delta}\right)\frac{1-r^2}{1+r^2-2r\cos(\vartheta_r-t)}P(t)\,dt.$$

As $r\to 1$, $|\vartheta_r|$ becomes and remains $<\delta/2$, so $(1-r^2)/(1+r^2-2r\cos(\vartheta_r-t))\to 0$ *uniformly* for $\delta<|t|<\pi$, and the *first integral* tends to *zero.*

The *second* is treated by partial integration. Writing $J(\vartheta)=\int_0^\vartheta P(t)\,dt$, that second integral becomes

$$\frac{1}{2\pi}\left[J(\delta)\frac{1-r^2}{1+r^2-2r\cos(\vartheta_r-\delta)}-J(-\delta)\frac{1-r^2}{1+r^2-2r\cos(\vartheta_r+\delta)}\right]$$

$$-\frac{1}{2\pi}\int_{-\delta}^{\delta}J(t)\frac{\partial}{\partial t}\left(\frac{1-r^2}{1+r^2-2r\cos(\vartheta_r-t)}\right)dt.$$

The two integrated terms in square brackets tend to zero as $r\to 1$. Since $J(t)=P(0)t+o(|t|)$, the *integral* equals

$$-\frac{1}{2\pi}\int_{-\delta}^{\delta}P(0)t\frac{\partial}{\partial t}\left(\frac{1-r^2}{1+r^2-2r\cos(\vartheta_r-t)}\right)dt$$

$$+\frac{1}{2\pi}\int_{-\delta}^{\delta}o(|t|)\frac{\partial}{\partial t}\left(\frac{1-r^2}{1+r^2-2r\cos(\vartheta_r-t)}\right)dt.$$

Here, the *first* term is readily seen (by *reverse* integration by parts!) to equal

$$o(1)+\frac{P(0)}{2\pi}\int_{-\delta}^{\delta}\frac{1-r^2}{1+r^2-2r\cos(\vartheta_r-t)}dt=o(1)+P(0)(1-o(1)),$$

which tends to $P(0)$ as $r\to 1$.

To estimate the *second* term, we have to use the fact that

$$\frac{1-r^2}{1+r^2-2r\cos(\vartheta_r - t)}$$

is a monotone function of t on each of the intervals $-\delta \leqslant t \leqslant \vartheta_r$ and $\vartheta_r \leqslant t \leqslant \delta$. (We are supposing that r is so close to 1 that $-\delta < \vartheta_r < \delta$.) Given any $\varepsilon > 0$, we can choose $\delta > 0$ so small to begin with that the *second term* is in absolute value

$$\leqslant \frac{\varepsilon}{2\pi}\int_{-\delta}^{\delta} |t| \left| \frac{\partial}{\partial t}\left(\frac{1-r^2}{1+r^2-2r\cos(\vartheta_r - t)}\right)\right| dt.$$

Writing $\tau = \vartheta_r - t$, this becomes

$$\frac{\varepsilon}{2\pi}\int_{\vartheta_r-\delta}^{\vartheta_r+\delta} |\tau - \vartheta_r| \left| \frac{\partial}{\partial \tau}\left(\frac{1-r^2}{1+r^2-2r\cos\tau}\right)\right| d\tau.$$

We break this up as

$$(*) \qquad \left(\frac{\varepsilon}{2\pi}\int_{\vartheta_r-\delta}^{0} + \frac{\varepsilon}{2\pi}\int_{0}^{\vartheta_r+\delta}\right) |\tau - \vartheta_r| \left| \frac{\partial}{\partial \tau}\left(\frac{1-r^2}{1+r^2-2r\cos\tau}\right)\right| d\tau;$$

in the *second integral*,

$$\frac{\partial}{\partial \tau}\left(\frac{1-r^2}{1+r^2-2r\cos\tau}\right) < 0,$$

so that second integral is $\leqslant$

$$-\frac{\varepsilon}{2\pi}\int_{0}^{\vartheta_r+\delta} \tau \frac{\partial}{\partial \tau}\left(\frac{1-r^2}{1+r^2-2r\cos\tau}\right) d\tau$$

$$-\frac{\varepsilon|\vartheta_r|}{2\pi}\int_{0}^{\vartheta_r+\delta} \frac{\partial}{\partial \tau}\left(\frac{1-r^2}{1+r^2-2r\cos\tau}\right) d\tau.$$

Here, the first term is $\varepsilon(\frac{1}{2} + o(1))$ (see above treatment of expression involving $P(0)$!), and the second is

$$\leqslant \frac{\varepsilon|\vartheta_r|}{2\pi}\cdot\frac{1+r}{1-r}.$$

This last, however, is $\leqslant (K/\pi)\varepsilon$ *in view of our condition on* ϑ_r. We see that the second integral in $(*)$ is $\leqslant (K/\pi + \frac{1}{2} + o(1))\varepsilon$ for r close enough to 1.

The first integral in $(*)$ is similarly treated, and seen to also be $\leqq (K/\pi + \frac{1}{2} + o(1))\varepsilon$ for r close to 1. In this way, we have found that the expression

$$\frac{1}{2\pi}\int_{-\delta}^{\delta} o(|t|) \frac{\partial}{\partial t}\left(\frac{1-r^2}{1+r^2-2r\cos(\vartheta_r - t)}\right) dt$$

is in absolute value $\leqslant (1 + 2K/\pi + o(1))\varepsilon$ if $\delta > 0$ is small enough to begin with, and r close enough to 1. However, according to the calculation at the beginning of this proof, the sum of the last expression and $P(0)$ differs by $o(1)$ from $U(re^{i\vartheta_0})$ when $r \to 1$. So, since $\varepsilon > 0$ is arbitrary, we have established the desired result.

Remark. Suppose that

$$U(re^{i\vartheta}) = \frac{1}{2\pi} \int_{-\pi}^{\pi} \frac{1 - r^2}{1 + r^2 - 2r\cos(\vartheta - t)} \, d\mu(t)$$

with a finite (complex valued) *measure* μ. Form the *primitive*

$$\mu(\vartheta) = \int_0^{\vartheta} d\mu(t).$$

Then it is *still true* that, wherever the *derivative* $\mu'(\vartheta)$ *exists and is finite*, we have $U(z) \to \mu'(\vartheta)$ for $z \mathrel{\not\angle\!\!\to} e^{i\vartheta}$. (Hence $\lim_{r\to 1} U(re^{i\vartheta})$ *exists and is finite* a.e. by Lebesgue's differentiation theorem.) The proof of this slightly more general result is *exactly the same* as that of the above one.

Problem 2

The purpose of this problem is to derive, from Szegő's theorem, the following result. *Let* $w(x) \geq 0$ *be in* $L_1(-\infty, \infty)$ *and let* $a > 0$. *There are finite sums* $S(x)$ *of the form* $S(x) = \sum_{\lambda \geq a} A_\lambda e^{i\lambda x}$ *with* $\int_{-\infty}^{\infty} |1 - S(x)| w(x) \, dx$ *arbitrarily small iff* $\int_{-\infty}^{\infty} (\log w(x)/(1 + x^2)) \, dx = -\infty$. *In case* $\int_{-\infty}^{\infty} (\log w(x)/(1 + x^2)) dx = -\infty$, *we can, given any bounded continuous function* $\varphi(x)$, *find finite sums* $S(x)$ *of the above mentioned form with* $\int_{-\infty}^{\infty} |\varphi(x) - S(x)| w(x) \, dx$ *arbitrarily small.* Establishment of this result is in a series of steps.

(a) Let $\alpha > 0$ and let p be a positive integer. There are numbers A_n with

$$\left(\frac{x}{1 - i\alpha x}\right)^p = \sum_0^{\infty} A_n \left(\frac{i - x}{i + x}\right)^n,$$

the series on the right being *uniformly convergent* for $-\infty < x < \infty$. (Hint: Put $w = (i - z)/(i + z)$ and look at where $f(w) = z/(1 - i\alpha z)$ is regular in the w-plane. *Little or no computation is used in doing* (a).)

(b) Let $\lambda > 0$. There are *finite sums* $S_k(x)$, each of the form $\sum_{n \geq 0} C_n((i - x)/(i + x))^n$, such that $|S_k(x)| \leq 2$ on $\mathbb{R}$ and $S_k(x) \underset{k}{\to} e^{i\lambda x}$ u.c.c.* on $\mathbb{R}$. (Hint: $e^{i\lambda x} = \lim_{\alpha \to 0+} e^{i\lambda x/(1 - i\alpha x)}$. For each $\alpha > 0$ the series for $\exp(i\lambda x/(1 - i\alpha x))$ is *uniformly convergent* for $-\infty < x < \infty$. *Little or no computation here.*)

(c) Given any integer $n > 0$ there are finite sums $T_k(x)$ of the form $\sum_{\lambda > 0} A_\lambda e^{i\lambda x}$ with $|T_k(x)| \leq C$ independent of k on $\mathbb{R}$ and $T_k(x) \to 1/(i + x)^n$ u.c.c.

* u.c.c means *uniform convergence on compacta.*

on $\mathbb{R}$. (Hint: Start from the integral formula

$$\frac{i}{i+x} = \int_0^\infty e^{-\lambda} e^{i\lambda x}\, d\lambda,$$

$$\frac{i}{(i+x)^2} = -\int_0^\infty i\lambda e^{-\lambda} e^{i\lambda x}\, d\lambda,$$

&c.)

(d) Given $w \geqslant 0$ in $L_1(\mathbb{R})$, denote by $\mathscr{A}$ the class of bounded continuous φ defined on $\mathbb{R}$ such that $\int_{-\infty}^{\infty} |\varphi(x) - S(x)| w(x)\, dx$ can be made arbitrarily small with suitable finite sums $S(x)$ of the form $\sum_{n \geqslant 0} A_n((i-x)/(i+x))^n$. Call $\mathscr{E}$ the set of bounded continuous φ for which finite sums $T(x) = \sum_{\lambda \geqslant 0} A_\lambda e^{i\lambda x}$ exist making $\int_{-\infty}^{\infty} |\varphi(x) - T(x)| w(x)\, dx$ arbitrarily small. *Prove that* $\mathscr{A} = \mathscr{E}$.

(e) Let $a > 0$, and denote by $\mathscr{F}$ the set of bounded continuous φ such that there are finite sums $\sigma(x) = \sum_{\lambda \geqslant a} A_\lambda e^{i\lambda x}$ with $\int_{-\infty}^{\infty} |\varphi(x) - \sigma(x)| w(x)\, dx$ arbitrarily small. Prove that if $1 \in \mathscr{F}$ then $\mathscr{F}$ contains *all* bounded continuous φ, and this happens iff $\int_{-\infty}^{\infty} (\log w(x)/(1+x^2))\, dx = -\infty$. (Hint: If $1 \in \mathscr{F}$, then $\mathscr{E}$ (of part (d)) includes all $e^{i\lambda x}$ with $\lambda \geqslant -a$, hence, by iteration, all $e^{i\lambda x}$ with $\lambda \geqslant -2a$, with $\lambda \geqslant -3a$, &c. So $\mathscr{E}$ includes *all integral* powers $((i-x)/(i+x))^n$ with *positive and negative n*. These are enough to approximate $e^{-iax}\varphi(x)$ for any bounded continuous φ.)

III

Entire functions of exponential type

An entire function $f(z)$ is said to be *of exponential type* if there is a constant A such that

$$|f(z)| \leqslant \text{const.}e^{A|z|}$$

everywhere. The *infimum* of the set of A for which such an inequality holds (with the constant in front on the right depending on A) is called the *type* of $f(z)$.

Entire functions of exponential type come up in various branches of analysis, partly on account of the evident fact that integrals of the form

$$\int_K e^{i\lambda z}\, d\mu(\lambda)$$

are equal to such functions whenever K is a compact subset of $\mathbb{C}$. In this chapter we establish some of the most important results concerning them, which find application throughout the rest of the book. We are not of course attempting to give a complete treatment of the subject. Fuller accounts are contained in the books by Boas and by Levin.

A. Hadamard factorization

As in Chapter I, we denote the *number* of *zeros* of $f(z)$ having modulus $\leqslant r$ by $n(r)$ (each zero being counted according to its multiplicity). We sometimes write $n_f(r)$ instead of $n(r)$ when several functions are being dealt with.

Theorem. *If $f(z)$ is entire and of exponential type, $n(r) \leqslant Cr + O(1)$.*

Proof. See Problem 1(a), Chapter I. If $|f(z)| \leqslant \text{const.}e^{A|z|}$ we can take $C = eA$.

Theorem (Hadamard factorization). *Let $f(z)$ be entire, of exponential type, and denote by $\{z_n\}$ the sequence of its zeros $\neq 0$ (multiplicities counted by repetition), so arranged that*

$$0 < |z_1| \leqslant |z_2| \leqslant |z_3| \leqslant \cdots.$$

Then

$$\boxed{f(z) = Cz^k e^{bz} \prod_n \left(1 - \frac{z}{z_n}\right) e^{z/z_n},}$$

the product being uniformly convergent on compact subsets of $\mathbb{C}$.

Terminology. Henceforth we abbreviate the last phrase as 'u.c.c. convergent on $\mathbb{C}$'.

Proof of theorem. By working with $f(z)/z^k$ instead of $f(z)$ (if necessary), we first reduce the situation to one where $f(0) \neq 0$. Then $n(r) \leqslant Kr$ for some K.

If, with a zero z_n of f, we have $|z_n| > 2R$, then, for $|z| \leqslant R$,

$$\log\left\{\left(1 - \frac{z}{z_n}\right)e^{z/z_n}\right\} = -\frac{z}{z_n} - \frac{1}{2}\left(\frac{z}{z_n}\right)^2 - \cdots + \frac{z}{z_n}$$

$$= -\frac{1}{2}\left(\frac{z}{z_n}\right)^2 - \frac{1}{3}\left(\frac{z}{z_n}\right)^3 - \cdots$$

$$= -\frac{1}{2}\left(\frac{z}{z_n}\right)^2 (1 + O(1))$$

(We are using the branch of the logarithm which is *zero* at 1). Therefore

$$\left|\log\left\{\left(1 - \frac{z}{z_n}\right)e^{z/z_n}\right\}\right| \leqslant \frac{1}{2}\left|\frac{z}{z_n}\right|^2 (1 + O(1)),$$

whence (assuming always that $|z| \leqslant R$),

$$\sum_{|z_n| \geqslant 2R} \left|\log\left\{\left(1 - \frac{z}{z_n}\right)e^{z/z_n}\right\}\right| \leqslant \frac{1 + O(1)}{2} \sum_{|z_n| \geqslant 2R} \left|\frac{R}{z_n}\right|^2$$

$$= \frac{1 + O(1)}{2} R^2 \int_{2R}^{\infty} \frac{dn(t)}{t^2}$$

$$= \frac{1 + O(1)}{2} R^2 \left\{\frac{-n(2R)}{4R^2} + 2\int_{2R}^{\infty} \frac{n(t)}{t^3} dt\right\}$$

$$\leqslant \frac{1+O(1)}{2}R^2 \cdot \int_{2R}^{\infty} \frac{2K}{t^2}dt = \frac{1+O(1)}{2}KR.$$

This inequality establishes absolute and uniform convergence of

$$\sum_{|z_n|\geqslant 2R} \log\left\{\left(1-\frac{z}{z_n}\right)e^{z/z_n}\right\}$$

for $|z|\leqslant R$, and hence the uniform convergence of

$$\prod_{|z_n|\geqslant 2R}\left(1-\frac{z}{z_n}\right)e^{z/z_n}$$

for such values of z.

Write $P(z)=\prod_n(1-z/z_n)e^{z/z_n}$; according to what has just been shown, $P(z)$ is an entire function of z. Since $f(0)\neq 0$, $f(z)/P(z)$ *is entire and has no zeros in* $\mathbb{C}$. There is thus an entire function $\varphi(z)$ with

$$\frac{f(z)}{P(z)} = e^{\varphi(z)},$$

and *it is claimed that* $\varphi(z)=a+bz$ with constants a and b.

To show that $\varphi(z)$ has the asserted form, we use the fact that $f(z)$ is of exponential type in conjunction with the inequality $n(r)\leqslant Kr$ in order to get some control on $|\Re\varphi(z)|$ for large $|z|$. For $|z|\leqslant R$,

$$e^{\Re\varphi(z)} = \left|\frac{f(z)}{\prod_{|z_n|<2R}\left(1-\frac{z}{z_n}\right)e^{z/z_n}}\right| \cdot \left|\frac{1}{\prod_{|z_n|\geqslant 2R}\left(1-\frac{z}{z_n}\right)e^{z/z_n}}\right| = \text{I}\cdot\text{II, say.}$$

The computation made above shows that $|\log \text{II}|\leqslant CR$ with some constant C (we estimated $|\log\{1-z/z_n\}e^{z/z_n}|$!), and it suffices to estimate I.

For I we use a *trick*. The ratio $\psi(z)=f(z)/\prod_{|z_n|<2R}(1-z/z_n)e^{z/z_n}$ is *entire*, so, by the *principle of maximum*, $\sup_{|z|\leqslant R}|\psi(z)|\leqslant \sup_{|z|=4R}|\psi(z)|$. Here, estimation of the quantity on the *right* will furnish an upper bound for I, which is at most equal to the left-hand side. We have, for $|z|=4R$, $|z/z_n|=4R/|z_n|$, whence

$$\prod_{|z_n|<2R}|e^{z/z_n}| \geqslant \exp\left\{-4R\sum_{|z_n|\leqslant 2R}1/|z_n|\right\}$$

$$= \exp\left\{-4R\int_0^{2R}\frac{dn(t)}{t}\right\} = \exp\left\{-2n(2R)-4R\int_0^{2R}\frac{n(t)}{t^2}dt\right\}.$$

Since $n(t)=0$ for $0\leqslant t<|z_1|$, a quantity >0, and $n(t)/t\leqslant K$, the last expression is

$$\geqslant e^{-4R(K\log R+O(1))}.$$

At the same time, for $|z| = 4R$ and $|z_n| \leqslant 2R$,

$$\left|1 - \frac{z}{z_n}\right| \geqslant 1,$$

so

$$\left|\prod_{|z_n|<2R}\left(1 - \frac{z}{z_n}\right)e^{z/z_n}\right| \geqslant e^{-4R(K\log R + O(1))}$$

when $|z| = 4R$. Because $f(z)$ is of exponential type we therefore have

$$|\psi(z)| \leqslant e^{4KR(\log R + O(1))}, \quad |z| = 4R,$$

whence

$$I \leqslant e^{4KR(\log R + O(1))},$$

and finally, since $e^{\Re\varphi(z)} = I \cdot II$,

$$\Re\varphi(z) \leqslant 4KR\log R + O(R) \quad \text{for} \quad |z| \leqslant R$$

in view of the fact that $\log II \leqslant CR$.

At this point, we use a device already applied to the study of $\log|F(re^{i\vartheta})|$ near the end of Chapter I. By analyticity of $\varphi(z)$,

$$\Re\varphi(0) = \frac{1}{2\pi}\int_{-\pi}^{\pi} \Re\varphi(Re^{i\vartheta})\,d\vartheta$$

$$= \frac{1}{2\pi}\int_{-\pi}^{\pi} [\Re\varphi(Re^{i\vartheta})]_+\,d\vartheta - \frac{1}{2\pi}\int_{-\pi}^{\pi} [\Re\varphi(Re^{i\vartheta})]_-\,d\vartheta$$

(with self-evident notation). Therefore,

$$\int_{-\pi}^{\pi} |\Re\varphi(Re^{i\vartheta})|\,d\vartheta = \int_{-\pi}^{\pi} \{[\Re\varphi(Re^{i\vartheta})]_+ + [\Re\varphi(Re^{i\vartheta})]_-\}\,d\vartheta$$

$$= 2\int_{-\pi}^{\pi} [\Re\varphi(Re^{i\vartheta})]_+\,d\vartheta - 2\pi\Re\varphi(0).$$

By the one-sided inequality just found for $\Re\varphi(z)$, $|z| \leqslant R$, this last expression is $\leqslant 16\pi KR\log R + O(R) + O(1)$.

Now we can conclude the proof. Since $\varphi(z)$ is entire,

$$\varphi(z) = \sum_0^\infty \gamma_n z^n,$$

so

$$2\Re\varphi(Re^{i\vartheta}) = \sum_1^\infty \bar{\gamma}_n R^n e^{-in\vartheta} + 2\Re\gamma_0 + \sum_1^\infty \gamma_n R^n e^{in\vartheta}$$

Using the relations $\int_{-\pi}^{\pi} e^{ik\vartheta}e^{-il\vartheta}\,d\vartheta = \begin{cases} 0, k \neq l \\ 2\pi, k = l \end{cases}$, we get, for $n \geqslant 2$,

$$\gamma_n = \frac{1}{\pi R^n}\int_{-\pi}^{\pi} \Re\varphi(Re^{i\vartheta})e^{-in\vartheta}\,d\vartheta.$$

Therefore

$$|\gamma_n| \leqslant \frac{1}{\pi R^n}\int_{-\pi}^{\pi} |\Re\varphi(Re^{i\vartheta})|\,d\vartheta$$

which, by the above work, is $\leqslant R^{-n}(16KR\log R + O(R) + O(1))$. Making $R \to \infty$, we see that $\gamma_n = 0$ for $n \geqslant 2$. Our power series for $\varphi(z)$ thus reduces to the *linear* expression $\gamma_0 + \gamma_1 z$, and finally

$$f(z) = e^{\varphi(z)}P(z) = e^{\gamma_0}e^{\gamma_1 z}\prod_n\left(1 - \frac{z}{z_n}\right)e^{z/z_n},$$

the required representation. We are done.

B. Characterization of the set of zeros of an entire function of exponential type. Lindelöf's theorems

While establishing the Hadamard factorization in the preceding § we found that

$$\mathrm{II} \leqslant e^{CR}$$

which was to be expected (having *started* with a function of exponential growth), but we could only show that

$$\mathrm{I} \leqslant e^{O(R\log R)}.$$

This, however, forced $\varphi(z)$ to be a first degree polynomial, whence, *in fact*,

$$\mathrm{I} \leqslant e^{O(R)},$$

because the method used to estimate II showed at the same time that

$$\mathrm{II} \geqslant e^{-O(R)}.$$

The refinement on our estimate of I from $e^{O(R\log R)}$ to $e^{O(R)}$ is due to the fact that $|f(z)| \leqslant e^{O(|z|)}$ for large $|z|$. Otherwise, the $R\log R$ growth is *best possible*, and if we *only know* that $n(r) \leqslant Kr$, *we can only conclude that*

$$\left|\prod_n\left(1 - \frac{z}{z_n}\right)e^{z/z_n}\right| \leqslant e^{O(|z|\log|z|)}$$

for $|z|$ large, most of the contribution coming from the factors with $|z_n| < 2|z|$.

The fact that $f(z)$ is of *exponential* type imposes *not only* the growth condition $n(r) \leqslant O(r)$ (for large r), but also *a certain symmetry* in the *distribution of the zeros* z_n. This symmetry is a deeper property of that set than the growth condition.

Theorem (Lindelöf). *Let $f(z)$ be entire, of exponential type, with $f(0) \neq 0$, and denote by $\{z_n\}$ the sequence of zeros of $f(z)$, with, as usual, each zero repeated therein according to its multiplicity. Put*

$$S(r) = \sum_{|z_n| \leqslant r} \frac{1}{z_n}.$$

Then $|S(r)|$ is bounded as $r \to \infty$.

Proof. By double integration. Since $n(r) \leqslant Kr$, f being of exponential type, we clearly have

$$|S(r) - S(R)| \leqslant 2K \quad \text{for} \quad R \leqslant r \leqslant 2R,$$

whence

$$\int_R^{2R} S(r) r\,dr = \tfrac{3}{2}R^2 S(R) + O(R^2).$$

We proceed to calculate the integral on the left.

Provided that $f(z)$ has no zeros for $|z| = r$, we have, by the calculus of residues,

$$S(r) = \sum_{|z_n| < r} \frac{1}{z_n} = \frac{1}{2\pi i}\int_0^{2\pi} \frac{f'(re^{i\vartheta})}{f(re^{i\vartheta})} \cdot \frac{ire^{i\vartheta}d\vartheta}{re^{i\vartheta}} - \frac{f'(0)}{f(0)}.$$

By the Cauchy–Riemann equations,

$$\frac{f'(z)}{f(z)} = \left(\frac{\partial}{\partial x} - i\frac{\partial}{\partial y}\right)\log|f(z)|,$$

whence, putting $z = re^{i\vartheta}$,

$$S(r) = \frac{1}{2\pi}\int_0^{2\pi}\left(\frac{\partial}{\partial x} - i\frac{\partial}{\partial y}\right)\log|f(re^{i\vartheta})|\,d\vartheta - \frac{f'(0)}{f(0)}.$$

This holds for all save a finite number of values of r on the interval $[R, 2R]$. Multiply by rdr and integrate from R to $2R$. We find

$$\tfrac{3}{2}R^2 S(R) = \int_R^{2R} S(r) r dr + O(R^2)$$

$$= \frac{1}{2\pi} \iint_{R \leqslant |z| \leqslant 2R} \left(\frac{\partial}{\partial x} - i\frac{\partial}{\partial y}\right)\log|f(z)|\,dx\,dy + O(R^2).$$

Since $S(r) = S(r+)$, there is no loss of generality in assuming that $f(z)$ has no zeros on $|z| = R$ or on $|z| = 2R$. We may then apply Green's theorem to the double integral on the right (this is justified by first excising a small disk of radius ρ, say, about each of the z_n in the annulus $R < |z| < 2R$, and then making $\rho \to 0$), obtaining for it the value

$$\frac{1}{2\pi}\int_0^{2\pi} (2R\log|f(2Re^{i\vartheta})| - R\log|f(Re^{i\vartheta})|)e^{-i\vartheta}\,d\vartheta,$$

whence, by the previous relation,

$$\tfrac{3}{2}R^2|S(R)| \leqslant \frac{R}{2\pi}\int_0^{2\pi} (2|\log|f(2Re^{i\vartheta})|| + |\log|f(Re^{i\vartheta})||)\,d\vartheta + O(R^2).$$

Here, by Jensen's inequality (see Chapter I),

$$\int_0^{2\pi} |\log|f(re^{i\vartheta})||\,d\vartheta \leqslant 2\int_0^{2\pi} \log^+|f(re^{i\vartheta})|\,d\vartheta - 2\pi\log|f(0)|,$$

which is $\leqslant 4\pi Ar + O(1)$ if $|f(z)| \leqslant \text{const.}e^{A|z|}$. Combined with the preceding, this yields

$$\tfrac{3}{2}R^2|S(R)| \leqslant O(R^2) + 8AR^2 + 2AR^2 + O(R),$$

and

$$|S(R)| \leqslant O(1) \quad \text{for} \quad R \to \infty.$$

Q.E.D.

The result just proven has an easy converse.

Theorem (also due to Lindelöf!). *Let*

$$0 < |z_1| \leqslant |z_2| \leqslant |z_3| \leqslant \cdots,$$

denote by $n(r)$ the number of z_k having modulus $\leqslant r$ (taking account of multiplicities, as usual), and suppose that $n(r) \leqslant Kr$. Suppose also that the sums

$$\sum_{|z_n| \leqslant r} \frac{1}{z_n}$$

remain bounded in absolute value as $r \to \infty$. Then the product

$$C(z) = \prod_n \left(1 - \frac{z}{z_n}\right)e^{z/z_n}$$

is equal to an entire function of exponential type.

Terminology. $C(z)$ is frequently called a *canonical* product.

Proof of theorem. Uniform convergence of the product on compact subsets of $\mathbb{C}$ has already been shown during the establishment of the Hadamard factorization (§A).

Let R be given, and, for $|z| = R$, write

$$|C(z)| = \left|\prod_{|z_n|<2R}\left(1-\frac{z}{z_n}\right)e^{z/z_n}\right|\cdot\left|\prod_{|z_n|\geqslant 2R}\left(1-\frac{z}{z_n}\right)e^{z/z_n}\right| = \mathrm{I}\cdot\mathrm{II},$$

say. It has already been shown that $\mathrm{II} \leqslant e^{O(R)}$ while we were deriving the Hadamard factorization, so we need only consider I. Clearly,

$$\mathrm{I} \leqslant \prod_{|z_n|\leqslant 2R}\left(1+\frac{R}{|z_n|}\right)\cdot\exp\left\{R\left|\sum_{|z_n|\leqslant 2R}\frac{1}{z_n}\right|\right\}.$$

By hypothesis, the exponential factor on the right is $\leqslant e^{O(R)}$, and we need only estimate the *product*.

The logarithm of that product is

$$\sum_{|z_n|\leqslant 2R}\log\left(1+\frac{R}{|z_n|}\right) = \int_0^{2R}\log\left(1+\frac{R}{t}\right)dn(t)$$

$$= n(R)\log\tfrac{3}{2} + \int_0^{2R}\frac{R}{R+t}\,\frac{n(t)}{t}\,dt,$$

since $n(t)$ is zero for t near 0. Plugging in $n(t) \leqslant Kt$, we see that the last expression is $\leqslant KR\log\frac{3}{2} + 2KR$ so that, finally,

$$\log \mathrm{I} \leqslant KR\log\tfrac{3}{2} + 2KR + O(R) = O(R).$$

Since II has a similar estimate, we see that $|C(z)| \leqslant e^{O(R)}$ for $|z| = R$.

We're done.

Here is an important consequence of the above results.

Theorem. *Let $f(z)$ and $g(z)$ be entire and of exponential type. If the ratio $f(z)/g(z)$ is also entire, it is of exponential type.*

Proof. Combine the Hadamard factorization theorem with the two Lindelöf theorems.

Problem 3

Let p be an integer > 1; suppose that $f(z)$ is entire with $f(0) \neq 0$ and that

$|f(z)| \leqslant Ce^{A|z|^p}$.

Prove that $n_f(r) \leqslant Kr^p$ and that the sums

$$T(r) = \sum_{|z_n|\leqslant r}\frac{1}{z_n^p}$$

are bounded. (Hint: In studying $T(r)$, express $f'(re^{i\vartheta})/f(re^{i\vartheta})$ in terms of $(\partial/\partial r)\log|f(re^{i\vartheta})|$ and $(\partial/\partial\vartheta)\log|f(re^{i\vartheta})|$, assuming, of course, that f has no zeros on $|z| = r$.)

C. Phragmén–Lindelöf theorems

The entire functions of exponential type one meets with in the following chapters (and, for that matter, in many parts of analysis where they find application) have their *size on the real axis* subject to some *restriction*. During the remainder of this chapter we will be concerned with such functions, and we start here by seeing what it means to impose *boundedness* on $\mathbb{R}$. Some of the following material is contained in textbooks on elementary complex variable theory; we include it for completeness.

Theorem (extended maximum principle). *Let $\mathscr{D}$ be a domain in $\mathbb{C}$ not equal to all of $\mathbb{C}$, and suppose that $f(z)$ is analytic and bounded in $\mathscr{D}$. Assume that, for each $\zeta \in \partial\mathscr{D}$, $\limsup_{\substack{z\to\zeta \\ z\in\mathscr{D}}} |f(z)| \leqslant m$. Then $|f(z)| \leqslant m$ in $\mathscr{D}$.*

Remark. If $\mathscr{D}$ is a *bounded* domain, this is the ordinary maximum principle, and then the assumption that $f(z)$ is bounded in $\mathscr{D}$ is superfluous. When $\mathscr{D}$ is *unbounded*, however, this assumption is *really necessary*, as the simplest examples show.

Proof of theorem. Wlog, say that $0 \in \partial\mathscr{D}$. Pick any $\eta > 0$ and fix it. According to the hypothesis, we can find a $\rho > 0$ such that $|f(z)| \leqslant m + \eta$ for $z \in \mathscr{D}$ and $|z| \leqslant \rho$; we fix such a ρ and write

$$\mathscr{D}_\rho = \mathscr{D} \cap \{|z| > \rho\}.$$

The open set $\mathscr{D}_\rho$ may not be connected, but that doesn't matter; its boundary consists of part of $\partial\mathscr{D}$ and the arcs of $\{|z| = \rho\}$ lying in $\mathscr{D}$.

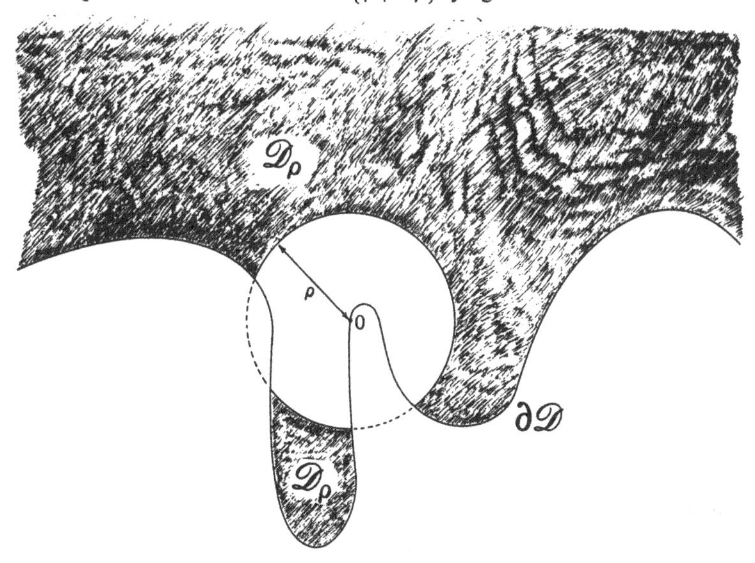

Figure 3

By choice of ρ, $\limsup_{\substack{z\to\zeta \\ z\in\mathscr{D}_\rho}} |f(z)| \leqslant m+\eta$ for $\zeta\in\partial\mathscr{D}_\rho$.

Take now a small $\varepsilon>0$ and consider, in $\mathscr{D}_\rho$, the subharmonic function

$$v_\varepsilon(z) = \log|f(z)| - \varepsilon\log|z|.$$

The right side is $\leqslant \log|f(z)| + \varepsilon\log(1/\rho)$ for $z\in\mathscr{D}_\rho$, and this is in turn $\leqslant \log|f(z)|+\eta$ if ε is chosen sufficiently small, which we assume henceforth. Referring to the previous relation, we see that

$$(*)\qquad \limsup_{\substack{z\to\zeta \\ z\in\mathscr{D}_\rho}} v_\varepsilon(z) \leqslant \log(m+\eta)+\eta$$

for each $\zeta\in\partial\mathscr{D}_\rho$.

Let $z_0\in\mathscr{D}_\rho$. *Since* $f(z)$ *is bounded in* $\mathscr{D}$, *say* $|f(z)|\leqslant M$ *there, we can find an* $R>|z_0|$ (depending of course on ε) *so large that* $v_\varepsilon(z)\leqslant \log M - \varepsilon\log|z|$ is $\leqslant \log m$ *for* $z\in\mathscr{D}_\rho$ and $|z|=R$. (This is the crucial step in the proof.) Denoting by $\mathscr{D}_{\rho,R}$ the *bounded* open set $\mathscr{D}_\rho\cap\{|z|<R\}$, we see that ( ) holds for *every* $\zeta\in\partial\mathscr{D}_{\rho,R}$, because any such ζ which is not on $\partial\mathscr{D}_\rho$ lies in the intersection of $\mathscr{D}_\rho$ with the circle $|\zeta|=R$.

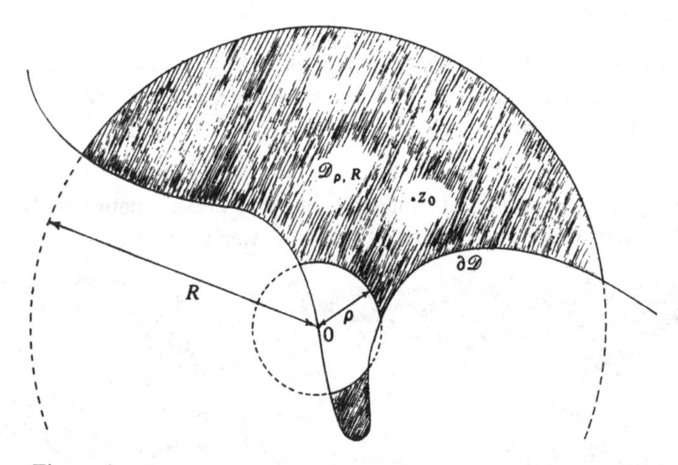

Figure 4

Since $\mathscr{D}_{\rho,R}$ is a bounded open set, we therefore have, by the (ordinary) maximum principle, $v_\varepsilon(z)\leqslant\log(m+\eta)+\eta$, $z\in\mathscr{D}_{\rho,R}$. This holds in particular for $z=z_0$, so

$$\log|f(z_0)| \leqslant \log(m+\eta)+\eta+\varepsilon\log|z_0|.$$

However, $\varepsilon>0$ could be chosen *as small as we pleased*. Therefore,

$$\log|f(z_0)| \leqslant \log(m+\eta)+\eta$$

and, since $\eta>0$ was arbitrary,

$$|f(z_0)|\leqslant m.$$

Q.E.D.

Remark. The peculiar reasoning followed in the above proof is called a *Phragmén–Lindelöf argument.* Most Phragmén–Lindelöf theorems are proved in the same way. Note the special rôle played by the harmonic function $\varepsilon \log|z|$; a function used in this way is called a *Phragmén–Lindelöf function.*

Theorem (Phragmén–Lindelöf). *Let $f(z)$ be analytic in a sector S of opening 2γ, and suppose that*

$$|f(z)| \leqslant Ce^{A|z|^\alpha}$$

in S, where $\alpha < \pi/2\gamma$. If, for every $\zeta \in \partial S$, $\limsup\limits_{\substack{z\to\zeta\\ z\in S}} |f(z)| \leqslant m$, *then $|f(z)| \leqslant m$ in S.*

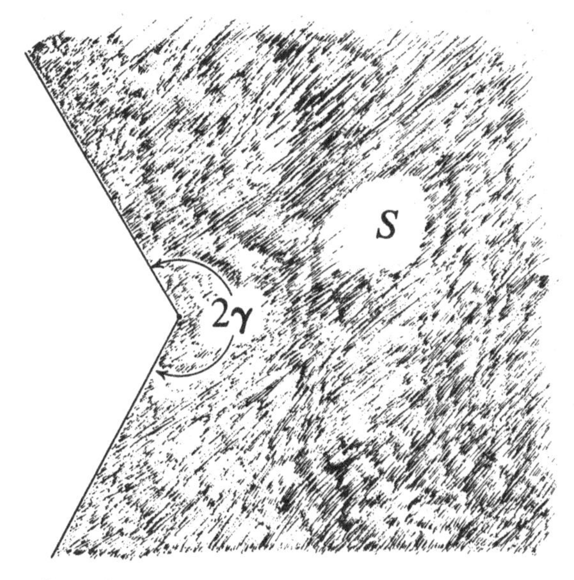

Figure 5

Proof. By making a change of variable, we may reduce our situation to the case where S is the sector

$$\{z: -\gamma < \arg z < \gamma\}$$

with vertex at the origin. Pick any number β, $\alpha < \beta < \pi/2\gamma$, and, with $\varepsilon > 0$ *fixed* but *arbitrary*, consider, in S, the subharmonic function

$$v_\varepsilon(z) = \log|f(z)| - \varepsilon\Re(z^\beta).$$

(Note: z^β is certainly analytic and single valued in S.) For $z = re^{i\vartheta}$ in S, we have

$$\Re(z^\beta) = r^\beta \cos\beta\vartheta > r^\beta \cos\beta\gamma,$$

and $\cos\beta\gamma > 0$ *since* $0 < \beta\gamma < \pi/2$. Therefore, in the first place, $v_\varepsilon(z) \leqslant \log|f(z)|$ in S, so, for $\zeta\in\partial S$,

$$\limsup_{\substack{z\to\zeta \\ z\in S}} v_\varepsilon(z) \leqslant \log m.$$

In the second place, since

$$\log|f(z)| \leqslant O(1) + A|z|^\alpha$$

in S and $\beta > \alpha$, we have

$$v_\varepsilon(z) \leqslant \log m$$

for $z\in S$ whenever $|z|$ is *large enough* (*how* large depends on ε!).

Suppose now that $z_0\in S$. With our fixed $\varepsilon > 0$, choose an $R > |z_0|$ so large that $v_\varepsilon(z) \leqslant \log m$ for $z\in S$ and $|z| = R$. Then

$$\limsup_{\substack{z\to\zeta \\ z\in S}} v_\varepsilon(z) \quad \text{is} \leqslant \log m$$

for any ζ on the boundary of the *bounded region* $S\cap\{|z| < R\}$,

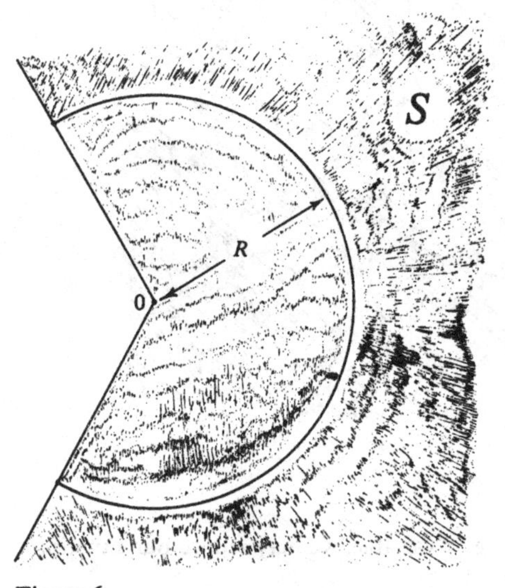

Figure 6

so, by the principle of maximum, $v_\varepsilon(z) \leqslant \log m$ throughout that region. In particular, $v_\varepsilon(z_0) \leqslant \log m$, so $\log|f(z_0)| \leqslant \log m + \varepsilon\Re(z_0^\beta)$. Now, keeping z_0 fixed, squeeze ε. Get $|f(z_0)| \leqslant m$, as required.

Important remark. The preceding two theorems *remain valid* if we merely suppose that $\log|f(z)|$ is *subharmonic* instead of taking $f(z)$ to be *analytic*. The proofs are exactly the same.

In the hypothesis of the *second* of the above two results we required $\alpha < \pi/2\gamma$ (with strict inequality); this in fact *cannot be relaxed* to the condition $\alpha \leqslant \pi/2\gamma$. What happens when $\alpha = \pi/2\gamma$ is seen from the following result, which, for simplicity, is stated for the case where $2\gamma = \pi$ (the only one which will arise on our work). We give its version for subharmonic functions.

Theorem. *Let $u(z)$ be subharmonic for $\Im z > 0$ with $u(z) \leqslant A|z| + o(|z|)$ there when $|z|$ is large. Suppose that, for each real x,*

$$\limsup_{\substack{z \to x \\ \Im z > 0}} u(z) \leqslant 0.$$

Then $u(z) \leqslant A\Im z$ for $\Im z > 0$.

Proof. Take any $\varepsilon > 0$. The function $v_\varepsilon(z) = u(z) - (A + \varepsilon)\Im z$ is subharmonic in the *first quadrant and* $v_\varepsilon(z) \leqslant O(|z|)$ there when $|z|$ is large. If ζ lies on the *boundary of the first quadrant*, we clearly have

$$\limsup_{\substack{z \to \zeta \\ \Im z > 0}} v_\varepsilon(z) \leqslant M$$

for some M since $u(iy) \leqslant Ay + o(y)$ for $y > 0$ and large.

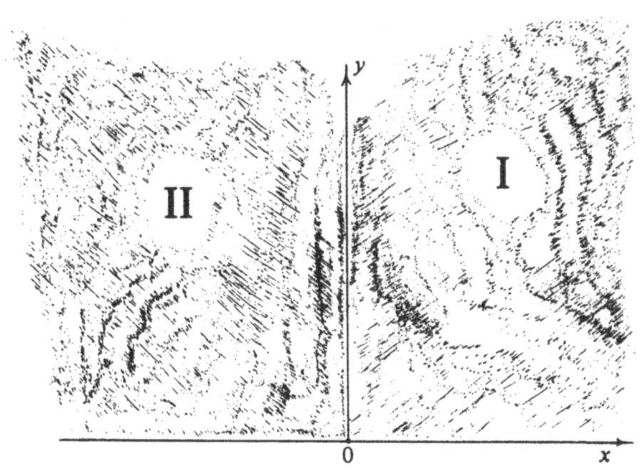

Figure 7

The first quadrant has opening $< \pi$, so by the preceding theorem (or rather by its version for subharmonic functions), $v_\varepsilon(z) \leqslant M$ throughout that region.

We see in like manner that $v_\varepsilon(z)$ is bounded above in the *second* quadrant, so, finally, $v_\varepsilon(z)$ is *bounded above for* $\Im z > 0$.

However, for x real,

$$\limsup_{\substack{z \to x \\ \Im z > 0}} v_\varepsilon(z) \leqslant 0.$$

Therefore, by the version for subharmonic functions of the first theorem in this §,

$$v_\varepsilon(z) \leqslant 0 \quad \text{for} \quad \Im z > 0.$$

That is,

$$u(z) \leqslant (A+\varepsilon)\Im z \quad \text{for} \quad \Im z > 0.$$

Squeeze ε. Get $u(z) \leqslant A\Im z$, $\Im z > 0$. Q.E.D.

Corollary. *Let $f(z)$ be analytic in $\Im z > 0$, continuous up to the real axis, and satisfy*

$$|f(z)| \leqslant Ce^{A|z|}$$

for $\Im z > 0$. If $|f(x)| \leqslant M$ for real x, then

$$|f(z)| \leqslant Me^{A\Im z}$$

when $\Im z > 0$.

Proof. Apply the theorem to $u(z) = \log|f(z)/M|$.

Remark. The example $f(z) = e^{-iAz}$ shows that the inequality furnished by the corollary cannot be improved. (Note also the relation between this particular function – or rather $\log|f(z)|$ – and the Phragmén–Lindelöf function $(A+\varepsilon)\Im z$ used in proving the theorem. That's no accident!)

The preceding theorem has an extension with a more elaborate statement, but the same proof. We give the version for analytic functions.

Theorem. *Let $f(z)$ be analytic in $\Im z > 0$ and continuous in $\Im z \geqslant 0$. Suppose that*

(i) $\log|f(z)| \leqslant O(|z|)$ *for large* $|z|$, $\Im z > 0$,
(ii) $|f(x)| \leqslant M$, $-\infty < x < \infty$,
(iii) $\limsup_{y\to\infty}(\log|f(iy)|)/y = A$.

Then, for $\Im z \geqslant 0$,

$$|f(z)| \leqslant Me^{A\Im z}.$$

Remark. The *growth of f on the imaginary axis* is thus *enough* to control the exponential furnished by the conclusion, as long as $|f(z)|$ has at most *some* finite exponential growth in $\Im z > 0$.

The *proof* of this result is *exactly* like that of the preceding one. It is enough to put $u(z) = \log|f(z)/M|$ and then *copy* the preceding argument word for word.

Any sector of the form $0 < \arg z < \alpha$ or $\alpha < \arg z < \pi$ has opening $< \pi$. Looking at the reasoning used to establish the above two theorems, *we see that we can even replace* (iii) *in the hypothesis of the preceding one* by

(iii)′ $\limsup_{R\to\infty}(\log|f(Re^{i\alpha})|)/R\sin\alpha = A$ for some α, $0 < \alpha < \pi$,

and the same conclusion holds good.

Theorem. *Let $f(z)$ be analytic for $\Im z > 0$ and continuous for $\Im z \geqslant 0$. Suppose that $|f(z)| \leqslant Ce^{A|z|}$ for $\Im z > 0$, that $|f(x)|$ is bounded on the real axis, and that*

$$f(x) \to 0 \quad \text{as} \quad x \to \infty.$$

Then $f(x + iy) \to 0$ uniformly in each strip $0 \leqslant y \leqslant L$ as $x \to \infty$.

Proof. If, say, $|f(x)| \leqslant M$ on $\mathbb{R}$, we have $|f(z)| \leqslant Me^{A\Im z}$ for $\Im z \geqslant 0$ by the corollary preceding the above theorem. Take any $B > A$ and some large K, and look at the function

$$g_K(z) = \frac{z}{z + iK} e^{iBz} f(z)$$

in $\Im z > 0$. Since $B > A$ and $K > 0$, we have $|g_K(z)| \leqslant M$, $\Im z \geqslant 0$. We can, however, do better than this.

Given $\varepsilon > 0$, we can find a Y so large that $e^{-(B-A)Y} < \varepsilon/M$; take such a Y and fix it. Then,

$$|g_K(z)| \;\leqslant\; \left|\frac{z}{z + iK}\right| e^{-(B-A)\Im z} M \;<\; \varepsilon$$

for $\Im z \geqslant Y$ as long as $K > 0$, Choose now $X > 0$ so large that $|f(x)| < \varepsilon$ for $x \geqslant X$; this we can do because $f(x) \to 0$ as $x \to \infty$.

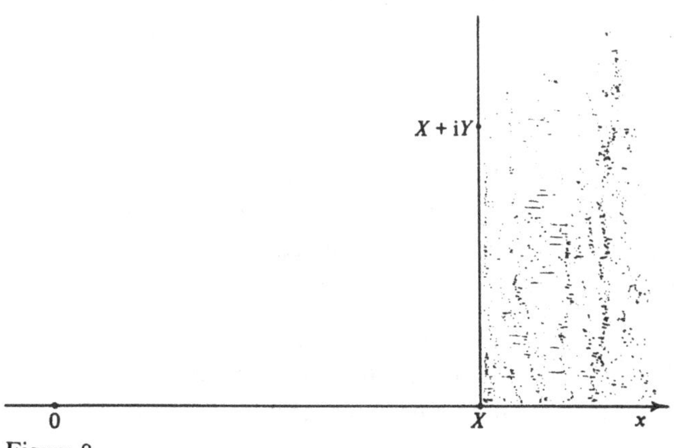

Figure 8

Having fixed X and Y, we now take K so large that $|(X+iy)/(X+iy+iK)|M < \varepsilon$ for $0 \leqslant y \leqslant Y$; fixing this K we will then have

$$|g_K(X+iy)| \leqslant \left|\frac{X+iy}{X+iy+iK}\right| e^{-(B-A)y} M < \varepsilon$$

for $0 \leqslant y \leqslant Y$. By choice of Y the same inequality also holds if $y \geqslant Y$. Finally, $|g_K(x)| \leqslant |f(x)| < \varepsilon$ for $x > X$.

We see that $|g_K(z)| < \varepsilon$ on the boundary of the quadrant $\{\Re z > X, \Im z > 0\}$. However, $|g_K(z)| \leqslant M$ *in* that quadrant, so, by the first theorem of this §, $|g_K(z)| < \varepsilon$ *throughout* it. Let then $\Re z > X$ and $\Im z > 0$. We have

$$|f(z)| = \left|\frac{z+iK}{z}\right| e^{B\Im z} |g_K(z)| < \left|\frac{z+iK}{z}\right| e^{B\Im z} \varepsilon.$$

Suppose that $0 \leqslant \Im z \leqslant L$. Then, if $\Re z > \max(X, K)$ we have, by the previous relation,

$$|f(z)| < 2e^{BL} \varepsilon.$$

Here, $\varepsilon > 0$ is arbitrary. Therefore $f(x+iy) \to 0$ uniformly for $0 \leqslant y \leqslant L$ as $x \to \infty$.

We are done.

D. The Paley–Wiener theorem

Theorem. *Let $f(z)$ be entire and of exponential type A. Suppose that*

$$\int_{-\infty}^{\infty} |f(x)|^2 \, dx < \infty.$$

Then there is a function $\varphi(\lambda) \in L_2(-A, A)$ with

$$f(z) = \frac{1}{2\pi} \int_{-A}^{A} e^{-iz\lambda} \varphi(\lambda) \, d\lambda.$$

Remark. If $f(z)$ is *given* by such an integral, it is obviously of exponential type A and belongs to $L_2(-\infty, \infty)$ on account of Plancherel's theorem. So the *converse of the theorem is evident*. The two results (the *theorem* and its *converse*) taken together constitute the celebrated and much used *Paley–Wiener theorem*.

Proof of theorem. Is essentially based on Plancherel's theorem, combined with contour integration and the third and fifth results of the previous §. An easy but rather fussy preliminary reduction is necessary.

Plancherel's theorem says that

$$\varphi(\lambda) = \underset{M\to\infty}{\text{l.i.m.}} \int_{-M}^{M} e^{i\lambda x} f(x)\,dx$$

exists and belongs to $L_2(-\infty,\infty)$, and that, for $x \in \mathbb{R}$,

$$f(x) = \frac{1}{2\pi} \underset{M\to\infty}{\text{l.i.m.}} \int_{-M}^{M} e^{-ix\lambda} \varphi(\lambda)\,d\lambda.$$

(Here, 'l.i.m.' stands for 'limit in mean (square)', and denotes a limit in $L_2(-\infty,\infty)$.) Our main task is to show that $\varphi(\lambda) \equiv 0$ a.e. for $\lambda > A$ and $\lambda < -A$.

To this end, let us introduce the function

$$f_h(z) = \frac{1}{2h} \int_{-h}^{h} f(z+t)\,dt;$$

$f_h(z)$ is clearly entire. Because $f(z)$ is of exponential type A, we have, for any $A' > A$,

$$|f(z)| \leqslant \text{const.}e^{A'|z|},$$

and from this it is clear that also

$$|f_h(z)| \leqslant \text{const.}e^{A'|z|}$$

(with a different constant).

By Schwarz' inequality we also have

$$|f_h(x)| \leqslant \sqrt{\left(\frac{1}{2h}\int_{x-h}^{x+h} |f(t)|^2\,dt\right)},$$

so, since $f(x) \in L_2(-\infty,\infty)$, $f_h(x)$ is *bounded on* $\mathbb{R}$ *and in fact* $f_h(x) \to 0$ for $x \to \pm\infty$. (This is the *main reason* for doing $(1/2h)\int_{-h}^{h}$ on f!) If we call $\sup_{x\in\mathbb{R}} |f_h(x)| = \tilde{M}_h$, we see by the previous inequality for $|f_h(z)|$ and the *third* theorem (p. 27) of the previous § (applied in *each* half plane $\Im z > 0$ and $\Im z < 0$), that

$$|f_h(z)| \leqslant \tilde{M}_h e^{A'|\Im z|}.$$

Here, A' can be *any* number $> A$, so in fact

$$(*) \qquad |f_h(z)| \leqslant \tilde{M}_h e^{A|\Im z|}.$$

The *fifth* theorem of § C (p. 29) shows moreover that

$(\ast\ast)$ $\qquad f_h(x+iy) \to 0$ uniformly for $-L \leqslant y \leqslant L$ when $x \to \pm\infty$.

In order to prove that

$$\varphi(\lambda) = \underset{M\to\infty}{\text{l.i.m.}} \int_{-M}^{M} e^{i\lambda x} f(x)\,dx$$

vanishes a.e. for $\lambda > A$ and for $\lambda < -A$ it is more than sufficient to show the same thing with $f(x)$ replaced by $f_h(x)$ in the right-hand integral, $h > 0$ being arbitrary. That's because

$$\underset{M\to\infty}{\text{l.i.m.}} \int_{-M}^{M} e^{i\lambda x} f_h(x)\,dx = \frac{\sin \lambda h}{\lambda h}\,\varphi(\lambda),$$

which we can check using Fubini's theorem and the fact that

$$\frac{1}{2h}\int_{-h}^{h} e^{i\lambda x}\,dx = \frac{\sin \lambda h}{\lambda h}.*$$

Taking a large M, look at $\int_{-M}^{M} e^{i\lambda x} f_h(x)\,dx$, assuming that $\lambda > A$. Let γ consist of the three *upper* pieces of the rectangular contour shown.

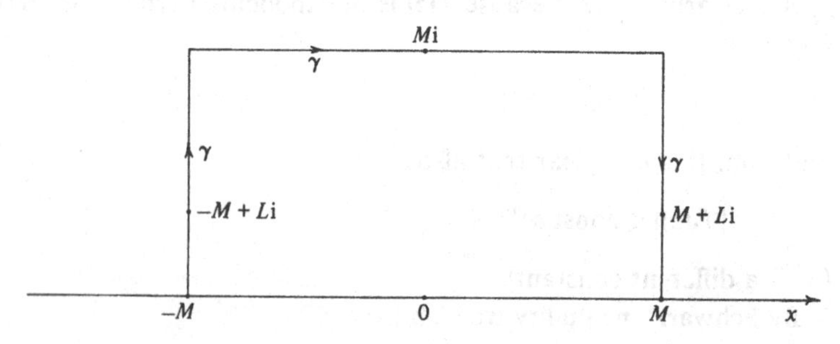

Figure 9

By Cauchy's theorem,

$$\int_{-M}^{M} e^{i\lambda x} f_h(x)\,dx = \int_{\gamma} e^{i\lambda z} f_h(z)\,dz.$$

The contribution to $\int_\gamma$ from the *top horizontal portion* of γ has absolute value

$$\left|\int_{-M}^{M} e^{i\lambda(x+iM)} f_h(x+iM)\,dx\right|,$$

and this, by (*), is

$$\leqslant 2M\cdot \tilde{M}_h e^{-(\lambda-A)M},$$

a quantity tending to zero as $M \to \infty$, since $\lambda > A$.

Fix any large number L. If $M > L$, we write the contribution to $\int_\gamma$ from

* To do this, one should start from the *second* formula at the top of p. 31 and conclude by applying Plancherel's theorem.

the *right-hand vertical portion* of γ as

$$-\left(\int_0^L + \int_L^M\right) e^{i\lambda(M+iy)} f_h(M+iy)\cdot i\,dy.$$

The *second* integral is in modulus

$$\leqslant \tilde{M}_h \int_L^\infty e^{-(\lambda-A)y}\,dy = \frac{e^{-(\lambda-A)L}}{\lambda-A}\tilde{M}_h,$$

again by (∗), and we can make the quantity on the right as small as we like by taking *L large*. The *first* integral, however, has modulus

$$\leqslant \int_0^L e^{-\lambda y}|f_h(M+iy)|\,dy$$

and this, for any fixed L, tends to 0 as $M\to\infty$ according to (⁑). We see that the contribution from the *right vertical portion* of γ *tends to zero* as $M\to\infty$; that of the *left vertical portion does the same*, as a similar argument shows.

In fine, $\int_\gamma e^{i\lambda z} f_h(z)\,dz \longrightarrow 0$ as $M\to\infty$, i.e.,

$$\int_{-M}^M e^{i\lambda x} f_h(x)\,dx \longrightarrow 0 \quad \text{as} \quad M\to\infty$$

when $\lambda > A$. For $\lambda < -A$ we establish the same result using a similar argument and this contour:

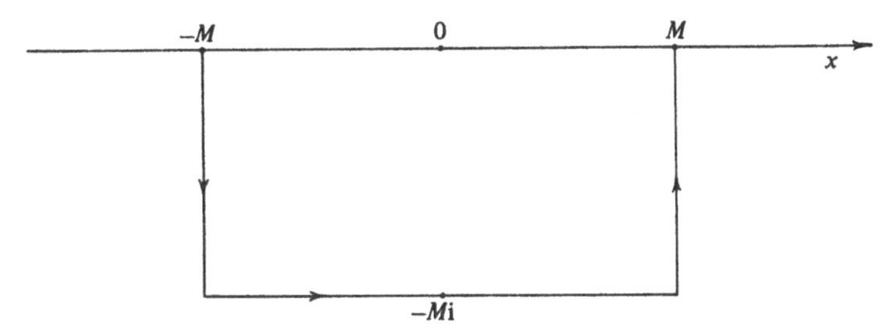

Figure 10

Thus, $\int_{-M}^M e^{i\lambda x} f_h(x)\,dx \longrightarrow 0$ *pointwise* in λ for $|\lambda| > A$ as $M\to\infty$. However, for *some sequence of Ms* tending to ∞, the integrals in question must tend a.e. to

$$\underset{M\to\infty}{\text{l.i.m.}} \int_{-M}^M e^{i\lambda x} f_h(x)\,dx = \frac{\sin\lambda h}{\lambda h}\varphi(\lambda).$$

(L_2 convergence of a *sequence* implies the a.e. *pointwise* convergence of some *subsequence to the same limit.*) This means that $(\sin\lambda h/\lambda h)\varphi(\lambda) = 0$

a.e. for $|\lambda| > A$, whence $\varphi(\lambda) = 0$ a.e. for $|\lambda| > A$. ($\sin \lambda h/\lambda h$ vanishes only on a countable set of points!)

The Fourier–Plancherel inversion formula now gives us, for $x \in \mathbb{R}$,

$$f(x) = \underset{M\to\infty}{\text{l.i.m.}} \frac{1}{2\pi}\int_{-M}^{M} e^{-ix\lambda}\varphi(\lambda)\,d\lambda$$

$$= \frac{1}{2\pi}\int_{-A}^{A} e^{-ix\lambda}\varphi(\lambda)\,d\lambda \quad \text{a.e.}$$

In fact, we have

$$f(z) = \frac{1}{2\pi}\int_{-A}^{A} e^{-iz\lambda}\varphi(\lambda)\,d\lambda$$

for all complex z. That's because *each* of the two sides is an *entire* function of z. Since these two entire functions coincide a.e. on $\mathbb{R}$, they must be everywhere equal by the identity theorem for analytic functions. Our theorem is completely proved.

If we refer to the *fourth* theorem of §C (p. 28), we see that we can give the result just proved a more general formulation. The statement thus obtained, which we give as a corollary, also goes under the names of Paley and Wiener.

Corollary. *Let $f(z)$ be entire and of (some) exponential type, with $f(x) \in L_2(\mathbb{R})$, and let*

$$\limsup_{y\to\infty} \frac{\log|f(iy)|}{y} = b,$$

$$\limsup_{y\to-\infty} \frac{\log|f(iy)|}{|y|} = -a.$$

Then

$$f(z) = \frac{1}{2\pi}\int_{a}^{b} e^{-iz\lambda}\varphi(\lambda)\,d\lambda,$$

where $\varphi \in L_2(a, b)$.

Proof. If f is of exponential type A, say, we certainly have

$$f(z) = \frac{1}{2\pi}\int_{-A}^{A} e^{-iz\lambda}\varphi(\lambda)\,d\lambda$$

by the theorem, so, if $x \in \mathbb{R}$,

$$|f(x)| \leqslant \frac{1}{2\pi}\sqrt{\left(2A\int_{-A}^{A} |\varphi(\lambda)|^2\,d\lambda\right)},$$

a *finite quantity*, i.e., f is *bounded on* $\mathbb{R}$ if we only assume $f \in L_2(\mathbb{R})$, *provided* that it is *entire* and of *exponential type*. Applying the fourth theorem of §C in each of the half planes $\Im z > 0$, $\Im z < 0$, we now see that

$$|f(z)| \leqslant \text{const.}e^{b\Im z}, \qquad \Im z > 0;$$
$$|f(z)| \leqslant \text{const.}e^{-a|\Im z|}, \quad \Im z < 0.$$

Symmetrize by taking $g(z) = e^{i\gamma z}f(z)$ with $\gamma = (b+a)/2$. Then $g(z)$ is also entire, $g(x) \in L_2(\mathbb{R})$, and, by the previous relations,

$$|g(z)| \leqslant \text{const.}e^{(b-a)|\Im z|/2}$$

in *both* upper and lower half planes, i.e., $g(z)$ is of *exponential type* $(b-a)/2$. (We see at this point that $(b-a)/2$ *cannot be* $\leqslant 0$ *unless* $f(z) \equiv 0$ – the reader is urged to think out why this is so.) Use the above theorem once more, this time for $g(z)$. We find

$$g(z) = \frac{1}{2\pi}\int_{-(1/2)(b-a)}^{(1/2)(b-a)} e^{-iz\lambda}\psi(\lambda)\,d\lambda$$

with a certain $\psi \in L_2$. Going back to $f(z) = e^{-i\gamma z}g(z)$, we have

$$f(z) = \frac{1}{2\pi}\int_a^b e^{-iz\lambda}\psi(\lambda-\gamma)\,d\lambda,$$

establishing the result with $\varphi(\lambda) = \psi(\lambda - \gamma)$.

Scholium. The Paley–Wiener theorem has more content than meets the eye. Suppose that $\varphi \in L_2(a,b)$; then

$$f(z) = \frac{1}{2\pi}\int_a^b e^{-iz\lambda}\varphi(\lambda)\,d\lambda$$

is entire, of exponential type, and belongs to L_2 on $\mathbb{R}$. We can also easily verify directly that

$$\limsup_{y\to\infty}\frac{\log|f(x+iy)|}{y} \leqslant b$$

and

$$\limsup_{y\to-\infty}\frac{\log|f(x+iy)|}{|y|} \leqslant -a$$

for each real x.

These inequalities remain true as long as φ vanishes a.e. outside $[a,b]$. If, however, we take for $[a,b]$ the *smallest closed interval containing* φ*'s support – the so-called supporting interval for* φ *– the inequalities become equalities!*

Without loss of generality, take $x = 0$, and suppose, for instance, that

$$\limsup_{y\to\infty} \frac{\log|f(iy)|}{y} = b' < b.$$

The above corollary then shows that

$$\varphi(\lambda) = \underset{M\to\infty}{\text{l.i.m.}} \int_{-M}^{M} e^{i\lambda x} f(x)\,dx$$

in fact vanishes a.e. for $\lambda > b'$. The support of φ would thus be contained in $[a, b']$, so $[a, b]$ would not be φ's supporting interval, and we have a contradiction.

If $[a, b]$ *is* the supporting interval of φ, we must therefore have

$$\limsup_{y\to\infty} \frac{\log|f(iy)|}{y} = b.$$

It is clear that this b can only come from the portions

$$\frac{1}{2\pi}\int_{b-\varepsilon}^{b} e^{-iz\lambda}\varphi(\lambda)\,d\lambda$$

of the integral giving $f(z)$, $\varepsilon > 0$. We *know* that $\varphi(\lambda)$ cannot vanish identically a.e. on any interval of the form $[b - \varepsilon, b]$, but it is *still quite conceivable that*

$$\left|\int_{b-\varepsilon}^{b} e^{y\lambda}\varphi(\lambda)\,d\lambda\right|$$

could come out much smaller than e^{by} *for large y on account of cancellation.* The Paley–Wiener theorem teaches us that *such cancellation cannot take place.* This is a remarkable and deep property of (square) integrable functions $\varphi(\lambda)$.

There are versions of the Paley–Wiener theorem for other spaces besides $L_2(\mathbb{R})$. The following is frequently used.

Theorem. *Let* $\varphi(\lambda) \in L_1(\mathbb{R})$ *have compact support, put*

$$f(z) = \frac{1}{2\pi}\int_{-\infty}^{\infty} e^{-iz\lambda}\varphi(\lambda)\,d\lambda,$$

and suppose that

$$\limsup_{y\to\infty} \frac{\log|f(iy)|}{y} = b, \quad \limsup_{y\to-\infty} \frac{\log|f(iy)|}{|y|} = -a.$$

Then $\varphi(\lambda)$ *vanishes a.e. outside* $[a, b]$.

Proof. This would be part of the corollary to the Paley–Wiener theorem, save that $f(x)$ is not necessarily in $L_2(\mathbb{R})$. For $h>0$, put

$$\varphi_h(\lambda)=\frac{1}{2h}\int_{-h}^{h}\varphi(\lambda+\tau)\,\mathrm{d}\tau;$$

then $\|\varphi_h-\varphi\|_1\to 0$ as $h\to 0$ and we need only show that $\varphi_h(\lambda)\equiv 0$ for $\lambda\notin[a-h,b+h]$.

Write

$$f_h(z)=\frac{1}{2\pi}\int_{-\infty}^{\infty}\mathrm{e}^{-\mathrm{i}z\lambda}\varphi_h(\lambda)\,\mathrm{d}\lambda;$$

then

$$f_h(z)=\frac{\sin hz}{hz}f(z),$$

so, since $f(x)$ is clearly *bounded* on $\mathbb{R}$ (φ being in $L_1(\mathbb{R})$), $f_h(x)\in L_2(\mathbb{R})$. By the hypothesis we now have

$$\limsup_{y\to\infty}\frac{\log|f_h(\mathrm{i}y)|}{y}=b+h,$$

$$\limsup_{y\to-\infty}\frac{\log|f_h(\mathrm{i}y)|}{|y|}=-a+h.$$

Therefore $\varphi_h(\lambda)\equiv 0$ outside $[a-h,b+h]$, by Paley–Wiener. We're done.

Remark. The same result holds (with almost the same proof) if we replace $\varphi(\lambda)\,\mathrm{d}\lambda$ (with $\varphi\in L_1$ and of compact support) by $\mathrm{d}\mu(\lambda)$, μ being any *finite signed measure* of compact support.

E. Introduction to the condition $\int_{-\infty}^{\infty}(\log^+|f(x)|/(1+x^2))\,\mathrm{d}x<\infty$

The entire functions of exponential type considered in the previous § certainly satisfy this condition, as do those arising in the study of many questions in analysis. We will meet repeatedly with such functions in the following chapters of this book, and the rest of the present chapter is mainly concerned with them. It turns out that the boundedness condition

$$\int_{-\infty}^{\infty}\frac{\log^+|f(x)|}{1+x^2}\mathrm{d}x \quad<\quad\infty$$

implies many results for entire functions f of exponential type.

The following simple result is very useful, and all that one needs for many investigations.

Theorem. *Let $f(z)$ be regular in $\Im z > 0$ and continuous up to the real axis. Suppose that $\log|f(z)| \leqslant O(|z|)$ for $|z|$ large when $\Im z > 0$, that*

$$\limsup_{y\to\infty} \frac{\log|f(iy)|}{y} = A,$$

and that

$$\int_{-\infty}^{\infty} \frac{\log^+|f(x)|}{1+x^2}dx < \infty.$$

Then, for $\Im z > 0$,

$$\boxed{\log|f(z)| \leqslant A\Im z + \frac{1}{\pi}\int_{-\infty}^{\infty} \frac{\Im z \log^+|f(t)|}{|z-t|^2}dt.}$$

Proof. $\Im z/|z-t|^2 = \Re(i/(z-t))$ is, for each $t\in\mathbb{R}$, a *positive and harmonic function* in $\Im z > 0$. For fixed z with positive real part we have, by calculus,

$$\frac{1}{\pi}\int_{-\infty}^{\infty} \frac{\Im z}{|z-t|^2}dt = 1,$$

and, if $z\to x_0\in\mathbb{R}$,

$$\sup_{|t-x_0|\geqslant\delta}\left\{\frac{\Im z}{|z-t|^2}\bigg/\frac{1}{t^2+1}\right\}\longrightarrow 0$$

for each $\delta > 0$. Therefore, if $P(t)$ is any positive continuous function with

$$\int_{-\infty}^{\infty} \frac{P(t)}{1+t^2}dt < \infty,$$

we have by the usual elementary approximate identity argument (no need to refer to Chapter II, §B, here!),

$$\frac{1}{\pi}\int_{-\infty}^{\infty} \frac{\Im z}{|z-t|^2}P(t)dt \longrightarrow P(x_0)$$

for $z\to x_0\in\mathbb{R}$.

In our present situation $P(t) = \log^+|f(t)|$ is continuous on $\mathbb{R}$, so if we put

$$U(z) = \frac{1}{\pi}\int_{-\infty}^{\infty} \frac{\Im z \log^+|f(t)|}{|z-t|^2}dt$$

for $\Im z > 0$, $U(z)$ is *positive and harmonic* in the upper half plane and

$U(z) \to \log^+|f(x_0)|$ for $z \to x_0 \in \mathbb{R}$. We see that in $\Im z > 0$, $\log|f(z)| - U(z)$ is *subharmonic*, is $\leqslant O(|z|)$ for large $|z|$, and *has boundary values* $\leqslant 0$ *everywhere on* $\mathbb{R}$. Moreover,

$$\log|f(\mathrm{i}y)| - U(\mathrm{i}y) \leqslant Ay + o(y)$$

for $y \to \infty$. The *fourth* theorem of §C (p. 28) (or rather its version for subharmonic functions) now yields without further ado

$$\log|f(z)| - U(z) \leqslant A\Im z, \quad \Im z > 0,$$

that is,

$$\log|f(z)| \leqslant A\Im z + U(z), \quad \Im z > 0.$$

We are done.

Later on we will give some refined versions of this result. Their derivation requires more effort.

F. Representation of positive harmonic functions as Poisson integrals

In order to proceed further with the discussion begun in §E, it is simplest to apply the Riesz–Evans–Herglotz representation for positive harmonic functions, although its use can in fact be avoided. We explain that representation here, together with some of its function-theoretic consequences.

1. The representation

Theorem. *Let $V(w)$ be positive and harmonic for $|w| < 1$. There is a finite positive measure ν on $[-\pi, \pi]$ with*

$$V(w) = \frac{1}{2\pi}\int_{-\pi}^{\pi} \frac{1 - |w|^2}{|w - \mathrm{e}^{\mathrm{i}t}|^2} \mathrm{d}\nu(t), \quad |w| < 1.$$

Sketch of Proof. By the ordinary Poisson formula, if $R < 1$, we have, for $|w| < R$,

$$V(w) = \frac{1}{2\pi}\int_{-\pi}^{\pi} \frac{R^2 - |w|^2}{|w - R\mathrm{e}^{\mathrm{i}t}|^2} V(R\mathrm{e}^{\mathrm{i}t})\mathrm{d}t,$$

that is, for $|w| < 1$,

$$(*) \qquad V(Rw) = \frac{1}{2\pi}\int_{-\pi}^{\pi} \frac{1 - |w|^2}{|w - \mathrm{e}^{\mathrm{i}t}|^2} V(R\mathrm{e}^{\mathrm{i}t})\mathrm{d}t.$$

In particular,

$$(\ddagger)\qquad \frac{1}{2\pi}\int_{-\pi}^{\pi} V(Re^{it})dt = V(0) < \infty,$$

no matter how close $R < 1$ is to 1.

We must now use some version of Tychonoff's theorem in order to obtain the measure ν.

Take any sequence $\{R_n\}$ tending monotonically to 1, for example $R_n = 1 - 1/n$. The functions $\varphi_n(t) = V(R_n e^{it})$ are all $\geqslant 0$ and have bounded integrals over $[-\pi, \pi]$ by ($\ddagger$); we can therefore (by using Cantor's diagonal process) extract a *subsequence of these functions*, which we also denote by $\{\varphi_n\}$ (so as not to write subscripts of subscripts!), having the property that

$$L(G) = \lim_{n\to\infty}\int_{-\pi}^{\pi} G(t)\varphi_n(t)dt$$

exists and is finite for G ranging over a *countable dense subset* of $\mathscr{C}(-\pi, \pi)$.

If, however, G and $G' \in \mathscr{C}(-\pi, \pi)$ and $\|G - G'\| < \varepsilon$, we have, for *every* n,

$$\left|\int_{-\pi}^{\pi} G(t)\varphi_n(t)dt - \int_{-\pi}^{\pi} G'(t)\varphi_n(t)dt\right| \leqslant \int_{-\pi}^{\pi} \varepsilon\varphi_n(t)dt = 2\pi\varepsilon V(0),$$

so *in fact* $L(G) = \lim_{n\to\infty}\int_{-\pi}^{\pi} G(t)\varphi_n(t)dt$ exists for *every* $G \in \mathscr{C}(-\pi, \pi)$, and $|L(G)| \leqslant 2\pi V(0)\|G\|$.

L is thus a bounded linear functional on $\mathscr{C}(-\pi, \pi)$; it is moreover positive because, if $G \in \mathscr{C}(-\pi, \pi)$ and $G \geqslant 0$, $L(G) \geqslant 0$ since $\varphi_n(t) \geqslant 0$ for each n. By the Riesz representation theorem there is thus a positive finite measure ν on $[-\pi, \pi]$ with

$$L(G) = \int_{-\pi}^{\pi} G(t)d\nu(t), \quad G \in \mathscr{C}(-\pi, \pi).$$

Taking in particular $G(t) = (1/2\pi)(1 - |w|^2)/|w - e^{it}|^2$ with a fixed w, $|w| < 1$, we obtain

$$\frac{1}{2\pi}\int_{-\pi}^{\pi}\frac{1 - |w|^2}{|w - e^{it}|^2}d\nu(t) = L(G)$$

$$= \lim_{n\to\infty}\frac{1}{2\pi}\int_{-\pi}^{\pi}\frac{1 - |w|^2}{|w - e^{it}|^2}\varphi_n(t)dt$$

$$= \lim_{n\to\infty}\frac{1}{2\pi}\int_{-\pi}^{\pi}\frac{1 - |w|^2}{|w - e^{it}|^2}V(R_n e^{it})dt.$$

Referring to ($*$), we see that the last expression equals $\lim_{n\to\infty} V(R_n w) = V(w)$, V being certainly *continuous* for $|w| < 1$.

This completes the proof.

Scholium. Once we *know* that the measure ν giving rise to the desired representation *exists*, we see that the *passage to a subsequence* of the $\varphi_n(t)$ in the above proof was not really *necessary* (although we are only able to see this once the proof has been carried out!).

Suppose $G(t)$ is any continuous function on $[-\pi,\pi]$ with $G(-\pi)=G(\pi)$; then, by the elementary approximate identity property of the Poisson kernel

$$\frac{1}{2\pi}\frac{1-|w|^2}{|w-e^{it}|^2},$$

$$\frac{1}{2\pi}\int_{-\pi}^{\pi}\frac{1-R^2}{|e^{it}-Re^{i\vartheta}|^2}G(\vartheta)d\vartheta \longrightarrow G(t)$$

uniformly for $-\pi\leqslant t\leqslant\pi$ as $R\to 1$, so, by Fubini's theorem,

$$\int_{-\pi}^{\pi}G(\vartheta)V(Re^{i\vartheta})d\vartheta = \frac{1}{2\pi}\int_{-\pi}^{\pi}\int_{-\pi}^{\pi}\frac{1-R^2}{|e^{it}-Re^{i\vartheta}|^2}G(\vartheta)d\vartheta d\nu(t)$$
$$\longrightarrow \int_{-\pi}^{\pi}G(t)\,d\nu(t)$$

as $R\to 1$. This simple fact can frequently be used to get information about the measure ν.

The reader should think through what happens with the argument just given when $G\in\mathscr{C}(-\pi,\pi)$ but $G(-\pi)\neq G(\pi)$. Here is a hint: we at least have

$$\frac{1}{2\pi}\int_{-\pi}^{\pi}\frac{1-R^2}{|e^{it}-Re^{i\vartheta}|^2}G(\vartheta)d\vartheta \longrightarrow \begin{cases} G(t), & t\neq -\pi,\pi,\\ \dfrac{G(\pi)+G(-\pi)}{2}, & t=-\pi,\pi,\end{cases}$$

as $R\to 1$, although the convergence is no longer uniform. The integrals on the left are, however, bounded.

Terminology. The situation of the scholium is frequently described by saying that

$$V(Re^{i\vartheta})d\vartheta \longrightarrow d\nu(\vartheta) \quad w^*$$

for $R\to 1$, or by writing

$$\text{‘}V(Re^{i\vartheta}) \rightharpoonup d\nu(\vartheta) \quad \text{as} \quad R\to 1\text{’}$$

(with a *half arrow*).

Theorem. *Let $v(z)$ be positive and harmonic in $\Im z>0$. There is a positive number α and a positive measure μ on $\mathbb{R}$ with*

$$\int_{-\infty}^{\infty}\frac{d\mu(t)}{1+t^2}<\infty$$

such that

$$v(z) = \alpha \Im z + \frac{1}{\pi}\int_{-\infty}^{\infty} \frac{\Im z}{|z-t|^2}\,d\mu(t) \quad \textit{for} \quad \Im z > 0.$$

Proof. From the previous theorem by making the change of variable

$$z \to w = \frac{i - z}{i + z}$$

which takes $\Im z > 0$ conformally onto the open unit disk.

Everybody should do this calculation at least once in his or her life, so let us give the good example. The conformal mapping just described takes $v(z)$ to a positive harmonic function $V(w) = v(z)$ defined for $|w| < 1$, so we have

$$V(w) = \frac{1}{2\pi}\int_{-\pi}^{\pi} \frac{1 - |w|^2}{|e^{i\tau} - w|^2}\,d\nu(\tau)$$

with a positive measure ν according to the result just proved. We write τ here because t will denote a variable *running along the real axis.*

We have $w = (i - z)/(i + z)$, and the real t corresponds in a similar way to

$$e^{i\tau} = \frac{i - t}{i + t}.$$

Therefore

$$\frac{1 - |w|^2}{|e^{i\tau} - w|^2} = \frac{1 - \left|\dfrac{i - z}{i + z}\right|^2}{\left|\dfrac{i - t}{i + t} - \dfrac{i - z}{i + z}\right|^2}$$

$$= \frac{(|i + z|^2 - |i - z|^2)|i + t|^2}{|(i - t)(i + z) - (i + t)(i - z)|^2} = \frac{4\Im z(t^2 + 1)}{|2i(z - t)|^2}$$

$$= \frac{\Im z}{|z - t|^2}(1 + t^2).$$

Since $e^{\pm i\pi} = -1$ corresponds to $t = \infty$, we see that

$$\frac{1}{2\pi}\int_{(-\pi,\pi)} \frac{1 - |w|^2}{|e^{i\tau} - w|^2}\,d\nu(\tau) = \frac{1}{\pi}\int_{-\infty}^{\infty} \frac{\Im z}{|z - t|^2}\,d\mu(t),$$

where $d\mu(t) = \frac{1}{2}(1 + t^2)\,d\nu(\tau)$.

We are finally left with the (possible) point masses coming from ν at $-\pi$ and π; their contribution gives us the term $\alpha\Im z$ with $\alpha \geqslant 0$. Recalling that $v(z) = V(w)$, we see that the proof is complete.

Remark. If $\Phi(t)$ is a continuous function of compact support,

$$\int_{-\infty}^{\infty} \Phi(t)\mathrm{d}\mu(t) = \lim_{y \to 0+} \int_{-\infty}^{\infty} \Phi(t)v(t+\mathrm{i}y)\mathrm{d}t.$$

To see this, just use the approximate identity property of $(1/\pi)(\Im z/|z-t|^2)$ (§E) – compare with the above scholium.

2. Digression on the a.e. existence of boundary values

The representation derived in the preceding article can be combined with the result in Chapter II, §B, to obtain some theorems about the a.e. existence of (non-tangential) finite boundary values for certain classes of harmonic and analytic functions defined in $\{|w|<1\}$ or in $\{\Im z > 0\}$. Although this is *not* a book about boundary behaviour or H_p spaces, it is perhaps a good idea to show here how such results are deduced, especially since that can be done with so little additional effort.

Theorem. *Let $V(w)$ be positive and harmonic in $\{|w|<1\}$. Then, for almost every ϑ, the non-tangential boundary value*

$$\lim_{w \not\angle\to e^{i\vartheta}} V(w)$$

exists and is finite.

Proof. By the previous article,

$$V(w) = \frac{1}{2\pi}\int_{-\pi}^{\pi} \frac{1-|w|^2}{|w-\mathrm{e}^{\mathrm{i}\tau}|^2}\mathrm{d}\nu(\tau)$$

where ν is a finite positive measure. A theorem of Lebesgue says that

$$\nu'(\vartheta) = \frac{\mathrm{d}}{\mathrm{d}\vartheta}\left(\int_0^{\vartheta} \mathrm{d}\nu(\tau)\right)$$

exists and is finite a.e. And by the remark at the end of Chapter 2, §B, $V(w) \to \nu'(\vartheta)$ as $w \not\angle\to \mathrm{e}^{\mathrm{i}\vartheta}$ wherever $\nu'(\vartheta)$ exists and is finite. We're done.

Corollary (Fatou). *Let $F(w)$ be analytic and bounded for $|w|<1$. Then*

$$\lim_{w \not\angle\to e^{i\vartheta}} F(w)$$

exists for almost all ϑ.

Proof. If $|F(w)| < M$ in $\{|w|<1\}$, $M + \Re F(w)$ and $M + \Im F(w)$ are both positive and harmonic there.

Notation. Let $F(w)$ be analytic and bounded for $\{|w|<1\}$. The *non-tangential limit*

$$\lim_{w \not\angle \to e^{i\vartheta}} F(w)$$

(which, by the corollary, exists a.e.) *is denoted by* $F(e^{i\vartheta})$. The function $F(e^{i\vartheta})$, thus defined a.e., is Lebesgue measurable (and, of course, bounded).

Theorem. *Let $F(w)$ be analytic and bounded for $|w|<1$. Then*

$$F(w) = \frac{1}{2\pi}\int_{-\pi}^{\pi} \frac{F(e^{i\vartheta})e^{i\vartheta}d\vartheta}{e^{i\vartheta}-w}, \quad |w|<1.$$

Remark. Thus, the boundary values $F(e^{i\vartheta})$ (which are defined a.e.) serve to recover $F(w)$.

Proof. For each $R<1$, we have, by Cauchy's theorem,

$$F(Rw) = \frac{1}{2\pi}\int_{-\pi}^{\pi} \frac{F(Re^{i\vartheta})e^{i\vartheta}d\vartheta}{e^{i\vartheta}-w}, \quad |w|<1.$$

Fix w, and take $R=R_n$ with $R_n \xrightarrow[n]{} 1$. We have $|F(R_n e^{i\vartheta})| \leqslant M$, say, and $F(R_n e^{i\vartheta}) \xrightarrow[n]{} F(e^{i\vartheta})$ a.e. by the corollary. The result follows by Lebesgue's dominated convergence theorem.

Lemma. *Let $F(w)$ be analytic for $|w|<1$ and suppose that $|F(w)| \leqslant 1$ there. Let $|\alpha|=1$, and take*

$$E = \{\vartheta : F(e^{i\vartheta}) = \alpha, \ -\pi < \vartheta \leqslant \pi\}.$$

*Then, unless $F(w) \equiv \alpha$, $|E| = 0$.**

Remark. The result is also true when $|\alpha|<1$. But then the proof is more difficult.

Proof of lemma. Take, wlog, $\alpha = 1$. We must then prove that $F(w) \equiv 1$ if $|E|>0$.

The function $((F(w)+1)/2)^n$ is analytic in $\{|w|<1\}$ and in modulus $\leqslant 1$ there, so, by the above theorem applied to it,

$$\left(\frac{F(0)+1}{2}\right)^n = \frac{1}{2\pi}\int_{-\pi}^{\pi}\left(\frac{F(e^{i\vartheta})+1}{2}\right)^n d\vartheta.$$

Here, $(F(e^{i\vartheta})+1)/2 = 1$ if $\vartheta \in E$, and, if ϑ, $-\pi<\vartheta\leqslant\pi$, is *not* in E, $|(F(e^{i\vartheta})+1)/2|<1$, so $((F(e^{i\vartheta})+1)/2)^n \xrightarrow[n]{} 0$. We see by bounded conver-

* We follow the customary practice of denoting the Lebesgue measure of $E \subseteq \mathbb{R}$ by $|E|$.

gence that

$$\frac{1}{2\pi}\int_{-\pi}^{\pi}\left(\frac{F(e^{i\vartheta})+1}{2}\right)^n d\vartheta = \frac{|E|}{2\pi} + o(1)$$

for $n \to \infty$. Suppose $|E| > 0$. Then the last relation combines with the previous to yield

$$\frac{F(0)+1}{2} = \sqrt[n]{\frac{|E|}{2\pi} + o(1)}, \quad n \to \infty,$$

after extracting an nth root. Since the right side tends to 1 for $n \to \infty$ we have finally $F(0) = 1$.

However, $|F(w)| \leqslant 1$, $|w| < 1$. Therefore $F(w) \equiv 1$ there by the strong maximum principle, Q.E.D.

Theorem. *Let $f(t) \in L_1(-\pi, \pi)$, and put, for $|w| < 1$,*

$$G(w) = \frac{1}{2\pi}\int_{-\pi}^{\pi} \frac{e^{it}+w}{e^{it}-w} f(t)dt.$$

Then $\lim_{w \not\to e^{i\vartheta}} G(w)$ exists and is finite a.e.

Proof. Wlog, $f(t) \geqslant 0$. Notice that $G(w)$ is certainly analytic in $\{|w| < 1\}$, and that

$$\Re G(w) = \frac{1}{2\pi}\int_{-\pi}^{\pi} \frac{1-|w|^2}{|w-e^{it}|^2} f(t)dt$$

is $\geqslant 0$ there. (Compare Chapter II, §A!)

The function

$$F(w) = \frac{G(w)-1}{G(w)+1}$$

is therefore analytic and in modulus $\leqslant 1$ for $|w| < 1$. So, by a previous corollary,

$$F(e^{i\vartheta}) = \lim_{w \not\to e^{i\vartheta}} F(w)$$

exists a.e. It follows that, whenever this limit exists,

$$\lim_{w \not\to e^{i\vartheta}} G(w)$$

must *also* exist and equal the finite quantity

$$\frac{1+F(e^{i\vartheta})}{1-F(e^{i\vartheta})}$$

unless $F(e^{i\vartheta}) = 1$.

However, $F(e^{i\vartheta})$ *can equal* 1 *only on a set of measure zero by the lemma–otherwise* $G(w)$ would equal ∞ everywhere in $\{|w|<1\}$, which is absurd. So $\lim_{w \not\to e^{i\vartheta}} G(w)$ exists and is finite a.e., as required.

Scholium. Write $w = re^{i\vartheta}$ and *suppose that* $f \in L_1(-\pi,\pi)$ *is real-valued.* Then we have

$$\frac{1}{2\pi}\int_{-\pi}^{\pi}\frac{e^{it}+w}{e^{it}-w}f(t)dt$$
$$= \frac{1}{2\pi}\int_{-\pi}^{\pi}\frac{1-r^2}{1+r^2-2r\cos(\vartheta-t)}f(t)dt$$
$$+ \frac{i}{2\pi}\int_{-\pi}^{\pi}\frac{2r\sin(\vartheta-t)}{1+r^2-2r\cos(\vartheta-t)}f(t)dt.$$

We see that both

$$U(re^{i\vartheta}) = \frac{1}{2\pi}\int_{-\pi}^{\pi}\frac{1-r^2}{1+r^2-2r\cos(\vartheta-t)}f(t)dt$$

and

$$\tilde{U}(re^{i\vartheta}) = \frac{1}{2\pi}\int_{-\pi}^{\pi}\frac{2r\sin(\vartheta-t)}{1+r^2-2r\cos(\vartheta-t)}f(t)dt$$

are *harmonic* in $\{|w|<1\}$, $U(w)$ being equal to $\Re G(w)$ there, and $\tilde{U}(w)$ equal to $\Im G(w)$, with $G(w)$ the analytic function considered in the above theorem.

$\tilde{U}(w)$ is frequently called a *harmonic conjugate* to $U(w)$; it has the property that $U(w) + i\tilde{U}(w)$ is *analytic* in $\{|w|<1\}$. It is an easy exercise to see that any two harmonic conjugates to $U(w)$ must differ by a constant; the particular one we are considering has the property that

$$\tilde{U}(0) = 0.$$

By Chapter II, §B, we *already know* that

$$\lim_{w \not\to e^{i\vartheta}} U(w)$$

exists and is finite a.e.; it is *in fact equal to* $f(\vartheta)$ almost everywhere. *The above theorem now tells us that* $\lim_{w \not\to e^{i\vartheta}} \tilde{U}(w)$ *also exists and is finite a.e., indeed, under the present circumstances,* $\tilde{U}(w) = \Im G(w)$. This conclusion is so important that it should be stated as a separate

Theorem. *Let $f \in L_1(-\pi, \pi)$. Then, for almost every φ, the limit of*

$$\frac{1}{2\pi}\int_{-\pi}^{\pi} \frac{2r\sin(\vartheta - t)}{1 + r^2 - 2r\cos(\vartheta - t)} f(t)\,dt$$

exists and is finite for $re^{i\vartheta} \xrightarrow{\not\,\angle} e^{i\varphi}$

Notation. The non-tangential limit in question is frequently denoted by $\tilde{f}(\varphi)$; for obvious reasons we often call $\tilde{f}$ the *harmonic conjugate* of f. It is also called the *Hilbert transform* of f. We will come back to the consideration of $\tilde{f}$ later on in this chapter.

G. Return to the subject of §E

1. Functions without zeros in $\Im z > 0$

Theorem. *Let $f(z)$ be analytic in $\Im z > 0$ and at the points of the real axis. Suppose that*

$$\log|f(z)| \leqslant O(|z|)$$

for $\Im z \geqslant 0$ and $|z|$ large, and that

$$\int_{-\infty}^{\infty} \frac{\log^+|f(x)|}{1 + x^2}\,dx \;<\; \infty.$$

Then, if $f(z)$ has no zeros in $\Im z > 0$,

$$\log|f(z)| = A\Im z + \frac{1}{\pi}\int_{-\infty}^{\infty} \frac{\Im z \log|f(t)|}{|z - t|^2}\,dt$$

there, where

$$A = \limsup_{y \to \infty} \frac{\log|f(iy)|}{y}.$$

Remark. $f(z)$ is *allowed* to have zeros *on* $\mathbb{R}$.

Proof of theorem. With

$$U(z) = \frac{1}{\pi}\int_{-\infty}^{\infty} \frac{\Im z \log^+|f(t)|}{|z - t|^2}\,dt$$

we have by §E

$$\log|f(z)| - A\Im z - U(z) \;\leqslant\; 0$$

for $\Im z > 0$. Since $f(z)$ has no zeros in $\Im z > 0$, $v(z) = \log|f(z)| - A\Im z - U(z)$ is *harmonic* there, and we have, by §F.1,

$$v(z) = -\alpha\Im z - \frac{1}{\pi}\int_{-\infty}^{\infty} \frac{\Im z}{|z-t|^2}\,\mathrm{d}\mu(t)$$

for $\Im z > 0$, where $\alpha \geqslant 0$ and μ is a positive measure on $\mathbb{R}$.

We use the remark at the end of §F.1 to obtain the description of μ. According to that remark, if $\Phi(t)$ is *continuous and of compact support*,

$$-\int_{-\infty}^{\infty} \Phi(t)\,\mathrm{d}\mu(t) = \lim_{y\to 0+}\int_{-\infty}^{\infty} \Phi(t)v(t+\mathrm{i}y)\,\mathrm{d}t.$$

In view of the formula for $U(z)$, we also have

$$\int_{-\infty}^{\infty} \Phi(t)\log^+|f(t)|\mathrm{d}t = \lim_{y\to 0+}\int_{-\infty}^{\infty} \Phi(t)U(t+\mathrm{i}y)\,\mathrm{d}t.$$

Therefore

$$\int_{-\infty}^{\infty} \Phi(t)(\log^+|f(t)|\mathrm{d}t - \mathrm{d}\mu(t))$$

$$= \lim_{y\to 0+}\int_{-\infty}^{\infty} \Phi(t)(U(t+\mathrm{i}y) + v(t+\mathrm{i}y))\,\mathrm{d}t$$

$$= \lim_{y\to 0+}\int_{-\infty}^{\infty} \Phi(t)(\log|f(t+\mathrm{i}y)| - Ay)\,\mathrm{d}t.$$

Under the hypothesis of the present theorem (analyticity *up to* and *on* $\mathbb{R}$), the last limit is just

$$\int_{-\infty}^{\infty} \Phi(t)\log|f(t)|\,\mathrm{d}t.$$

Indeed, we easily verify directly (using dominated convergence) that

$$\int_J \big|\log|f(t+\mathrm{i}y)| - \log|f(t)|\big|\,\mathrm{d}t \longrightarrow 0$$

as $y \to 0$ for any finite interval J on $\mathbb{R}$. (The argument is essentially the same as that used in the proof of Jensen's formula, Chapter I.) We thus have

$$\int_{-\infty}^{\infty} \Phi(t)(\log^+|f(t)|\mathrm{d}t - \mathrm{d}\mu(t)) = \int_{-\infty}^{\infty} \Phi(t)\log|f(t)|\,\mathrm{d}t$$

for each continuous function Φ of compact support, and hence

$$\log^+|f(t)|\mathrm{d}t - \mathrm{d}\mu(t) = \log|f(t)|\,\mathrm{d}t.$$

Therefore, for $\Im z > 0$,

$$\log|f(z)| = A\Im z + U(z) + v(z)$$
$$= (A-\alpha)\Im z + \frac{1}{\pi}\int_{-\infty}^{\infty}\frac{\Im z \log|f(t)|}{|z-t|^2}\mathrm{d}t,$$

by the formulas for $U(z)$ and $v(z)$.

In order to complete the proof, we must show that $\alpha = 0$. To see this, recall that by §F.1 the positive measure μ introduced above satisfies

$$\int_{-\infty}^{\infty}\frac{\mathrm{d}\mu(t)}{1+t^2} < \infty.$$

(We are already tacitly using this property – without it the formulas for $v(z)$ and especially for $\log|f(z)|$ make no sense!) Therefore, by the evaluation of $\mathrm{d}\mu(t)$ just made,

$$(*) \qquad \int_{-\infty}^{\infty}\frac{|\log|f(t)||}{1+t^2}\mathrm{d}t \ < \ \infty.$$

The formula for $\log|f(z)|$ just obtained now yields

$$\frac{\log|f(\mathrm{i}y)|}{y} = A - \alpha + \frac{1}{\pi}\int_{-\infty}^{\infty}\frac{\log|f(t)|}{t^2+y^2}\mathrm{d}t.$$

Making $y \to \infty$, we see from $(*)$ that $\log|f(\mathrm{i}y)|/y \to A - \alpha$. Since we called

$$A = \limsup_{y\to\infty}\frac{\log|f(\mathrm{i}y)|}{y},$$

we have $\alpha = 0$, and the theorem is proved.

Remark. Under the conditions of the present theorem, we see that $\limsup_{y\to\infty}(\log|f(\mathrm{i}y)|/y)$ is actually a *limit.*

2. Convergence of $\int_{-\infty}^{\infty}(\log^{-}|f(x)|/(1+x^2))\,\mathrm{d}x$

We are going to extend the work of the previous article to functions $f(z)$ *having* zeros in $\Im z > 0$. For this purpose, we need some preparatory material.

Lemma. *Let $S(w)$ be positive and superharmonic in $\{|w| < 1\}$ and suppose that* $\lim_{r\to 1} S(re^{\mathrm{i}\tau}) = S(e^{\mathrm{i}\tau})$ *exists a.e. Then, if* $|w| < 1$,

$$\frac{1}{2\pi}\int_{-\pi}^{\pi}\frac{1-|w|^2}{|w-e^{\mathrm{i}\tau}|^2}S(e^{\mathrm{i}\tau})\,\mathrm{d}\tau \leqslant S(w).$$

Remark. The assumption on the a.e. existence of the radial limit $S(e^{\mathrm{i}\tau})$ is superfluous. This is a consequence of a difficult theorem of Littlewood, which can be found in the books of Tsuji and Garnett. In our applications, this existence will, however, be manifest, so we may as well require it in the hypothesis of the lemma.

Proof of lemma. Let $|w_0| < 1$. By superharmonicity (mean value property) $\liminf_{w \to w_0} S(w) \leqslant S(w_0)$. If r_n increases towards 1 we can therefore find a sequence $\{w_n\}$ with $|w_n| < 1$ and $w_n \underset{n}{\longrightarrow} w_0$ such that $\liminf_{n \to \infty} S(r_n w_n) \leqslant S(w_0)$.

By superharmonicity of $S(w)$ for $|w| < 1$ we have, for each $r_n < 1$,

$$\frac{1}{2\pi} \int_{-\pi}^{\pi} \frac{1 - |w_n|^2}{|w_n - e^{i\tau}|^2} S(r_n e^{i\tau})\, d\tau \leqslant S(r_n w_n).$$

Also, $((1 - |w_n|^2)/|w_n - e^{i\tau}|^2) S(r_n e^{i\tau})$ is $\geqslant 0$ and tends to $((1 - |w_0|^2)/|w_0 - e^{i\tau}|^2) S(e^{i\tau})$ for almost all τ as $n \to \infty$ (separate convergence of the two factors!). Therefore, by Fatou's lemma,

$$\frac{1}{2\pi} \int_{-\pi}^{\pi} \frac{1 - |w_0|^2}{|w_0 - e^{i\tau}|^2} S(e^{i\tau})\, d\tau$$

$$\leqslant \liminf_{n \to \infty} \frac{1}{2\pi} \int_{-\pi}^{\pi} \frac{1 - |w_n|^2}{|w_n - e^{i\tau}|^2} S(r_n e^{i\tau})\, d\tau$$

$$\leqslant \liminf_{n \to \infty} S(r_n w_n) \leqslant S(w_0),$$

and we are done.

Lemma. *If $v(z)$ is subharmonic and $\leqslant 0$ in $\Im z > 0$ and $v(t) = \lim_{z \not\to t} v(z)$ exists a.e. on $\mathbb{R}$, then, for $\Im z > 0$,*

$$v(z) \;\leqslant\; \frac{1}{\pi} \int_{-\infty}^{\infty} \frac{\Im z}{|z - t|^2} v(t)\, dt.$$

Proof. Apply the previous lemma to $S(w)$, given by the formula

$$S\left(\frac{i - z}{i + z}\right) = -v(z),$$

and use the calculation made to establish the second theorem of §F.1.

From this lemma we have first of all the very important and much used

Theorem. *Let $f(z)$ be analytic for $\Im z > 0$ and continuous up to $\mathbb{R}$. Suppose also that $\log|f(z)| \leqslant O(|z|)$ for $|z|$ large, $\Im z \geqslant 0$, and that*

$$\int_{-\infty}^{\infty} \frac{\log^+ |f(x)|}{1 + x^2} dx \;<\; \infty.$$

Then, unless $f(z) \equiv 0$,

$$\int_{-\infty}^{\infty} \frac{\log^-|f(x)|}{1+x^2}\mathrm{d}x \;<\; \infty.$$

Proof. Without loss of generality, $f(\mathrm{i}) \neq 0$; otherwise work with

$$\left(\frac{z+\mathrm{i}}{z-\mathrm{i}}\right)^k f(z)$$

instead of $f(z)$ if f should have a zero of multiplicity k at i. By the theorem of §E, if we write $A = \limsup_{y\to\infty}(\log|f(\mathrm{i}y)|/y)$, the function

$$v(z) = \log|f(z)| - A\Im z - \frac{1}{\pi}\int_{-\infty}^{\infty} \frac{\Im z \log^+|f(t)|}{|z-t|^2}\mathrm{d}t$$

is $\leqslant 0$ for $\Im z > 0$. It is not, however, *harmonic* there as in the previous subsection, but merely *subharmonic.*

For $z \to t \in \mathbb{R}$, the right-hand integral in the previous formula tends to $\log^+|f(t)|$, since $\log^+|f(x)|$ is continuous on $\mathbb{R}$. Therefore, when $z \to t$,

$$v(z) \to \log|f(t)| - \log^+|f(t)| = -\log^-|f(t)|.$$

We may now apply the preceding lemma with $v(t) = -\log^-|f(t)|$. Since $f(\mathrm{i}) \neq 0$, we find that

$$-\infty \;<\; v(\mathrm{i}) \;<\; -\frac{1}{\pi}\int_{-\infty}^{\infty} \frac{\log^-|f(t)|}{1+t^2}\mathrm{d}t.$$

We are done.

Theorem. *Under the hypothesis of the preceding result,*

$$\log|f(z)| \leqslant A\Im z + \frac{1}{\pi}\int_{-\infty}^{\infty} \frac{\Im z \log|f(t)|}{|z-t|^2}\mathrm{d}t$$

for $\Im z > 0$, *where*

$$A = \limsup_{y\to\infty} \frac{\log|f(\mathrm{i}y)|}{y},$$

the integral on the right being absolutely convergent.

Remark. This is an improvement of the result in §E, where we have $\log^+|f(t)| \geqslant 0$ instead of the (signed) quantity $\log|f(t)|$ in the right-hand integral.

Proof. Taking $v(z)$ as in the proof of the preceding theorem we have, by the discussion there,

$$v(z) \leqslant -\frac{1}{\pi}\int_{-\infty}^{\infty} \frac{\Im z \log^-|f(t)|}{|z-t|^2}\,dt,$$

$\Im z > 0$, according to the above lemma. Adding

$$A\Im z + \frac{1}{\pi}\int_{-\infty}^{\infty} \frac{\Im z \log^+|f(t)|}{|z-t|^2}\,dt$$

to both sides of this inequality gives the desired result.

3. Taking the zeros in $\Im z > 0$ into account. Use of Blaschke products

Theorem. *Let $f(z)$ be analytic in $\{\Im z > 0\}$ and continuous up to $\mathbb{R}$, and suppose that $\log|f(z)| \leqslant O(|z|)$ for $|z|$ large, $\Im z > 0$, and that*

$$\int_{-\infty}^{\infty} \frac{\log^+|f(x)|}{1+x^2}\,dx < \infty.$$

► *Assume also that $f(0) \neq 0$.*

Denote by $\{z_n\}$ the sequence of zeros of $f(z)$ in $\Im z > 0$ (with repetitions according to multiplicities). Then

$$\sum_n \frac{\Im z_n}{|z_n|^2} < \infty.$$

Remark. The requirement that $f(0) \neq 0$ is *essential.*

Proof of theorem. Since $f(0) \neq 0$ and $f(z)$ is continuous up to $\mathbb{R}$, the z_n cannot accumulate at 0, i.e., $|z_n| > c$ for some $c > 0$. We may, wlog, take $c = 3$, for, if c is smaller than 3, we can work with $f((3/c)z)$ instead of $f(z)$.

The integral

$$\frac{i}{\pi}\int_{-\infty}^{\infty}\left(\frac{1}{z-t}+\frac{t}{t^2+1}\right)\log^+|f(t)|\,dt$$

is absolutely convergent for $\Im z > 0$ because of our condition

$$\int_{-\infty}^{\infty}\frac{\log^+|f(t)|}{1+t^2}\,dt \;<\; \infty;$$

it therefore represents a function *analytic* in that half plane whose *real part* is none other than

$$\frac{1}{\pi}\int_{-\infty}^{\infty}\frac{\Im z}{|z-t|^2}\log^+|f(t)|\,dt.$$

From this observation and the result of §E we see that

$$g(z) \;=\; \frac{e^{iAz}f(z)}{\exp\left\{\dfrac{i}{\pi}\displaystyle\int_{-\infty}^{\infty}\left(\frac{1}{z-t}+\frac{t}{t^2+1}\right)\log^+|f(t)|\,dt\right\}}$$

is *analytic and in modulus* $\leqslant 1$ for $\Im z > 0$, where the constant A is defined in the usual fashion. We have $g(i) \neq 0$, since (here) all the $|z_n|$ are > 3. For each N, apply the principle of maximum to

$$g(z)\Big/\prod_1^N\left(\frac{z-z_n}{z-\bar{z}_n}\right)$$

in $\Im z > 0$. Since $|g(z)| \leqslant 1$ there, we find

$$0 \;<\; |g(i)| \;\leqslant\; \prod_1^N\left|\frac{i-z_n}{i-\bar{z}_n}\right|.$$

For each n, we have

$$\left|\frac{i-z_n}{i-\bar{z}_n}\right| \;=\; \left|\frac{1-i/z_n}{1-i/\bar{z}_n}\right| \;=\; \left|1-\frac{2}{1-i/\bar{z}_n}\cdot\frac{\Im z_n}{|z_n|^2}\right|.$$

Here, however, $|z_n| > 3$, so the last expression is certainly

$$\leqslant\; 1-\frac{2}{1+1/|z_n|}\cdot\frac{\Im z_n}{|z_n|^2},$$

as is evident if we look at the image of the circle $|\omega| = 1/|z_n|$ $(<\frac{1}{3})$ under the linear fractional transformation

$$\omega \longrightarrow \frac{2}{1+\omega}\frac{\Im z_n}{|z_n|^2}:$$

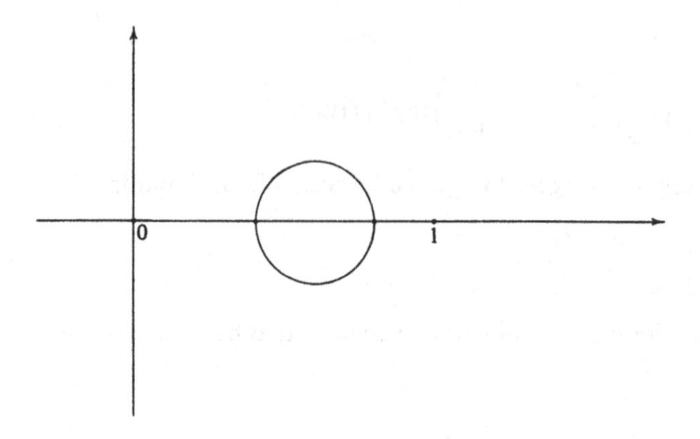

Figure 11

Substituting into the previous inequality and taking logarithms, we get

$$-\infty \;<\; \log|g(\mathrm{i})| \;\leqslant\; \sum_1^N \log\left(1 - \frac{2|z_n|}{|z_n|+1}\cdot\frac{\Im z_n}{|z_n|^2}\right)$$

and this is in turn

$$\leqslant\; -\sum_1^N \frac{2|z_n|}{|z_n|+1}\,\frac{\Im z_n}{|z_n|^2} \;\leqslant\; -\frac{3}{2}\sum_1^N \frac{\Im z_n}{|z_n|^2}$$

since $|z_n| > 3$.

We thus have

$$\sum_1^N \frac{\Im z_n}{|z_n|^2} \;\leqslant\; \frac{2}{3}\log\frac{1}{|g(\mathrm{i})|} \;<\; \infty$$

for all N, and our theorem follows on making $N \to \infty$.

Theorem. *Let $f(z)$ be an entire function of exponential type with*

$$\int_{-\infty}^{\infty} \frac{\log^+|f(x)|}{1+x^2}\,\mathrm{d}x \;<\; \infty.$$

There is an entire function $g(z)$ of exponential type with no zeros in $\Im z > 0$ and $|g(x)| = |f(x)|$ for $-\infty < x < \infty$.

Proof. Let $\{\lambda_n\}$ denote the set of zeros of $f(z)$ in $\Im z > 0$, and $\{\mu_n\}$ *all the other zeros* of $f(z)$ (repetitions according to multiplicities, as usual). Wlog, $f(0) \neq 0$, otherwise work with $f(z)/z^k$ instead of $f(z)$. The Hadamard factorization of $f(z)$ can then be written

$$f(z) = C\mathrm{e}^{bz} \lim_{R\to\infty} \left\{ \prod_{|\lambda_n|\leqslant R} \left(1 - \frac{z}{\lambda_n}\right)\mathrm{e}^{z/\lambda_n} \cdot \prod_{|\mu_n|\leqslant R} \left(1 - \frac{z}{\mu_n}\right)\mathrm{e}^{z/\mu_n} \right\}.$$

By the previous theorem, the sums $\sum_{|\lambda_n|\leqslant R}(\Im\lambda_n/|\lambda_n|^2)$, and by Lindelöf's theorem (§B), the sums

$$\sum_{|\lambda_n|\leqslant R}\frac{1}{\lambda_n} + \sum_{|\mu_n|\leqslant R}\frac{1}{\mu_n}$$

are bounded in absolute value.

The sums

$$\sum_{|\lambda_n|\leqslant R}\frac{1}{\bar{\lambda}_n} + \sum_{|\mu_n|\leqslant R}\frac{1}{\mu_n}$$

are therefore also bounded in absolute value, so, by the (easy) converse of Lindelöf's theorem, the products

$$Ce^{bz}\prod_{|\lambda_n|\leqslant R}\left(1-\frac{z}{\bar{\lambda}_n}\right)e^{z/\bar{\lambda}_n}\cdot\prod_{|\mu_n|\leqslant R}\left(1-\frac{z}{\mu_n}\right)e^{z/\mu_n}$$

converge u.c.c. in the complex plane to an *entire function $g(z)$ of exponential type* as $R\to\infty$.

For real x,

$$\frac{g(x)}{f(x)} = \lim_{R\to\infty}\prod_{|\lambda_n|\leqslant R}\left(\frac{1-x/\bar{\lambda}_n}{1-x/\lambda_n}\right)\exp\left(2i\frac{\Im\lambda_n}{|\lambda_n|^2}x\right),$$

the product being u.c.c. convergent on $\mathbb{R}$ in view of the condition

$$\sum_n\frac{\Im\lambda_n}{|\lambda_n|^2} < \infty.$$

The right side of the above expression is clearly of modulus 1 on $\mathbb{R}$, so $|f(x)| = |g(x)|$ there. And $g(z)$ has no zeros in $\Im z > 0$.

Theorem (Riesz–Fejér). *Let $F(z)$ be entire and of exponential type, let $F(x)\geqslant 0$ on $\mathbb{R}$, and suppose that*

$$\int_{-\infty}^{\infty}\frac{\log^+ F(x)}{1+x^2}dx < \infty.$$

Then there is an entire function $f(z)$ of exponential type without zeros in $\Im z > 0$ such that $F(z) = f(z)\cdot\overline{f(\bar{z})}$. In particular, $F(x) = |f(x)|^2$, $x\in\mathbb{R}$.

Proof. Since $F(x)$ is *real* on $\mathbb{R}$, $F(\bar{z}) = \overline{F(z)}$ by the Schwarz reflection principle, so, if $\Im\lambda > 0$, λ is a *zero* of $F(z)$ iff $\bar{\lambda}$ is also a zero thereof, with the same multiplicity. Because $F(x)\geqslant 0$ on $\mathbb{R}$, any *real* zero of F must have *even* multiplicity.

Denote by $\{\lambda_n\}$ the set of zeros of $F(z)$ in $\Im z > 0$ (repetitions according to multiplicities); then, by the observations just made, the Hadamard factorization of F must take the form

$$F(z) = Ce^{bz}\lim_{R\to\infty}\left\{\prod_{|\lambda_n|\leqslant R}\left(1-\frac{z}{\lambda_n}\right)e^{z/\lambda_n}\cdot\prod_{|\lambda_n|\leqslant R}\left(1-\frac{z}{\bar{\lambda}_n}\right)e^{z/\bar{\lambda}_n}\right.$$
$$\left.\cdot\prod_{|\alpha_n|\leqslant R}\left(1-\frac{z}{\alpha_n}\right)^2 e^{2z/\alpha_n}\right\}.^*$$

Here, the α_n are certain *real* numbers corresponding to the possible real zeros of $F(z)$ – of course, there may not really *be* any α_n. Using the fact that $F(x)\geqslant 0$ we easily check that $C\geqslant 0$ and that b is *real*.

As in the proof of the preceding theorem, we have $\sum_n \Im\lambda_n/|\lambda_n|^2<\infty$, and this, together with the Lindelöf theorems of §B, implies that the products

$$\sqrt{C}e^{bz/2}\prod_{|\lambda_n|\leqslant R}\left(1-\frac{z}{\bar{\lambda}_n}\right)e^{z/\bar{\lambda}_n}\prod_{|\alpha_n|\leqslant R}\left(1-\frac{z}{\alpha_n}\right)e^{z/\alpha_n}$$

converge u.c.c. in $\mathbb{C}$ to a certain entire function $f(z)$ of *exponential type* as $R\to\infty$. We clearly have

$$F(z) = f(z)\overline{f(\bar{z})},$$

and $f(z)$ has no zeros in $\Im z>0$. As required.

Theorem. *Let $f(z)$ be entire and of exponential type and suppose that*

$$\int_{-\infty}^{\infty}\frac{\log^+|f(x)|}{1+x^2}\,dx < \infty.$$

Denote by $\{\lambda_n\}$ the set of zeros of $f(z)$ in $\Im z>0$ (repetitions according to multiplicities), and put

$$A = \limsup_{y\to\infty}\frac{\log|f(iy)|}{y}.$$

Then, for $\Im z>0$,

$$\boxed{\log|f(z)| = A\Im z + \sum_n \log\left|\frac{1-z/\lambda_n}{1-z/\bar{\lambda}_n}\right| + \frac{1}{\pi}\int_{-\infty}^{\infty}\frac{\Im z}{|z-t|^2}\log|f(t)|\,dt.}$$

Proof. With the entire function $g(z)$ used in proving the theorem before the last one, put

$$G(z) = g(z)\cdot\exp\left\{-2i\sum_n\frac{\Im\lambda_n}{|\lambda_n|^2}z\right\}.$$

* As long as $F(0)\neq 0$. Otherwise, an additional factor z^{2k} appears on the right, and the description of $f(z)$ following in text must be modified accordingly.

We have, of course,

$$(*) \qquad \sum_n \frac{\Im \lambda_n}{|\lambda_n|^2} < \infty,$$

so $G(z)$ is an entire function of exponential type because $g(z)$ is, and, for $x \in \mathbb{R}$, $|G(x)| = |g(x)| = |f(x)|$. $G(z)$ is without zeros for $\Im z > 0$, and, in view of the description of $g(z)$ given where it was introduced,

$$G(z) = f(z) \cdot \lim_{R \to \infty} \prod_{|\lambda_n| \leqslant R} \left(\frac{1 - z/\bar{\lambda}_n}{1 - z/\lambda_n} \right).$$

Using $(*)$ and the fact that $\lambda_n \underset{n}{\longrightarrow} \infty$, we readily verify directly that the infinite product

$$\prod_n \left(\frac{1 - z/\lambda_n}{1 - z/\bar{\lambda}_n} \right)$$

is u.c.c. convergent for $\Im z \geqslant 0$; in the upper half plane, $G(z)$ evidently equals $f(z)$ *divided* by this infinite product.

For $\Im z > 0$, $|(1 - z/\bar{\lambda}_n)/(1 - z/\lambda_n)| > 1$, therefore $|G(z)| > |f(z)|$. Hence, if we call

$$A' = \limsup_{y \to \infty} \frac{\log |G(\mathrm{i}y)|}{y},$$

we have

$$A' \geqslant \limsup_{y \to \infty} \frac{\log |f(\mathrm{i}y)|}{y} = A.$$

Apply now the theorem of article 1 to $G(z)$, which has no zeros in $\Im z > 0$. Since $|G(t)| = |f(t)|$ for real t, we find that

$$\log |G(z)| = A' \Im z + \frac{1}{\pi} \int_{-\infty}^{\infty} \frac{\Im z}{|z - t|^2} \log |f(t)| \, \mathrm{d}t$$

for $\Im z > 0$. In view of the relation between $G(z)$ and $f(z)$, this yields

$$\log |f(z)| = A' \Im z + \sum_n \log \left| \frac{1 - z/\lambda_n}{1 - z/\bar{\lambda}_n} \right|$$

$$+ \frac{1}{\pi} \int_{-\infty}^{\infty} \frac{\Im z}{|z - t|^2} \log |f(t)| \, \mathrm{d}t, \quad \Im z > 0.$$

We will be done when we show that $A' = A$. Indeed, we have already seen that $A' \geqslant A$ so it is only necessary to prove that $A' \leqslant A$.

To this end, consider the functions

$$G_N(z) = f(z) \Big/ \prod_1^N \left(\frac{1 - z/\lambda_n}{1 - z/\bar{\lambda}_n} \right).$$

For each fixed N, $|G_N(x)| = |f(x)|$ on $\mathbb{R}$, and $G_N(\mathrm{i}y)/f(\mathrm{i}y) \to 1$ for $y \to \infty$, since the product on the right has only a finite number of factors. Because of this,

$$\limsup_{y\to\infty} \frac{\log|G_N(\mathrm{i}y)|}{y} = \limsup_{y\to\infty} \frac{\log|f(\mathrm{i}y)|}{y} = A,$$

whence, *by the second theorem of article* 2,

$$\log|G_N(z)| \leqslant A\Im z + \frac{1}{\pi}\int_{-\infty}^{\infty} \frac{\Im z}{|z-t|^2}\log|f(t)|\,\mathrm{d}t$$

for $\Im z > 0$. Make now $N \to \infty$; then $G_N(z) \xrightarrow[N]{} G(z)$ u.c.c. in $\Im z > 0$, so finally

$$\log|G(z)| \leqslant A\Im z + \frac{1}{\pi}\int_{-\infty}^{\infty} \frac{\Im z}{|z-t|^2}\log|f(t)|\,\mathrm{d}t$$

there. However, the *left* side *equals*

$$A'\Im z + \frac{1}{\pi}\int_{-\infty}^{\infty} \frac{\Im z}{|z-t|^2}\log|f(t)|\,\mathrm{d}t.$$

Therefore $A' \leqslant A$, whence finally $A' = A$, and the theorem is proved.

Remark. The expression

$$\prod_n \left(\frac{1 - z/\lambda_n}{1 - z/\bar{\lambda}_n}\right)$$

is called a *Blaschke product.*

Problem 4

Let φ and $\psi \in L_1(\mathbb{R})$ be functions of compact support. Let $[a,b]$ be the *supporting interval* for φ (that's the *smallest* closed interval *outside* of which $\varphi \equiv 0$ a.e.), and denote by $[a',b']$ the supporting interval for ψ. *Prove* that the supporting interval for $\varphi * \psi$ is precisely $[a+a', b+b']$. (Note: By $\varphi * \psi$ we mean the *convolution*

$$(\varphi * \psi)(\lambda) = \int_{-\infty}^{\infty} \varphi(\lambda - \tau)\psi(\tau)\,\mathrm{d}\tau.)$$

H. Levinson's theorem on the density of the zeros

We are going to close this chapter by proving a version of Levinson's theorem on the distribution of the zeros of entire functions of exponential type for such functions f which also satisfy the condition

$$\int_{-\infty}^{\infty} \frac{\log^+|f(x)|}{1+x^2}\,\mathrm{d}x \;<\; \infty.$$

This version is in fact due to Miss Cartwright and, although sufficient for most applications in analysis, is not the most general form of Levinson's theorem. For the latter one should consult the books by Boas or Levin, or, for that matter, the one by Levinson himself.

The proof to be given here depends on Kolmogorov's theorem on the *harmonic conjugates* of integrable functions, so we turn first to the establishment of that result.

1. Kolmogorov's theorem on the harmonic conjugate

Let $f(z)$ be analytic and *bounded* for $\Im z > 0$. An obvious application of a result of Fatou (the corollary in §F.2) shows that the boundary value

$$f(t) = \lim_{y \to 0+} f(t + \mathrm{i}y)$$

exists for almost every real t. *In the application of the following lemma to be given below, this fact can also be verified directly*; the reader interested in economy of thought may therefore include it in the hypothesis if he or she wants to.

Lemma. *Let $f(z)$ be bounded and analytic in $\Im z > 0$. Then*

$$f(z) = \frac{1}{\pi} \int_{-\infty}^{\infty} \frac{\Im z}{|z - t|^2} f(t) \mathrm{d}t$$

there.

Proof. If z lies inside the contour Γ shown below, we have, by Cauchy's theorem,

$$f(z) = \frac{1}{2\pi \mathrm{i}} \int_{\Gamma} \frac{f(\zeta) \mathrm{d}\zeta}{\zeta - z}.$$

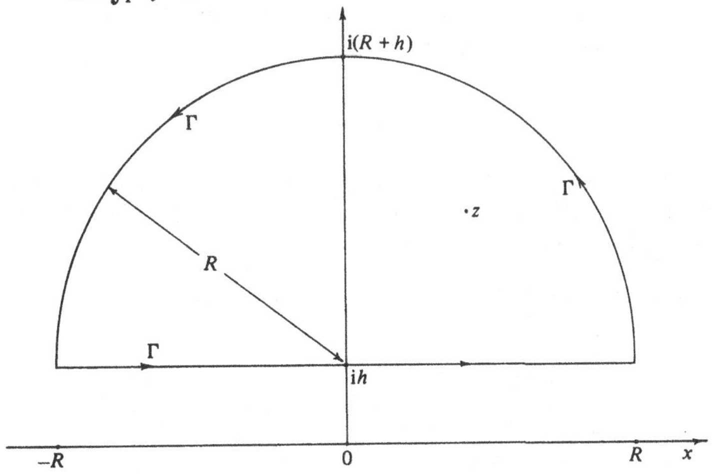

Figure 12

We fix R and make $h \to 0$ through some sequence of values; since $|f(t+ih)| \leqslant C$ say and $f(t+ih) \to f(t)$ a.e. for $h \to 0$, we have, by dominated convergence,

$$\int_{-R}^{R} \frac{f(t+ih)}{t+ih-z}\,dt \longrightarrow \int_{-R}^{R} \frac{f(t)}{t-z}\,dt.$$

Similarly,

$$\int_0^{\pi} \frac{f(ih+Re^{i\vartheta})}{ih+Re^{i\vartheta}-z}\, iRe^{i\vartheta}\,d\vartheta \longrightarrow \int_0^{\pi} \frac{f(Re^{i\vartheta})}{Re^{i\vartheta}-z}\, iRe^{i\vartheta}\,d\vartheta$$

as $h \to 0$. We thus see that

$$(*) \qquad f(z) = \frac{1}{2\pi i}\int_{-R}^{R} \frac{f(t)}{t-z}\,dt + \frac{1}{2\pi}\int_0^{\pi} \frac{f(Re^{i\vartheta})}{Re^{i\vartheta}-z}\, Re^{i\vartheta}\,d\vartheta.$$

Taking the relation

$$0 = \frac{1}{2\pi i}\int_{\Gamma} \frac{f(\zeta)}{\zeta-\bar{z}}\,d\zeta$$

and making $h \to 0$, we see in the same way that

$$0 = \frac{1}{2\pi i}\int_{-R}^{R} \frac{f(t)\,dt}{t-\bar{z}} + \frac{1}{2\pi}\int_0^{\pi} \frac{f(Re^{i\vartheta})}{Re^{i\vartheta}-\bar{z}}\, Re^{i\vartheta}\,d\vartheta.$$

Subtract this equation from (*). We get

$$f(z) = \frac{1}{\pi}\int_{-R}^{R} \frac{\Im z}{|t-z|^2} f(t)\,dt + \frac{i}{\pi}\int_0^{\pi} \frac{\Im z\, f(Re^{i\vartheta})Re^{i\vartheta}}{(Re^{i\vartheta}-z)(Re^{i\vartheta}-\bar{z})}\,d\vartheta.$$

Since f is bounded, the second term on the right is $O(1/R)$ for large R, so, making $R \to \infty$, we end with

$$f(z) = \frac{1}{\pi}\int_{-\infty}^{\infty} \frac{\Im z}{|z-t|^2} f(t)\,dt. \qquad \text{Q.E.D.}$$

Scholium. The reader is invited to obtain the lemma from the second theorem of §F.1 and the remark thereto (on representation of positive harmonic functions in $\Im z > 0$).

Suppose now that $u(t)$ is real valued and that

$$\int_{-\infty}^{\infty} \frac{|u(t)|}{1+t^2}\,dt \;<\; \infty.$$

Then the integral

$$\frac{i}{\pi}\int_{-\infty}^{\infty} \left(\frac{1}{z-t} + \frac{t}{t^2+1} \right) u(t)\,dt$$

converges absolutely for $\Im z > 0$ and equals an analytic function of z – call it $F(z)$ – there. $(\Re F)(z)$ is simply the by now familiar harmonic function

$$\frac{1}{\pi}\int_{-\infty}^{\infty}\frac{\Im z}{|z-t|^2}u(t)\mathrm{d}t;$$

$(\Im F)(z)$ is equal to

$$\frac{1}{\pi}\int_{-\infty}^{\infty}\left(\frac{\Re z-t}{|z-t|^2}+\frac{t}{t^2+1}\right)u(t)\,\mathrm{d}t.$$

Let us call the former expression $U(z)$ and the latter $\tilde{U}(z)$. Both are real valued and harmonic in $\{\Im z > 0\}$, and the latter is a *harmonic conjugate* of the former for that region since $U(z)+\mathrm{i}\tilde{U}(z)$ is equal to the function $F(z)$, *analytic* therein.

In order to examine the *boundary behaviour* of $U(z)$ and $\tilde{U}(z)$ we may first map $\{\Im z > 0\}$ onto $\{|w|<1\}$ by taking $w=(\mathrm{i}-z)/(\mathrm{i}+z)$ and then appeal to the results in Chapter II §B and in §F.2 of this chapter. From the first of those §§ we see that one simply has

$$U(t+\mathrm{i}y)\to u(t)$$

at almost every $t\in\mathbb{R}$ when $y\to 0$. The behaviour of $\tilde{U}(t+\mathrm{i}y)$ for $y\to 0$ is less transparent. According to the last theorem and scholium following it in §F.2, $\tilde{U}(t+\mathrm{i}y)$ *must, however, tend to a definite finite limit for almost every* $t\in\mathbb{R}$ as $y\to 0$. It is not very easy to see how that limit is related to our original function u; we get around this difficulty by denoting the limit by $\tilde{u}(t)$ and calling $\tilde{u}$ the *Hilbert transform* (or 'harmonic conjugate') of u.

Under certain circumstances one *can* in fact write a formula for $\tilde{u}(t)$ and verify almost *by inspection* that $\tilde{U}(t+\mathrm{i}y)$ tends to $\tilde{u}(t)$ (as given by the formula) when $y\to 0$. When this happens, *we do not need to use the general result of* §F.2 *to establish existence of* $\lim_{y\to 0}\tilde{U}(t+\mathrm{i}y)$. *That will indeed be the case in the application we make here*; the reader who is merely interested in arriving at Levinson's result may therefore include *existence* of the appropriate Hilbert transforms in the *hypothesis* of Kolmogorov's theorem, to be given below. It *is*, however, *true* that the Hilbert transforms in question *do* always exist a.e.

Here is a situation in which the *existence* of $\lim_{y\to 0}\tilde{U}(t+\mathrm{i}y)$ is *elementary*. Suppose that the integral

$$\int_0^1\frac{u(x_0-\tau)-u(x_0+\tau)}{\tau}\,\mathrm{d}\tau$$

is absolutely convergent; this will *certainly* be the case, for instance, if $u(t)$ is

Lip 1 (or even Lip α, $\alpha > 0$) at x_0. Then, if we write

$$\tilde{u}(x_0) = \frac{1}{\pi}\int_0^1 \frac{u(x_0 - \tau) - u(x_0 + \tau)}{\tau}\,d\tau$$

$$+ \frac{1}{\pi}\int_{x_0-1}^{x_0+1} \frac{tu(t)}{t^2+1}\,dt + \frac{1}{\pi}\int_{|t-x_0|>1} \left(\frac{1}{x_0 - t} + \frac{t}{t^2+1}\right) u(t)\,dt,$$

we easily verify that

$$\tilde{U}(x_0 + iy) = \frac{1}{\pi}\int_{-\infty}^{\infty} \left(\frac{x_0 - t}{(x_0 - t)^2 + y^2} + \frac{t}{t^2+1}\right) u(t)\,dt$$

tends to $\tilde{u}(x_0)$ as $y \to 0$. Just break up the integral on the right into two pieces, one with $x_0 - 1 \leqslant t \leqslant x_0 + 1$ and the other with $|t - x_0| > 1$. The *first* piece is readily seen to tend to the sum of the *first two* right-hand terms in the formula for $\tilde{u}(x_0)$ when $y \to 0$, and the *second* piece tends to the *third* term. In proving Levinson's theorem, the functions $u(t)$ which concern us are of the form $u(t) = \log^+|f(t)|$ or $u(t) = \log^-|f(t)|$ with $f(z)$ *entire* and such that

$$\int_{-\infty}^{\infty} \frac{|\log|f(t)||}{1+t^2}\,dt \;<\; \infty.$$

The function $\log^+|f(t)|$ is *certainly* Lip 1 at *every point* of $\mathbb{R}$ – $f(z)$ is *analytic*! And $\log^-|f(t)|$ is Lip 1 at all the points of $\mathbb{R}$ save those *isolated* ones where f has a *zero*. In *either* case, then, $\tilde{u}(x_0)$ is defined by the elementary procedure just described for all $x_0 \in \mathbb{R}$ except those belonging to some *countable* set of isolated points.

Our purpose in dwelling on the above matter at such length has been to explain that the proof of Levinson's theorem to be given below does *not really depend* on deep theorems about the *existence* of the Hilbert transform. The question of that existence is, however, *close enough* to the subject at hand to require our giving it *some* attention. The reader who wants to learn *more* about this question should consult the books of Zygmund or Garnett (*Bounded Analytic Functions*) or my own (on H_p spaces). There is also a beautiful real-variable treatment in Garsia's book on almost everywhere convergence.

Without further ado, let us now give

Kolmogorov's theorem. *Let $u(t)$ be real valued, let*

$$\int_{-\infty}^{\infty} \frac{|u(t)|}{1+t^2}\,dt \;<\; \infty,$$

and put

$$\tilde{u}(x) = \lim_{y \to 0+} \frac{1}{\pi} \int_{-\infty}^{\infty} \left(\frac{x-t}{(x-t)^2 + y^2} + \frac{t}{t^2+1} \right) u(t) \mathrm{d}t,$$

the limit on the right existing a.e. Then, if $\lambda > 0$,

$$\int_{\{|\tilde{u}(t)| > \lambda\}} \frac{\mathrm{d}t}{t^2+1} \leqslant \frac{4}{\lambda} \int_{-\infty}^{\infty} \frac{|u(t)|}{t^2+1} \mathrm{d}t.$$

Proof. Consider first the *special case where* $u(t) \geqslant 0$; this is actually where most of the work has to be done. The following argument was first published by Katznelson, and is due to Carleson.

Wlog,

$$\frac{1}{\pi} \int_{-\infty}^{\infty} \frac{u(t)}{1+t^2} \mathrm{d}t = 1.$$

Put

$$F(z) = \frac{\mathrm{i}}{\pi} \int_{-\infty}^{\infty} \left(\frac{1}{z-t} + \frac{t}{t^2+1} \right) u(t) \mathrm{d}t;$$

this function is analytic in $\Im z > 0$ *and has positive real part there.* Also, $F(\mathrm{i}) = 1$. For almost all $t \in \mathbb{R}$,

$$\Re F(t + \mathrm{i}y) \longrightarrow u(t)$$

and

$$\Im F(t + \mathrm{i}y) \longrightarrow \tilde{u}(t)$$

as $y \to 0+$.

Fix now any $\lambda > 0$, and take

$$f(z) = 1 + \frac{F(z) - \lambda}{F(z) + \lambda}$$

for $\Im z > 0$; $f(z)$ is analytic there and has modulus at most 2. For almost every $t \in \mathbb{R}$, $f(t) = \lim_{y \to 0+} f(t + \mathrm{i}y)$ exists, and can be expressed in terms of $u(t) + \mathrm{i}\tilde{u}(t)$.

By the lemma, for $\Im z > 0$,

$$f(z) = \frac{1}{\pi} \int_{-\infty}^{\infty} \frac{\Im z}{|z-t|^2} f(t) \mathrm{d}t,$$

therefore

$$(*) \qquad \frac{1}{\pi}\int_{-\infty}^{\infty} \frac{\Re f(t)}{1+t^2}\,dt \;=\; \Re f(i) \;=\; \Re\left(1+\frac{1-\lambda}{1+\lambda}\right) \;=\; \frac{2}{1+\lambda}.$$

The transformation $F \to f = 1 + (F-\lambda)/(F+\lambda)$ makes the half-plane $\Re F > 0$ correspond to the circle $|f-1| < 1$ *and takes the two portions of the imaginary axis where* $|F| > \lambda$ *onto the right half of the circle* $|f-1| = 1$.

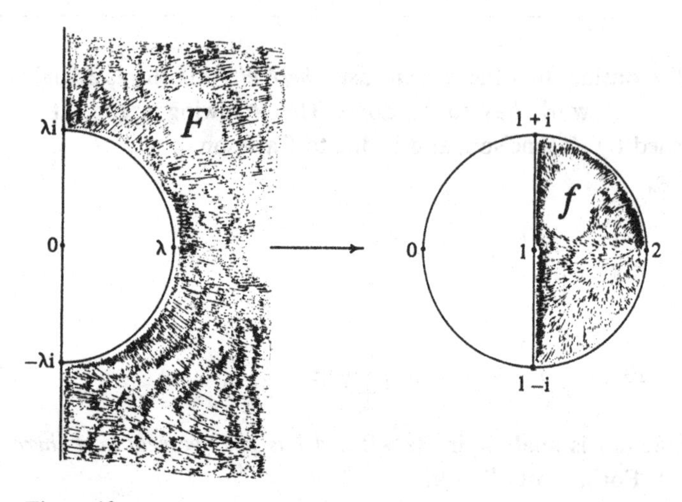

Figure 13

From the picture we therefore see that $\Re f(t) \geqslant 1$ whenever $|F(t)| = |u(t) + i\tilde{u}(t)| \geqslant \lambda$, hence, *surely whenever* $|\tilde{u}(t)| \geqslant \lambda$. Since we always have $\Re f(t) \geqslant 0$, we get from $(*)$,

$$\frac{1}{\pi}\int_{\{|\tilde{u}(t)|\geqslant\lambda\}} \frac{dt}{1+t^2} \;\leqslant\; \frac{2}{1+\lambda} \;<\; \frac{2}{\lambda}\cdot 1 \;=\; \frac{2}{\pi\lambda}\int_{-\infty}^{\infty}\frac{u(t)dt}{1+t^2},$$

since we assumed that the *integral alone* is equal to π. By homogeneity, we therefore get

$$\int_{\{|\tilde{u}(t)|\geqslant\lambda\}} \frac{dt}{1+t^2} \;\leqslant\; \frac{2}{\lambda}\int_{-\infty}^{\infty}\frac{u(t)dt}{1+t^2}$$

for the case where $u(t) \geqslant 0$.

In the general case of real u, write $u(t) = u_+(t) - u_-(t)$ and observe that $\tilde{u}(t) = \tilde{u}_+(t) - \tilde{u}_-(t)$, whence

$$\{t: |\tilde{u}(t)| \geqslant \lambda\} \subseteq \{t: |\tilde{u}_+(t)| \geqslant \lambda/2\} \cup \{t: |\tilde{u}_-(t)| \geqslant \lambda/2\}.$$

The inequality just established may be applied to each of the functions u_+, u_- and we obtain the desired result by adding, since

$$\int_{-\infty}^{\infty} \frac{u_+(t)}{1+t^2}\mathrm{d}t + \int_{-\infty}^{\infty} \frac{u_-(t)}{1+t^2}\mathrm{d}t = \int_{-\infty}^{\infty} \frac{|u(t)|\mathrm{d}t}{1+t^2}.$$

Kolmogorov's theorem is proved.

When $\lambda \to \infty$, the *right-hand side* of the inequality furnished by Kolmogorov's theorem may be replaced by $\mathrm{o}(1/\lambda)$.

Corollary. *Let $u(t)$ be real-valued and $\int_{-\infty}^{\infty}(|u(t)|/(1+t^2))\mathrm{d}t < \infty$. Then*

$$\int_{\{|\tilde{u}(t)|>\lambda\}} \frac{\mathrm{d}t}{1+t^2} = \mathrm{o}\left(\frac{1}{\lambda}\right) \quad \textit{for } \lambda \to \infty.$$

Proof. Take any $\varepsilon > 0$. We can find a *continuously differentiable function* φ *of compact support with*

$$\int_{-\infty}^{\infty} \frac{|u(t)-\varphi(t)|}{1+t^2}\mathrm{d}t < \varepsilon.$$

Referring to the discussion preceding the statement of the above theorem, we see that $\tilde{\varphi}$ can be readily computed; in fact

$$\tilde{\varphi}(x) = \frac{1}{\pi}\int_0^1 \frac{\varphi(x-\tau)-\varphi(x+\tau)}{\tau}\mathrm{d}\tau$$

$$+ \frac{1}{\pi}\int_{x-1}^{x+1} \frac{t\varphi(t)\mathrm{d}t}{t^2+1} + \frac{1}{\pi}\int_{|t-x|>1}\left(\frac{1}{x-t} + \frac{t}{t^2+1}\right)\varphi(t)\mathrm{d}t.$$

Because $\varphi'(t)$ is bounded and of compact support, the expression on the right is *bounded*; it is even $\mathrm{O}(1/|x|)$ for large x. So $|\tilde{\varphi}(x)| \leqslant M$, say ($M$, of course, *depends* on φ, hence on ε!), and, if $\lambda > 2M$, the set $\{t: |\tilde{u}(t)| > \lambda\}$ is included in $\{t: |\tilde{u}(t) - \tilde{\varphi}(t)| > \lambda/2\}$. Applying the theorem to the function $u - \varphi$, we therefore find that

$$\int_{\{|\tilde{u}(t)|>\lambda\}} \frac{\mathrm{d}t}{1+t^2} < \frac{8\varepsilon}{\lambda}$$

for $\lambda > 2M$. ε, however, was arbitrary. We're done.

2. Functions with only real zeros

If we want to study the *distribution of the zeros* of an entire function $f(z)$ of exponential type with

$$\int_{-\infty}^{\infty} \frac{\log^+|f(x)|}{1+x^2}\mathrm{d}x < \infty,$$

and we put

$$\limsup_{y\to\infty}\frac{\log|f(\mathrm{i}y)|}{y} = A,$$

$$\limsup_{y\to-\infty}\frac{\log|f(\mathrm{i}y)|}{|y|} = A',$$

there is *no loss of generality in assuming that* $A = A'$. The latter situation may always be arrived at by working with

$$\mathrm{e}^{\mathrm{i}(A-A')z/2}f(z)$$

instead of $f(z)$; here, the new function has the *same* zeros as $f(z)$ and *equals* $f(z)$ *in modulus on the real axis.*

We begin by looking at such functions f which have only real zeros.

Theorem. *Let $f(z)$ be entire and of exponential type, have only real zeros, and satisfy the condition*

$$\int_{-\infty}^{\infty}\frac{\log^+|f(x)|}{1+x^2}\mathrm{d}x \;<\; \infty.$$

Suppose that

$$\limsup_{y\to\infty}\frac{\log|f(\mathrm{i}y)|}{y} \;=\; \limsup_{y\to-\infty}\frac{\log|f(\mathrm{i}y)|}{|y|} \;=\; A.$$

For $t \geqslant 0$, let $\nu(t)$ be the number of zeros of *f on* $[0, t]$, *and, if $t < 0$, take $\nu(t)$ as* minus *the* number of zeros *of f in* $[t, 0)$. *(In both cases, multiplicities are counted.) Then $\nu(t)/t \to A/\pi$ for $t \to \infty$ and for $t \to -\infty$.*

Proof. By the theorem of §G.1,

$$(*)\qquad \log|f(z)| = A\Im z + \frac{1}{\pi}\int_{-\infty}^{\infty}\frac{\Im z}{|z-t|^2}\log|f(t)|\,\mathrm{d}t$$

for $\Im z > 0$, and by the same token, for $\Im z < 0$,

$$\log|f(z)| = A|\Im z| + \frac{1}{\pi}\int_{-\infty}^{\infty}\frac{|\Im z|}{|z-t|^2}\log|f(t)|\,\mathrm{d}t.$$

From these two relations, we see that the function $f(z)/\overline{f(\bar{z})}$, *analytic* in $\{\Im z > 0\}$, has constant modulus equal to 1 there. Therefore *in fact* $f(z)/\overline{f(\bar{z})} \equiv \beta$, a constant of modulus 1, for $\Im z > 0$. Making $z \to$ any point x of the real axis where $f(x) \neq 0$, we see that $f(x)/\overline{f(x)} = \beta$. This means that any continuous determination of $\arg f(x)$ on a *zero-free interval* (for f) is *constant* on *that interval.*

Since $f(z) \neq 0$ in $\{\Im z > 0\}$, we can define a (single valued) analytic branch of $\log f(z)$ in that half plane, and then take $\arg f(z)$ as the harmonic function $\Im \log f(z)$ there. For $x \in \mathbb{R}$ such that $f(x) \neq 0$, *define* $\arg f(x)$ as $\lim_{y \to 0+} \arg f(x + iy)$; as we have just seen, this function $\arg f(x)$ is *constant* on each interval of $\mathbb{R}$ where $f(x) \neq 0$. If x *increases* and passes through a zero x_0 of f, $\arg f(x)$ clearly *jumps down* by $\pi \times$ *the multiplicity of the zero* x_0. Therefore

$$\arg f(x) = -\pi\nu(x) + \text{const.}$$

for real x with $f(x) \neq 0$.

From (*) we see that, in $\{\Im z > 0\}$, the harmonic function $\log|f(z)|$ is the *real part* of the *analytic* one

$$-iAz + \frac{i}{\pi}\int_{-\infty}^{\infty}\left(\frac{1}{z-t} + \frac{t}{t^2+1}\right)\log|f(t)|\,dt.$$

It is, at the same time, the real part of $\log f(z)$ there. The *imaginary part* of the *latter analytic function* must therefore differ by a *constant* from that of the *former* one in $\{\Im z > 0\}$, and we have

$$\arg f(z) = -A\Re z + \frac{1}{\pi}\int_{-\infty}^{\infty}\left(\frac{\Re z - t}{|z-t|^2} + \frac{t}{t^2+1}\right)\log|f(t)|\,dt + \text{const.}$$

there. Taking $z = x + iy$ with x not a zero of f and making $y \to 0$, we obtain

$$\nu(x) = \frac{A}{\pi}x - \frac{1}{\pi^2}\lim_{y\to 0+}\int_{-\infty}^{\infty}\left(\frac{x-t}{(x-t)^2+y^2} + \frac{t}{t^2+1}\right)\log|f(t)|\,dt$$
$$+ \text{const.},$$

in view of the relation between $\arg f(x)$ and $\nu(x)$.

Write

$$\Delta(x) = \lim_{y\to 0+}\frac{1}{\pi^2}\int_{-\infty}^{\infty}\left(\frac{x-t}{(x-t)^2+y^2} + \frac{t}{t^2+1}\right)\log|f(t)|\,dt,$$

so that $\nu(x) = (A/\pi)x - \Delta(x) + \text{const.}$, save perhaps when $f(x) = 0$. (The limit in question certainly exists if $f(x) \neq 0$; see the discussion just preceding Kolmogorov's theorem in the previous article.)

In the course of proving the theorem of §G.1 we showed that the condition

$$\int_{-\infty}^{\infty}\frac{\log^+|f(x)|}{1+x^2}\,dx \;<\; \infty$$

(which is part of our hypothesis) actually *implies* that

$$\int_{-\infty}^{\infty}\frac{|\log|f(t)||}{1+t^2}\,dt \;<\; \infty;$$

we have of course been tacitly using the latter relation all along, since without it, (∗) and the formulas following therefrom would not make much sense. We can therefore apply Kolmogorov's theorem, and especially its *corollary*, to $u(t) = \log|f(t)|$. We have $\Delta(x) = (1/\pi)\tilde{u}(x)$ with this u, and therefore, by the corollary,

$$(\ast\ast) \qquad \int_{\{|\Delta(x)|>\lambda\}} \frac{dx}{1+x^2} = o\left(\frac{1}{\lambda}\right)$$

for large λ.

In order to prove that

$$\frac{v(x)}{x} \longrightarrow \frac{A}{\pi} \quad \text{for } x \to \pm\infty,$$

it is enough to show that $\Delta(x)/x \to 0$, $x \to \pm\infty$, and we restrict ourselves to the situation where $x \to \infty$, since the other one is treated in the same manner.

Pick any $\gamma > 1$, as close to 1 as we please, and any $\varepsilon > 0$. For large n, we have

$$\int_{\gamma^n}^{\gamma^{n+1}} \frac{dx}{1+x^2} \sim \frac{\gamma - 1}{\gamma^{n+1}},$$

so, taking $\lambda = \varepsilon\gamma^n$ in (∗∗) and making $n \to \infty$, we see that there must be *some* $x_n \in [\gamma^n, \gamma^{n+1}]$ with $|\Delta(x_n)| \leqslant \varepsilon\gamma^n$ if n is *large enough*. Since $v(x) = (A/\pi)x - \Delta(x) + \text{const.}$ is *increasing* (by its definition!), we have, for $\gamma^n \leqslant x \leqslant \gamma^{n+1}$ with n large,

$$-\Delta(x_{n-1}) - \frac{A}{\pi}(\gamma^{n+1} - \gamma^{n-1}) \leqslant -\Delta(x) \leqslant -\Delta(x_{n+1}) + \frac{A}{\pi}(\gamma^{n+2} - \gamma^n).$$

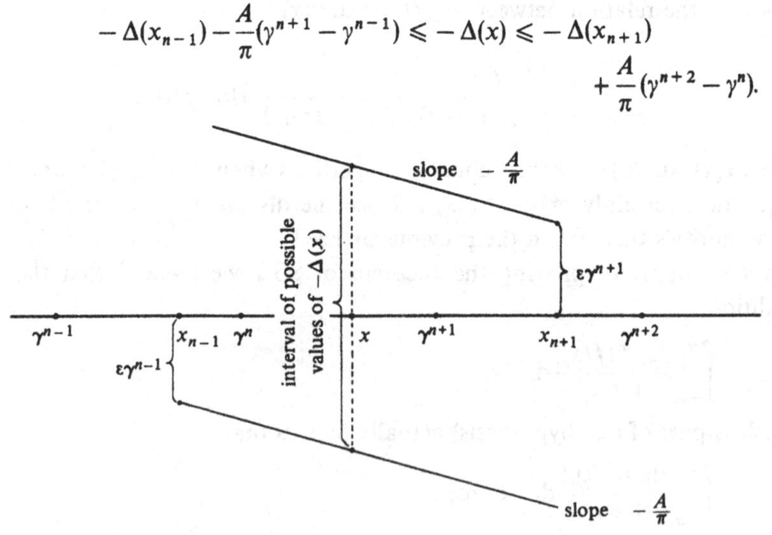

Figure 14

Thus, if n is large enough and $\gamma^n \leqslant x \leqslant \gamma^{n+1}$,

$$|\Delta(x)| \leqslant \varepsilon\gamma^{n+1} + \frac{A}{\pi}(\gamma^2 - 1)\gamma^n,$$

and

$$\left|\frac{\Delta(x)}{x}\right| \leqslant (\gamma^2 - 1)\frac{A}{\gamma} + \varepsilon\gamma.$$

Since $\gamma > 1$ and $\varepsilon > 0$ are arbitrary, we see that $\Delta(x)/x \to 0$ when $x \to \infty$, as required.

Our proof is now complete.

3. The zeros not necessarily real

Given an entire function $f(z)$, let us denote by $n_+(r)$ the number of its zeros with real part $\geqslant 0$ having modulus $\leqslant r$, and by $n_-(r)$ the number of its zeros with real part < 0 having modulus $\leqslant r$. As usual,

$$n(r) = n_+(r) + n_-(r)$$

is the *total number* of zeros of f with modulus $\leqslant r$, and multiple zeros of f are counted according to their multiplicities in reckoning the quantities $n_+(r)$, $n_-(r)$ and $n(r)$.

Theorem (Levinson). *Let the entire function $f(z)$ of exponential type be such that*

$$\int_{-\infty}^{\infty} \frac{\log^+|f(x)|}{1+x^2}\mathrm{d}x \;<\; \infty,$$

and suppose that

$$\limsup_{y\to\infty} \frac{\log|f(\mathrm{i}y)|}{y} \;=\; \limsup_{y\to-\infty} \frac{\log|f(\mathrm{i}y)|}{|y|} \;=\; A.$$

Then

$$\frac{n_+(r)}{r} \longrightarrow \frac{A}{\pi}$$

and

$$\frac{n_-(r)}{r} \longrightarrow \frac{A}{\pi}$$

as $r \to \infty$. Given any $\delta > 0$, the number of zeros of f with modulus $\leqslant r$ lying outside both of the two sectors $|\arg z| < \delta$, $|\arg z - \pi| < \delta$ is $\mathrm{o}(r)$ for large r.

Proof. Without loss of generality, $f(0) \neq 0$, for if $f(0) = 0$ we can work with a suitable quotient $f(z)/z^k$ instead of $f(z)$. *We may thus just as well take $f(0) = 1$ in what follows.*

Denote by $\{\lambda_n\}$ the sequence of zeros of $f(z)$, each zero being repeated in that sequence according to its multiplicity. By the first theorem of §G.3,

$$\sum_{\Im\lambda_n > 0} \frac{\Im\lambda_n}{|\lambda_n|^2} < \infty,$$

and similarly (referring to the *lower* half plane),

$$\sum_{\Im\lambda_n < 0} \frac{\Im\lambda_n}{|\lambda_n|^2} > -\infty.$$

Hence,

$$\sum_n \frac{|\Im\lambda_n|}{|\lambda_n|^2} < \infty.$$

Take any $\delta > 0$. From the previous relation, we have

$$\sum_{\delta \leqslant |\arg\lambda_n| \leqslant \pi - \delta} \frac{1}{|\lambda_n|} < \infty.$$

This certainly implies that the number of λ_n with $\delta \leqslant |\arg \lambda_n| \leqslant \pi - \delta$ and $|\lambda_n| \leqslant r$ is $o(r)$ for $r \to \infty$; the *last* affirmation of the theorem is thus established.

To get the rest (and main part) of the theorem, we follow an idea due to Levinson himself and compare $f(z)$ with another entire function *having only real zeros* to which the conclusion of the theorem in the previous subsection can be applied.

The Hadamard factorization of our function f has the form

$$f(z) = e^{cz} \prod_n \left(1 - \frac{z}{\lambda_n}\right) e^{z/\lambda_n}.$$

Corresponding to each λ_n we now compute a real number λ_n' according to the formula

$$\frac{1}{\lambda_n'} = \Re\left(\frac{1}{\lambda_n}\right);$$

if perchance λ_n is pure imaginary we put $\lambda_n' = \infty$. Let us now write

$$\varphi(z) = e^{cz} \prod_n \left(1 - \frac{z}{\lambda_n'}\right) e^{z/\lambda_n}.$$

We must, first of all, show that the product just written converges u.c.c. in $\mathbb{C}$. But

$$\left(1-\frac{z}{\lambda_n'}\right)e^{z/\lambda_n} = \left(1-\frac{z}{\lambda_n'}\right)e^{z/\lambda_n'}\cdot e^{-iz\Im\lambda_n/|\lambda_n|^2}$$

Here, $\sum_n(|\Im\lambda_n|/|\lambda_n|^2)<\infty$. Also $|\lambda_n'|\geqslant|\lambda_n|$ with $n(r)$, the number of λ_n having modulus $\leqslant r$, at most $O(r)$ (see §A), so $\prod_n(1-z/\lambda_n')e^{z/\lambda_n'}$ does converge u.c.c. in $\mathbb{C}$, by §A. The product defining $\varphi(z)$ therefore converges u.c.c. in $\mathbb{C}$, and $\varphi(z)$ *is an entire function* whose zeros are the real numbers λ_n'.

We want to show that $\varphi(z)$ is of exponential type. This can be done most easily by appealing to the Lindelöf theorems of §B, and the reader is invited to see how that goes. One can also make a direct verification without resorting to the Lindelöf theorems by proceeding as follows.

In the first place, we clearly have

$$|\varphi(x)|\leqslant|f(x)|$$

for $x\in\mathbb{R}$, so, since $f(z)$ *is* of exponential type, $|\varphi(x)|$ is at most $e^{O(|x|)}$ on the real axis. Consider now $z=iy$; here,

$$|\varphi(iy)| = e^{-y\Im c}\prod_n\left(1+\left(\frac{y}{\lambda_n'}\right)^2\right)^{1/2}e^{y\Im\lambda_n/|\lambda_n|^2}.$$

The right side is easily seen to be $\leqslant e^{O(|y|)}$; the only place where calculation is required is in the evaluation of $\sum_n\log(1+(y/\lambda_n')^2)$. To compute this sum, write it as a Stieltjes integral and integrate by parts, using the fact that the number of λ_n' with absolute value $\leqslant r$ is $O(r)$; we find without trouble that the sum is $O(|y|)$. (A very similar calculation was made in proving the second (easy) Lindelöf theorem of §B.)

Having seen that $|\varphi(x)|\leqslant e^{O(|x|)}$ and $|\varphi(iy)|\leqslant e^{O(|y|)}$, we must examine $|\varphi(z)|$ for general complex z. According to the discussion at the beginning of §B, from the fact that the number of λ_n' with modulus $\leqslant r$ is $O(r)$ we can only deduce an inequality of the form

$$\log|\varphi(z)| \leqslant O(|z|\log|z|),$$

valid for large $|z|$. At this point, however, we can apply the *second* Phragmén–Lindelöf theorem of §C. Look at $\varphi(z)$ in each of the quadrants I, II, III and IV. Take, say, the *first* one. For proper choice of the *complex* (!) constant γ,

$$e^{\gamma z}\varphi(z)$$

will be *bounded* on both the *positive real* and *positive imaginary* axes. For

z lying in the first quadrant and $|z|$ large, we will certainly have

$$|e^{\gamma z}\varphi(z)| \leqslant e^{O(|z|)}e^{O(|z|\log|z|)} \leqslant e^{|z|^{3/2}},$$

say.

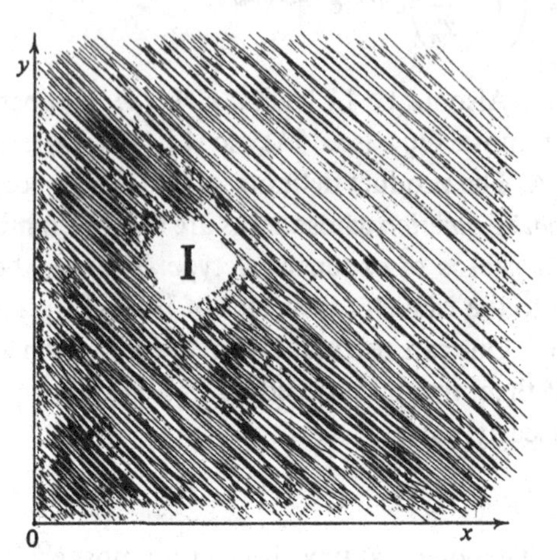

Figure 15

Therefore, because the opening of the quadrant, 90°, is *less than* $\frac{2}{3}\pi$ *radians*, the function $e^{\gamma z}\varphi(z)$, bounded on the *sides* of that quadrant, is *bounded in its interior*, and we see that

$$|\varphi(z)| \leqslant \text{const.}e^{|\gamma||z|}$$

in I. A similar argument works in each of the remaining three quadrants, and $\varphi(z)$ *is* therefore of exponential type.

Knowing φ to be of exponential type, we refer once more to the property $|\varphi(x)| \leqslant |f(x)|$, $x \in \mathbb{R}$, in order to obtain the condition

$$\int_{-\infty}^{\infty} \frac{\log^+|\varphi(x)|}{1+x^2}\,dx < \infty,$$

from the similar one assumed for f in the hypothesis. *We are in a position to apply the theorem of the previous article with our function* φ.

Write

$$B = \limsup_{y\to\infty} \frac{\log|\varphi(iy)|}{y}, \quad B' = \limsup_{y\to-\infty} \frac{\log|\varphi(iy)|}{|y|},$$

and call $\nu(t)$ the number of points λ'_n (counting multiplicities) in $[0, t]$ if $t \geqslant 0$. For $t < 0$, let $\nu(t)$ be *minus* that number of points λ'_n in $[t, 0)$. The theorem of the previous article is directly applicable to the function

$$e^{i(B-B')z/2}\varphi(z),$$

and tells us that

$$\frac{v(t)}{t} \longrightarrow \frac{B+B'}{2\pi} \quad \text{for} \quad t \to \pm\infty.$$

It is now claimed that $(B+B')/2 = A$, the common value of $\limsup_{y\to\infty} \log|f(iy)|/y$ and $\limsup_{y\to-\infty} \log|f(iy)|/|y|$.

First of all, for real y,

$$\left|1 - \frac{iy}{\lambda_n}\right|^2 = 1 - 2y\frac{\Im\lambda_n}{|\lambda_n|^2} + \frac{y^2}{|\lambda_n|^2}$$

$$= \left(1 + \frac{y^2}{|\lambda_n|^2}\right)\left(1 - \frac{2y\Im\lambda_n}{|\lambda_n|^2 + y^2}\right).$$

Noting that

$$\frac{1}{|\lambda_n|^2} = \frac{1}{|\lambda_n'|^2} + \left(\frac{\Im\lambda_n}{|\lambda_n|^2}\right)^2,$$

we see that the last expression is in turn

$$\leqslant \left|1 - \frac{iy}{\lambda_n'}\right|^2\left(1 + \left(\frac{y\Im\lambda_n}{|\lambda_n|^2}\right)^2\right)\left(1 - \frac{2y\Im\lambda_n}{|\lambda_n|^2 + y^2}\right).$$

Since $(1-s)e^s \leqslant 1$ for real s we also have

$$1 - \frac{2y\Im\lambda_n}{|\lambda_n|^2 + y^2} \leqslant \exp\left(-\frac{2y\Im\lambda_n}{|\lambda_n|^2 + y^2}\right).$$

Comparing the product representations for $f(iy)$ and $\varphi(iy)$, we see from the inequalities just written that

$$|f(iy)| \leqslant |\varphi(iy)|\prod_n\left(1 + \left(\frac{y\Im\lambda_n}{|\lambda_n|^2}\right)^2\right)^{1/2} \cdot \exp\left(-\sum_n \frac{y\Im\lambda_n}{|\lambda_n|^2 + y^2}\right).$$

Because $\sum_n(|\Im\lambda_n|/|\lambda_n|^2) < \infty$,

$$\left|\sum_n \frac{y\Im\lambda_n}{|\lambda_n|^2 + y^2}\right| = o(|y|) \quad \text{for} \quad y \to \pm\infty.$$

The same is true for the logarithm of the *product* on the right side of the previous relation. To verify that, denote by $N(t)$ the *number* of the quantities $|\lambda_n|^2/|\Im\lambda_n|$ lying between 0 and t (counting repetitions in the usual way), and rewrite

$$\sum_n \log\left(1 + \left(\frac{y\Im\lambda_n}{|\lambda_n|^2}\right)^2\right)^{1/2}$$

as

$$\frac{1}{2}\int_0^\infty \log\left(1+\frac{y^2}{t^2}\right)\mathrm{d}N(t).$$

Since $\sum_n |\Im\lambda_n|/|\lambda_n|^2 < \infty$, we have $N(t) = \mathrm{o}(t)$ for $t \to \infty$, from which the integral just written is easily seen to be $\mathrm{o}(|y|)$ for $y \to \pm\infty$ after an integration by parts.

In view of these facts, the above relation between $|f(\mathrm{i}y)|$ and $|\varphi(\mathrm{i}y)|$ shows that

$$\log|f(\mathrm{i}y)| \leqslant \log|\varphi(\mathrm{i}y)| + \mathrm{o}(|y|)$$

for $y \to \pm\infty$. Therefore

$$A = \limsup_{y\to\infty}\frac{\log|f(\mathrm{i}y)|}{y} \leqslant \limsup_{y\to\infty}\frac{\log|\varphi(\mathrm{i}y)|}{y} = B,$$

and, in like manner, $A \leqslant B'$. We have thus proved that $A \leqslant (B + B')/2$.

We wish now to prove the reverse inequality. Take any $\delta > 0$. We showed at the very beginning of this demonstration that, for *large* r, *all but* $\mathrm{o}(r)$ *of the original zeros* λ_n *of* $f(z)$ *with modulus* $\leqslant r$ *lie in one of the two sectors*

$$|\arg z| < \delta, \quad |\arg z - \pi| < \delta.$$

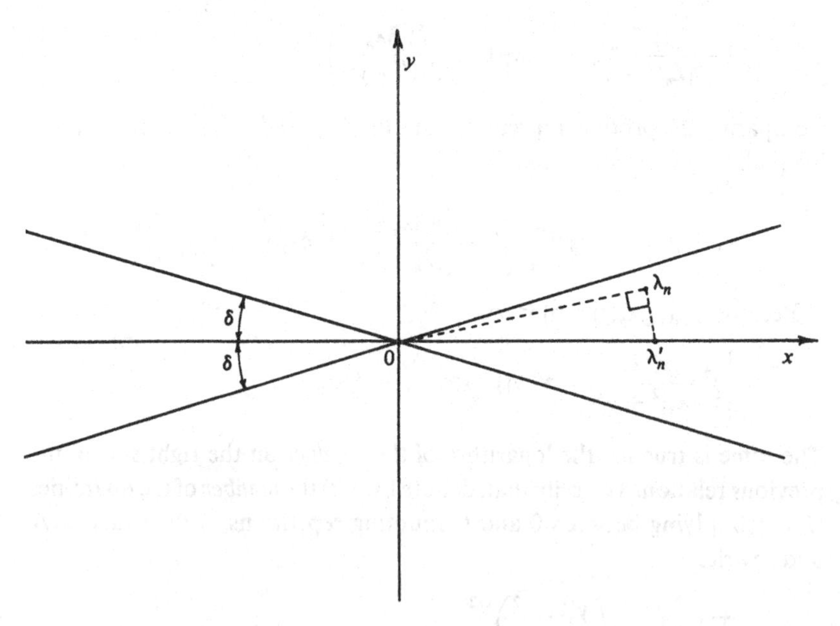

Figure 16

For λ_n in *either* of those two sectors,

$$|\lambda_n| \leqslant |\lambda_n'| = |\lambda_n|^2/|\Re\lambda_n| \leqslant |\lambda_n|\sec\delta,$$

so, for $r>0$ and large,

$$\nu(r) \leqslant n_+(r) \leqslant \nu(r\sec\delta)+o(r),$$
$$-\nu(-r) \leqslant n_-(r) \leqslant -\nu(-r\sec\delta)+o(r).$$

(Recall that $\nu(t)$ is by its definition *negative* for $t<0$!) Since $\delta>0$ is arbitrary, these relations and the known asymptotic behaviour of $\nu(t)$ for $t\to\pm\infty$ imply that

$$\frac{n_+(r)}{r} \longrightarrow \frac{B+B'}{2\pi}$$

and

$$\frac{n_-(r)}{r} \longrightarrow \frac{B+B'}{2\pi}$$

for $r\to\infty$.

According to the theorem of §E (the first and simplest one of its kind!),

$$\log|f(z)| \leqslant A\Im z + \frac{1}{\pi}\int_{-\infty}^{\infty}\frac{\Im z}{|z-t|^2}\log^+|f(t)|\,dt$$

for $\Im z>0$, and similarly, referring to the *lower* half plane,

$$\log|f(z)| \leqslant A|\Im z| + \frac{1}{\pi}\int_{-\infty}^{\infty}\frac{|\Im z|}{|z-t|^2}\log^+|f(t)|\,dt$$

for $\Im z<0$. Taking any $\delta>0$, we see from these two formulas that

$$\log|f(Re^{i\vartheta})| \leqslant AR|\sin\vartheta| + o(R)$$

holds *uniformly* in each of the two sectors

$$\delta<\vartheta<\pi-\delta, \quad \pi+\delta<\vartheta<2\pi-\delta$$

when $R\to\infty$ on account of the finiteness of $\int_{-\infty}^{\infty}(\log^+|f(t)|/(1+t^2))dt$.* In the remaining sectors $|\vartheta|\leqslant\delta$, $|\vartheta-\pi|\leqslant\delta$ we surely have

$$\log|f(Re^{i\vartheta})| \leqslant KR.$$

Therefore, for large R,

$$\frac{1}{2\pi}\int_0^{2\pi}\log|f(Re^{i\vartheta})|\,d\vartheta \leqslant \frac{2A}{\pi}R + o(R) + \frac{2\delta}{\pi}KR,$$

* In the integrals figuring in the two preceding relations the denominator of the integrand, $|z-t|^2$, is, for $z=Re^{i\vartheta}$, equal to $R^2+t^2-2Rt\cos\vartheta$. When $\delta<\vartheta<\pi-\delta$ or $\pi+\delta<\vartheta<2\pi-\delta$, this is $\geqslant(1-\cos\delta)(t^2+R^2)$.

or, since $\delta > 0$ is arbitrary,

$$\frac{1}{2\pi}\int_0^{2\pi} \log|f(Re^{i\vartheta})|\,d\vartheta \;\leqslant\; \frac{2A}{\pi}R + o(R)$$

for $R \to \infty$.

Now apply Jensen's formula, recalling that $n(r) = n_+(r) + n_-(r)$. By what has already been proved, we know that

$$\frac{n(r)}{r} \longrightarrow \frac{B+B'}{\pi}$$

for $r \to \infty$. Therefore

$$\int_0^R \frac{n(r)}{r}\,dr = \frac{B+B'}{\pi}R + o(R)$$

for large R. The left-hand side is, however, equal to

$$\frac{1}{2\pi}\int_0^{2\pi} \log|f(Re^{i\vartheta})|\,d\vartheta$$

which, as we have just seen, is $\leqslant (2A/\pi)R + o(R)$ for large R. Therefore $(B+B')/2 \leqslant A$.

The reverse inequality has already been shown. Therefore, $(B+B')/2 = A$, which means that

$$\frac{n_+(r)}{r} \longrightarrow \frac{A}{\pi} \quad\text{and}\quad \frac{n_-(r)}{r} \longrightarrow \frac{A}{\pi}$$

as $r \to \infty$. The first (principal) affirmation of our theorem is thus established, and we are done.

Remark. The above proof of the Cartwright version of Levinson's theorem depends on the elementary material of §§A and C, Kolmogorov's result, the formulas in §§E and G.1, and the first (easy) theorem in §G.3. The more delicate results in §§G.2 and G.3 (the one involving Blaschke products, in particular) are not used, nor are Lindelöf's theorems.

Problem 5

Let $f(z)$ be entire and of exponential type, with

$$\int_{-\infty}^{\infty} \frac{\log^+|f(x)|}{1+x^2}\,dx < \infty$$

and suppose that

$$\limsup_{y\to\infty} \frac{\log|f(iy)|}{y} = A.$$

(a) Show that

$$\int_0^{\pi}\left|\frac{\log|f(Re^{i\vartheta})|}{R} - A\sin\vartheta\right|\mathrm{d}\vartheta \longrightarrow 0$$

as $R \to \infty$.

(b) Let $\{\lambda_n\}$ be the sequence of zeros of $f(z)$ in $\{\Im z > 0\}$ (repetitions according to multiplicities), and put

$$B(z) = \prod_n \left(\frac{1 - z/\lambda_n}{1 - z/\bar{\lambda}_n}\right)$$

for $\Im z > 0$. Show that

$$\int_0^{\pi} \frac{1}{R}\log|B(Re^{i\vartheta})|\,\mathrm{d}\vartheta \longrightarrow 0$$

for $R \to \infty$.

IV

Quasianalyticity

One of the first applications of the relation

$$\int_{-\infty}^{\infty} \frac{\log^- |f(x)|}{1+x^2} dx < \infty$$

established in §G.2 of the previous chapter was to the study of quasianalyticity. This subject may now be treated by purely real-variable methods (thanks, in particular, to the work of Bang); the older function-theoretic approach of Carleman and Ostrowski is still, however, an excellent illustration of the power of complex-variable technique when applied in the investigation of real-variable phenomena, and it will be outlined here.

The material in the present chapter is due to Denjoy, Carleman, Ostrowski, Mandelbrojt and H. Cartan. We are only presenting an introduction to the subject of quasianalyticity; there are, for instance, other notions of that concept besides the one adopted here. One such, due to Beurling, will be taken up in Chapter VII, but for others, the reader should consult the books of Mandelbrojt and, regarding more recent work, Kahane's thesis published in the *Annales de l'Institut Fourier* in the early 1950s. Mandelbrojt's books on quasianalyticity also contain, of course, more elaborate treatments of the material given below. His 1952 book is the most complete, but his two earlier ones are easier to read.

A. Quasianalyticity. Sufficiency of Carleman's criterion

1. Definition of the classes $\mathscr{C}_I(\{M_n\})$

Suppose that $f(x)$ is infinitely differentiable on $\mathbb{R}$. The familiar example

$$f(x)=\begin{cases}\exp(-1/x^2), & x\neq 0,\\ 0, & x=0,\end{cases}$$

shows that, if, at some x_0, f and all of its derivatives *vanish*, we *cannot* conclude that $f(x)\equiv 0$. Under certain restrictions on f and its derivatives, however, such a conclusion may become legitimate. Consider, for example, functions f subject to the inequalities

$$|f^{(n)}(x)|\leqslant K^n n!,\quad x\in\mathbb{R},$$

on their successive derivatives. By looking, for instance, at Taylor's formula with Lagrange's form of the remainder, we see that

$$f(x)=f(x_0)+\sum_{n=1}^{\infty}\frac{f^{(n)}(x_0)}{n!}(x-x_0)^n$$

for $|x-x_0|<1/K$, x_0 being any point of $\mathbb{R}$, and this means that f is in fact the restriction to $\mathbb{R}$ of a function *analytic* in $|\Im z|<K$. Such a function cannot vanish together with all its derivatives at any point of $\mathbb{R}$ without being identically zero.

Are there perhaps some *other* systems of inequalities which, imposed on the successive derivatives of f, will *imply* the uniqueness property in question *without*, however, forcing f to be actually *analytic*? This question (which, like so many others in analysis, comes from mathematical physics) was raised at the beginning of the present century. The answer turns out to be *yes*, and the classes of functions thus obtained which, without necessarily being *themselves* analytic, share with the latter ones the property of being uniquely determined by their values and those of their successive derivatives at any point, are called *quasianalytic*. (Note: There are also *pseudoanalytic functions* in analysis. Those have *nothing to do* with the present discussion.)

Definition. Given any interval $I\subseteq\mathbb{R}$ and a sequence of numbers $M_n>0$, we say that a function f, infinitely differentiable on I, belongs to the class

$$\mathscr{C}_I(\{M_n\})$$

if there are two numbers c and ρ, depending on f, such that

$$|f^{(n)}(x)|\leqslant c\rho^n M_n\quad\text{for}\quad x\in I$$

and $n=0,1,2,3,\ldots$.

Remarks. The number c is introduced *mainly for convenience*, because we want $\mathscr{C}_I(\{M_n\})$ to be a *vector space*. The number ρ is introduced because, in the case $I=\mathbb{R}$, we want $f(\rho x)$ to belong to $\mathscr{C}_{\mathbb{R}}(\{M_n\})$ when f belongs to that class.

Scholium. Suppose I has a finite endpoint, say a, but that we are *not* taking a to be in I. Then, by requiring

$$|f^{(n)}(x)| \leqslant c\rho^n M_n$$

for $x \in I$, we obtain the existence of

$$\lim_{\substack{x \to a \\ x \in I}} f^{(n)}(x)$$

for $n = 0, 1, 2, \ldots$, so that f *and all its derivatives* may be *defined by continuity* at a. We will *then still have*

$$|f^{(n)}(a)| \leqslant c\rho^n M_n$$

for $n = 0, 1, 2, \ldots$. This means that for the classes $\mathscr{C}_I(\{M_n\})$ as we have defined them, we may *always assume that the intervals I are closed.*

Definition. A class $\mathscr{C}_I(\{M_n\})$ is called *quasianalytic* if, given *any* $x_0 \in I$, *the only* $f \in \mathscr{C}_I(\{M_n\})$ such that

$$f^{(n)}(x_0) = 0, \quad n = 0, 1, 2, \ldots,$$

has $f(x) \equiv 0$, $x \in I$.

Now we have the problem: which classes $\mathscr{C}_I(\{M_n\})$ are quasianalytic and which are not?

2. The function $T(r)$. Carleman's criterion

The quasianalyticity of the class $\mathscr{C}_I(\{M_n\})$ turns out to be governed by the function

$$T(r) = \sup_{n \geqslant 0} \frac{r^n}{M_n}$$

defined for $r > 0$, whose use is due to Ostrowski.

Theorem. *If* $\int_0^\infty (\log T(r)/(1 + r^2))dr = \infty$, *the class* $\mathscr{C}_I(\{M_n\})$ *is quasianalytic.*

Proof. Suppose that $x_0 \in I$ and that $f \in \mathscr{C}_I(\{M_n\})$ and $f^{(n)}(x_0) = 0$ for $n = 0, 1, 2, \ldots$. To prove that $f(x) \equiv 0$ on I it is enough to show that $f(x) \equiv 0$ *on any interval* $J \subseteq I$ *having* x_0 *as an endpoint.*

Without loss of generality, take $J = [0, 1]$ and suppose that x_0 is 1, i.e., that $f^{(n)}(1) = 0$, $n = 0, 1, 2, \ldots$. (Having an interval J of length $\neq 1$ only means that the parameter ρ in the bounds on $\sup_{x\in J}|f^{(n)}(x)|$ gets changed, while the M_n remain *unchanged*.) We have

$$f^{(n)}(x) \leqslant c\rho^n M_n, \quad n=0,1,2,\ldots \quad \text{and} \quad 0\leqslant x\leqslant 1.$$

For $\Re\sigma \geqslant 0$, put

$$\varphi(\sigma) = \int_0^1 t^\sigma f(t)\,\mathrm{d}t.$$

$\varphi(\sigma)$ is clearly *analytic* for $\Re\sigma > 0$ and *continuous* for $\Re\sigma \geqslant 0$, and

$$|\varphi(\sigma)| \leqslant cM_0, \quad \Re\sigma \geqslant 0.$$

We are going to show that $\varphi(\sigma) \equiv 0$. By the theorem of §G.2 in the preceding chapter (applied, of course to the *right* half plane instead of the *upper* half plane), this will certainly follow from the relation

$$\int_{-\infty}^{\infty} \frac{\log|\varphi(\mathrm{i}\tau)|}{1+\tau^2}\,\mathrm{d}\tau = -\infty,$$

which we now set out to establish.

Since $f(1)=0$, we have, when $\Re\sigma \geqslant 0$,

$$\varphi(\sigma) = f(t)\frac{t^{\sigma+1}}{\sigma+1}\bigg|_0^1 - \frac{1}{\sigma+1}\int_0^1 t^{\sigma+1}f'(t)\,\mathrm{d}t$$

$$= -\frac{1}{\sigma+1}\int_0^1 t^{\sigma+1}f'(t)\,\mathrm{d}t.$$

Again, $f'(1)=0$, so a similar integration by parts gives us

$$\varphi(\sigma) = \frac{1}{(\sigma+1)(\sigma+2)}\int_0^1 t^{\sigma+2}f''(t)\,\mathrm{d}t.$$

Repeating this process yields

$$\varphi(\sigma) = \frac{(-1)^n}{(\sigma+1)(\sigma+2)\ldots(\sigma+n)}\int_0^1 t^{\sigma+n}f^{(n)}(t)\,\mathrm{d}t.$$

Therefore, for $\Re\sigma \geqslant 0$,

$$|\varphi(\sigma)| \leqslant \frac{1}{|\sigma+1||\sigma+2|\cdots|\sigma+n|}\cdot c\rho^n M_n, \quad n=1,2,3,\ldots.$$

We have already seen that an analogous inequality holds for $n=0$.

Putting $\sigma = \mathrm{i}\tau$ with τ real, we get

$$|\varphi(\mathrm{i}\tau)| \leqslant \frac{c\rho^n M_n}{|\tau|^n}, \qquad n=0,1,2,\ldots,$$

that is,

$$\frac{1}{M_n}\left(\frac{|\tau|}{\rho}\right)^n |\varphi(i\tau)| \leqslant c, \quad n = 0, 1, 2, \ldots .$$

Since by definition

$$T\left(\frac{|\tau|}{\rho}\right) = \sup_{n\geqslant 0} \frac{1}{M_n}\left(\frac{|\tau|}{\rho}\right)^n,$$

we see that

$$|\varphi(i\tau)|\, T\left(\frac{|\tau|}{\rho}\right) \leqslant c,$$

so that

$$\log|\varphi(i\tau)| \leqslant \log c - \log T\left(\frac{|\tau|}{\rho}\right).$$

Since $T(r) \geqslant 1/M_0$ is *bounded below* (wlog $M_0 > 0$, for otherwise *surely* $f \equiv 0$), the relation

$$\int_0^\infty \frac{\log T(r)}{1+r^2}\,dr = \infty$$

implies that

$$\int_0^\infty \frac{\log T(r)}{1+\rho^2 r^2}\,dr = \infty.$$

Therefore

$$\int_{-\infty}^\infty \frac{\log|\varphi(i\tau)|}{1+\tau^2}\,d\tau \leqslant \pi \log c - \int_{-\infty}^\infty \frac{\log T(|\tau|/\rho)}{1+\tau^2}\,d\tau$$

$$= \pi \log c - 2\rho \int_0^\infty \frac{\log T(r)}{1+\rho^2 r^2}\,dr = -\infty,$$

as claimed above.

For this reason, $\varphi(\sigma) \equiv 0$ in $\Re\sigma \geqslant 0$ and in particular $\varphi(0) = \varphi(1) = \varphi(2) = \cdots = 0$. In other words,

$$\int_0^1 t^k f(t)\,dt = 0, \quad k = 0, 1, 2, \ldots .$$

By Weierstrass' theorem on polynomial approximation, this makes the (continuous!) function $f(t)$ vanish identically on $[0, 1]$. That's what we had to prove. We're done.

B. Convex logarithmic regularization of $\{M_n\}$ and the necessity of Carleman's criterion

We are going to see that the *converse* of the theorem at the end of the previous § is true. This requires us to make a preliminary study of the geometrical relationship between the sequence $\{M_n\}$ and the function $T(r)$.

1. Definition of the sequence $\{\underline{M}_n\}$. Its relation to $\{M_n\}$ and $T(r)$

By the definition of $T(r)$ given at the beginning of §A.2, we have

$$\log T(r) = \sup_{n \geqslant 0}(n \log r - \log M_n);$$

moreover (unless $\mathscr{C}_I(\{M_n\})$ consists *only* of the function *identically zero* on I, which situation we henceforth *exclude* from consideration),

$$\log T(r) \geqslant -\log M_0 > -\infty.$$

The function $\log T(r)$ clearly *increases* with $\log r$. This description of $\log T(r)$ is conveniently shown by the following diagram:

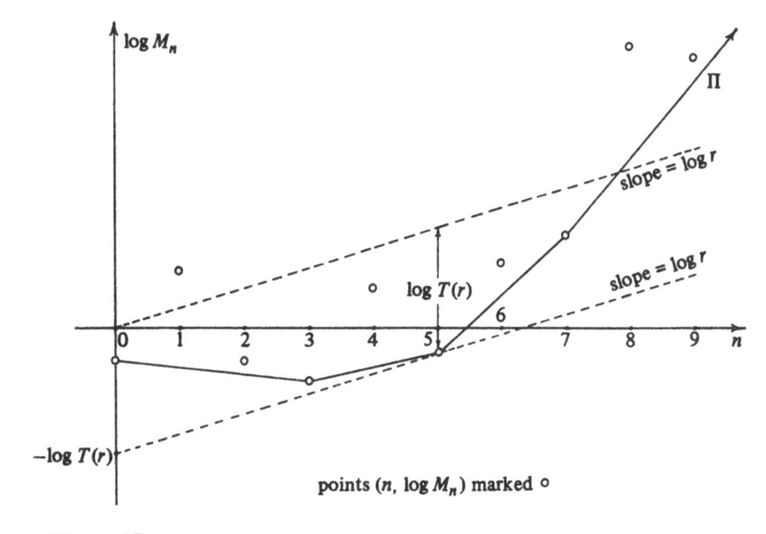

Figure 17

We see that $-\log T(r)$ is the y-intercept of the *highest* straight line of slope $\log r$ that lies *under all the points* $(n, \log M_n)$. It is convenient at this time to introduce the *highest convex curve*, Π, *lying under all the points* $(n, \log M_n)$. Π is the so-called *Newton polygon* of that collection of points (first applied by Isaac Newton in the computation of power series expansions of algebraic functions!). The straight line of ordinate

$n \log r - \log T(r)$ on the above diagram is, for each fixed $r > 0$, the *supporting line to* Π *having slope* $\log r$.

At any abscissa n, the *ordinate* of Π is simply the supremum of the ordinates of all of its supporting lines, since Π is convex. Therefore, the *ordinate* of Π at n is

$$\sup_{r>0}(n \log r - \log T(r)).$$

This quantity is henceforth denoted by $\log \underline{M}_n$; *and* $\{\underline{M}_n\}$ *is called the convex logarithmic regularization of the sequence* $\{M_n\}$; $\log \underline{M}_n$ is clearly a *convex function of* n. The following diagram shows the relation between $\log M_n$ and $\log \underline{M}_n$:

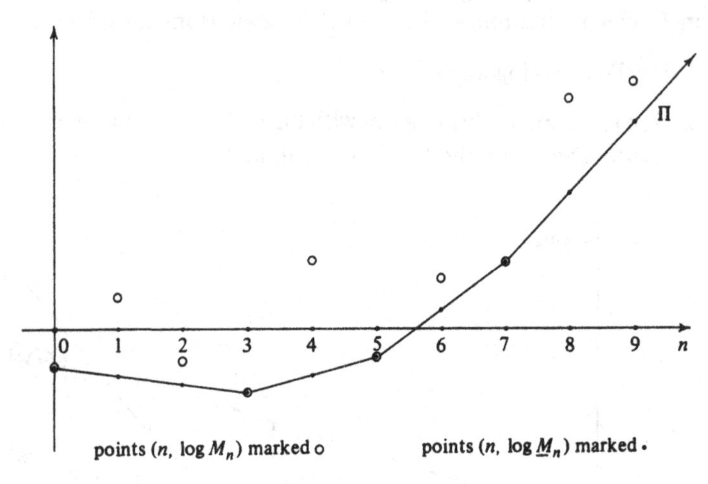

Figure 18

It is evident that $\underline{M}_n \leqslant M_n$ for each n.

The Newton polygon Π is *also* the highest convex curve lying under all the points $(n, \log \underline{M}_n)$, as the figures show. Therefore, by the above geometric characterization of $\log T(r)$ in terms of supporting lines, we also have

$$\log T(r) = \sup_{n \geqslant 0}[n \log r - \log \underline{M}_n],$$

i.e.,

$$T(r) = \sup_{n \geqslant 0} \frac{r^n}{\underline{M}_n}.$$

We see that the function $T(r)$ *cannot distinguish between the sequence* $\{M_n\}$ *and its convex logarithmic regularization* $\{\underline{M}_n\}$.

It will be convenient to consider the ordinates of the Newton polygon

Π at *non-integer* abscissae ν. We denote this ordinate simply by $\Pi(\nu)$; $\Pi(\nu)$ is a convex, piecewise linear function.

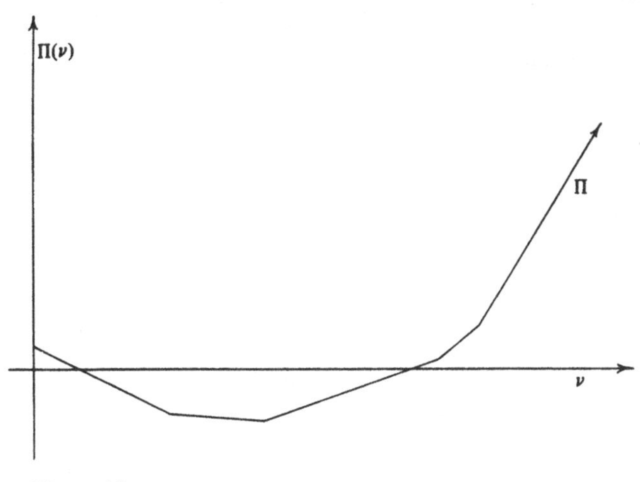

Figure 19

Lemma. *If $T(r)$ is finite for all finite r, then $\underline{M}_n^{1/n}$ is an increasing function of n when $n \geqslant$ some n_0.*

Proof. Under the hypothesis, the slope $\Pi'(\nu)$ of Π must tend to ∞ as $\nu \to \infty$. Otherwise, for some $r_0 < \infty$, $\Pi'(\nu)$ would remain $\leqslant \log r_0$ for all ν, and *certainly no straight line of slope* $\log r$, *with* $r > r_0$, *could lie below* Π.

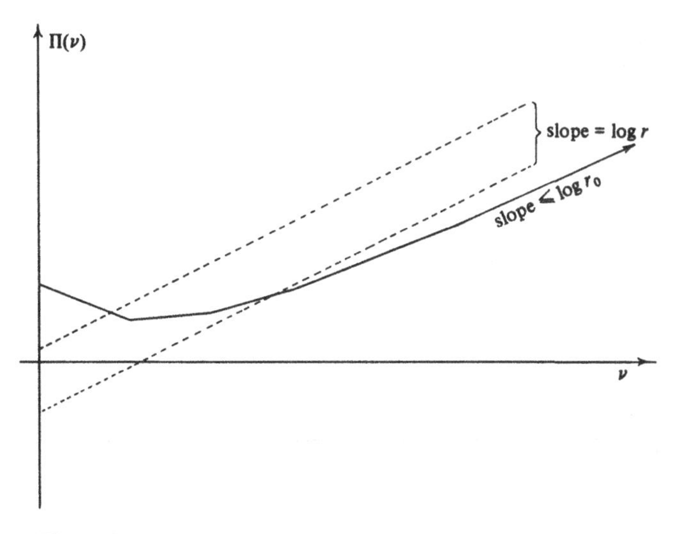

Figure 20

This means that $T(r) = \infty$ for $r > r_0$, contrary to hypothesis.

Because $\Pi'(\nu) \to \infty$ as $\nu \to \infty$ we certainly have $\Pi(\nu) \to \infty$. Pick any ν_0 with $\Pi(\nu_0)/\nu_0 > 0$, and draw a line $\mathscr{L}$ through the origin with slope *bigger than* $\Pi(\nu_0)/\nu_0$. Then $\mathscr{L}$ certainly passes *above* the point $(\nu_0, \Pi(\nu_0))$, lying on the convex curve Π.

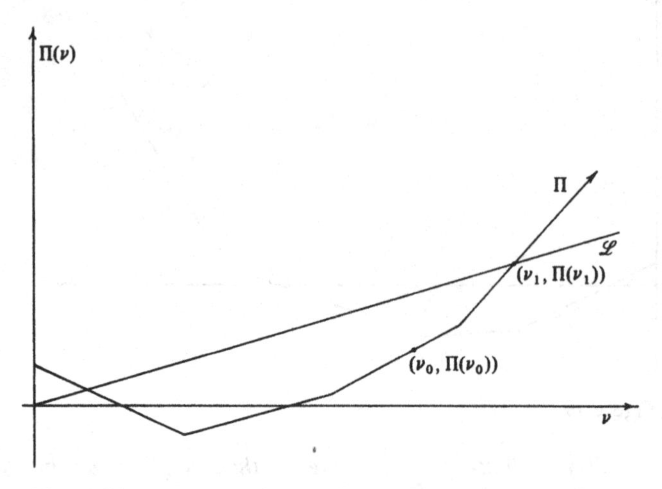

Figure 21

However, since $\Pi'(\nu) \to \infty$ as $\nu \to \infty$, $\mathscr{L}$ cannot lie above Π *forever*. Let ν_1 be the *last abscissa to the right* where Π cuts $\mathscr{L}$ – there *is* such a last abscissa because Π is convex. The figure shows that, at ν_1, Π *cuts* $\mathscr{L}$ *from below*. This means that

$$\frac{\Pi(\nu_1)}{\nu_1} < \Pi'(\nu_1).*$$

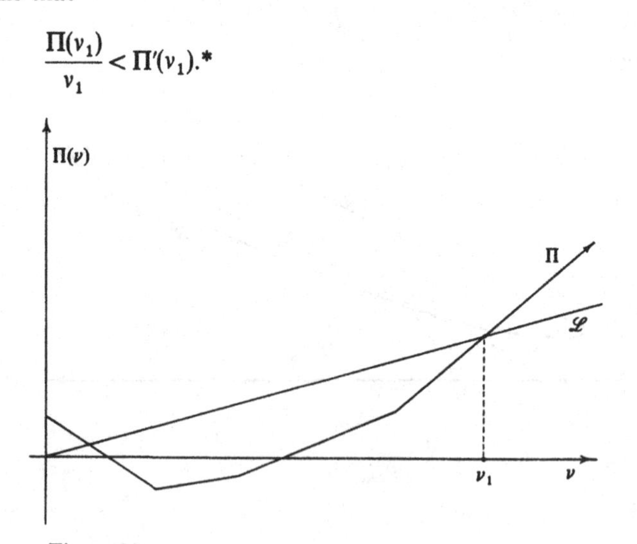

Figure 22

* If necessary, turn $\mathscr{L}$ slightly about O to ensure that $(\nu_1, \Pi(\nu_1))$ is not a vertex of Π.

It is claimed that $\Pi(\nu)/\nu$ *increases for* $\nu \geqslant \nu_1$.

To see this, observe that

$$\frac{d}{d\nu}\left(\frac{\Pi(\nu)}{\nu}\right) = \frac{\nu\Pi'(\nu) - \Pi(\nu)}{\nu^2};$$

so it is enough to verify that

$$\nu\Pi'(\nu) - \Pi(\nu) > 0 \quad \text{for} \quad \nu \geqslant \nu_1.$$

However, since $\Pi(\nu)$ is *convex*, $d(\nu\Pi'(\nu) - \Pi(\nu)) = \nu d\Pi'(\nu)$ is certainly $\geqslant 0$, so $\nu\Pi'(\nu) - \Pi(\nu)$ *increases*. And $\nu_1\Pi'(\nu_1) - \Pi(\nu_1)$ *is* > 0 by our construction. So $\nu\Pi'(\nu) - \Pi(\nu)$ remains > 0 for $\nu \geqslant \nu_1$, and therefore $\Pi(\nu)/\nu$ increases for such ν.

Let n_0 be any *integer* $\geqslant \nu_1$. Then, if $n \geqslant n_0$,

$$\frac{\log \underline{M}_n}{n} = \frac{\Pi(n)}{n} \leqslant \frac{\Pi(n+1)}{n+1} = \frac{\log \underline{M}_{n+1}}{n+1}.$$

This proves the lemma.

Lemma. *Suppose that* $\underline{M}_n^{1/n} \to \infty$ *for* $n \to \infty$. *Then, for sufficiently large* n, $\underline{M}_n^{1/n}$ *is increasing as a function of* n, *and*

$$\underline{M}_n^{1/n} \leqslant \frac{\underline{M}_{n+1}}{\underline{M}_n}.$$

Proof. Since $(\log \underline{M}_n)/n = \Pi(n)/n$ tends to ∞ with n, the *slope* of the convex curve Π *cannot remain bounded* as $\nu \to \infty$, and $\Pi'(\nu) \to \infty$, $\nu \to \infty$. We are therefore back in the situation of the previous lemma, and the argument used there shows that $\Pi(\nu)/\nu$ *increases* for large ν, and in particular $(\log \underline{M}_n)/n$ increases for large n.

The reasoning used in the above proof also showed that $\Pi'(\nu) > \Pi(\nu)/\nu$ if ν is large. Therefore, for large n,

$$\Pi'(n) > \frac{\Pi(n)}{n}.$$

Since, for $\nu \geqslant n$, $\Pi'(\nu) \geqslant \Pi'(n)$, we thus have, by the mean value theorem,*

$$\Pi(n+1) - \Pi(n) > \frac{\Pi(n)}{n},$$

i.e.,

$$\log \underline{M}_{n+1} - \log \underline{M}_n > \frac{\log \underline{M}_n}{n}$$

for large n. We're done.

* Note that all the *vertices* of Π have *integer* abscissae.

Corollary. *If* $\sum_{n=1}^{\infty} \underline{M}_n^{-1/n} < \infty$, *then*

$$\sum_{n=1}^{\infty} \frac{\underline{M}_{n-1}}{\underline{M}_n} < \infty.$$

Proof. Clear.

Theorem. *If* $\int_0^\infty (\log T(r)/(1+r^2))dr < \infty$, *then*

$$\sum_{n=1}^{\infty} \underline{M}_n^{-1/n} < \infty.$$

Proof. (Rudin). Since $T(r)$ is increasing, convergence of $\int_0^\infty (\log T(r)/(1+r^2))dr$ certainly implies that $T(r) < \infty$ for all finite r, so, by the first of the above lemmas, $\underline{M}_n^{1/n}$ is *increasing* for $n \geqslant n_0$, say.

As we saw during the discussion preceding the above two lemmas, $T(r) = \sup_{k \geqslant 0}(r^k/\underline{M}_k)$, and this is $\geqslant e^n$ when $r \geqslant e\underline{M}_n^{1/n}$. If $n \geqslant n_0$ so that $\underline{M}_n^{1/n} \leqslant \underline{M}_{n+1}^{1/(n+1)}$, we therefore get

$$\int_{e\underline{M}_n^{1/n}}^{e\underline{M}_{n+1}^{1/(n+1)}} \frac{\log T(r)}{r^2} dr \geqslant \frac{n}{e}\left[\frac{1}{\underline{M}_n^{1/n}} - \frac{1}{\underline{M}_{n+1}^{1/(n+1)}}\right].$$

Similarly,

$$\int_{e\underline{M}_n^{1/n}}^{\infty} \frac{\log T(r)}{r^2} dr \geqslant \frac{n}{e}\underline{M}_n^{-1/n}.$$

Using these inequalities and taking an arbitrary $m > n_0$, we find that

$$\begin{aligned}\int_{e\underline{M}_{n_0}^{1/n_0}}^{\infty} \frac{\log T(r)}{r^2} dr &= \sum_{n=n_0}^{m-1} \int_{e\underline{M}_n^{1/n}}^{e\underline{M}_{n+1}^{1/(n+1)}} \frac{\log T(r)}{r^2} dr \\ &\quad + \int_{e\underline{M}_m^{1/m}}^{\infty} \frac{\log T(r)}{r^2} dr \\ &\geqslant \frac{1}{e}\sum_{n=n_0}^{m-1} n(\underline{M}_n^{-1/n} - \underline{M}_{n+1}^{-1/(n+1)}) + \frac{m}{e}\underline{M}_m^{-1/m} \\ &= \frac{n_0}{e}\underline{M}_{n_0}^{-1/n_0} + \sum_{n=n_0+1}^{m} \frac{1}{e\underline{M}_n^{1/n}}.\end{aligned}$$

We see that

$$\sum_{n=n_0+1}^{m} \underline{M}_n^{-1/n} \leqslant e\int_{e\underline{M}_{n_0}^{1/n_0}}^{\infty} \frac{\log T(r)}{r^2} dr.$$

Since $\log T(r)$ is bounded below, the hypothesis makes the integral on the right finite. Therefore, making $m \to \infty$, we get

$$\sum_{n_0+1}^{\infty} \underline{M}_n^{-1/n} < \infty,$$

Q.E.D.

Corollary. *If* $\int_0^\infty (\log T(r)/(1+r^2))\mathrm{d}r < \infty$,

$$\sum_{n=1}^{\infty} \frac{\underline{M}_{n-1}}{\underline{M}_n} < \infty.$$

Proof. By the theorem and the corollary just before it.

2. Necessity of Carleman's criterion and the characterization of quasianalytic classes

Using the work of the preceding article, we can now establish the

Theorem. *If* $\int_0^\infty (\log T(r)/(1+r^2))\mathrm{d}r < \infty$, $\mathscr{C}_I(\{M_n\})$ *is not quasianalytic for any interval* I *of positive length.*

Proof. Take the convex logarithmic regularization $\{\underline{M}_n\}$ of $\{M_n\}$; then, by the corollary at the end of the last article, we have

$$\sum_1^{\infty} \frac{\underline{M}_{n-1}}{\underline{M}_n} < \infty.$$

The following construction works with the ratios $\mu_n = \underline{M}_{n-1}/\underline{M}_n$. Picking any $\varepsilon > 0$, we fix an $n_0 > 1$ such that

$$\sum_{n_0}^{\infty} \mu_n < \varepsilon.$$

For each n, $\sin \mu_n z/\mu_n z$ is entire, of exponential type μ_n, and bounded by 1 in absolute value on the real axis. Hence, by Phragmén–Lindelöf (§C of Chapter III),

$$\left| \frac{\sin \mu_n z}{\mu_n z} \right| \leqslant e^{\mu_n |\Im z|}.$$

The product $\prod_{n \geqslant n_0} (\sin \mu_n z/\mu_n z)$ will therefore equal a non-zero entire function $\varphi(z)$ with $|\varphi(z)| \leqslant e^{\varepsilon |\Im z|}$, if only it is convergent. However, if $|\mu_n z|$ is *small*,

$$\frac{\sin \mu_n z}{\mu_n z} = 1 - \frac{1}{3!}\mu_n^2 z^2 + O(\mu_n^4 |z|^4).$$

Since $\sum_n \mu_n < \infty$, we also have $\sum_n \mu_n^2 < \infty$; the product in question is therefore u.c.c. convergent in the complex plane.

Put

$$f(z) = \underline{M}_{n_0-1}\left(\frac{\sin(\varepsilon/n_0)z}{(\varepsilon/n_0)z}\right)^{2n_0}\prod_{n\geqslant n_0}\frac{\sin\mu_n z}{\mu_n z};$$

$f(z)$ is a non-zero entire function with

$$|f(z)| \leqslant \underline{M}_{n_0-1}e^{3\varepsilon|\Im z|}$$

and

$$\int_{-\infty}^{\infty}|f(x)|\,dx < \infty.$$

This last relation and the boundedness of f on the real axis certainly make $\int_{-\infty}^{\infty}|f(x)|^2\,dx < \infty$, and $f(z)$ is obviously of exponential type $\leqslant 3\varepsilon$. We therefore conclude by the Paley–Wiener theorem (Chapter II, §D) that

$$(*)\qquad F(\lambda) = \int_{-\infty}^{\infty} e^{i\lambda x}f(x)\,dx$$

vanishes identically for $\lambda < -3\varepsilon$ and for $\lambda > 3\varepsilon$. (Vanishes identically there and not just a.e., because here, $f(x)$ being in $L_1(\mathbb{R})$, $F(\lambda)$ is *continuous*.) Because $f(z) \not\equiv 0$, $F(\lambda)$ *cannot* be everywhere zero.

As we just remarked, $F(\lambda)$ is continuous on $\mathbb{R}$. It is even infinitely differentiable there. Indeed, if $k < n_0$,

$$|x^k f(x)| \leqslant \underline{M}_{n_0-1}\left(\frac{n_0}{\varepsilon}\right)^k\left|\frac{\sin(\varepsilon/n_0)x}{(\varepsilon/n_0)x}\right|^{2n_0-k}$$

certainly belongs to $L_1(\mathbb{R})$, so we can differentiate $(*)$ k times with respect to λ under the integral sign. In this way we see that

$$F^{(k)}(\lambda) = \int_{-\infty}^{\infty}(ix)^k e^{i\lambda x}f(x)\,dx$$

is in absolute value

$$\leqslant \left(\frac{n_0}{\varepsilon}\right)^k \underline{M}_{n_0-1}\int_{-\infty}^{\infty}\left|\frac{\sin(\varepsilon/n_0)x}{(\varepsilon/n_0)x}\right|^{n_0+1}dx,$$

a *finite constant*, for $k < n_0$. (Remember, we took $n_0 > 1$.) When $k \geqslant n_0$, we can start to use products of the factors $(\sin\mu_n x)/\mu_n x$, $n \geqslant n_0$, to absorb powers of x:

$$|x^k f(x)| \leqslant \left(\frac{n_0}{\varepsilon}\right)^{n_0-1}\left|\frac{\sin(\varepsilon/n_0)x}{(\varepsilon/n_0)}\right|^{n_0+1}\cdot\underline{M}_{n_0-1}\cdot\frac{1}{\mu_{n_0}}\cdot\cdots\cdot\frac{1}{\mu_k}$$

$$= \left(\frac{n_0}{\varepsilon}\right)^{n_0 - 1} \left|\frac{\sin(\varepsilon/n_0)x}{(\varepsilon/n_0)x}\right|^{n_0+1} \cdot \underline{M}_{n_0-1} \cdot \frac{\underline{M}_{n_0}}{\underline{M}_{n_0-1}} \cdots \frac{\underline{M}_k}{\underline{M}_{k-1}}$$

$$= \underline{M}_k \left(\frac{n_0}{\varepsilon}\right)^{n_0 - 1} \left|\frac{\sin(\varepsilon/n_0)x}{(\varepsilon/n_0)x}\right|^{n_0+1}$$

Therefore $|F^{(k)}(\lambda)| \leqslant \underline{M}_k (n_0/\varepsilon)^{n_0-1} \int_{-\infty}^{\infty} |\sin(\varepsilon/n_0)x/(\varepsilon/n_0)x|^{n_0+1}\, dx$ for $k \geqslant n_0$. We see that we can choose c in such a way that

$$|F^{(k)}(\lambda)| \leqslant c\underline{M}_k \quad \text{on} \quad \mathbb{R}$$

for *all* k (including the finite number of values from 0 to $n_0 - 1$).

We know, however, that $\underline{M}_k \leqslant M_k$ for each k. Therefore our function F belongs to $\mathscr{C}_{\mathbb{R}}(\{M_n\})$, does not vanish identically, but *is* identically zero outside the interval $[-3\varepsilon, 3\varepsilon]$. This means that F and all of its derivatives must *vanish* at both points 3ε and -3ε. Since $F(\lambda) \not\equiv 0$ on $[-3\varepsilon, 3\varepsilon]$, the class $\mathscr{C}_I(\{M_n\})$ cannot be quasianalytic when $I = [-3\varepsilon, 3\varepsilon]$. By translating F, we see that the same is true when I is any interval of length 6ε. Here, $\varepsilon > 0$ is arbitrary, so we're done.

The work just done can now be combined with that in §A.2 to give a *complete characterization* of the quasianalytic clases $\mathscr{C}_I(\{M_n\})$.

Theorem. *Given any interval I of positive length, the class $\mathscr{C}_I(\{M_n\})$ is quasianalytic iff any one of the following equivalent relations holds:*

(a) $\int_0^{\infty} (\log T(r)/(1 + r^2))dr = \infty$
(b) $\sum_n \underline{M}_n^{-1/n} = \infty$
(c) $\sum_n (\underline{M}_{n-1}/\underline{M}_n) = \infty$.

Proof. By the preceding material and logic-chopping.

We know that (a) implies quasianalyticity of $\mathscr{C}_I(\{M_n\})$ for any I by the theorem of §A.2. Also, *not*-(a) implies that $\sum_n \underline{M}_n^{-1/n} < \infty$ by the theorem of the preceding subsection, and this last relation *by itself* implies that

$$(\dagger) \qquad \sum_n (\underline{M}_{n-1}/\underline{M}_n) < \infty$$

according to the corollary immediately preceding that theorem.

Now, the *proof* of the preceding theorem is entirely based on the condition (†). Therefore, (†) *by itself* implies that $\mathscr{C}_I(\{M_n\})$ is *not* quasianalytic (for any I). So *quasianalyticity* of $\mathscr{C}_I(\{M_n\})$ *implies* (c) (the negation of (†)) which implies (b) which implies (a), which, however, *itself* implies the quasianalyticity of $\mathscr{C}_I(\{M_n\})$ as we have already seen. So complete equivalence of the latter property with any of the conditions (a), (b) and (c) is fully established. Q.E.D.

C. Scholium, Direct establishment of the equivalence between the three conditions $\int_0^\infty (\log T(r)/(1+r^2))\,dr < \infty$, $\sum_n \underline{M}_n^{-1/n} < \infty$ and $\sum_n \underline{M}_{n-1}/\underline{M}_n < \infty$.

These three conditions are of course equivalent according to the theorem at the end of the preceding §. In §B.1 we gave direct arguments, based upon the geometric properties of the Newton polygon Π of the set of points $(n, \log M_n)$, to show that the *first* condition implied the *second*, and the *second* the *third*. The above establishment of the *reverse* implications depended, however, on the theorem of §A.2 as well as on the construction used to prove the first result of §B.2; in other words, on lots of complex variable theory. Since the relationship between $T(r)$ and the convex logarithmic regularization $\{\underline{M}_n\}$ is *strictly graphical*, i.e., *geometric*, we should also give a *direct proof* of the reverse implications. Let's do that. The reasoning used involves a peculiar change of variable and a summation by parts.

Theorem. $\sum_{n=1}^\infty (\underline{M}_{n-1}/\underline{M}_n) < \infty$ *implies that*

$$\int_0^\infty \frac{\log T(r)}{1+r^2}\,dr \;<\; \infty.$$

Proof. Look at the Newton polygon Π, denoting, as in §B.1, its ordinate corresponding to the (real) abscissa ν by $\Pi(\nu)$:

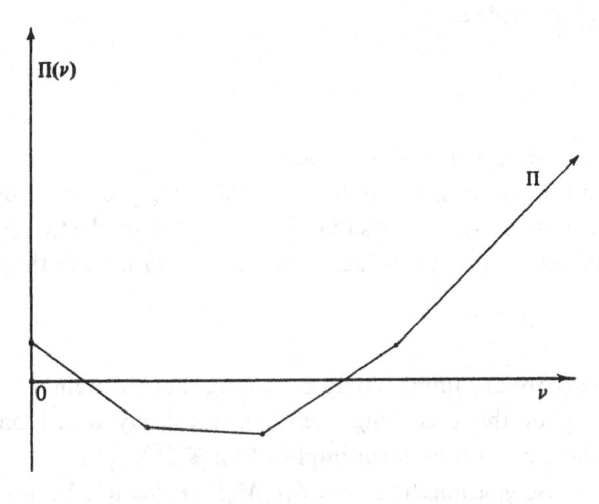

Figure 23

Π consists of certain straight segments, and each vertex of Π has a certain *integer* abscissa. Therefore, on each segment $[n-1, n]$, Π is a *straight line with the constant slope*

$$\Pi'(v) = \Pi(n) - \Pi(n-1) = \log \frac{\underline{M}_n}{\underline{M}_{n-1}}.$$

If, then,

$$\sum_n \frac{\underline{M}_{n-1}}{\underline{M}_n} < \infty,$$

we must have $\underline{M}_n/\underline{M}_{n-1} \xrightarrow[n]{} \infty$, and so the slope $\Pi'(v)$ of Π must tend to ∞ as $v \to \infty$. Either, then, Π has an *infinite* number of vertices, or else, if it has only finitely many, its *last* side must be *vertical* and have *infinite slope.* We examine only the *first* of these situations; treatment of the *second* is similar (and easier).

We are dealing, then, with a Newton polygon Π having an infinite number of vertices and thus an infinite number of straight sides whose slopes increase without limit. Denote by v_1 the abscissa of the *first vertex* of Π *where two sides of positive slope meet*, and by v_2, v_3, etc. those of the successive vertices lying to *the right* of $(v_1, \Pi(v_1))$.

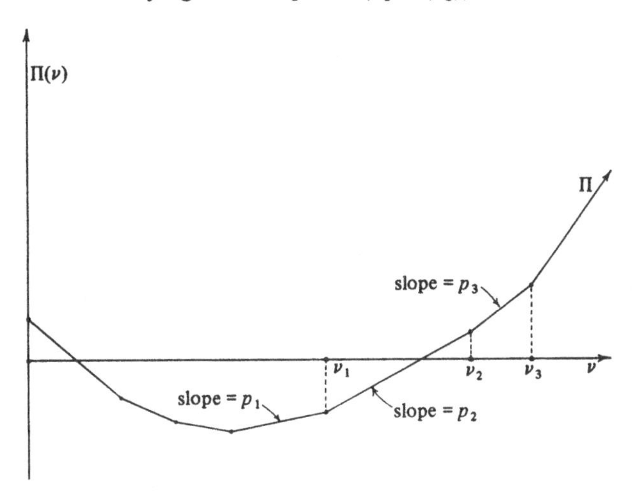

Figure 24

We call p_k the slope of the side of Π meeting the vertex $(v_k, \Pi(v_k))$ *from the left*; thus, $\Pi'(v) = p_k$ for $v_{k-1} < v < v_k$, and therefore $\underline{M}_{n-1}/\underline{M}_n = e^{-p_k}$ for $v_{k-1} < n \leqslant v_n$. (Keep in mind that the vertex abscissae v_k are *integers.*) Since $p_k \xrightarrow[k]{} \infty$ and $p_{k+1} > p_k$ (convexity of Π), we can break up

$$\int_{\exp p_1}^{\infty} \frac{\log T(r)}{r^2}\, dr$$

into a sum of integrals of the form

$$\int_{\exp p_k}^{\exp p_{k+1}} \frac{\log T(r)}{r^2}\,\mathrm{d}r.$$

Make the change of variable $\log r = p$ and put $\log T(r) = \tau(p)$. Integrating by parts between $r_k = \mathrm{e}^{p_k}$ and $r_{k+1} = \mathrm{e}^{p_{k+1}}$, we have

$$\int_{r_k}^{r_{k+1}} \frac{\log T(r)}{r^2}\,\mathrm{d}r = \frac{\log T(r_k)}{r_k} - \frac{\log T(r_{k+1})}{r_{k+1}} + \int_{r_k}^{r_{k+1}} \frac{1}{r}\,\mathrm{d}T(r)$$

$$= \mathrm{e}^{-p_k}\tau(p_k) - \mathrm{e}^{-p_{k+1}}\tau(p_{k+1}) + \int_{p_k}^{p_{k+1}} \mathrm{e}^{-p}\,\mathrm{d}\tau(p).$$

However, for $p_k \leqslant p \leqslant p_{k+1}$, $\tau(p) = \nu_k p - \Pi(\nu_k)$:

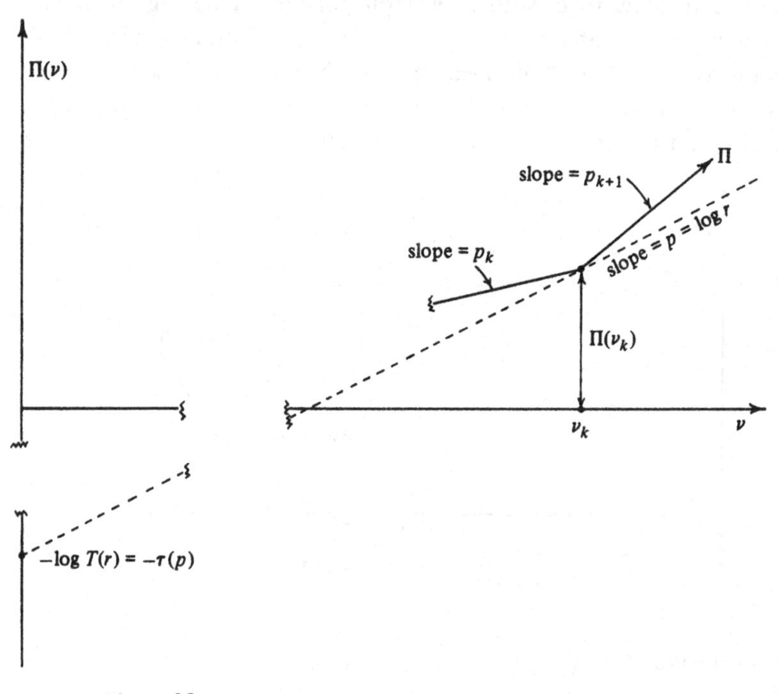

Figure 25

Therefore

$$\int_{p_k}^{p_{k+1}} \mathrm{e}^{-p}\,\mathrm{d}\tau(p) = \nu_k \int_{p_k}^{p_{k+1}} \mathrm{e}^{-p}\,\mathrm{d}p = \nu_k(\mathrm{e}^{-p_k} - \mathrm{e}^{-p_{k+1}}).$$

Adding, we thus get

$$\int_{r_1}^{r_{m+1}} \frac{\log T(r)}{r^2} \mathrm{d}r = \sum_{k=1}^{m} (\mathrm{e}^{-p_k}\tau(p_k) - \mathrm{e}^{-p_{k+1}}\tau(p_{k+1}))$$

$$+ \sum_{k=1}^{m} \nu_k(\mathrm{e}^{-p_k} - \mathrm{e}^{-p_{k+1}}).$$

The first sum on the right telescopes, and to the second we apply summation by parts. In this way, we see that the right side of the previous relation equals

$$\mathrm{e}^{-p_1}\tau(p_1) - \mathrm{e}^{-p_{m+1}}\tau(p_{m+1}) + \nu_1\mathrm{e}^{-p_1}$$

$$+ \sum_{k=2}^{m} (\nu_k - \nu_{k-1})\mathrm{e}^{-p_k} - \nu_m\mathrm{e}^{-p_{m+1}}.$$

Recall that $\log T(r)$ is increasing, so, if $\tau(p_{m+1}) = \log T(r_{m+1})$ remains <0 for $m \to \infty$,

$$\int_0^\infty \frac{\log T(r)}{1+r^2} \mathrm{d}r$$

will *certainly be* $<\infty$. There is thus no loss of generality in assuming $\tau(p_{m+1}) \geqslant 0$ for large m. If that is the case, we may drop the two terms prefixed by – signs from the previous expression and make $m \to \infty$, getting finally

$$\int_{r_1}^\infty \frac{\log T(r)}{r^2} \mathrm{d}r \;\leqslant\; \frac{\log T(r_1)}{r_1} + \nu_1\mathrm{e}^{-p_1} + \sum_{k=2}^{\infty} (\nu_k - \nu_{k-1})\mathrm{e}^{-p_k},$$

since, as we know, $r_{m+1} = \mathrm{e}^{p_{m+1}} \to \infty$ for $m \to \infty$.

As we saw at the beginning of this discussion, we have $\underline{M}_{n-1}/\underline{M}_n = \mathrm{e}^{-p_k}$ for $\nu_{k-1} < n \leqslant \nu_k$; there are clearly $\nu_k - \nu_{k-1}$ such values of n. Therefore the preceding inequality becomes

$$\int_{r_1}^\infty \frac{\log T(r)}{r^2} \mathrm{d}r \;\leqslant\; \frac{\log T(r_1)}{r_1} + \nu_1\mathrm{e}^{-p_1} + \sum_{n>\nu_1} \frac{\underline{M}_{n-1}}{\underline{M}_n}.$$

Hence $\int_{r_1}^\infty (\log T(r)/r^2)\,\mathrm{d}r < \infty$ if $\sum_n \underline{M}_{n-1}/\underline{M}_n$ converges, and thence $\int_0^\infty (\log T(r)/(1+r^2))\mathrm{d}r < \infty$, $T(r)$ being increasing.

Q.E.D.

This theorem, combined with that of §B.1, establishes equivalence of the two conditions

$$\int_0^\infty \frac{\log T(r)}{1+r^2} \mathrm{d}r \;<\; \infty \quad \text{and} \quad \sum_n \frac{\underline{M}_{n-1}}{\underline{M}_n} \;<\; \infty$$

since, as we saw in §B.1, the latter condition *is implied* by the inequality $\sum_n \underline{M}_n^{-1/n} < \infty$. In order to obtain full equivalence of our three conditions, it is still necessary to show that $\sum_n (\underline{M}_{n-1}/\underline{M}_n) < \infty$ *also implies* the relation $\sum_n \underline{M}_n^{-1/n} < \infty$. It was in order to establish such an implication that Carleman proved his celebrated inequality which says that

$$\sum_{n=1}^{\infty} (a_1 a_2 \cdots a_n)^{1/n} \leqslant c \sum_{n=1}^{\infty} a_n$$

with an absolute constant c for any sequence of numbers $a_k \geqslant 0$. Given this fact, we need only observe that

$$\frac{\underline{M}_0}{\underline{M}_1} \cdot \frac{\underline{M}_1}{\underline{M}_2} \cdots \frac{\underline{M}_{n-1}}{\underline{M}_n} = \frac{\underline{M}_0}{\underline{M}_n}.$$

Since $\underline{M}_0^{1/n} \underset{n}{\longrightarrow} 1$, the desired implication follows directly from the inequality.

Problem 6

Prove Carleman's inequality using Lagrange's method of undetermined multipliers. (Hint: Take $x_1, x_2, \ldots, x_N \geqslant 0$; the problem is to find the *maximum* of $\sum_{n=1}^{N} (x_1 x_2 \cdots x_n)^{1/n}$ subject to the condition that $\sum_{k=1}^{N} x_k = 1$. Show first that the maximum is attained at a place where *all* the x_k are *strictly positive* for $1 \leqslant k \leqslant M$, say, $M \leqslant N$, and the x_k with $M < k \leqslant N$ (if there are any) *all vanish*. The effect of this is to merely lower N, so we may always take the maximum to be an *internal* one, obtained for $x_k > 0$, $1 \leqslant k \leqslant N$.

Now apply Lagrange's method with the undetermined multiplier λ, and show that at the presumed maximum, $\sum_1^N (x_1 x_2 \cdots x_n)^{1/n} = \lambda$ by *adding equations*. The whole problem reduces to *getting a bound* on λ. To this end, write *each* of the N equations involving λ, and in them, make the substitutions

$$x_r = \frac{\xi_r}{r}, \quad \xi_r > 0.$$

Pick out the equation obtained by doing $\partial/\partial x_k$ in the Lagrange procedure, with k so chosen that ξ_k (at the sought maximum) is $\geqslant$ *all the other* ξ_r. This will give you the estimate

$$\lambda \leqslant \sum_{n \geqslant k} \frac{k}{n(n!)^{1/n}},$$

which yields a bound on λ independent of k.)

D. The Paley–Wiener construction of entire functions of small exponential type decreasing fairly rapidly along the real axis

Suppose we are given an *increasing* function $S(r)$, defined for $r \geqslant 0$ and (say) $\geqslant 1$ there. Do there exist any non-zero entire functions $\varphi(z)$ of exponential type such that

$$|\varphi(x)| \leqslant \frac{1}{S(|x|)}, \quad x \in \mathbb{R}?$$

According to the *first* theorem of Chapter III, §G.2, there are *no* such φ if

$$\int_0^\infty \frac{\log S(r)}{1+r^2} \mathrm{d}r = \infty.$$

If, however, the integral on the left is *convergent, we can use the construction in* §B.2 (applied in proving the first theorem of that article) to *obtain* such φ with, indeed, *arbitrarily small* exponential type. This application requires us to go a little further with the graphical work of §§B.1 and C.

As far as the problem taken up in this § is concerned, there is no loss of generality in assuming that $S(r) \equiv 1$ for $0 \leqslant r \leqslant 1$, say.

Lemma. *Let* $S(r)$ *be increasing on* $[0, \infty)$, *with* $S(r) \equiv 1$ *for* $0 \leqslant r \leqslant 1$, *and suppose that*

$$\int_0^\infty \frac{\log S(r)}{r^2} \mathrm{d}r < \infty.$$

Then there is an increasing function $T(r) \geqslant S(r)$ *with* $\log T(r)$ *a convex function of* $\log r$ *and also*

$$\int_0^\infty \frac{\log T(r)}{r^2} \mathrm{d}r < \infty.$$

Proof. Just put

$$\log T(r) = \int_0^{er} \frac{\log S(\rho)}{\rho} \mathrm{d}\rho \ !$$

Then, since $S(\rho)$ is increasing and $\geqslant 1$,

$$\log T(r) \geqslant \log S(r) \int_r^{er} \frac{\mathrm{d}\rho}{\rho} = \log S(r).$$

Again,

$$\frac{\mathrm{d} \log T(r)}{\mathrm{d} \log r} = r \frac{\mathrm{d} \log T(r)}{\mathrm{d}r} = \log S(er),$$

an increasing function of r, so $\log T(r)$ is convex in $\log r$. Finally,

$$\int_0^\infty \frac{\log T(r)}{r^2}\,dr = \int_0^\infty \int_0^{er} \frac{\log S(\rho)}{\rho r^2}\,d\rho\,dr$$

$$= \int_0^\infty \int_{\rho/e}^\infty \frac{\log S(\rho)}{\rho}\frac{dr}{r^2}\,d\rho = \int_0^\infty \frac{e\log S(\rho)}{\rho^2}\,d\rho < \infty.$$

Suppose now that we are given a function $S(r)$ satisfying the hypothesis of the lemma. We may, if we like, first obtain the function $T(r)$ and then search for entire functions φ satisfying the inequality

$$(*) \qquad |\varphi(x)| \leqslant \frac{1}{T(|x|)}.$$

Any such φ *will also satisfy*

$$|\varphi(x)| \leqslant \frac{1}{S(|x|)}$$

and hence solve our original problem. Our task thus reduces to the construction of entire functions φ of exponential type satisfying $(*)$, given that $T(r) \geqslant 1$, *that* $\log T(r)$ *is a convex function of* $\log r$, and that

$$\int_0^\infty \frac{\log T(r)}{r^2}\,dr < \infty.$$

Starting with such a function $T(r)$, put

$$M_n = \sup_{r>0} \frac{r^n}{T(r)} \quad \text{for} \quad n = 0, 1, 2, \ldots.$$

Since $\log M_n = \sup_{r>0}\{n\log r - \log T(r)\}$, *the sequence* $\{\log M_n\}$ *is already convex in* n. Now take

$$\tilde{T}(r) = \sup_{n\geqslant 0} \frac{r^n}{M_n}.$$

Lemma. *For* $r \geqslant 1$,

$$\log \tilde{T}(r) \leqslant \log T(r) \leqslant \log \tilde{T}(r) + \log r.$$

Proof. Uses graphs *dual* to the ones employed up to now to study the sequence $\{M_n\}$. Because $\log T(r)$ is a convex function of $\log r$, we have the following picture:

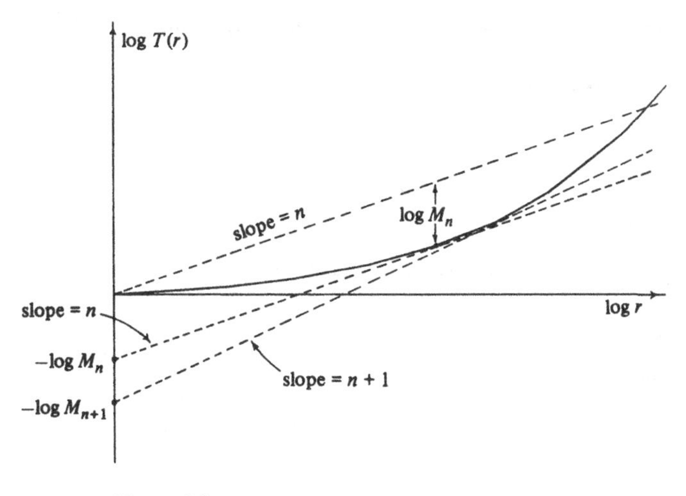

Figure 26

The supporting line to the graph of $\log T(r)$ vs $\log r$ *with integral slope n* has ordinate $n(\log r) - \log M_n$ at the abscissa $\log r$. It is therefore clear that $\log \tilde{T}(r)$, the *largest ordinate* of *those supporting lines with integral slope*, must lie *below* $\log T(r)$. This proves *one* of our desired inequalities.

To show the *other* one, take any $r > 1$, and look at any supporting line through the point $(\log r, \log T(r))$ of our graph. Since $\log T(r)$ is increasing, the *slope*, ν, of that supporting line must be $\geqslant 0$.

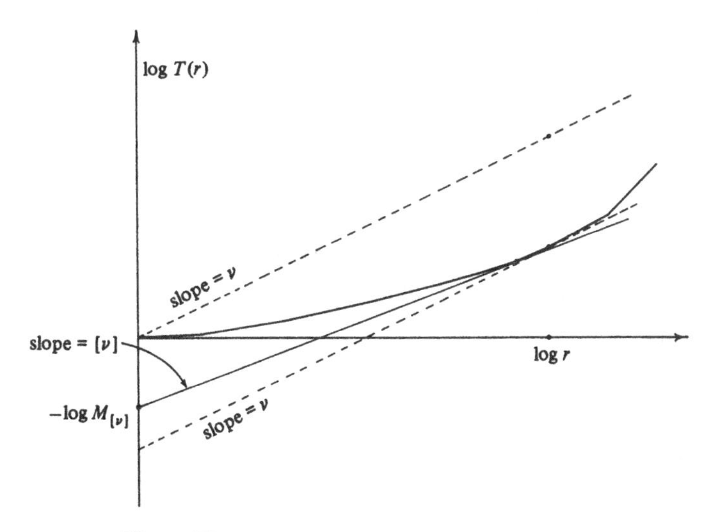

Figure 27

If $[\nu]$ denotes the largest integer $\leqslant \nu$, it is clear from the figure that

$$\nu \log r - \log M_{[\nu]} \geqslant \log T(r).$$

Therefore,

$$\begin{aligned}\log T(r) &\leqslant ([\nu]+1)\log r - \log M_{[\nu]} \\ &= [\nu]\log r - \log M_{[\nu]} + \log r \leqslant \log \tilde{T}(r) + \log r\end{aligned}$$

by definition of the function $\tilde{T}(r)$. We are done.

Theorem. *If* $T(r) \geqslant 1$ *is increasing, with* $T(r) \equiv 1$ *for* $0 \leqslant r \leqslant 1$ *and* $\log T(r)$ *a convex function of* $\log r$, *and if*

$$\int_0^\infty \frac{\log T(r)}{r^2}\,dr < \infty,$$

then, given any $\eta > 0$ *there is a non-zero entire function* $\varphi(z)$ *of exponential type* $< 2\eta$ *with*

$$|\varphi(x)| \leqslant \frac{1}{T(|x|)}, \quad x \in \mathbb{R}.$$

Proof. Form the sequence $\{M_n\}$ and then the function $\tilde{T}(r)$ in the manner described above. According to the preceding lemma, it is enough to find an entire function $\varphi \not\equiv 0$ of exponential type $< 2\eta$ with $|\varphi(x)| \leqslant 1$ on $\mathbb{R}$ and

$$|\varphi(x)| \leqslant \frac{1}{|x|\tilde{T}(|x|)} \quad \text{for} \quad |x| \geqslant 1, x \in \mathbb{R}.$$

The lemma and the hypothesis taken together tell us that

$$(\dagger) \qquad \int_1^\infty \frac{\log \tilde{T}(r)}{r^2}\,dr < \infty.$$

Also, since $\log M_n$ is here a convex function of n, the sequence $\{M_n\}$ is *identical* with its convex logarithmic regularization $\{\underline{M}_n\}$. Therefore $(\dagger)$ implies that

$$\sum_{n=1}^\infty \frac{M_{n-1}}{M_n} < \infty$$

by the corollary at the end of §B.1.

Write $\mu_n = M_{n-1}/M_n$ and take N so large that

$$\sum_{n=N}^\infty \mu_n < \eta.$$

Then put

$$f(z) = M_{N-1}\left(\frac{\sin(\eta/N)z}{(\eta/N)z}\right)^N \prod_{n=N}^{\infty} \frac{\sin \mu_n z}{\mu_n z}.$$

We see as in the proof of the first theorem of §B.2 that $f(z)$ is entire, not identically zero, and of exponential type $< \eta + \eta = 2\eta$. Arguing as in §B.2, we see also that $|x^k f(x)|$ is bounded on $\mathbb{R}$ for each positive integer k and moreover, when $k \geqslant N$, that

$$|x^k f(x)| \;=\; \frac{|x^{k+1} f(x)|}{|x|} \;\leqslant\; \left(\frac{N}{\eta}\right)^N \frac{M_{N-1}}{|x|\mu_N \mu_{N+1} \cdots \mu_k} \;=\; \left(\frac{N}{\eta}\right)^N \frac{M_k}{|x|},$$

whence

$$\frac{|x|^k}{M_k}|f(x)| \;\leqslant\; \frac{1}{|x|}\left(\frac{N}{\eta}\right)^N, \quad x \in \mathbb{R}.$$

Since similar inequalities hold also for $k = 0, 1, \ldots, N-1$, we have

$$\frac{|x|^k}{M_k}|f(x)| \;\leqslant\; \frac{C}{|x|}$$

for $x \in \mathbb{R}$ and $k = 0, 1, 2, \ldots$, C being a certain constant. Given $x \in \mathbb{R}$, take the supremum of the left-hand side for $k = 0, 1, 2, \ldots$. Referring to the definition of $\tilde{T}(r)$, we get $\tilde{T}(|x|)|f(x)| \leqslant C/|x|$, i.e.

$$|f(x)| \;\leqslant\; \frac{C}{|x|\tilde{T}(|x|)}, \quad x \in \mathbb{R}.$$

It is now evident that we can take φ as a suitable constant multiple of f, and φ will satisfy the required conditions. The theorem is proved.

Now we may refer to the lemma at the beginning of this §, and to the discussion given there. In that manner, we deduce from the result just established the following.

Corollary. *Let $S(r) \geqslant 1$ be increasing. A necessary and sufficient condition for there to exist entire functions $\varphi \not\equiv 0$ of exponential type with*

$$|\varphi(x)| \;\leqslant\; \frac{1}{S(|x|)} \quad \textit{on} \quad \mathbb{R}$$

is that

$$\int_0^\infty \frac{\log S(r)}{1+r^2} dr \;<\; \infty.$$

If that condition is met, there are entire $\varphi \not\equiv 0$ of arbitrarily small exponential type satisfying the inequality in question.

This result, which is due to Paley and Wiener, has found extensive use. Generalizations of it will be taken up in Chapters X and XI.

E. Theorem of Cartan and Gorny on equality of $\mathscr{C}_{\mathbb{R}}(\{M_n\})$ and $\mathscr{C}_{\mathbb{R}}(\{\underline{M}_n\})$. $\mathscr{C}_{\mathbb{R}}(\{M_n\})$ an algebra

The criteria for quasianalyticity of the class $\mathscr{C}_I(\{M_n\})$ given in §B.2 all depend on the *convex logarithmic regularization* $\{\underline{M}_n\}$ of $\{M_n\}$ rather than on the latter sequence itself. This makes it seem plausible that our initial consideration of classes $\mathscr{C}_I(\{M_n\})$ with *completely general* sequences $\{M_n\}$ was in fact *superfluous*, and that any such class is *in reality equal* to one of the form $\mathscr{C}_I(\{M'_n\})$ with a sequence $\{M'_n\}$ having fairly regular behaviour.

We are going to verify this hunch for the case where $I = \mathbb{R}$ by proving that $\mathscr{C}_{\mathbb{R}}(\{M_n\})$ always equals $\mathscr{C}_{\mathbb{R}}(\{\underline{M}_n\})$. A similar result holds when the interval I is *not* the whole real line, but then the regularized sequence $\{M'_n\}$ such that $\mathscr{C}_I(\{M_n\}) = \mathscr{C}_I(\{M'_n\})$ is no longer necessarily $\{\underline{M}_n\}$. In that circumstance, which we do not treat here, the regularization process used to pass from $\{M_n\}$ to $\{M'_n\}$ is more complicated than the one yielding $\{\underline{M}_n\}$. Interested readers may find a discussion of this and other related matters in Mandelbrojt's 1952 book.

Lemma (S. Bernstein). *Let $P(z)$ be a polynomial of degree n, and suppose that $|P(x)| \leqslant M$ for $-R \leqslant x \leqslant R$. If $a > 1$, we have $|P(z)| \leqslant Ma^n$ for all z of the form $\frac{1}{2}R(w + w^{-1})$ with $1 \leqslant |w| \leqslant a$.*

Remark. For fixed a, the set of z in question fills out a certain *ellipse* with foci at $\pm R$.

Proof of lemma Under the conformal mapping $w \longrightarrow z = \frac{1}{2}R(w + w^{-1})$, the region $|w| > 1$ goes onto the complement of the segment $[-R, R]$, and the unit circumference is taken onto that segment.

With z related to w in the manner described, put, for $|w| > 1$,

$$f(w) = \frac{P(z)}{w^n};$$

$f(w)$ is then certainly analytic outside the unit circle, and continuous up to it.

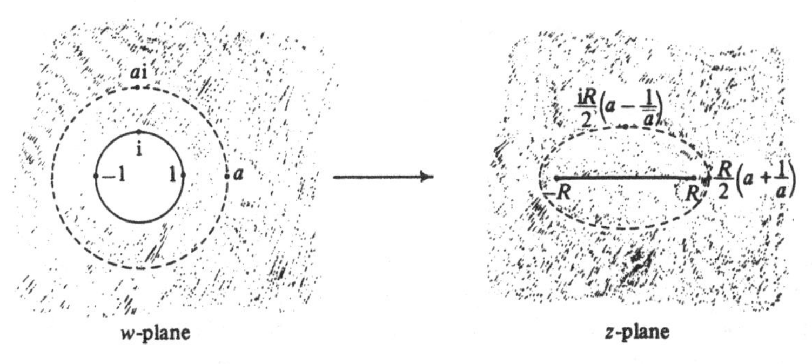

Figure 28

For large $|w|$, $z \sim Rw/2$. Since $P(z)$ is of degree n, we see that $f(w)$ is *bounded* in $\{|w| > 1\}$. Therefore, by the extended principle of maximum (Chapter III, §C!), we have, in that region,

$$|f(w)| \leqslant \sup_{|\omega|=1} |f(\omega)|.$$

However, for $|\omega| = 1$, $z = (R/2)(\omega + 1/\omega)$ is a *real number* x *on the segment* $[-R, R]$, so $|f(\omega)| = |P(x)|$ is $\leqslant M$. Hence $|f(w)| \leqslant M$ for $|w| \geqslant 1$, i.e. $|P(z)| = |f(w)|\,|w|^n \leqslant M|w|^n$. For $1 \leqslant |w| \leqslant a$, the right side is $\leqslant Ma^n$. We are done.

Problem 7

(a) Let P be a polynomial of degree $n-1$ with $|P(x)| \leqslant M$ for $-R \leqslant x \leqslant R$. Show that

$$|P'(0)| \leqslant \frac{en}{R} M.$$

(Hint: Apply Cauchy's inequality, using a circle of suitably chosen radius with its center at 0.)

(b) Let $f(x)$ be infinitely differentiable on $\mathbb{R}$, and *bounded* thereon. Suppose that each of f's derivatives is also bounded on $\mathbb{R}$, and write

$$B_k = \sup_x |f^{(k)}(x)|.$$

Show that there is a constant C independent of n such that

$$B_1 \leqslant C B_0^{(n-1)/n} B_n^{1/n}$$

for $n = 1, 2, 3, \ldots$.
(Hint: To show that

$$|f'(0)| \leqslant C B_0^{(n-1)/n} B_n^{1/n},$$

take

$$P(x) = f(0) + x f'(0) + \cdots + \frac{x^{n-1}}{(n-1)!} f^{(n-1)}(0),$$

and apply (a) to $P(x)$ with a suitably chosen R, using Lagrange's formula for the remainder to estimate $\sup_{-R\leqslant x\leqslant R}|P(x)|$.)

(c) By iterating the result found in (b), show that

$$B_k \leqslant C^k B_0^{(n-k)/n} B_n^{k/n} \quad \text{for } 1 \leqslant k \leqslant n-1.$$

(Hint. $f''(x)$ is $\mathrm{d}f'(x)/\mathrm{d}x$, and so forth.)

Remarks on problem 7. In the result of (a), the factor e is not necessary. The inequality without e requires a more sophisticated proof. The result of (c) was first established (independently) by Gorny and by Cartan. In it, the factor C^k may be replaced by 2. This improvement, due to Kolmogorov, is quite a bit deeper. A discussion of it is found in Mandelbrojt's 1952 book. Another treatment is in the complements near the end of Akhiezer's book on the theory of approximation.

The final result of the last problem is used to establish the following

Theorem (Cartan, Gorny). *Let $M_n > 0$, and let $\{\underline{M}_n\}$ be the convex logarithmic regularization of $\{M_n\}$. Then*

$$\mathscr{C}_{\mathbb{R}}(\{M_n\}) = \mathscr{C}_{\mathbb{R}}(\{\underline{M}_n\}).$$

Proof. Since $\underline{M}_n \leqslant M_n$, it is manifest that $\mathscr{C}_{\mathbb{R}}(\{\underline{M}_n\}) \subseteq \mathscr{C}_{\mathbb{R}}(\{M_n\})$, so our real task is to prove the opposite inclusion.

For each $N = 2, 3, 4, \ldots$, put

$$M_n(N) = \begin{cases} M_n & 0 \leqslant n \leqslant N, \\ \infty, & n > N, \end{cases}$$

and form the convex logarithmic regularization $\{\underline{M}_n(N)\}$ of $\{M_n(N)\}$ in the usual fashion:

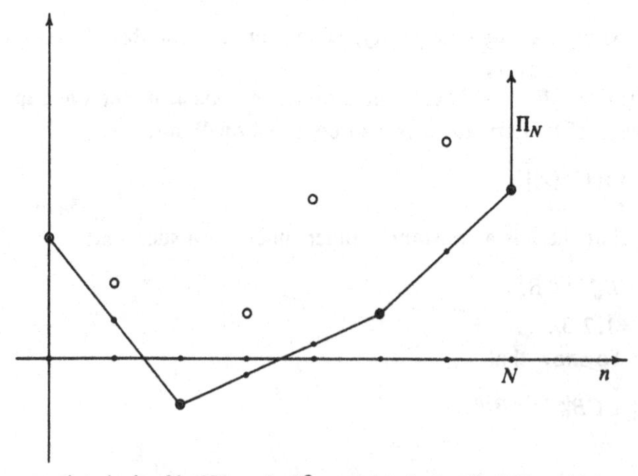

points $(n, \log M_n(N))$ marked ○ points $(n, \log \underline{M}_n(N))$ marked •

Figure 29

For $0 \leqslant n \leqslant N$, $\log \underline{M}_n(N)$ is the ordinate at n of Π_N, the *highest convex polygon* lying under the first $N+1$ points $(n, \log M_n)$, $n = 0, 1, \ldots, N$. As $N \to \infty$, the polygons Π_N *go down* towards Π, the Newton polygon of the set of *all* the points $(n, \log M_n)$, $n = 0, 1, 2, \ldots$. This means that $\underline{M}_n(N) \to \underline{M}_n$ for each n as $N \to \infty$.

Let $f \in \mathscr{C}_{\mathbb{R}}(\{M_n\})$. In order to show that $f \in \mathscr{C}_{\mathbb{R}}(\{\underline{M}_n\})$ (which will complete the proof of our theorem) it is enough, according to the observation just made, to show that there are constants a and σ independent of N with

$$|f^{(n)}(x)| \leqslant a\sigma^n \underline{M}_n(N)$$

for every x and $n = 0, 1, \ldots, N$.

Pick any N, which we fix for the moment, and denote by n_k, $0 = n_0 < n_1 < n_2 < \cdots < n_p = N$, the *abscissae of the vertices* of Π_N. Since $f \in \mathscr{C}_{\mathbb{R}}(\{M_n\})$, we have $|f^{(k)}(x)| \leqslant b\rho^k M_k$ with some constants b and ρ for all $x \in \mathbb{R}$ and each $k = 0, 1, 2, 3, \ldots$. Therefore, *if n is one of the n_j, we already have*

$$|f^{(n)}(x)| \leqslant b\rho^n \underline{M}_n(N), \quad x \in \mathbb{R},$$

since in that case $\underline{M}_n(N) = \underline{M}_{n_j}(N) = M_{n_j}(N) = M_n$. Suppose, then, that $n_j < n < n_{j+1}$. We at least have

$$(*) \qquad |f^{(n_j)}(x)| \leqslant b\rho^{n_j} M_{n_j}, \quad x \in \mathbb{R},$$

and

$$(\ast\!\ast) \qquad |f^{(n_{j+1})}(x)| \leqslant b\rho^{n_{j+1}} M_{n_{j+1}}, \quad x \in \mathbb{R}.$$

Now apply result (c) of problem 7 to the function

$$g(x) = f^{(n_j)}(x).$$

With $(*)$ and $(\ast\!\ast)$, that result yields

$$|f^{(n)}(x)| = |g^{(n-n_j)}(x)| \leqslant C^{n-n_j} \{b\rho^{n_j} M_{n_j}\}^{(n_{j+1}-n)/(n_{j+1}-n_j)} \times \{b\rho^{n_{j+1}} M_{n_{j+1}}\}^{(n-n_j)/(n_{j+1}-n_j)}$$

for $x \in \mathbb{R}$. Here, C is an *absolute constant which we can wlog take* $\geqslant 1$. However,

$$M_{n_j}^{(n_{j+1}-n)/(n_{j+1}-n_j)} M_{n_{j+1}}^{(n-n_j)/(n_{j+1}-n_j)} = \underline{M}_n(N):$$

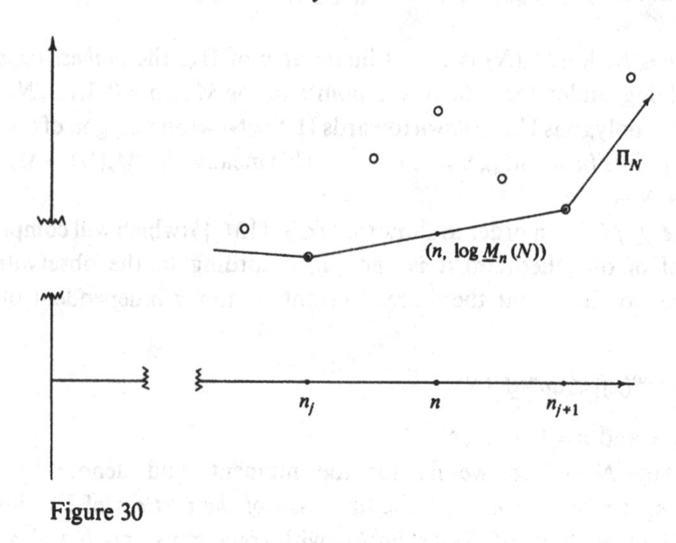

Figure 30

So the preceding relation becomes

$$|f^{(n)}(x)| \leqslant C^{n-n_j} b\rho^n \underline{M}_n(N), \quad x \in \mathbb{R},$$

or, since $C \geqslant 1$,

$$|f^{(n)}(x)| \leqslant b(C\rho)^n \underline{M}_n(N), \quad x \in \mathbb{R}.$$

We have thus established the desired inequality for $n = 0, 1, 2, \ldots, N$ with $a = b$ and $\sigma = C\rho$. As remarked above, this is enough to prove the theorem, and we are done.

The result just established has an important theoretical consequence.

Theorem. *$\mathscr{C}_{\mathbb{R}}(\{M_n\})$ is an algebra, i.e., if f and g belong to that class, so does $f \cdot g$.*

Proof. Let f and g belong to $\mathscr{C}_{\mathbb{R}}(\{M_n\})$; since $\underline{M}_n \leqslant M_n$ it is enough to prove that $f \cdot g \in \mathscr{C}_{\mathbb{R}}(\{\underline{M}_n\})$.

By the above theorem, we certainly have f and g in $\mathscr{C}_{\mathbb{R}}(\{\underline{M}_n\})$, so, for $n \geqslant 0$,

$$|f^{(n)}(x)| \leqslant a\rho^n \underline{M}_n, \quad x \in \mathbb{R},$$

and

$$|g^{(n)}(x)| \leqslant a\sigma^n \underline{M}_n, \quad x \in \mathbb{R},$$

if the constant a is chosen sufficiently large. According to Leibniz' formula,

$$\left(\frac{\mathrm{d}}{\mathrm{d}x}\right)^n (f(x)g(x)) = \sum_{k=0}^{n} \binom{n}{k} f^{(k)}(x) g^{(n-k)}(x),$$

and, by the inequalities just given, the sum on the right is in modulus

$$\leqslant \sum_{k=0}^{n} \binom{n}{k} a\rho^k \underline{M}_k \cdot a\sigma^{n-k} \underline{M}_{n-k}.$$

However, since $\log \underline{M}_n$ is a *convex function* of m,

$$\log \underline{M}_n - \log \underline{M}_{n-k} \geqslant \log \underline{M}_k - \log \underline{M}_0,$$

so $\underline{M}_k \underline{M}_{n-k} \leqslant \underline{M}_0 \underline{M}_n$.

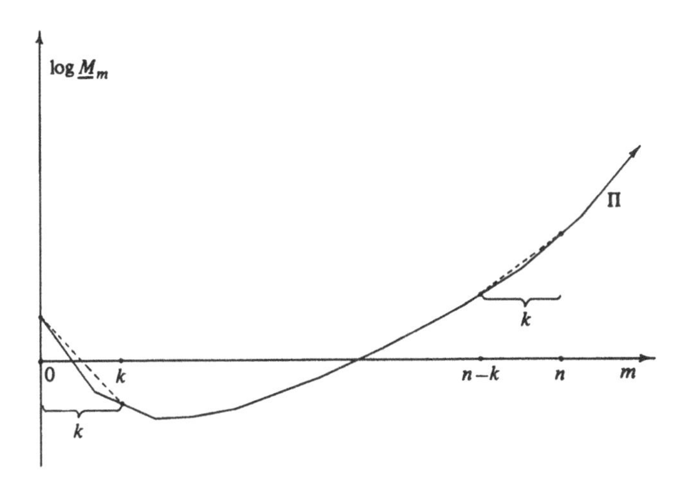

Figure 31

The preceding sum is therefore

$$\leqslant a^2 \sum_{k=0}^{n} \binom{n}{k} \rho^k \sigma^{n-k} \underline{M}_0 \underline{M}_n = a^2 \underline{M}_0 (\rho + \sigma)^n \underline{M}_n,$$

in other words,

$$\left| \left(\frac{\mathrm{d}}{\mathrm{d}x} \right)^n \Big(f(x) g(x) \Big) \right| \leqslant a^2 \underline{M}_0 (\rho + \sigma)^n \underline{M}_n, \quad x \in \mathbb{R},$$

when $n = 1, 2, 3, \ldots$.

We also clearly have $|f(x)g(x)| \leqslant a^2 M_0$ on $\mathbb{R}$. Therefore $f \cdot g \in \mathscr{C}_{\mathbb{R}}(\{\underline{M}_n\})$, and the theorem is proved.

Here is a good place to end our elementary discussion of quasianalyticity. Several ideas introduced in this chapter find applications in other parts of analysis, and will be met with again in this book. The Paley–Wiener construction in §D has various uses, and is the starting point of some important further investigations, which we will take up in Chapters X and XI. The whole notion of convex regularization, which played such a big rôle in this chapter, turns out to be especially important. A similar kind of regularization is used in some work of Beurling, and in the proof of Volberg's theorem. Those matters will be studied in Chapter VII.

V

The moment problem on the real line

The moment problem on $\mathbb{R}$, known also as the Hamburger moment problem, consists of two questions:

1. Given a numerical sequence $S_0, S_1, S_2, \ldots$, when is there a *positive* Radon measure μ on $\mathbb{R}$ with all the integrals $\int_{-\infty}^{\infty} |x|^k \mathrm{d}\mu(x)$, $k \geqslant 0$, convergent, such that

$$S_k = \int_{-\infty}^{\infty} x^k \,\mathrm{d}\mu(x), \quad k = 0, 1, 2, 3, \ldots?$$

2. If the answer to 1 is *yes*, is there *only one positive measure* μ on $\mathbb{R}$ with

$$S_k = \int_{-\infty}^{\infty} x^k \,\mathrm{d}\mu(x), \quad k = 0, 1, 2, 3, \ldots?$$

When the answer to 1 is *yes*, $\{S_k\}$ is called a *moment sequence* and the numbers S_k are called *the moments* of the measure μ. If, for a moment sequence $\{S_k\}$, the answer to 2 is *yes*, we say that $\{S_k\}$ is *determinate*. If the answer to 2 is *no*, the moment sequence $\{S_k\}$ is said to be *indeterminate*.

The study of various kinds of moment problems goes back to the second half of the last century, when Tchebyshev and Stieltjes investigated the moment problem on the *half-line* $[0, \infty)$. Stieltjes' research thereon led him to invent the integral bearing his name. A lot of familiar ideas and notions in analysis did in fact originate in work on the moment problem, and the subject as we now know it has many of its roots in such work. The following discussion will perhaps give the reader some perceptions of this relationship.

It is really only question 2 (the one involving *uniqueness* of μ) that has to do with the subject of this book, mainly through its connection with the material in the previous and next chapters. It would not, however, make much sense to discuss 2 without at first dealing with 1 (on the *existence* of μ). We give what is essentially M. Riesz' treatment of 1 in §A. Most of the rest of the material in this chapter is also based on M. Riesz' work.

The reader should not believe that our discussion of the moment problem reflects its real scope. We do not even touch on some very important approaches to it. There is, for instance, a vast formal structure involved with the recurrence relations of orthogonal polynomials and the algebra of continued fractions which is relevant to other parts of analysis such as Sturm–Liouville theory and the problem of interpolation by bounded analytic functions, as well as to the moment problem. Our subject is also connected in various ways with the theory of operators in Hilbert space, Krein's work being especially important in this regard. It would require a whole book to deal with all of these matters.

There is a very good book, namely, the one by Akhiezer (*The Classical Moment Problem*). That book has been translated into English; unfortunately, both the Russian and the English versions are now out of print and very hard to find – at present there is *one* copy that I know of in the city of Montreal! The older work of Tamarkin and Shohat is somewhat more accessible. The reader may also be interested in looking at the original papers by M. Riesz; they are in the *Arkiv för Matematik, Astronomi och Fysik*, and appeared around 1922–3. I am indebted to Professor R. Vermes for showing me those papers.

A. Characterization of moment sequences. Method based on extension of positive linear functionals

Theorem. *There is a positive measure μ on $\mathbb{R}$ with*

$$S_k = \int_{-\infty}^{\infty} x^k \, d\mu(x), \quad k = 0, 1, 2, 3, \ldots,$$

if and only if

$$(*) \qquad \sum_{j=0}^{N} \sum_{i=0}^{N} S_{i+j} \xi_i \xi_j \geqslant 0$$

for any N and any choice of the real numbers $\xi_0, \xi_1, \ldots, \xi_N$.

Proof. (M. Riesz) The condition $(*)$ is certainly *necessary*, for if

$$P(x) = \sum_{i=0}^{N} \xi_i x^i$$

is any real polynomial of degree N, we certainly have

$$\int_{-\infty}^{\infty} (P(x))^2 \, d\mu(x) \geqslant 0;$$

the integral is, however, clearly equal to the left side of (*). The real work is to prove (*) sufficient.

Denote by $\mathscr{P}$ the set of *real polynomials*, and for $P(x) \in \mathscr{P}$ put

$$L(P) = \sum_{k=0}^{N} S_k a_k,$$

where $\sum_{k=0}^{N} a_k x^k = P(x)$. L is then a *real linear form* on the *vector space* $\mathscr{P}$; it is claimed that L is *positive* on $\mathscr{P}$, i.e., that if $P \in \mathscr{P}$ and $P(x) \geqslant 0$ on $\mathbb{R}$, we have $L(P) \geqslant 0$. *Take* a real polynomial $P(x)$ which is non-negative on $\mathbb{R}$. By Schwarz' reflection principle, $\overline{P(\bar{z})} = P(z)$, so if $\alpha \notin \mathbb{R}$ is a root of P, so is $\bar{\alpha}$, and $\bar{\alpha}$ has the same multiplicity as α. Again, every *real* root of P must have *even* multiplicity. Factoring $P(x)$ completely, we see that $P(x)$ must be of the form $|g(x)|^2$ (for real x), where $g(x)$ is a certain polynomial with *complex coefficients*. We can write

$$g(x) = R(x) + iS(x)$$

where R and S are *polynomials* with *real coefficients*, and then we will have $P(x) = (R(x))^2 + (S(x))^2$, so that

$$L(P) = L(R^2) + L(S^2).$$

However, if, for example,

$$R(x) = \sum_i \xi_i x^i,$$

the coefficient of x^k in $(R(x))^2$ is $\sum_{i+j=k} \xi_i \xi_j$, whence

$$L(R^2) = \sum_k S_k \left(\sum_{i+j=k} \xi_i \xi_j \right) = \sum_i \sum_j S_{i+j} \xi_i \xi_j$$

which is $\geqslant 0$ by (*). In the same way, we see that $L(S^2) \geqslant 0$, so finally $L(P) \geqslant 0$ as asserted.

In order to prove the sufficiency of (*), we have to obtain a positive measure μ on $\mathbb{R}$ with $S_k = \int_{-\infty}^{\infty} x^k \, d\mu(x)$ for $k \geqslant 0$; for this purpose, M. Riesz brought into play the rather peculiar space $\mathscr{E} = \mathscr{P} \dotplus \mathscr{C}_0$ consisting of *all sums* $P(x) + \varphi(x)$, where $P \in \mathscr{P}$ and $\varphi \in \mathscr{C}_0$, the set of *real continuous functions* on $\mathbb{R}$ *tending to zero as* $x \to \pm\infty$. Our linear form L is *defined and positive* on the vector subspace $\mathscr{P}$ of $\mathscr{E}$, and the idea is to *extend* L *to all of* $\mathscr{E}$, in such fashion that *it remain positive on this larger space.* The extension is carried out inductively.

Let us take any fixed countable subset $\{\varphi_n: n=1,2,3,\ldots\}$ of $\mathscr{C}_0$, *dense therein with respect to the usual sup-norm* $\|\ \ \|_\infty$, and, for $n=1,2,\ldots$, call $\mathscr{E}_n$ the *vector subspace* of $\mathscr{E}$ generated by $\mathscr{P}$ and $\varphi_1,\ldots,\varphi_n$. In order to have a uniform notation, we write $\mathscr{E}_0=\mathscr{P}$; then $\mathscr{E}_0\subset\mathscr{E}_1\subset\mathscr{E}_2\subset\cdots$, so the union $\mathscr{E}_\infty=\bigcup_{n=0}^\infty\mathscr{E}_n$ is also a vector subspace of $\mathscr{E}$. We first extend L from $\mathscr{E}_0=\mathscr{P}$ to $\mathscr{E}_\infty$ so as to keep it positive on $\mathscr{E}_\infty$.

Suppose, for $n\geqslant 0$, that L has already been extended to $\mathscr{E}_n$ and is positive thereon. We show how to extend it to $\mathscr{E}_{n+1}$ so that it stays positive. In case $\varphi_{n+1}\in\mathscr{E}_n$, we have $\mathscr{E}_{n+1}=\mathscr{E}_n$, and then nothing need be done. We must examine the situation where $\varphi_{n+1}\notin\mathscr{E}_n$. Because L is already defined on $\mathscr{E}_n$, the two quantities

$$A=\sup\{Lf:\ f\in\mathscr{E}_n \text{ and } f\leqslant\varphi_{n+1}\}$$

and

$$B=\inf\{Lg:\ g\in\mathscr{E}_n \text{ and } g\geqslant\varphi_{n+1}\}$$

are available. (If peradventure there were *no* $f\in\mathscr{E}_n$ with $f\leqslant\varphi_{n+1}$ we would put $A=-\infty$. And if there were no $g\in\mathscr{E}_n$ with $g\geqslant\varphi_{n+1}$, we'd take $B=\infty$. Neither of these possibilities can, however, occur in the present situation as we shall see immediately.) It is important to verify that

$$-\infty<A\leqslant B<\infty.$$

We do this by using the function 1, which, *as a polynomial* (!), *belongs to* $\mathscr{P}=\mathscr{E}_0$, *hence to* $\mathscr{E}_n$. Because $\varphi_{n+1}\in\mathscr{C}_0$, we have the evident inequality

$$-\|\varphi_{n+1}\|_\infty\cdot 1\ \leqslant\ \varphi_{n+1}(x)\ \leqslant\ \|\varphi_{n+1}\|_\infty\cdot 1,$$

so $A\geqslant-\|\varphi_{n+1}\|_\infty L(1)$ and $B\leqslant\|\varphi_{n+1}\|_\infty L(1)$. Moreover, if f and $g\in\mathscr{E}_n$ are such that $f\leqslant\varphi_{n+1}\leqslant g$ (*note that we have just seen that there* ***are*** *such functions f and g!*), we have $g-f\in\mathscr{E}_n$ and $g-f\geqslant 0$, so $L(g-f)\geqslant 0$ *by positivity of L on $\mathscr{E}_n$*, i.e.,

$$L(f)\leqslant L(g).$$

This shows that $A\leqslant B$.

Take any number c with

$$A\leqslant c\leqslant B,$$

and put

$$L(\varphi_{n+1})=c,$$

which, taking L as linear, gives an extension of L from $\mathscr{E}_n$ to $\mathscr{E}_{n+1}$. It is claimed that L as thus extended is still positive on $\mathscr{E}_{n+1}$.

Any element of $\mathscr{E}_{n+1}$ can be expressed as $h + a\varphi_{n+1}$ with $h \in \mathscr{E}_n$ and $a \in \mathbb{R}$. If $h + a\varphi_{n+1} \geqslant 0$ and $a = 0$, then we already know $L(h + a\varphi_{n+1}) = L(h)$ is $\geqslant 0$. If $h + a\varphi_{n+1} \geqslant 0$ and $a > 0$, $(1/a)h + \varphi_{n+1} \geqslant 0$, i.e., $-(1/a)h \leqslant \varphi_{n+1}$. Since the left side belongs to $\mathscr{E}_n$, we have by the definition of A that

$$L\left(-\frac{1}{a}h\right) \leqslant A \leqslant c = L(\varphi_{n+1}),$$

whence $0 \leqslant L((1/a)h) + L(\varphi_{n+1})$ and finally $L(h + a\varphi_{n+1}) \geqslant 0$. There remains the case where $h + a\varphi_{n+1} \geqslant 0$ and $a < 0$. Here $(1/a)h + \varphi_{n+1} \leqslant 0$ and $\varphi_{n+1} \leqslant -(1/a)h \in \mathscr{E}_n$. Therefore $L(-(1/a)h) \geqslant B \geqslant c = L(\varphi_{n+1})$, $L(-(1/a)h) - L(\varphi_{n+1}) \geqslant 0$, and finally $L(h + a\varphi_{n+1}) \geqslant 0$, since $-a > 0$. Taking $L(\varphi_{n+1}) = c$ thus produces a positive extension of L from $\mathscr{E}_n$ to $\mathscr{E}_{n+1}$, as asserted.

Indefinite continuation of this process yields a positive extension of L from $\mathscr{E}_0 = \mathscr{P}$ to $\mathscr{E}_\infty$. Once this has been done, however, we may extend L from $\mathscr{E}_\infty$ to all of $\mathscr{E}$ *by continuity*.

Take any $f \in \mathscr{E}$; we have $f = P + \varphi$ where $P \in \mathscr{P}$ and $\varphi \in \mathscr{C}_0$. Since the functions φ_n used in forming the $\mathscr{E}_n$ are *dense* in $\mathscr{C}_0$, there is some subsequence of them, $\{\varphi_{n_k}\}$, with

$$\|\varphi - \varphi_{n_k}\|_\infty \longrightarrow 0 \quad \text{for} \quad k \to \infty.$$

Then $\lim_{k\to\infty} L(\varphi_{n_k})$ exists. Indeed, since L is positive on $\mathscr{E}_\infty$, which includes all the φ_n, we have

$$|L(\varphi_{n_k}) - L(\varphi_{n_l})| \leqslant \|\varphi_{n_k} - \varphi_{n_l}\|_\infty \cdot L(1),$$

because

$$-\|\varphi_{n_k} - \varphi_{n_l}\|_\infty \leqslant \varphi_{n_k}(x) - \varphi_{n_l}(x) \leqslant \|\varphi_{n_k} - \varphi_{n_l}\|_\infty.$$

Here, $\|\varphi_{n_k} - \varphi_{n_l}\|_\infty \xrightarrow[k,l]{} 0$, so the limit in question does exist. If $\{\psi_k\}$ is *any other* sequence of functions in $\mathscr{E}_\infty$ with $\|\varphi - \psi_k\|_\infty \xrightarrow[k]{} 0$, we have $\|\varphi_{n_k} - \psi_k\|_\infty \xrightarrow[k]{} 0$, so, by the argument just given,

$$L(\psi_k) - L(\varphi_{n_k}) \xrightarrow[k]{} 0,$$

and $\lim_{k\to\infty} L(\psi_k)$ exists, equalling $\lim_{k\to\infty} L(\varphi_{n_k})$. We see in this way that the latter limit is independent of the choice of the particular sequence, $\{\varphi_{n_k}\}$, of φ_n used to approximate φ in norm $\| \quad \|_\infty$, so it makes sense to define

$$L(\varphi) = \lim_{k\to\infty} L(\varphi_{n_k}).$$

We can then put $L(f) = L(P) + L(\varphi)$, and L is in this way extended so as to be a linear form on $\mathscr{E}$.

L as thus extended is *positive* on $\mathscr{E}$. Suppose that $f \in \mathscr{E}$ is non-negative on $\mathbb{R}$. Writing as above $f = P + \varphi$ with $P \in \mathscr{P}$ and $\varphi \in \mathscr{C}_0$, we can take a sequence $\{\varphi_{n_k}\}$ of the φ_n with $\| \varphi - \varphi_{n_k} \|_\infty \xrightarrow[k]{} 0$, and we'll have

$$L(f) = L(P) + \lim_{k \to \infty} L(\varphi_{n_k}).$$

Since $P + \varphi \geqslant 0$, we see that

$$P + \| \varphi - \varphi_{n_k} \|_\infty + \varphi_{n_k} \geqslant 0;$$

this function, however, belongs to $\mathscr{E}_\infty$ (the sum of the first two terms is a polynomial!). Therefore, L being positive on $\mathscr{E}_\infty$,

$$L(P) + \| \varphi - \varphi_{n_k} \|_\infty \cdot L(1) + L(\varphi_{n_k}) \geqslant 0,$$

and, making $k \to \infty$, we get $L(f) \geqslant 0$, as claimed.

The linear form L *is in particular positive on* $\mathscr{C}_0$. Therefore, *by F. Riesz' representation theorem*, there is a *positive measure* μ on $\mathbb{R}$ with

$$L(\varphi) = \int_{-\infty}^{\infty} \varphi(x)\, \mathrm{d}\mu(x)$$

for $\varphi \in \mathscr{C}_0$. We would like now to show that in fact

$$L(f) = \int_{-\infty}^{\infty} f(x)\, \mathrm{d}\mu(x)$$

for all f *in* $\mathscr{E}$. This seems at first unlikely, because there is *so little connection* between the two vector spaces $\mathscr{P}$ and $\mathscr{C}_0$ used to make up $\mathscr{E}$ – there doesn't seem to be much hope of relating L's behaviour on $\mathscr{P}$ to that on $\mathscr{C}_0$. The formula in question turns out nevertheless to be correct.

In order to accomplish the passage from $\mathscr{C}_0$ to $\mathscr{P}$, M. Riesz *used a trick* (which was later codified by Choquet into the so-called 'method of adapted cones'). Let us start with an even power x^{2k} of x, and show that

$$\int_{-\infty}^{\infty} x^{2k}\, \mathrm{d}\mu(x)$$

is *finite and equal to* $L(x^{2k})$. For each large N, take the function $\varphi_N \in \mathscr{C}_0$ defined thus:

$$\varphi_N(x) = \begin{cases} x^{2k}, & |x| \leqslant N, \\ 0, & |x| \geqslant 2N, \\ \textit{a linear function} \text{ on } [N, 2N] \\ \textit{and on } [-2N, -N]. \end{cases}$$

There is clearly a quantity ε_N, tending to zero as $N \to \infty$, such that

$$(\dagger) \qquad x^{2k} \leqslant \varphi_N(x) + \varepsilon_N x^{2k+2}, \quad x \in \mathbb{R}.$$

(That's the *main idea* here!) We also have

$$\varphi_N(x) \leqslant x^{2k}, \quad x \in \mathbb{R},$$

with equality for $-N \leqslant x \leqslant N$, and of course $\varphi_N \geqslant 0$.

Since the measure μ is positive,

$$\int_{-N}^{N} x^{2k}\,\mathrm{d}\mu(x) = \int_{-N}^{N} \varphi_N(x)\,\mathrm{d}\mu(x)$$

$$\leqslant \int_{-\infty}^{\infty} \varphi_N(x)\,\mathrm{d}\mu(x) = L(\varphi_N) \leqslant L(x^{2k}),$$

the last inequality holding because L is positive on $\mathscr{E}$. At the same time, by (†) and the positivity of L,

$$L(x^{2k}) \leqslant L(\varphi_N) + \varepsilon_N L(x^{2k+2}),$$

whilst

$$L(\varphi_N) = \int_{-2N}^{2N} \varphi_N(x)\,\mathrm{d}\mu(x) \leqslant \int_{-2N}^{2N} x^{2k}\,\mathrm{d}\mu(x).$$

Combining these two relations with the preceding one, we get

$$\int_{-N}^{N} x^{2k}\,\mathrm{d}\mu(x) \leqslant L(x^{2k}) \leqslant \int_{-2N}^{2N} x^{2k}\,\mathrm{d}\mu(x) + \varepsilon_N L(x^{2k+2}).$$

Making $N \to \infty$, we see that

$$L(x^{2k}) = \int_{-\infty}^{\infty} x^{2k}\,\mathrm{d}\mu(x),$$

since $\varepsilon_N \to 0$. Because $|x|^k \leqslant 1 + x^{2k}$ and L is finite (!), this reasoning also shows that all the integrals

$$\int_{-\infty}^{\infty} |x|^k\,\mathrm{d}\mu(x)$$

are *convergent*.

We must still treat the odd powers of x. This can be done by going through an argument like the one just made, working with $x^{2k} + x^{2k+1} + x^{2k+2}$ (a *non-negative* function of x!) instead of with x^{2k}. In that way, we can conclude that

$$L(x^{2k} + x^{2k+1} + x^{2k+2}) = \int_{-\infty}^{\infty} (x^{2k} + x^{2k+1} + x^{2k+2})\,\mathrm{d}\mu(x),$$

whence, using what we already know for the even powers of x,

$$L(x^{2k+1}) = \int_{-\infty}^{\infty} x^{2k+1}\,\mathrm{d}\mu(x).$$

The relation $L(x^n) = \int_{-\infty}^{\infty} x^n\,\mathrm{d}\mu(x)$ is now established for $n = 0, 1, 2, 3, \ldots$. *However, $L(x^n) = S_n$ according to our original definition of the linear form L!* So

$$S_n = \int_{-\infty}^{\infty} x^n\,\mathrm{d}\mu(x),$$

and $\{S_n\}$ *is* a moment sequence. We are done.

Remark. The argument at the beginning of the above proof can be followed so as to establish a general theorem about the extension of positive linear functionals on real linear spaces $\mathscr{E}$ with positive cones. (A *positive cone* in $\mathscr{E}$ is a cone $\mathscr{K}$ with vertex at 0 such that $\mathscr{E} = \mathscr{K} - \mathscr{K}$, and a functional on $\mathscr{E}$ is called *positive* if it takes non-negative values on $\mathscr{K}$.) The reader is invited to formulate and prove such a theorem.

Remark. The reader's attention is directed to the similarity of the inductive extension procedure used in the above proof and the inductive step in the proof of the Hahn–Banach theorem. What is the relation of the general theorem alluded to in the previous remark and Hahn–Banach? Can either one be obtained from the other?

B. Scholium. Determinantal criterion for $\{S_n\}$ to be a moment sequence

The necessary and sufficient condition for $\{S_k\}$ to be a moment sequence furnished by the theorem of the preceding §, namely, that the forms

$$\sum_{i=0}^{N} \sum_{j=0}^{N} S_{i+j}\xi_i\xi_j$$

be *positive*, is equivalent to another one involving the principal determinants of the infinite matrix

$$\begin{bmatrix} S_0, & S_1, & S_2, & S_3, & \cdots \\ S_1, & S_2, & S_3, & S_4, & \cdots \\ S_2, & S_3, & S_4, & S_5, & \cdots \\ S_3, & S_4, & & & \\ \vdots & \vdots & \vdots & \vdots & \end{bmatrix}.$$

This determinantal condition played an important rôle in the older investigations on the moment problem, and we give it here for the sake of completeness. Matrices like the one just written are called *Hankel matrices* and were extensively studied towards the end of the last century, in particular, by Frobenius.

We need two lemmas from linear algebra.

Lemma. *Given the symmetric matrix*

$$S = \begin{bmatrix} s_{0,0} & s_{0,1} & \cdots & s_{0,N+1} \\ s_{1,0} & s_{1,1} & \cdots & s_{1,N+1} \\ \vdots & & & \\ s_{N+1,0} & s_{N+1,1} & \cdots & s_{N+1,N+1} \end{bmatrix}$$

where $N \geqslant -1$ (*sic*!), *the form*

$$\sum_{i=0}^{N+1} \sum_{j=0}^{N+1} s_{i,j}\xi_i\xi_j$$

is strictly (*sic*!) *positive definite if and only if all the principal determinants*

$$\det \begin{bmatrix} s_{0,0} & s_{0,1} & \cdots & s_{0,M} \\ s_{1,0} & s_{1,1} & \cdots & s_{1,M} \\ \vdots & & & \\ s_{M,0} & s_{M,1} & \cdots & s_{M,M} \end{bmatrix}$$

are *strictly positive* for $M = 0, 1, \ldots, N + 1$.

▶ ***Remark and warning.*** If we replace '*strictly* positive definite' by 'positive definite' and merely require the principal determinants to be $\geqslant 0$, the corresponding statement is *false*. Example:

$$S = \begin{bmatrix} 1 & 0 & 0 \\ 0 & 0 & 0 \\ 0 & 0 & -1 \end{bmatrix}.$$

The danger of this pitfall (in which I myself landed during one of my lectures!) was pointed out to me by Professor G. Schmidt.

Proof of lemma. If the quadratic form in question *is* strictly positive definite, then so is each of the forms

$$\sum_{i=0}^{M} \sum_{j=0}^{M} s_{i,j}\xi_i\xi_j$$

for $M = 0, 1, \ldots, N + 1$. This means that the *characteristic values* of the *matrix* of any such form are all *strictly positive*. But the *product* of the characteristic values of the form just written is equal to the *determinant* figuring in the lemma's statement. So, in *one* direction, the lemma is *clear*.

To go in the *opposite* direction, we argue by *induction* on N. For $N = -1$ we have the quadratic form $s_{0,0}\zeta_0^2$ whose determinant is just $s_{0,0}$. In this case, the desired result is manifest.

Let us therefore *assume* that the lemma is *true* with N standing in place of $N+1$, and then *prove* that it is *also true* as stated, with $N+1$. We are given that the determinants in question are all >0 for $M = 0, 1, \ldots, N+1$. In particular, then, we have $s_{0,0} > 0$ (*this is the place in the proof where strict positivity of the principal determinants is used!*) so we may wlog take $s_{0,0} = 1$, since multiplication of the quadratic form by a constant >0 does not affect its strict positivity. With this normalization, our $(N+2) \times (N+2)$ matrix S take the form

$$S = \left[\begin{array}{c|cccc} 1 & \sigma_1 & \sigma_2 & \cdots & \sigma_{N+1} \\ \hline \sigma_1 & & & & \\ \sigma_2 & & & & \\ \vdots & & & S' & \\ \sigma_{N+1} & & & & \end{array}\right]$$

where S' is a certain $(N+1) \times (N+1)$ symmetric matrix.

To show that S is strictly positive definite, it is enough to show that the matrix T congruent to it equal to

$$(\ast)\qquad \begin{bmatrix} 1 & 0 & 0 & \cdots & 0 \\ -\sigma_1 & 1 & 0 & \cdots & 0 \\ -\sigma_2 & 0 & 1 & 0\cdots & 0 \\ \vdots & & & & \\ -\sigma_{N+1} & 0 & 0 & \cdots & 1 \end{bmatrix} \times S \times \begin{bmatrix} 1 & -\sigma_1 & -\sigma_2 & \cdots & -\sigma_{N+1} \\ 0 & 1 & 0 & \cdots & 0 \\ 0 & 0 & 1 & \cdots & 0 \\ \vdots & & & & \\ 0 & 0 & \cdots & \cdots & 1 \end{bmatrix}$$

is strictly positive definite. Observe first of all that

$$T = \left[\begin{array}{c|cccc} 1 & 0 & 0 & \cdots & 0 \\ \hline 0 & & & & \\ 0 & & & T' & \\ \vdots & & & & \\ 0 & & & & \end{array}\right]$$

with a certain $(N+1) \times (N+1)$ symmetric matrix T'. It is therefore clear that T will be strictly positive definite if T' is. *On account of the particular triangular forms* of the matrices standing on each side of S in $(\ast)$, we have, however, for any *principal minor*

$$\begin{bmatrix} 1 & 0 & 0 & \cdots & 0 \\ 0 & t_{1,1} & t_{1,2} & \cdots & t_{1,R} \\ 0 & t_{2,1} & t_{2,2} & \cdots & t_{2,R} \\ \vdots & & & & \\ 0 & t_{R,1} & t_{R,2} & \cdots & t_{R,R} \end{bmatrix}$$

of T $(1 \leqslant R \leqslant N+1)$:

$$\begin{bmatrix} 1 & 0 & \cdots & 0 \\ 0 & t_{1,1} & \cdots & t_{1,R} \\ \vdots & & & \\ 0 & t_{R,1} & \cdots & t_{R,R} \end{bmatrix} = \begin{bmatrix} 1 & 0 & \cdots & 0 \\ -\sigma_1 & 1 & \cdots & 0 \\ \vdots & & & \\ -\sigma_R & 0 & \cdots & 1 \end{bmatrix}$$

$$\times \begin{bmatrix} s_{0,0} & \cdots & s_{0,R} \\ \vdots & & \\ s_{R,0} & \cdots & s_{R,R} \end{bmatrix} \begin{bmatrix} 1 & -\sigma_1 & \cdots & -\sigma_R \\ 0 & 1 & \cdots & 0 \\ \vdots & & & \\ 0 & 0 & \cdots & 1 \end{bmatrix}.$$

The determinant of the matrix on the left is therefore equal to

$$\det \begin{bmatrix} s_{0,0} & \cdots & s_{0,R} \\ s_{R,0} & \cdots & s_{R,R} \end{bmatrix}$$

which by hypothesis is >0 for $1 \leqslant R \leqslant N+1$. The determinant of the left-hand matrix is, however, just

$$\det \begin{bmatrix} t_{1,1} & \cdots & t_{1,R} \\ \vdots & & \\ t_{R,1} & \cdots & t_{R,R} \end{bmatrix},$$

i.e., *the determinant of the Rth principal minor* of the $(N+1) \times (N+1)$ *symmetric matrix* T'. Those determinants are therefore all >0, so, since T' has one row and one column *less* than T, it is strictly positive definite by our induction hypothesis. So, therefore, is T, and hence S, as we wished to show. The lemma is proved.

Kronecker's lemma. *Let a sequence* $s_0, s_1, s_2, \ldots$ *be given, and denote, for* $n \geqslant 0$, *the matrix*

$$\begin{bmatrix} s_0 & s_1 & \cdots & s_n \\ s_1 & s_2 & \cdots & s_{n+1} \\ \vdots & & & \\ s_n & s_{n+1} & \cdots & s_{2n} \end{bmatrix}$$

by Δ_n. *Suppose there is a number* $m \geqslant 1$ *such that* $\det \Delta_n \neq 0$ *for* $n = 0, 1, \ldots, m-1$ *while* $\det \Delta_n = 0$ *for all* $n \geqslant m$. *Then there are numbers* $\alpha_0, \alpha_1, \ldots, \alpha_{m-1}$ *such that, for all* $p \geqslant 0$,

$$s_{m+p} + \alpha_{m-1} s_{m-1+p} + \cdots + \alpha_0 s_p = 0.$$

Proof. Since $\det \Delta_m = 0$, there is a non-trivial relation of linear dependence

$$\alpha_0 \begin{bmatrix} s_0 \\ s_1 \\ \vdots \\ s_m \end{bmatrix} + \alpha_1 \begin{bmatrix} s_1 \\ s_2 \\ \vdots \\ s_{m+1} \end{bmatrix} + \cdots + \alpha_m \begin{bmatrix} s_m \\ s_{m+1} \\ \vdots \\ s_{2m} \end{bmatrix} = 0$$

between the columns of Δ_m. Since $\det \Delta_{m-1} \neq 0$, we *cannot have* $\alpha_m = 0$, and may as well take $\alpha_m = 1$. Then the desired relation clearly holds for $p = 0, 1, 2, \ldots, m$, and we want to show that it holds for $p > m$. This we do by induction.

Write, for $p \geqslant 0$,

$$\Sigma_{m+p} = s_{m+p} + \alpha_{m-1} s_{m-1+p} + \cdots + \alpha_0 s_p,$$

and *assume* that $\Sigma_{m+p} = 0$ for $p = 0, 1, \ldots, r-1$, with $r - 1 \geqslant m$, i.e. $r \geqslant m+1$. Let us then *prove* that $\Sigma_{m+r} = 0$.

We have $\det \Delta_r = 0$. Since $r \geqslant m+1$, we can write

$$\Delta_r = \left[\begin{array}{ccc|cccc} & & & s_m & s_{m+1} & \cdots & s_r \\ & \Delta_{m-1} & & s_{m+1} & & \cdots & \\ & & & \vdots & & & \vdots \\ & & & s_{2m-1} & & \cdots & s_{m+r-1} \\ \hline s_m & \cdots & \cdots & s_{2m} & & \cdots & s_{m+r} \\ \vdots & & & \vdots & & & \vdots \\ s_r & & & s_{m+r} & & \cdots & s_{2r} \end{array}\right]$$

Denote by σ_k the kth *column* of this matrix, whose *initial* column is called the *zeroth* one. The *determinant* of the matrix is then *unchanged* if, for each $k \geqslant m$ we *add* to σ_k the linear combination

$$\alpha_{m-1}\sigma_{k-1} + \alpha_{m-2}\sigma_{k-2} + \cdots + \alpha_0 \sigma_{k-m}$$

of the m columns preceding it. These column operations convert Δ_r to the matrix

$$\left[\begin{array}{ccc|ccc} & & & \Sigma_m & \cdots & \Sigma_r \\ & \Delta_{m-1} & & \vdots & & \\ & & & \Sigma_{2m-1} & \cdots & \Sigma_{m+r-1} \\ \hline s_m & \cdots & s_{2m-1} & \Sigma_{2m} & \cdots & \Sigma_{m+r} \\ \vdots & & & \vdots & & \\ s_r & \cdots & s_{m+r-1} & \Sigma_{m+r} & \cdots & \Sigma_{2r} \end{array}\right]$$

which, by our induction hypothesis, equals

$$\left[\begin{array}{ccc|cccc} & & & 0 & 0 & \cdots & 0 \\ & & & 0 & 0 & \cdots & 0 \\ & \Delta_{m-1} & & \vdots & \vdots & & \vdots \\ & & & 0 & 0 & \cdots & 0 \\ \hline s_m & \cdots & s_{2m-1} & 0 & \cdots & 0 & \Sigma_{m+r} \\ \vdots & & \vdots & \vdots & & & \vdots \\ & & & 0 & \Sigma_{m+r} & & \\ s_r & \cdots & s_{m+r-1} & \Sigma_{m+r} & \cdots & & \Sigma_{2r} \end{array}\right]$$

The determinant of this latter matrix is, however, just

$$\det \Delta_{m-1} \cdot \det \begin{bmatrix} 0 & 0 & \cdots & \Sigma_{m+r} \\ \vdots & & & \\ 0 & \Sigma_{m+r} & & \\ \Sigma_{m+r} & \Sigma_{m+r+1} & \cdots & \Sigma_{2r} \end{bmatrix}$$

$= \pm \det \Delta_{m-1}(\Sigma_{m+r})^{r-m+1}$. This quantity, then, is equal to $\det \Delta_r$ which we know must be *zero* since $r \geqslant m+1$. But, according to the hypothesis of the lemma, $\det \Delta_{m-1} \neq 0$. Hence $\Sigma_{m+r} = 0$, which is what we wanted to prove. The lemma is established.

Now we are able to prove the main result of this §.

Theorem. *Given a sequence of numbers* $s_0, s_1, s_2, \ldots$, *form the matrices*

$$\Delta_n = \begin{bmatrix} s_0 & s_1 & \cdots & s_n \\ s_1 & s_2 & \cdots & \\ \vdots & & & \\ s_n & s_{n+1} & \cdots & s_{2n} \end{bmatrix}.$$

A necessary and sufficient condition for the s_k *to be the moments of a non-zero positive measure* μ *is that either*

(i) *all the quantities* $\det \Delta_n$ *are* > 0 (*sic!*) *for* $n = 0, 1, 2, \ldots$,

or else

(ii) *for some* $m \geqslant 1$, $\det \Delta_n > 0$ *for* $n = 0, 1, \ldots, m-1$, *while* $\det \Delta_n = 0$ *for all* $n \geqslant m$.

Remarks. The condition that $\det \Delta_n \geqslant 0$ for $n \geqslant 0$ is *necessary, but not sufficient* for $\{s_k\}$ to be a moment sequence. Case (ii) of the theorem is *degenerate* and, as we shall see, happens iff the s_k are the moments of a positive measure *supported on a finite set of points.*

Proof of theorem. Suppose, in the first place, that we *have* a positive non-zero measure μ on $\mathbb{R}$ with

$$s_k = \int_{-\infty}^{\infty} x^k \, \mathrm{d}\mu(x), \quad k = 0, 1, 2, \ldots.$$

Then, as we observed at the beginning of the proof of the theorem in §A,

$$\sum_{i=0}^{n} \sum_{j=0}^{n} s_{i+j}\xi_i\xi_j = \int_{-\infty}^{\infty} \left(\sum_{k=0}^{n} \xi_k x^k \right)^2 \mathrm{d}\mu(x).$$

If μ is *not* supported on a finite set of points, the integral on the right can only vanish when

$$\xi_0 = \xi_1 = \xi_2 = \cdots = \xi_n = 0,$$

so in this case all the forms $\sum_{i=0}^{n}\sum_{j=0}^{n} s_{i+j}\xi_i\xi_j$ are *strictly positive definite.* Here, $\det \Delta_n > 0$ for all $n \geqslant 0$ by the *first* of the above two lemmas.

Suppose now that our positive measure μ is supported on m points, call them $x_1, x_2, \ldots, x_m$. If $n < m$, the polynomial

$$\sum_{k=0}^{n} \xi_k x^k$$

vanishes at *each* of those points only when $\xi_0 = \xi_1 = \cdots = \xi_n = 0$, so, if $\mu(\{x_p\}) > 0$ for $1 \leqslant p \leqslant m$, the form

$$\sum_{i=0}^{n}\sum_{j=0}^{n} s_{i+j}\xi_i\xi_j$$

is *strictly positive definite* when $0 \leqslant n < m$. By the first lemma, then, $\det \Delta_n > 0$, $0 \leqslant n < m$. Consider now a value of n which is $\geqslant m$. We can then take the polynomial

$$x^{n-m}(x - x_1)(x - x_2)\ldots(x - x_m)$$

which vanishes on the support of μ. Rewriting that polynomial as

$$\sum_{k=0}^{n} \xi_k x^k$$

we must therefore have

$$\sum_{i=0}^{n}\sum_{j=0}^{n} s_{i+j}\xi_i\xi_j = 0,$$

although here $\xi_n = 1 \neq 0$. For such n, our quadratic form, although *positive definite,* is *not strictly so,* and hence at least *one* characteristic value of the matrix Δ_n must be *zero.* This makes $\det \Delta_n = 0$ whenever $n \geqslant m$.

Our theorem is proved in one direction.

Going the other way, suppose, first of all, that we are in case (i). Then, by the first lemma, *all* the quadratic forms

$$\sum_{i=0}^{n}\sum_{j=0}^{n} s_{i+j}\xi_i\xi_j$$

are (strictly) positive definite, so $\{s_k\}$ *is* a moment sequence by the theorem of §A. The argument just given shows that, *here, no* positive measure of which the s_k are the moments *can be supported* on a *finite set of points.*

It remains for us to treat case (ii). By the theorem of §A, we will be *through* when we show that *all the forms*

$$\sum_{i=0}^{n}\sum_{j=0}^{n} s_{i+j}\xi_i\xi_j$$

are positive definite. For $0 \leqslant n < m$ we do have $\det \Delta_n > 0$, so we can by the *first* lemma conclude that those forms *are* positive definite for such n.

To handle the forms with $n \geqslant m$ we must apply *Kronecker's lemma.* According to that result, we have some quantities $\alpha_0, \alpha_1, \ldots, \alpha_{m-1}$ such that

$$\alpha_0 s_p + \alpha_1 s_{p+1} + \cdots + \alpha_{m-1} s_{m-1+p} + s_{m+p} = 0$$

for $p \geqslant 0$. For $n \geqslant m$, our matrix Δ_n takes the form

$$\left[\begin{array}{ccc|ccc} & & & s_m & \cdots & s_n \\ & \Delta_{m-1} & & & & \\ & & & s_{2m-1} & \cdots & s_{n+m-1} \\ \hline s_m & \cdots & \cdots & s_{2m} & \cdots & s_{n+m} \\ \vdots & & & \vdots & & \\ s_n & \cdots & \cdots & s_{n+m} & \cdots & s_{2n} \end{array}\right]$$

The $(n+1) \times (n+1)$ matrix

$$\left[\begin{array}{cccc|ccccc} 1 & 0 & \cdots & 0 & 0 & & \cdots & \cdots & 0 \\ 0 & 1 & \cdots & 0 & 0 & & \cdots & \cdots & 0 \\ \vdots & & & & \vdots & & & & \vdots \\ 0 & 0 & \cdots & 1 & 0 & & 0 & \cdots & 0 \\ \hline \alpha_0 & \alpha_1 & \cdots & \alpha_{m-1} & 1 & & 0 & \cdots & 0 \\ 0 & \alpha_0 & \cdots & \alpha_{m-2} & \alpha_{m-1} & & 1 & & \vdots \\ \vdots & & & & & & & & \\ 0 & \cdots & \alpha_0 & & & & \cdots & \alpha_{m-1} & 1 \end{array}\right]$$

is non-singular. Therefore, positive definiteness of Δ_n is *implied* by that of the product

$$\left[\begin{array}{cccc|ccc} 1 & 0 & \cdots & 0 & 0 & \cdots & 0 \\ 0 & 1 & \cdots & 0 & & & \\ \vdots & & & & \vdots & & \vdots \\ 0 & & \cdots & 1 & 0 & \cdots & 0 \\ \hline \alpha_0 & & \cdots & \alpha_{m-1} & 1 & & \\ \vdots & & & & & & \\ 0 & \cdots & \alpha_0 & & \cdots & \alpha_{m-1} & 1 \end{array}\right] \left[\begin{array}{c|ccc} \Delta_{m-1} & s_m & \cdots & s_n \\ & \vdots & & \vdots \\ \hline s_m \quad \cdots & s_{2m} & \cdots & s_{m+n} \\ & \vdots & & \vdots \\ s_n \quad \cdots & s_{n+m} & \cdots & s_{2n} \end{array}\right]$$

$$\times \left[\begin{array}{cccc|ccc} 1 & 0 & \cdots & 0 & \alpha_0 & \cdots & 0 \\ 0 & 1 & & & \vdots & & \vdots \\ & & & & & & \alpha_0 \\ 0 & & \cdots & 1 & \alpha_{m-1} & & \vdots \\ \hline 0 & & \cdots & 0 & 1 & & \\ & & & & & & \vdots \\ & & & & & & \alpha_{m-1} \\ 0 & & \cdots & 0 & & \cdots & 1 \end{array}\right].$$

Using the relation furnished by Kronecker's lemma, we see that the product is just

$$\left[\begin{array}{ccc|ccc} & & & 0 & \cdots & 0 \\ & \Delta_{m-1} & & \vdots & & \vdots \\ & & & 0 & \cdots & 0 \\ \hline 0 & \cdots & 0 & 0 & \cdots & 0 \\ \vdots & & \vdots & & & \\ 0 & \cdots & 0 & 0 & \cdots & 0 \end{array}\right].$$

This matrix is *certainly* positive definite (although, of course, not *strictly* so!), because Δ_{m-1} is, as we already know. So Δ_n is positive definite (*not* strictly) also for $n \geqslant m$, and the proof is finished for case (ii). We are all done.

Remark. Since large determinants are hard to compute, the theorem just proved may not seem to be of much use. It does, in any event, furnish the complete answer to a rather interesting question.

Suppose we *lift* the requirement that the measures considered be positive in our statement of the moment problem. Let us, in other words, ask *which real sequences* $\{A_k\}$ *can be represented in the form*

$$A_k = \int_{-\infty}^{\infty} x^k \, d\tau(x), \quad k = 0, 1, 2, \ldots,$$

with *real signed measures* τ such that $\int_{-\infty}^{\infty} |x|^k |d\tau(x)| < \infty$ for every $k \geqslant 0$. The rather surprising answer turns out to be that *every real sequence* $\{A_k\}$ *can be so represented.*

In order to establish this fact, it is enough to show that, *given any real sequence* $\{A_k\}$, *two moment sequences* $\{S_k\}$ and $\{S'_k\}$ can be found with

$$A_k = S'_k - S_k, \quad k = 0, 1, 2, \ldots.$$

We use an inductive procedure to do this.

Take first $S_0 > 0$, and sufficiently large so that $S'_0 = A_0 + S_0$ is *also* > 0. Put $S_1 = 0$ and $S'_1 = A_1$. It is clear that if $S_2 > 0$ is *large enough*, and $S'_2 = A_2 + S_2$, *both the determinants*

$$\det\begin{bmatrix} S_0 & S_1 \\ S_1 & S_2 \end{bmatrix}, \quad \det\begin{bmatrix} S'_0 & S'_1 \\ S'_1 & S'_2 \end{bmatrix}$$

will be strictly positive.

Now just keep going. We can take $S_3 = 0$ and $S'_3 = A_3$. *Because the above two determinants are* > 0, we can find $S_4 > 0$ *large enough* so that

$$\det\begin{bmatrix} S_0 & S_1 & S_2 \\ S_1 & S_2 & S_3 \\ S_2 & S_3 & S_4 \end{bmatrix} \quad \text{and} \quad \det\begin{bmatrix} S'_0 & S'_1 & S'_2 \\ S'_1 & S'_2 & S'_3 \\ S'_2 & S'_3 & S'_4 \end{bmatrix}$$

are both > 0, where $S'_4 = A_4 + S_4$. There is clearly nothing to stop the continuation of this process. For each *odd* k we take $S_k = 0$ and $S'_k = A_k$. If $k = 2m + 2$, we can adjust $S_k > 0$ so as to make the corresponding $(m + 2) \times (m + 2)$ determinants involving the S_l and S'_l, $0 \leqslant l \leqslant k$, both > 0 (with, of course, $S'_k = A_k + S_k$) by *merely taking account* of the S_l and S'_l *already gotten* for $0 \leqslant l \leqslant k - 1$. This is because the *preceding step* has *already ensured* that

$$\det\begin{bmatrix} S_0 & S_1 & \cdots & S_m \\ S_1 & & & \\ \vdots & & & \\ S_m & \cdots & & S_{2m} \end{bmatrix} > 0.$$

and

$$\det\begin{bmatrix} S'_0 & S'_1 & \cdots & S'_m \\ S'_1 & & & \\ \vdots & & & \\ S'_m & \cdots & & S'_{2m} \end{bmatrix} > 0.$$

The sequences $\{S_k\}$ and $\{S'_k\}$ arrived at by following this procedure indefinitely are *moment sequences* according to the above theorem, and their construction is such that $A_k = S'_k - S_k$ for $k = 0, 1, 2, \ldots$. That is what we needed.

The result just found should have some applications. I do not know of any.

C. Determinacy. Two conditions, one sufficient and the other necessary

Having discussed the circumstances under which $\{S_k\}$ is a moment sequence, we come to the second question: if it *is*, when is the positive measure with moments S_k *unique*? In this §, we derive some simple partial answers to this question from earlier results.

1. Carleman's sufficient condition

Theorem (Carleman). *A moment sequence* $\{S_k\}$ *is determinate provided that*

$$\sum_{k=0}^{\infty} \frac{1}{S_{2k}^{1/2k}} = \infty.$$

Proof. Suppose we have *two* positive measures, μ and ν, with

$$S_k = \int_{-\infty}^{\infty} x^k \, \mathrm{d}\mu(x) = \int_{-\infty}^{\infty} x^k \, \mathrm{d}\nu(x), \quad k = 0, 1, 2, \ldots.$$

We have to show that $\mu = \nu$, and, as is well known, this will be the case if the Fourier–Stieltjes transform

$$f(\lambda) = \frac{1}{2}\int_{-\infty}^{\infty} \mathrm{e}^{\mathrm{i}\lambda x}(\mathrm{d}\mu(x) - \mathrm{d}\nu(x))$$

vanishes identically on $\mathbb{R}$.

It is now claimed that $f(\lambda)$ is infinitely differentiable on $\mathbb{R}$ and in fact belongs to a *quasianalytic class* thereon (see previous chapter). Observe that

$$\frac{1}{2}\int_{-\infty}^{\infty} x^{2k}(\mathrm{d}\mu(x) + \mathrm{d}\nu(x)) = S_{2k} < \infty;$$

therefore all the integrals

$$\frac{1}{2}\int_{-\infty}^{\infty} (\mathrm{i}x)^k \mathrm{e}^{\mathrm{i}\lambda x}(\mathrm{d}\mu(x) - \mathrm{d}\nu(x))$$

are *absolutely convergent* (at least, first of all, for *even* $k \geqslant 0$ and hence for *all* $k \geqslant 0$), since the measures μ and ν are *positive* (*here* is where we *use* their positivity!). This means that $f(\lambda)$ is infinitely differentiable on $\mathbb{R}$, and that

$$f^{(k)}(\lambda) = \frac{1}{2}\int_{-\infty}^{\infty} (\mathrm{i}x)^k \mathrm{e}^{\mathrm{i}\lambda x}(\mathrm{d}\mu(x) - \mathrm{d}\nu(x)).$$

For $\lambda \in \mathbb{R}$ we have

$$|f^{(k)}(\lambda)| \leqslant \frac{1}{2}\int_{-\infty}^{\infty} |x|^k(\mathrm{d}\mu(x) + \mathrm{d}\nu(x)),$$

a *finite quantity* independent of λ.

Denote $\sup_{\lambda \in \mathbb{R}} |f^{(n)}(\lambda)|$ by M_n. Then,

$$M_{2k} \leqslant \frac{1}{2}\int_{-\infty}^{\infty} x^{2k}(\mathrm{d}\mu(x) + \mathrm{d}\nu(x)) = S_{2k}.$$

Bringing in, as in §B.1 of the previous chapter, the *convex logarithmic regularization* $\{\underline{M}_n\}$ of the sequence $\{M_n\}$, we see that

$$\underline{M}_{2k} \leqslant S_{2k},$$

so

$$\sum_{n=0}^{\infty} \underline{M}_n^{-1/n} \geqslant \sum_{k=0}^{\infty} \underline{M}_{2k}^{-1/2k} \geqslant \sum_{k=0}^{\infty} S_{2k}^{-1/2k}.$$

The last sum on the right is, however, *infinite* by hypothesis. Therefore, by the *second* theorem of §B.2, Chapter IV, the class $\mathscr{C}_{\mathbb{R}}(\{M_n\})$ is *quasianalytic*.

However, $f(\lambda) \in \mathscr{C}_{\mathbb{R}}(\{M_n\})$ and

$$f^{(k)}(0) = \frac{1}{2}\int_{-\infty}^{\infty} (\mathrm{i}x)^k(\mathrm{d}\mu(x) - \mathrm{d}\nu(x)) = \mathrm{i}^k(S_k - S_k) = 0$$

for $k = 0, 1, 2, \ldots$ according to our initial supposition. Therefore $f(\lambda) \equiv 0$ on $\mathbb{R}$, as required, and we are done.

Scholium. If $\{S_k\}$ is a moment sequence, $\log S_{2k}$ is a *convex function of* k. This is an elementary consequence of Hölder's inequality. Taking, namely, a positive measure μ with

$$S_k = \int_{-\infty}^{\infty} x^k \mathrm{d}\mu(x),$$

we have, for $r, s \geqslant 0$ and $0 < \lambda < 1$,

$$\int_{-\infty}^{\infty} |x|^{\lambda r + (1-\lambda)s} \mathrm{d}\mu(x) \leqslant \left(\int_{-\infty}^{\infty} |x|^r \mathrm{d}\mu(x)\right)^{\lambda} \cdot \left(\int_{-\infty}^{\infty} |x|^s \mathrm{d}\mu(x)\right)^{1-\lambda},$$

so, if $r = 2k$, $s = 2l$, and $\lambda r + (1-\lambda)s$ is an *even integer*, $2m$ say, with (say) $2k < 2m < 2l$, we find that

$$S_{2m} \leqslant S_{2k}^{(2l-2m)/(2l-2k)} \cdot S_{2l}^{(2m-2k)/(2l-2k)},$$

the asserted convexity of $\log S_{2k}$.

An obvious adaptation of the work in Chapter IV, §§B.1 and C now shows that this convexity has the following consequence:

Theorem. *Let* $S(r) = \sup_{k\geq 0}(r^{2k}/S_{2k})$ *for* $r > 0$. *Then* $\sum_k S_{2k}^{-1/2k} = \infty$ *iff*

$$\int_0^\infty \frac{\log S(r)}{1+r^2}\,dr \;=\; \infty.$$

Corollary. *If* $\int_0^\infty (\log S(r)/(1+r^2))\,dr = \infty$ *with* $S(r)$ *as defined in the theorem, then the moment sequence* $\{S_k\}$ *is determinate.*

Proof. Combine the preceding theorem with that of Carleman.

2. A necessary condition

Theorem. *Let* $w(x) \geq 0$, *suppose that* $\int_{-\infty}^{\infty} |x|^k w(x)\,dx < \infty$ *for* $k = 0, 1, 2, \ldots$, *and put*

$$S_k = \int_{-\infty}^{\infty} x^k w(x)\,dx, \quad k = 0, 1, 2, \ldots.$$

If

$$\int_{-\infty}^{\infty} \frac{1}{1+x^2} \log\left(\frac{1}{w(x)}\right) dx \;<\; \infty,$$

the moment sequence $\{S_k\}$ *is indeterminate.*

Remark. Since $\int_{-\infty}^{\infty} (w(x)/(1+x^2))\,dx \leq \int_{-\infty}^{\infty} w(x)\,dx < \infty$, we have

$$\int_{-\infty}^{\infty} \frac{\log^+ w(x)}{1+x^2}\,dx \;<\; \infty$$

in any case, by the inequality between arithmetic and geometric means.

Proof of theorem. According to Problem 2 (at the end of §B, Chapter II!), if $\int_{-\infty}^{\infty} (\log w(x)/(1+x^2))\,dx > -\infty$, the infimum of $\int_{-\infty}^{\infty} |e^{-ix} - \sum_{\lambda\geq 0} A_\lambda e^{i\lambda x}|\,w(x)\,dx$, taken over *all finite sums* $\sum_{\lambda\geq 0} A_\lambda e^{i\lambda x}$, is *strictly positive*. By the Hahn–Banach theorem and the known form of linear functionals on $L_1(\mu)$ for σ-finite measures μ, we get a Borel function $\varphi(x)$, defined on

$$\{x\colon\ w(x) > 0\}$$

and *essentially bounded on that set*, with

$$\int_{-\infty}^{\infty} \varphi(x)\cdot e^{-ix} w(x)\,dx \neq 0$$

(hence φw is *not* almost everywhere zero!), whilst

$$(*)\qquad \int_{-\infty}^{\infty} \varphi(x)\cdot e^{i\lambda x} w(x)\,dx = 0, \quad \lambda \geqslant 0.$$

Under the conditions of this theorem, $w(x) > 0$ a.e., so $\varphi(x)$ is in fact defined a.e. on $\mathbb{R}$ and essentially bounded there, i.e., $\varphi \in L_\infty(\mathbb{R})$. Without loss of generality,

$$|\varphi(x)| \leqslant \tfrac{1}{2} \quad \text{a.e.}, \quad x \in \mathbb{R}.$$

Differentiating $(*)$ successively with respect to λ (which we can *do*, since the integrals $\int_{-\infty}^{\infty} |x|^k w(x)\,dx$ are all finite for $k \geqslant 0$) and looking at the resulting derivatives at $\lambda = 0$, we find that

$$\int_{-\infty}^{\infty} x^k \varphi(x) w(x)\,dx = 0, \quad k = 0, 1, 2, \ldots.$$

The functions $\Re\varphi(x)$ and $\Im\varphi(x)$ *can't both be zero* a.e.; say, wlog, that $\Re\varphi(x)$ isn't zero a.e. Then, from the preceding relation, we have

$$\int_{-\infty}^{\infty} x^k \Re\varphi(x) w(x)\,dx = 0, \quad k = 0, 1, 2, \ldots,$$

so that

$$S_k = \int_{-\infty}^{\infty} x^k (1 - \Re\varphi(x)) w(x)\,dx, \quad k = 0, 1, 2, \ldots,$$

as well as

$$S_k = \int_{-\infty}^{\infty} x^k w(x)\,dx, \quad k = 0, 1, 2, \ldots.$$

Here, $|\varphi(x)| \leqslant \frac{1}{2}$ a.e. but $\Re\varphi(x)$ is *not* a.e. equal to zero; therefore $(1 - \Re\varphi(x))w(x)\,dx$ is the *differential of a certain positive measure on* $\mathbb{R}$, *different from the positive measure with differential* $w(x)\,dx$, but having the *same moments*, S_k, as the latter. The moment sequence $\{S_k\}$ is thus indeterminate. Q.E.D.

Corollary. *Let $T(r)$ be $\geqslant 1$ for $r \geqslant 0$, and bounded near 0. Suppose that* $\log T(r)$ *is a convex function of* $\log r$, *and that*

$$\int_0^{\infty} \frac{r^k}{T(r)}\,dr \;<\; \infty \quad \textit{for} \quad k \geqslant 0.$$

The moment sequence

$$S_k \;=\; \int_{-\infty}^{\infty} \frac{x^k}{T(|x|)}\,dx, \quad k = 0, 1, 2, \ldots,$$

is determinate iff

$$\sum_{k=0}^{\infty} S_{2k}^{-1/2k} = \infty.$$

Remark. Here, of course, the S_k with *odd* k are all *zero*.

Proof of corollary. The *if* part follows by Carleman's theorem (preceding article).

To do *only if*, suppose that

$$\sum_k S_{2k}^{-1/2k} < \infty.$$

By the formula for the S_k, we have first of all

$$(\ast)\qquad S_{2k} = \mathrm{o}(1) + 2\int_1^{\infty} \frac{x^{2k}}{(T(x)/x^2)}\cdot\frac{\mathrm{d}x}{x^2}.$$

Put

$$M_n = \sup_{x\geqslant 0} \frac{x^n}{(T(x)/x^2)};$$

$\log M_n$ is then a *convex function* of n, and we proceed to apply to it and to $T(x)$ some of the work on convex logarithmic regularization from the preceding chapter.

From ($\ast$), we see that

$$S_{2k} \leqslant \mathrm{o}(1) + 2M_{2k},$$

whence surely $\sum_k M_{2k}^{-1/2k} < \infty$. This certainly implies that $M_{2k}^{1/2k} \xrightarrow[k]{} \infty$, so, since $\log M_n$ is a convex function of n, the proof of the *second* lemma in §B.1 of Chapter IV shows that the expression $M_n^{1/n}$ is *eventually increasing*. Therefore the convergence of $\sum_k M_{2k}^{-1/2k}$ implies that

$$\sum_n M_n^{-1/n} < \infty.$$

Taking, for $x > 0$,

$$P(x) = \sup_{n\geqslant 0} \frac{x^n}{M_n},$$

we will then get

$$(\dagger)\qquad \int_1^{\infty} \frac{\log P(x)}{x^2}\,\mathrm{d}x < \infty$$

by the second lemma of §B.1, Chapter IV and the theorem of §C in that chapter.

Since $\log T(x)$ is a convex function of $\log x$, so is $\log(T(x)/x^2)$. The *second* lemma of §D, Chapter IV therefore shows that, for all sufficiently large x.

$$\log(T(x)/x^2) \leqslant \log P(x) + \log x,$$

in view of the relations between $T(x)/x^2$, M_n and $P(x)$. (The convex function $\log T(x)$ of $\log x$ must *eventually be increasing*, since all the integrals $\int_0^\infty (x^k/T(x))\,dx$, $k \geqslant 0$, converge!) Referring to (†), we see that

$$\int_1^\infty \frac{\log T(x)}{x^2}\,dx \;<\; \infty,$$

i.e.,

$$\int_{-\infty}^\infty \frac{1}{1+x^2}\log T(|x|)\,dx \;<\; \infty.$$

Indeterminateness of $\{S_k\}$ now follows by the above theorem.

Example. The sequence of moments

$$S_k(\alpha) = \int_{-\infty}^\infty x^k e^{-|x|^\alpha}\,dx$$

is *determinate* for $\alpha \geqslant 1$ and *indeterminate* if $0 < \alpha < 1$. (Note: In applying the above results it is better to work directly with $T(x) = e^{x^\alpha}$ for $\alpha \geqslant 1$ as well as for $0 < \alpha < 1$. Otherwise one should express $S_k(\alpha)$ in terms of the Γ-function and use Stirling's formula.)

Problem 8

The moment sequence

$$S_k = \int_1^\infty x^k e^{-x/\log x}\,dx$$

is *determinate*, but the Taylor series $\sum_0^\infty (S_n/n!)(i\lambda)^n$ of $\int_1^\infty e^{i\lambda x} e^{-x/\log x}\,dx$ *does not converge for any* $\lambda \neq 0$. (Hint: To see that the Taylor series can't converge for $\lambda \neq 0$, *estimate* S_n *from below* for large n. To do this, write

$$S_n = \int_1^\infty e^{-\varphi_n(x)}\,dx$$

with $\varphi_n(x) = x/\log x - n\log x$, and use *Laplace's method* to estimate the integral. To a first approximation, the *zero* x_0 of $\varphi_n'(x)$ has $x_0 \sim n\log n$, and this yields a good enough approximation to $\varphi_n''(x_0)$. To get a lower bound for $e^{-\varphi_n(x_0)}$, compute $\varphi_n(x)$ for $x = n\log n + n\log\log n$.)

D. M. Riesz' general criterion for indeterminacy

Let $\{S_k\}$ be a moment sequence. If we put

$$S(r) = \sup_{k\geq 0} \frac{r^{2k}}{S_{2k}} \quad \text{for} \quad r > 0,$$

then, according to the corollary at the end of §C.1, $\{S_k\}$ is *determinate* when

$$\int_0^\infty \frac{\log S(r)}{1+r^2} dr = \infty.$$

If, on the other hand, there is a density $w(x) \geq 0$ with

$$S_k = \int_{-\infty}^\infty x^k w(x)\,dx, \quad k = 0, 1, 2, \ldots,$$

$\{S_k\}$ is *indeterminate* provided that

$$\int_{-\infty}^\infty \frac{1}{1+x^2} \log\left(\frac{1}{w(x)}\right) dx < \infty,$$

as we have seen in §C.2.

Both conditions involve integrals of *the same form*, containing, however, different functions. This leads one to think that they might both be reflections of some general *necessary and sufficient condition* expressed in terms of the integral which is the subject of this book. As we shall now see, that turns out to be the case.

1. The criterion with Riesz' function $R(z)$

Given a moment sequence $\{S_k\}$, we take a positive measure μ on $\mathbb{R}$ having the moments S_k, and, for $z \in \mathbb{C}$, put

$$R(z) = \sup\left\{ |P(z)|^2 \colon P \text{ a } \textit{polynomial} \text{ with } \int_{-\infty}^\infty |P(x)|^2\,d\mu(x) \leq 1 \right\}.$$

It is only the sequence of S_k which is needed to get $R(z)$ and not the measure μ itself of which they are the moments; indeed, if

$$P(z) = \sum_{k=0}^N c_k z^k$$

with the $c_k \in \mathbb{C}$,

$$\int_{-\infty}^\infty |P(x)|^2\,d\mu(x) = \sum_{i=0}^N \sum_{j=0}^N S_{i+j} c_i \bar{c}_j.$$

Thus, $R(z)$ (which may be infinite at some points) *depends just on the sequence* $\{S_k\}$; it turns out to govern that moment sequence's *determinacy*. Marcel Riesz worked with the *reciprocal* $\rho(z) = 1/R(z)$ instead of with $R(z)$, and the reader should note that, in literature on the moment problem, results are usually stated in terms of $\rho(z)$.

Theorem (M. Riesz). *Given a moment sequence* $\{S_k\}$ *and its associated function* $R(z)$, $\{S_k\}$ *is indeterminate if* $R(x) < \infty$ *on a non-denumerable subset of* $\mathbb{R}$. *Conversely, if* $\{S_k\}$ *is indeterminate,* $R(x) < \infty$ *everywhere on* $\mathbb{R}$ *and*

$$\int_{-\infty}^{\infty} \frac{\log^+ R(x)}{1+x^2}\,dx \;<\; \infty.$$

Proof. For the first (and longest) part of the proof, let us suppose that $R(x) < \infty$ for all x belonging to some *non-denumerable* subset E of $\mathbb{R}$. We must establish indeterminacy of $\{S_k\}$.

Take any positive measure μ with $S_k = \int_{-\infty}^{\infty} x^k\,d\mu(x)$, $k = 0, 1, 2, \ldots$, and let us first show that μ *cannot be supported on a finite set of points.* Suppose, on the contrary, that μ *were* supported on $\{x_1, x_2, \ldots, x_N\}$, say. Put $P_M(x) = M(x-x_1)(x-x_2)\ldots(x-x_N)$; then,

$$\int_{-\infty}^{\infty} |P_M(x)|^2\,d\mu(x) = 0,$$

but, if $x \neq x_1,\ x_2, \ldots$ or x_N, $P_M(x) \to \infty$ as $M \to \infty$, so $R(x) = \infty$. In that case, $R(x)$ could *not* be finite on the non-denumerable set E.

Having established that μ is not supported on a finite set, let us apply Schmidt's orthogonalization procedure to the sequence $1, x, x^2, \ldots$ and the measure μ, obtaining, one after the other, the real polynomials $p_n(x)$, $n \geqslant 0$, with $p_n(x)$ of degree n such that

$$\int_{-\infty}^{\infty} x^k p_n(x)\,d\mu(x) = 0 \quad \text{for} \quad k = 0, \ldots, n-1,$$

when $n \geqslant 1$. Of course, the construction of the $p_n(x)$ really only depends on the S_k, and not on the particular positive measure μ of which they are the moments. Since no such μ can be supported on a finite set of points, the *orthogonalization process never stops,* and we obtain a non-zero p_n for *each n.* These orthogonal polynomials will be used presently.

Pick any x_0 with $R(x_0) < \infty$. We are going to construct a positive measure ν on $\mathbb{R}$ *having the moments* S_k, *but such that*

$$\nu(\{x_0\}) \geqslant 1/R(x_0) \quad (\textit{sic!}).$$

In order to obtain ν, let us take any large N, and try to find M points $x_1, x_2, \ldots, x_M$ *different from* x_0, with $M = N-1$ or N (it turns out that *either* possibility may occur) such that the *Gauss quadrature formula*

$$(*) \qquad \int_{-\infty}^{\infty} P(x)\,d\mu(x) = \sum_{k=0}^{M} \mu_k P(x_k)$$

holds for all (complex) polynomials P of degree $\leqslant M+N$; here, the μ_k are supposed to be certain coefficients independent of P.

Assume for the time being that we can obtain a quadrature formula $(*)$ for every large N, and consider the situation for *any given fixed* N. In the first place, *the coefficients* μ_k *are all* > 0. To see this, pick any k, $0 \leqslant k \leqslant M$, and write

$$Q_k(x) = \prod_{\substack{i \neq k \\ 0 \leqslant i \leqslant M}} (x - x_i).$$

The polynomial Q_k is of degree M, so $P(x) = [Q_k(x)]^2$ is of degree $2M \leqslant M+N$, and we can apply $(*)$ to it, getting

$$(Q_k(x_k))^2 \mu_k = \int_{-\infty}^{\infty} [Q_k(x)]^2\,d\mu(x).$$

The right side is surely > 0, for μ is not supported on any finite set of points. Therefore $\mu_k > 0$.

Using the polynomial

$$q(x) = \frac{Q_0(x)}{\sqrt{\mu_0}\,Q_0(x_0)}$$

of degree M, we have, by $(*)$ applied to $(q(x))^2$,

$$\int_{-\infty}^{\infty} (q(x))^2\,d\mu(x) = 1,$$

whilst $(q(x_0))^2 = 1/\mu_0$. Therefore, since $q(x)$ is a real polynomial, surely

$$R(x_0) \geqslant \frac{1}{\mu_0}$$

by definition of $R(z)$, i.e.,

$$\mu_0 \geqslant 1/R(x_0).$$

Let ν_N be the discrete positive measure supported on the set $x_0, x_1, \ldots, x_M$ defined by the relations

$$\nu_N(\{x_k\}) = \mu_k, \quad k = 0, 1, \ldots, M;$$

according to (*) we will then have

$$(\substack{*\\*}) \qquad S_k = \int_{-\infty}^{\infty} x^k \, d\mu(x) = \int_{-\infty}^{\infty} x^k \, d\nu_N(x)$$

for $0 \leqslant k \leqslant M + N$, hence certainly for $k = 0, 1, 2, \ldots, 2N - 1$. And, as we have just seen, $\nu_N(\{x_0\}) \geqslant 1/R(x_0)$.

Given any fixed k, the integrals

$$\int_{-\infty}^{\infty} (x^2 + 1)^k \, d\nu_N(x)$$

can, according to what has just been shown, be expressed in obvious fashion in terms of the S_n as soon as $N > k$. *They are hence bounded*, and this means we can find *some sequence of N's tending to* ∞, and *finite positive measures* $\nu^{(k)}$ on $\mathbb{R}$, $k = 0, 1, 2, \ldots$, with, for *each* k,

$$(x^2 + 1)^k \, d\nu_N(x) \longrightarrow d\nu^{(k)}(x) \quad w^*$$

as $N \to \infty$ through that sequence. (See Chapter III, §F.1). Let $l \geqslant 0$; then, since $(x^2 + 1)^{-l}$ is bounded and continuous on $\mathbb{R}$, the w^* convergence just mentioned certainly implies that

$$(x^2 + 1)^{k-l} d\nu_N(x) \longrightarrow (x^2 + 1)^{-l} d\nu^{(k)}(x),$$

so, if $l = 0, 1, \ldots, k$,

$$d\nu^{(k-l)}(x) = (x^2 + 1)^{-l} d\nu^{(k)}(x)$$

and thus

$$d\nu^{(k)}(x) = (x^2 + 1)^l \, d\nu^{(k-l)}(x).$$

Put $\nu^{(0)} = \nu$. By the preceding relation, $(x^2 + 1)^k d\nu(x) = d\nu^{(k)}(x)$ for $k = 0, 1, 2, \ldots$, so, since the measures $\nu^{(k)}$ are all finite, we have

$$\int_{-\infty}^{\infty} (x^2 + 1)^k \, d\nu(x) \; < \; \infty$$

for $k \geqslant 0$. *It is now claimed that the* S_n *are the moments of the measure* ν.

Fixing any n, take a $k > n$. Then

$$\int_{-\infty}^{\infty} x^n \, d\nu(x) = \int_{-\infty}^{\infty} \frac{x^n}{(x^2 + 1)^k} \, d\nu^{(k)}(x).$$

By the above mentioned w^* convergence, the integral on the right is just the limit of

$$\int_{-\infty}^{\infty} \frac{x^n}{(x^2 + 1)^k} (x^2 + 1)^k \, d\nu_N(x)$$

as N goes to ∞ through its special sequence of values. Each of the latter integrals, however, $= \int_{-\infty}^{\infty} x^n \mathrm{d}\nu_N(x)$ which, by ($\ast\!\ast$), is just S_n as soon as $2N-1 \geqslant n$. Therefore

$$\int_{-\infty}^{\infty} x^n \mathrm{d}\nu(x) = S_n$$

for any $n \geqslant 0$, as claimed.

We have, moreover, $\nu_N(\{x_0\}) \geqslant 1/R(x_0)$ by our construction. Therefore, since the ν_N are positive measures, of which a subsequence tends w^* to ν, we *certainly have*

$$(\dagger) \qquad \nu(\{x_0\}) \geqslant 1/R(x_0).$$

In this way, we have obtained a *positive measure* ν *having the moments* S_k *and satisfying* ($\dagger$), where x_0 is *any one of the points in the non-denumerable set* E on which $R(x) < \infty$.

From this fact it follows, however, *that* $\{S_k\}$ *cannot be determinate*. We have, indeed, a positive measure ν with moments S_k satisfying ($\dagger$) for *each* $x_0 \in E$, and, *in the case of determinacy, those measures* ν *would have to be all the same*. In other words, there would be a *single measure* ν with $\nu(\{x_0\}) > 0$ *for a non-denumerable set of points* x_0. But that is nonsense. So, if $R(x)$ is finite on a non-denumerable set, we can establish indeterminacy using the quadrature formula ($\ast$). Everything turns, then, on the establishment of that formula, to which we will immediately direct our attention.

There is, however, one remark which should be made at this point, even though it has no bearing on the proof, namely, *that in* ($\dagger$) *we in fact have equality*,

$$\nu(\{x_0\}) = 1/R(x_0).$$

To see this, suppose that $\nu(\{x_0\}) > 1/R(x_0)$. We can get a polynomial P with

$$\int_{-\infty}^{\infty} |P(x)|^2 \mathrm{d}\mu(x) = \int_{-\infty}^{\infty} |P(x)|^2 \mathrm{d}\nu(x) = 1$$

but $|P(x_0)|^2$ *as close as we like to* $R(x_0)$. Then, however,

$$\int_{-\infty}^{\infty} |P(x)|^2 \mathrm{d}\nu(x) \geqslant |P(x_0)|^2 \nu(\{x_0\})$$

would be $> |P(x_0)|^2/R(x_0)$, *and hence* > 1, a contradiction. We see that the function $R(x)$ *gives the solution to a certain extremal problem*:

$$\boxed{1/R(x_0) = \max\{\mu(\{x_0\}) : \mu \text{ a positive measure with the moments } S_k\}.}$$

We have now to prove the quadrature formula (*). For this purpose we use the orthogonal polynomials $p_n(x)$ described at the beginning of the present demonstration; the idea goes back to Gauss. Take any $x_0 \in \mathbb{R}$ and any positive integer N. We can surely find *two real numbers* α and β, *not both zero*, such that

$$Q(x) = \alpha p_N(x) + \beta p_{N+1}(x)$$

vanishes at x_0. The polynomial Q is certainly not identically zero, and in fact it is of degree N or $N+1$, depending on whether $p_N(x_0) = 0$ or not. It is this uncertainty in the degree of Q which forces us to bring in the number M; we take $M = (\text{degree of } Q) - 1$; thus, $M = N-1$ or N.

$Q(x)$, being of degree $M+1$, vanishes at x_0 *and at* M *other points*; it is claimed that these points are *real and distinct*. This statement will be seen to rest entirely on the relation

$$(\S) \qquad \int_{-\infty}^{\infty} P(x)Q(x)\,\mathrm{d}\mu(x) = 0,$$

valid for any polynomial P *of degree* $\leqslant N-1$, which is an obvious consequence of the formula for Q and the orthogonality property of the polynomials $p_n(x)$.

Suppose, to begin with, that $Q(x)$ *has the real zeros* $x_0, \ldots, x_r$ (with repetitions according to multiplicities, as in Chapter III), *and no others, and that* $r < M-1$. Then, if

$$P(x) = (x-x_0)(x-x_1)\ldots(x-x_r),$$

$P(x)Q(x)$ *will not change sign on* $\mathbb{R}$, so, for a suitable constant $c \neq 0$, $cP(x)Q(x) \geqslant 0$ on $\mathbb{R}$. Therefore $\int_{-\infty}^{\infty} cP(x)Q(x)\mathrm{d}\mu(x) > 0$, for *$\mu$ is not supported on a finite set of points.* However $P(x)$ has degree $r+1 < M \leqslant N-1$, so $\int_{-\infty}^{\infty} cP(x)Q(x)\mathrm{d}\mu(x) = 0$ by (§). We have reached a contradiction, showing that $Q(x)$ must have *at least* M *real zeros* (counting multiplicities), including x_0. However, $Q(x)$ is of degree $M+1$. Therefore Q can have *at most one* non-real zero. The coefficients of Q are *real*, however, like those of p_N and p_{N+1}. Hence, *non-real zeros* of Q *must occur in pairs*, and Q *cannot have just one such zero.* This shows that *all the zeros* of Q *are real.*

The real zeros of Q are *distinct*. Suppose, for instance, that Q has at least a *double zero* at a_0; denote the remaining (real) zeros of Q by $a_1, \ldots, a_{M-1}$; it is, of course, not excluded that some of them coincide with a_0. Put

$$P(x) = (x-a_1)(x-a_2)\ldots(x-a_{M-1});$$

since $Q(x)$ has the factor $(x-a_0)^2$, $P(x)Q(x)$ *does not change sign* on $\mathbb{R}$. Thus, for a suitable constant c,

$$\int_{-\infty}^{\infty} cP(x)Q(x)\,d\mu(x) > 0$$

as before, and this contradicts (§) since P is of degree $M-1 < N$.

Denote now the *real and distinct zeros* of Q by $x_0, x_1, \ldots, x_M$, and let us complete the proof of the quadrature formula (∗). Take any polynomial $P(x)$ of degree $\leqslant M+N$. Then, long division of $P(x)$ by $Q(x)$ yields

$$P(x) = D(x)Q(x) + R(x)$$

where, since degree of $Q = M+1$, the degree of R is $\leqslant M$ and the degree of D is $\leqslant N-1$. This last fact implies, by (§), that

$$\int_{-\infty}^{\infty} D(x)Q(x)\,d\mu(x) = 0,$$

so

$$\int_{-\infty}^{\infty} P(x)\,d\mu(x) = \int_{-\infty}^{\infty} R(x)\,d\mu(x).$$

Now, since degree of $R \leqslant M$, Lagrange's interpolation formula gives us

$$R(x) = \sum_{k=0}^{M} \frac{R(x_k)}{Q'(x_k)(x-x_k)} Q(x),$$

i.e.

$$R(x) = \sum_{k=0}^{M} \frac{P(x_k)}{Q'(x_k)(x-x_k)} Q(x),$$

since $R(x_k) = P(x_k)$ at each zero x_k of Q. Therefore

$$\int_{-\infty}^{\infty} P(x)\,d\mu(x) = \int_{-\infty}^{\infty} R(x)\,d\mu(x)$$

$$= \int_{-\infty}^{\infty} \left(\sum_{k=0}^{M} \frac{P(x_k)}{Q'(x_k)(x-x_k)} Q(x) \right) d\mu(x) = \sum_{k=0}^{M} P(x_k)\mu_k,$$

where

$$\mu_k = \int_{-\infty}^{\infty} \frac{Q(x)}{Q'(x_k)(x-x_k)}\,d\mu(x), \quad k = 0, 1, \ldots, M.$$

Our quadrature formula (∗) is thus established, and therewith, the *first part of the theorem.*

Proof of the *second part* of the theorem is quite a bit shorter. Here, we suppose that $\{S_k\}$ is an *indeterminate moment sequence*, and use that property to obtain information about $R(z)$.

We have, then, *two different* positive measures μ and ν with

$$S_k = \int_{-\infty}^{\infty} x^k \, \mathrm{d}\mu(x) = \int_{-\infty}^{\infty} x^k \, \mathrm{d}\nu(x), \quad k = 0, 1, 2, \ldots.$$

Denote by σ the positive measure $\frac{1}{2}(\mu + \nu)$ and by τ the *real signed measure* $\frac{1}{2}(\mu - \nu)$. Then also

$$S_k = \int_{-\infty}^{\infty} x^k \, \mathrm{d}\sigma(x), \quad k = 0, 1, 2, \ldots,$$

so that, if $p(x)$ is *any polynomial*,

$$(\dagger\dagger) \qquad \int_{-\infty}^{\infty} |p(x)|^2 \, \mathrm{d}\mu(x) = \int_{-\infty}^{\infty} |p(x)|^2 \, \mathrm{d}\sigma(x).$$

For the signed measure τ,

$$\int_{-\infty}^{\infty} x^k \, \mathrm{d}\tau(x) = 0, \quad k = 0, 1, 2, \ldots.$$

There is a *trick* based on this identity which, according to M. Riesz, goes back to Markov who used it in studying the moment problem around 1890. The same idea was used by Riesz himself and then, around 1950, by Pollard in the study of weighted polynomial approximation (see next chapter). Take any polynomial P and any $z_0 \notin \mathbb{R}$. Then

$$\frac{P(x) - P(z_0)}{x - z_0}$$

is also a polynomial in x, so, by the identity just written,

$$\int_{-\infty}^{\infty} \frac{P(x) - P(z_0)}{x - z_0} \, \mathrm{d}\tau(x) = 0.$$

From this we have

$$(\S\S) \qquad P(z) \int_{-\infty}^{\infty} \frac{\mathrm{d}\tau(t)}{t - z} = \int_{-\infty}^{\infty} \frac{P(t)\mathrm{d}\tau(t)}{t - z}$$

whenever $z \notin \mathbb{R}$.

The function

$$F(z) = \int_{-\infty}^{\infty} \frac{\mathrm{d}\tau(t)}{t - z}$$

is clearly analytic for $\Im z > 0$; moreover, *it cannot be identically zero there.*

Indeed,

$$\Im F(z) = \int_{-\infty}^{\infty} \frac{\Im z}{|z-t|^2} \, d\tau(t),$$

τ being real, so by the remark at the end of §F.1, Chapter III,

$$\Im F(x+iy)\,dx \longrightarrow \pi d\tau(x) \quad w^*$$

for $y \to 0+$. Therefore $F(z) \equiv 0$ for $\Im z > 0$ would make $\tau = 0$, which is, however, contrary to the initial assumption that $\mu \neq \nu$.

Since $F(z) \not\equiv 0$ in $\{\Im z > 0\}$, we can use (§§) to get a formula for $P(z)$ in that half plane:

$$P(z) = \frac{1}{F(z)} \int_{-\infty}^{\infty} \frac{P(t)\,d\tau(t)}{t - z}.$$

In particular, if $z = \xi + i$ with ξ real,

$$\begin{aligned} |P(\xi + i)| \;&\leqslant\; \frac{1}{|F(\xi+i)|} \int_{-\infty}^{\infty} |P(t)|\,|d\tau(t)| \\ &\leqslant\; \frac{1}{|F(\xi+i)|} \int_{-\infty}^{\infty} |P(t)|\,d\sigma(t), \end{aligned}$$

since $|d\tau(t)| \leqslant d\sigma(t)$. Let now $P = p^2$, where p is any polynomial. By the preceding relation and (††), $|p(\xi+i)|^2 \leqslant (1/|F(\xi+i)|)\int_{-\infty}^{\infty} |p(t)|^2 d\mu(t)$, so, by definition of $R(z)$,

$$R(\xi + i) \;\leqslant\; \frac{1}{|F(\xi+i)|}.$$

The analytic function $F(z)$ is clearly *bounded* in $\{\Im z \geqslant 1\}$ and continuous up to the line $\Im z = 1$. Since, as we have seen, $F(z) \not\equiv 0$ there, we have, applying the first theorem of §G.2, Chapter III to the half plane $\{\Im z \geqslant 1\}$,

$$\int_{-\infty}^{\infty} \frac{\log^- |F(\xi+i)|}{\xi^2 + 1} \, d\xi \;<\; \infty.$$

Combined with the previous inequality, this yields

$$(\ddagger) \qquad \int_{-\infty}^{\infty} \frac{\log^+ R(\xi+i)}{\xi^2+1} \, d\xi \;<\; \infty.$$

Using this result, we can now estimate $R(x)$ *on the real axis.*

Let $p(z)$ be any polynomial with

$$\int_{-\infty}^{\infty} |p(t)|^2 d\mu(t) \leqslant 1;$$

then, *by definition* (!),

$$|p(\xi + \mathrm{i})|^2 \leqslant R(\xi + \mathrm{i}).$$

On the other hand, by the theorem of §E, Chapter III, applied to the half plane $\Im z \leqslant 1$,

$$\log|p(x)| \;\leqslant\; \frac{1}{\pi}\int_{-\infty}^{\infty} \frac{\log^+|p(\xi+\mathrm{i})|}{(x-\xi)^2+1}\,\mathrm{d}\xi$$

for $x \in \mathbb{R}$. (Note that $p(z)$ is an *entire function of exponential type zero*! The reader who does not wish to resort to the result from Chapter III may of course easily verify the inequality for *polynomials* $p(z)$ directly.) These two relations yield

$$2\log|p(x)| \;\leqslant\; \frac{1}{\pi}\int_{-\infty}^{\infty} \frac{\log^+ R(\xi+\mathrm{i})}{(x-\xi)^2+1}\,\mathrm{d}\xi, \quad x \in \mathbb{R},$$

whence, taking the supremum of $2\log|p(x)|$ for such polynomials p,

$$\log R(x) \;\leqslant\; \frac{1}{\pi}\int_{-\infty}^{\infty} \frac{\log^+ R(\xi+\mathrm{i})}{(x-\xi)^2+1}\,\mathrm{d}\xi, \quad x \in \mathbb{R},$$

(by definition again!). We can, of course, replace $\log R(x)$ by $\log^+ R(x)$ in this inequality, since the right-hand side is $\geqslant 0$.

We see from (‡) that the integral on the right in the relation just obtained is $< \infty$ for each $x \in \mathbb{R}$. That is, $R(x) < \infty$ *for every real* x *if our moment sequence is indeterminate*, this is part of what we wanted to prove. Again,

$$\int_{-\infty}^{\infty} \frac{\log^+ R(x)}{1+x^2}\,\mathrm{d}x \;\leqslant\; \frac{1}{\pi}\int_{-\infty}^{\infty}\int_{-\infty}^{\infty} \frac{\log^+ R(\xi+\mathrm{i})}{(x-\xi)^2+1}\cdot\frac{1}{x^2+1}\,\mathrm{d}\xi\,\mathrm{d}x$$

$$= \frac{1}{\pi}\int_{-\infty}^{\infty} \log^+ R(\xi+\mathrm{i}) \int_{-\infty}^{\infty} \frac{\mathrm{d}x}{((\xi-x)^2+1)(x^2+1)}\,\mathrm{d}\xi$$

$$= \int_{-\infty}^{\infty} \frac{2\log^+ R(\xi+\mathrm{i})}{\xi^2+4}\,\mathrm{d}\xi,$$

and the last integral is finite by (‡). This shows that

$$\int_{-\infty}^{\infty} \frac{\log^+ R(x)}{x^2+1}\,\mathrm{d}x \;<\; \infty,$$

and the *second part* of our theorem is completely proved.

Corollary. *The moment sequence* $\{S_k\}$ *is determinate iff, for the function* $R(z)$ *associated with it,*

$$\int_{-\infty}^{\infty} \frac{\log^+ R(x)}{1+x^2}\,\mathrm{d}x \;=\; \infty.$$

Remark. The corollary does not give the full story. What the theorem really says is that there is an *alternative* for the function $R(x)$: *either* $R(x) = \infty$ *everywhere on* $\mathbb{R}$ *save, perhaps, on a countable set of points, or else* $R(x) < \infty$ *everywhere on* $\mathbb{R}$ *and*

$$\int_{-\infty}^{\infty} \frac{\log^+ R(x)}{x^2+1} \mathrm{d}x \ < \ \infty.$$

Scholium. Take the *normalized orthogonal polynomials* $P_n(x)$ corresponding to a positive measure μ with the moments S_k. Like the $p_n(x)$ used in the first part of the proof of the above theorem, the P_n are gotten by applying Schmidt's orthogonalization procedure (with the measure μ) to the successive powers $1, x, x^2, x^3, \ldots$; here, however, one also imposes the supplementary conditions

$$\int_{-\infty}^{\infty} [P_n(x)]^2 \, \mathrm{d}\mu(x) = 1,$$

making each $P_n(x)$ a *constant multiple* of $p_n(x)$. One of course needs only the S_k to compute the successive P_n.

It is easy to express $R(x)$ in terms of these P_n; we have, in fact,

$$R(x) \ = \ \sum_{n=0}^{\infty} (P_n(x))^2.$$

Proof of this relation may be left to the reader – first work out

$$R_N(x) = \max\{|p(x)|^2 \colon p \text{ a polynomial of degree } \leqslant N \text{ with } \int_{-\infty}^{\infty} |p(t)|^2 \, \mathrm{d}\mu(t) = 1\}$$

by writing $p(t) = \sum_{n=0}^{N} \alpha_n P_n(t)$ and using Lagrange's method; then make $N \to \infty$. The boxed formula seems at first sight to break down if any μ with the moments S_k is supported on a finite number of points, say M. In that case, the formula can, however, be saved by taking $P_n(x) = \infty$ for $n \geqslant M$ and x lying outside the support of μ. This makes sense, because the *only polynomial of degree* $n \geqslant M$ orthogonal to the powers $1, x, \ldots, x^{M-1}$ with respect to a measure supported on M points is ***zero**, hence can't be normalized.* The vain attempt to normalize it gives us the form $0/0$, which we are of course at liberty to take as ∞ outside the support of that measure.

2. Derivation of the results in §C from the above one

Let us first deduce Carleman's theorem in §C.1 from that of M. Riesz. According to the *second* theorem of §C.1 (in the scholium of that article), Carleman's theorem is *equivalent* to the following statement: *the moment sequence* $\{S_k\}$ *is determinate provided that*

$$(*) \qquad \int_0^\infty \frac{\log S(r)}{1+r^2}\,\mathrm{d}r \;=\; \infty,$$

where

$$S(r) \;=\; \sup_{k\geqslant 0} \frac{r^{2k}}{S_{2k}} \quad \text{for } r>0.$$

To verify this, observe that, if the S_k are the moments of a positive measure μ, the polynomials

$$q_k(x) \;=\; x^k/\sqrt{S_{2k}}$$

satisfy

$$\int_{-\infty}^{\infty} (q_k(x))^2\,\mathrm{d}\mu(x) = 1,$$

so surely $R(x) \geqslant (q_k(x))^2$ for each k, by definition of $R(z)$. Therefore

$$R(x) \geqslant S(|x|).$$

Also, $S(|x|) \;\geqslant\; 1/S_0 \;>\; 0$, so $\log S(|x|)$ is *bounded below*. It is thus clear that $(*)$ implies

$$\int_{-\infty}^{\infty} \frac{\log^+ R(x)}{1+x^2}\,\mathrm{d}x \;=\; \infty.$$

The moment sequence $\{S_k\}$ is therefore *determinate* by the corollary to Riesz' theorem.

Consider now the theorem of §C.2. We are given a positive integrable function $w(x)$ with

$$\int_{-\infty}^{\infty} \frac{\log w(x)}{1+x^2}\,\mathrm{d}x \;>\; -\infty,$$

and want to prove that the moment sequence $S_k = \int_{-\infty}^{\infty} x^k w(x)\mathrm{d}x$ is *indeterminate* using Riesz' theorem.

Observe that the integrability of $w(x)$ makes $\int_{-\infty}^{\infty}(w(x)/(1+x^2))\mathrm{d}x < \infty$, so, surely, $\int_{-\infty}^{\infty}(\log^+ w(x)/(1+x^2))\mathrm{d}x < \infty$. Our other assumption on w therefore implies that

$$(\ast\ast) \qquad \int_{-\infty}^{\infty} \frac{\log^- w(x)}{1+x^2}\,\mathrm{d}x \;<\; \infty.$$

Take any polynomial p with $\int_{-\infty}^{\infty}|p(x)|^2 w(x)\mathrm{d}x \leqslant 1$. Then, surely,

$$\frac{1}{\pi}\int_{-\infty}^{\infty}\frac{|p(x)|^2 w(x)}{(x-\xi)^2+1}\mathrm{d}x \;\leqslant\; \frac{1}{\pi}$$

for any real ξ, whence, by the inequality between arithmetic and geometric means,

$$\frac{1}{\pi}\int_{-\infty}^{\infty}\frac{2\log|p(x)|+\log w(x)}{(x-\xi)^2+1}\mathrm{d}x \;\leqslant\; \log\frac{1}{\pi} \;\leqslant\; 0,$$

i.e.

$$\frac{1}{\pi}\int_{-\infty}^{\infty}\frac{2\log|p(x)|}{(x-\xi)^2+1}\mathrm{d}x \;\leqslant\; -\frac{1}{\pi}\int_{-\infty}^{\infty}\frac{\log w(x)}{(x-\xi)^2+1}\mathrm{d}x.$$

The left-hand integral is, however, $\geqslant 2\log|p(\xi+\mathrm{i})|$ by the second theorem of §G.2, Chapter III. ($p(z)$ is entire, of exponential type *zero*. For polynomials, the fact in question may also be easily verified directly.) We therefore have

$$\log|p(\xi+\mathrm{i})|^2 \;\leqslant\; \frac{1}{\pi}\int_{-\infty}^{\infty}\frac{\log^- w(x)}{(x-\xi)^2+1}\mathrm{d}x,$$

and, taking the supremum over such polynomials p,

$$\log R(\xi+\mathrm{i}) \;\leqslant\; \frac{1}{\pi}\int_{-\infty}^{\infty}\frac{\log^- w(x)}{(x-\xi)^2+1}\mathrm{d}x.$$

Here, one may, of course, replace $\log R(\xi+\mathrm{i})$ by $\log^+ R(\xi+\mathrm{i})$ on the left.

For $x_0\in\mathbb{R}$,

$$\log R(x_0) \;\leqslant\; \frac{1}{\pi}\int_{-\infty}^{\infty}\frac{\log^+ R(\xi+\mathrm{i})}{(x_0-\xi)^2+1}\mathrm{d}\xi,$$

just as in the proof of the second part of Riesz' theorem. Substituting in the previous inequality on the right and changing the order of integration, we get finally

$$\log R(x_0) \;\leqslant\; \frac{1}{\pi}\int_{-\infty}^{\infty}\frac{2\log^- w(x)}{(x_0-x)^2+4}\mathrm{d}x, \quad x_0\in\mathbb{R}.$$

But the integral on the right is *finite* by (*). Therefore $R(x_0)<\infty$ for each $x_0\in\mathbb{R}$, so the moment sequence $\{S_k\}$ is *indeterminate* by Riesz' theorem. We are done.

VI

Weighted approximation on the real line

In the study of weighted approximation on $\mathbb{R}$, we start with a function $W(x) \geqslant 1$, henceforth called a *weight*, defined for $-\infty < x < \infty$. We usually suppose that $W(x) \to \infty$ for $x \to \pm\infty$, but *do not always assume W continuous*, and frequently allow it to be *infinite* on some large sets.

Given a weight W, we take the space $\mathscr{C}_W(\mathbb{R})$ consisting of continuous functions $\varphi(x)$ defined on $\mathbb{R}$ with $\varphi(x)/W(x) \longrightarrow 0$ for $x \to \pm\infty$, and write $\|\varphi\|_W = \sup_x |\varphi(x)/W(x)|$ for $\varphi \in \mathscr{C}_W(\mathbb{R})$. Being presented with a certain subset $\mathscr{E}$ of $\mathscr{C}_W(\mathbb{R})$, we then ask whether $\mathscr{E}$ is *dense* in $\mathscr{C}_W(\mathbb{R})$ in the norm $\| \ \|_W$ – this is the so-called *weighted approximation problem.*

The following preliminary observation will be used continually.

Lemma. *$\mathscr{E}$ is $\| \ \|_W$-dense in $\mathscr{C}_W(\mathbb{R})$ iff, for some $c \notin \mathbb{R}$, all the functions*

$$\frac{1}{(x-c)^n W(x)} \quad \textit{and} \quad \frac{1}{(x-\bar{c})^n W(x)}, \qquad n = 1, 2, 3, \ldots,$$

can be approximated uniformly on $\mathbb{R}$ by functions of the form $f(x)/W(x)$ with $f \in \mathscr{E}$.

Proof. *Only if is* manifest. For *if*, take any function $\varphi \in \mathscr{C}_W(\mathbb{R})$ and *first* construct a continuous function ψ of *compact support* such that $|(\varphi(x) - \psi(x))/W(x)| < \varepsilon/3$ on $\mathbb{R}$. We can, for instance, put

$$\psi(x) = \begin{cases} \varphi(x), & |x| \leqslant A, \\ \dfrac{2A - |x|}{A}\varphi(x), & A \leqslant |x| \leqslant 2A, \\ 0, & |x| \geqslant 2A; \end{cases}$$

the desired relation will then hold if A is taken *large enough*, since $\varphi(x)/W(x) \longrightarrow 0$ for $x \to \pm\infty$.

By the appropriate version of Weierstrass' theorem, linear combinations of the functions $(x-c)^{-n}$, $(x-\bar{c})^{-n}$, $n = 1,2,3,\ldots$, can now be used to approximation $\psi(x)$ *uniformly* on $\mathbb{R}$, so we can get such a linear combination $\sigma(x)$ with $|\psi(x)-\sigma(x)| < \varepsilon/3$ on $\mathbb{R}$, whence (since $W(x) \geqslant 1!$), $|(\psi(x)-\sigma(x))/W(x)| < \varepsilon/3$ there.

If, now, we can find an $f \in \mathscr{E}$ with $|\sigma(x)/W(x) - f(x)/W(x)| < \varepsilon/3$ for $x \in \mathbb{R}$, we'll have $\|\sigma - f\|_W \leqslant \varepsilon/3$. Then, since $\|\varphi - \psi\|_W \leqslant \varepsilon/3$ and $\|\psi - \sigma\|_W \leqslant \varepsilon/3$, we obtain $\|\varphi - f\|_W \leqslant \varepsilon$. This establishes the *if* part of the lemma.

In the first weighted approximation problem we consider, the so-called *Bernstein approximation problem*, $\mathscr{E}$ consists of *polynomials*. *Then, of course,* ▶ *we must impose on the weight W the supplementary requirement that*

$$\frac{x^n}{W(x)} \longrightarrow 0 \quad \text{as } x \to \pm\infty$$

for all $n \geqslant 0$. In another problem, whose treatment is similar to that of Bernstein's, $\mathscr{E}$ consists of all finite linear combinations of the exponentials $e^{i\lambda x}$ with $-a \leqslant \lambda \leqslant a$ for some given positive a. If a weight W satisfies the supplementary condition, it is natural to compare the solution of Bernstein's problem with those of the latter one for different values of a. One may also study approximation using *weighted* L_p *norms* instead of the weighted uniform norm $\| \quad \|_W$.

These questions are taken up in the present chapter. Some of the methods applied in studying them resemble closely the one used to prove the *second* part of Riesz' theorem (previous chapter, §D.1). There is indeed a relation between the material of this chapter and the determinacy problem discussed in the preceding one, and results obtained in the study of either subject may sometimes be applied to the other.

I know of no book entirely devoted to the matters mentioned above; the one by Nachbin has very little concerning them and is really about something else. One who wishes to go into the subject should first read Mergelian's *Uspekhi* paper and then Akhiezer's; both have been translated into English. There is material in the complements at the end of the second edition of Akhiezer's book on approximation theory, and also in de Branges' book. It is worthwhile to study S. Bernstein's original investigations on weighted polynomial approximation; most of his papers on this are in volume two of his collected works. Some of the results given near the end of the present chapter are from a paper of mine published around 1964.

A. Mergelian's treatment of weighted polynomial approximation

Let W be a weight such that

$$\frac{x^n}{W(x)} \longrightarrow 0 \quad \text{as } x \to \pm\infty \quad \text{for } n = 1, 2, 3, \ldots;$$

the solution of Bernstein's approximation problem for W turns out to be governed by the quantity

$$\Omega(z) = \sup\left\{ |P(z)| \colon\; P \text{ a polynomial and } \left|\frac{P(t)}{(t-\mathrm{i})W(t)}\right| \leqslant 1 \quad \text{on } \mathbb{R} \right\}$$

introduced by Mergelian.

Note the similarity between the definition of this quantity and that of the Riesz function $R(z)$ given in §D.1 of the preceding chapter.

Note especially that the condition $|P(t)/(t-\mathrm{i})W(t)| \leqslant 1$ *and not the seemingly more natural one* $|P(t)/W(t)| \leqslant 1$ *is used in defining* $\Omega(z)$. About this, more later.

1. Criterion in terms of finiteness of $\Omega(z)$

Theorem (Mergelian). *The polynomials are* $\|\ \|_W$*-dense in* $\mathscr{C}_W(\mathbb{R})$ *iff* $\Omega(z_0) = \infty$ *for one non-real* z_0, *and, if this happens, then* $\Omega(z) = \infty$ *for all non-real* z.

Proof (Mergelian).

Only if: Suppose the polynomials are dense in $\mathscr{C}_W(\mathbb{R})$. Then, given any $z_0 \notin \mathbb{R}$, we can find polynomials $Q_n(t)$ such that the quantities

$$\delta_n \;=\; \sup_{t\in\mathbb{R}} \left| \frac{(t-z_0)^{-1} - Q_n(t)}{W(t)} \right|$$

tend to *zero* as $n \to \infty$. Put

$$P_n(t) \;=\; \frac{1-(t-z_0)Q_n(t)}{\delta_n};$$

$P_n(t)$ is for each n a polynomial, $P_n(z_0) = 1/\delta_n \xrightarrow[n]{} \infty$, and $|P_n(t)/(t-z_0)W(t)| \leqslant 1$ for $t\in\mathbb{R}$. There is obviously a number $K(z_0) > 0$ *depending only on* z_0 such that

$$\frac{1}{K(z_0)} \;\leqslant\; \left|\frac{t-z_0}{t-\mathrm{i}}\right| \;\leqslant\; K(z_0) \quad \text{for } t\in\mathbb{R}.$$

Therefore $|P_n(t)/(t-\mathrm{i})W(t)| \leqslant K(z_0)$ on $\mathbb{R}$, so, since $P_n(z_0) \underset{n}{\longrightarrow} \infty$, we see that $\Omega(z_0) = \infty$, establishing the *only if* part.

If: When $z_0 \notin \mathbb{R}$, it is convenient to work with the quantity

$$M(z_0) = \sup\left\{|P(z_0)|:\ P \text{ a polynomial and } \left|\frac{P(t)}{(t-z_0)W(t)}\right| \leqslant 1 \text{ on } \mathbb{R}\right\}.$$

One sees by using the number $K(z_0)$ brought in during the above argument that $\Omega(z_0) = \infty$ iff $M(z_0) = \infty$, as long as $z_0 \notin \mathbb{R}$.

One advantage of introducing $M(z)$ lies in its *continuity property*:

$$(*) \qquad \left|\frac{1}{M(z)} - \frac{1}{M(\zeta)}\right| \leqslant \frac{|\zeta - z|}{|\Im z|\,|\Im \zeta|}.$$

To verify this, take any polynomial P with $|P(t)/(t-\zeta)W(t)| \leqslant 1$ on $\mathbb{R}$ and $|P(\zeta)|$ *close to* $M(\zeta)$. For the polynomial in t

$$Q(t) = \frac{P(t) - P(\zeta)}{t - \zeta}(t - z) + P(\zeta)$$

we have $Q(z) = P(\zeta)$, whilst, for $t \in \mathbb{R}$, $|Q(t)/(t-z)W(t)| \leqslant$ $|P(t)/(t-\zeta)W(t)| + |P(\zeta)|\,|(z-\zeta)/(t-z)(t-\zeta)W(t)|$ and this is $\leqslant 1 + |P(\zeta)|\,|z-\zeta|/|\Im z|\,|\Im \zeta|$, since, as we are always assuming, $W(x) \geqslant 1$. Put now

$$R(t) = \left(1 + |P(\zeta)|\frac{|z-\zeta|}{|\Im z|\,|\Im \zeta|}\right)^{-1} \cdot Q(t);$$

R is a polynomial in t, and $|R(t)/(t-z)W(t)| \leqslant 1$ on $\mathbb{R}$. Therefore $|R(z)| \leqslant M(z)$; however,

$$|R(z)| = \frac{|P(\zeta)|}{1 + |P(\zeta)|\dfrac{|z-\zeta|}{|\Im z|\,|\Im \zeta|}}.$$

Thence,

$$\frac{1}{M(z)} \leqslant \frac{1}{|R(z)|} = \frac{1}{|P(\zeta)|} + \frac{|z-\zeta|}{|\Im z|\,|\Im \zeta|},$$

so, since we can have $|P(\zeta)|$ as close as we like to $M(\zeta)$, we get

$$\frac{1}{M(z)} \leqslant \frac{1}{M(\zeta)} + \frac{|z-\zeta|}{|\Im z|\,|\Im \zeta|}$$

This relation and the similar one obtained by reversing the rôles of z and ζ in the argument just made give us $(*)$.

Armed with $(*)$, we proceed with the *if* part of our proof. Suppose then that $z_0 \notin \mathbb{R}$ and that $\Omega(z_0) = \infty$; this means that $M(z_0) = \infty$, so we can find polynomials $P_n(t)$ with $|P_n(t)/(t - z_0)W(t)| \leqslant 1$ on $\mathbb{R}$ whilst $|P_n(z_0)| \underset{n}{\longrightarrow} \infty$. For the polynomials

$$Q_n(t) \;=\; \frac{P_n(z_0) - P_n(t)}{(t - z_0)P_n(z_0)},$$

we will have, for $t \in \mathbb{R}$,

$$\sup_{t\in\mathbb{R}} \left| \frac{1}{(t - z_0)W(t)} - \frac{Q_n(t)}{W(t)} \right| \;\leqslant\; \frac{1}{|P_n(z_0)|},$$

a quantity tending to 0 as $n \to \infty$.

Therefore the function $(t - z_0)^{-1}$ can be approximated as closely as we like, in the norm $\| \; \|_W$, by polynomials.

It is now claimed that

$$(\ast\!\!\ast) \qquad \int_0^{2\pi} \Omega(z_0 + \rho e^{i\vartheta})d\vartheta = \infty$$

for each $\rho > 0$. Since $\Omega(z_0) = \infty$, there are polynomials $q_n(t)$ with $|q_n(t)/(t - i)W(t)| \leqslant 1$ on $\mathbb{R}$ (N.B. Here it *is* $t - i$ in the denominator and *not* $t - z_0$!), and $|q_n(z_0)| \underset{n}{\longrightarrow} \infty$. For each n, $|q_n(z)| \leqslant \Omega(z)$ *by definition*, therefore

$$\int_0^{2\pi} \Omega(z_0 + \rho e^{i\vartheta})d\vartheta \;\geqslant\; \int_0^{2\pi} |q_n(z_0 + \rho e^{i\vartheta})|d\vartheta \;\geqslant\; 2\pi|q_n(z_0)|.$$

Since the quantity on the right $\to \infty$ with n, we have $(\ast\!\!\ast)$.

If $0 < \rho < |\Im z_0|$, there is a z_ρ, $|z_\rho - z_0| = \rho$, with $\Omega(z_\rho) = \infty$. Indeed, $(\ast\!\!\ast)$ implies the existence of a sequence of points ζ_k with $|\zeta_k - z_0| = \rho$ and $\Omega(\zeta_k) \underset{k}{\longrightarrow} \infty$. Suppose, wlog, that $\zeta_k \underset{k}{\longrightarrow} z_\rho$. Comparison of the definitions of $\Omega(\zeta)$ and $M(\zeta)$ shows immediately that

$$\Omega(\zeta) \;\leqslant\; \sup_{t\in\mathbb{R}} \left| \frac{t - i}{t - \zeta} \right| \cdot M(\zeta).$$

Here, the supremum is *clearly bounded above* for $|\zeta - z_0| = \rho$ since $|\Im z_0| - \rho > 0$; therefore our choice of the sequence $\{\zeta_k\}$ makes $M(\zeta_k) \underset{k}{\longrightarrow} \infty$. Because $\zeta_k \underset{k}{\longrightarrow} z_\rho$, *we then get* $M(z_\rho) = \infty$ *by* $(*)$, since $|\Im\zeta| \geqslant |\Im z_0| - \rho > 0$ on the circle $|\zeta - z_0| = \rho$. Thus, $\Omega(z_\rho) = \infty$. (A mistake I made here while lecturing was pointed out to me by Dr Raymond Couture.)

We thus have points z_ρ for which $\Omega(z_\rho) = \infty$ with $|z_\rho - z_0| = \rho$, when $\rho > 0$ is sufficiently small.

For each such z_ρ, $1/(t - z_\rho)$ can be approximated in $\| \; \|_W$-norm by polynomials in t; this is shown by the argument used above for $1/(t - z_0)$. We can therefore obtain a sequence of points $z_n \neq z_0$ tending to z_0, such that each of the functions

$$\frac{1}{(t - z_n)W(t)}$$

is the *uniform limit*, on $\mathbb{R}$, of polynomials in t divided by $W(t)$. This fact makes it possible for us to show (by taking limits of difference quotients of successively higher order) that *each* of the expressions

$$\frac{1}{(t - z_0)^m W(t)}, \quad m = 1, 2, 3, \ldots,$$

can be uniformly approximated on $\mathbb{R}$ by polynomials in t divided by $W(t)$ $(\geqslant 1)$.

Now we have $\Omega(\bar{z}_0) = \Omega(z_0)$, for, if P is a polynomial with $|P(t)/(t - \mathrm{i})W(t)| \leqslant 1$ on $\mathbb{R}$, the polynomial $P^*(t)$ whose coefficients are the *complex conjugates* of the corresponding ones of $P(t)$ also satisfies $|P^*(t)/(t - \mathrm{i})W(t)| \leqslant 1$, $t \in \mathbb{R}$. Therefore in the present case $\Omega(\bar{z}_0) = \infty$, so, by the above discussion, each of the functions $1/(t - \bar{z}_0)^m W(t)$, $m = 1, 2, 3, \ldots$, can be uniformly approximated on $\mathbb{R}$ by polynomials in t divided by $W(t)$. As we just saw, the same is true for the functions $1/(t - z_0)^m W(t)$. *According to the general lemma given at the very beginning of this chapter, polynomials must hence be* $\| \ \|_W$*-dense in* $\mathscr{C}_W(\mathbb{R})$. The *if* part of our theorem is thus established.

We are done.

2. A computation

In the next article and later on, as well, we will need a *formula* for

$$\sup_{t \in \mathbb{R}} \left| \frac{t - \mathrm{i}}{t - z} \right|.$$

Lemma. *When* $\Im z > 0$,

$$\boxed{\sup_{t \in \mathbb{R}} \left| \frac{t - \mathrm{i}}{t - z} \right| = \frac{|z + \mathrm{i}| + |z - \mathrm{i}|}{2\Im z}}$$

Proof. $|(t - \mathrm{i})/(t - z)| = |1 - (z - \mathrm{i})/(t - \mathrm{i})|^{-1}$. In order to simplify the writing, put $z - \mathrm{i} = \zeta$; then we have to calculate $\inf_{t \in \mathbb{R}} |1 - \zeta/(t - \mathrm{i})|$. The linear fractional transformation $t \to 1/(t - \mathrm{i})$ takes the real axis into the *circle* having the segment $[0, \mathrm{i}]$ as *diameter*:

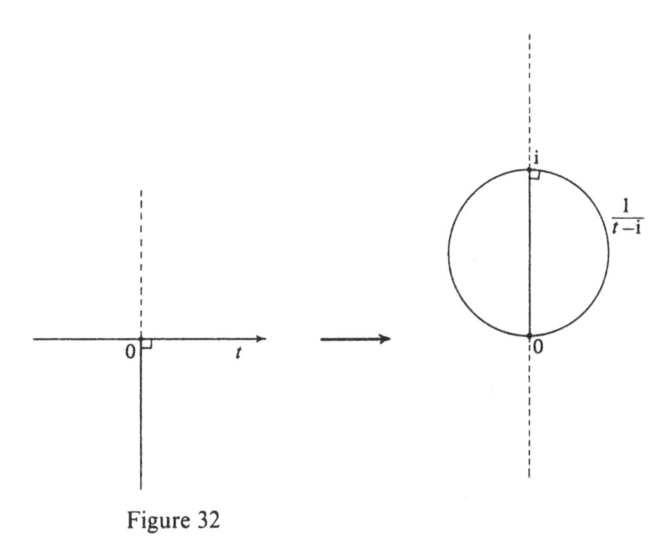

Figure 32

Therefore, as t ranges over the real axis, $\zeta/(t - \mathrm{i})$ ranges over the circle γ_ζ with segment $[0, \zeta\mathrm{i}]$ as diameter:

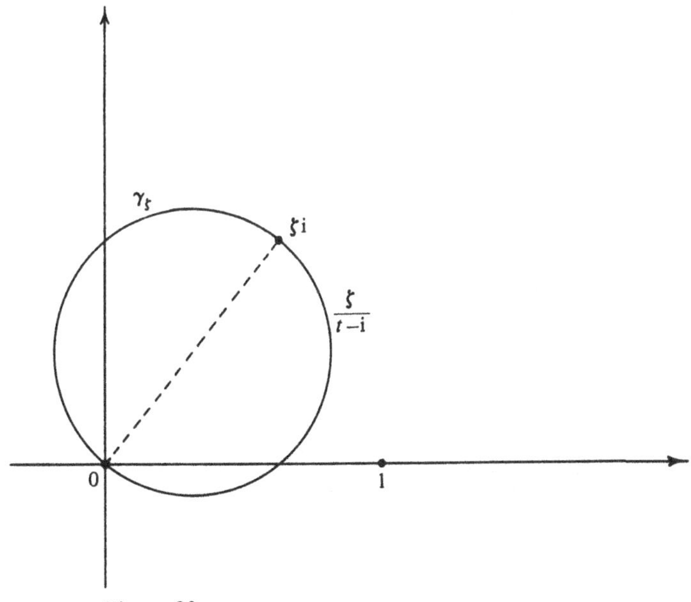

Figure 33

Since $\Im z > 0$, $\Im\zeta > -1$, so $\Re(\zeta \mathrm{i}) < 1$. Therefore the point 1 must lie *outside* the circle γ_ζ:

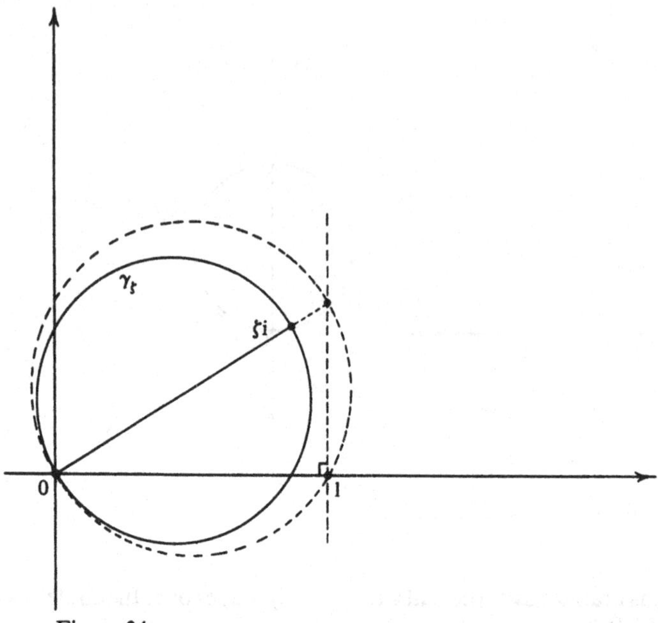

Figure 34

Our quantity $\inf_{t\in\mathbb{R}} |1 - \zeta/(t - \mathrm{i})|$, which is simply the distance from 1 to γ_ζ, can thus be read off from the diagram:

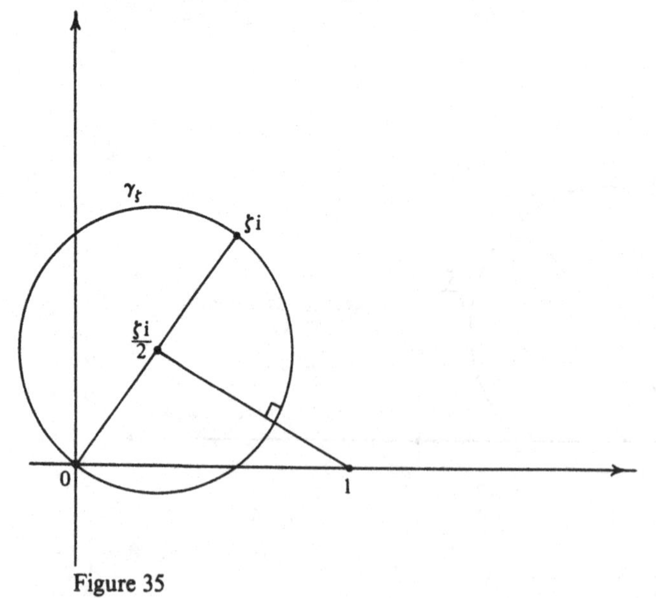

Figure 35

We see that

$$\inf_{t\in\mathbb{R}}\left|1-\frac{\zeta}{t-\mathrm{i}}\right| = \left|1-\frac{\zeta\mathrm{i}}{2}\right| - \text{radius of } \gamma_\zeta$$

$$= \left|1-\frac{\zeta\mathrm{i}}{2}\right| - \left|\frac{\zeta\mathrm{i}}{2}\right| = \left|\frac{1}{2}-\frac{\mathrm{i}z}{2}\right| - \left|\frac{1}{2}+\frac{\mathrm{i}z}{2}\right|.$$

Finally,

$$\sup_{t\in\mathbb{R}}\left|\frac{t-\mathrm{i}}{t-z}\right| = \frac{1}{\left|\frac{1}{2}-\frac{\mathrm{i}z}{2}\right| - \left|\frac{1}{2}+\frac{\mathrm{i}z}{2}\right|} = \frac{|z+\mathrm{i}|+|z-\mathrm{i}|}{2\Im z},$$

proving the lemma.

Corollary. *For $\Im z > 0$,*

$$\sup_{t\in\mathbb{R}}\left|\frac{t-\mathrm{i}}{t-z}\right| \leqslant \frac{1+|z|}{\Im z}.$$

This inequality will be sufficient for our purposes.

3. Criterion in terms of $\int_{-\infty}^{\infty}(\log\Omega(t)/(1+t^2))\,\mathrm{d}t$

We return to the consideration of the quantity $\Omega(t)$ introduced at the beginning of this §, and to its connection with weighted polynomial approximation.

Theorem (Mergelian). *Polynomials are dense in $\mathscr{C}_W(\mathbb{R})$ iff*

$$(*) \qquad \int_{-\infty}^{\infty}\frac{\log\Omega(t)}{1+t^2}\mathrm{d}t = \infty.$$

Remark. Since $W(t)\geqslant 1$ we *always have* $\log\Omega(z)\geqslant 0$, i.e., $\Omega(z)\geqslant 1$, because 1 is a *polynomial* (!), and $|1/(t-\mathrm{i})W(t)|\leqslant 1$ on $\mathbb{R}$.

Proof of theorem

Only if: We must show that, if (∗) *fails*, the polynomials *can't be dense* in $\mathscr{C}_W(\mathbb{R})$.

Assume, then, that

$$\int_{-\infty}^{\infty}\frac{\log\Omega(t)}{1+t^2}\mathrm{d}t < \infty,$$

and take *any polynomial* $P(t)$ with $|P(t)/(t-\mathrm{i})W(t)|\leqslant 1$ on $\mathbb{R}$. Then, by a very simple version of the second theorem of §G.2, Chapter III (which, for *polynomials*, can be easily verified directly),

$$\log|P(\mathrm{i})| \leq \frac{1}{\pi}\int_{-\infty}^{\infty} \frac{\log|P(t)|}{1+t^2}\,\mathrm{d}t.$$

Here, *by definition* (compare with the proof of the second part of M. Riesz' theorem in §D.1, Chapter V),

$$|P(t)| \leq \Omega(t),$$

so

$$\log|P(\mathrm{i})| \leq \frac{1}{\pi}\int_{-\infty}^{\infty} \frac{\log \Omega(t)}{1+t^2}\,\mathrm{d}t,$$

and, taking the supremum of $\log|P(\mathrm{i})|$ over all polynomials P subject to the condition given above, we get

$$\log \Omega(\mathrm{i}) \leq \frac{1}{\pi}\int_{-\infty}^{\infty} \frac{\log \Omega(t)}{1+t^2}\,\mathrm{d}t < \infty.$$

The quantity $\Omega(\mathrm{i})$ is thus finite. Therefore polynomials *cannot be dense* in $\mathscr{C}_W(\mathbb{R})$ by the first Mergelian theorem of article 1.

If: Supposing that polynomials *are not dense* in $\mathscr{C}_W(\mathbb{R})$, we must show that (*) *is false*.

If polynomials are *not dense* in $\mathscr{C}_W(\mathbb{R})$, the Hahn–Banach theorem (whose validity does *not*, by the way, depend on $\mathscr{C}_W(\mathbb{R})$'s being complete!) furnishes us with a *bounded linear functional* on $\mathscr{C}_W(\mathbb{R})$ which is *not identically zero thereon*, but *is zero at each of the polynomials*. It is convenient to denote the *value* of this linear functional at a member φ of $\mathscr{C}_W(\mathbb{R})$ by the expression

$$L\left(\frac{\varphi(t)}{W(t)}\right);$$

the reason for this is that *then* we will simply have

$$(*\!*)\qquad \left|L\left(\frac{\varphi(t)}{W(t)}\right)\right| \leq C \sup_{t\in\mathbb{R}}\left|\frac{\varphi(t)}{W(t)}\right|$$

with some constant C, for $\varphi \in \mathscr{C}_W(\mathbb{R})$.

▶ (N.B. We are NOT writing $L(\varphi(t)/W(t))$ as $\int_{-\infty}^{\infty}(\varphi(t)/W(t))\,\mathrm{d}\mu(t)$ with a Radon measure μ. That's because we are *not assuming any continuity* of $W(t)$ here, so the existence of such a measure μ is problematical.)

Let us continue with this part of the proof. We have our linear functional L, such that

$$(\dagger)\qquad L\left(\frac{P(t)}{W(t)}\right) = 0$$

for every polynomial P, whilst

$$(\S)\qquad L\left(\frac{\varphi_0(t)}{W(t)}\right)\neq 0$$

for some $\varphi_0\in\mathscr{C}_W(\mathbb{R})$.

Consider the function

$$F(z) \;=\; L\left(\frac{1}{(t-z)W(t)}\right),$$

defined whenever $z\notin\mathbb{R}$. In the first place, $F(z)$ *cannot vanish identically for both* $\Im z>0$ and $\Im z<0$. Suppose it *did*. A simple modification of the general lemma given at the beginning of this chapter (whose verification is left to the reader) guarantees, for each $\varepsilon>0$, the existence of a finite linear combination $\varphi_\varepsilon(t)$ of the fractions $1/(t-c)$, $c\notin\mathbb{R}$, such that $\|\varphi_0-\varphi_\varepsilon\|_W<\varepsilon$ for the function φ_0 figuring in (§). If, then, $F(c)=L(1/(t-c)W(t))=0$ for every $c\notin\mathbb{R}$, we'd have $L(\varphi_\varepsilon(t)/W(t))=0$, whence $|L(\varphi_0(t)/W(t))|<C\varepsilon$ by (⁑). Squeezing ε, we get a contradiction with (§).

Wlog, $F(z)$ is not identically zero in $\{\Im z>0\}$. *It is analytic there.* To see this, observe that if $z\notin\mathbb{R}$, the difference quotient

$$\frac{(t-z-\Delta z)^{-1}-(t-z)^{-1}}{\Delta z} \;=\; \frac{1}{(t-z)(t-z-\Delta z)}$$

tends to $(t-z)^{-2}$ *uniformly for* $t\in\mathbb{R}$ as $\Delta z\to 0$. Therefore, by the linearity of L and (⁑),

$$\frac{F(z+\Delta z)-F(z)}{\Delta z} \;\longrightarrow\; L\left(\frac{1}{(t-z)^2W(t)}\right)$$

as $\Delta z\to 0$, since $W(t)\geqslant 1$ on $\mathbb{R}$. This shows that $F'(z)$ exists at every $z\notin\mathbb{R}$ and establishes analyticity of $F(z)$ in $\{\Im z>0\}$.

From (⁑), we get

$$|F(z)|\leqslant C \quad \text{for } \Im z\geqslant 1.$$

Since $F(z)\not\equiv 0$ in the upper half plane, the first theorem of §G.2, Chapter III, shows that

$$(\dagger\dagger)\qquad \int_{-\infty}^{\infty}\frac{\log^-|F(x+i)|}{1+x^2}\,\mathrm{d}x \;<\; \infty.$$

We can now bring in the Markov–Riesz–Pollard trick already used in proving the second part of Riesz' theorem in §D.1 of the previous chapter.

Take any polynomial $P(t)$ and any fixed z, $\Im z > 0$. Then

$$\frac{P(t) - P(z)}{t - z}$$

is *also* a polynomial in t, so, applying (†) to *it*, we get

$$L\left(\frac{P(t) - P(z)}{(t - z)W(t)}\right) = 0,$$

i.e., in terms of $F(z) = L(1/(t - z)W(t))$,

$$F(z)P(z) = L\left(\frac{P(t)}{(t - z)W(t)}\right).$$

We can thus write

$$(\S\S) \qquad P(z) = \frac{1}{F(z)} L\left(\frac{P(t)}{(t - z)W(t)}\right)$$

for $\Im z > 0$, provided z is not a zero of the analytic function $F(z)$. The idea now is to use (§§) together with (††) in order to show that

$$\int_{-\infty}^{\infty} \frac{\log \Omega(t)}{1 + t^2} dt < \infty.$$

Take any polynomial $P(t)$ such that

$$\left|\frac{P(t)}{(t - \mathrm{i})W(t)}\right| \leqslant 1 \quad \text{on} \quad \mathbb{R}.$$

Then, $|P(t)/(t - z)W(t)| \leqslant \sup_{t\in\mathbb{R}} |(t - \mathrm{i})/(t - z)|$ *which, by the previous article, is* $\leqslant (1 + |z|)/\Im z$ for $\Im z > 0$ (see the corollary there). Putting $z = x + \mathrm{i}$, $x \in \mathbb{R}$, we thus get, from $(\ast\!\!\ast)$,

$$\left|L\left(\frac{P(t)}{(t - x - \mathrm{i})W(t)}\right)\right| \leqslant C(1 + \sqrt{(x^2 + 1)}).$$

Referring to (§§), we see that

$$|P(x + \mathrm{i})| \leqslant \frac{C(1 + \sqrt{(x^2 + 1)})}{|F(x + \mathrm{i})|}$$

for any polynomial P with $|P(t)/(t - \mathrm{i})W(t)| \leqslant 1$ on $\mathbb{R}$. Taking the supremum of $|P(x + \mathrm{i})|$ over such P, we find that

$$\Omega(x + \mathrm{i}) \leqslant \frac{C(1 + \sqrt{(x^2 + 1)})}{|F(x + \mathrm{i})|},$$

that is, writing $C' = \log C$,

$$(\ddagger) \qquad \log\Omega(x+\mathrm{i}) \;\leqslant\; C' + \log(1+\sqrt{(x^2+1)}) + \log^-|F(x+\mathrm{i})|.$$

We use the last relation in conjunction with (††) in order to get a grip on $\log\Omega(t)$ for real t. The procedure being followed here is like the one used in proving the second part of Riesz' theorem (§D.1, previous chapter). I call it a *hall of mirrors* argument because it consists in our *first going up to the line* $\Im z = 1$ from the *real axis* and *then going back down to the real axis again*. Our reason for engaging in this roundabout manoeuvre is that we do not have any simple way of controlling $|F(z)|$ when z gets near $\mathbb{R}$ (unless we bring in H_p-spaces, whose use we are avoiding as much as possible!). Let P be a polynomial. By the second theorem of §G.2, Chapter III,

$$\log|P(t)| \;\leqslant\; \frac{1}{\pi}\int_{-\infty}^{\infty}\frac{\log|P(x+\mathrm{i})|}{(x-t)^2+1}\mathrm{d}x.$$

If also $|P(t)/(t-\mathrm{i})W(t)| \leqslant 1$, we have, of course, $|P(x+\mathrm{i})| \leqslant \Omega(x+\mathrm{i})$, so, taking the supremum of $\log|P(t)|$ for such P,

$$\log\Omega(t) \;\leqslant\; \frac{1}{\pi}\int_{-\infty}^{\infty}\frac{\log\Omega(x+\mathrm{i})}{(t-x)^2+1}\mathrm{d}x.$$

Plug in (‡) on the right, multiply by $1/(t^2+1)$, and integrate t from $-\infty$ to ∞. After changing the order of integration and using the identity

$$\frac{1}{\pi}\int_{-\infty}^{\infty}\frac{\mathrm{d}t}{((t-x)^2+1)(t^2+1)} \;=\; \frac{2}{x^2+4},$$

we obtain

$$\int_{-\infty}^{\infty}\frac{\log\Omega(t)}{1+t^2}\mathrm{d}t \;\leqslant\; \pi C' + \int_{-\infty}^{\infty}\frac{2\log(1+\sqrt{(x^2+1)})}{x^2+4}\mathrm{d}x$$
$$+ \int_{-\infty}^{\infty}\frac{2\log^-|F(x+\mathrm{i})|}{x^2+4}\mathrm{d}x.$$

The first integral on the right is obviously finite. The *second is also finite by* (††). The integral

$$\int_{-\infty}^{\infty}\frac{\log\Omega(t)}{1+t^2}\mathrm{d}t$$

is therefore *finite*, contradicting (∗). This completes the proof of the *if* part of the theorem, and we are done.

B. Akhiezer's method, based on use of $W_*(z)$

The function $\Omega(z)$ introduced by Mergelian, which, as we have seen, indicates by its size whether or not the polynomials are dense in $\mathscr{C}_W(\mathbb{R})$, is equal to $\sup\{|P(z)|: P \text{ a polynomial and } |P(t)| \leqslant |t-\mathrm{i}|W(t) \text{ for } t\in\mathbb{R}\}$. The presence of the multiplier $|t-\mathrm{i}|$ in front of $W(t)$ is disconcerting, and it would seem more natural to work with the quantity

$$W_*(z) = \sup\{|P(z)|: P \text{ a polynomial and } |P(t)| \leqslant W(t) \text{ on } \mathbb{R}\}.$$

On the real axis, $W_*(x) \leqslant W(x)$, and so $W_*(x)$ is a kind of *lower regularization* of $W(x)$ by *polynomials*. (Recall that the idea of using *some* kind of lower regularization occurred already in the study of quasianalyticity (Chapter IV); the *convex logarithmic regularization* which turned out to be useful there is *not the same*, however, as the *regularization by polynomials* dealt with here.)

We are always assuming that $W(x) \geqslant 1$. Therefore, since 1 is a *polynomial* (!), we *certainly* have $W_*(z) \geqslant 1$.

1. Criterion in terms of $\int_{-\infty}^{\infty}(\log W_*(x)/(1+x^2))\mathrm{d}x$

The following theorem, due to Akhiezer, is implicity contained in the work of S. Bernstein, who was in possession of all the elements of the proof. Bernstein, who devoted much effort to the study of the problem bearing his name, was apparently unable to see that a solution was within his reach, and never formulated this next result.

Theorem (Akhiezer). *Let $W(x)$ be continuous. Then the polynomials are dense in $\mathscr{C}_W(\mathbb{R})$ iff*

$$\int_{-\infty}^{\infty} \frac{\log W_*(x)}{1+x^2}\mathrm{d}x = \infty.$$

Remark. As we shall see from the proof, the *continuity* requirement on $W(x)$ can be much relaxed. What is really needed here is that $W(x)$ be *finite on a set of points which is not too sparse.*

Proof of Theorem.

If: Comparison of the definitions of $W_*(z)$ and $\Omega(z)$ shows that $W_*(x) \leqslant \Omega(x)$. Therefore, if $\int_{-\infty}^{\infty}(\log W_*(x)/(1+x^2))\mathrm{d}x = \infty$, we certainly have $\int_{-\infty}^{\infty}(\log \Omega(x)/(1+x^2))\mathrm{d}x = \infty$, so polynomials *are* dense in $\mathscr{C}_W(\mathbb{R})$ by Mergelian's second theorem (§A.3). Note that the continuity of W plays no role here.

Only if: Assuming that $\int_{-\infty}^{\infty}(\log W_*(x)/(1+x^2))\mathrm{d}x$ is *finite*, we show that *any collection of polynomials P with $\|P\|_W \leqslant 1$ forms a normal family in the complex plane.* For this, a *hall of mirrors* argument like the one at the end of §A.3 is used. If P is any polynomial with $\|P\|_W \leqslant 1$, the second theorem of

§G.2, Chapter III and the very definition of W_* give, for real ξ,

$$\log|P(\xi+\mathrm{i})| \leqslant \frac{1}{\pi}\int_{-\infty}^{\infty}\frac{\log|P(t)|}{(t-\xi)^2+1}\mathrm{d}t \leqslant \frac{1}{\pi}\int_{-\infty}^{\infty}\frac{\log W_*(t)}{(t-\xi)^2+1}\mathrm{d}t.$$

Taking the supremum of $\log|P(\xi+\mathrm{i})|$ for such P, we find, as usual,

$$\log W_*(\xi+\mathrm{i}) \leqslant \frac{1}{\pi}\int_{-\infty}^{\infty}\frac{\log W_*(t)}{(t-\xi)^2+1}\mathrm{d}t.$$

Now suppose that $\Im z\leqslant 0$. Using again the second theorem of §G.2, Chapter III, but this time in the half plane $\Im z\leqslant 1$, we see that for any polynomial P,

$$\log|P(z)| \leqslant \frac{1}{\pi}\int_{-\infty}^{\infty}\frac{(1+|\Im z|)\log|P(\xi+\mathrm{i})|}{|\xi+\mathrm{i}-z|^2}\mathrm{d}\xi.$$

If also $\|P\|_W\leqslant 1$, we have $|P(\xi+\mathrm{i})|\leqslant W_*(\xi+\mathrm{i})$, so, by the inequality just found for the latter function (which, by the way, is $\geqslant 1$),

$$\log|P(z)| \leqslant \frac{1}{\pi^2}\int_{-\infty}^{\infty}\int_{-\infty}^{\infty}\frac{(1+|\Im z|)\log W_*(t)}{|\xi+\mathrm{i}-z|^2|\xi+\mathrm{i}-t|^2}\mathrm{d}t\,\mathrm{d}\xi.$$

Changing the order of integration, and using the identity

$$\frac{2+|\Im z|}{|t+2\mathrm{i}-z|^2} = \frac{1}{\pi}\int_{-\infty}^{\infty}\frac{(1+|\Im z|)\mathrm{d}\xi}{|\xi+\mathrm{i}-z|^2|t-\xi-\mathrm{i}|^2},$$

valid for $\Im z\leqslant 0$, we find that

$$\log|P(z)| \leqslant \frac{1}{\pi}\int_{-\infty}^{\infty}\frac{(2+|\Im z|)\log W_*(t)}{|t+2\mathrm{i}-z|^2}\mathrm{d}t.$$

Apply now the corollary from §A.2. In the present situation, where $\Im z\leqslant 0$, we get

$$\sup_{t\in\mathbb{R}}\left|\frac{t-\mathrm{i}}{t+2\mathrm{i}-z}\right|^2 \leqslant \left(\frac{1+|z-2\mathrm{i}|}{2+|\Im z|}\right)^2,$$

whence, by the preceding, for $\Im z\leqslant 0$,

$$(\ast)\qquad \log|P(z)| \leqslant \frac{(3+|z|)^2}{2+|\Im z|}\cdot\frac{1}{\pi}\int_{-\infty}^{\infty}\frac{\log W_*(t)}{1+t^2}\mathrm{d}t$$

whenever P is a polynomial with $\|P\|_W\leqslant 1$.

For such polynomials P, however, $(\ast)$ is also valid for $\Im z\geqslant 0$. This is seen by an argument *just* like the above one, working *first* with $\log|P(\xi-\mathrm{i})|$ instead of $\log|P(\xi+\mathrm{i})|$, and *then* using the second theorem of §G.2, Chapter III in the half plane $\Im z\geqslant -1$ instead of the half plane $\Im z\leqslant 1$.

The polynomials P with $\|P\|_W\leqslant 1$ thus satisfy $(\ast)$ in the whole complex plane. Since $\int_{-\infty}^{\infty}(\log W_*(t)/(1+t^2))dt$ is *finite*, such polynomials *form* a *normal family* in $\mathbb{C}$.

Once we know that the polynomials P with $\|P\|_W \leqslant 1$ *do* form a *normal family* in $\mathbb{C}$, it is *manifest* that polynomials *cannot* be $\|\quad\|_W$-dense in $\mathscr{C}_W(\mathbb{R})$. Suppose, indeed, that $\varphi \in \mathscr{C}_W(\mathbb{R})$ and that we have polynomials P_n with $\|P_n - \varphi\|_W \underset{n}{\longrightarrow} 0$. We may wlog take $\|\varphi\|_W < 1$, then $\|P_n\|_W \leqslant 1$ for all sufficiently large n, hence, wlog, for all n.

These $P_n(z)$ therefore form a normal family in $\mathbb{C}$, so a *subsequence of them* must tend u.c.c. in $\mathbb{C}$ to *some entire function* $\Phi(z)$. *At every* $x \in \mathbb{R}$, $\Phi(x)$ *and* $\varphi(x)$ *must coincide, since* $W(x)$, *being continuous, must be finite on* $\mathbb{R}$ (!). The function $\varphi \in \mathscr{C}_W(\mathbb{R})$ which is $\|\quad\|_W$-approximable by polynomials *must thus coincide on* $\mathbb{R}$ *with some entire function.* Since *lots* of continuous $\varphi \in \mathscr{C}_W(\mathbb{R})$ *don't do that*, we see that polynomials *cannot be* $\|\quad\|_W$-dense in $\mathscr{C}_W(\mathbb{R})$.

We have finished the *only if* part of Akhiezer's theorem, which is now completely proved.

Remark. We see already from the argument at the very end of the above proof that *we need merely assume* $W(x) < \infty$ *on some closed subset of* $\mathbb{R}$ *with a finite limit point, instead of the continuity* of W on $\mathbb{R}$, and *then* the property

$$\int_{-\infty}^{\infty} \frac{\log W_*(t)}{1+t^2}\,dt \;<\; \infty$$

will surely imply that the polynomials are not $\|\quad\|_W$-dense in $\mathscr{C}_W(\mathbb{R})$. Even this assumption on $W(x)$ can be very much weakened, as we shall see in the next article.

2. Description of $\|\quad\|_W$ limits of polynomials when $\int_{-\infty}^{\infty}(\log W_*(t)/(1+t^2))dt < \infty$

A small refinement of the calculations made in proving the *only if* part of the previous theorem yields an elegant result.

Theorem (Akhiezer). *If* $\int_{-\infty}^{\infty}(\log W_*(t)/(1+t^2))dt < \infty$, *every function in* $\mathscr{C}_W(\mathbb{R})$ *which can be* $\|\quad\|_W$*-approximated by polynomials coincides, on the subset of* $\mathbb{R}$ *where* $W(x) < \infty$, *with some entire function of zero exponential type.*

Proof. We start from the estimate of $\log|P(z)|$ found in the preceding article for polynomials P with $\|P\|_W \leqslant 1$. As we saw there, if $\Im z \leqslant 0$ and P is a polynomial with $\|P\|_W \leqslant 1$,

$$\log|P(z)| \;\leqslant\; \frac{1}{\pi}\int_{-\infty}^{\infty} \frac{(2+|\Im z|)\log W_*(t)}{|t+2i-z|^2}\,dt.$$

Take any $\varepsilon > 0$. Since $\int_{-\infty}^{\infty}(\log W_*(t)/(1+t^2))dt < \infty$, there is a finite M_ε

such that

$$(\dagger) \qquad \frac{1}{\pi}\int_{\{\log W_*(t)>M_\varepsilon\}} \frac{\log W_*(t)}{1+t^2}\,dt \;<\; \varepsilon;$$

we then break up the integral of the preceding relation into the sum of *two*, *one* over

$$\{t\in\mathbb{R}:\quad \log W_*(t)\leqslant M_\varepsilon\}$$

and the *other* over the set where $\log W_*(t) > M_\varepsilon$. We obtain in this way

$$\log|P(z)| \;\leqslant\; M_\varepsilon + \frac{1}{\pi}\int_{\{\log W_*(t)>M_\varepsilon\}} \frac{(2+|\Im z|)\log W_*(t)}{|t+2i-z|^2}\,dt.$$

Apply now the corollary from §A.2. We find, by virtue of (†), that

$$\log|P(z)| \;\leqslant\; M_\varepsilon + \frac{(1+|z-2i|)^2}{(2+|\Im z|)}\varepsilon;$$

this holds whenever $\Im z \leqslant 0$ if P is a polynomial with $\|P\|_W \leqslant 1$.

One can, of course, use exactly the same kind of reasoning for the half plane $\Im z \geqslant 0$. We see in this way that if P is any polynomial with $\|P\|_W \leqslant 1$, the relation

$$(\dagger\dagger) \qquad \log|P(z)| \;\leqslant\; M_\varepsilon + \frac{(3+|z|)^2}{2+|\Im z|}\varepsilon$$

holds in the entire complex plane.

This inequality we refine still more by use of a Phragmén–Lindelöf argument.

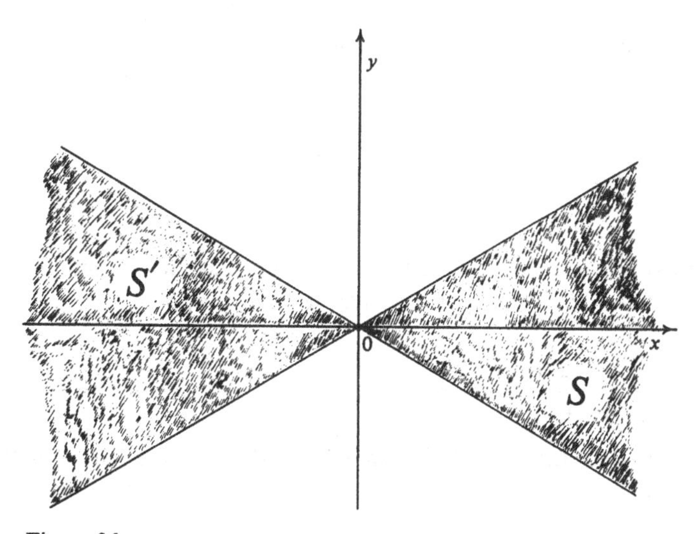

Figure 36

Take the two sectors S and S' where $|y| < \frac{1}{2}|x|$; what is important here is that S and S' *have opening* $< 90°$. *Outside* both S and S', $|x| \leqslant 2|y|$, so $|z| \leqslant 3|y|$, and (††) gives

(§) $$\log|P(z)| \leqslant M_\varepsilon + 9\varepsilon(1 + |\Im z|).$$

This also holds on the *boundaries* of S and S', where it can be rewritten thus:

$$\log|P(z)| \leqslant M_\varepsilon + 9\varepsilon + \tfrac{9}{2}\varepsilon|\Re z|.$$

Let us consider the sector S. Inside S, $\log|P(z)| - \frac{9}{2}\varepsilon\Re z$ is *subharmonic*, and $\leqslant \text{const.}|z|^2$ for large $|z|$ by (††). On the *boundary* of S, $\log|P(z)| - \frac{9}{2}\varepsilon\Re z \leqslant M_\varepsilon + 9\varepsilon$ as we have just seen. So, since the opening of S is $< 90°$, *this last inequality must in fact hold throughout* S, by the *second* Phragmén–Lindelöf theorem of §C, Chapter III. We thus get

$$\log|P(z)| \leqslant M_\varepsilon + 9\varepsilon + \tfrac{9}{2}\varepsilon|\Re z|$$

in S.

The same reasoning applies to S'. Referring to (§), which holds *outside* S and S', and contenting ourselves with a result *slightly worse* than what we *actually have*, we see that

$$\log|P(z)| \leqslant M_\varepsilon + 9\varepsilon(1 + |z|)$$

throughout $\mathbb{C}$, whenever $\|P\|_W \leqslant 1$.

If, now, $\Phi(z)$ is *any* u.c.c. *limit* of polynomials P with $\|P\|_W \leqslant 1$, we must *also* have

$$\log|\Phi(z)| \leqslant M_\varepsilon + 9\varepsilon(1 + |z|)$$

for all complex z. Since $\varepsilon > 0$ is *arbitrary* (with M_ε depending on ε through (†)), we see that the entire function $\Phi(z)$ must be of *exponential type zero.* Any $\varphi \in \mathscr{C}_W(\mathbb{R})$ with $\|\varphi\|_W < 1$ which is the $\|\ \ \|_W$-limit of polynomials must, *on the set of* x *where* $W(x) < \infty$, *coincide* with *such an entire function* $\Phi(z)$, as we saw at the end of the preceding subsection.

We are done.

Remark. Let $\varphi \in \mathscr{C}_W(\mathbb{R})$ be such that there exist polynomials P_n with $\|\varphi - P_n\|_W \underset{n}{\longrightarrow} 0$. Then, as the above theorem shows, if

$$\int_{-\infty}^{\infty} \frac{\log W_*(t)}{1+t^2}\,dt \ < \ \infty,$$

$\varphi(x)$ must, on $\{x\colon W(x) < \infty\}$, *coincide* with an entire function $\Phi(z)$ of *exponential type zero.* Sometimes it is useful to know that $\Phi(z)$ satisfies the more precise condition

$$\log|\Phi(z)| \leqslant \log\|\Phi\|_W + M_\varepsilon + 9\varepsilon(1+|z|), \quad z\in\mathbb{C}.$$

Here, $\varepsilon > 0$ is *arbitrary and* M_ε *depends only on* ε (through (†)), *and not on* Φ. This fact follows immediately from the proof just given – we need only note that $\|\varphi\|_W = \|\Phi\|_W$, so that $\|K^{-1}\varphi\|_W < 1$ for every $K > \|\Phi\|_W$.

Remark. Given that $\int_{-\infty}^{\infty}(\log W_*(t)/(1+t^2))\,\mathrm{d}t < \infty$, is it true that for *every* entire function $\Psi(z)$ of *exponential type zero* whose restriction, $\Psi(x)$, to $\mathbb{R}$ belongs to $\mathscr{C}_W(\mathbb{R})$, we *do have* a sequence of polynomials P_n with $\|\Psi - P_n\|_W \xrightarrow[n]{} 0$? As we shall see later on, the *answer to this question is no* for some weights $W(x)$ with seemingly rather regular behaviour.

Theorem. *Suppose that* $W(x_k) < \infty$ *for a sequence of points* x_k *going to* ∞, *and, if* $n(t)$ *denotes the number of the points* x_k *in* $[0,t]$, *suppose that*

$$\limsup_{r\to\infty}\frac{n(t)}{t} > 0.$$

Then, if $\int_{-\infty}^{\infty}(\log W_*(t)/(1+t^2))\,\mathrm{d}t < \infty$, *the polynomials are not dense in* $\mathscr{C}_W(\mathbb{R})$.

Proof. Take any function $\varphi\in\mathscr{C}_W(\mathbb{R})$ such that $\varphi(x_0) = 1$ but $\varphi(x_k) = 0$ for $k \geqslant 1$. Then φ *cannot* be $\|\ \ \|_W$-approximated by polynomials.

If, indeed, it *could* be so approximated, the preceding theorem would furnish an entire function $\Phi(z)$ of exponential type *zero* with $\Phi(x_k) = \varphi(x_k)$ for $k \geqslant 0$. Then in particular $\Phi(x_0) = 1$, so $\Phi(z) \not\equiv 0$. At the same time $\Phi(x_k) = 0$ for $k \geqslant 1$, so, if $N(r)$ denotes the number of zeros of $\Phi(z)$ with modulus $\leqslant r$, $N(r) \geqslant n(r) - 1$, and $\limsup_{r\to\infty}(N(r)/r)$ would have to be > 0 by hypothesis.

This, however, is *impossible*. For, $\Phi(z)$ being of *exponential type zero* and $\not\equiv 0$, we must have $N(r) = o(r)$ for $r \to \infty$ by an easy application of Jensen's formula (see problem 1(a) in Chapter 1!).

The theorem is proved.

Remark. We shall soon see that the condition $\limsup_{t\to\infty}(n(t)/t) > 0$ in this theorem cannot be relaxed much.

3. Strengthened version of Akhiezer's criterion. Pollard's theorem

The Bernstein approximation problem was also studied by the American mathematician Pollard, whose work was largely independent of Akhiezer's and Mergelian's. Pollard published one of the first solutions, I think in fact *before* the appearance of the other two mathematicians' articles. *In one direction*, the criterion given by him strengthens that furnished by the Akhiezer theorem of article 1. The way this happens is shown by the following

Theorem (due, essentially, to Pollard). *If*

$$\sup\left\{\int_{-\infty}^{\infty}\frac{\log|P(x)|}{1+x^2}\,dx : P \text{ a polynomial and } \|P\|_W \leqslant 1\right\}$$

is finite, then $\int_{-\infty}^{\infty}(\log W_*(x)/(1+x^2))\,dx < \infty$.

Proof. As $x \to \pm\infty$, $W(x) \to \infty$ *faster than any power of* x. So, if we take a suitable constant C,

$$\tilde{W}(x) \;=\; C\frac{W(x)}{|x-i|}$$

is $\geqslant 1$ on $\mathbb{R}$. $\tilde{W}(x)$ obviously grows faster than any power of x as $x \to \pm\infty$, and we may consider *weighted polynomial approximation with the weight* $\tilde{W}$. To this situation we apply the Mergelian theorems in §§A.1 and A.3.

Put, for $z \in \mathbb{C}$,

$$\tilde{\Omega}(z) = \sup\left\{|P(z)|: P \text{ a polynomial and } \left|\frac{P(t)}{(t-i)\tilde{W}(t)}\right| \leqslant 1 \text{ on } \mathbb{R}\right\};$$

note that $\tilde{\Omega}(z)$ is just $CW_*(z)$. According to the theorem of §A.1, $\tilde{\Omega}(i) < \infty$ implies that polynomials are *not* $\|\ \ \|_{\tilde{W}}$-dense in $\mathscr{C}_{\tilde{W}}(\mathbb{R})$, and, by §A.3, the latter fact makes

$$\int_{-\infty}^{\infty}\frac{\log\tilde{\Omega}(t)}{1+t^2}\,dt \;<\; \infty,$$

i.e.,

$$\int_{-\infty}^{\infty}\frac{\log W_*(t)}{1+t^2}\,dt \;<\; \infty.$$

In order to show this last relation, it is therefore enough to verify that $\tilde{\Omega}(i) < \infty$, or, what comes to the same thing, that $W_*(i) < \infty$.

Take a sequence of polynomials $\{P_n(z)\}$ with $\|P_n\|_W \leqslant 1$ and $|P_n(i)| \xrightarrow[n]{} W_*(i)$; by §G.2 of Chapter III,

$$\log|P_n(i)| \;\leqslant\; \frac{1}{\pi}\int_{-\infty}^{\infty}\frac{\log|P_n(t)|}{1+t^2}\,dt.$$

Under the hypothesis, however, the integrals on the right are *bounded above*. Therefore $W_*(i) < \infty$, which is what we needed. We are done.

In the course of the argument just given, we established a subsidiary result, important in its own right. We state it as a

Corollary. *If* $W_*(\mathrm{i}) < \infty$, *then* $\int_{-\infty}^{\infty} (\log W_*(t)/(1+t^2))\,\mathrm{d}t < \infty$.

Remark. Sometimes it is easier to get an upper bound for

$$\int_{-\infty}^{\infty} \frac{\log|P(t)|}{1+t^2}\,\mathrm{d}t$$

when P is a polynomial with $\|P\|_W \leqslant 1$ than to try to directly obtain good estimates on $W_*(x)$. If we can show that the upper bound is finite, the description of functions $\|\ \ \|_W$-approximable by polynomials given in article 2 is available, and hence the consequences of that description. For this reason, the result proved here is quite useful.

C. Mergelian's criterion really more general in scope than Akhiezer's. Example

Given a weight $W(x) \geqslant 1$ on $\mathbb{R}$ which goes to ∞ faster than any power of x as $x \to \pm\infty$, we can form the two functions

$$\Omega(z) = \sup\left\{|P(z)|\colon P \text{ a polynomial and } \left|\frac{P(t)}{(t-\mathrm{i})W(t)}\right| \leqslant 1 \text{ on } \mathbb{R}\right\},$$

and

$$W_*(z) = \sup\left\{|P(z)|\colon P \text{ a polynomial and } \left|\frac{P(t)}{W(t)}\right| \leqslant 1 \text{ on } \mathbb{R}\right\}.$$

Mergelian's second theorem (§A.3) says that the condition

$$\int_{-\infty}^{\infty} \frac{\log \Omega(t)}{1+t^2}\,\mathrm{d}t = \infty$$

is necessary and sufficient for polynomials to be $\|\ \ \|_W$-dense in $\mathscr{C}_W(\mathbb{R})$. Akhiezer's theorem (§B.1) says that the condition

$$\int_{-\infty}^{\infty} \frac{\log W_*(t)}{1+t^2}\,\mathrm{d}t = \infty$$

is *always sufficient* for the $\|\ \ \|_W$-density of polynomials in $\mathscr{C}_W(\mathbb{R})$, and *necessary* for that density to hold *provided that* $W(x)$ *has a certain regularity*. As we saw in §B.1, *continuity* of W is *enough* here; it suffices in fact that $W(x)$ be finite on an infinite closed set with a finite point of accumulation. The work of §B.2 shows that the *set on which* $W(x)$ *is finite need not even have a finite point of accumulation*; it is enough that the set be *infinite and not too sparse*. As long as

$$\limsup_{r\to\infty} \frac{\text{number of points in the set and in } [-r,r]}{r}$$

is *positive*, the necessity of Akhiezer's criterion (involving $\log W_*(t)$) holds good.

These successive relaxations in the regularity required of $W(x)$ for Akhiezer's criterion to hold make us hope that perhaps *all* restrictions on W's regularity may be dispensed with. Maybe the lower polynomial regularization $W_*(x)$ of $W(x)$ is all we need for the study of Bernstein's problem, no matter what the behaviour of the latter function is, and we can *forget* about $\Omega(x)$ altogether. Do we *really need* $\Omega(x)$ at all in order to have a completely general test for $\| \ \|_W$-density of polynomials in $\mathscr{C}_W(\mathbb{R})$?

We do. Here is an *example of a weight* $W(x)$ *such that*

$$\int_{-\infty}^{\infty} \frac{\log W_*(t)}{1+t^2} dt \ < \ \infty,$$

but nevertheless $\Omega(\mathrm{i}) = \infty$, *making the polynomials* $\| \ \|_W$*-dense in* $\mathscr{C}_W(\mathbb{R})$.

Our construction is based on use of the entire function

$$S(z) = \prod_{n=1}^{\infty} \left(1 - \frac{z^2}{4^n}\right)$$

of *exponential type zero*. The weight $W(x)$ will be *identically infinite* outside the set of points

$$x_k = \operatorname{sgn} k \cdot 2^{|k|}; \quad k = \pm 1, \pm 2, \ldots;$$

these are just the *zeros* of $S(z)$. *On that set* we take

$$W(x_k) = C\sqrt{|x_k| \cdot |S'(x_k)|}$$

with a constant C chosen so as to make $W(x_k) \geqslant 1$ for all k.

We start with the asymptotic evaluation of $S'(x_k)$ for large $|k|$; on account of symmetry we need only consider positive values of k. For $k \geqslant 1$, then,

$$S'(2^k) \ = \ -\frac{2}{2^k} \prod_{1 \leqslant n < k} \left(1 - \frac{4^k}{4^n}\right) \cdot \prod_{n > k} \left(1 - \frac{4^k}{4^n}\right).$$

Here is a *trick* which can be used to good effect in many calculations of this kind. *Factor* each ratio $4^k/4^n$ with $k > n$ *out of the first product*. One finds that

$$S'(2^k) \ = \ -\frac{2}{2^k}(-1)^{k-1} \prod_{1 \leqslant n < k} 4^{k-n} \cdot \prod_{1 \leqslant n < k} \left(1 - \frac{4^n}{4^k}\right) \prod_{n > k} \left(1 - \frac{4^k}{4^n}\right)$$

$$= \ \frac{2(-1)^k}{2^k} \cdot 2^{(k-1)k} \prod_{l=1}^{k-1} \left(1 - \frac{1}{4^l}\right) \prod_{m=1}^{\infty} \left(1 - \frac{1}{4^m}\right).$$

For large k, this is

$$\sim \ (-1)^k 2^{(k-1)^2} (S(1))^2,$$

and we see that $|S'(x_k)|$ behaves like a constant multiple of $|x_k|^{(|k|-2)}$ for $k \to \pm\infty$. This evidently tends to ∞ faster than any power of x_k as $k \to \pm\infty$; the same is then true of $W(x_k)$.

We need also to consider the partial products

$$S_N(z) = \prod_{n=1}^{N} \left(1 - \frac{z^2}{4^n}\right)$$

of $S(z)$. For $1 \leqslant k \leqslant N$, we have

$$S'_N(2^k) = -\frac{2}{2^k} \prod_{1 \leqslant n < k} \left(1 - \frac{4^k}{4^n}\right) \prod_{k < n \leqslant N} \left(1 - \frac{4^k}{4^n}\right);$$

comparison of this with the *first* of the above formulas for $S'(2^k)$ shows that

$$|S'_N(x_k)| \geqslant |S'(x_k)| \quad \text{for } 1 \leqslant k \leqslant N.$$

On account of this fact and of the *growth* of $|S'(x_k)|$ for $k \to \pm\infty$, we have, for any *polynomial* P,

$$\sum_{-N}^{N}{}' \frac{P(x_k)}{S'_N(x_k)(z - x_k)} \xrightarrow[N]{} \sum_{-\infty}^{\infty}{}' \frac{P(x_k)}{S'(x_k)(z - x_k)}$$

as long as z is different from all the x_k.

▶ N.B. *A prime next to a summation sign means that there is no term corresponding to the value zero of the summation index.* (This convention is fairly widespread, by the way.)

Let us fix any polynomial P. As soon as $2N >$ degree of P, we have, by Lagrange's interpolation formula,

$$P(z) = S_N(z) \sum_{-N}^{N}{}' \frac{P(x_k)}{S'_N(x_k)(z - x_k)},$$

provided that z is different from all the x_k. Fixing such a z, and making $N \to \infty$, we get, by virtue of the previous relation,

$$(*) \qquad P(z) = S(z) \sum_{-\infty}^{\infty}{}' \frac{P(x_k)}{S'(x_k)(z - x_k)}.$$

This formula is valid, then, for any polynomial P and any z different from all the x_k.

We estimate $W_*(\mathrm{i})$. Take any polynomial P with $\|P\|_W \leqslant 1$, i.e., with

$$|P(x_k)| \leqslant C\sqrt{|x_k|} \cdot |S'(x_k)|.$$

Substituting into $(*)$, we find that

$$|P(\mathrm{i})| \leqslant CS(i) \sum_{-\infty}^{\infty}{}' \frac{\sqrt{|x_k|}}{|\mathrm{i} - x_k|} \leqslant 2CS(\mathrm{i}) \sum_{1}^{\infty} 2^{-k/2} = \frac{2CS(\mathrm{i})}{\sqrt{2} - 1}.$$

Taking the supremum of $|P(\mathrm{i})|$ over all such P, we see that

$$W_*(\mathrm{i}) \leqslant \frac{2CS(\mathrm{i})}{\sqrt{2}-1} < \infty.$$

According to the corollary in §B.3, this implies that

$$\int_{-\infty}^{\infty} \frac{\log W_*(t)}{1+t^2}\,dt < \infty.$$

It is now claimed that $\Omega(\mathrm{i}) = \infty$. To see this, consider the polynomials

$$P_N(x) = \sqrt{x_N}\cdot S_N(x);$$

since $P_N(\mathrm{i})/\sqrt{x_N} \underset{N}{\longrightarrow} S(\mathrm{i}) > 0$, it is clear that $P_N(\mathrm{i}) \underset{N}{\longrightarrow} \infty$. It is therefore enough to show that

$$\left|\frac{P_N(x)}{(x-\mathrm{i})W(x)}\right| \leqslant \frac{1}{2CS(1)}$$

on $\mathbb{R}$ in order to conclude that $\Omega(\mathrm{i}) = \infty$. We have, in other words, to verify that

$$|P_N(x_k)| \leqslant \frac{|x_k - \mathrm{i}|\,W(x_k)}{2CS(1)}$$

for $k = \pm 1, \pm 2, \ldots$. This is true for $1 \leqslant |k| \leqslant N$ because then $P_N(x_k) = 0$. Suppose, therefore, that $k > N$. Then

$$|P_N(x_k)| = \sqrt{x_N}\cdot \prod_{1\leqslant n\leqslant N}\left|\frac{4^k}{4^n}-1\right| \leqslant \sqrt{x_N}\cdot \prod_{1\leqslant n<k}\left|\frac{4^k}{4^n}-1\right|$$

$$= \frac{\sqrt{x_N}\cdot x_k|S'(x_k)|}{2\prod_{n>k}\left(1-\frac{4^k}{4^n}\right)}$$

$$\doteq \frac{\sqrt{x_N}\cdot x_k|S'(x_k)|}{2S(1)}.$$

Taking symmetry into account, we see that, for $|k| > N$,

$$|P_N(x_k)| \leqslant \frac{|x_k|}{2S(1)}\sqrt{x_N}\cdot|S'(x_k)|$$

$$\leqslant \frac{1}{2S(1)}|x_k - \mathrm{i}|\sqrt{|x_k|}\,|S'(x_k)| = \frac{|x_k-\mathrm{i}|\,W(x_k)}{2CS(1)}.$$

We thus have $|P_N(x_k)| \leqslant |x_k - \mathrm{i}|\,W(x_k)/2CS(1)$ for all k and every N, and this, as we have seen, ensures that $\Omega(\mathrm{i}) = \infty$.

Because $\Omega(\mathrm{i}) = \infty$, polynomials are $\| \ \|_W$-dense in $\mathscr{C}_W(\mathbb{R})$ by the first Mergelian theorem (§A.1). *However, as we already have shown,*

$$\int_{-\infty}^{\infty} \frac{\log W_*(t)}{1+t^2} \mathrm{d}t \ < \ \infty.$$

Application of Akhiezer's criterion to $\mathscr{C}_W(\mathbb{R})$ with the weight W considered here would therefore lead to a false result.

D. Some partial results involving the weight W explicitly

Let us see how much information about the $\| \ \|_W$-density of polynomials in $\mathscr{C}_W(\mathbb{R})$ can be obtained by direct examination of the weight $W(x)$ itself. Here, first of all, is an easy negative result.

Theorem (T. Hall). *If $\int_{-\infty}^{\infty} (\log W(x)/(1+x^2))\mathrm{d}x < \infty$, the polynomials are not $\| \ \|_W$-dense in $\mathscr{C}_W(\mathbb{R})$.*

Proof. Trivially, $\Omega(x) \leqslant |x - \mathrm{i}|\, W(x)$ for $x \in \mathbb{R}$, so, if the above integral with $\log W(x)$ converges, so is

$$\int_{-\infty}^{\infty} \frac{\log \Omega(x)}{1+x^2} \mathrm{d}x \ < \ \infty.$$

The desired result now follows from Mergelian's second theorem, §A.3.

One is, naturally, very interested in finding simple conditions on W which will guarantee $\| \ \|_W$-*density* of polynomials in $\mathscr{C}_W(\mathbb{R})$. In this direction, we begin with a very old result.

Theorem (S. Bernstein). *Let $W(x) = \sum_{n=0}^{\infty} A_n x^{2n}$ where the $A_n \geqslant 0$ and the series is everywhere convergent. If*

$$\int_{-\infty}^{\infty} \frac{\log W(x)}{1+x^2} \mathrm{d}x \ = \ \infty,$$

polynomials are $\| \ \|_W$-dense in $\mathscr{C}_W(\mathbb{R})$.

Proof. The polynomials

$$P_N(x) \ = \ \sum_{n=0}^{N} A_n x^{2n}$$

satisfy $\|P_N\|_W \leqslant 1$, because all the A_n are $\geqslant 0$. Clearly $P_N(x) \xrightarrow[N]{} W(x)$ for each x, so *here* $W_*(x) = W(x)$. Our result now follows from Akhiezer's theorem (§B.1).

Corollary. *The polynomials are $\| \ \|_W$-dense in $\mathscr{C}_W(\mathbb{R})$ for $W(x) = \mathrm{e}^{x^2}$.*

Corollary. *The polynomials are* $\| \ \|_W$*-dense in* $\mathscr{C}_W(\mathbb{R})$ *for* $W(x) = e^{|x|}$.

Proof. Use the fact that $\frac{1}{2}e^{|x|} \leqslant \cosh x \leqslant e^{|x|}$ and work with the weight

$$\cosh x = 1 + \frac{x^2}{2!} + \frac{x^4}{4!} + \cdots.$$

Theorem. *Let* $W(x) \geqslant 1$ *be even, and suppose that, for* $x > 0$, $\log W(x)$ *is a convex function of* $\log x$. *Then, if*

$$\int_{-\infty}^{\infty} \frac{\log W(x)}{1 + x^2} dx = \infty,$$

polynomials are $\| \ \|_W$*-dense in* $\mathscr{C}_W(\mathbb{R})$.

Proof. Starts out like that of the corollary in §C.2, Chapter V, with the use of some material from convex logarithmic regularization. Write

$$S_n = \sup_{r>0} \frac{r^n}{W(r)} \quad \text{for } n = 0, 1, 2, \ldots,$$

and then put, for $r > 0$,

$$T(r) = \sup_{n \geqslant 0} \frac{r^{2n}}{S_{2n}}.$$

Since $\log W(r)$ is a convex function of $\log r$, we have, by the *proof* of the second lemma in Chapter IV, §D, that

$$(*) \qquad \frac{W(r)}{r^2} \leqslant T(r) \leqslant W(r)$$

whenever r is sufficiently large. (It's $W(r)/r^2$ on the left and not $W(r)/r$ because we use the *even* powers of r in forming $T(r)$.)

Take any fixed number λ between 0 and 1, and form the function

$$S(x) = (1 - \lambda^2) \sum_{n=0}^{\infty} \lambda^{2n} \frac{x^{2n}}{S_{2n}}.$$

We see from the definition of $T(r)$ and $(*)$ that $S(x) \leqslant W(x)$ for $|x|$ sufficiently large, since W is even. Therefore, as in the proof of the above theorem of Bernstein (the numbers S_{2n} are all positive!), we at least have

$$W_*(x) \geqslant CS(x)$$

for some $C > 0$ chosen so as to make $CS(x) \leqslant W(x)$ for *all* x.

Referring again to $(*)$, we see, however, that

$$S(x) \geqslant (1 - \lambda^2) T(\lambda|x|) \geqslant (1 - \lambda^2) \frac{W(\lambda|x|)}{\lambda^2 x^2}$$

for $|x|$ sufficiently large. Hence, taking a big enough,

$$\int_a^\infty \frac{\log W_*(x)}{x^2}\mathrm{d}x \geqslant \frac{\log C}{a} + \int_a^\infty \frac{\log S(x)}{x^2}\mathrm{d}x$$

$$\geqslant \frac{\log C}{a} + \frac{1}{a}\log\left(\frac{1-\lambda^2}{\lambda^2}\right) - 2\int_a^\infty \frac{\log x}{x^2}\mathrm{d}x + \int_a^\infty \frac{\log W(\lambda x)}{x^2}\mathrm{d}x.$$

The last integral on the right equals $\lambda\int_{\lambda a}^\infty(\log W(\xi)/\xi^2)\mathrm{d}\xi$ which is clearly infinite if $\int_{-\infty}^\infty(\log W(x)/(1+x^2))\mathrm{d}x = \infty$.
So $\int_{-\infty}^\infty(\log W_*(x)/(1+x^2))\mathrm{d}x = \infty$, and the result follows by Akhiezer's theorem.

Remark. Is the theorem still true if the even function $W(x)$ is merely required to be *increasing* for $x > 0$? An example to be given in Chapter VII shows that the answer to this question is *no.*

E. Weighted approximation by sums of imaginary exponentials with exponents from a finite interval

Let $W(x) \geqslant 1$ be a weight which is now merely assumed to tend to ∞ as $x \longrightarrow \pm\infty$. We fix some $A > 0$ and ask whether the collection of *finite sums* of the form

$$\sum_{-A\leqslant\lambda\leqslant A} C_\lambda e^{i\lambda x}$$

is $\| \ \|_W$-dense in $\mathscr{C}_W(\mathbb{R})$. It turns out that the theorems of Mergelian and Akhiezer given in §§A and B above have *complete analogues* in the present situation. We will be able to see this in the present § without having to repeat most of the details from the preceding discussion.

1. Equivalence with weighted approximation by certain entire functions of exponential type. The collection $\mathscr{E}_A$

If $\sigma(t)$ is a finite sum of the form $\sum_{-A\leqslant\lambda\leqslant A}C_\lambda e^{i\lambda t}$ and z_0 is a complex number, the ratio $(\sigma(t)-\sigma(z_0))/(t-z_0)$ is no longer expressible as such a sum. Therefore the useful Markov–Riesz–Pollard trick applied, for instance, in the proof of the second Mergelian theorem (§A.3) is not available for such sums. For this reason, the following result is very important.

Lemma. *If $W(x) \geqslant 1$ and $W(x) \to \infty$ for $x \to \pm\infty$, every entire function of exponential type $\leqslant A$, bounded on the real axis, can be $\| \ \|_W$-approximated by finite sums of the form*

$$\sum_{-A\leqslant\lambda\leqslant A} C_\lambda e^{i\lambda x}.$$

Proof. Take any entire function $f(z)$ of exponential type $\leqslant A$, bounded on $\mathbb{R}$. Since $W(x) \to \infty$ for $x \to \pm\infty$,

$$\sup_{x \in \mathbb{R}} \left| \frac{f(x) - f(\rho x)}{W(x)} \right| \longrightarrow 0$$

for $\rho \to 1$ by *continuity* of f. Given $\varepsilon > 0$, fix a $\rho < 1$ such that the above supremum is $< \varepsilon$. If $h > 0$ is small enough, we will also have

$$\sup_{x \in \mathbb{R}} \left| \frac{1}{W(x)} \left(f(\rho x) - f(\rho x) \frac{\sin hx}{hx} \right) \right| \;<\; \varepsilon;$$

we take such an h so small that

$$g(z) = f(\rho z) \frac{\sin hz}{hz}$$

is *of exponential type* $\leqslant A$, and fix it.

We thus have $\| f - g \|_W < 2\varepsilon$. However, g, besides being of exponential type $\leqslant A$, is *also* in L_2 on the *real axis*. We can therefore apply the Paley–Wiener theorem (Chapter III, §D) to g, obtaining

$$g(x) \;=\; \int_{-A}^{A} \mathrm{e}^{\mathrm{i}\lambda x} G(\lambda) \mathrm{d}\lambda$$

with some $G \in L_2(-A, A)$. This property of G also makes $G \in L_1(-A, A)$, by Schwarz' inequality.

For large integers N, put

$$g_N(x) \;=\; \sum_{k=0}^{N-1} \mathrm{e}^{\mathrm{i}(-A + (2k/N)A)x} \int_{-A+(2k/N)A}^{-A+((2k+2)/N)A} G(\lambda) \mathrm{d}\lambda.$$

For each N, $|g_N(x)| \leqslant \| G \|_1$ for $x \in \mathbb{R}$, and we clearly have $g_N(x) \underset{N}{\longrightarrow} g(x)$ u.c.c. on $\mathbb{R}$. Therefore $\| g - g_N \|_W \underset{N}{\longrightarrow} 0$, so, taking N large enough, we get

$$\| f - g_N \|_W \;\leqslant\; \| f - g \|_W + \| g - g_N \|_W \;<\; 3\varepsilon.$$

Since $g_N(x)$ is a finite sum of the form

$$\sum_{-A \leqslant \lambda \leqslant A} C_\lambda \mathrm{e}^{\mathrm{i}\lambda x},$$

we are done.

Definition. $\mathscr{E}_A$ denotes the set of entire functions of exponential type $\leqslant A$, bounded on the real axis.

Since every finite sum $\sum_{-A \leqslant \lambda \leqslant A} C_\lambda \mathrm{e}^{\mathrm{i}\lambda x}$ certainly *belongs* to $\mathscr{E}_A$, the above lemma has the obvious

Corollary. *Let $\varphi \in \mathscr{C}_W(\mathbb{R})$. There are finite sums $\sigma(x)$ of the form*

$$\sum_{-A \leqslant \lambda \leqslant A} C_\lambda \mathrm{e}^{\mathrm{i}\lambda x}$$

making $\|\varphi - \sigma\|_W$ arbitrarily small if and only if there are $f_n \in \mathscr{E}_A$ with $\|\varphi - f_n\|_W \xrightarrow[n]{} 0$.

Remark. What is important here is that, if $f \in \mathscr{E}_A$ and $z_0 \in \mathbb{C}$, the ratio $(f(t) - f(z_0))/(t - z_0)$ *also* belongs to $\mathscr{E}_A$.

2. The functions $\Omega_A(z)$ and $W_A(z)$. Analogues of Mergelian's and Akhiezer's theorems

In analogy with the definition of $\Omega(z)$ (beginning of §A), we put

$$\Omega_A(z) = \sup\left\{ |f(z)|\colon\ f \in \mathscr{E}_A \text{ and } \left|\frac{f(t)}{(t - \mathrm{i})W(t)}\right| \leqslant 1 \text{ on } \mathbb{R} \right\}.$$

Remark. A slight extension of the argument used to prove the lemma in the preceding subsection shows that $\Omega_A(z)$ is already obtained if we use only finite sums $f(t)$ of the form $\sum_{-A \leqslant \lambda \leqslant A} C_\lambda \mathrm{e}^{\mathrm{i}\lambda t}$ in taking the supremum on the right. Verification of this fact is left to the reader.

Observe that, if $f \in \mathscr{E}_A$, so is the function f^* defined by the formula $f^*(z) = \overline{f(\bar{z})}$. This makes $\Omega_A(z) = \Omega_A(\bar{z})$. As we have already noted, when $f(t) \in \mathscr{E}_A$, the quotient $(f(t) - f(z_0))/(t - z_0)$ also belongs to $\mathscr{E}_A$. So, by the way, does

$$\frac{f(t) - f(z_0)}{t - z_0}(t - z_1)$$

belong to $\mathscr{E}_A$ then. These evident facts make it possible for us to virtually *copy* the proof of the first Mergelian theorem as given in §A.1, replacing the collection of *polynomials* by $\mathscr{E}_A$. Keeping the lemma from the previous subsection in mind, we obtain, in this way, the

Theorem. *If $\Omega_A(z) = \infty$ for one non-real z, the collection of finite sums of the form*

$$\sum_{-A \leqslant \lambda \leqslant A} C_\lambda \mathrm{e}^{\mathrm{i}\lambda x}$$

is $\|\ \ \|_W$-dense in $\mathscr{C}_W(\mathbb{R})$. Conversely, if such sums are $\|\ \ \|_W$-dense in $\mathscr{C}_W(\mathbb{R})$, $\Omega_A(z) = \infty$ for all non-real z.

The second theorem in §G.2 of Chapter III applies to the functions $f \in \mathscr{E}_A$, and we have, for them,

$$(*) \qquad \log|f(z)| \leq A|\Im z| + \frac{1}{\pi}\int_{-\infty}^{\infty} \frac{|\Im z|\log|f(t)|}{|z-t|^2}\,dt.$$

Using this relation we can copy the proof of Mergelian's *second* theorem (§A.3), to get

Theorem. *The finite sums of the form* $\sum_{-A\leq\lambda\leq A} C_\lambda e^{i\lambda x}$ *are* $\|\ \|_W$*-dense in* $\mathscr{C}_W(\mathbb{R})$ *if and only if*

$$\int_{-\infty}^{\infty} \frac{\log \Omega_A(x)}{1+x^2}\,dx = \infty.$$

To obtain analogues of Akhiezer's theorems (§§B.1 and B.2), we write

$$W_A(z) = \sup\{|f(z)|: f\in\mathscr{E}_A \text{ and } \|f\|_W \leq 1\}.$$

As in the formation of $\Omega_A(z)$, we can limit the set of functions f occurring on the right to the ones expressible as finite sums $\sum_{-A\leq\lambda\leq A} C_\lambda e^{i\lambda x}$. $W_A(x)$ is thus a *lower regularization* of $W(x)$ by such finite sums.

Arguing as in §§B.1, B.2, with use of $(*)$ in the appropriate places, we find:

Theorem. *For continuous W, finite sums of the form*

$$\sum_{-A\leq\lambda\leq A} C_\lambda e^{i\lambda x}$$

are $\|\ \|_W$*-dense in* $\mathscr{C}_W(\mathbb{R})$ *if and only if*

$$\int_{-\infty}^{\infty} \frac{\log W_A(x)}{1+x^2}\,dx = \infty.$$

Theorem. *If* $\int_{-\infty}^{\infty}(\log W_A(x)/(1+x^2))\,dx < \infty$, *any function in* $\mathscr{C}_W(\mathbb{R})$ *which can be* $\|\ \|_W$*-approximated by finite sums of the form* $\sum_{-A\leq\lambda\leq A} C_\lambda e^{i\lambda x}$ *coincides, on the set of points where* $W(x) < \infty$, *with an entire function* $\Phi(z)$ *satisfying, for each* $\varepsilon > 0$, *an inequality of the form*

$$|\Phi(z)| \leq \|\Phi\|_W M_\varepsilon \exp(A|\Im z| + \varepsilon|z|).$$

Here, M_ε *depends only on* ε, *and is independent of the particular function* Φ *arising in this manner.*

Corollary. *Let* $\int_{-\infty}^{\infty}(\log W_A(x)/(1+x^2))\,dx < \infty$, *and denote by E the set of points on* $\mathbb{R}$ *where* $W(x) < \infty$. *If either*

$$\limsup_{r\to\infty} \frac{\text{number of points in } E\cap[0,r]}{r} > \frac{A}{\pi}$$

or

$$\limsup_{r\to\infty} \frac{\text{number of points in } E\cap[-r,0]}{r} > \frac{A}{\pi},$$

then $\mathscr{E}_A$ *cannot be* $\|\ \ \|_W$*-dense in* $\mathscr{C}_W(\mathbb{R})$.

Proof. Is based on a result much deeper than the one needed for the corresponding proposition about weighted polynomial approximation (end of §B.2).

Suppose, wlog, that $W(x_k)<\infty$ where $0\leqslant x_0<x_1<x_2<\cdots$, and that $\limsup_{n\to\infty} n/x_n > A/\pi$. (If the set E has a *finite* limit point, one can give a much simpler argument.) Take any continuous bounded φ (belonging thus to $\mathscr{C}_W(\mathbb{R})$) with $\varphi(x_0)=1$ and $\varphi(x_k)=0$ for $k\geqslant 1$; it is claimed that such a function φ cannot be $\|\ \ \|_W$-approximated by functions in $\mathscr{E}_A$.

If it could, we would, by the theorem, get an entire function $\Phi(z)$ with $\Phi(x_k)=\varphi(x_k)$, $k\geqslant 0$ (hence $\Phi(x_0)=1$ so that $\Phi\not\equiv 0$) satisfying, for every $\varepsilon>0$, an inequality of the form

$$|\Phi(z)| \ \leqslant\ C_\varepsilon \exp(A|\Im z|+\varepsilon|z|).$$

This certainly makes Φ of exponential type $\leqslant A$. We also have

$$(\dagger)\qquad \int_{-\infty}^{\infty} \frac{\log^+|\Phi(x)|}{1+x^2}\,dx \ <\ \infty.$$

Indeed, there is a sequence of functions $f_n\in\mathscr{E}_A$ with

$$\|\varphi - f_n\|_W \xrightarrow[n]{} 0$$

(hence, wlog, $\|f_n\|_W\leqslant 1$), and $f_n(z)\xrightarrow[n]{}\Phi(z)$ u.c.c. (That's how one shows there *is* such a function Φ – see §§B.1 and B.2!) Since $\|f_n\|_W\leqslant 1$ we have *by definition* $|f_n(z)|\leqslant W_A(z)$, and thus finally $|\Phi(z)|\leqslant W_A(z)$. We are, however, *assuming* that $\int_{-\infty}^{\infty}(\log W_A(x)/(1+x^2))dx<\infty$, and, in the last integral, we may replace log by $\log^+$, because $W_A(z)\geqslant 1$. (Note that $1\in\mathscr{E}_A$!) Therefore $(\dagger)$ holds.

The *hypothesis of Levinson's theorem*, from §H.3 of Chapter III *is thus satisfied*. If $n_+(r)$ denotes the number of zeros of $\Phi(z)$ in the right half plane having modulus $\leqslant r$, that theorem says that $\lim_{r\to\infty} n_+(r)/r$ *exists*, and *here* has a value $\leqslant A/\pi$. However, $\Phi(x_k)=\varphi(x_k)=0$ for $k\geqslant 1$, so certainly $n_+(x_k)\geqslant k$. Our assumption that $\limsup_{k\to\infty} k/x_k > A/\pi$ therefore leads to a contradiction. The corollary is proved.

3. Scholium. Pólya's maximum density

We have not really used the full strength of Levinson's theorem in proving the corollary at the end of the preceding article. One can in fact replace the assumption that

$$\limsup_{r\to\infty} \frac{\text{number of points in } E\cap[0,r]}{r} > \frac{A}{\pi}$$

by a *weaker* one, and the *corollary's conclusion will still apply.*

Suppose we have any increasing sequence of points $x_k \geqslant 0$, some of which may be repeated. For $r>0$, denote the number of those points on $[0,r]$ (counting repetitions) by $N(r)$, and, for each positive $\lambda<1$, put

$$D_\lambda = \limsup_{r\to\infty} \frac{N(r)-N(\lambda r)}{(1-\lambda)r}$$

Note that if $\limsup_{r\to\infty} N(r)/r = \bar{D}$ is *finite*, we certainly have $\bar{D} \leqslant D_\lambda$ for each $\lambda<1$, as simple verification shows.

Lemma. $\lim_{\lambda\to 1} D_\lambda$ *exists (it may be infinite).*

Proof. Let $0<\lambda<\lambda'<1$, Writing $\lambda/\lambda' = \mu$, we have the identity

$$\frac{N(r)-N(\lambda r)}{(1-\lambda)r} = \frac{1-\lambda'}{1-\lambda}\frac{N(r)-N(\lambda' r)}{(1-\lambda')r} + \frac{\lambda'-\lambda}{1-\lambda}\frac{N(\lambda' r)-N(\mu\lambda' r)}{(1-\mu)\lambda' r},$$

whence

$$(*) \qquad D_\lambda \leqslant \frac{1-\lambda'}{1-\lambda}D_{\lambda'} + \frac{\lambda'-\lambda}{1-\lambda}D_{\lambda/\lambda'}.$$

Since $N(r)$ is increasing we also have, for $0<\lambda<\lambda'<1$,

$$\frac{N(r)-N(\lambda r)}{(1-\lambda)r} \geqslant \frac{1-\lambda'}{1-\lambda}\frac{N(r)-N(\lambda' r)}{(1-\lambda')r},$$

so

$$D_\lambda \geqslant \frac{1-\lambda'}{1-\lambda}D_{\lambda'}.$$

Suppose first of all that $\limsup_{\lambda\to 1} D_\lambda = \infty$. Then, if we have $D_{\lambda_0} \geqslant M$, say, for some λ_0, $0<\lambda_0<1$, the previous relation shows that $D_\lambda \geqslant (1/(1+\lambda_0))D_{\lambda_0} > \frac{1}{2}M$ for $\lambda_0^2 \leqslant \lambda \leqslant \lambda_0$. However, substituting $\lambda=\lambda_0$ and $\lambda'=\sqrt{\lambda_0}$ in $(*)$, we get $D_{\lambda_0} \leqslant D_{\sqrt{\lambda_0}}$, so also $D_{\sqrt{\lambda_0}} \geqslant M$. Then, by the reasoning just given, $D_\lambda \geqslant M/2$ for $\lambda_0 \leqslant \lambda \leqslant \sqrt{\lambda_0}$. This same argument can evidently be repeated indefinitely, getting $D_{\lambda_0^{1/4}} \geqslant M$, $D_\lambda \geqslant M/2$ for $\sqrt{\lambda_0} \leqslant \lambda \leqslant \lambda_0^{1/4}$, and so forth. Hence $D_\lambda \geqslant M/2$ for $\lambda_0^2 \leqslant \lambda < 1$, so, since M was arbitrary, $D_\lambda \to \infty$ as $\lambda \to 1$.

Consider now the case where $\limsup_{\lambda\to\infty} D_\lambda = L$ is finite, and, picking any $\varepsilon>0$, take any λ_0, $0<\lambda_0<1$, such that $D_{\lambda_0} > L-\varepsilon$ but $D_\lambda < L+\varepsilon$

for $\lambda_0 \leqslant \lambda < 1$. Putting $\lambda = \lambda_0$ in (*), we find, for $\lambda_0 \leqslant \lambda' < 1$,

$$L - \varepsilon \;\leqslant\; \frac{1-\lambda'}{1-\lambda_0}D_{\lambda'} \;+\; \frac{\lambda'-\lambda_0}{1-\lambda_0}(L+\varepsilon),$$

that is,

$$\frac{1-\lambda'}{1-\lambda_0}L \;-\; \varepsilon \;-\; \frac{\lambda'-\lambda_0}{1-\lambda_0}\varepsilon \;\leqslant\; \frac{1-\lambda'}{1-\lambda_0}D_{\lambda'},$$

and

$$D_{\lambda'} \;\geqslant\; L - \frac{1+\lambda'-2\lambda_0}{1-\lambda'}\varepsilon.$$

For $\lambda_0 \leqslant \lambda' \leqslant \sqrt{\lambda_0}$, the right-hand side is $\geqslant\; L-(1+2\sqrt{\lambda_0})\varepsilon \;>\; L-3\varepsilon$, so we see that in fact

$$L - 3\varepsilon \;<\; D_{\lambda'} \;<\; L+\varepsilon \quad \text{for } \lambda_0 \leqslant \lambda' \leqslant \sqrt{\lambda_0}.$$

As we already saw, $D_{\sqrt{\lambda_0}} \geqslant D_{\lambda_0}$. Therefore $D_{\sqrt{\lambda_0}} > L - \varepsilon$, and we may repeat the last argument with $\sqrt{\lambda_0}$ instead of λ_0 to conclude that

$$L - 3\varepsilon \;<\; D_{\lambda'} \;<\; L+\varepsilon \quad \text{for } \sqrt{\lambda_0} \leqslant \lambda' \leqslant \lambda_0^{1/4}.$$

Continuing in this way, we see that

$$L - 3\varepsilon \;<\; D_{\lambda'} \;<\; L+\varepsilon \quad \text{for } \lambda_0 \leqslant \lambda' < 1,$$

so, since $\varepsilon > 0$ was arbitrary, $D_\lambda \to L$ for $\lambda \to 1$. The lemma is proved.

Definition. $D^* = \lim_{\lambda\to 1} D_\lambda = \lim_{\lambda \to 1-} (\limsup_{r\to\infty} (N(r) - N(\lambda r))/(1-\lambda)r)$ is called the *maximum density* of the sequence $\{x_k\}$.

Since, as we have already remarked,

$$D_\lambda \;\geqslant\; \bar{D} \;=\; \limsup_{r\to\infty} \frac{N(r)}{r},$$

we *certainly* have $\bar{D} \leqslant D^*$. Simple examples (furnished by sequences with *large gaps*) show that D^* may be *much larger* than $\bar{D}$. Therefore, if, in any theorem whose hypothesis requires $\bar{D}$ to *exceed* some value, we can replace $\bar{D}$ by D^*, we obtain a stronger result thereby. This observation applies to the corollary at the end of the preceding article.

Theorem. *Given a weight $W(x) \geqslant 1$ tending to ∞ as $x \longrightarrow \pm\infty$ and a number $A > 0$, suppose that*

$$\int_{-\infty}^{\infty} \frac{\log W_A(x)}{1+x^2}\,dx \;<\; \infty$$

and that $W(x_k) < \infty$ on a strictly increasing sequence of points $x_k \geqslant 0$. If the maximum density D^* of the sequence $\{x_k\}$ is $> A/\pi$, then $\mathscr{E}_A$ is not *$\| \ \|_W$-dense in $\mathscr{C}_W(\mathbb{R})$.*

Proof. Taking the index k of the sequence $\{x_k\}$ to start from the value $k = 0$, we begin as in the proof of the corollary by choosing a $\varphi \in \mathscr{C}_W(\mathbb{R})$ with $\varphi(x_0) = 1$ and $\varphi(x_k) = 0$ for $k \geqslant 1$, and argue that, if φ *could* be $\| \ \|_W$-approximated by functions in $\mathscr{C}_W(\mathbb{R})$, there would be an entire function $\Phi(z)$ of exponential type $\leqslant A$ with

$$\int_{-\infty}^{\infty} \frac{\log^+|\Phi(x)|}{1+x^2} dx < \infty$$

and $\Phi(x_k) = \varphi(x_k)$, $k \geqslant 0$. Letting $n_+(r)$ be the number of zeros of $\Phi(z)$ with real part $\geqslant 0$ having modulus $\leqslant r$, we have, by Levinson's theorem (§H.3, Chapter III), that

$$\frac{n_+(r)}{r} \longrightarrow \text{some } D \leqslant \frac{A}{\pi}$$

as $r \to \infty$.

The x_k with $k \geqslant 1$ are zeros of $\Phi(z)$; therefore, if $N(r)$ denotes the he number of such x_k in $[0, r]$, we have, for each $\lambda < 1$, $N(r) - N(\lambda r) \leqslant n_+(r) - n_+(\lambda r)$. In view of the limit relation just written, the quantity on the right equals $(1 - \lambda)Dr + o(r)$ for large r, so we get

$$D_\lambda \leqslant D$$

for each $\lambda < 1$. Therefore $D^* = \lim_{\lambda \to 1} D_\lambda$ is *also* $\leqslant D \leqslant A/\pi$, contradicting our assumption that $D^* > A/\pi$. We are done.

Remark. Towards the end of Chapter IX, we will see that the theorem remains true when we replace the *maximum density* D^* of the sequence $\{x_k\}$ by a *still larger* density associated with that sequence.

The maximum density D^* associated with an increasing sequence of positive numbers $\{x_k\}$ has an elegant geometric interpretation.

Definition. Let $\{\xi_n\}$ be an increasing sequence of positive numbers, some of which may be repeated, and let $\nu(r)$ denote the number of points ξ_n in the interval $[0, r]$, counting repeated ξ_n according to their multiplicities as usual. The sequence $\{\xi_n\}$ is called *measurable* if $\lim_{r \to \infty} \nu(r)/r$ *exists and is finite*. The *value* of that *limit* is called the *density* of $\{\xi_n\}$.

We have then the

Theorem (Pólya). *Let the maximum density D^* of the increasing sequence of positive numbers x_k be finite. Then any measurable sequence of positive* *numbers containing all the* x_k *has density* $\geqslant D^*$, *and there* is *such a measurable sequence whose density is* exactly D^*.

If $D^* = \infty$, *there is* no *measurable sequence* (*of finite density*) *containing all the* x_k.

Proof. If $\{x_k\}$ is contained in an increasing sequence of numbers $\xi_n \geqslant 0$, and if, with $\nu(r)$ denoting the number of points ξ_n in $[0, r]$, $\nu(r)/r \to D$ for $r \to \infty$, we see, just as in the proof of the preceding theorem, that $D^* \leqslant D$. The *first* and *last* statements of the present theorem are therefore true. To complete the proof we must, when $D^* < \infty$, show how to *construct* a measurable sequence of density D^* containing the points x_k. The idea here is transparent enough, but the details are a bit fussy.

Call $N(r)$ the number of points x_k (counting repetitions) in $[0, r]$, and write $\lambda_n = 2^{-1/2^n}$. We have $\lambda_n \underset{n}{\uparrow} 1$, so, if we put

$$\varepsilon_n = \sup\{|D^* - D_\lambda|:\ \lambda_n \leqslant \lambda < 1\},$$

ε_n *decreases monotonically* to zero as $n \to \infty$, since, *according to the lemma* at the beginning of this subsection, $D_\lambda \to D^*$ for $\lambda \to 1$. Referring to the definition of D_λ and taking $\lambda = \lambda_n$, we see that for each n there is an r_n with

$$(\dagger) \qquad \frac{N(r) - N(\lambda_n r)}{(1 - \lambda_n)r} < D^* + 2\varepsilon_n \quad \text{for} \quad r \geqslant r_n.$$

It is convenient in what follows to write $\Lambda_n = 1/\lambda_n = 2^{1/2^n}$. For each n, take an *integral power* R_n of Λ_{n-1} (*sic*!) which is $> r_n$ and large enough so that

$$(\ddagger) \qquad (\Lambda_n - 1)\varepsilon_n R_n > 1.$$

We require also that $R_n > R_{n-1}$ if $n > 1$ so as to make the sequence $\{R_n\}$ *increasing*.

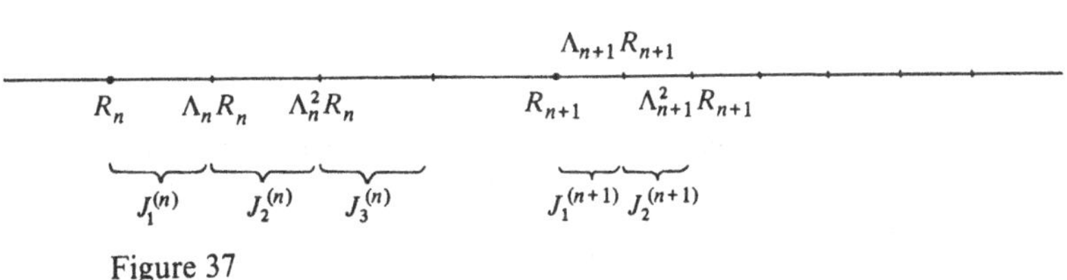

Figure 37

Using the numbers R_n and Λ_n we construct certain intervals, in the following manner. Given n, we have, from R_n up to R_{n+1}, the intervals $(R_n,\ \Lambda_n R_n]$, $(\Lambda_n R_n,\ \Lambda_n^2 R_n], \ldots, (\Lambda_n^{K-1} R_n,\ \Lambda_n^K R_n]$, say, with $\Lambda_n^K R_n = R_{n+1}$. From R_{n+1} onwards, each of the intervals $(\Lambda_n^l R_n,\ \Lambda_n^{l+1} R_n]$ splits into *two*, both of the form $(\Lambda_{n+1}^m R_{n+1},\ \Lambda_{n+1}^{m+1} R_{n+1}]$. After R_{n+2}, each of *those* splits further into two, and so forth. *We denote the intervals of the form* $(\Lambda_n^{p-1}R_n, \Lambda_n^p R_n]$ *lying between* R_n *and* R_{n+1} *by* $J_p^{(n)}$.

Consider any of the intervals $J_p^{(n)}$. Since $\Lambda_n = 1/\lambda_n$, we have, by (†) and the choice of R_n, the inequality

$$\frac{N(\Lambda_n^p R_n) - N(\Lambda_n^{p-1} R_n)}{\Lambda_n^p R_n - \Lambda_n^{p-1} R_n} < D^* + 2\varepsilon_n$$

for the number $N(\Lambda_n^p R_n) - N(\Lambda_n^{p-1} R_n)$ of points x_k in $J_p^{(n)}$. *If the ratio on the left is* $< D^*$, *let us throw new points into* $J_p^{(n)}$ *until we arrive at a total number of such points (including the* x_k *already* $\in J_p^{(n)}$*) lying between*

$$(\Lambda_n^p R_n - \Lambda_n^{p-1} R_n)D^* \quad \textit{and} \quad (\Lambda_n^p R_n - \Lambda_n^{p-1} R_n)(D^* + 2\varepsilon_n).$$

This we can do, thanks to ($\ast\!\ast$).

In this manner we adjoin points to the sequence $\{x_k\}$ in each of the intervals $J_p^{(n)}$ lying between R_n and R_{n+1}, to the extent necessary. We do that for every n. When finished, we have a *new sequence of points containing all the original* x_k. It is claimed that this new sequence is *measurable*, and of density D^*.

For $r > 0$, call $\nu(r)$ the number of points *of our new sequence* in $[0, r]$. Suppose that $R > R_n$; then R lies in one of the intervals $J_p^{(m)}$ with $m \geqslant n$, and, since the ε_l *decrease monotonically*,

$$(R - R_n)D^* - (D^* + 2\varepsilon_n)|J_p^{(m)}| \leqslant \nu(R) - \nu(R_m)$$
$$\leqslant (R - R_n)(D^* + 2\varepsilon_n) + (D^* + 2\varepsilon_n)|J_p^{(m)}|,$$

as is evident from our construction. Because $\Lambda_m \underset{m}{\longrightarrow} 1$ and $|J_p^{(m)}| \leqslant (\Lambda_m - 1)R\Lambda_m \leqslant (\Lambda_m - 1)\Lambda_n R$, the last relation shows that

$$D^* - \varepsilon_n \leqslant \frac{\nu(R) - \nu(R_n)}{R - R_n} \leqslant D^* + 3\varepsilon_n$$

as soon as R is sufficiently large $> R_n$.* This means that we have

$$D^* - 2\varepsilon_n \leqslant \frac{\nu(R)}{R} \leqslant D^* + 4\varepsilon_n$$

for R large enough, so, since ε_n can be taken as small as we like,

$$\frac{\nu(R)}{R} \longrightarrow D^* \quad \text{for} \quad R \to \infty.$$

Our new sequence is thus measurable and of density D^*, which is what was needed. We are done.

4. The analogue of Pollard's theorem

Returning from the above digression to the main subject of the

* The upper index m of the interval $J_p^{(m)}$ containing R tends to ∞ with R.

present §, let us complete our exposition of the parallel between weighted approximation by linear combinations of the $e^{i\lambda x}$, $-A \leqslant \lambda \leqslant A$, and that by polynomials. To do this, we need the analogue of the Pollard theorem in §C.3.

In the present situation, we cannot just *copy* the proof given for $W_*(x)$ in §C.3. That's because we now suppose merely that $W(x) \to \infty$ for $x \to \pm\infty$, and *no longer* assume the growth of $W(x)$ to be more rapid than that of any power of x as $x \to \pm\infty$. This means that we no longer necessarily have $W(x)/|x - i| \longrightarrow \infty$ for $x \to \pm\infty$, or even $W(x)/|x - i| \geqslant$ const. > 0 on $\mathbb{R}$.

The *method* of the proof in §B.3 can, however, be *adapted* to the treatment of the present case.

Theorem. *Let $W(x) \geqslant 1$ and $W(x) \to \infty$ for $x \longrightarrow \pm\infty$, and suppose $A > 0$. If $W_A(i) < \infty$, then*

$$\int_{-\infty}^{\infty} \frac{\log W_A(t)}{1 + t^2} dt \ < \ \infty.$$

Proof. As in §C.3, put $\tilde{W}(x) = W(x)/|x - i|$. For each fixed $z_0 \notin \mathbb{R}$, the ratio $1/(t - z_0)\tilde{W}(t)$ is bounded above on $\mathbb{R}$ and $\to 0$ as $t \to \pm\infty$.

Let us define $\mathscr{C}_{\tilde{W}}(\mathbb{R})$ as the set of functions φ continuous on $\mathbb{R}$ for which $|\varphi(x)/\tilde{W}(x)|$ is bounded and tends to zero as $x \to \pm\infty$ (just as in the situation where $\tilde{W}(x) \geqslant 1$), and put

$$\|\varphi\|_{\tilde{W}} \ = \ \sup_{x \in \mathbb{R}} \left| \frac{\varphi(x)}{\tilde{W}(x)} \right|$$

for such φ. As we have just seen, all the functions $1/(x - z_0)$, $z_0 \notin \mathbb{R}$, *do* belong to $\mathscr{C}_{\tilde{W}}(\mathbb{R})$.

Denote by $\tilde{\mathscr{E}}_A$ the set of functions $f(t)$ in $\mathscr{E}_A$ such that $tf(t)$ *also* belongs to $\mathscr{E}_A$; $\tilde{\mathscr{E}}_A$ is just the set of entire functions f of exponential type $\leqslant A$ with $f(t)$ and $tf(t)$ both bounded on $\mathbb{R}$. There are plenty of such functions; $\sin A(t - z_0)/(t - z_0)$ is one for each complex z_0.

We have $\tilde{\mathscr{E}}_A \subseteq \mathscr{C}_{\tilde{W}}(\mathbb{R})$. It is claimed that, if $W_A(i) < \infty$, $\tilde{\mathscr{E}}_A$ is *not* $\| \ \|_{\tilde{W}}$-dense in $\mathscr{C}_{\tilde{W}}(\mathbb{R})$. To see this, it is enough to verify that the function $1/(t - i)$ (which belongs to $\mathscr{C}_{\tilde{W}}(\mathbb{R})$) is not the $\| \ \|_{\tilde{W}}$-limit of functions in $\tilde{\mathscr{E}}_A$.

Suppose, for $\eta > 0$, that we *had* an $f \in \tilde{\mathscr{E}}_A$ with

$$\left\| \frac{1}{t - i} - f(t) \right\|_{\tilde{W}} \ \leqslant \ \eta.$$

Then

$$\left| \frac{1 - (t - i)f(t)}{W(t)} \right| \ = \ \left| \frac{1 - (t - i)f(t)}{(t - i)\tilde{W}(t)} \right| \ \leqslant \ \eta$$

for $t \in \mathbb{R}$, so, putting $G(t) = (1 - (t-\mathrm{i})f(t))/\eta$, we would have a $G \in \mathscr{E}_A$ (because $f \in \tilde{\mathscr{E}}_A$) with $\|G\|_W \leqslant 1$ and $G(\mathrm{i}) = 1/\eta$. This means that $1/\eta$ would have to be $\leqslant W_A(\mathrm{i})$, so η cannot be smaller than $1/W_A(\mathrm{i})$, and our assertion holds.

Assuming henceforth that $W_A(\mathrm{i}) < \infty$, we see by the Hahn–Banach theorem (same application as in §A.3) that there is a linear form L on the functions of the form $\varphi/\tilde{W}$, $\varphi \in \mathscr{C}_{\tilde{W}}(\mathbb{R})$, with

$$\left| L\left(\frac{\varphi}{\tilde{W}}\right)\right| \leqslant \|\varphi\|_{\tilde{W}}, \quad \varphi \in \mathscr{C}_{\tilde{W}}(\mathbb{R}),$$

and $L(f/\tilde{W}) = 0$ for all $f \in \tilde{\mathscr{E}}_A$, whilst $L(1/(t-\mathrm{i})\tilde{W}(t)) \neq 0$.

Let now $G \in \mathscr{E}_A$ (*sic*!), and take any fixed $z_0 \notin \mathbb{R}$. Since the function

$$\frac{G(t) - G(z_0)}{t - z_0}$$

belongs to $\tilde{\mathscr{E}}_A$ (!), we have

$$L\left(\frac{G(t) - G(z_0)}{(t - z_0)\tilde{W}(t)}\right) = 0,$$

whence (the Markov–Riesz–Pollard trick again!),

$$(\S) \qquad G(z_0) = L\left(\frac{G(t)}{(t - z_0)\tilde{W}(t)}\right) \Big/ L\left(\frac{1}{(t - z_0)\tilde{W}(t)}\right),$$

provided that the denominator on the right is different from zero.

If $\|G\|_W \leqslant 1$, we have, since

$$\frac{G(t)}{(t - z_0)\tilde{W}(t)} = \frac{|t - \mathrm{i}|}{t - z_0}\cdot\frac{G(t)}{W(t)},$$

that

$$\left\|\frac{G(t)}{t - z_0}\right\|_{\tilde{W}} \leqslant \sup_{t \in \mathbb{R}} \left|\frac{t - \mathrm{i}}{t - z_0}\right|.$$

The quantity on the right was worked out in §A.2 and seen there (in the corollary) to be $\leqslant (1 + |z_0|)/|\Im z_0|$. Therefore, if $\|G\|_W \leqslant 1$,

$$\left| L\left(\frac{G(t)}{(t - z_0)\tilde{W}(t)}\right)\right| \leqslant \left\|\frac{G(t)}{t - z_0}\right\|_{\tilde{W}} \leqslant \frac{1 + |z_0|}{|\Im z_0|}.$$

Calling

$$\Phi(z) = L\left(\frac{1}{(t - z)\tilde{W}(t)}\right) \quad \text{for} \quad z \notin \mathbb{R}$$

we see from this last relation and from (§) that, for $G \in \mathscr{E}_A$,

$$(\S\S) \qquad |G(z)| \leqslant \frac{1+|z|}{|\Im z||\Phi(z)|} \quad \text{if} \quad \|G\|_W \leqslant 1.$$

We know that $\Phi(\mathrm{i}) \neq 0$. Also, $\Phi(z)$ is analytic for $\Im z > 0$. That's because

$$|t-\mathrm{i}|\left\{\frac{(t-z-\Delta z)^{-1}-(t-z)^{-1}}{\Delta z} - \frac{1}{(t-z)^2}\right\} = \frac{|t-\mathrm{i}|\Delta z}{(t-z)^2(t-z-\Delta z)}$$

tends to zero *uniformly* for $-\infty < t < \infty$ as $\Delta z \to 0$, provided that $z \notin \mathbb{R}$. Since $|t-\mathrm{i}|\tilde{W}(t) = W(t)$ *is* $\geqslant 1$ on $\mathbb{R}$, this implies that

$$\frac{\Phi(z+\Delta z)-\Phi(z)}{\Delta z} \longrightarrow L\left(\frac{1}{(t-z)^2\tilde{W}(t)}\right)$$

when $\Delta z \to 0$ as long as $z \notin \mathbb{R}$, and thus establishes analyticity of $\Phi(z)$ in the upper and lower half planes.

The function $\Phi(z)$, analytic and not $\equiv 0$ for $\Im z > 0$, is not quite bounded in $\{\Im z \geqslant 1\}$; it is, however, not far from being bounded in the latter region. (*Here*, by the way, lies the main difference between our present situation and the one discussed in §A.3.) We have, for $t \in \mathbb{R}$,

$$\left|\frac{1}{(t-z)\tilde{W}(t)}\right| = \left|\frac{t-\mathrm{i}}{(t-z)W(t)}\right| \leqslant \left|\frac{t-\mathrm{i}}{t-z}\right|$$

since $W(t) \geqslant 1$, whence, by §A.2, $\|1/(t-z)\|_{\tilde{W}} \leqslant (1+|z|)/|\Im z|$, so

$$|\Phi(z)| = \left|L\left(\frac{1}{(t-z)\tilde{W}(t)}\right)\right| \leqslant \frac{1+|z|}{|\Im z|}.$$

The function $\Phi(z)/(z+\mathrm{i})$ is thus analytic and bounded in $\{\Im z > 1\}$ and continuous in the closure of that half plane; it is certainly not identically zero there because $\Phi(\mathrm{i}) \neq 0$. Therefore, by §G.2 of Chapter III,

$$(\dagger\dagger) \qquad \int_{-\infty}^{\infty} \frac{1}{1+x^2} \log^- \left|\frac{\Phi(x+\mathrm{i})}{x+\mathrm{i}}\right| \mathrm{d}x < \infty.$$

By the definition of W_A and (§§) we now obtain

$$W_A(x+\mathrm{i}) = \sup\{|G(x+\mathrm{i})|: G \in \mathscr{E}_A \text{ and } \|G\|_W \leqslant 1\}$$
$$\leqslant \frac{1+|x+\mathrm{i}|}{|\Phi(x+\mathrm{i})|} \leqslant 2\frac{|x+\mathrm{i}|}{|\Phi(x+\mathrm{i})|},$$

and

$$\log W_A(x+\mathrm{i}) \leqslant \log 2 + \log^- \left|\frac{\Phi(x+\mathrm{i})}{x+\mathrm{i}}\right|.$$

Now we are in the hall of mirrors again! Take any $G \in \mathscr{E}_A$ with $\|G\|_W \leqslant 1$. Then, on the one hand,

$$\log|G(x+i)| \;\leqslant\; \log W_A(x+i),$$

while, on the other,

$$\log|G(\xi)| \;\leqslant\; A + \frac{1}{\pi}\int_{-\infty}^{\infty} \frac{\log|G(x+i)|}{(\xi-x)^2+1}\,dx$$

for $\xi \in \mathbb{R}$, according to the second theorem of §G.2, Chapter III, applied in the half plane $\Im z \leqslant 1$. Substituting into this last inequality the two preceding it, we find

$$\log|G(\xi)| \;\leqslant\; A + \log 2 + \frac{1}{\pi}\int_{-\infty}^{\infty} \frac{\log^-|\Phi(x+i)/(x+i)|}{(\xi-x)^2+1}\,dx,$$

and, since $\log W_A(\xi)$ is the supremum of $\log|G(\xi)|$ for such G, we see that

$$\log W_A(\xi) \;\leqslant\; A + \log 2 + \frac{1}{\pi}\int_{-\infty}^{\infty} \frac{\log^-|\Phi(x+i)/(x+i)|}{(\xi-x)^2+1}\,dx.$$

Using Fubini's theorem in the usual way together with this relation, we finally get

$$\int_{-\infty}^{\infty} \frac{\log W_A(\xi)}{\xi^2+1}\,d\xi$$

$$\leqslant\; \pi(A+\log 2) + \frac{1}{\pi}\int_{-\infty}^{\infty} \frac{2}{x^2+4}\log^-\left|\frac{\Phi(x+i)}{x+i}\right|dx.$$

The integral on the right is, however, *finite* by (††). The theorem is thus proved, and we are done.

Remark. Thus, if $W_A(i) < \infty$, the fourth theorem of article 2 (Akhiezer's description) and the discussion in article 3 related to it apply.

F. L. de Branges' description of extremal unit measures orthogonal to the $e^{i\lambda x}/W(x)$, $-A \leqslant \lambda \leqslant A$, when $\mathscr{E}_A$ is not dense in $\mathscr{C}_W(\mathbb{R})$.

We have now about finished with the individual treatment of uniform weighted approximation by polynomials or by functions in $\mathscr{E}_A$. There remains one thing, however.

In his study of the situation where linear combinations of the $e^{i\lambda x}$, $-A \leqslant \lambda \leqslant A$, (or polynomials) are *not* $\|\ \|_W$-dense in $\mathscr{C}_W(\mathbb{R})$, Louis de Branges obtained a beautiful description (valid for weights which are not too irregular) of the *extremal unit measures orthogonal to the functions*

$e^{i\lambda x}/W(x)$, $-A \leqslant \lambda \leqslant A$ (or to the polynomials divided by W). We should not end the present discussion without giving it.

> For the treatment of de Branges' description, we assume that $W(x)$ is continuous (and finite) on a certain closed unbounded subset E of $\mathbb{R}$, and that $W(x) \equiv \infty$ on $\mathbb{R} \sim E$. As always, $W(x) \geqslant 1$.

In this circumstance, the ratios $\varphi(x)/W(x)$ with $\varphi \in \mathscr{C}_W(\mathbb{R})$ are *continuous* on the *locally compact set* E, and tend to *zero* whenever $x \to \pm\infty$ in E. We can write

$$\|\varphi\|_W = \sup_{t\in E}\left|\frac{\varphi(t)}{W(t)}\right| \quad \text{for} \quad \varphi \in \mathscr{C}_W(\mathbb{R}),$$

and we see that the correspondence $\varphi \leftrightarrow \varphi/W$ is an isometric isomorphism between $\mathscr{C}_W(\mathbb{R})$ and $\mathscr{C}_0(E)$, the usual Banach space of functions continuous and zero at ∞ on the locally compact Hausdorff space E. The bounded linear functionals on $\mathscr{C}_0(E)$ are given by the Riesz representation theorem. Therefore the $\|\ \ \|_W$-bounded linear functionals on $\mathscr{C}_W(\mathbb{R})$ are all of the form

$$\int_E \frac{\varphi(t)}{W(t)}\,d\mu(t) = \int_{-\infty}^{\infty} \frac{\varphi(t)}{W(t)}\,d\mu(t)$$

with *totally finite complex Radon measures* μ *supported on E.*

We consider in the following discussion the case where linear combinations of the $e^{i\lambda x}$, $-A \leqslant \lambda \leqslant A$, are *not* $\|\ \ \|_W$-dense in $\mathscr{C}_W(\mathbb{R})$. We could also treat the situation where *polynomials* are not $\|\ \ \|_W$-dense in $\mathscr{C}_W(\mathbb{R})$ and obtain a result analogous to the one to be found for approximation by exponentials; here, of course, one needs to make the supplementary assumption that $x^n/W(x) \longrightarrow 0$ for every $n \geqslant 0$ as $x \to \pm\infty$ in E.

Granted, then, that linear combinations of the $e^{i\lambda x}$, $-A \leqslant \lambda \leqslant A$, are not $\|\ \ \|_W$-dense in $\mathscr{C}_W(\mathbb{R})$, there must, by the Hahn–Banach theorem, be some *non-zero* $\|\ \ \|_W$-bounded linear functional on $\mathscr{C}_W(\mathbb{R})$ which is *orthogonal to* (i.e., *annihilates*) all the $e^{i\lambda x}$, $-A \leqslant \lambda \leqslant A$, or, what comes to the same thing (lemma of §E.1), to all the functions $f \in \mathscr{E}_A$. According to the above description of such linear functionals, there is thus a *non-zero totally finite Radon measure* μ on E with

$$(*) \qquad \int_E \frac{f(x)}{W(x)}\,d\mu(x) = \int_{-\infty}^{\infty} \frac{f(x)}{W(x)}\,d\mu(x) = 0 \quad \text{for} \quad f \in \mathscr{E}_A.$$

The idea now is to try to obtain a description of the non-zero measures μ on E satisfying $(*)$.

In the first place, if a complex measure μ satisfies $(*)$, *so do its real and* *imaginary parts.* That's because $u = \frac{1}{2}(f + f^*)$ and $v = (1/2i)(f - f^*)$ *both belong to* $\mathscr{E}_A$ *if* f *does* (and conversely). However, $u(x)$ and $v(x)$ are both *real-valued* on $\mathbb{R}$ (recall that $f^*(z) = \overline{f(\bar{z})}$), so, if μ is any *complex* measure on E satisfying $(*)$, we have, given $f \in \mathscr{E}_A$,

$$\int_E \frac{u(x)}{W(x)} d\mu(x) = \int_E \frac{v(x)}{W(x)} d\mu(x) = 0,$$

whence (taking real and imaginary parts)

$$\int_E \frac{u(x)}{W(x)} d\Re\mu(x) = \int_E \frac{u(x)}{W(x)} d\Im\mu(x) = 0,$$

$$\int_E \frac{v(x)}{W(x)} d\Re\mu(x) = \int_E \frac{v(x)}{W(x)} d\Im\mu(x) = 0,$$

and thus

$$\int_E \frac{f(x)}{W(x)} d\Re\mu(x) = \int_E \frac{f(x)}{W(x)} d\Im\mu(x) = 0.$$

This means that a description of the *real-valued* (signed) μ on E satisfying $(*)$ provides us, at the same time, with one for all such *complex* measures μ. Our investigation thus reduces to the study of the *real signed measures* μ *on* E satisfying $(*)$.

Notation. Call Σ the set of *finite real-valued Radon measures* μ *on* E *satisfying* $(*)$ *and such that*

$$\|\mu\| = \int_E |d\mu(x)| \leqslant 1.$$

The set Σ is *convex* and *w*-compact* (over $\mathscr{C}_0(E)$). We can therefore apply to it the celebrated *Krein–Millman theorem,* which says that Σ *is the w*-closed convex hull of its extreme points.* (Recall: an *extreme point* μ of Σ is a member thereof which *cannot be written* as $\lambda\mu_1 + (1-\lambda)\mu_2$ with $0 < \lambda < 1$ and measures μ_1 and μ_2 in Σ *different* from μ.) More explicitly, we can, given any $\mu \in \Sigma$, find a sequence of *finite convex combinations*

$$\mu_N = \sum_k \lambda_k(N) \nu_k(N)$$

of *extreme points* $\nu_k(N)$ of Σ (the $\lambda_k(N) > 0$ and $\sum_k \lambda_k(N) = 1$ for each N) with

$$d\mu_N(x) \longrightarrow d\mu(x) \quad w^* \quad \text{as} \quad N \to \infty.$$

We can, in fact, even do *better* – *Choquet's theorem* furnishes a representation for μ as a kind of *integral over the set of extreme points of* Σ. (The book of Phelps is an excellent introduction to these matters. In the present situation, Σ is *metrizable*, so a particularly simple and elegant form of Choquet's theorem applies to it.) The point here is that a good description of the *extreme points* of Σ will already tell us a *great deal* about all the members of Σ and thus, in turn, about the *complex*-valued μ satisfying $(*)$. Knowledge of those extreme points therefore takes us a long way towards a complete description of the measures μ which satisfy $(*)$.

What Louis de Branges found is an *explicit description of the extreme points* of Σ. We now set out to explain his work.

1. Three lemmas

The main idea behind the following development is contained in

De Branges' lemma. *Let μ be an extreme point of Σ and h a* bounded *real-valued Borel function. Suppose that*

$$\int_{-\infty}^{\infty} \frac{f(x)}{W(x)} h(x)\,\mathrm{d}\mu(x) \;=\; 0$$

for all $f \in \mathscr{E}_A$. Then $h(x)$ is a.e. $(|\mathrm{d}\mu|)$ equal to a constant.

Remark. It is enough to assume that h is *essentially* $(|\mathrm{d}\mu|)$ *bounded*, as will be clear in the proof.

Proof. We start out by *observing once and for all that an extreme point μ of Σ is never the zero measure.* That's because

$$0 = \tfrac{1}{2}\nu + \tfrac{1}{2}(-\nu)$$

where, for ν, we can take *any non-zero member of* Σ. In the situation of this §, Σ *has* non-zero elements.

In view of this fact, we must have $\int_{-\infty}^{\infty} |\mathrm{d}\mu(t)| = 1$ for any extreme point μ of Σ. Otherwise we could write

$$\mu \;=\; (1 - \|\mu\|)\cdot 0 + \|\mu\| \left(\frac{\mu}{\|\mu\|} \right),$$

with $0 < \|\mu\| < 1$ and the measures 0 and $\mu/\|\mu\|$ both belonging to Σ.

Now we take a function h as in the hypothesis, and an extreme point μ of Σ. The relation $(*)$ holds, therefore

$$\int_E \frac{f(x)}{W(x)} (h(x) + C)\,\mathrm{d}\mu(x) \;=\; 0$$

for every constant C. Since h is bounded, we will have $h(x) + C \geqslant 0$ for suitable C; *we may therefore just as well assume that* $h(x)$ *is positive to begin with*, since *otherwise* we would *only need to replace* h *by* $h+C$ in the following argument.

We have, then, a *positive* bounded h satisfying the hypothesis. Unless $h(x) \equiv 0$ a.e. ($|\mathrm{d}\mu|$) (in which case the lemma is already proved), we have

$$\int_E h(t)|\mathrm{d}\mu(t)| \;>\; 0.$$

Multiplication of h by a positive constant will then give us a new positive function like the h in the hypothesis, *which we henceforth also denote by* h, fulfilling the condition

$$\int_E h(t)|\mathrm{d}\mu(t)| \;=\; 1.$$

Since h is bounded, there is a λ, $0 < \lambda < 1$, with $0 \leqslant \lambda h(x) \leqslant 1$. Picking such a λ, we have

$$\int_E \left|\frac{1-\lambda h(t)}{1-\lambda}\right| |\mathrm{d}\mu(t)| \;=\; \int_E \frac{1-\lambda h(t)}{1-\lambda}|\mathrm{d}\mu(t)|$$
$$=\; \frac{1}{1-\lambda}\int_E |\mathrm{d}\mu(t)| - \frac{\lambda}{1-\lambda}\int_E h(t)|\mathrm{d}\mu(t)| \;=\; 1,$$

since $\int_E |\mathrm{d}\mu(t)| = 1$.

Also,

$$\int_E \frac{f(x)}{W(x)}\frac{1-\lambda h(x)}{1-\lambda}\mathrm{d}\mu(x) \;=\; 0$$

for all $f \in \mathscr{E}_A$ by the hypothesis and the property (*). In view of the previous relation, we see that the measure μ_2 on E such that

$$\mathrm{d}\mu_2(t) \;=\; \frac{1-\lambda h(t)}{1-\lambda}\mathrm{d}\mu(t)$$

belongs to Σ.

The same is true for the measure μ_1 on E with

$$\mathrm{d}\mu_1(t) = h(t)\mathrm{d}\mu(t).$$

However,

$$\mathrm{d}\mu(t) = \lambda \mathrm{d}\mu_1(t) + (1-\lambda)\mathrm{d}\mu_2(t),$$

and we assumed that μ *was an extreme point of* Σ. Since $0 < \lambda < 1$, we therefore must have $\mathrm{d}\mu_1(t) = \mathrm{d}\mu_2(t)$, i.e., $(1-\lambda h(t))/(1-\lambda) = h(t)$ a.e. ($|\mathrm{d}\mu|$), and finally $h(t) \equiv 1$ a.e. ($|\mathrm{d}\mu|$). The lemma is proved.

Lemma. *Let μ be an extreme point of Σ, let $F \in L_1(|d\mu|)$, and suppose that $\int_E F(t)d\mu(t) = 0$. Then there are $f_n \in \mathscr{E}_A$ with*

$$\int_E \left| \frac{f_n(t)}{W(t)} - F(t) \right| |d\mu(t)| \underset{n}{\longrightarrow} 0.$$

Proof. By duality. The usual application of the Hahn–Banach theorem shows that the *infimum*, for f ranging over $\mathscr{E}_A$, of

$$\int_E \left| F(t) - \frac{f(t)}{W(t)} \right| |d\mu(t)|$$

is equal to the *supremum* of

$$\left| \int_E F(t)h(t)d\mu(t) \right|$$

for Borel functions h such that $|h(t)| \leqslant 1$ a.e. ($|d\mu|$) and

$$(\dagger) \qquad \int_E \frac{f(t)}{W(t)} h(t)\,d\mu(t) \;=\; 0$$

whenever $f \in \mathscr{E}_A$.

Since μ is a real measure and any $f \in \mathscr{E}_A$ can be written as $u + iv$ with u and v in $\mathscr{E}_A$ and *real* on $\mathbb{R}$, we see that, if h satisfies ($\dagger$) for all $f \in \mathscr{E}_A$, so do $\Re h$ and $\Im h$. De Branges' lemma now shows that these latter functions are *constant* a.e. ($|d\mu|$) if h is bounded; in other words, the functions h over which the above mentioned supremum is taken *are all constant* a.e. ($|d\mu|$).

But then $\int_E h(t)F(t)d\mu(t) = 0$ for such functions h, according to our assumption on F. So the supremum in question is zero, and the infimum is also zero. Done.

Lemma. *Let $\mu \neq 0$ belong to Σ. The functions $f(z)$ in $\mathscr{E}_A$ with*

$$\int_E \left| \frac{f(t)}{W(t)} \right| |d\mu(t)| \;\leqslant\; 1$$

form a normal family *in the complex plane. The limit $F(z)$ of any u.c.c. convergent sequence of such functions f is an entire function of exponential type $\leqslant A$ with*

$$\int_{-\infty}^{\infty} \frac{\log^+|F(t)|}{1+t^2}\,dt \;<\; \infty.$$

Proof. Since μ is real, the function $\Phi(z) = \int_E (d\mu(t)/(t-z)W(t))$, which is analytic in both half planes $\Im z > 0$, $\Im z < 0$, *cannot vanish identically in either* (otherwise μ would be 0). $\Phi(z)$ is *bounded* for $\Im z \geqslant 1$ and for $\Im z \leqslant -1$.

If now $f \in \mathscr{E}_A$, the function of t, $(f(t) - f(z))/(t - z)$, *also* belongs to $\mathscr{E}_A$, making $\int_E ((f(t) - f(z))/(t - z))(\mathrm{d}\mu(t)/W(t)) = 0$. Therefore, if $z \notin \mathbb{R}$,

$$f(z) \;=\; \frac{1}{\Phi(z)} \int_E \frac{f(t)\mathrm{d}\mu(t)}{(t-z)W(t)}$$

(the Markov–Riesz–Pollard trick again!). When $\int_E |f(t)/W(t)||\mathrm{d}\mu(t)| \leqslant 1$, this yields, for $z = x \pm \mathrm{i}$, $|f(x \pm \mathrm{i})| \leqslant |1/\Phi(x \pm \mathrm{i})|$, and, by §E of Chapter III,

$$\log|f(\zeta)| \;\leqslant\; A|\Im\zeta \mp 1| + \frac{1}{\pi}\int_{-\infty}^{\infty} \frac{|\Im\zeta \mp 1| \log^+ |f(x \pm \mathrm{i})|}{|\zeta - x \mp \mathrm{i}|^2}\mathrm{d}x$$

$$\leqslant\; A|\Im\zeta \mp 1| + \frac{1}{\pi}\int_{-\infty}^{\infty} \frac{|\Im\zeta \mp 1| \log^- |\Phi(x \pm \mathrm{i})|}{|\zeta - x \mp \mathrm{i}|^2}\mathrm{d}x.$$

Since $\Phi(z)$ is $\not\equiv 0$ both in $\Im z > 0$ and in $\Im z < 0$, we have

$$\int_{-\infty}^{\infty} (1/(1+x^2)) \log^- |\Phi(x + \mathrm{i})| \mathrm{d}x \;<\; \infty$$

and

$$\int_{-\infty}^{\infty} (1/(1+x^2)) \log^- |\Phi(x - \mathrm{i})| \mathrm{d}x \;<\; \infty.$$

From here on, the proof is like that of Akhiezer's second theorem (§B.2 – see also §E.2, especially the proof of the corollary at the end of that article).

2. De Branges' theorem

Lemma. *Let μ be an extreme point of Σ. Then μ is supported on a countably infinite subset of $\mathbb{R}$ without finite limit point.*

Proof. As we saw in proving de Branges' lemma (previous article), an extreme point μ of Σ *cannot be the zero measure.*

Such a measure μ *cannot have compact support.* Suppose, indeed, that μ were supported on the compact set $K \subseteq E$; then we would have

$$\int_K \frac{\mathrm{e}^{\mathrm{i}\lambda x}}{W(x)} \mathrm{d}\mu(x) \;=\; 0, \quad -A \leqslant \lambda \leqslant A.$$

Here, since K is compact, we can differentiate with respect to λ under the integral sign as many times as we wish, obtaining (for $\lambda = 0$)

$$\int_K \frac{x^n}{W(x)} \mathrm{d}\mu(x) \;=\; 0, \quad n = 0, 1, 2, \ldots.$$

From this, Weierstrass' theorem would give us

$$\int_K \frac{g(x)}{W(x)} \mathrm{d}\mu(x) \;=\; 0$$

for all continuous functions g on K. Then, however, μ would have to be *zero*, since $K \subseteq E$, and $W(x)$ is continuous (and $< \infty$) on E.

The fact that μ does *not* have compact support implies the existence of a *finite interval J containing two disjoint open intervals*, I_1 and I_2, with

$$\int_{I_1} |\mathrm{d}\mu(t)| > 0 \quad \text{and} \quad \int_{I_2} |\mathrm{d}\mu(t)| > 0.$$

This means that we can find a Borel function φ, *identically zero outside* $I_1 \cup I_2$ (hence identically zero outside J) with $|\varphi(t)|$ *equal to non-zero constants* on *each* of the intervals I_1, I_2, such that

$$\int_{-\infty}^{\infty} \varphi(t)\mathrm{d}\mu(t) = 0.$$

On account of this relation we have, by the *second* lemma of the preceding article, a sequence of functions $f_n \in \mathscr{E}_A$ with

$$\int_{-\infty}^{\infty} \left| \frac{f_n(t)}{W(t)} - \varphi(t) \right| |\mathrm{d}\mu(t)| \;\xrightarrow[n]{}\; 0.$$

The *third* lemma of the above article now shows that a subsequence of the $f_n(z)$ converges u.c.c. in $\mathbb{C}$ to some entire function $F(z)$ of exponential type $\leqslant A$, and we see by Fatou's lemma that

$$\int_{-\infty}^{\infty} \left| \frac{F(t)}{W(t)} - \varphi(t) \right| |\mathrm{d}\mu(t)| \;=\; 0,$$

i.e.,

$$\frac{F(t)}{W(t)} \;=\; \varphi(t) \text{ a.e. } (|\mathrm{d}\mu|).$$

By its construction, φ is *not a.e.* ($|\mathrm{d}\mu|$) *equal to zero*, hence $F(z) \not\equiv 0$. The function φ *does*, however, vanish identically outside the finite interval J. Therefore

$$\frac{F(t)}{W(t)} \;\equiv\; 0 \text{ a.e. } (|\mathrm{d}\mu|),\ t \notin J.$$

Since $F(z) \not\equiv 0$ is entire, F *can only* vanish on a certain countable set without finite limit point. We see that μ, *outside* J, must be supported on this countable set, consisting of zeros of F.

Because the support of μ is not compact, there is a *finite interval J', disjoint from J*, and containing two disjoint open intervals I'_1 and I'_2 with

$$\int_{I'_1} |d\mu(t)| > 0, \quad \int_{I'_2} |d\mu(t)| > 0.$$

Repetition of the argument just made, with J' playing the rôle of J, shows now that μ, outside J' (and hence in particular *in J*!) is also supported on a countable set without finite limit point. Therefore the *whole support* of μ in E must be such a set, which is what we had to prove.

Remark. The support of μ must *really be infinite*. Otherwise it would be compact, and this, as we have seen, is impossible.

Now we are ready to establish the

Theorem (Louis de Branges). *Let $W(x) \geqslant 1$ be a weight having the properties stated at the beginning of this §, and let E be the associated closed set on which $W(x)$ is finite.*

Suppose that $\mathscr{E}_A$ is not *$\| \ \|_W$-dense in $\mathscr{C}_W(\mathbb{R})$, and let μ be an* extreme point *of the set Σ of real signed measures ν on E such that*

$$\int_E |d\nu(t)| \leqslant 1$$

and

$$\int_E \frac{f(t)}{W(t)} d\nu(t) = 0 \quad \textit{for all } f \in \mathscr{E}_A.$$

Then μ is supported on an infinite sequence $\{x_n\}$ without finite limit point, lying in E. There is an entire function $S(z)$ of exponential type A having a simple zero *at each point x_n and* no other zeros, *with*

$$\boxed{\mu(\{x_n\}) = \frac{W(x_n)}{S'(x_n)}.}$$

Moreover,

$$\int_{-\infty}^{\infty} \frac{\log^+ |S(x)|}{1+x^2} dx < \infty$$

and

$$\lim_{y \to \infty} \frac{\log |S(iy)|}{y} = \lim_{y \to -\infty} \frac{\log |S(iy)|}{|y|} = A.$$

Proof. Let us begin by first establishing an auxiliary proposition.

Given an extreme point μ of Σ, suppose that we have a sequence of functions $f_n \in \mathscr{E}_A$ such that

$$\int_{-\infty}^{\infty} |f_n(x) - f_m(x)| \frac{|d\mu(x)|}{W(x)} \underset{n,m}{\longrightarrow} 0.$$

We know from the third lemma of the preceding article that a subsequence of the f_n tends u.c.c. in $\mathbb{C}$ to some entire function F of exponential type $\leqslant A$, and here it is clear by Fatou's lemma that

$$(*) \qquad \int_{-\infty}^{\infty} |f_n(x) - F(x)| \frac{|d\mu(x)|}{W(x)} \underset{n}{\longrightarrow} 0,$$

whence surely

$$\int_{-\infty}^{\infty} \frac{F(x)}{W(x)} d\mu(x) = 0$$

since $\int_{-\infty}^{\infty}(f_n(x)/W(x))d\mu(x) = 0$ for all n.

Our auxiliary proposition says that the relations

$$\int_{-\infty}^{\infty} \frac{F(x) - F(a)}{x - a} \frac{d\mu(x)}{W(x)} = 0, \qquad \int_{-\infty}^{\infty} (x-b) \frac{F(x) - F(a)}{x - a} \frac{d\mu(x)}{W(x)} = 0$$

also hold for the limit function F; here, a and b are arbitrary complex numbers. Both of these formulas are proved in the same way, and it is enough to deal here only with the *second* one.

Wlog, $f_n(z) \underset{n}{\longrightarrow} F(z)$ u.c.c., whence $f_n(a) \longrightarrow F(a)$ and thence, by $(*)$,

$$\int_{-\infty}^{\infty} \frac{|f_n(x) - f_n(a) - (F(x) - F(a))|}{W(x)} |d\mu(x)| \underset{n}{\longrightarrow} 0,$$

since $W(x) \geqslant 1$. From this is clear that

$$\int_{|x-a|\geqslant 1} \left| \frac{x-b}{x-a} \left(f_n(x) - f_n(a) - F(x) + F(a) \right) \right| \frac{|d\mu(x)|}{W(x)} \underset{n}{\longrightarrow} 0.$$

Also, the u.c.c. convergence of $f_n(z)$ to $F(z)$ makes

$$\frac{f_n(z) - f_n(a)}{z - a}(z - b) \underset{n}{\longrightarrow} \frac{F(z) - F(a)}{z - a}(z - b)$$

u.c.c., by the elementary theory of analytic functions (Cauchy's formula!). Therefore we also have

$$\int_{|x-a|\leqslant 1} \left| (x-b) \left(\frac{f_n(x) - f_n(a)}{x - a} - \frac{F(x) - F(a)}{x - a} \right) \right| \frac{|d\mu(x)|}{W(x)} \underset{n}{\longrightarrow} 0.$$

and finally

$$\int_{-\infty}^{\infty} \left| \frac{f_n(x) - f_n(a)}{x-a}(x-b) - \frac{F(x) - F(a)}{x-a}(x-b) \right| \frac{|d\mu(x)|}{W(x)} \xrightarrow[n]{} 0.$$

Since, however, the $f_n \in \mathscr{E}_A$, we have $(x-b)(f_n(x) - f_n(a))/(x-a) \in \mathscr{E}_A$, whence

$$\int_{-\infty}^{\infty} \frac{f_n(x) - f_n(a)}{x-a}(x-b)\frac{d\mu(x)}{W(x)} = 0$$

for each n. Referring to the previous relation, we see that

$$\int_{-\infty}^{\infty} \frac{F(x) - F(a)}{x-a}(x-b)\frac{d\mu(x)}{W(x)} = 0,$$

as we set out to show.

Now we turn to the theorem itself. According to the lemma at the beginning of this article, our measure μ is supported on a countable set $\{x_n\} \subseteq E$ without finite limit point, and, by the remark following that lemma, $\mu(\{x_n\}) \neq 0$ *for infinitely many* of the points x_n. There is thus no loss of generality in supposing that $\mu(\{x_n\}) \neq 0$ *for each n*.

Take any two points from among the x_n, say x_0 and x_1, and put

$$\varphi(x_0) = \frac{1}{\mu(\{x_0\})(x_0 - x_1)},$$

$$\varphi(x_1) = \frac{1}{\mu(\{x_1\})(x_1 - x_0)},$$

and $\varphi(x) = 0$ for $x \neq x_0$ or x_1. Then

$$\int_{-\infty}^{\infty} \varphi(x) d\mu(x) = 0,$$

so, as in the proof of the preceding lemma, there is a sequence of $f_n \in \mathscr{E}_A$ with

$$\int_{-\infty}^{\infty} \left| \frac{f_n(x)}{W(x)} - \varphi(x) \right| |d\mu(x)| \xrightarrow[n]{} 0$$

and, wlog, $f_n(z) \xrightarrow[n]{} F(z)$ u.c.c., F being some entire function of exponential type $\leqslant A$. We see that

$$(*) \qquad \frac{F(x)}{W(x)} = \varphi(x) \text{ a.e. } (|d\mu|),$$

whence

$$(\dagger) \qquad \int_{-\infty}^{\infty} \frac{|f_n(x) - F(x)|}{W(x)} |d\mu(x)| \xrightarrow[n]{} 0.$$

From (⁑) and the definition of φ, we have $F(x_0) \neq 0$, $F(x_1) \neq 0$, so $F(z) \not\equiv 0$. For the same reasons, however, *$F(x)$ vanishes at all the other points x_n, $n \neq 0, 1$.*

Put

$$S(z) = F(z)(z - x_0)(z - x_1).$$

Then S, like F, is an entire function of exponential type $\leqslant A$. $S(z)$ vanishes at each of the points x_n in the support of μ. Finally,

$$\int_{-\infty}^{\infty} \frac{\log^+ |S(x)|}{1 + x^2} dx \ < \ \infty,$$

since, by the third lemma of the preceding article, the function F has this property.

Let us compute the quantities $S'(x_n)$. We already know that

$$S'(x_0) \ = \ F(x_0)(x_0 - x_1) \ = \ W(x_0)\varphi(x_0)(x_0 - x_1) \ = \ \frac{W(x_0)}{\mu(\{x_0\})},$$

and similarly $S'(x_1) = W(x_1)/\mu(\{x_1\})$. Take any other point x_n, $n \neq 0, 1$, and form the function

$$\frac{S(x)}{(x - x_0)(x - x_n)} \ = \ \frac{F(x)}{x - x_n}(x - x_1).$$

Since $F(x_n) = 0$, ($\dagger$) implies, *by our auxiliary proposition*, that

$$\int_{-\infty}^{\infty} \frac{F(x)}{x - x_n}(x - x_1)\frac{d\mu(x)}{W(x)} \ = \ 0.$$

The function $S(x)/(x - x_0)(x - x_n)$ vanishes at all the x_k, save x_0 and x_n. The previous relation therefore reduces to

$$\frac{S'(x_0)\mu(\{x_0\})}{(x_0 - x_n)W(x_0)} + \frac{S'(x_n)\mu(\{x_n\})}{(x_n - x_0)W(x_n)} \ = \ 0,$$

i.e.,

$$\frac{S'(x_n)}{W(x_n)}\mu(\{x_n\}) \ = \ 1,$$

and finally $S'(x_n) = W(x_n)/\mu(\{x_n\})$.

The function $S(z)$ can have no zeros apart from the x_n. Suppose, indeed, that $S(a)=0$ with a different from all the x_n; then we would also have $F(a)=0$, so, in the identity

$$\int_{-\infty}^{\infty} \frac{S(x)}{(x-x_0)(x-a)} \frac{d\mu(x)}{W(x)} = \int_{-\infty}^{\infty} \frac{F(x)}{x-a}(x-x_1)\frac{d\mu(x)}{W(x)},$$

the *right-hand integral would have to vanish* by our auxiliary proposition. The quantity $(F(x)/(x-a))(x-x_1)$ is, however, *different from* 0 *at only one of the points* x_n *in* μ's *support, namely, at* x_0, where

$$\frac{F(x_0)}{x_0-a}(x_0-x_1) = \frac{S'(x_0)}{x_0-a}.$$

We would thus get

$$\frac{S'(x_0)}{x_0-a}\cdot\frac{\mu(\{x_0\})}{W(x_0)} = 0,$$

i.e., in view of the computation made in the previous paragraph,

$$\frac{1}{x_0-a} = 0,$$

which is absurd.

The function $S(z)$ thus *vanishes once at each* x_n, and *only at those points.* As we have already seen, $\mu(\{x_n\}) = W(x_n)/S'(x_n)$, $S(z)$ is of exponential type $\leqslant A$, and

$$\int_{-\infty}^{\infty} \frac{\log^+|S(x)|}{1+x^2}dx < \infty.$$

To complete our proof, we have to show that $\log|S(iy)|/|y| \longrightarrow A$ for $y\to\pm\infty$.

In order to do this, let us first derive the partial fraction decomposition

$$\frac{1}{S(z)} = \sum_n \frac{1}{(z-x_n)S'(x_n)}.$$

Note that $\sum_n |1/S'(x_n)|$ is *surely convergent* because $\mu(\{x_n\}) = W(x_n)/S'(x_n)$, μ is a *finite* measure, and $W(x)\geqslant 1$. Take the function $G(t) = F(t)(t-x_1) = S(t)/(t-x_0)$, and, for *fixed* z, observe that

$$\frac{G(t)-G(z)}{t-z} = \frac{F(t)(t-x_1)-F(z)(z-x_1)}{t-z}$$

$$= \frac{F(t)-F(z)}{t-z}(t-x_1) + F(z).$$

By our auxiliary proposition,

$$\int_{-\infty}^{\infty} \frac{F(t) - F(z)}{t - z}(t - x_1)\frac{\mathrm{d}\mu(t)}{W(t)} = 0,$$

and of course

$$\int_{-\infty}^{\infty} F(z)\cdot\frac{\mathrm{d}\mu(t)}{W(t)} = 0$$

since $1 \in \mathscr{E}_A$. Therefore

$$\int_{-\infty}^{\infty} \frac{G(t) - G(z)}{t - z}\frac{\mathrm{d}\mu(t)}{W(t)} = 0,$$

or, since $G(t)$ vanishes at *all the* x_n *save* x_0,

$$\frac{G(x_0)}{x_0 - z}\frac{\mu(\{x_0\})}{W(x_0)} - G(z)\sum_n \frac{1}{(x_n - z)S'(x_n)} = 0.$$

This is the same as

$$\frac{S'(x_0)}{x_0 - z}\cdot\frac{1}{S'(x_0)} = \frac{S(z)}{z - x_0}\sum_n \frac{1}{(x_n - z)S'(x_n)},$$

or

$$S(z)\sum_n \frac{1}{(z - x_n)S'(x_n)} = 1,$$

the desired relation.

From the result just found we derive a more general *interpolation formula*. Let $-A \leqslant \lambda \leqslant A$. Then $(\mathrm{e}^{\mathrm{i}\lambda t} - \mathrm{e}^{\mathrm{i}\lambda z})/(t - z)$ belongs, as a function of t, to $\mathscr{E}_A$, so

$$\int_{-\infty}^{\infty} \frac{\mathrm{e}^{\mathrm{i}\lambda t} - \mathrm{e}^{\mathrm{i}\lambda z}}{t - z}\frac{\mathrm{d}\mu(t)}{W(t)} = 0.$$

In other words,

$$\sum_n \frac{\mathrm{e}^{\mathrm{i}\lambda x_n}}{(x_n - z)S'(x_n)} = \mathrm{e}^{\mathrm{i}\lambda z}\sum_n \frac{1}{(x_n - z)S'(x_n)}.$$

According to our previous result, the right-hand side is just $-\mathrm{e}^{\mathrm{i}\lambda z}/S(z)$. Therefore

$$\boxed{\frac{\mathrm{e}^{\mathrm{i}\lambda z}}{S(z)} = \sum_n \frac{\mathrm{e}^{\mathrm{i}\lambda x_n}}{(z - x_n)S'(x_n)} \quad \text{for} \quad -A \leqslant \lambda \leqslant A.}$$

(An analogous formula with $e^{i\lambda t}$ replaced by any $f(t) \in \mathscr{E}_A$ also holds, by the way – the proof is the same.)

In the boxed relation, put $\lambda = -A$ and take $z = iy$, $y > 0$. We get

$$\frac{e^{Ay}}{S(iy)} = \sum_n \frac{e^{-iAx_n}}{(iy - x_n)S'(x_n)}.$$

Since $\sum_n |1/S'(x_n)| < \infty$ and the x_n are real, the right side tends to 0 for $y \to \infty$. Thus,

$$\liminf_{y\to\infty} \frac{\log|S(iy)|}{y} \geqslant A.$$

But $\limsup_{y\to\infty} \log|S(iy)|/y \leqslant A$ since S is of exponential type $\leqslant A$. Therefore $\log|S(iy)|/y \to A$ for $y \to \infty$.

On taking $\lambda = A$ in the above boxed formula and making $y \to -\infty$, we see in like manner that

$$\frac{\log|S(iy)|}{|y|} \longrightarrow A \quad \text{for} \quad y \to -\infty.$$

De Branges' theorem is now completely proved. We are done.

3. Discussion of the theorem

De Branges' description of the extreme points of Σ is a most beautiful result; I still do not understand the full meaning of it.

Since $\int_{-\infty}^{\infty} (\log^+|S(x)|/(1+x^2))\,dx < \infty$ and $S(z)$ is of exponential type, the set of zeros $\{x_n\}$ of S, on which the extremal measure μ corresponding to S is supported, has a distribution governed by *Levinson's theorem* (Chapter III; here the version in §H.2 suffices). Because

$$\frac{\log|S(iy)|}{|y|} \longrightarrow A \quad \text{for} \quad y \longrightarrow \pm\infty,$$

we see by that theorem that

$$\frac{\text{number of } x_n \text{ in } [0,t]}{t} \longrightarrow \frac{A}{\pi} \quad \text{as} \quad t \to \infty$$

and

$$\frac{\text{number of } x_n \text{ in } [-t,0]}{t} \longrightarrow \frac{A}{\pi} \quad \text{as} \quad t \to \infty.$$

The zeros of $S(z)$ *are distributed roughly* (very *roughly!*) *like the points*

$$\frac{\pi}{A}n, \quad n = 0, \pm 1, \pm 2, \pm 3, \ldots.$$

(We shall see towards the end of Chapter IX that a certain refinement of this description is possible; we cannot, however obtain much more information about the actual *position* of the points x_n.)

De Branges' result is an *existence theorem*. It says that, if W is a weight of the kind considered in this § such that the $e^{i\lambda x}$, $-A \leqslant \lambda \leqslant A$, are *not* $\| \quad \|_W$-dense in $\mathscr{C}_W(\mathbb{R})$, *then* there *exists* an entire function $\Phi(z)$ of exponential type A with

$$\frac{\log|\Phi(iy)|}{|y|} \longrightarrow A \quad \text{for} \quad y \to \pm\infty, \qquad \int_{-\infty}^{\infty} \frac{\log^+|\Phi(x)|}{1+x^2} dx < \infty,$$

and $|\Phi(x_n)| \geqslant W(x_n)$ on a set of points x_n with $x_n \sim (\pi/A)n$ for $n \to \pm\infty$.

It suffices to take $\Phi(x) = S'(x)$ with one of the functions $S(z)$ furnished by the theorem. (There *will be* such a function S because here Σ is not reduced to $\{0\}$, and *will have* extreme points by the Krein–Millman theorem!) If $\{x_n\}$ is the set of zeros of S, we have

$$\sum_n \frac{W(x_n)}{|S'(x_n)|} = \int_{-\infty}^{\infty} |d\mu(x)| = 1,$$

so $|S'(x_n)| \geqslant W(x_n)$. Let us verify that

$$\int_{-\infty}^{\infty} \frac{\log^+|S'(x)|}{1+x^2} dx < \infty.$$

Our function $S(z)$ is of exponential type; *therefore, so is* $S'(z)$. The desired relation will hence follow in now familiar fashion via Fubini's theorem and §E of Chapter III from the inequality

$$\int_{-\infty}^{\infty} \frac{\log^+|S'(x+i)|}{1+x^2} dx < \infty,$$

which we proceed to establish (cf. the hall of mirrors argument at the end of §E.4).

Since $S(z)$ is free of zeros in $\Im z > 0$, we have there, by §G.1 of Chapter III,

$$\begin{aligned} \log|S(z)| &= A\Im z + \frac{1}{\pi}\int_{-\infty}^{\infty} \frac{\Im z \log|S(t)|}{|z-t|^2} dt \\ &= A\Im z + \frac{1}{\pi}\int_{-\infty}^{\infty} \Im\left(\frac{1}{t-z}\right) \log|S(t)| dt. \end{aligned}$$

For the same reason one can define an analytic function $\log S(z)$ in $\Im z > 0$. Using the previous relation together with the Cauchy–Riemann equations we thus find that

$$\frac{S'(z)}{S(z)} = \frac{\mathrm{d}\log S(z)}{\mathrm{d}z} = \left(\frac{\partial}{\partial x} - \mathrm{i}\frac{\partial}{\partial y}\right)\log|S(z)|$$

$$= -\mathrm{i}A - \frac{\mathrm{i}}{\pi}\int_{-\infty}^{\infty}\frac{\log|S(t)|}{(z-t)^2}\mathrm{d}t, \quad \Im z > 0,$$

whence, taking $z = x + \mathrm{i}$,

$$\left|\frac{S'(x+\mathrm{i})}{S(x+\mathrm{i})}\right| \leqslant A + \frac{1}{\pi}\int_{-\infty}^{\infty}\frac{|\log|S(t)||}{(x-t)^2+1}\mathrm{d}t.$$

Here, since

$$\int_{-\infty}^{\infty}\frac{\log^+|S(t)|}{t^2+1}\mathrm{d}t < \infty,$$

we of course have

$$\int_{-\infty}^{\infty}\frac{\log^-|S(t)|}{t^2+1}\mathrm{d}t < \infty,$$

(Chapter III, §G.2) so

$$\frac{1}{\pi}\int_{-\infty}^{\infty}\frac{|\log|S(t)||}{1+t^2}\mathrm{d}t = \text{say } C,$$

a finite quantity. By §B.2 we also have

$$\left|\frac{t-\mathrm{i}}{t-\mathrm{i}-x}\right|^2 \leqslant (|x|+2)^2, \quad t\in\mathbb{R},$$

so

$$\frac{1}{\pi}\int_{-\infty}^{\infty}\frac{|\log|S(t)||}{(t-x)^2+1}\mathrm{d}t \leqslant C(|x|+2)^2$$

and thence, by the previous relation,

$$\frac{|S'(x+\mathrm{i})|}{|S(x+\mathrm{i})|} \leqslant A + C(|x|+2)^2.$$

This means, however, that

$$\log|S'(x+\mathrm{i})| \leqslant \log(A + C(|x|+2)^2) + \log|S(x+\mathrm{i})|,$$

from which

$$\int_{-\infty}^{\infty}\frac{\log^+|S'(x+\mathrm{i})|}{x^2+1}\mathrm{d}t \leqslant \int_{-\infty}^{\infty}\frac{\log^+|S(x+\mathrm{i})|}{1+x^2}\mathrm{d}x$$
$$+ \int_{-\infty}^{\infty}\frac{\log^+(A + C(|x|+2)^2)}{x^2+1}\mathrm{d}x.$$

Both integrals on the right are *finite*, however, the *first* because

$$\int_{-\infty}^{\infty} \frac{\log^+ |S(x)|}{1+x^2} dx \ < \ \infty,$$

and the *second* by inspection. Therefore

$$\int_{-\infty}^{\infty} \frac{\log^+ |S'(x+i)|}{1+x^2} dx \ < \ \infty,$$

which is what we needed to show.

We still have to check that

$$\frac{\log |S'(iy)|}{|y|} \longrightarrow A \quad \text{for} \quad y \longrightarrow \pm\infty.$$

There are several ways of doing this; one goes as follows. Since the limit relation in question is *true* for S, we have, for each $\varepsilon > 0$,

$$|S(z)| \ \leqslant \ M_\varepsilon \exp(A|\Im z| + \varepsilon|z|)$$

(see discussion at end of §B.2). Using Cauchy's formula (for the derivative) with circles of radius 1 centered on the imaginary axis, we see from this relation that

$$|S'(iy)| \ \leqslant \ \text{const.}e^{(A+\varepsilon)|y|},$$

so, since $\varepsilon > 0$ is arbitrary,

$$\limsup_{y \to \pm\infty} \frac{\log |S'(iy)|}{|y|} \ \leqslant \ A.$$

However, $S(iy) = S(0) + i\int_0^y S'(i\eta)\,d\eta$. Therefore the above limit superior along either direction of the imaginary axis must *be* A, otherwise $\log|S(iy)|/|y|$ could not tend to A as $y \to \pm\infty$. By a remark at the end of §G.1, Chapter III, it will follow from this fact that the ratio $\log|S'(iy)|/|y|$ actually *tends* to A as $y \to \pm\infty$, if we can verify that $S'(z)$ *has only real zeros.*

To see this, write the Hadamard factorization (Chapter III, §A) for S:

$$S(z) \ = \ Ae^{cz}\prod_n \left(1 - \frac{z}{x_n}\right)e^{z/x_n}.$$

(We are assuming that none of the zeros x_n of S is equal to 0; if one of them is, a slight modification in this formula is necessary.) Here, as we know, all the x_n are real, therefore

$$\left|\frac{S(iy)}{S(-iy)}\right| \ = \ e^{-2y\Im c}.$$

Since $\log|S(\mathrm{i}y)|/y$ and $\log|S(-\mathrm{i}y)|/y$ both tend to the same limit, A, as $y \to \infty$, we must have $\Im c = 0$, i.e., c is *real*. Logarithmic differentiation of the above Hadamard product now yields

$$\frac{S'(z)}{S(z)} = c + \sum_n \left(\frac{1}{z - x_n} + \frac{1}{x_n} \right),$$

whence

$$\Im\left(\frac{S'(z)}{S(z)} \right) = -\sum_n \frac{\Im z}{|z - x_n|^2}.$$

The expression on the right is <0 for $\Im z > 0$ and >0 for $\Im z < 0$; $S'(z)$ can hence *not vanish* in *either of those half planes*. This argument (which goes back to Gauss, by the way), shows that *all the zeros of* $S'(z)$ *must be real*, as required.

We have now finished showing that the function $\Phi(z) = S'(z)$ has all the properties claimed for it. As an observation of general interest, let us just mention one more fact: *the zeros of* $S'(z)$ *are simple and lie between the zeros* x_n *of* $S(z)$. To see that, differentiate the above formula for $S'(z)/S(z)$ one more time, getting

$$\frac{\mathrm{d}}{\mathrm{d}z}\left(\frac{S'(z)}{S(z)} \right) = -\sum_n \frac{1}{(z - x_n)^2}.$$

From this it is clear that $S'(x)/S(x)$ decreases strictly from ∞ to $-\infty$ on each open interval with endpoints at two successive points x_n, and hence *vanishes precisely once therein*. $S'(z)$ therefore has exactly *one* zero in each such interval, and, since all its zeros are *real*, *no others*.

This property implies that the (real) zeros of $S'(z)$ *have the same asymptotic distributions as the* x_n. From that it is easy to obtain another proof of the limit relation

$$\frac{\log|S'(\mathrm{i}y)|}{|y|} \longrightarrow A, \quad y \longrightarrow \pm\infty.$$

Just use the Hadamard factorization of $S'(z)$ to write $\log|S'(\mathrm{i}y)|$ as a *Stieltjes integral*, then perform an integration by parts in the latter. The desired result follows without difficulty (see a similar computation in §H.3, Chapter III).

Let us summarize. If, for a weight $W(x)$, the $\mathrm{e}^{\mathrm{i}\lambda x}$, $-A \leqslant \lambda \leqslant A$, are *not* $\|\ \|_W$-dense in $\mathscr{C}_W(\mathbb{R})$, Louis de Branges' theorem furnishes entire functions of *precise exponential type* A having *convergent* $\log^+$ *integrals* which are at the same time *large* ($\geqslant W$ in absolute value) *fairly often*, namely on a set of points x_n with $x_n \sim (\pi/A)n$ for $n \to \pm\infty$. These points

x_n are of course located in the set E where $W(x) < \infty$; the theorem, unfortunately, does not provide much more information about their position, even though some refinement in the description of their asymptotic distribution is possible (Chapter IX). One would like to know more about the location of the x_n.

4. Scholium. Krein's functions

Entire functions whose reciprocals have partial fraction decompositions like the one for $1/S(z)$ figuring in the proof of de Branges' theorem arise in the study of various questions. They were investigated by M.G. Krein, in connection, I believe, with the inverse Sturm–Liouville problem. We give some results about such functions here, limiting the discussion to those with *real zeros*. More material on Krein's work (he allowed complex zeros) can be found, together with references, in Levin's book.

Theorem. *Let $S(z)$ be entire, of exponential type, and have only the real simple zeros $\{x_n\}$. Suppose that $S(z) \to \infty$ as $z \to \infty$ along each of four rays*

$$\arg z = \alpha_k,$$

with

$$0 < \alpha_1 < \frac{\pi}{2} < \alpha_2 < \pi < \alpha_3 < \frac{3\pi}{2} < \alpha_4 < 2\pi,$$

and that also

$$\sum_n |1/S'(x_n)| < \infty.$$

Then

$$\frac{1}{S(z)} = \sum_n \frac{1}{(z - x_n)S'(x_n)}.$$

Proof. The function

$$L(z) = \sum_n \frac{S(z)}{(z - x_n)S'(x_n)}$$

is entire, since $S(x_n) = 0$ for each n and $\sum_n |1/S'(x_n)| < \infty$.

I claim that *$L(z)$ is of exponential type.* Clearly,

$$|L(z)| \leqslant \text{const.} \frac{|S(z)|}{|\Im z|},$$

so the growth of $L(z)$ is dominated by that of $S(z)$ outside the strip $|\Im z| \leqslant 1$. For $|\Im z| \leqslant 1$, one may use the following trick. The function $\sqrt{|L(z)|}$ is *subharmonic*, therefore

$$\sqrt{|L(z)|} \leqslant \frac{1}{2\pi}\int_{-\pi}^{\pi} \sqrt{|L(z+2e^{i\theta})|}\,d\theta.$$

Substituting the preceding inequality into the integral on the right, we obtain for it a bound of the form const.$e^{K|z|/2}\int_{-\pi}^{\pi} d\theta/\sqrt{|\Im z + 2\sin\theta|}$, and this is clearly $\leqslant$ const.$e^{K|z|/2}$ for $|\Im z| \leqslant 1$. We see in this way that $|L(z)| \leqslant Ce^{K|z|}$ for all z.

We have $L(x_k) = 1$ at each x_k. Therefore

$$\frac{L(z)-1}{S(z)} = \sum_n \frac{1}{(z-x_n)S'(x_n)} - \frac{1}{S(z)}$$

is *entire; as the ratio of two entire functions of exponential type it is also of exponential type by Lindelöf's theorem.* (See *third* theorem of §B, Chapter III.)

Since $\sum_n |1/S'(x_n)| < \infty$,

$$\sum_n \frac{1}{(z-x_n)S'(x_n)} \longrightarrow 0$$

as $z \to \infty$ along each of the rays $\arg z = \alpha_k$, $k = 1, 2, 3$ and 4, and by hypothesis $1/S(z) \to 0$ for $z \to \infty$ along each of those rays. Therefore $(L(z)-1)/S(z)$ is certainly *bounded* on each of those rays, so, since it is *entire and of exponential type*, it is *bounded in each of the four sectors separated by them* (and having opening $< 180°$) according to the *second Phragmén–Lindelöf theorem* of §C, Chapter III. The entire function $(L(z)-1)/S(z)$ is thus *bounded in* $\mathbb{C}$, hence *equal to a constant*, by Liouville's theorem. Since, as we have seen, it tends to *zero* for z tending to ∞ along certain rays, *the constant must be zero.* Hence

$$\sum_n \frac{1}{(z-x_n)S'(x_n)} - \frac{1}{S(z)} \equiv 0,$$

Q.E.D.

Remark. The hypothesis of the theorem just proved is very ungainly, and one would *like* to be able to affirm the following more general result:

Let $S(z)$, of exponential type, have only the real simple zeros x_n, let $\sum_n |1/S'(x_n)| < \infty$, and suppose that $S(iy) \to \infty$ for $y \to \pm\infty$. Then

$$\frac{1}{S(z)} = \sum_n \frac{1}{(z-x_n)S'(x_n)}.$$

One can waste much time attempting to prove this statement, all in vain, because it is false! In order to lay this ghost for good, here is a *counter example.*

Take

$$S(z) \;=\; \prod_1^\infty \left(1 - \frac{z}{2^n}\right) e^{z/2^n};$$

since $\sum_1^\infty 2^{-n} < \infty$, $S(z)$ is of exponential type. One readily computes $|S'(2^n)|$ by the method used in §C, and finds that

$$|S'(2^n)| \;\sim\; \frac{1}{e} 2^{(n(n-3)/2)} e^{2^n} (S(1))^2$$

for $n \to \infty$; we thus *certainly have*

$$\sum_1^\infty |1/S'(2^n)| < \infty.$$

It is also true that

$$|S(iy)|^2 \;=\; \prod_1^\infty \left(1 + \frac{y^2}{4^n}\right) \;\longrightarrow\; \infty$$

for $y \to \pm\infty$. However, $\prod_1^\infty (1 + |z|/2^n) \leqslant e^{o(|z|)}$ for $z \to \infty$, again by convergence of $\sum_1^\infty 2^{-n}$ (see calculations in §A, Chapter III!). So, for x *real and negative,*

$$S(x) \;=\; e^x \prod_1^\infty \left(1 + \frac{|x|}{2^n}\right) \;\leqslant\; e^{-|x| + o(|x|)},$$

and, for $x \to -\infty$, $1/S(x)$ tends to ∞ like an exponential (!). Therefore $1/S(x)$ certainly *cannot* equal

$$\sum_1^\infty \frac{1}{(x - 2^n)S'(2^n)}$$

which *tends to zero* as $x \to -\infty$.

Theorem (Krein). *Let an entire function $S(z)$ have only the real simple zeros x_n; suppose that $\sum_n |1/S'(x_n)| < \infty$ and that*

$$\frac{1}{S(z)} \;=\; \sum_n \frac{1}{(z - x_n)S'(x_n)}.$$

Then $S(z)$ is of exponential type, *and*

$$\int_{-\infty}^{\infty} \frac{\log^+ |S(x)|}{1 + x^2} dx \;<\; \infty.$$

Remark. In particular, for functions $S(z)$ satisfying the hypothesis of the *previous* theorem, we have

$$\int_{-\infty}^{\infty} \frac{\log^+ |S(x)|}{1+x^2} dx \;<\; \infty.$$

The reader who wants only *this* result may skip *all but the last paragraph* of the following demonstration.

Proof of theorem. Without loss of generality, $\sum_n |1/S'(x_n)| = 1$, whence, by the assumed representation for $1/S(z)$,

$$\left|\frac{1}{S(z)}\right| \;\leqslant\; \frac{1}{|\Im z|}.$$

Given any $h > 0$, the reciprocal $1/S(z)$ is thus *bounded and non-zero* in each of the half-planes $\{\Im z \geqslant h\}$, $\{\Im z \leqslant -h\}$, as well as being *analytic* in slightly larger open half planes containing them. The representation of §G.1, Chapter III, therefore applies in each of those half planes, and we find that in fact

$$\int_{-\infty}^{\infty} \frac{\log^- |1/S(t+ih)|}{1+t^2} dt \;<\; \infty,$$

and that

$$\log\left|\frac{1}{S(z)}\right| \;=\; -A_h(\Im z - h) + \frac{1}{\pi}\int_{-\infty}^{\infty} \frac{(\Im z - h)\log|1/S(t+ih)|}{|z-t-ih|^2} dt$$

for $\Im z > h$, while $\int_{-\infty}^{\infty} (\log^- |1/S(t-ih)|/(1+t^2))\, dt < \infty$ and

$$\log\left|\frac{1}{S(z)}\right| \;=\; -B_h(|\Im z| - h) + \frac{1}{\pi}\int_{-\infty}^{\infty} \frac{(|\Im z| - h)\log|1/S(t-ih)|}{|z-t+ih|^2} dt$$

when $\Im z < -h$.

Here, A_h and B_h are constants which, *a priori*, depend on h. *In fact, they do not*, because, by the remark at the end of §G.1, Chapter III, $\lim_{y\to\infty}(1/y)\log|1/S(iy)|$ *exists* and equals $-A_h$, with a similar relation involving B_h for $y \to -\infty$. All the numbers $-A_h$ for $h > 0$ are thus equal to the limit just mentioned, say to $-A$, and all the B_h are similarly equal to some number B.

For each $h > 0$ we thus have, for $\Im z > h$,

$$\log|S(z)| \;=\; A(\Im z - h) + \frac{1}{\pi}\int_{-\infty}^{\infty} \frac{(\Im z - h)\log|S(t+ih)|}{|z-t-ih|^2} dt$$

$$\leqslant\; A(\Im z - h) + \frac{1}{\pi}\int_{-\infty}^{\infty} \frac{(\Im z - h)\log^+ |S(t+ih)|}{|z-t-ih|^2} dt.$$

Similar relations involving B, which we do not bother to write down, hold for $\Im z < -h$. Let us fix some value of h, say $h = 1$. We have

$$\int_{-\infty}^{\infty} \frac{\log^+ |S(t \pm \mathrm{i})|}{1+t^2} \, \mathrm{d}t \;=\; \int_{-\infty}^{\infty} \frac{\log^- |1/S(t \pm \mathrm{i})|}{1+t^2} \, \mathrm{d}t,$$

both of which are *finite*. Knowing this we can, by using the two inequalities for $\log|S(z)|$ involving integrals with $\log^+$, in $\{\Im z > 1\}$ and in $\{\Im z < -1\}$, verify immediately that $S(z)$ is *of exponential growth at most* in *each* of the two sectors $\delta < \arg z < \pi - \delta$, $\pi + \delta < \arg z < 2\pi - \delta$, $\delta > 0$ being arbitrary. This verification proceeds in the same way as the corresponding one made while proving Akhiezer's *second* theorem, §B.2.

It remains to show that $S(z)$ is of at most exponential growth in each of the two sectors $|\arg z| < \delta$, $|\arg z - \pi| < \delta$. This can be done by choosing $\delta < \pi/4$ and then following the Phragmén–Lindelöf procedure used at the end of the proof of Akhiezer's second theorem, *provided* that we know that $|S(z)| \leqslant \exp(O(|z|^2))$ for large $|z|$ in *each* of those two sectors. This property we now proceed to establish.

The method followed here is like that used to discuss $L(z)$ in the proof of the previous theorem. For $h > 0$, we have $|S(t + \mathrm{i}h)| \geqslant h$, so

$$\frac{1}{\pi} \int_{-\infty}^{\infty} \frac{\log^- |S(t+\mathrm{i}h)|}{1+t^2} \, \mathrm{d}t \;\leqslant\; \log^+ \frac{1}{h}.$$

At the same time,

$$A + \frac{1}{\pi} \int_{-\infty}^{\infty} \frac{\log |S(t+\mathrm{i}h)|}{t^2+1} \, \mathrm{d}t \;=\; \log |S(\mathrm{i} + \mathrm{i}h)|,$$

and, since $\mathrm{i} + \mathrm{i}h$ lies on the positive imaginary axis, we *already know* that $\log |S(\mathrm{i} + \mathrm{i}h)| \leqslant C(h+1)$ for $h > 0$, for the positive imaginary axis lies in the sector $\delta < \arg z < \pi - \delta$ where (at most) exponential growth of $S(z)$ *is* clear. Because $\log^+ |S(t+\mathrm{i}h)| = \log |S(t+\mathrm{i}h)| + \log^- |S(t+\mathrm{i}h)|$, the above relations yield, for $h > 0$.

$$\frac{1}{\pi} \int_{-\infty}^{\infty} \frac{\log^+ |S(t+\mathrm{i}h)|}{t^2+1} \, \mathrm{d}t \;\leqslant\; -A + C(h+1) + \log^+ \frac{1}{h}.$$

In like manner,

$$\frac{1}{\pi} \int_{-\infty}^{\infty} \frac{\log^+ |S(t-\mathrm{i}h)|}{t^2+1} \, \mathrm{d}t \;\leqslant\; -B + C'(h+1) + \log^+ \frac{1}{h}$$

when $h > 0$.

From these two inequalities we now find, for large (!) R, that

$$\frac{1}{\pi} \int_{-R}^{R} \int_{-\infty}^{\infty} \frac{\log^+ |S(t+\mathrm{i}y)|}{1+t^2} \, \mathrm{d}t \, \mathrm{d}y \;\leqslant\; \text{const.} R^2$$

(note that $\int_{-R}^{R} \log^+ (1/|y|)\,dy \leqslant$ const!). This inequality yields, in turn

$$\int_{-R}^{R}\int_{-R}^{R} \log^+ |S(z)|\,dx\,dy \;\leqslant\; \text{const.}R^4$$

for large R.

Let z_0 be given. Since $\log^+ |S(z)|$ is *subharmonic*,

$$\log^+ |S(z_0)| \;\leqslant\; \frac{1}{\pi|z_0|^2} \iint\limits_{|z-z_0|\leqslant|z_0|} \log^+ |S(z)|\,dx\,dy$$

$$\leqslant\; \frac{4}{\pi R^2}\int_{-R}^{R}\int_{-R}^{R} \log^+ |S(z)|\,dx\,dy,$$

where $R = 2|z_0|$. By what we have just seen, the expression on the *right* is $\leqslant (1/R^2)O(R^4) = O(|z_0|^2)$ for large values of $|z_0|$, i.e., $|S(z_0)| \leqslant \exp(O(|z_0|^2))$ when $|z_0|$ is large. This is what we wanted to show; as explained above, it implies that $S(z)$ is actually of exponential growth in the two sectors $|\arg z| < \delta$, $|\arg z - \pi| < \delta$, and hence, finally, that the entire function $S(z)$ *is of exponential type.*

We still have to show that

$$\int_{-\infty}^{\infty} \frac{\log^+ |S(x)|}{1+x^2}\,dx \;<\; \infty.$$

That is, however, immediate. In the course of the argument just completed, we had (taking, for instance, $h = 1$) the relation

$$\int_{-\infty}^{\infty} \frac{\log^+ |S(t+i)|}{1+t^2}\,dt \;<\; \infty.$$

Because S is of *exponential type*, the desired inequality follows from this one by §E of Chapter III (applied in the half plane $\Im z < 1$) and Fubini's theorem, in the usual fashion (hall of mirrors). We are done.

Remark. We remind the reader that, since the functions $S(z)$ considered here have no zeros either in $\Im z > 0$ or in $\Im z < 0$, the representation of §G.1, Chapter III holds for them in each of those half planes. That is,

$$\log|S(z)| \;=\; A\Im z + \frac{1}{\pi}\int_{-\infty}^{\infty} \frac{\Im z \log|S(t)|}{|z-t|^2}\,dt \quad \text{for} \quad \Im z > 0,$$

and

$$\log|S(z)| \;=\; B|\Im z| + \frac{1}{\pi}\int_{-\infty}^{\infty} \frac{|\Im z| \log|S(t)|}{|z-t|^2}\,dt \quad \text{for} \quad \Im z < 0.$$

Problem 9

Let $x_{-n} = -x_n$, let $\sum_1^\infty 1/x_n^2 < \infty$, and suppose that $\sum_{-\infty}^{\prime\infty} |1/S'(x_n)| < \infty$, where

$$S(z) = \prod_1^\infty \left(1 - \frac{z^2}{x_n^2}\right).$$

The x_n are assumed to be *real*.

Show that

$$\frac{1}{S(z)} = \sum_{-\infty}^{\infty}{}' \frac{1}{(z - x_n)S'(x_n)},$$

and hence that $S(z)$ is of *exponential type*, and that

$$\int_{-\infty}^{\infty} \frac{\log^+ |S(x)|}{1 + x^2} dx < \infty.$$

(Hint: First put $S_R(z) = \prod_{0 < x_n \leqslant R}(1 - z^2/x_n^2)$, and show that one can make $R \to \infty$ in the Lagrange formula

$$1 = \sum_{|x_n| \leqslant R} \frac{S_R(z)}{(z - x_n)S_R'(x_n)}$$

so as to obtain

$$1 = \sum_{-\infty}^{\infty}{}' \frac{S(z)}{(z - x_n)S'(x_n)}.$$

At this point, one may either invoke Krein's theorem, or else look at the Poisson representation of the (negative) harmonic function $\log|1/S(z)|$ in a suitable half-plane $\{\Im z > H\}$, noting that here $|S(z)| \leqslant S(\mathrm{i}|z|)$.)

Problem 10

Let $S(z)$ be entire, of exponential type, and satisfy the rest of the hypothesis of the *first* theorem of this article. That is, S has only the real simple zeros x_n, $\sum_n |1/S'(x_n)| < \infty$, and $S(z) \to \infty$ for z tending to ∞ along *four rays*, *one* in the *interior* of *each* of the *four quadrants*. Suppose also that the *two limits* (which *exist* by the above discussion) of $\log|S(\mathrm{i}y)|/|y|$, for $y \to \infty$ and for $y \to -\infty$, *are equal*, say to $A > 0$. The purpose of this problem is to prove that

$$\sum_n \frac{e^{\mathrm{i}\lambda x_n}}{S'(x_n)} = 0 \quad \text{for} \quad -A \leqslant \lambda \leqslant A.$$

(a) If

$$F(z) = \sum_n \frac{S(z)e^{\mathrm{i}\lambda x_n}}{(z - x_n)S'(x_n)},$$

show that $F(z)$ is entire and of exponential type A, and that

$$\int_{-\infty}^{\infty} \frac{\log^+|F(x)|}{1+x^2}\,dx \;<\; \infty.$$

(Hint: Refer to the trick with $L(z)$, pp. 203–4.)

(b) Show that $Q(z) = (F(z) - e^{i\lambda z})/S(z)$ is *entire and of exponential type*, and that

$$\int_{-\infty}^{\infty} \frac{\log^+|Q(x)|}{1+x^2}\,dx \;<\; \infty.$$

(c) If $-A < \lambda < A$ and $Q(z)$ is the function constructed in (b), show that $Q(z) \equiv 0$. (Hint: First show that $Q(iy) \to 0$ for $y \to \pm\infty$ when $-A < \lambda < A$. Use this fact and the result proved in (b) to show that

$$\limsup_{R\to\infty} \frac{\log|Q(Re^{i\varphi})|}{R} \;\leqslant\; 0$$

if $\varphi \neq 0$ or π. Then use boundedness of Q on the imaginary axis and apply the Poisson representation for $\log|Q(z)|$ (or else Phragmén–Lindelöf) in the *right* and *left* half-planes.)

(d) $\sum_{-\infty}^{\infty}(e^{i\lambda x_n}/S'(x_n)) = 0$ for $-A \leqslant \lambda \leqslant A$. (Hint: Show this for $-A < \lambda < A$ and argue by continuity. Here, one may observe that

$$\sum_{-\infty}^{\infty} \frac{e^{i\lambda x_n}}{S'(x_n)} \;=\; \lim_{y\to\infty}\left(iy\cdot\sum_{-\infty}^{\infty} \frac{e^{i\lambda x_n}}{(iy - x_n)S'(x_n)}\right),$$

and use the result of (c).)

G. Weighted approximation with L_p norms

The results established in §§A–E apply to *uniform* weighted approximation, i.e., to approximation using the norm

$$\|\varphi\|_W \;=\; \sup_{t\in\mathbb{R}} \left|\frac{\varphi(t)}{W(t)}\right|.$$

One may ask what happens if, instead of *this* norm, we use a weighted L_p one, viz.

$$\|\varphi\|_{W,p} = \sqrt[p]{\int_{-\infty}^{\infty} \left|\frac{\varphi(x)}{W(x)}\right|^p dx};$$

here, p is some number $\geqslant 1$. The answer is that *all the results except for de Branges' theorem* (§F) *carry over with hardly any change, not even in the proofs*. Here, we of course have to assume that, for $x \to \pm\infty$, $W(x) \to \infty$ rapidly enough to make

$$\int_{-\infty}^{\infty} \left(\frac{1}{W(x)}\right)^p dx \;<\; \infty.$$

(some weakening of this restriction is possible; compare with the discussion in §E.4.)

It is enough to merely peruse the proofs of Mergelian's and Akhiezer's theorems, whether for approximation by polynomials or by functions in $\mathscr{E}_A$, to see that they are applicable *as is* with the norms $\| \quad \|_{W,p}$. Here the functions $\Omega(z)$, $\Omega_A(z)$, $W_*(z)$ and $W_A(z)$ have evidently to be defined using the appropriate norm $\| \quad \|_{W,p}$ instead of $\| \quad \|_W$. And it is no longer necessarily true that $W_*(x) \leqslant W(x)$.

Verification of all this is left to the reader. In general, in the *kind* of approximation problem considered here (that of the *density* of a certain simple class of functions in the whole space), *it makes very little difference which L_p norm is chosen.* If the proofs vary in difficulty, they are hardest for the L_1 norm or for the uniform one. Here, the *continuous functions* (with the uniform norm) play the rôle of '$\lim_{p\to\infty} L_p$', and *not* L_∞, which is not even separable.

H. Comparison of weighted approximation by polynomials and by functions in $\mathscr{E}_A$

We now turn to the examination of the *relations* between the $\| \quad \|_W$-closed subspaces of $\mathscr{C}_W(\mathbb{R})$ generated by the *polynomials* and by the *linear combinations of the* $e^{i\lambda x}$, $-A \leqslant \lambda \leqslant A$, for $A > 0$.

▶ *In order to consider the former subspace, it is of course necessary to assume that*

$$x^n/W(x) \longrightarrow 0 \quad \textit{for } x \longrightarrow \pm\infty$$

when $n \geqslant 0$. This we do throughout the present §.

We also use systematically the following

Notation. $\mathscr{C}_W(0)$ is the $\| \quad \|_W$-closure of the set of polynomials in $\mathscr{C}_W(\mathbb{R})$. For $A > 0$, $\mathscr{C}_W(A)$ is the $\| \quad \|_W$-closure of the set of finite linear combinations of the $e^{i\lambda x}$; $-A \leqslant \lambda \leqslant A$. (*Equivalently, $\mathscr{C}_W(A)$ is the $\| \quad \|_W$-closure of $\mathscr{E}_A$*; see §E.1.)

It also turns out to be useful to introduce some intersections:

Definition. For $A \geqslant 0$ (*sic!*),

$$\mathscr{C}_W(A+) = \bigcap_{A' > A} \mathscr{C}_W(A').$$

In this §, we shall be especially interested in $\mathscr{C}_W(0+)$, the set of functions in $\mathscr{C}_W(\mathbb{R})$ *which can be* $\| \ \|_W$*-approximated by entire functions of arbitrarily small exponential type.*

We clearly have $\mathscr{C}_W(A) \subseteq \mathscr{C}_W(A+)$ for $A > 0$. But also:

Lemma. $\mathscr{C}_W(0) \subseteq \mathscr{C}_W(0+)$.

Proof. We have to show that $\mathscr{C}_W(0) \subseteq \mathscr{C}_W(A)$ for every $A > 0$. *Fix* any such A.

We have $x/W(x) \longrightarrow 0$ for $x \to \pm\infty$. Therefore, for the functions

$$f_h(x) \;=\; \frac{e^{i(\lambda+h)x} - e^{i\lambda x}}{h}, \qquad h > 0,$$

we have $\|f_h\|_W \leqslant \text{const.}$, $h > 0$, and $f_h(x)/W(x) \longrightarrow 0$ *uniformly* for $h > 0$ as $x \to \pm\infty$. Since $f_h(x) \longrightarrow xe^{i\lambda x}$ u.c.c. in x for $h \to 0$, we thus have $\|f_h(x) - xe^{i\lambda x}\|_W \longrightarrow 0$ as $h \to 0$, and $xe^{i\lambda x} \in \mathscr{C}_W(A)$ if $-A < \lambda < A$.

By iterating this procedure, we find that $x^n e^{i\lambda x} \in \mathscr{C}_W(A)$ for $n = 0, 1, 2, 3, \ldots$ if $-A < \lambda < A$. In particular, then, all the *powers* x^n, $n = 0, 1, 2, \ldots$, belong to $\mathscr{C}_W(A)$, so $\mathscr{C}_W(0) \subseteq \mathscr{C}_W(A)$, as required.

Remark. This *justifies* the notation $\mathscr{C}_W(0)$ for the $\| \ \|_W$-closure of *polynomials* in $\mathscr{C}_W(\mathbb{R})$.

Once we know that $\mathscr{C}_W(0) \subseteq \mathscr{C}_W(0+)$, it is natural to ask whether $\mathscr{C}_W(0) = \mathscr{C}_W(0+)$ for the weights considered in this §, and, if the equality does not hold for *all* such weights, for *which ones* it *is* true. In other words, if a given function can be $\| \ \|_W$-approximated by *entire functions of arbitrarily small exponential type*, can it be $\| \ \|_W$-approximated by *polynomials*? This question, which interested some probabilists around 1960, was studied by Levinson and McKean who used the quadratic norm $\| \ \|_{W,2}$ (§G) instead of $\| \ \|_W$, and, simultaneously and independently, by me, in terms of the uniform norm $\| \ \|_W$. I learned later, around 1967, that *I.O. Khachatrian* had done some of the same work that I had a couple of years before me, in a somewhat different way. He has a paper in the Kharkov University Mathematics and Mechanics Faculty's *Uchonye Zapiski* for 1964, and a short note in the (more accessible) 1962 *Doklady* (vol. 145).

The remainder of this § is concerned with the question of equality of the subspaces $\mathscr{C}_W(0)$ and $\mathscr{C}_W(0+)$. It turns out that *in general* they are *not equal*, but that they *are* equal when the weight $W(x)$ enjoys a certain *regularity*.

1. Characterization of the functions in $\mathscr{C}_W(A+)$

Akhiezer's second theorem (§§B.2 and E.2) generally furnishes only a partial description of the functions in $\mathscr{C}_W(A)$ when that subspace does not coincide with $\mathscr{C}_W(\mathbb{R})$. One important reason for introducing the intersections $\mathscr{C}_W(A+)$ is that we can give a *complete* description of the functions belonging to any one of them which is properly contained in $\mathscr{C}_W(\mathbb{R})$.

Lemma. *Suppose that $f(z)$ is an entire function of exponential type with*

$$|f(z)| \leqslant C_\varepsilon \exp(A|\Im z| + \varepsilon|z|)$$

for each $\varepsilon > 0$. Then, if $\delta > 0$, the Fourier transform

$$F_\delta(\lambda) = \int_{-\infty}^{\infty} e^{-\delta|x|} e^{i\lambda x} f(x)\, dx$$

belongs to $L_1(\mathbb{R})$, and, if $A' > A$,

$$(*) \qquad \int_{|\lambda| > A'} |F_\delta(\lambda)|\, d\lambda \longrightarrow 0 \quad \textit{for} \quad \delta \to 0.$$

Proof. For each $\delta > 0$, $e^{-\delta|x|} f(x)$ is in $L_1(\mathbb{R})$ (choose $\varepsilon < \delta$ in the given condition on $f(x)$), so $F_\delta(\lambda)$ is *continuous* and *therefore* integrable on $[-A', A']$. The whole lemma will thus follow as soon as we prove $(*)$.

Fix $A' > A$, and suppose for the moment that $\delta > 0$ is also fixed. Take an $\varepsilon > 0$ *less than* both $\delta/2$ and $(A' - A)/2$. If $\lambda \geqslant A'$, we then have, for $y = \Im z \geqslant 0$,

$$|e^{-\delta z} e^{i\lambda z} f(z)| \leqslant C_\varepsilon e^{(A - A')y - \delta x + \varepsilon|z|},$$

and, for $x = \Re z \geqslant 0$, this is in turn $< C_\varepsilon e^{-\varepsilon x - \varepsilon y}$.

Let us now apply Cauchy's theorem using the following contour Γ_R:

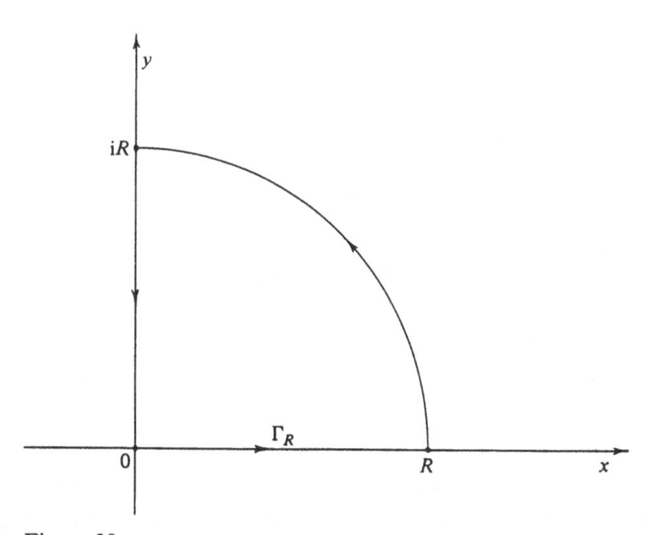

Figure 38

We have $\int_{\Gamma_R} e^{-\delta z} e^{i\lambda z} f(z) dx = 0$. For large R, $|e^{-\delta z} e^{i\lambda z} f(z)|$ is, by the preceding inequality, $< C_\varepsilon e^{-\varepsilon R/\sqrt{2}}$ on the *circular* part of Γ_R. Therefore the portion of our integral taken along this circular part tends to *zero* as $R \to \infty$, and we see that

$$\int_0^\infty e^{-\delta x} e^{i\lambda x} f(x) dx \quad = \quad i \int_0^\infty e^{-i\delta y} e^{-\lambda y} f(iy) dy.$$

This formula is valid whenever $\lambda \geqslant A' > A$ and $\delta > 0$. By integrating around the following contour

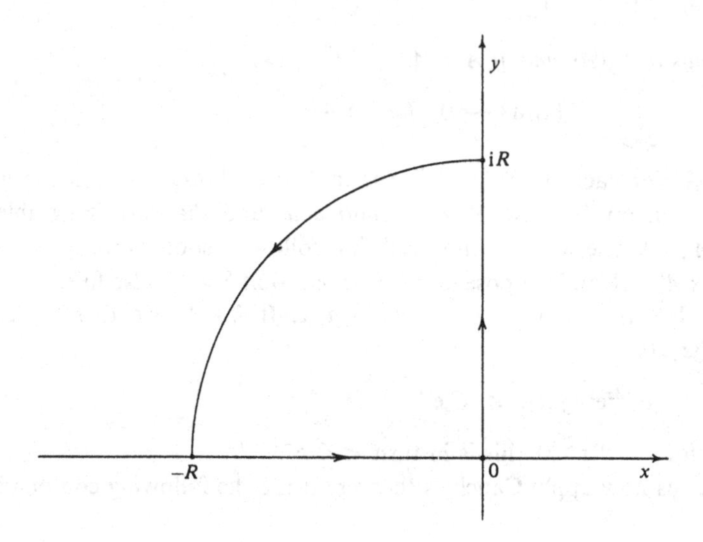

Figure 39

we see in like manner, on making $R \to \infty$, that

$$\int_{-\infty}^0 e^{\delta x} e^{i\lambda x} f(x) dx \quad = \quad -i \int_0^\infty e^{i\delta y} e^{-\lambda y} f(iy) dy,$$

whenever $\delta > 0$ and $\lambda \geqslant A' > A$. Combining this with the previous formula we get

$$F_\delta(\lambda) \quad = \quad \int_{-\infty}^\infty e^{-\delta|x|} e^{i\lambda x} f(x) dx \quad = \quad 2 \int_0^\infty e^{-\lambda y} \sin \delta y f(iy) dy;$$

this holds whenever $\lambda \geqslant A' > A$ and $\delta > 0$.

Take now any η, $0 < \eta < (A' - A)/2$, and *fix* it for the following computation. Since, for $y \geqslant 0$, $|f(iy)| \leqslant C_\eta e^{(A+\eta)y}$, the formula just derived yields, for $\lambda \geqslant A'$,

$$|F_\delta(\lambda)| \leqslant 2\int_0^\infty C_\eta e^{(A+\eta-\lambda)y}|\sin\delta y|\,dy.$$

By Schwarz' inequality, the right-hand side is in turn

$$\leqslant 2C_\eta\sqrt{\left(\int_0^\infty e^{-(\lambda-A-\eta)y}\,dy\int_0^\infty e^{-(\lambda-A-\eta)y}\sin^2\delta y\,dy\right)}$$

For the second integral under the radical we have

$$\int_0^\infty e^{-(\lambda-A-\eta)y}\sin^2\delta y\,dy = \tfrac{1}{2}\Re\int_0^\infty e^{-(\lambda-A-\eta)y}(1-e^{2i\delta y})\,dy$$

$$= \frac{2\delta^2}{(\lambda-A-\eta)|\lambda-A-\eta-2i\delta|^2}.$$

Therefore, for $\lambda \geqslant A'$,

$$|F_\delta(\lambda)| \leqslant \frac{2\sqrt{2}C_\eta\delta}{(\lambda-A-\eta)|\lambda-A-\eta-2i\delta|} \leqslant \frac{2\sqrt{2}C_\eta\delta}{(\lambda-A-\eta)^2}.$$

And

$$\int_{A'}^\infty |F_\delta(\lambda)|\,d\lambda \leqslant \frac{2\sqrt{2}C_\eta\delta}{A'-A-\eta} \leqslant \frac{2\sqrt{2}C_\eta\delta}{\eta}.$$

We see now that

$$\int_{A'}^\infty |F_\delta(\lambda)|\,d\lambda \longrightarrow 0 \quad \text{for} \quad \delta\to 0.$$

Working with contours in the *lower* half plane, we see in the same way that

$$\int_{-\infty}^{-A'} |F_\delta(\lambda)|\,d\lambda \longrightarrow 0 \quad \text{as} \quad \delta\to 0.$$

We have proved (*), and are done.

Theorem (de Branges). *Let $f(z)$ be entire, and suppose that*

$$|f(z)| \leqslant C_\varepsilon e^{A|\Im z|+\varepsilon|z|}$$

for each $\varepsilon > 0$. Then,

$$\text{if} \quad f\in\mathscr{C}_W(\mathbb{R}), \quad f\in\mathscr{C}_W(A+).$$

Proof. We have to show that, if $f\in\mathscr{C}_W(\mathbb{R})$, then in fact $f\in\mathscr{C}_W(A')$ for each $A' > A$; this we do by duality.

Fix any $A' > A$. According to the Hahn–Banach theorem it is enough to show that if L is any bounded linear functional on functions of the form $\varphi(t)/W(t)$ with $\varphi \in \mathscr{C}_W(\mathbb{R})$, and if

$$L\left(\frac{e^{i\lambda t}}{W(t)}\right) = 0 \quad \text{for} \quad -A' \leqslant \lambda \leqslant A',$$

then

$$L\left(\frac{f(t)}{W(t)}\right) = 0.$$

To see this, observe in the first place that $\| f(t) - e^{-\delta|t|}f(t) \|_W \longrightarrow 0$ for $\delta \to 0$, so surely

$$L\left(\frac{f(t)}{W(t)}\right) = \lim_{\delta \to 0} L\left(\frac{e^{-\delta|t|}f(t)}{W(t)}\right).$$

Our task thus reduces to showing that the limit on the right is zero; this we do with the help of the above lemma.

Writing, as in the lemma,

$$F_\delta(\lambda) = \int_{-\infty}^{\infty} e^{-\delta|x|} e^{i\lambda x} f(x)\, dx,$$

we have $F_\delta \in L_1(\mathbb{R})$ as we have seen. Hence, by the Fourier inversion formula,

$$e^{-\delta|t|}f(t) = \frac{1}{2\pi}\int_{-\infty}^{\infty} e^{-i\lambda t} F_\delta(\lambda)\, d\lambda.$$

In order to bring the functional L into play, we approximate the integral on the right by *finite sums*.

Put

$$S_N(t) = \frac{1}{2\pi}\sum_{k=-N^2}^{N^2-1} e^{-i(k/N)t} \int_{k/N}^{(k+1)/N} F_\delta(\lambda)\, d\lambda;$$

since $F_\delta \in L_1(\mathbb{R})$,

$$S_N(t) \longrightarrow \frac{1}{2\pi}\int_{-\infty}^{\infty} e^{-i\lambda t} F_\delta(\lambda)\, d\lambda = e^{-\delta|t|}f(t)$$

u.c.c. in t as $N \to \infty$, and, at the same time, $|S_N(t)| \leqslant \| F_\delta \|_1$ on $\mathbb{R}$ for all N. Therefore, since $W(t) \to \infty$ for $t \to \pm\infty$,

$$\| e^{-\delta|t|}f(t) - S_N(t) \|_W \underset{N}{\longrightarrow} 0,$$

so, by the *boundedness* of L,

$$L\left(\frac{e^{-\delta|t|}f(t)}{W(t)}\right) = \lim_{N\to\infty} L\left(\frac{S_N(t)}{W(t)}\right).$$

However, $\|e^{-i\lambda t} - e^{-i\lambda' t}\|_W \longrightarrow 0$ when $|\lambda - \lambda'| \to 0$, so $L(e^{-i\lambda t}/W(t))$ is a *continuous function* of λ on $\mathbb{R}$ as well as being *bounded* there (note that $|e^{i\lambda t}| = 1$!). Hence, since $F_\delta(\lambda) \in L_1(\mathbb{R})$, we have

$$L\left(\frac{S_N(t)}{W(t)}\right) = \frac{1}{2\pi}\sum_{k=-N^2}^{N^2-1} L\left(\frac{e^{-i(k/N)t}}{W(t)}\right)\int_{k/N}^{(k+1)/N} F_\delta(\lambda)\,d\lambda$$

$$\longrightarrow \frac{1}{2\pi}\int_{-\infty}^{\infty} L\left(\frac{e^{-i\lambda t}}{W(t)}\right)F_\delta(\lambda)\,d\lambda$$

for $N \to \infty$. In view of the previous relation, we thus get

$$L\left(\frac{e^{-\delta|t|}f(t)}{W(t)}\right) = \frac{1}{2\pi}\int_{-\infty}^{\infty} L\left(\frac{e^{-i\lambda t}}{W(t)}\right)F_\delta(\lambda)\,d\lambda.$$

We are assuming that

$$L\left(\frac{e^{-i\lambda t}}{W(t)}\right) = 0 \quad \text{for} \quad -A' \leqslant \lambda \leqslant A'.$$

The integral on the right thus reduces to

$$\frac{1}{2\pi}\int_{|\lambda|\geqslant A'} L\left(\frac{e^{-i\lambda t}}{W(t)}\right)F_\delta(\lambda)\,d\lambda.$$

Here, as already noted,

$$|L(e^{-i\lambda t}/W(t))| \leqslant \text{const.}, \quad \lambda \in \mathbb{R},$$

so the last integral is bounded in absolute value by

$$\text{const.}\int_{|\lambda|\geqslant A'} |F_\delta(\lambda)|\,d\lambda.$$

This, however, tends to 0 by the lemma as $\delta \to 0$. We see that

$$L\left(\frac{e^{-\delta|t|}f(t)}{W(t)}\right) \longrightarrow 0$$

for $\delta \to 0$, which is what was needed. The theorem is proved.

Remark. Since we are not supposing anything about *continuity* of $W(t)$, we are *not* in general permitted to write

$$L\left(\frac{f(t)}{W(t)}\right) \quad \text{as} \quad \int_{-\infty}^{\infty} \frac{f(t)}{W(t)}\,d\mu(t)$$

with a finite (complex-valued) Radon *measure* on $\mathbb{R}$. This makes the above proof *appear* a little more involved than in the case where use of such a measure is allowed. The difference is only in the *appearance*, however. The argument with a measure *is the same*, and only *looks* simpler.

Thanks to the above result, we can strengthen Akhiezer's second theorem so as to arrive at the following *characterization of the subspaces* $\mathscr{C}_W(A+)$. Recall the definition (§E.2):

$$W_A(z) = \sup\{|f(z)|: \ f \in \mathscr{E}_A \text{ and } \|f\|_W \leqslant 1\}.$$

Then we have the

Theorem. *Let $A \geqslant 0$. Either*

$$\int_{-\infty}^{\infty} \frac{\log W_{A'}(x)}{1+x^2}\,dx \;=\; \infty$$

for every $A' > A$, in which case $\mathscr{C}_W(A+)$ is equal to $\mathscr{C}_W(\mathbb{R})$, or else $\mathscr{C}_W(A+)$ consists precisely *of all the entire functions f such that $f(x)/W(x) \longrightarrow 0$ for $x \to \pm\infty$ and*

$$|f(z)| \;\leqslant\; C_\varepsilon e^{A|\Im z| + \varepsilon|z|}$$

for each $\varepsilon > 0$.

► ***Remark.*** *In the second case, $\mathscr{C}_W(A+)$ may* still *coincide with $\mathscr{C}_W(\mathbb{R})$. (If, for example, the set of points x where $W(x) < \infty$ is sufficiently sparse. See* §C *and end of* §E.2)

Proof. For the Mergelian function $\Omega_A(z)$ defined in §E.2, we have $\Omega_A(z) \geqslant W_A(z)$, so, if the *first alternative holds*,

$$\int_{-\infty}^{\infty} \frac{\log \Omega_{A'}(x)}{1+x^2}\,dx \;=\; \infty$$

for every $A' > A$. Then, by Mergelian's second theorem, $\mathscr{C}_W(A') = \mathscr{C}_W(\mathbb{R})$ for each $A' > A$, so $\mathscr{C}_W(A+) = \mathscr{C}_W(\mathbb{R})$.

The supremum $W_{A'}(z)$ is an *increasing* function of A' for each fixed z by virtue of the obvious inclusion of $\mathscr{E}_{A'}$ in $\mathscr{E}_{A''}$ when $A' \leqslant A''$. Therefore, if the *second alternative holds*, we have

$$(\dagger) \qquad \int_{-\infty}^{\infty} \frac{\log W_{A'}(x)}{1+x^2}\,dx \;<\; \infty$$

for each $A' \leqslant A_0$, some number *larger* than A.

Let $\varepsilon > 0$ be given, wlog $\varepsilon < A_0 - A$, and put $\delta = \varepsilon/2$. Then, if $f \in \mathscr{C}_W(A+)$, surely $f \in \mathscr{C}_W(A')$, where $A' = A + \delta$. For *this* A', (†) holds, so, by Akhiezer's *second theorem* (§E.2), we have

$$|f(z)| \leqslant K_\delta e^{A'|\Im z| + \delta|z|}.$$

Therefore

$$|f(z)| \leqslant K_{\varepsilon/2} e^{A|\Im z| + \varepsilon|z|}.$$

Saying that $f(x)/W(x) \longrightarrow 0$ for $x \to \pm\infty$ is simply another way of expressing the fact that $f \in \mathscr{C}_W(\mathbb{R})$. Thus, in the event of the *second alternative, all* the functions f in $\mathscr{C}_W(A+)$ *have* the two asserted properties.

However, *any* entire function f with those two properties *does* belong to $\mathscr{C}_W(A+)$. For such a function will be in $\mathscr{C}_W(\mathbb{R})$, and then *must* belong to $\mathscr{C}_W(A+)$ by the preceding theorem. The subspace $\mathscr{C}_W(A+)$ thus consists *precisely* of the functions having the two properties in question (and no others) when the second alternative holds. We are done.

Corollary. *For the intersections* $\mathscr{C}_W(A+)$ *the following alternative holds: Either* $\mathscr{C}_W(A+) = \mathscr{C}_W(\mathbb{R})$, *or, if* $\mathscr{C}_W(A+) \neq \mathscr{C}_W(\mathbb{R})$, *the former space consists* precisely *of the entire functions* $f(z)$ *belonging to* $\mathscr{C}_W(\mathbb{R})$ *with*

$$|f(z)| \leqslant C_\varepsilon e^{A|\Im z| + \varepsilon|z|}$$

for each $\varepsilon > 0$.

Remark. Even when $\mathscr{C}_W(A+) = \mathscr{C}_W(\mathbb{R})$, *all* the functions in $\mathscr{C}_W(A+)$ *may* have the form described in the *second* clause of this statement. That happens when $\mathscr{C}_W(\mathbb{R})$ consists entirely of the restrictions of such functions to the set of real x where $W(x) < \infty$. See remark following the statement of the preceding theorem.

2. Sufficient conditions for equality of $\mathscr{C}_W(0)$ and $\mathscr{C}_W(0+)$

Lemma. *Let* $w(z) = c\prod_1^N (z - a_k)$, *where the* a_k *are distinct, with* $\Im a_k < 0$. *Let* $g(z)$ *be an entire function of exponential type* $\leqslant A$ *with* $|(x + i)g(x)|$ *bounded for real* x. *Then, for* $x \in \mathbb{R}$,

$$\left| \frac{e^{-iAx} g(x)}{w(x)} - \sum_{k=1}^{N} \frac{g(a_k) e^{-iAa_k}}{w'(a_k)(x - a_k)} \right|$$
$$\leqslant \frac{e}{\pi} \int_{-\infty}^{\infty} \left| \frac{g(t + (i/A))}{w(t + (i/A))} \right| \left| \frac{\sin A(x - t)}{x - t - (i/A)} \right| dt.$$

Proof. $(z + i)g(z)$ is of exponential type $\leqslant A$ and is bounded on $\mathbb{R}$, hence has modulus $\leqslant \text{const.} e^{A|\Im z|}$ by the third Phragmén–Lindelöf theorem of §C,

Chapter III. Hence

$$(*) \qquad |g(z)| \leqslant \text{const.}\frac{e^{A|\Im z|}}{|z+i|}.$$

We are going to use (∗) together with some contour integration.

Fix $b > 0$, take any large R, and let Γ_R be the following contour:

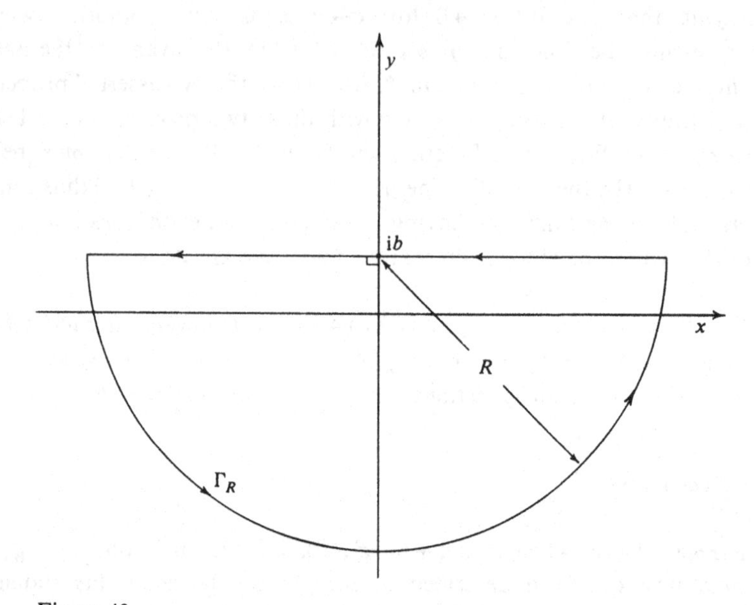

Figure 40

If R is large enough for Γ_R to encircle all the a_k and the real point x, the calculus of residues gives

$$\frac{1}{2\pi i}\int_{\Gamma_R}\frac{g(\zeta)e^{iA(x-\zeta)}}{W(\zeta)(x-\zeta)}d\zeta = \sum_{k=1}^{N}\frac{g(a_k)e^{iA(x-a_k)}}{W'(a_k)(x-a_k)} - \frac{g(x)}{W(x)}.$$

By (∗), $|g(\zeta)e^{-iA\zeta}|$ is $O(1/(|\zeta|-1)) = O(1/R)$ on the semi-circular part of Γ_R, so, as $R \to \infty$, the portion of the integral taken along that part of the contour tends to zero. Therefore

$$\frac{1}{2\pi i}\int_{-\infty}^{\infty}\frac{g(t+ib)e^{iA(x-t-ib)}}{W(t+ib)(x-t-ib)}dt = \frac{g(x)}{W(x)} - \sum_k\frac{g(a_k)e^{iA(x-a_k)}}{W'(a_k)(x-a_k)}.$$

We rewrite this relation as follows:

$$(\ast\ast) \qquad \frac{e^{Ab}}{2\pi i}\int_{-\infty}^{\infty}\frac{g(t+ib)e^{iA(x-t)}}{W(t+ib)(x-t-ib)}dt = \frac{g(x)}{W(x)} - \sum_k\frac{g(a_k)e^{iA(x-a_k)}}{W'(a_k)(x-a_k)}.$$

Let now Γ'_R be the contour obtained by reflecting Γ_R in the line $\Im z = b$:

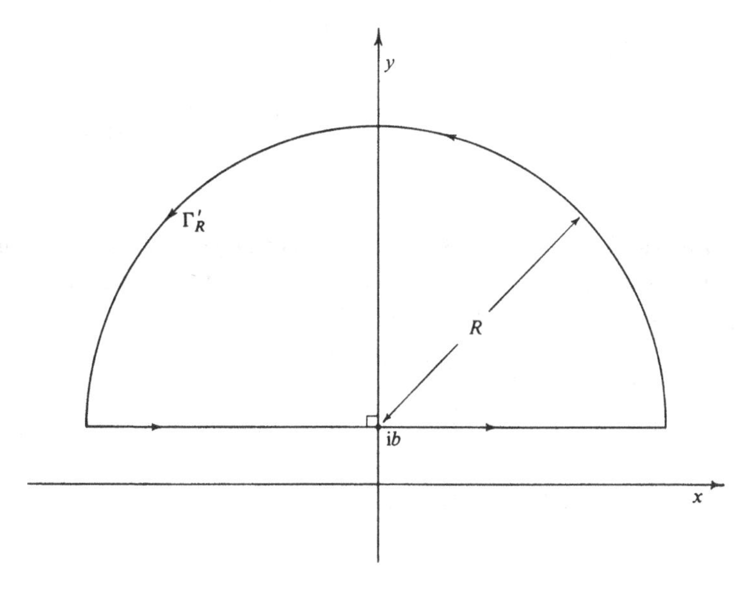

Figure 41

We have

$$\int_{\Gamma'_R} \frac{g(\zeta)e^{-iA(x-\zeta)}}{w(\zeta)(x-\zeta)} d\zeta \;=\; 0.$$

Here, $|g(\zeta)e^{iA\zeta}| = O(1/R)$ on the semi-circular part of Γ'_R, so, making $R \to \infty$, we get

$$\int_{-\infty}^{\infty} \frac{g(t+ib)e^{-iA(x-t-ib)}}{w(t+ib)(x-t-ib)} dt \;=\; 0,$$

that is,

$$\int_{-\infty}^{\infty} \frac{g(t+ib)e^{-iA(x-t)}}{w(t+ib)(x-t-ib)} dt \;=\; 0.$$

Multiplying the last relation by $e^{Ab}/2\pi i$ and subtracting the result from the left side of ($\ast\ast$), we find

$$\frac{e^{Ab}}{\pi}\int_{-\infty}^{\infty} \frac{g(t+ib)\sin A(x-t)}{w(t+ib)(x-t-ib)} dt \;=\; \frac{g(x)}{w(x)} \;-\; \sum_k \frac{g(a_k)e^{iA(x-a_k)}}{w'(a_k)(x-a_k)}.$$

Now put $b = 1/A$ and multiply what has just been written by e^{-iAx}. After taking absolute values, we see that

$$\left|\frac{g(x)e^{-iAx}}{w(x)} - \sum_k \frac{g(a_k)e^{-iAa_k}}{w'(a_k)(x-a_k)}\right|$$

$$\leqslant \frac{e}{\pi}\int_{-\infty}^{\infty}\left|\frac{g(t+(i/A))}{w(t+(i/A))}\right|\left|\frac{\sin A(x-t)}{x-t-(i/A)}\right| dt,$$

Q.E.D.

Corollary. *Let $w(z)$ be as in the lemma, and suppose that $f(z)$ is entire, of exponential type $\leqslant A/2$, and bounded on $\mathbb{R}$. Then there is a polynomial $P(z)$ of degree less than that of $w(z)$, such that*

$$\left|\frac{P(x) - e^{-iAx}\dfrac{\sin(Ax/2)}{(Ax/2)}f(x)}{w(x)}\right| \leqslant \frac{Ke}{\pi}\sup_{t\in\mathbb{R}}\left|\frac{f(t+(i/A))}{w(t+(i/A))}\right| \quad \text{for } x\in\mathbb{R}.$$

Here K is an absolute numerical constant, whose value we do not bother to calculate.

Proof. Put $g(z) = (\sin(Az/2)/(Az/2))f(z)$; then $g(z)$ satisfies the hypothesis of the previous lemma, so, with the polynomial

$$P(x) = \sum_{k=1}^{N}\frac{w(x)g(a_k)e^{-iAa_k}}{w'(a_k)(x-a_k)},$$

we get, for $x\in\mathbb{R}$,

$$(\dagger) \qquad \left|\frac{P(x)-e^{-iAx}g(x)}{w(x)}\right| \leqslant \frac{e}{\pi}\sup_{t\in\mathbb{R}}\left|\frac{f(t+(i/A))}{w(t+(i/A))}\right|$$

$$\times\int_{-\infty}^{\infty}\left|\frac{\sin\dfrac{A}{2}\left(t+\dfrac{i}{A}\right)}{\dfrac{A}{2}\left(t+\dfrac{i}{A}\right)}\right|\left|\frac{\sin A(x-t)}{x-t-(i/A)}\right| dt.$$

In the integral on the right, make the substitutions $At/2 = \tau$, $Ax/2 = \xi$. That integral then becomes

$$2\int_{-\infty}^{\infty}\left|\frac{\sin(\tau+(i/2))}{\tau+(i/2)}\right|\left|\frac{\sin 2(\xi-\tau)}{2(\xi-\tau)-i}\right| d\tau.$$

By Schwarz, this last is

$$\leqslant 2\sqrt{\left(\int_{-\infty}^{\infty}\left|\frac{\sin(\tau+(i/2))}{\tau+(i/2)}\right|^2 d\tau\cdot\int_{-\infty}^{\infty}\frac{\sin^2 2(\tau-\xi)}{4(\tau-\xi)^2}d\tau\right)},$$

a *finite quantity* – call it K – *independent of* ξ, hence (clearly) *independent* of A and x.

The right-hand side of (†) is thus bounded above by

$$K\cdot\frac{\mathrm{e}}{\pi}\sup_{t\in\mathbb{R}}\left|\frac{f(t+(\mathrm{i}/A))}{w(t+(\mathrm{i}/A))}\right|,$$

and the corollary is established.

Theorem. *Let* $W(x)=\sum_0^\infty a_{2k}x^{2k}$, where the a_{2k} are all $\geqslant 0$, *with* $a_0\geqslant 1$ *and* $a_{2k}>0$ *for infinitely many values of* k. *Then* $\mathscr{C}_W(0)=\mathscr{C}_W(0+)$.

Remark. We require $a_0\geqslant 1$ because our weights $W(x)$ are supposed to be $\geqslant 1$. We require $a_{2k}>0$ for infinitely many k because $W(x)$ is supposed to go to ∞ faster than any polynomial as $x\to\pm\infty$.

Proof of theorem. Let $\varphi\in\mathscr{C}_W(0+)$. Then there are finite sums

$$f_n(x) \;=\; \sum_{-1/2n\leqslant\lambda\leqslant 1/2n} a_n(\lambda)\mathrm{e}^{\mathrm{i}\lambda x}$$

with

$$\|f_n-\varphi\|_W \;\xrightarrow[n]{}\; 0.$$

We put $g_n(x)=(\sin(x/2n)/(x/2n))f_n(x)$, and set out to apply the above corollary with $f=f_n$ and suitable polynomials w. Note that f_n is entire, of exponential type $1/2n$, and bounded on the real axis. Since $\|f_n-\varphi\|_W \xrightarrow[n]{} 0$, we also have

$$\|\mathrm{e}^{-\mathrm{i}x/n}g_n(x)-\varphi(x)\|_W \;\xrightarrow[n]{}\; 0,$$

in view of the fact that $W(x)\to\infty$ for $x\to\pm\infty$.*

Choose any $\varepsilon>0$. The norms $\|f_n\|_W$ must be bounded; wlog $|f_n(x)/W(x)|\leqslant 1$, say, for $x\in\mathbb{R}$ and every n. For the function $\varphi\in\mathscr{C}_W(\mathbb{R})$ we of course have $\varphi(x)/W(x)\xrightarrow[n]{}0$ for $x\to\pm\infty$, so, since $\|f_n-\varphi\|_W\xrightarrow[n]{}0$, *there must be an* L (depending on ε) such that

$$\left|\frac{f_n(x)}{W(x)}\right| \;\leqslant\; \varepsilon \quad\text{for}\quad |x|\geqslant L$$

whenever n *is sufficiently large*. Take such an L and *fix* it.

Pick any n large enough for the previous relation to be true, and *fix* it *for the moment*. Our individual function $f_n(x)$ is *bounded* on $\mathbb{R}$ (true, with perhaps an enormous bound!), so, for some N_0, we will surely have

$$\left|f_n(x)\Big/\sum_0^{N_0}a_{2k}x^{2k}\right| \;<\; \varepsilon \quad\text{for}\quad |x|>A,\ \text{say},$$

* Note that $\|\mathrm{e}^{-\mathrm{i}x/n}(\sin(x/2n)/(x/2n))\varphi(x) \;-\; \varphi(x)\|_W \;\xrightarrow[n]{}\; 0$ for any $\varphi\in\mathscr{C}_W(\mathbb{R})$.

where, wlog, $A > L$, the number chosen above. Also,

$$1 \leqslant \sum_0^N a_{2k}x^{2k} \underset{N}{\longrightarrow} W(x),$$

the sums on the left being monotone increasing with N ($a_{2k} \geqslant 0$!). Therefore,

$$\left| f_n(x) \Big/ \sum_0^N a_{2k}x^{2k} \right| \underset{N}{\longrightarrow} |f_n(x)/W(x)|$$

uniformly for $-A \leqslant x \leqslant A$, and, if $N \geqslant N_0$ is large enough, we have, in view of the previous inequality,

$$(\S) \qquad \left| f_n(x) \Big/ \sum_0^N a_{2k}x^{2k} \right| \leqslant 2 \quad \text{for} \quad x \in \mathbb{R}$$

(since $\| f_n \|_W \leqslant 1$), and also

$$(\dagger\dagger) \qquad \left| f_n(x) \Big/ \sum_0^N a_{2k}x^{2k} \right| \leqslant 2\varepsilon \quad \text{for} \quad |x| \geqslant L.$$

Fix such an N for the moment (it depends of course on n which we have already fixed!), and call

$$V(x) = \sum_0^N a_{2k}x^{2k}.$$

Because $V(x) \geqslant 1$ on $\mathbb{R}$, we can find another polynomial $w(x)$, *with all its zeros in* $\Im z < 0$, such that $|w(x)| = V(x)$, $x \in \mathbb{R}$.

There is no loss of generality in supposing that the zeros of w are *distinct*. There are, in any case, a finite number ($2N$) of them, lying in the open lower half plane. Separating each *multiple* zero (if there are any*) into a cluster of *simple* ones, *very close together*, will change $w(x)$ to a polynomial $\tilde{w}(x)$ having the new zeros, and such that

$$(1-\delta)|w(x)| \leqslant |\tilde{w}(x)| \leqslant (1+\delta)|w(x)|$$

on $\mathbb{R}$, with $\delta > 0$ *as small as we like*. One may then run through the following argument with $\tilde{w}$ in place of w; the effect of this will merely be to render the final inequality worse by a harmless factor of $(1+\delta)/(1-\delta)$.

Let us proceed, then, assuming that the zeros of w *are* simple. Desiring, as we do, to use the above corollary, we need an estimate for

$$\sup_{t\in\mathbb{R}} \left| \frac{f_n(t+in)}{w(t+in)} \right|.$$

The function $e^{iz/2n} f_n(z)/w(z)$ is *analytic and bounded* for $\Im z > 0$, and continuous up to $\mathbb{R}$. Therefore we can use Poisson's formula (lemma of

* and there *are*! – *all* zeros of $w(z)$ are of *even order*!

§H.1, Chapter III), getting

$$\frac{e^{i(t+in)/2n}f_n(t+in)}{w(t+in)} = \frac{1}{\pi}\int_{-\infty}^{\infty}\frac{n}{(t-x)^2+n^2}\cdot\frac{e^{ix/2n}f_n(x)}{w(x)}dx.$$

Since $|w(x)| = V(x)$, we see, by (§) and (††), that the integral on the right is in absolute value

$$\leqslant \frac{2}{\pi}\int_{-L}^{L}\frac{n}{(t-x)^2+n^2}dx + \frac{2\varepsilon}{\pi}\int_{|x|\geqslant L}\frac{n}{(t-x)^2+n^2}dx.$$

This is in turn

$$\leqslant \frac{2}{\pi}\left(\arctan\frac{t+L}{n} - \arctan\frac{t-L}{n}\right) + 2\varepsilon \leqslant \frac{4L}{\pi n} + 2\varepsilon,$$

so we have

$$\left|\frac{e^{-1/2}f_n(t+in)}{w(t+in)}\right| \leqslant \frac{4L}{\pi n} + 2\varepsilon, \quad t\in\mathbb{R}.$$

Apply now the corollary with $f = f_n$ and $A = 1/n$. According to it and to the inequality just proved, there is a *polynomial* $P_n(x)$ (depending, of course, partly on our $w(x)$ whose choice *also* depended on the n we have taken!) such that, for $x\in\mathbb{R}$,

$$\left|\frac{P_n(x) - e^{-ix/n}g_n(x)}{w(x)}\right| \leqslant \frac{Ke}{\pi}\cdot e^{1/2}\left(\frac{4L}{\pi n} + 2\varepsilon\right),$$

where (as we recall),

$$g_n(x) = \frac{\sin(x/2n)}{(x/2n)}f_n(x).$$

Therefore, since $|w(x)| = V(x) \leqslant W(x)$ (!),

$$\left|\frac{P_n(x) - e^{-ix/n}g_n(x)}{W(x)}\right| \leqslant \frac{Ke^{3/2}}{\pi}\left(\frac{4L}{\pi n} + 2\varepsilon\right), \quad x\in\mathbb{R}.$$

Our number L depended *only* on ε, and the intermediate partial sum

$$V(x) = \sum_{0}^{N} a_{2k}x^{2k}$$

(which depended on n) is now *gone. The only restriction on n* (which was kept fixed during the above argument) was that it be *sufficiently large* (*how* large depended on L). For *fixed L*, then, there is, for *each* sufficiently large n, a polynomial $P_n(x)$ satisfying the above relation. If such an n is also $> 2L/\pi\varepsilon$, we will thus certainly have

$$\left|\frac{P_n(x) - e^{-ix/n}g_n(x)}{W(x)}\right| \leqslant 4\varepsilon\frac{Ke^{3/2}}{\pi}$$

for $x \in \mathbb{R}$.

Let us return to our function $\varphi \in \mathscr{C}_W(0+)$. For each *sufficiently large n*, we have a *polynomial* $P_n(x)$ with

$$\|P_n - \varphi\|_W \leqslant \|\varphi(x) - e^{-ix/n}g_n(x)\|_W + \|e^{-ix/n}g_n(x) - P_n(x)\|_W,$$

which, according to the inequality we have finally established, is

$$\leqslant 4\varepsilon\frac{Ke^{3/2}}{\pi} + \|e^{-ix/n}g_n(x) - \varphi(x)\|_W.$$

However, $\|e^{-ix/n}g_n(x) - \varphi(x)\|_W \underset{n}{\longrightarrow} 0$ by choice of our functions f_n. Therefore $\|P_n - \varphi\|_W \leqslant 8\varepsilon(Ke^{3/2}/\pi)$ for all sufficiently large n. Since $\varepsilon > 0$ was arbitrary, we have, then, $\|P_n - \varphi\|_W \underset{n}{\longrightarrow} 0$, and $\varphi \in \mathscr{C}_W(0)$.

This proves that $\mathscr{C}_W(0+) \subseteq \mathscr{C}_W(0)$. Since the reverse inclusion is always true, we are done.

Remark. An analogous result holds for approximation in the norms $\|\ \ \|_{W,p}$, $1 < p < \infty$. There, a much *easier* proof can be given, based on duality and the fact that the Hilbert transform is a bounded operator on $L_p(\mathbb{R})$ for $1 < p < \infty$. The reader is encouraged to try to work out such a proof.

We can apply the technique of convex logarithmic regularisation developed in Chapter IV together with the theorem just proved so as to obtain another result in which a *regularity condition* on $W(x)$ replaces the explicit representation for it figuring above.

Theorem. *Let* $W(x) \geqslant 1$ *be* even, *with* $\log W(x)$ *a* convex *function of* $\log x$ *for* $x > 0$. *Suppose that for* each $\Lambda > 1$ *there is a constant* C_Λ *such that*

$$x^2 W(x) \leqslant C_\Lambda W(\Lambda x), \quad x \in \mathbb{R}.$$

Then $\mathscr{C}_W(0) = \mathscr{C}_W(0+)$.

Remark. Speaking, as we are, of $\mathscr{C}_W(0)$, we of course require that $x^n/W(x) \longrightarrow 0$ for $x \to \pm\infty$ and all $n \geqslant 0$, so $W(x)$ must tend to ∞ fairly rapidly as $x \to \pm\infty$. But one *cannot derive* the condition involving numbers $\Lambda > 1$ from this fact and the convexity of $\log W(x)$ in $\log|x|$. Nor have I been able to *dispense* with that ungainly condition.

Proof of theorem. Let us first show that, if $\varphi \in \mathscr{C}_W(\mathbb{R})$ and we write $\varphi_\lambda(x) = \varphi(\lambda^2 x)$ for $\lambda < 1$, then $\|\varphi - \varphi_\lambda\|_W \longrightarrow 0$ as $\lambda \to 1$.

We know that $\log W(x)$ tends to ∞ as $x \to \pm\infty$. Hence, since that function is *convex* in $\log x$ for $x > 0$, it must be *increasing* in x for all *sufficiently large* x. Take any $\varphi \in \mathscr{C}_W(\mathbb{R})$; since φ is continuous on $\mathbb{R}$ we certainly have $|\varphi(x) - \varphi_\lambda(x)| \longrightarrow 0$ uniformly on *any* interval $[-M, M]$ as $\lambda \to 1$. Also, $|\varphi(x)/W(x)| < \varepsilon$ for $|x|$ sufficiently large. Choose M big enough so that this inequality holds for $|x| \geqslant M/4$ and also $W(x)$ increases for $x \geqslant M/4$. Then, if $\frac{1}{2} < \lambda < 1$ and $|x| \geqslant M$,

$$\left|\frac{\varphi_\lambda(x)}{W(x)}\right| = \left|\frac{\varphi(\lambda^2 x)}{W(x)}\right| \leqslant \left|\frac{\varphi(\lambda^2 x)}{W(\lambda^2 x)}\right| < \varepsilon,$$

as well as $|\varphi(x)/W(x)| < \varepsilon$, so

$$\left|\frac{\varphi(x) - \varphi_\lambda(x)}{W(x)}\right| < 2\varepsilon$$

for $|x| \geqslant M$ and $\frac{1}{2} < \lambda < 1$. Making λ close enough to 1, we get the quantity on the left $< 2\varepsilon$ for $-M \leqslant x \leqslant M$ also, so $\|\varphi - \varphi_\lambda\|_W < 2\varepsilon$.

Take now any $\varphi \in \mathscr{C}_W(0+)$. We have to show that φ also belongs to $\mathscr{C}_W(0)$, and, by what we have just proved, this will follow if we establish that $\varphi_\lambda \in \mathscr{C}_W(0)$ for *each* $\lambda < 1$. We proceed to verify that fact.

We may, wlog, assume that $W(x) \equiv 1$ for $|x| \leqslant 1$ and *increases* for $x \geqslant 1$. For $n = 0, 1, 2, \ldots$, put

$$S_n = \sup_{r>0} \frac{r^n}{W(r)}$$

and, then, for $r > 0$, write

$$T(r) = \sup_{n \geqslant 0} \frac{r^{2n}}{S_{2n}}.$$

Since $\log W(r)$ increases for $r \geqslant 1$, the *proof* of the *second* lemma from §D of Chapter IV shows that

$$\frac{W(r)}{r^2} \leqslant T(r) \leqslant W(r) \quad \text{for} \quad r \geqslant 1$$

(cf. proof of second theorem in §D, this chapter). Take now

$$(\S\S) \qquad S(x) = 1 + \sum_{n=0}^{\infty} \frac{x^{2n+2}}{S_{2n}}.$$

Then, by the preceding inequalities, for $|x| \geqslant 1$,

$$S(x) \geqslant x^2 T(|x|) \geqslant W(x)$$

whilst, for any λ, $0<\lambda<1$,

$$S(x) \;=\; 1+x^2\sum_0^\infty \lambda^{2n}\frac{(x/\lambda)^{2n}}{S_{2n}} \;\leqslant\; 1+\frac{x^2}{1-\lambda^2}T\left(\frac{|x|}{\lambda}\right)$$

$$\leqslant\; 1+\frac{x^2}{1-\lambda^2}W\left(\frac{x}{\lambda}\right).$$

The first of these relations* clearly also holds for $|x|\leqslant 1$, because $W(x)\equiv 1$ there. *So does the second.* For, the inequality between its last two members is true for $|x|\geqslant\lambda$, while $T(|x|/\lambda)$ is, by its definition, *increasing* when $0<|x|<\lambda$, and $W(x/\lambda)$ *constant* for such x. We thus have

$$W(x) \;\leqslant\; S(x) \;\leqslant\; 1+\frac{x^2}{1-\lambda^2}W\left(\frac{x}{\lambda}\right)$$

for all x.

According to the hypothesis, there is a constant K_λ for each $\lambda<1$ with

$$\frac{x^2}{1-\lambda^2}W\left(\frac{x}{\lambda}\right) \;\leqslant\; K_\lambda W\left(\frac{x}{\lambda^2}\right).$$

We may of course take $K_\lambda\geqslant 1$, and thus get finally

$$(\ddagger) \qquad W(x) \;\leqslant\; S(x) \;\leqslant\; 2K_\lambda W\left(\frac{x}{\lambda^2}\right), \quad x\in\mathbb{R}.$$

Given our function $\varphi\in\mathscr{C}_W(0+)$, we have a sequence of functions f_n, $f_n\in\mathscr{E}_{1/n}$, with

$$\|\varphi-f_n\|_W \;\underset{n}{\longrightarrow}\; 0.$$

Thence, by $(\ddagger)$, *a fortiori*,

$$\|\varphi-f_n\|_S \;\underset{n}{\longrightarrow}\; 0,$$

so $\varphi\in\mathscr{C}_S(0+)$ as well. Now, however, $S(x)$ has the form (§§), so *we may apply the previous theorem, getting* $\varphi\in\mathscr{C}_S(0)$. There is thus a sequence of *polynomials* $P_n(x)$ with

$$\|\varphi-P_n\|_S \;\underset{n}{\longrightarrow}\; 0.$$

From this we see, by $(\ddagger)$ again, that

$$\sup_{x\in\mathbb{R}}\left|\frac{\varphi(x)-P_n(x)}{W(x/\lambda^2)}\right| \;\underset{n}{\longrightarrow}\; 0,$$

* i.e., that between $S(x)$ and $W(x)$

i.e.,

$$\sup_{x\in\mathbb{R}} \left| \frac{\varphi(\lambda^2 x) - P_n(\lambda^2 x)}{W(x)} \right| \xrightarrow[n]{} 0$$

for each λ, $0<\lambda<1$.

But this means that $\varphi_\lambda \in \mathscr{C}_W(0)$ for each such λ, the fact we had to verify. The theorem is proved.

Remark. *Some* regularity in $W(x)$ is *necessary* for the equality of $\mathscr{C}_W(0)$ and $\mathscr{C}_W(0+)$; exactly *what kind* is not yet known. In the next article we give an example showing that the behaviour of $W(x)$ *cannot depart too much* from that required in the above two theorems if we are to have $\mathscr{C}_W(0) = \mathscr{C}_W(0+)$. In the first part of the next chapter we will give another example, of an even weight $W(x)$, *increasing* for $x>0$, such that $\mathscr{C}_W(0) \neq \mathscr{C}_W(0+) = \mathscr{C}_W(\mathbb{R})$.

3. Example of a weight W with $\mathscr{C}_W(0) \neq \mathscr{C}_W(0+) \neq \mathscr{C}_W(\mathbb{R})$

The idea for this example comes from a letter J.-P. Kahane sent me in 1963.

Take

$$S(z) = \prod_1^\infty \left(1 - \frac{z^2}{4^n}\right),$$

pick any fixed number λ_1, $1<\lambda_1<2$, and write

$$C(z) = \left(1 - \frac{z^2}{\lambda_1^2}\right) \prod_2^\infty \left(1 - \frac{z^2}{4^n}\right).$$

This function $S(z)$ is the same as the one used in §C, and $C(z)$ differs from it *only* in that the two zeros, -2 and 2, of $S(z)$ *closest to the origin* have been *moved* slightly, the first towards -1 and the second towards 1.

Let us write $\lambda_{-1} = -\lambda_1$, and, for $|n| \geqslant 2$, $\lambda_n = (\operatorname{sgn} n) 2^{|n|}$. Then

$$C(z) = \prod_1^\infty \left(1 - \frac{z^2}{\lambda_n^2}\right),$$

and

$$\sum_{-\infty}^{\infty}{}' \frac{\lambda_n S(\lambda_n)}{C'(\lambda_n)} = 2\frac{\lambda_1 S(\lambda_1)}{C'(\lambda_1)} < 0.$$

For large n, we clearly have

$$C'(\lambda_n) \sim \frac{4}{\lambda_1^2} S'(\lambda_n),$$

where $S'(\lambda_n)$ was studied in §C. There we found that

$$|S'(\lambda_n)| \;=\; |S'(2^n)| \;\sim\; \text{const.}2^{(n-1)^2}$$

for large n, so surely (in view of the evenness of $C(z)$),

$$\sum_{-\infty}^{\infty}{}' \frac{|\lambda_n|^p}{|C'(\lambda_n)|} \;<\; \infty \quad \text{for} \quad p = 0, 1, 2, 3, \ldots$$

Use of the Lagrange interpolation formula now shows, as in §C (where an analogous result was proved with $S'(\lambda_n)$ in place of $C'(\lambda_n)$), that

$$P(z) \;=\; C(z) \sum_{-\infty}^{\infty}{}' \frac{P(\lambda_n)}{(z-\lambda_n)C'(\lambda_n)}$$

for any *polynomial* P. Taking $P(z) = z^{p+1}$ and then putting $z = 0$ gives us

$$\sum_{-\infty}^{\infty}{}' \frac{\lambda_n^p}{C'(\lambda_n)} \;=\; 0, \quad p = 0, 1, 2, \ldots.$$

We are ready to construct our weight W. Taking a large constant K (chosen so as to make $W(x)$ come out $\geqslant 1$), put $W(x) = KS(1)$ for $|x| \leqslant 1$. For $|x| \geqslant 1$, make $W(x) = Kx^2|S(x)|$ when x lies *outside all the intervals*

$$[2^n(1-2^{-4n}),\quad 2^n(1+2^{-4n})].$$

Finally, if $2^n(1-2^{-4n}) \;\leqslant\; |x| \;\leqslant\; 2^n(1+2^{-4n})$ for some $n \geqslant 1$, define $W(x)$ as

$$\sup\{K\xi^2|S(\xi)|:\quad |\xi - 2^n| \leqslant 2^{-3n}\}.$$

We see first of all that $xS(x)/W(x) \longrightarrow 0$ for $x \to \pm\infty$, so $xS(x) \in \mathscr{C}_W(\mathbb{R})$. Hence, since $S(z)$, and therefore $zS(z)$, is of exponential type *zero* we have $xS(x) \in \mathscr{C}_W(0+)$ by the *first* theorem of article 1.

We need some information about the asymptotic behaviour of $S(x)$ for $x \to \infty$. This may be obtained by the method followed in §C. Suppose that $x = 2^n\alpha$ with $1/\sqrt{2} \;\leqslant\; \alpha \;\leqslant\; \sqrt{2}$. Then we have

$$\begin{aligned} |S(x)| \;&=\; \prod_{k=1}^{n-1}\left|1 - \frac{4^n\alpha^2}{4^k}\right| \times |1-\alpha^2| \times \prod_{k=n+1}^{\infty}\left|1 - \frac{4^n\alpha^2}{4^k}\right| \\ &=\; |1-\alpha^2| \prod_{k=1}^{n-1}\left(\frac{4^n\alpha^2}{4^k}\right) \prod_{l=1}^{n-1}\left(1 - \frac{1}{4^l\alpha^2}\right) \prod_{l=1}^{\infty}\left(1 - \frac{\alpha^2}{4^l}\right), \end{aligned}$$

and this last is

$$\sim\; |1-\alpha^2| \frac{(4^n\alpha^2)^{n-1}}{2^{n(n-1)}} S\left(\frac{1}{\alpha}\right) S(\alpha) \;=\; |1-\alpha^2|(2^n\alpha^2)^{n-1} S\left(\frac{1}{\alpha}\right) S(\alpha).$$

Thence, for large n,

$$\sup\{|S(\xi)|: \ |\xi - 2^n| \leqslant 2^{-3n}\} \ \sim \ \text{const.}2^{n^2-5n},$$

so, for $2^n(1-2^{-4n}) \ \leqslant \ |x| \ \leqslant \ 2^n(1+2^{-4n})$,

$$W(x) \ \sim \ \text{const.}2^{n^2-3n},$$

and, in particular,

$$W(\lambda_n) \ \sim \ \text{const.}2^{n^2-3n}.$$

Comparing this with the relation $|C'(\lambda_n)| \sim \text{const.}2^{(n-1)^2}$, valid for large n, which we already know, we see that

$$\sum_{-\infty}^{\infty}{}' \frac{W(\lambda_n)}{|C'(\lambda_n)|} \ < \ \infty.$$

This permits us to define a *finite* signed Radon measure μ on the set of points λ_n, $n = \pm 1, \pm 2, \ldots$, by putting

$$\mu(\{\lambda_n\}) \ = \ \frac{W(\lambda_n)}{C'(\lambda_n)}.$$

Then, for $p = 0, 1, 2, \ldots$,

$$\int_{-\infty}^{\infty} \frac{x^p}{W(x)} \mathrm{d}\mu(x) \ = \ \sum_{-\infty}^{\infty}{}' \frac{\lambda_n^p}{C'(\lambda_n)},$$

which is *zero*, as we have seen, whilst

$$\int_{-\infty}^{\infty} \frac{xS(x)}{W(x)} \mathrm{d}\mu(x) \ = \ \sum_{-\infty}^{\infty}{}' \frac{\lambda_n S(\lambda_n)}{C'(\lambda_n)} \ < \ 0.$$

So $xS(x) \in \mathscr{C}_W(0+)$ *can't* be in $\mathscr{C}_W(0)$, and $\mathscr{C}_W(0) \neq \mathscr{C}_W(0+)$.

In the present example, $\mathscr{C}_W(0+)$ is a *proper subspace* of $\mathscr{C}_W(\mathbb{R})$. Indeed, this is almost immediate. By the above asymptotic computation of $S(x)$ we clearly have

$$W(x) \ = \ \text{const.}|x|^{\log_2|x| + \theta(x)}$$

with a quantity $\theta(x)$ *varying between two constants.* Therefore,

$$\int_{-\infty}^{\infty} \frac{\log W(x)}{1+x^2} \mathrm{d}x \ < \ \infty,$$

so we surely have $\mathscr{C}_W(A) \neq \mathscr{C}_W(\mathbb{R})$ for each A by an obvious extension of

T. Hall's theorem (p. 169). Hence $\mathscr{C}_W(0+) \neq \mathscr{C}_W(\mathbb{R})$. The construction of our example is finished.*

Remark. Let $\Omega(x) = \prod_1^\infty (1 + x^2/4^n)$. The asymptotic evaluation of $\Omega(x)$ for $x \to \infty$ can be made in fashion similar to that for $S(x)$, and is, in fact, *easier* than the latter. As is clear after a moment's thought, here one also obtains

$$\Omega(x) \sim \text{const.}|x|^{\log_2|x| + \varphi(x)} \quad \text{for} \quad |x| \to \infty$$

with a certain $\varphi(x)$ varying between two constants. Thus,

$$\frac{\Omega(x)}{W(x)} = \text{const.}|x|^{\psi(x)}, \quad x \in \mathbb{R},$$

where $A \leqslant \psi(x) \leqslant B$, say.

However, $\mathscr{C}_\Omega(0) = \mathscr{C}_\Omega(0+)$. This follows from the first theorem of the previous article, in view of the evident fact that $\Omega(x) = 1 + a_2x^2 + a_4x^4 + \cdots$ with $a_{2k} > 0$.

The difference in behaviour of $\Omega(x)$ *and* $W(x)$ *is small in comparison to their size, and yet* $\mathscr{C}_\Omega(0) = \mathscr{C}_\Omega(0+)$ *although* $\mathscr{C}_W(0) \neq \mathscr{C}_W(0+)$.

The question of *how* a weight *W*'s *local behaviour* is related to the *equality* of $\mathscr{C}_W(0)$ and $\mathscr{C}_W(0+)$ *merits further study.*

* $W(x)$ has jump discontinuities among the points $\pm(2^n \pm 2^{-3n})$, $n \geqslant 1$, but a continuous weight with the same properties as W is furnished by an evident elaboration of the procedure in the text.

VII

How small can the Fourier transform of a rapidly decreasing non-zero function be?

Let us consider functions $F(x) \in L_1(\mathbb{R})$ whose modulus goes to zero *rapidly* as $x \to \infty$, in such fashion that

$$\int_{-\infty}^{\infty} \frac{1}{1+x^2} \log^-\left(\int_x^{\infty} |F(t)|\,dt\right) dx \quad = \quad \infty.$$

The general theme of this chapter is that, for such a function F, the Fourier transform

$$\hat{F}(\lambda) \quad = \quad \int_{-\infty}^{\infty} e^{i\lambda x} F(x)\,dx$$

cannot be too small anywhere unless F vanishes identically.

The first result of this kind (obtained by Levinson) said that if (for such an F) $\hat{F}$ *vanishes throughout an interval of positive length*, then $F \equiv 0$. This was refined by Beurling, who proved that $\hat{F}(\lambda)$ *cannot even vanish on a set of positive measure unless* $F \equiv 0$. Analogues of these theorems hold for measures as well as functions F; they, and the methods used to establish them, have various important consequences, some of which apply to material already taken up in the present book.

These things have been known for more than 20 years. Until recently, the only developments since the sixties in the subject matter of this chapter had to do mainly with aspects of its presentation. That state of affairs was changed in 1982 by the appearance of a remarkable result, due to A.L. Volberg, which says that if

$$f(\vartheta) \quad = \quad \sum_{-\infty}^{\infty} \hat{f}(n)^{in\vartheta}$$

has

$$|\hat{f}(n)| \;\leqslant\; e^{-M(n)} \quad \text{for} \quad n > 0$$

with $M(n)$ sufficiently regular and increasing, and if

$$\sum_1^\infty \frac{M(n)}{n^2} = \infty,$$

then

$$\int_{-\pi}^{\pi} \log|f(\vartheta)|\,d\vartheta > -\infty$$

unless $f(\vartheta) \equiv 0$. The proof of this uses new ideas (coming from the study of weighted *planar* approximation by polynomials) and is very long; its inclusion has necessitated a considerable extension of the present chapter. I still do not completely understand the result's meaning; it applies to the *unit circle* and seems to *not have* a natural analogue for the *real line* which would generalize Levinson's and Beurling's theorems.

There are not too many easily accessible references for this chapter. The earliest results are in Levinson's book; material relating to them can also be found in the book by de Branges (some of it being set as problems). The main source for the first two §§ of this chapter consists, however, of the famous mimeographed notes for Beurling's Standford lectures prepared by P. Duren; those notes came out around 1961. Volberg published his theorem in a 6-page (!) *Doklady* note at the beginning of 1982. That paper is quite difficult to get through on account of its being so condensed.

A. The Fourier transform vanishes on an interval. Levinson's result

Levinson originally proved his theorem by means of a complicated argument, involving contour integration, which figured later on as one of the main ingredients in Beurling's proof of his deeper result. Beurling observed that Levinson's theorem (and others related to it) could be obtained more easily by the use of test functions, and then de Branges simplified that treatment by bringing Akhiezer's first theorem from §E.2 of Chapter VI into it. I follow this procedure in the present §. The particularly convenient and elegant test function used here (which has several other applications, by the way) was suggested to me by my reading of a paper of H. Widom.

1. Some shop math

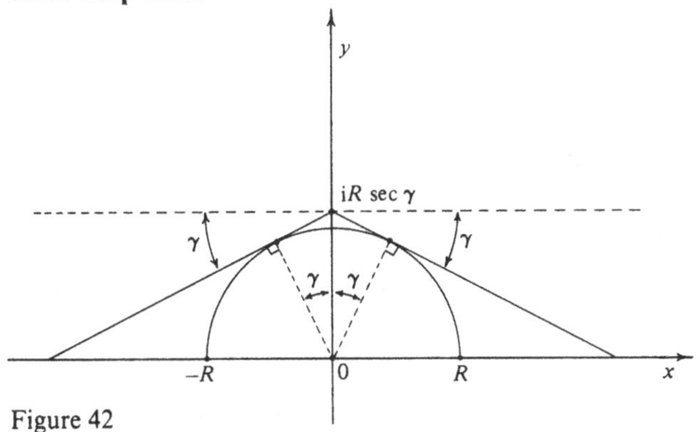

Figure 42

The circle of radius R about 0 lies *under* the two straight lines of slopes $\pm \tan\gamma$ passing through the point $iR\sec\gamma$. Therefore, if $A > 0$,

$$A\sqrt{(R^2 - x^2)} \leqslant AR\sec\gamma - (A\tan\gamma)|x|, \quad -R \leqslant x \leqslant R.$$

Consider any function $\omega(x) \geqslant 0$ such that

$$|\omega(x) - \omega(x')| \leqslant (A\tan\gamma)|x - x'|.$$

If we adjust R so as to make $AR\sec\gamma = \omega(0)$, we have, for $-R \leqslant x \leqslant R$,

$$\omega(x) \geqslant \omega(0) - (A\tan\gamma)|x|,$$

which, by the above, is $\geqslant A\sqrt{(R^2 - x^2)}$. The function $\cos(A\sqrt{(x^2 - R^2)})$ is, however, in modulus $\leqslant 1$ for $|x| > R$, and for $-R \leqslant x \leqslant R$ it equals $\cosh(A\sqrt{(R^2 - x^2)}) \leqslant \exp(A\sqrt{(R^2 - x^2)})$. Therefore, for $x \in \mathbb{R}$,

$$\omega(x) \geqslant \log|\cos(A\sqrt{(x^2 - R^2)})|$$

when

$$R = \frac{\omega(0)}{A\sec\gamma}.$$

Let us apply these considerations to a function $W(x) \geqslant 1$ defined on $\mathbb{R}$ and satisfying

$$|\log W(x) - \log W(x')| \leqslant C|x - x'|$$

there. Taking any fixed $A > 0$, we determine an acute angle γ such that $A\tan\gamma = C$. Suppose $x_0 \in \mathbb{R}$ is given. Then we translate x_0 to the origin, using the above calculation with

$$\omega(x - x_0) = \log W(x).$$

We see that

$$|\cos(A\sqrt{((x - x_0)^2 - R^2)})| \leqslant W(x) \quad \text{for} \quad x \in \mathbb{R},$$

where

$$R = \frac{\log W(x_0)}{A \sec \gamma} = \frac{\log W(x_0)}{\sqrt{(A^2+C^2)}}.$$

Here, $\cos(A\sqrt{((z-x_0)^2 - R^2)})$ *is an entire function of* z because the Taylor development of $\cos w$ about the origin contains only *even powers* of w. It is clearly *of exponential type* A, and, for $z = x_0$, has the value

$$\cosh AR \geq \tfrac{1}{2}e^{AR} = \tfrac{1}{2}(W(x_0))^{A/\sqrt{(A^2+C^2)}}.$$

Recall now the definition of the Akhiezer function $W_A(x)$ given in Chapter VI, §E.2, namely

$$W_A(x) = \sup\{|f(x)|\colon f \text{ entire of exponential type} \leq A, \\ \text{bounded on } \mathbb{R} \text{ and } |f(t)/W(t)| \leq 1 \text{ on } \mathbb{R}\}.$$

In terms of W_A, we have, by the computation just made, the

Theorem. *Let* $W(x) \geq 1$ *on* $\mathbb{R}$, *with*

$$|\log W(x) - \log W(x')| \leq C|x - x'|$$

for x *and* $x' \in \mathbb{R}$. *Then, if* $A > 0$,

$$W_A(x) \geq \tfrac{1}{2}(W(x))^{A/\sqrt{(A^2+C^2)}}, \quad x \in \mathbb{R}.$$

Corollary. *Let* $W(x) \geq 1$, *with* $\log W(x)$ *uniformly* Lip 1 *on* $\mathbb{R}$. *Then, if*

$$\int_{-\infty}^{\infty} \frac{\log W(x)}{1+x^2}\,dx = \infty,$$

we have

$$\int_{-\infty}^{\infty} \frac{\log W_A(x)}{1+x^2}\,dx = \infty$$

for each $A > 0$.

According to Akhiezer's first theorem (Chapter VI, §E.2), this in turn implies the

Theorem. *Let* $W(x) \geq 1$, *with* $\log W(x)$ *uniformly* Lip 1 *on* $\mathbb{R}$, *and* $W(x)$ *tending to* ∞ *as* $x \to \pm\infty$. *If*

$$\int_{-\infty}^{\infty} \frac{\log W(x)}{1+x^2}\,dx = \infty,$$

linear combinations of $e^{i\lambda x}$, $-A \leq \lambda \leq A$, *are, for each* $A > 0$, $\|\ \|_W$*-dense in* $\mathscr{C}_W(\mathbb{R})$.

2. Beurling's gap theorem

As a first application of the above fairly easy result, let us prove the following beautiful proposition of Beurling:

Theorem. *Let μ be a totally finite complex Radon measure on $\mathbb{R}$ with $|d\mu(t)| = 0$ on each of the disjoint intervals (a_n, b_n), $0 < a_1 < b_1 < a_2 < b_2 < \cdots$, and suppose that*

$$(*) \qquad \sum_1^\infty \left(\frac{b_n - a_n}{a_n}\right)^2 \;=\; \infty.$$

If $\hat{\mu}(\lambda) = \int_{-\infty}^{\infty} e^{i\lambda x} d\mu(x)$ vanishes identically on some real interval of positive length, then $\mu \equiv 0$.

Remark. This is not the only time we shall encounter the condition $(*)$ in the present book.

Proof of theorem (de Branges). We start by taking an *even* function $T(x) \geqslant 1$ whose logarithm is uniformly Lip 1 on $\mathbb{R}$, and which *increases to ∞ so slowly as $|x| \to \infty$*, that

$$\int_{-\infty}^{\infty} T(x)|d\mu(x)| \;<\; \infty.$$

(Construction of such a function T is in terms of the given measure μ, and is left to the reader as an easy exercise.)

For each n, let b_n' be *the lesser* of b_n and $2a_n$. Then, given that $(*)$ holds, we also have

$$\sum_n \left(\frac{b_n' - a_n}{a_n}\right)^2 \;=\; \infty.$$

Indeed, this sum certainly diverges if the one in $(*)$ does, when $(b_n' - a_n)/a_n$ differs from $(b_n - a_n)/a_n$ for only *finitely many n*. But the sum in question *also* diverges when *infinitely many* of its terms differ from the corresponding ones in $(*)$, since $(b_n' - a_n)/a_n = 1$ when $b_n' = 2a_n$.

Let $\omega(x)$ be *zero* outside the intervals (a_n, b_n'), and *on each one of those intervals let the graph of $\omega(x)$ vs x be a $45°$ triangle with base on (a_n, b_n').*

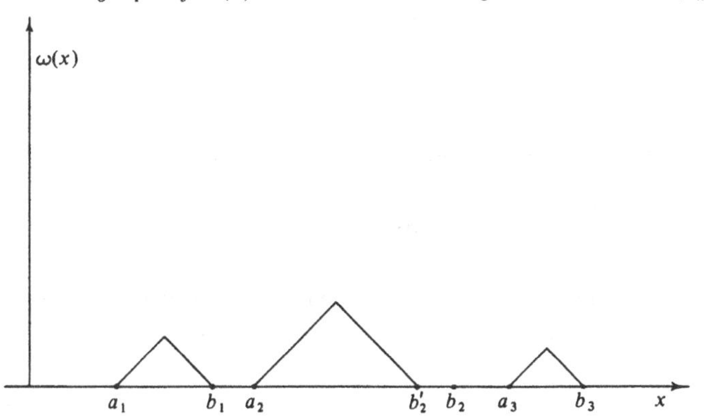

Figure 43

The function $\omega(x)$ is clearly uniformly Lip 1 on $\mathbb{R}$.

Put $W(x) = e^{\omega(x)}T(x)$. Then $W(x) \geqslant 1$ and $\log W(x)$ is uniformly Lip 1 on $\mathbb{R}$; also, $W(x) \to \infty$ for $x \to \pm\infty$. Since $|d\mu(x)| = 0$ throughout each interval (a_n, b'_n) and $\omega(x)$ is *zero* outside those intervals,

$$\int_{-\infty}^{\infty} W(x)|d\mu(x)| \;=\; \int_{-\infty}^{\infty} T(x)|d\mu(x)| \;<\; \infty.$$

The complex Radon measure ν with

$$d\nu(x) = W(x)d\mu(x)$$

is therefore *totally finite*.

Suppose now that $\hat{\mu}(\lambda)$ vanishes on some *interval*; say, wlog, that

$$\int_{-\infty}^{\infty} e^{i\lambda x}d\mu(x) \;=\; 0 \quad\text{for}\quad -A \leqslant \lambda \leqslant A.$$

This can be rewritten as

$$(*) \qquad \int_{-\infty}^{\infty} \frac{e^{i\lambda x}}{W(x)}d\nu(x) \;=\; 0, \qquad -A \leqslant \lambda \leqslant A.$$

However, $\log W(x) \;=\; \omega(x) + \log T(x) \;\geqslant\; \omega(x)$, and

$$\int_{a_1}^{\infty} \frac{\omega(x)}{x^2}dx \;\geqslant\; \sum_1^{\infty}\left(\frac{1}{b'_n}\right)^2\left(\frac{b'_n - a_n}{2}\right)^2 \;\geqslant\; \frac{1}{4}\sum_1^{\infty}\left(\frac{b'_n - a_n}{2a_n}\right)^2,$$

which is *infinite*, as we saw above. Therefore

$$\int_{-\infty}^{\infty} \frac{\log W(x)}{1+x^2}dx \;=\; \infty,$$

and *linear combinations* of the $e^{i\lambda x}$, $-A \leqslant \lambda \leqslant A$, are $\|\ \|_W$-*dense* in $\mathscr{C}_W(\mathbb{R})$ by the *second* theorem of the preceding article.

Referring to $(*)$, we thence see that $\nu \equiv 0$, i.e., $d\nu(x) = W(x)d\mu(x) \equiv 0$ and $\mu = 0$. Q.E.D.

Problem 11

Let μ be a finite complex measure on $\mathbb{R}$, and put

$$e^{-\sigma(x)} \;=\; \int_{-\infty}^{\infty} e^{-|x-t|}|d\mu(t)|.$$

Suppose that $\int_{-\infty}^{\infty}(\sigma(x)/(1+x^2))dx = \infty$. Then, if $\hat{\mu}(\lambda)$ vanishes identically on any interval, $\mu \equiv 0$ (Beurling). (Hint. Wlog, $\int_{-\infty}^{\infty}|d\mu(t)| \leqslant 1$ so that

$\sigma(x) \geqslant 0$. Assuming that $\hat{\mu}(\lambda) \equiv 0$ for $-A \leqslant \lambda \leqslant A$, write the relation

$$\int_{-\infty}^{\infty} e^{\sigma(x)-|x-t|}|d\mu(t)| \;=\; 1,$$

and use the picture

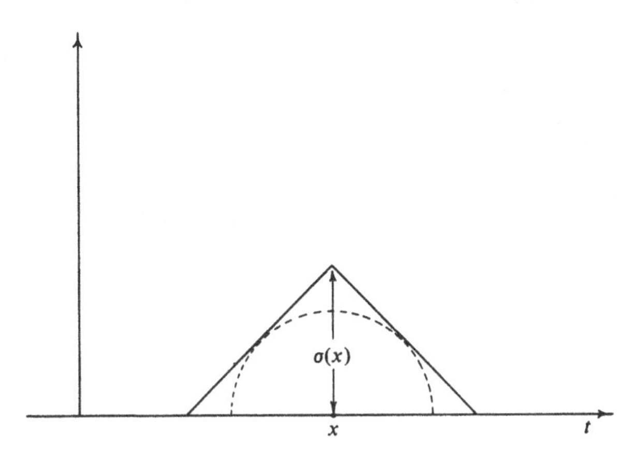

Figure 44

to estimate the supremum of $|f(x)|$ for entire functions f of exponential type $\leqslant A$, bounded on $\mathbb{R}$, and such that

$$\int_{-\infty}^{\infty} |f(t)|\,|d\mu(t)| \;\leqslant\; 1.)$$

Remark. Beurling generalized the result of problem 11 to complex Radon measures μ *which are not necessarily totally finite.* This extension will be taken up in Chapter X.

3. Weights which increase along the positive real axis

Lemma. *Let* $T(x) \geqslant 1$ *be defined and* increasing *for* $x \geqslant 0$, *and denote by* $\mathcal{T}(x)$ *the* largest minorant *of* $T(x)$ *with the property that* $|\log \mathcal{T}(x) - \log \mathcal{T}(x')| \leqslant |x - x'|$ *for* x *and* $x' \geqslant 0$. *If* $\int_1^\infty (\log T(x)/x^2)dx = \infty$, *then also* $\int_1^\infty (\log \mathcal{T}(x)/x^2)dx = \infty$.

Proof. The graph of $\log \mathcal{T}(x)$ vs x is obtained from that of $\log T(x)$ by means of the following construction:

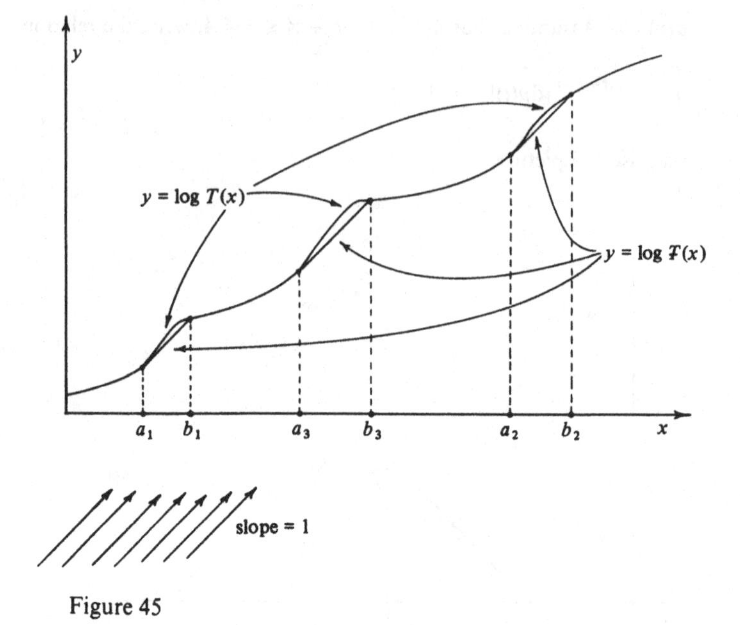

Figure 45

One imagines rays of light of slope 1 shining upwards *underneath* the graph of $\log T(x)$ vs x. The graph of $\log \mathcal{T}(x)$ is made up of the portions of the *former* one which are *illuminated* by those rays of light and some straight segments of slope 1. Those segments lie over certain intervals $[a_n, b_n]$ on the x-axis, of which there are generally countably many, that cannot necessarily be indexed in such fashion that $b_n \leqslant a_{n+1}$ for all n. The *open* intervals (a_n, b_n) are *disjoint*, and on *any one of them* we have

$$\log \mathcal{T}(x) \;=\; \log T(a_n) + (x - a_n).$$

On $[0, \infty) \sim \bigcup_n (a_n, b_n)$, $\mathcal{T}(x)$ and $T(x)$ are *equal*.

In order to prove the lemma, let us *assume* that $\int_1^\infty (\log \mathcal{T}(x)/x^2)\mathrm{d}x < \infty$ and then *show* that $\int_1^\infty (\log T(x)/x^2)\mathrm{d}x < \infty$. If, in the first place, (a_n, b_n) is any of the aforementioned intervals with $1 \leqslant a_n < b_n/2$, we have, since $\log T(a_n) \geqslant 0$,

$$\int_{a_n}^{b_n} \frac{\log \mathcal{T}(x)}{x^2}\mathrm{d}x \;\geqslant\; \int_{a_n}^{b_n} \frac{x - a_n}{x^2}\mathrm{d}x \;>\; \int_1^2 \frac{\xi - 1}{\xi^2}\mathrm{d}\xi$$

$$= \;\log 2 - \tfrac{1}{2}. > \;0.$$

We can therefore only have *finitely many intervals* (a_n, b_n) with $b_n > 2a_n$ and $a_n \geqslant 1$ if $\int_1^\infty (\log \mathcal{T}(x)/x^2)\mathrm{d}x$ is *finite.*

This being granted, consider any *other* of the intervals (a_n, b_n) with $a_n \geqslant 1$.

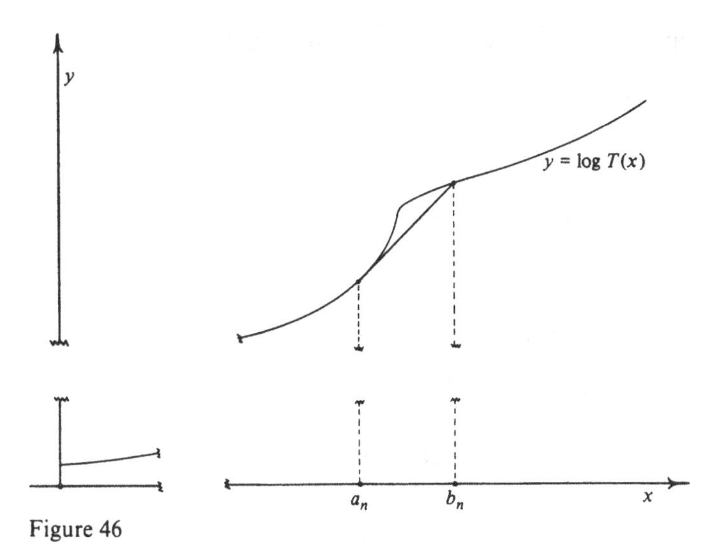

Figure 46

By shop math,

$$\int_{a_n}^{b_n} \frac{1}{x^2} \log \text{₮}(x) \mathrm{d}x \;\geqslant\; \frac{1}{b_n^2} \int_{a_n}^{b_n} \log \text{₮}(x) \mathrm{d}x$$

$$= \frac{1}{b_n^2} \cdot \frac{\log T(a_n) + \log T(b_n)}{2} \cdot (b_n - a_n) \;\geqslant\; \frac{(b_n - a_n) \log T(b_n)}{2b_n^2}.$$

At the same time, since $T(x)$ *increases*,

$$\int_{a_n}^{b_n} \frac{1}{x^2} \log T(x) \mathrm{d}x \;\leqslant\; \frac{(b_n - a_n) \log T(b_n)}{a_n^2} \;\leqslant\; 8 \cdot \frac{(b_n - a_n) \log T(b_n)}{2b_n^2}$$

when $b_n \leqslant 2a_n$. Therefore, for all the intervals (a_n, b_n) with $a_n \geqslant 1$ and $b_n \leqslant 2a_n$, hence, certainly, *for all save a finite number of the* (a_n, b_n) *contained in* $[1, \infty)$, we have

$$\int_{a_n}^{b_n} \frac{1}{x^2} \log T(x) \mathrm{d}x \;\leqslant\; 8 \int_{a_n}^{b_n} \frac{1}{x^2} \log \text{₮}(x) \mathrm{d}x.$$

The *sum* of the integrals $\int_{a_n}^{b_n} (1/x^2) \log T(x) \mathrm{d}x$ for the *remaining finite number of* (a_n, b_n) in $[1, \infty)$ *is surely finite* – note that *none* of those intervals can have *infinite length*, for such a one would be of the form (a_l, ∞), and in that case we would have

$$\int_{a_l}^{\infty} \frac{1}{x^2} \log \text{₮}(x) \mathrm{d}x \;\geqslant\; \int_{a_l}^{\infty} \frac{x - a_l}{x^2} \mathrm{d}x \;=\; \infty,$$

contrary to our assumption on $\mathcal{F}(x)$. We see that

$$\sum_{a_n \geq 1} \int_{a_n}^{b_n} \frac{\log T(x)}{x^2} dx \ < \ \infty,$$

since

$$\sum_{\substack{a_n \geq 1 \\ 2b_n \leq a_n}} 8 \int_{a_n}^{b_n} \frac{\log \mathcal{F}(x)}{x^2} dx$$

is finite.

On the complement

$$E = [1, \infty) \cap \sim \bigcup_n (a_n, b_n),$$

$T(x) = \mathcal{F}(x)$ by our construction. Hence

$$\int_E \frac{\log T(x)}{x^2} dx \ = \ \int_E \frac{\log \mathcal{F}(x)}{x^2} dx \ < \ \infty.$$

The whole half line $[1, \infty)$ can differ from the union of E and the (a_n, b_n) with $a_n \geq 1$ by *at most* an interval of the form $[1, b_m)$, which happens when there is an m such that $a_m < 1 < b_m$. If there *is* such an m, however, b_m must be *finite* (see above), and then

$$\int_1^{b_m} \frac{\log T(x)}{x^2} dx \ < \ \infty.$$

Putting everything together, we see that

$$\int_1^{\infty} \frac{\log T(x)}{x^2} dx \ < \ \infty,$$

which is what we had to show. We are done.

Corollary. *Let $W(x) \geq 1$ be defined on $\mathbb{R}$ and* increasing *for $x \geq 0$. If*

$$\int_1^{\infty} \frac{\log W(x)}{x^2} dx \ = \ \infty,$$

we have

$$\int_1^{\infty} \frac{\log W_A(x)}{x^2} dx \ = \ \infty$$

for each of the Akhiezer functions W_A, $A > 0$ (Chapter VI, §E.2).

Proof. Let, for $x \geq 0$, $\mathcal{F}(x)$ be the *largest minorant* of $W(x)$ on $[0, \infty)$ with

$$|\log \mathcal{F}(x) - \log \mathcal{F}(x')| \ \leq \ |x - x'|$$

there, and put $\mathcal{F}(x) = \mathcal{F}(0)$ for $x < 0$. By the lemma, $\int_1^\infty (\log W(x)/x^2)\mathrm{d}x = \infty$ implies that $\int_1^\infty (\log \mathcal{F}(x)/x^2)\mathrm{d}x = \infty$. Here, $\log \mathcal{F}(x)$ is certainly uniformly Lip 1 (and $\geqslant 0$) on $\mathbb{R}$, so, by the corollary of article 1, we see that

$$\int_1^\infty \frac{\log \mathcal{F}_A(x)}{x^2}\,\mathrm{d}x \;=\; \infty$$

for each $A > 0$.

We have $\mathcal{F}(x) \leqslant W(x) + \mathcal{F}(0)$ (the term $\mathcal{F}(0)$ on the right being perhaps needed for negative x). Therefore

$$\mathcal{F}_A(x) \;\leqslant\; (1 + \mathcal{F}(0))W_A(x),$$

and

$$\int_1^\infty \frac{\log W_A(x)}{x^2}\,\mathrm{d}x \;=\; \infty$$

for each $A > 0$ by the previous relation. Q.E.D.

From this, Akhiezer's first theorem (Chapter VI, §E.2) gives, without further ado, the following

Theorem. *Let $W(x) \geqslant 1$ on $\mathbb{R}$, with $W(x) \to \infty$ for $x \longrightarrow \pm\infty$. Suppose that $W(x)$ is* monotone *on* one *of the two half lines $(-\infty, 0]$, $[0, \infty)$, and that the* integral *of $\log W(x)/(1 + x^2)$, taken over whichever of those half lines on which monotoneity holds,* diverges. *Then $\mathscr{C}_W(A) = \mathscr{C}_W(\mathbb{R})$ for* every *$A > 0$, so $\mathscr{C}_W(0+) = \mathscr{C}_W(\mathbb{R})$.*

Remark. The notation is that of §E.2, Chapter VI. This result is due to Levinson. It is remarkable because only the monotoneity of $W(x)$ on a *half line* figures in it.

4. Example on the comparison of weighted approximation by polynomials and that by exponential sums

If $W(x) \geqslant 1$ tends to ∞ as $x \to \pm\infty$, we know that $\mathscr{C}_W(A)$ is *properly contained* in $\mathscr{C}_W(\mathbb{R})$ for each $A > 0$ in the case that

$$\int_{-\infty}^\infty \frac{\log W(x)}{1 + x^2}\,\mathrm{d}x \;<\; \infty.$$

(See Chapter VI, §E.2 and also the beginning of §D.) The theorem of the previous article shows that *mere monotoneity of $W(x)$ on $[0, \infty)$ without any additional regularity*, when accompanied by the condition

$$\int_0^\infty \frac{\log W(x)}{1 + x^2}\,\mathrm{d}x \;=\; \infty,$$

already guarantees the equality of $\mathscr{C}_W(A)$ and $\mathscr{C}_W(\mathbb{R})$ for each $A > 0$.

The question arises as to whether this also works for $\mathscr{C}_W(0)$, the $\| \ \|_W$-closure of the *polynomials* in $\mathscr{C}_W(\mathbb{R})$. (Here, of course we must assume that $x^n/W(x) \longrightarrow 0$ as $x \longrightarrow \pm\infty$ for all $n \geqslant 0$.) *The following example will show that the answer to this question is NO.*

We start with a *very rapidly increasing* sequence of numbers λ_n. It will be sufficient to take

$$\lambda_1 = 2,$$
$$\lambda_2 = \mathrm{e}^{\lambda_1},$$

and, in general, $\lambda_n = \mathrm{e}^{\lambda_{n-1}}$. Let us check that $\lambda_n > \lambda_{n-1}^2$ for $n > 1$. We have $\mathrm{e}^2 > 2^2 = 4$, and $(\mathrm{d}/\mathrm{d}x)(\mathrm{e}^x - x^2) = \mathrm{e}^x - 2x$ is > 0 for $x = 2$. Also $(\mathrm{d}^2/\mathrm{d}x^2)(\mathrm{e}^x - x^2) = \mathrm{e}^x - 2 > 0$ for $x \geqslant 2$, so $\mathrm{e}^x - x^2$ continues to increase strictly on $[2, \infty)$. Therefore $\mathrm{e}^x > x^2$ for $x \geqslant 2$, so $\lambda_n = \mathrm{e}^{\lambda_{n-1}} > \lambda_{n-1}^2$. We note that λ_{n-1}^2 is turn $\geqslant 2\lambda_{n-1}$, since the numbers λ_{n-1} are $\geqslant 2$.

We proceed to the construction of the weight W. For $0 \leqslant x \leqslant \lambda_1$, take $\log W(x) = 0$, and for $2\lambda_{n-1} \leqslant x \leqslant \lambda_n$ with $n > 1$ put $\log W(x) = n\lambda_{n-1}/2$ (by the computation just made we do have $2\lambda_{n-1} < \lambda_n$). We then specify $\log W(x)$ on the segments $[\lambda_{n-1}, 2\lambda_{n-1}]$ by making it *linear on each of them*, and finally define $W(x)$ for negative x by putting $W(-x) = W(x)$.

Here is the picture:

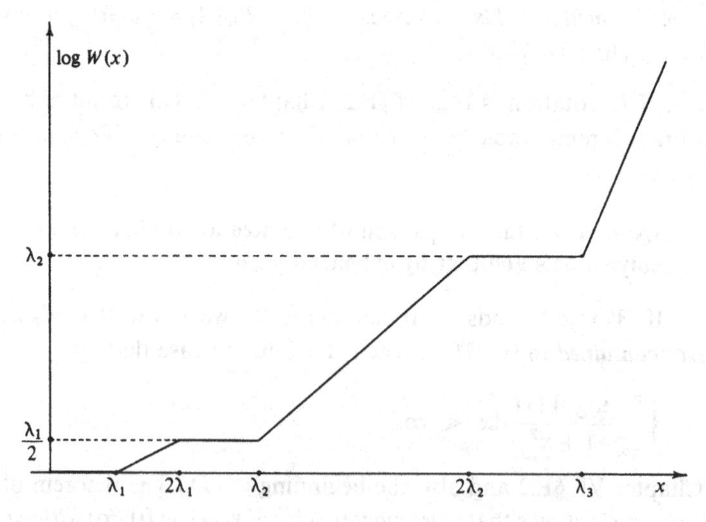

Figure 47

$W(x)$ is $\geqslant 1$ and *increasing* for $x \geqslant 0$, and, for large n,

$$\int_{2\lambda_n}^{\lambda_{n+1}} \frac{\log W(x)}{x^2}\,\mathrm{d}x \;=\; \frac{(n+1)\lambda_n}{2}\left(\frac{1}{2\lambda_n} - \frac{1}{\lambda_{n+1}}\right)$$

is $\geqslant (n+1)\lambda_n/8\lambda_n = \frac{1}{8}(n+1)$. Therefore $\int_0^\infty (\log W(x)/(1+x^2))dx = \infty$, so, *by the theorem of the previous article,* $\mathscr{C}_W(A) = \mathscr{C}_W(\mathbb{R})$ *for each* $A > 0$ *and* $\mathscr{C}_W(0+) = \mathscr{C}_W(\mathbb{R})$.

For $2\lambda_{n-1} \leqslant |x| \leqslant 2\lambda_n$,

$$W(x) \;\geqslant\; e^{n\lambda_{n-1}/2} \;=\; \lambda_n^{n/2} \;\geqslant\; |x/2|^{n/2}.$$

Hence

$$\frac{x^p}{W(x)} \longrightarrow 0 \quad \text{as} \quad x \longrightarrow \pm\infty$$

for every $p \geqslant 0$, and it makes sense to talk about the space $\mathscr{C}_W(0)$. *It is claimed that* $\mathscr{C}_W(0) \neq \mathscr{C}_W(\mathbb{R})$.

To see this, take the entire function

$$C(z) \;=\; \prod_1^\infty \left(1 - \frac{z^2}{\lambda_n^2}\right).$$

Because the λ_n go to ∞ so rapidly, $C(z)$ is of *zero exponential type*. For $n > 1$,

$$|C'(\lambda_n)| \;=\; \frac{2}{\lambda_n}\left(\frac{\lambda_n}{\lambda_1}\right)^2\left(\frac{\lambda_n}{\lambda_2}\right)^2 \cdots \left(\frac{\lambda_n}{\lambda_{n-1}}\right)^2$$

$$\times \prod_{k=1}^{n-1}\left(1 - \frac{\lambda_k^2}{\lambda_n^2}\right) \cdot \prod_{l=n+1}^{\infty}\left(1 - \frac{\lambda_n^2}{\lambda_l^2}\right).$$

Since the ratios λ_{j+1}/λ_j are always > 2 and $\to \infty$ as $j \to \infty$, the two products written with the sign $\prod$ on the right are both *bounded below* by *strictly positive constants* for $n > 1$ and indeed tend to 1 as $n \to \infty$. The product standing before them,

$$2\frac{\lambda_n}{\lambda_1^2}\cdot\frac{\lambda_n}{\lambda_2^2}\cdots\frac{\lambda_n}{\lambda_{n-1}^2}\cdot\lambda_n^{n-2},$$

far exceeds $2\lambda_n^{n-2}$ because $\lambda_j > \lambda_{j-1}^2$. Therefore we surely have

$$|C'(\lambda_n)| \;\geqslant\; \lambda_n^{n-2}$$

for large n.

At the same time, $W(\lambda_n) = e^{n\lambda_{n-1}/2} = \lambda_n^{n/2}$, whence, for large n,

$$\frac{W(\lambda_n)}{|C'(\lambda_n)|} \;\leqslant\; \frac{\lambda_n^{n/2}}{\lambda_n^{n-2}} \;=\; \frac{1}{\lambda_n^{(n/2)-2}}.$$

Since the sequence $\{\lambda_n\}$ tends to ∞, we thus have

$$\sum_1^\infty \frac{W(\lambda_n)}{|C'(\lambda_n)|} \;<\; \infty.$$

For $n = 1, 2, 3, \ldots$ it is convenient to put $\lambda_{-n} = -\lambda_n$. Let us then define a discrete measure μ supported on the points λ_n, $n = \pm 1, \pm 2, \ldots$, by putting

$$\mu(\{\lambda_n\}) = \frac{W(\lambda_n)}{C'(\lambda_n)}.$$

The functions $W(x)$ and $C(x)$ are even, hence

$$\int_{-\infty}^{\infty} |d\mu(x)| < \infty$$

by the calculation just made.

We can now verify, just as in §H.3 of Chapter VI, that

$$(\dagger) \qquad \int_{-\infty}^{\infty} \frac{x^p}{W(x)} d\mu(x) = 0 \quad \text{for} \quad p = 0, 1, 2, \ldots.$$

The integral on the right is just the (absolutely convergent) sum

$$\sum_{-\infty}^{\infty}{}' \frac{\lambda_n^p}{C'(\lambda_n)},$$

and we have to show that this is zero for $p \geqslant 0$. Taking

$$C_N(z) = \prod_{n=1}^{N} \left(1 - \frac{z^2}{\lambda_n^2}\right)$$

(cf. §C, Chapter VI), we have the Lagrange interpolation formula

$$z^l = \sum_{-N}^{N}{}' \frac{\lambda_n^l C_N(z)}{(z - \lambda_n) C_N'(\lambda_n)},$$

valid for $0 \leqslant l < 2N$. Fix l. Clearly, $|C_N'(\lambda_n)| \geqslant |C'(\lambda_n)|$ for $-N \leqslant n \leqslant N$. Therefore, since $\sum_{-\infty}^{\prime\infty} |\lambda_n^l / C'(\lambda_n)| < \infty$, we can make $N \to \infty$ in the preceding relation and use dominated convergence to obtain

$$z^l = \sum_{-\infty}^{\infty}{}' \frac{\lambda_n^l C(z)}{(z - \lambda_n) C'(\lambda_n)}.$$

Putting $l = p + 1$ and specializing to $z = 0$, the desired result follows, and we have (†).

Our measure μ is not zero. The strict inclusion of $\mathscr{C}_W(0)$ in $\mathscr{C}_W(\mathbb{R})$ is thus a consequence of (†), and the construction of our example is completed.

Let us summarize what we have. We have found *an even weight* $W(x) \geqslant 1$, *increasing on* $[0, \infty)$ *at a rate faster than that of any power of* x, *such that* $\mathscr{C}_W(0) \neq \mathscr{C}_W(\mathbb{R})$ *but* $\mathscr{C}_W(0+) = \mathscr{C}_W(\mathbb{R})$. This was promised at the end of §H.2, Chapter VI. In §H.3 of that chapter we constructed an even weight W with $\mathscr{C}_W(0) \neq \mathscr{C}_W(0+)$ and $\mathscr{C}_W(0+) \neq \mathscr{C}_W(\mathbb{R})$.

Scholium

As the work of Chapter VI shows, the condition

$$\int_{-\infty}^{\infty} \frac{\log W(x)}{1+x^2} \mathrm{d}x \;<\; \infty$$

is sufficient to guarantee proper inclusion in $\mathscr{C}_W(\mathbb{R})$ of each of the spaces $\mathscr{C}_W(0)$ and $\mathscr{C}_W(A)$, $A > 0$ (for $\mathscr{C}_W(0)$ see §D of that chapter). The question is, *how much regularity do we have to impose on* $W(x)$ *in order that the contrary property*

$$(\dagger\dagger) \qquad \int_{-\infty}^{\infty} \frac{\log W(x)}{1+x^2} \mathrm{d}x \;=\; \infty$$

should imply that $\mathscr{C}_W(0) = \mathscr{C}_W(\mathbb{R})$ *or that* $\mathscr{C}_W(A) = \mathscr{C}_W(\mathbb{R})$ *for* $A > 0$?

As we saw in the previous article, *monotoneity* of $W(x)$ on $[0, \infty)$ is *enough* for ($\dagger\dagger$) to make $\mathscr{C}_W(A) = \mathscr{C}_W(\mathbb{R})$ when $A > 0$, in the case of *even* weights W. In §D, Chapter VI, it was also shown that ($\dagger\dagger$) implies $\mathscr{C}_W(0) = \mathscr{C}_W(\mathbb{R})$ for *even weights* W *with* $\log W(x)$ *convex in* $\log|x|$. The example just given shows that *logarithmic convexity cannot be replaced by monotoneity when weighted polynomial approximation is involved*, even though the *later* is *good enough* when we deal with *weighted approximation by exponential sums.*

We have here a *qualitative difference* between weighted polynomial approximation and that by linear combinations of the $e^{i\lambda x}$, $-A \leqslant \lambda \leqslant A$, and in fact the *first real distinction* we have seen between these two kinds of approximation. In Chapter VI, the study of the latter paralleled that of the former in almost every detail.

The reason for this difference is that (for weights W which are finite reasonably often) the $\| \quad \|_W$-density of polynomials in $\mathscr{C}_W(\mathbb{R})$ is governed by the *lower polynomial regularization* $W_*(x)$ of W, whereas that of $\mathscr{E}_A$ is determined by the lower regularization $W_A(x)$ of W based on the use of entire functions of exponential type $\leqslant A$. The latter are better than polynomials for getting at $W(x)$ from underneath. As the example shows, they are qualitatively better.

5. Levinson's theorem

There is one other easy application of the material in article 1 which should be mentioned. Although the result obtained in that way has been superseded by a deeper (and more difficult) one of Beurling, to be given in the next §, it is still worthwhile, and serves as a basis for Volberg's very refined work presented in the last § of this chapter.

Theorem (Levinson). *Let μ be a finite Radon measure on $\mathbb{R}$, and suppose that*

$$\int_0^\infty \frac{1}{1+x^2}\log\left(\frac{1}{\int_x^\infty |d\mu(t)|}\right)dx = \infty.$$

Then the Fourier–Stieltjes transform

$$\hat{\mu}(\lambda) = \int_{-\infty}^{\infty} e^{i\lambda x}\,d\mu(x)$$

cannot vanish identically over any interval of positive length unless $\mu \equiv 0$.

Remark 1. Of course, the same result holds if

$$\int_{-\infty}^0 \frac{1}{1+x^2}\log\left(\frac{1}{\int_{-\infty}^x |d\mu(t)|}\right)dx = \infty.$$

Remark 2. Beurling's theorem, to be proved in the next §, says that under the stated condition on $\log(\int_x^\infty |d\mu(t)|)$, $\hat{\mu}(\lambda)$ cannot even vanish on a *set of positive measure* unless $\mu \equiv 0$.

Proof of theorem. It is enough, in the first place, to establish the result for *absolutely continuous* measures μ. Suppose, indeed, that μ is *any* measure satisfying the hypothesis; from it let us form the absolutely continuous measures μ_h, $h > 0$, having the densities

$$\frac{d\mu_h(x)}{dx} = \frac{1}{h}\int_x^{x+h} d\mu(t).$$

Then

$$\hat{\mu}_h(\lambda) = \frac{1-e^{-i\lambda h}}{i\lambda h}\,\hat{\mu}(\lambda),$$

so $\hat{\mu}_h(\lambda)$ *vanishes wherever $\hat{\mu}(\lambda)$ does.* Also,

$$\int_x^\infty |d\mu_h(t)| \leqslant \int_x^\infty |d\mu(t)|$$

for $x > 0$, so

$$\int_0^\infty \frac{1}{1+x^2}\log\left(\frac{1}{\int_x^\infty |d\mu_h(t)|}\right)dx = \infty$$

for each $h > 0$ by the hypothesis. Truth of our theorem for absolutely continuous measures would thus make the μ_h all zero if $\hat{\mu}(\lambda)$ vanishes on an interval of length > 0. But then $\mu \equiv 0$.

We may therefore take μ to be absolutely continuous. Assume, without loss of generality, that

$$\int_{-\infty}^{\infty} |d\mu(t)| \leqslant 1$$

and that $\hat{\mu}(\lambda) \equiv 0$ for $-A \leqslant \lambda \leqslant A$, $A > 0$.

For $x \geqslant 0$, write $W(x) = (\int_x^{\infty} |d\mu(t)|)^{-1/2}$, and, for $x < 0$, put

$$W(x) = \left(\int_{-\infty}^{x} |d\mu(t)| \right)^{-1/2}$$

The function $W(x)$ (perhaps discontinuous at 0) is $\geqslant 1$ and tends to ∞ as $x \to \pm\infty$. It is monotone on $(-\infty, 0)$ and on $[0, \infty)$, and *continuous* on each of those intervals (in the *extended sense*, as it *may take the value* ∞).

By integral calculus (!), we now find that

$$\int_0^{\infty} W(x)|d\mu(x)| = \int_0^{\infty} \frac{|d\mu(x)|}{\sqrt{\int_x^{\infty} |d\mu(t)|}} = 2\sqrt{\int_0^{\infty} |d\mu(t)|} < \infty,$$

and, in like manner,

$$\int_{-\infty}^{0} W(x)|d\mu(x)| < \infty.$$

The measure ν with $d\nu(x) = W(x)d\mu(x)$ is therefore *totally finite* on $\mathbb{R}$. (If $W(x)$ is infinite on any semi-infinite interval J, we of course must have $d\mu(x) \equiv 0$ on J, so $d\nu(x)$ is also zero there.) For $-A \leqslant \lambda \leqslant A$,

$$(\S) \qquad \int_{-\infty}^{\infty} \frac{e^{i\lambda x}}{W(x)} d\nu(x) = \hat{\mu}(\lambda) = 0.$$

However, by hypothesis,

$$\int_0^{\infty} \frac{\log W(x)}{1+x^2} dx = \frac{1}{2}\int_0^{\infty} \frac{1}{1+x^2} \log\left(\frac{1}{\int_x^{\infty} |d\mu(t)|} \right) dx = \infty,$$

so, since $W(x)$ is *increasing* on $[0, \infty)$, $\mathscr{C}_W(A)$ is $\| \quad \|_W$-dense in $\mathscr{C}_W(\mathbb{R})$ according to the theorem of article 3. Therefore, by (§),

$$\int_{-\infty}^{\infty} \varphi(x) d\mu(x) = \int_{-\infty}^{\infty} \frac{\varphi(x)}{W(x)} d\nu(x) = 0$$

for every continuous φ of compact support. This means that $\mu \equiv 0$. We are done.

The proof of Volberg's theorem uses the following

Corollary. Let $f(\vartheta) \sim \sum_{-\infty}^{\infty} \hat{f}(n) e^{in\vartheta}$ *belong to* $L_1(-\pi, \pi)$, *and suppose that* $f(\vartheta) = 0$ *a.e. on an interval* J *of positive length. If* $|\hat{f}(n)| \leqslant e^{-M(n)}$ *for*

$n > 0$ *with* $M(n)$ increasing, *and such that*

$$\sum_1^\infty \frac{M(n)}{n^2} = \infty,$$

then $f(\vartheta) \equiv 0, \ -\pi \leqslant \vartheta \leqslant \pi$.

Proof. Take any small $h > 0$ and form the convolution

$$f_h(\vartheta) = \frac{1}{h}\int_{-h}^{h}\left(1 - \frac{|t|}{h}\right) f(\vartheta - t)\,\mathrm{d}t.$$

If $h < \frac{1}{2}$ (length of J), $f_h(\vartheta)$ *also vanishes identically on an interval of positive length*

From the rudiments of Fourier series, we have

$$f_h(\vartheta) = \sum_{-\infty}^{\infty}\left(\frac{\sin(nh/2)}{nh/2}\right)^2 \hat{f}(n)\mathrm{e}^{in\vartheta}.$$

The sum on the right can be rewritten in evident fashion as $\int_{-\infty}^{\infty} \mathrm{e}^{i\vartheta x}\,\mathrm{d}\mu(x)$ with a (discrete) totally finite measure μ. Let $x > 0$ be given. If n is *the next integer* $\geqslant x$ we have, since $M(n)$ *increases,*

$$\int_x^\infty |\mathrm{d}\mu(t)| = \sum_{l \geqslant n}\left(\frac{\sin(lh/2)}{lh/2}\right)^2 |\hat{f}(l)|$$

$$\leqslant \mathrm{e}^{-M(n)} \sum_{l \geqslant n} \frac{4}{h^2 l^2} \leqslant \frac{\text{const.}}{h^2}\mathrm{e}^{-M(n)}.$$

Because $\sum_1^\infty M(n)/n^2 = \infty$, we see that

$$\int_0^\infty \frac{1}{1+x^2} \log^-\left(\int_x^\infty |\mathrm{d}\mu(t)|\right)\mathrm{d}x = \infty,$$

and conclude by the theorem that $f_h \equiv 0$. Making $h \to 0$, we see that $f \equiv 0$, Q.E.D.

B. The Fourier transform vanishes on a set of positive measure. Beurling's theorems

Beurling was able to extend considerably the theorem of Levinson given at the end of the preceding §. The main improvement in technique which made this extension possible involved the use of harmonic measure.

Harmonic measure will play an increasingly important rôle in the remaining chapters of this book. We therefore begin this § with a brief general discussion of what it is and what it does.

1. What is harmonic measure?

Suppose we have a finitely connected bounded domain $\mathscr{D}$ whose boundary, $\partial\mathscr{D}$, consists of several piecewise smooth Jordan curves. The *Dirichlet problem* for $\mathscr{D}$ requires us to find, for any given φ continuous on $\partial\mathscr{D}$, a function $U_\varphi(z)$ *harmonic in* $\mathscr{D}$ *and continuous up to* $\partial\mathscr{D}$ with $U_\varphi(\zeta) = \varphi(\zeta)$ for $\zeta \in \partial\mathscr{D}$. It is well known that the Dirichlet problem can always be solved for domains like those considered here. Many books on complex variable theory or potential theory contain proofs of this fact, which we henceforth take for granted.

Let us, however, tarry long enough to remind the reader of one particularly easy proof, available for *simply connected* domains $\mathscr{D}$. There, the Riemann mapping theorem provides us with a *conformal mapping* F of $\mathscr{D}$ onto the unit disk $\{|w| < 1\}$. Such a function F extends continuously up to $\partial\mathscr{D}$ and maps the latter in one–one fashion onto $\{|\omega| = 1\}$; this is true by a famous theorem of Carathéodory and can also be directly verified in many cases where $\partial\mathscr{D}$ has a simple explicit description (including all the ones to be met with in this book).

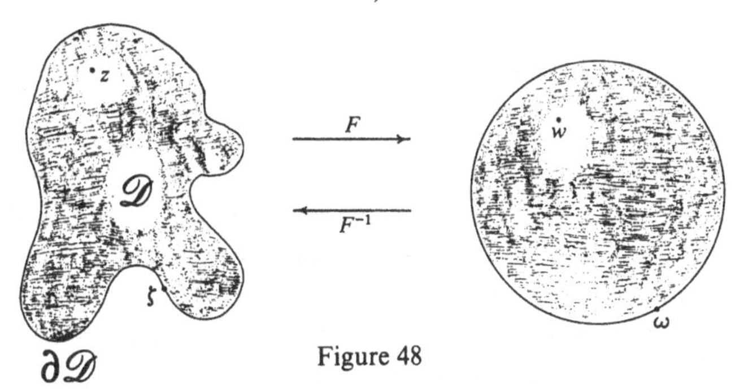

Figure 48

Denote by F^{-1} the inverse mapping to F. The function $\psi(\omega) = \varphi(F^{-1}(\omega))$ is then continuous on $\{|\omega| = 1\}$, and, if U_φ is the harmonic function sought which is to agree with φ on $\partial\mathscr{D}$, $V(w) = U_\varphi(F^{-1}(w))$ must be harmonic in $\{|w| < 1\}$ and continuous up to $\{|\omega| = 1\}$, where $V(\omega)$ must equal $\psi(\omega)$. A function V with these properties (there is only *one* such) can, however, be obtained from ψ by *Poisson's formula*:

$$V(w) \;=\; \frac{1}{2\pi}\int_{|\omega|=1} \frac{1-|w|^2}{|w-\omega|^2}\,\psi(\omega)|\mathrm{d}\omega|.$$

Going back to $\mathscr{D}$, and writing $z = F^{-1}(w)$, $\zeta = F^{-1}(\omega)$, we get

$$U_\varphi(z) \;=\; \frac{1}{2\pi}\int_{\partial\mathscr{D}} \frac{1-|F(z)|^2}{|F(z)-F(\zeta)|^2}\,\varphi(\zeta)|\mathrm{d}F(\zeta)|$$

for $z \in \mathscr{D}$. This is a *formula* for solving the Dirichlet problem for $\mathscr{D}$, based on the conformal mapping function F. Knowledge of this formula will help us later on to get general qualitative information about the behaviour near $\partial\mathscr{D}$ of certain functions harmonic in $\mathscr{D}$ but *not* continuous up to $\partial\mathscr{D}$, even when $\mathscr{D}$ is *not* simply connected.

Let us return to the multiply connected domains $\mathscr{D}$ of the kind considered here. If φ is real and continuous on $\partial\mathscr{D}$ and U_φ, harmonic in $\mathscr{D}$ and continuous on $\bar{\mathscr{D}}$, agrees with φ on $\partial\mathscr{D}$, we have, by the principle of maximum,

$$-\|\varphi\|_\infty \leqslant U_\varphi(z) \leqslant \|\varphi\|_\infty$$

for each $z \in \mathscr{D}$; here we are writing

$$\|\varphi\|_\infty = \sup_{\zeta \in \partial\mathscr{D}} |\varphi(\zeta)|.$$

This shows in the first place that *there can only be one function* U_φ corresponding to a *given* function φ. We see, secondly, that *there must be a* (signed) *measure* μ_z on $\partial\mathscr{D}$ (depending, of course, on z) with

$$(*) \qquad U_\varphi(z) = \int_{\partial\mathscr{D}} \varphi(\zeta) \mathrm{d}\mu_z(\zeta).$$

The latter statement is simply a consequence of the Riesz representation theorem applied to the space $\mathscr{C}(\partial\mathscr{D})$. Since U_φ can be found for *every* $\varphi \in \mathscr{C}(\partial\mathscr{D})$ (i.e., the Dirichlet problem for $\mathscr{D}$ can be solved!) and since, corresponding to each given φ, there is only *one* U_φ, there can, for any $z \in \mathscr{D}$, be *only one* measure μ_z on $\partial\mathscr{D}$ such that $(*)$ is true with every $\varphi \in \mathscr{C}(\partial\mathscr{D})$. The measure μ_z is thus *a function of* $z \in \mathscr{D}$, and we proceed to make a gross examination of its dependence on z.

If $\varphi(\zeta) \geqslant 0$ we must have $U_\varphi(z) \geqslant 0$ throughout $\mathscr{D}$ by the principle of maximum. Referring to $(*)$, we see that *the measures* μ_z *must be positive.* Also, 1 is a harmonic function (!), so, if $\varphi(\zeta) \equiv 1$, $U_\varphi(z) \equiv 1$. Therefore

$$\int_{\partial\mathscr{D}} \mathrm{d}\mu_z(\zeta) = 1$$

for every $z \in \mathscr{D}$. Let $\zeta_0 \in \partial\mathscr{D}$ and consider any small fixed neighborhood $\mathscr{V}$ of ζ_0. Take any continuous function φ on $\partial\mathscr{D}$ such that $\varphi(\zeta_0) = 0$, $\varphi(\zeta) \equiv 1$ for $\zeta \notin \mathscr{V} \cap \partial\mathscr{D}$, and $0 \leqslant \varphi(\zeta) \leqslant 1$ on $\mathscr{V} \cap \partial\mathscr{D}$.

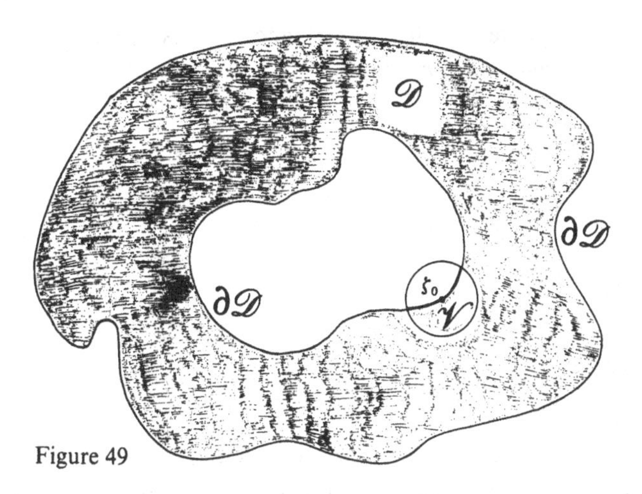

Figure 49

Since U_φ is a solution of the Dirichlet problem, we certainly have

$$U_\varphi(z) \longrightarrow \varphi(\zeta_0) = 0 \quad \text{for} \quad z \longrightarrow \zeta_0.$$

The positivity of the μ_z therefore makes

$$\int_{\zeta \notin \mathscr{V} \cap \partial\mathscr{D}} d\mu_z(\zeta) \longrightarrow 0$$

for $z \longrightarrow \zeta_0$. Because all the μ_z have total mass 1, we must also have

$$\int_{\mathscr{V} \cap \partial\mathscr{D}} d\mu_z(\zeta) \longrightarrow 1 \quad \text{for} \quad z \longrightarrow \zeta_0.$$

When z is near $\zeta_0 \in \partial\mathscr{D}$, μ_z has almost all of its total mass (1) *near* ζ_0 (on $\partial\mathscr{D}$). This is the so-called *approximate identity property* of the μ_z.

There is also a continuity property for the μ_z applying to variations of *z in the interior of $\mathscr{D}$*.

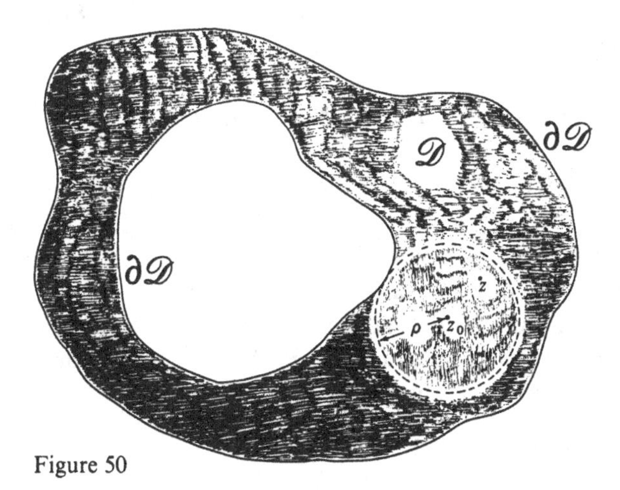

Figure 50

Take any $z_0 \in \mathscr{D}$, write $\rho = \operatorname{dist}(z_0, \partial\mathscr{D})$ and suppose that $|z - z_0| < \rho$. Then, if φ is *continuous and positive on* $\partial\mathscr{D}$,

$$\int_{\partial\mathscr{D}} \varphi(\zeta)\mathrm{d}\mu_z(\zeta)$$

lies between

$$\frac{\rho - |z - z_0|}{\rho + |z - z_0|}\int_{\partial\mathscr{D}} \varphi(\zeta)\mathrm{d}\mu_{z_0}(\zeta)$$

and

$$\frac{\rho + |z - z_0|}{\rho - |z - z_0|}\int_{\partial\mathscr{D}} \varphi(\zeta)\mathrm{d}\mu_{z_0}(\zeta).$$

This is nothing but *Harnack's inequality* applied to the circle $\{|z - z_0| < \rho\}$, $U_\varphi(z)$ being *harmonic and positive* in that circle. (The reader who does not recall Harnack's inequality may derive it very easily from the Poisson representation of positive harmonic functions for the unit disk given in Chapter III, §F.1.) These inequalities hold for *any* positive $\varphi \in \mathscr{C}(\partial\mathscr{D})$, so the signed measures

$$\mu_z \;-\; \frac{\rho - |z - z_0|}{\rho + |z - z_0|}\mu_{z_0},$$

$$\frac{\rho + |z - z_0|}{\rho - |z - z_0|}\mu_{z_0} \;-\; \mu_z$$

are in fact positive. This fact is usually expressed by the double inequality

$$\frac{\rho - |z - z_0|}{\rho + |z - z_0|}\mathrm{d}\mu_{z_0}(\zeta) \;\leqslant\; \mathrm{d}\mu_z(\zeta) \;\leqslant\; \frac{\rho + |z - z_0|}{\rho - |z - z_0|}\mathrm{d}\mu_{z_0}(\zeta).$$

What is important here is that we have a number $K(z, z_0)$, $0 < K(z, z_0) < \infty$, depending *only* on z and z_0 (and $\mathscr{D}$!), such that

$$\frac{1}{K(z, z_0)}\mathrm{d}\mu_{z_0}(\zeta) \;\leqslant\; \mathrm{d}\mu_z(\zeta) \;\leqslant\; K(z, z_0)\mathrm{d}\mu_{z_0}(\zeta).$$

Such an inequality in fact holds for *any* two points z, z_0 in $\mathscr{D}$; one needs only to join z to z_0 by a path lying in $\mathscr{D}$ and then take a chain of overlapping disks $\subseteq \mathscr{D}$ having their centres on that path, applying the previous special version of the inequality in each disk.

In order to indicate the dependence of the measures μ_z in (*) *on the domain* $\mathscr{D}$ *as well as on* $z \in \mathscr{D}$, we use a special notation for them which is now becoming standard. ***We write***

$$\mathrm{d}\omega_{\mathscr{D}}(\zeta, z) \quad \textit{for} \quad \mathrm{d}\mu_z(\zeta),$$

***so that* (*) *has this appearance*:**

$$U_\varphi(z) = \int_{\partial\mathscr{D}} \varphi(\zeta) d\omega_{\mathscr{D}}(\zeta, z).$$

We call $\omega_{\mathscr{D}}(\ , z)$ ***harmonic measure for*** $\mathscr{D}$ ***(or relative to*** $\mathscr{D}$***) as seen from z.*** $\omega_{\mathscr{D}}(\ ,z)$ is a positive Radon measure on $\partial\mathscr{D}$, of total mass 1, which serves to recover functions harmonic in $\mathscr{D}$ and continuous on $\bar{\mathscr{D}}$ from their boundary values on $\partial\mathscr{D}$ by means of the boxed formula. That formula is just the analogue of *Poisson's* for our domains $\mathscr{D}$.

If E is a Borel set on $\partial\mathscr{D}$,

$$\omega_{\mathscr{D}}(E, z) = \int_E d\omega_{\mathscr{D}}(\zeta, z)$$

is called *the harmonic measure of E relative to* $\mathscr{D}$ *(or in* $\mathscr{D}$*), seen from z.* We have, of course,

$$0 \leqslant \omega_{\mathscr{D}}(E, z) \leqslant 1.$$

Also, for fixed $E \subseteq \partial\mathscr{D}$, $\omega_{\mathscr{D}}(E, z)$ is a *harmonic function of z.* This almost obvious property may be verified as follows. Given $E \subseteq \partial\mathscr{D}$, take a sequence of functions $\varphi_n \in \mathscr{C}(\partial\mathscr{D})$ with $0 \leqslant \varphi_n \leqslant 1$ such that

$$\int_{\partial\mathscr{D}} |\chi_E(\zeta) - \varphi_n(\zeta)| d\omega_{\mathscr{D}}(\zeta, z_0) \underset{n}{\longrightarrow} 0$$

for the characteristic function χ_E of E. Here, z_0 is *any fixed point of* $\mathscr{D}$ which may be chosen at pleasure. Since $d\omega_{\mathscr{D}}(\zeta, z) \leqslant K(z, z_0) d\omega_{\mathscr{D}}(\zeta, z_0)$ as we have seen above, the previous relation makes

$$U_{\varphi_n}(z) = \int_{\partial\mathscr{D}} \varphi_n(\zeta) d\omega_{\mathscr{D}}(\zeta, z) \underset{n}{\longrightarrow} \omega_{\mathscr{D}}(E, z)$$

for every $z \in \mathscr{D}$; the convergence is even u.c.c. in $\mathscr{D}$ because $0 \leqslant U_{\varphi_n}(z) \leqslant 1$ there for each n. Therefore $\omega_{\mathscr{D}}(E, z)$ is harmonic in $z \in \mathscr{D}$ since the $U_{\varphi_n}(z)$ are.

Harmonic measure is also available for many *unbounded* domains $\mathscr{D}$. Suppose we have such a domain (perhaps of infinite connectivity) with a decent boundary $\partial\mathscr{D}$. The latter may consist of infinitely many pieces, but each individual piece should be nice, and they should not accumulate near any finite point in such a way as to cause trouble for the solution of the Dirichlet problem. In such case, $\partial\mathscr{D}$ is at least *locally compact* and, if $\varphi \in \mathscr{C}_0(\partial\mathscr{D})$ (the space of functions continuous on $\partial\mathscr{D}$ which *tend to zero* as one goes out towards ∞ thereon), there is one and *only* one function U_φ *harmonic and bounded* in $\mathscr{D}$, and continuous up to $\partial\mathscr{D}$, with $U_\varphi(\zeta) = \varphi(\zeta)$, $\zeta \in \partial\mathscr{D}$. (Here it is *absolutely necessary* to assume *boundedness* of U_φ in $\mathscr{D}$ in order to get *uniqueness*; look at the function y in $\Im z > 0$ which takes the value 0 on $\mathbb{R}$. Uniqueness of the *bounded* harmonic function with prescribed boundary values is a direct consequence of the *first* Phragmén–Lindelöf theorem in §C, Chapter III.) Riesz' representation theorem still holds in the present situation, and we will have $(*)$ for $\varphi \in \mathscr{C}_0(\partial\mathscr{D})$. The examination of the μ_z carried out above goes through almost without change, and we write $d\mu_z(\zeta) = d\omega_{\mathscr{D}}(\zeta, z)$ as before, calling $\omega_{\mathscr{D}}(\ ,z)$ the *harmonic measure for* $\mathscr{D}$, *as seen from* z. It serves to recover *bounded* functions harmonic in $\mathscr{D}$ and continuous up to $\partial\mathscr{D}$ from their boundary values, at least when the latter come from functions in $\mathscr{C}_0(\partial\mathscr{D})$.

Let us return for a moment to *bounded, finitely connected domains* $\mathscr{D}$. Suppose we are given a function $f(z)$, *analytic and bounded in* $\mathscr{D}$, and *continuous up to* $\partial\mathscr{D}$. An important problem in the theory of functions is to *obtain an upper bound for* $|f(z)|$ *when* $z \in \mathscr{D}$, *in terms of the boundary values* $f(\zeta)$, $\zeta \in \partial\mathscr{D}$. A very useful estimate is furnished by the

Theorem (on harmonic estimation). *For* $z \in \mathscr{D}$,

$$(\dagger) \qquad \log|f(z)| \leqslant \int_{\partial\mathscr{D}} \log|f(\zeta)|\, d\omega_{\mathscr{D}}(\zeta, z).$$

Proof. The result is really a generalization of Jensen's inequality. Take any $M > 0$. The function

$$V_M(z) = \max(\log|f(z)|, -M)$$

is continuous in $\bar{\mathscr{D}}$ and *subharmonic* in $\mathscr{D}$. Therefore the difference

$$V_M(z) - \int_{\partial\mathscr{D}} V_M(\zeta)\, d\omega_{\mathscr{D}}(\zeta, z)$$

is subharmonic in $\mathscr{D}$ and continuous up to $\partial\mathscr{D}$ where it takes the boundary value $V_M(\zeta) - V_M(\zeta) = 0$ everywhere. Hence that difference is $\leqslant 0$ throughout $\mathscr{D}$ by the principle of maximum, and

$$\log|f(z)| \leqslant V_M(z) \leqslant \int_{\partial\mathscr{D}} V_M(\zeta)\, d\omega_{\mathscr{D}}(\zeta, z)$$

for $z \in \mathscr{D}$. On making $M \to \infty$, the right side tends to

$$\int_{\partial\mathscr{D}} \log|f(\zeta)|\, d\omega_{\mathscr{D}}(\zeta, z)$$

by Lebesgue's monotone convergence theorem, since $\log|f(\zeta)|$, and hence the $V_M(\zeta)$, are *bounded above*, $|f(z)|$ being continuous and thus bounded on the compact set $\bar{\mathscr{D}}$. The proof is finished.

The result just established is true for *bounded* analytic functions in *unbounded* domains subject to the restrictions on such domains mentioned above. *Here the boundedness of $f(z)$ in $\mathscr{D}$ becomes crucial* (look at the functions e^{-inz} in $\Im z > 0$ with $n \to \infty$!). Verification of this proceeds very much as above, using the functions $V_M(\zeta)$. These are continuous and bounded (above *and* below) on $\partial\mathscr{D}$, so the functions

$$H_M(z) = \int_{\partial\mathscr{D}} V_M(\zeta) d\omega_{\mathscr{D}}(\zeta, z)$$

are harmonic and bounded in $\mathscr{D}$, and for each $\zeta_0 \in \mathscr{D}$ we can *check directly*, by using the *approximate identity* property of $\omega_{\mathscr{D}}(\ , z)$ established in the above discussion, that

$$H_M(z) \longrightarrow V_M(\zeta_0) \quad \text{for} \quad z \longrightarrow \zeta_0.$$

(It is *not necessary* that $V_M(\zeta)$ belong to $\mathscr{C}_0(\partial\mathscr{D})$ in order to draw this conclusion; only that it be *continuous and bounded* on $\partial\mathscr{D}$.) The difference

$$V_M(z) - H_M(z)$$

is thus *subharmonic and bounded above* in $\mathscr{D}$, and tends to 0 as z tends to *any* point of $\partial\mathscr{D}$. We can therefore conclude by the first Phragmén–Lindelöf theorem of §C, Chapter III (or, rather, by its analogue for *subharmonic* functions), that $V_M(z) - H_M(z) \leqslant 0$ in $\mathscr{D}$. The rest of the argument is as above.

The inequality (†) has one very important consequence, called the ***theorem on two constants***. *Let $f(z)$ be analytic and bounded in a domain $\mathscr{D}$ of the kind considered above, and continuous up to $\partial\mathscr{D}$. Suppose that $|f(\zeta)| \leqslant M$ on $\partial\mathscr{D}$, and that there is a Borel set $E \subseteq \partial\mathscr{D}$ with $|f(\zeta)| \leqslant$ some number m ($< M$) on E. Then, for $z \in \mathscr{D}$,*

$$|f(z)| \leqslant m^{\omega_{\mathscr{D}}(E,z)} M^{1-\omega_{\mathscr{D}}(E,z)}.$$

Deduction of this inequality from (†) is immediate.

Much of the importance of harmonic measure in analysis is due to this formula and to (†). For this reason, analysts have devoted (and continue to devote) considerable attention to the *estimation* of harmonic measure. We shall see some of this work later on in the present book. The systematic use of harmonic measure in analysis is mainly due to Nevanlinna, who also gave us the *name* for it. There are beautiful examples of its application in his book, *Eindeutige analytische Funktionen* (now translated into English), of which every analyst should own a copy.

Before ending our discussion of harmonic measure, let us describe a few more of its qualitative properties.

The first observation to be made is that the measures $\omega_{\mathscr{D}}(\ ,z)$ are *absolutely continuous with respect to arc length* on $\partial\mathscr{D}$ for the kind of domains considered here. This will follow if we can show that

$$\omega_{\mathscr{D}}(E_n, z_0) \underset{n}{\longrightarrow} 0 \quad \text{for} \quad z_0 \in \mathscr{D}$$

when the E_n *lie on any particular component* Γ of $\partial\mathscr{D}$ and

$$\int_\Gamma \chi_{E_n}(\zeta)|d\zeta| \underset{n}{\longrightarrow} 0.$$

(Here, χ_{E_n} denotes the *characteristic function* of E_n.) We do this by comparing $\omega_{\mathscr{D}}(\ ,z)$ with harmonic measure for a *simply connected domain*; the method is of independent interest and is frequently used.

Let $\mathscr{E}$ be the *simply connected domain on the Riemann sphere* (including perhaps ∞), *bounded by the component* Γ *of* $\partial\mathscr{D}$ *and including all the points of* $\mathscr{D}$.

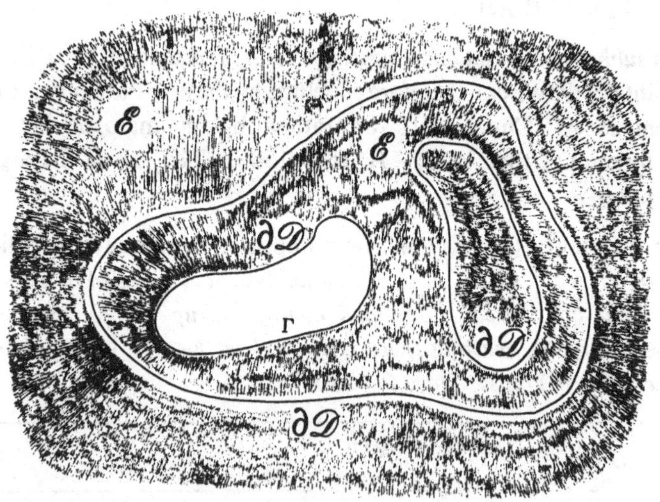

Figure 51

If $\varphi \in \mathscr{C}(\partial\mathscr{D})$ is *positive*, and *zero on all* the *components* of $\partial\mathscr{D}$ *save* Γ, we have

$$\int_\Gamma \varphi(\zeta) d\omega_{\mathscr{E}}(\zeta, z) \geqslant \int_{\partial\mathscr{D}} \varphi(\zeta) d\omega_{\mathscr{D}}(\zeta, z)$$

for $z \in \mathscr{D}$. Indeed, both integrals give us functions harmonic in $\mathscr{D}$ ($\subseteq \mathscr{E}$!), and continuous up to $\partial\mathscr{D}$. The right-hand function, $U_\varphi(z)$, equals $\varphi(\zeta)$ on Γ and *zero* on the *other* components of $\partial\mathscr{D}$. The left-hand one – call it $V(z)$ for the moment – *also* equals $\varphi(z)$ on Γ but is *surely* $\geqslant 0$ on the *other* components of $\partial\mathscr{D}$, because they lie in $\mathscr{E}$ and $\varphi \geqslant 0$. Therefore $V(z) \geqslant U_\varphi(z)$ throughout $\mathscr{D}$ by the principle of maximum, as claimed. This inequality holds for *every* function φ of the kind described above, whence, on Γ,

$$\boxed{\mathrm{d}\omega_{\mathscr{E}}(\zeta, z) \;\geqslant\; \mathrm{d}\omega_{\mathscr{D}}(\zeta, z) \quad \text{for} \quad z \in \mathscr{D}.}$$

This relation is an example of what Nevanlinna called ***the principle of extension of domain.***

Let us return to our sets

$$E_n \subseteq \Gamma \quad \text{with} \quad \int_\Gamma \chi_{E_n}(\zeta)|\mathrm{d}\zeta| \;\underset{n}{\longrightarrow}\; 0;$$

in order to verify that

$$\omega_{\mathscr{D}}(E_n, z) \underset{n}{\longrightarrow} 0, \quad z \in \mathscr{D},$$

it is enough, in virtue of the inequality just established, to check that $\omega_{\mathscr{E}}(E_n, z_0) \underset{n}{\longrightarrow} 0$ for each $z_0 \in \mathscr{E}$. Because $\mathscr{E}$ is *simply connected*, we may, however, use the formula derived near the beginning of the present article. Fixing $z_0 \in \mathscr{E}$, take a conformal mapping F of $\mathscr{E}$ onto $\{|w| < 1\}$ which sends z_0 to 0. From the formula just mentioned, it is clear that

$$\omega_{\mathscr{E}}(E_n, z_0) \;=\; \frac{1}{2\pi}\int_\Gamma \chi_{E_n}(\zeta)|\mathrm{d}F(\zeta)|.$$

The component Γ of $\partial\mathscr{D}$ is, however, *rectifiable*; a theorem of the brothers Riesz therefore guarantees that the mapping F from Γ onto the unit circumference is *absolutely continuous with respect to arc length.* For domains $\mathscr{D}$ whose boundary components are given explicitly and in fairly simple form (the sort we will be dealing with), that property can also be verified directly. We can hence write

$$\omega_{\mathscr{E}}(E_n, z_0) \;=\; \frac{1}{2\pi}\int_\Gamma \chi_{E_n}(\zeta)\left|\frac{\mathrm{d}F(\zeta)}{\mathrm{d}\zeta}\right||\mathrm{d}\zeta|$$

with

$$\left|\frac{\mathrm{d}F(\zeta)}{\mathrm{d}\zeta}\right| \quad \text{in} \quad L_1(\Gamma,\ |\mathrm{d}\zeta|),$$

and from this we see that $\omega_{\mathscr{E}}(E_n, z_0) \underset{n}{\longrightarrow} 0$ when

$$\int_\Gamma \chi_{E_n}(\zeta)|\mathrm{d}\zeta| \underset{n}{\longrightarrow} 0.$$

The absolute continuity of $\omega_{\mathscr{D}}(\ ,z)$ with respect to arc length on $\partial\mathscr{D}$ is thus verified.

The property just established makes it possible for us to write

$$\omega_{\mathscr{D}}(E, z) = \int_{\partial\mathscr{D}} \chi_E(\zeta) \frac{\mathrm{d}\omega_{\mathscr{D}}(\zeta, z)}{|\mathrm{d}\zeta|} \cdot |\mathrm{d}\zeta|$$

for $E \subseteq \partial\mathscr{D}$ and $z \in \mathscr{D}$. It is important for us to be able to majorize the integral on the right by one of the form

$$K_z \int_{\partial\mathscr{D}} \chi_E(\zeta)|\mathrm{d}\zeta|$$

(with K_z depending on z and, of course, on $\mathscr{D}$) when dealing with *certain kinds* of simply connected domains $\mathscr{D}$. In order to see for *which* kind, let us, for fixed $z_0 \in \mathscr{D}$, take a conformal mapping F of $\mathscr{D}$ onto $\{|w| < 1\}$ which sends z_0 to 0 and apply the formula used in the preceding argument, which here takes the form

$$\omega_{\mathscr{D}}(E, z_0) = \frac{1}{2\pi} \int_{\partial\mathscr{D}} \chi_E(\zeta) \left| \frac{\mathrm{d}F(\zeta)}{\mathrm{d}\zeta} \right| |\mathrm{d}\zeta|.$$

If the boundary $\partial\mathscr{D}$ is an *analytic curve*, or merely has a *differentiably turning tangent*, the derivative $F'(z)$ of the conformal mapping function *will be* continuous up to $\partial\mathscr{D}$; in such circumstances $|\mathrm{d}F(\zeta)/\mathrm{d}\zeta|$ is bounded on $\partial\mathscr{D}$ (the bound depends evidently on z_0), and we *have* a majorization of the desired kind. This is even true when $\partial\mathscr{D}$ *has a finite number of corners* and is sufficiently smooth *away* from them, *provided that all those corners stick out.*

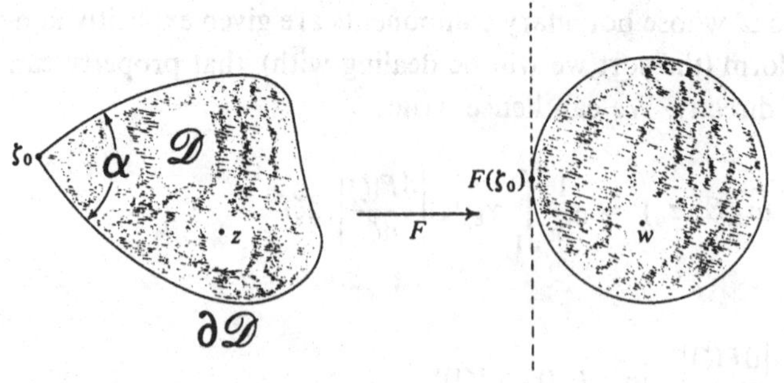

Figure 52

In this situation, where $\partial\mathscr{D}$ has a corner with internal angle α at ζ_0, $F(z) = F(\zeta_0) + (C + o(1))(z - \zeta_0)^{\pi/\alpha}$ for z in $\bar{\mathscr{D}}$ (*sic!*) *near* ζ_0; we see that $F'(\zeta_0) = 0$ if $\alpha < \pi$, and that $F'(\zeta)$ is *near* 0 if $\zeta \in \partial\mathscr{D}$ is *near* ζ_0 (sufficient smoothness of $\partial\mathscr{D}$ away from its corners is being assumed). In the present case, then, $|F'(\zeta)|$ *is* bounded on $\partial\mathscr{D}$, and an estimate

$$\omega_{\mathscr{D}}(E, z_0) \;\leqslant\; K_{z_0}\int_{\partial\mathscr{D}} \chi_E(\zeta)|d\zeta|$$

does hold good. It is *really necessary* that the corners *stick out*. If, for instance, $\alpha > \pi$, then $|F'(\zeta_0)| = \infty$, and $|F'(\zeta)|$ *tends* to ∞ for ζ on $\partial\mathscr{D}$ tending to ζ_0:

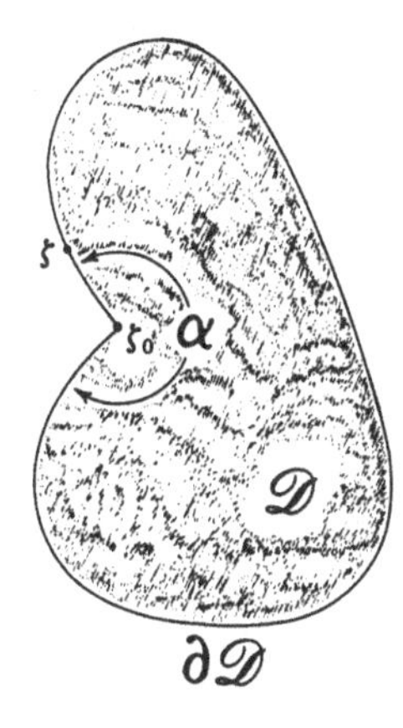

Figure 53

Here, we *do not* have

$$\omega_{\mathscr{D}}(E, z_0) \;\leqslant\; \text{const.}\int_{\partial\mathscr{D}} \chi_E(\zeta)|d\zeta|$$

for sets $E \subseteq \partial\mathscr{D}$ located near ζ_0.

Let us conclude with a general examination of the *boundary behaviour* of $\omega_{\mathscr{D}}(E, z)$ for $E \subseteq \partial\mathscr{D}$. Consider first of all the case where E is an *arc*, σ, *on one of the components* of $\partial\mathscr{D}$. Then the *simple approximate identity property* of $\omega_{\mathscr{D}}(\ ,z)$ established above immediately shows that

$$\omega_{\mathscr{D}}(\sigma, z) \longrightarrow 1 \quad \text{if} \quad z \longrightarrow \zeta \in \sigma$$

and ζ is not an endpoint of σ, while

$$\omega_{\mathscr{D}}(\sigma, z) \longrightarrow 0 \quad \text{if} \quad z \longrightarrow \zeta \in \partial\mathscr{D} \sim \sigma$$

and ζ is not an endpoint of σ. If $z \in \mathscr{D}$ *tends* to an endpoint of σ, we cannot say much (in general) about $\omega_{\mathscr{D}}(\sigma, z)$, save that it remains between 0 and

1. These properties, however, *already suffice to determine the harmonic function* $\omega_{\mathscr{D}}(\sigma, z)$ (defined in $\mathscr{D}$) *completely*. This may be easily verified by using the principle of maximum together with an evident modification of the first Phragmén–Lindelöf theorem from §C, Chapter III; such verification is left to the reader. One sometimes *uses* this characterization in order to compute or estimate harmonic measure. Of course, once $\omega_{\mathscr{D}}(\sigma, z)$ is *known for arcs* $\sigma \subseteq \partial\mathscr{D}$, we can *get* $\omega_{\mathscr{D}}(E, z)$ for Borel sets E by the standard construction applying to all positive Radon measures.

What about the boundary behaviour of $\omega_{\mathscr{D}}(E, z)$ for a *more general* set E? We only consider *closed sets E lying on a single component* Γ of $\partial\mathscr{D}$; knowledge about this situation is all that is needed in practice.

Take, then, a closed subset E of the component Γ of $\partial\mathscr{D}$. In the first place, $\omega_{\mathscr{D}}(E, z) \leqslant \omega_{\mathscr{D}}(\Gamma, z)$. When z tends to any point of a component Γ' of $\partial\mathscr{D}$ *different from* Γ, $\omega_{\mathscr{D}}(\Gamma, z)$ tends to *zero* by the previous discussion (Γ is an arc without endpoints!) Hence $\omega_{\mathscr{D}}(E, z) \longrightarrow 0$ for $z \longrightarrow \zeta$ if $\zeta \in \partial\mathscr{D}$ belongs to a *component of the* latter *other than* Γ.

Examination of the boundary of $\omega_{\mathscr{D}}(E, z)$ *for z near* Γ is more delicate.

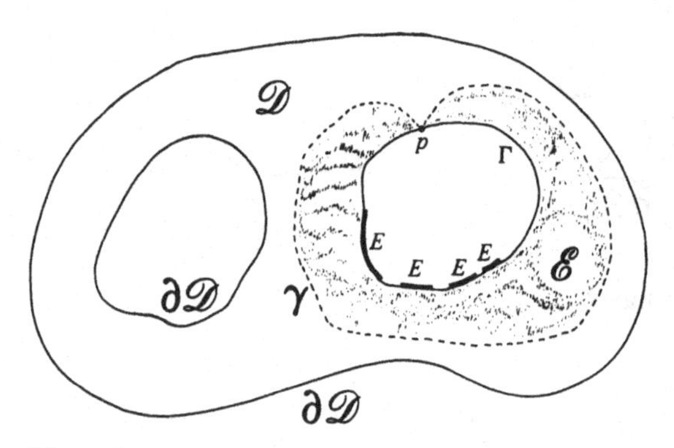

Figure 54

Take any point p on Γ lying outside the closed set E (if E were *all* of Γ, we could conclude by the case for *arcs* handled previously), and draw a curve γ lying in $\mathscr{D}$ like the one shown, with its two endpoints at p. Together, the curves γ and Γ bound a certain *simply connected* domain $\mathscr{E} \subseteq \mathscr{D}$.

We are going to derive the formula

$$\omega_{\mathscr{D}}(E, z) \;=\; \int_{\gamma} \omega_{\mathscr{D}}(E, \zeta)\, d\omega_{\mathscr{E}}(\zeta, z) \;+\; \omega_{\mathscr{E}}(E, z),$$

valid for $z \in \mathscr{E}$. Take any finite union $\mathscr{U}$ of arcs on Γ containing the closed set E but avoiding a whole neighborhood of the point p, and let ψ be any function continuous on Γ with $0 \leqslant \psi(\zeta) \leqslant 1$, $\psi(\zeta) \equiv 0$ outside $\mathscr{U}$, and $\psi(\zeta) \equiv 1$ on E. Since ψ is zero on a neighborhood of p, the function

$$U_\psi(z) = \int_\Gamma \psi(\zeta)\,d\omega_{\mathscr{D}}(\zeta, z)$$

tends to *zero* as $z \to p$. Write $U_\psi(\zeta) = \psi(\zeta)$ for $\zeta \in \Gamma$; the function $U_\psi(\zeta)$ then becomes continuous on $\Gamma \cup \gamma = \partial\mathscr{E}$, so

$$V(z) = \int_{\partial\mathscr{E}} U_\psi(\zeta)\,d\omega_{\mathscr{E}}(\zeta, z)$$

is harmonic in $\mathscr{E}$ and continuous up to $\partial\mathscr{E}$, where it takes the boundary value $U_\psi(z)$. For this reason, the function

$$U_\psi(z) - V(z),$$

harmonic in $\mathscr{E}$, is *identically zero* therein, and we have

$$\int_\gamma U_\psi(\zeta)\,d\omega_{\mathscr{E}}(\zeta, z) + \int_\Gamma \psi(\zeta)\,d\omega_{\mathscr{E}}(\zeta, z) = V(z) = U_\psi(z)$$
$$= \int_\Gamma \psi(\zeta)\,d\omega_{\mathscr{D}}(\zeta, z)$$

for $z \in \mathscr{E}$. Making the covering $\mathscr{U}$ shrink down to E, we end with

$$\omega_{\mathscr{D}}(E, z) = \int_\gamma \omega_{\mathscr{D}}(E, \zeta)\,d\omega_{\mathscr{E}}(\zeta, z) + \omega_{\mathscr{E}}(E, z),$$

our desired relation.

The function $\omega_{\mathscr{D}}(E, \zeta)$ is continuous on γ and zero at p, because $\omega_{\mathscr{D}}(E, z) \leqslant$ each of the functions $U_\psi(z)$ considered above. The function harmonic in $\mathscr{E}$ with boundary values equal to $\omega_{\mathscr{D}}(E, \zeta)$ on γ and to *zero* on Γ is therefore *continuous on* $\gamma \cup \Gamma = \partial\mathscr{E}$, so

$$\int_\gamma \omega_{\mathscr{D}}(E, \zeta)\,d\omega_{\mathscr{E}}(\zeta, z)$$

tends to zero when $z \in \mathscr{E}$ *tends to any point of* Γ. Referring to the previous relation, we see that

$$(\ast\ast) \qquad \omega_{\mathscr{D}}(E, z) - \omega_{\mathscr{E}}(E, z) \longrightarrow 0$$

whenever $z \in \mathscr{E}$ tends to any point of Γ. The *behaviour* of the *first term* on the left is thus *the same* as that of *the second*, for $z \longrightarrow \zeta_0 \in \Gamma$.

Because $\mathscr{E}$ is simply connected, we may use conformal mapping to study $\omega_{\mathscr{E}}(E, z)$'s boundary behaviour.

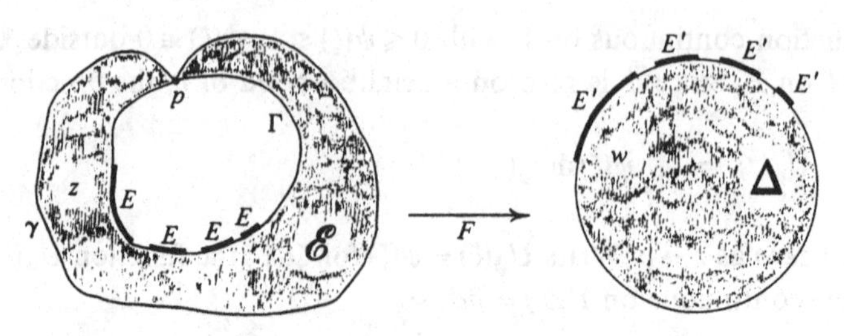

Figure 55

Let F map $\mathscr{E}$ conformally onto $\Delta = \{|w| < 1\}$; F takes $E \subseteq \Gamma$ onto a certain closed subset E' of the unit circumference, and we have

$$\omega_{\mathscr{E}}(E, z) = \omega_{\Delta}(E', F(z))$$

for $z \in \mathscr{E}$ (see the formula near the beginning of this article). *Assume that* Γ *is smooth, or at least that* E *lies on a smooth part of* Γ. Then it is a fact (easily verifiable directly in the cases which will interest us – the general result for curves with a tangent at every point being due to Lindelöf) that *F preserves angles right up to* Γ, as long as we stay away from p:

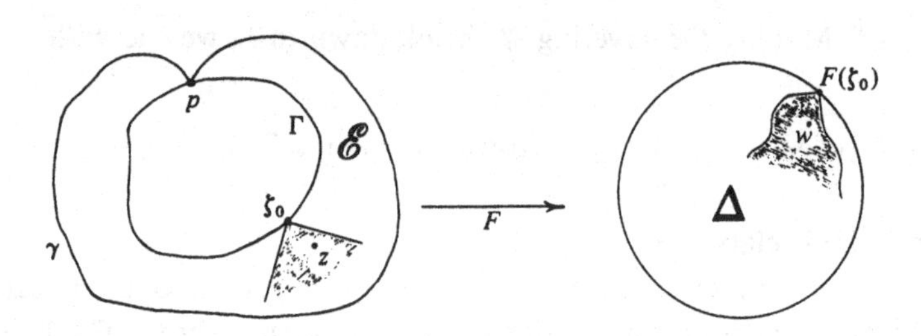

Figure 56

This means that if $z \in \mathscr{E}$ tends to any point ζ_0 of E *from within an acute angle with vertex at* ζ_0, *lying strictly in* $\mathscr{E}$ (we henceforth write this as '$z \xrightarrow{\angle} \zeta_0$'), *the image* $w = F(z)$ *will tend to* $F(\zeta_0) \in E'$ *from within such an angle lying in* Δ.

However,

$$\omega_{\Delta}(E', w) \;=\; \frac{1}{2\pi}\int_0^{2\pi} \frac{1 - |w|^2}{|w - e^{i\varphi}|^2}\chi_{E'}(e^{i\varphi})\,d\varphi.$$

A study of the boundary behaviour of the integral on the right was made in §B of Chapter II. According to the result proved there,

$$\omega_\Delta(E', w) \longrightarrow \chi_{E'}(\omega_0)$$

as $w \not\angle\!\!\!\longrightarrow \omega_0$, for almost every ω_0 on the unit circumference. In the present situation (E' *closed*) we even have

$$\omega_\Delta(E', w) \longrightarrow 0$$

whenever $w \to$ a point of the unit circumference *not* in E'. Under the conformal mapping F, sets of (arc length) measure zero on Γ *correspond precisely* to sets of measure zero on $\{|\omega| = 1\}$. (As before, one can verify this statement directly for the simple situations we will be dealing with. The general result is due to F. and M. Riesz.) In view of the angle preservation just described, we see, going back to $\mathscr{E}$, that, *for almost every* $\zeta_0 \in E$,

$$\omega_{\mathscr{E}}(E, z) \longrightarrow 1 \quad \text{as} \quad z \not\angle\!\!\!\longrightarrow \xi_0,$$

and that

$$\omega_{\mathscr{E}}(E, z) \longrightarrow 0 \quad \text{as} \quad z \longrightarrow \zeta_0$$

for $\zeta_0 \in \Gamma$ *not belonging to* E.

Now we bring in ($\ast$). According to what has just been shown, that relation tells us that

$$\omega_{\mathscr{D}}(E, z) \longrightarrow 1 \quad \text{as} \quad z \not\angle\!\!\!\longrightarrow \zeta_0$$

for *almost every* $\zeta_0 \in E$, whilst

$$\omega_{\mathscr{D}}(E, z) \longrightarrow 0 \quad \text{as} \quad z \longrightarrow \zeta_0$$

for $\zeta_0 \in \Gamma \sim E$, except possibly when $\zeta_0 = p$. By moving p slightly and taking a new curve γ (and new domain $\mathscr{E}$) we can, however, remove any doubt about that case. Referring to the already known boundary behaviour of $\omega_{\mathscr{D}}(E, z)$ at the *other* components of $\partial\mathscr{D}$, we have, finally,

$$\omega_{\mathscr{D}}(E, z) \longrightarrow \begin{cases} 0 & \text{as} \quad z \longrightarrow \zeta_0 \in \partial\mathscr{D} \sim E, \\ 1 & \text{as} \quad z \not\angle\!\!\!\longrightarrow \zeta_0 \text{ for } \textit{almost every } \zeta_0 \in E. \end{cases}$$

This completes our elementary discussion of harmonic measure.

2. Beurling's improvement of Levinson's theorem

We need two auxiliary results.

Lemma. *Let μ be a totally finite (complex) measure on $\mathbb{R}$, and put*

$$\hat{\mu}(\lambda) = \int_{-\infty}^{\infty} e^{i\lambda t}\, d\mu(t)$$

(as usual). Suppose, for some real λ_0, that

$$\int_{\lambda_0}^{\infty} e^{-Y\lambda} e^{iX\lambda} \hat{\mu}(\lambda)\, d\lambda \;\equiv\; 0$$

and

$$\int_{-\infty}^{\lambda_0} e^{Y\lambda} e^{iX\lambda} \hat{\mu}(\lambda)\, d\lambda \;\equiv\; 0$$

for all $X \in \mathbb{R}$ and all $Y > 0$. Then $\mu \equiv 0$.

Proof. If we write $d\mu_{\lambda_0}(t) = e^{i\lambda_0 t}\, d\mu(t)$, we have $\hat{\mu}(\tau + \lambda_0) = \hat{\mu}_{\lambda_0}(\tau)$, and the identical vanishing of μ_{λ_0} clearly implies that of μ. In terms of $\hat{\mu}_{\lambda_0}$, the two relations from the hypothesis reduce to

$$\int_0^{\infty} e^{-Y\tau} e^{iX\tau} \hat{\mu}_{\lambda_0}(\tau)\, d\tau \;\equiv\; 0,$$

$$\int_{-\infty}^{0} e^{Y\tau} e^{iX\tau} \hat{\mu}_{\lambda_0}(\tau)\, d\tau \;\equiv\; 0,$$

valid for $X \in \mathbb{R}$ and $Y > 0$. Therefore, *if we prove the lemma for the case where* $\lambda_0 = 0$, we will *have* $\mu \equiv 0$. We thus proceed under the assumption that $\lambda_0 = 0$.

By direct calculation (!), for $X \in \mathbb{R}$ and $Y > 0$,

$$\frac{Y}{(X+t)^2 + Y^2} \;=\; \frac{1}{2}\int_{-\infty}^{\infty} e^{-Y|\lambda|} e^{iX\lambda} e^{i\lambda t}\, d\lambda.$$

The integral on the right is absolutely convergent, so, multiplying it by $d\mu(t)$, integrating with respect to t, and changing the order of integration, we find

$$\int_{-\infty}^{\infty} \frac{Y}{(X+t)^2 + Y^2}\, d\mu(t) \;=\; \frac{1}{2}\int_{-\infty}^{\infty} e^{-Y|\lambda|} e^{iX\lambda} \hat{\mu}(\lambda)\, d\lambda.$$

Under our assumption, the integral on the right vanishes identically for $X \in \mathbb{R}$ and $Y > 0$. Calling the one on the left $J_Y(X)$, we have, however,

$$J_Y(-X)\, dX \;\longrightarrow\; \pi\, d\mu(X) \quad w^*$$

for $Y \longrightarrow 0$. Therefore $d\mu(X) \equiv 0$, and we are done.

Lemma. *Let $M(r) \geqslant 0$ be increasing on $[0, \infty)$, and put*

$$M_*(r) = \min(r, M(r))$$

for $r \geqslant 0$. Then, if

$$\int_0^{\infty} \frac{M(r)}{1 + r^2}\, dr \;=\; \infty$$

we also have

$$\int_0^\infty \frac{M_*(r)}{1+r^2}\,\mathrm{d}r \;=\; \infty.$$

Proof. Is like that of the lemma in §A.3. The following diagram shows that $M(r) = M_*(r)$ outside of a certain open set $\mathcal{O}$, the union of disjoint intervals (a_n, b_n), on which $M_*(r) = r$.

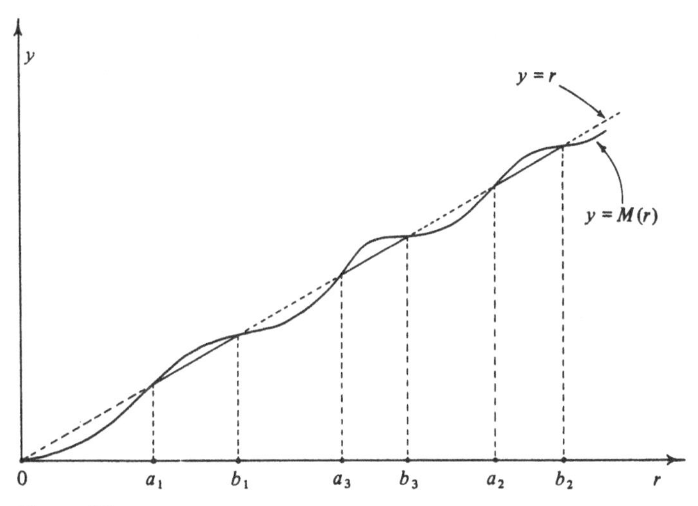

Figure 57

It is enough to show that

$$\int_1^\infty \frac{M_*(r)}{r^2}\,\mathrm{d}r \;=\; \infty.$$

If

$$\int_{\mathcal{O}\cap[1,\infty)} \frac{M_*(r)}{r^2}\,\mathrm{d}r \;=\; \infty,$$

we are already finished; let us therefore assume that this last integral is *finite*. We then sureiy have

$$\sum_{a_n\geqslant 1}\int_{a_n}^{b_n} \frac{M_*(r)}{r^2}\,\mathrm{d}r \;=\; \sum_{a_n\geqslant 1}\int_{a_n}^{b_n}\frac{\mathrm{d}r}{r} \;=\; \sum_{a_n\geqslant 1}\log\left(\frac{b_n}{a_n}\right) \;<\; \infty,$$

so $b_n/a_n \underset{n}{\longrightarrow} 1$ which, fed back into the last relation, gives us

$$\sum_{a_n\geqslant 1}\frac{b_n - a_n}{a_n} \;<\; \infty.$$

Since, however, $M(r)$ is increasing, we see from the picture that

$$\int_{a_n}^{b_n} \frac{M(r)}{r^2}\,dr \;\leqslant\; M(b_n)\int_{a_n}^{b_n}\frac{dr}{r^2} \;=\; b_n\cdot\frac{b_n - a_n}{a_n b_n} \;=\; \frac{b_n - a_n}{a_n}.$$

Therefore

$$\sum_{a_n \geqslant 1} \int_{a_n}^{b_n} \frac{M(r)}{r^2}\,dr \;<\; \infty$$

by the previous relation, so, since we are assuming

$$\int_0^{\infty} \frac{M(r)}{1+r^2}\,dr \;=\; \infty$$

which implies

$$\int_1^{\infty} \frac{M(r)}{r^2}\,dr \;=\; \infty$$

(M being increasing), we must have

$$\int_E \frac{M(r)}{r^2}\,dr \;=\; \infty,$$

where

$$E \;=\; [1,\infty) \sim \bigcup_{a_n \geqslant 1} (a_n, b_n).$$

The set E is either *equal* to the complement of $\mathcal{O}$ in $[1,\infty)$ or else *differs* therefrom by an interval of the form $[1, b_k)$ where (a_k, b_k) is a component of $\mathcal{O}$ straddling the point 1 (in case there is one). Since $M_*(r) = M(r)$ *outside* $\mathcal{O}$, we thus have

$$\int_E \frac{M_*(r)}{r^2}\,dr \;=\; \infty$$

(including in the possible situation where $b_k = \infty$), and therefore

$$\int_1^{\infty} \frac{M_*(r)}{r^2}\,dr \;=\; \infty$$

as required.

Theorem (Beurling). *Let μ be a finite complex measure on $\mathbb{R}$ such that*

$$\int_{-\infty}^{0} \frac{1}{1+x^2} \log\left(\frac{1}{\int_{-\infty}^{x} |d\mu(t)|}\right) dx \;=\; \infty.$$

If

$$\hat{\mu}(\lambda) = \int_{-\infty}^{\infty} e^{i\lambda t}\, d\mu(t)$$

vanishes *on a set* $E \subseteq \mathbb{R}$ *of* positive measure, *then* $\mu \equiv 0$.

Proof. In the *complex* λ-plane, let $\mathscr{D}$ be the strip

$$\{0 < \Im\lambda < 1\};$$

we work with harmonic measure $\omega_{\mathscr{D}}(\ ,\lambda)$ *for* $\mathscr{D}$ (see article 1).

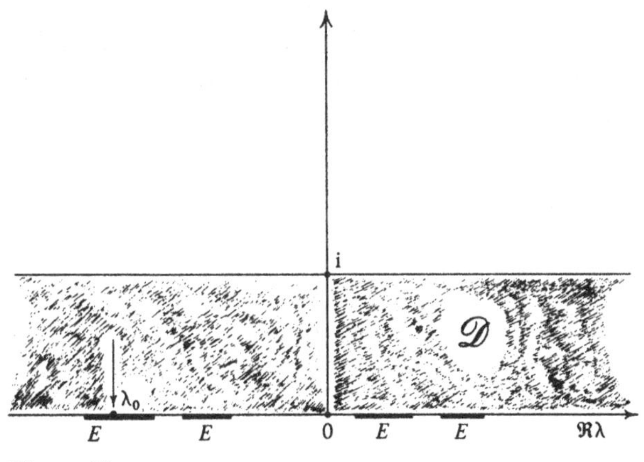

Figure 58

Because $|E| > 0$, E contains a compact set of *positive* Lebesgue measure; there is thus no loss of generality in assuming E *closed* and *bounded.* According to the discussion at the end of the previous article, we then have

$$\omega_{\mathscr{D}}(E, \lambda) \longrightarrow 1$$

as $\lambda \not\longrightarrow \lambda_0$ *for almost every* λ_0 *in the set* E (*of positive measure*). There is thus certainly a $\lambda_0 \in E$ with

$$\omega_{\mathscr{D}}(E, \lambda_0 + i\tau) \longrightarrow 1$$

for $\tau \to 0+$; *we take such a* λ_0 *and fix it* throughout the rest of the proof.

We are going to show that

$$\int_{-\infty}^{\lambda_0} e^{Y\lambda} e^{iX\lambda} \hat{\mu}(\lambda)\, d\lambda \equiv 0$$

and

$$\int_{\lambda_0}^{\infty} e^{-Y\lambda} e^{iX\lambda} \hat{\mu}(\lambda)\, d\lambda \equiv 0$$

for $Y > 0$ and $X \in \mathbb{R}$; by the *first* of the above lemmas we will then have $\mu \equiv 0$ which is what we want to establish. The argument here is the same for any value of λ_0. *In order not to burden the exposition with a proliferation of symbols, we give it for the case where $\lambda_0 = 0$, which we henceforth assume. We have, then, $\hat{\mu}(\lambda) = 0$ on the closed set E, $0 \in E$, and $\omega_{\mathscr{D}}(E, i\tau) \longrightarrow 1$ for $\tau \longrightarrow 0+$.*

Consider the *second* of the above two integrals. Under the present circumstances, it is equal to

$$\int_0^\infty e^{iZ\lambda}\hat{\mu}(\lambda)\,d\lambda \;=\; F(Z),$$

say, where $Z = X + iY$. The function $F(Z)$ defined in this fashion is *analytic* for $\Im Z > 0$ and *bounded* in each half plane of the form $\Im Z \geqslant h > 0$. By §G.2 of Chapter III, we will therefore have $F(Z) \equiv 0$ for $\Im Z > 0$ provided that

$$(*) \qquad \int_0^\infty \frac{\log|F(X+i)|}{1+X^2}\,dX \;=\; -\infty.$$

This relation we now proceed to establish.

Take a number $A > 0$ (later on, A will be made to depend on X), and write

$$\hat{\mu}(\lambda) \;=\; \hat{\mu}_A(\lambda) + \hat{\rho}_A(\lambda),$$

with

$$\hat{\mu}_A(\lambda) \;=\; \int_{-A}^\infty e^{i\lambda t}\,d\mu(t)$$

and

$$\hat{\rho}_A(\lambda) \;=\; \int_{-\infty}^{-A} e^{i\lambda t}\,d\mu(t).$$

The function $\hat{\mu}_A(\lambda)$ is actually defined for $\Im\lambda \geqslant 0$ and *analytic* when $\Im\lambda > 0$. $\hat{\rho}_A(\lambda)$ is not, in general, defined for $\Im\lambda > 0$; when A is large, it is, however, *very small* on the real axis in view of our assumption on

$$\int_{-\infty}^x |d\mu(t)|$$

in the hypothesis. We think of $\hat{\rho}_A(\lambda)$ as a *correction* to $\hat{\mu}_A(\lambda)$ on $\mathbb{R}$.

Wlog,

$$\int_{-\infty}^\infty |d\mu(t)| \;\leqslant\; 1.$$

Then, writing

$$\int_{-\infty}^{-A} |\mathrm{d}\mu(t)| = \mathrm{e}^{-M(A)},$$

we have

$$|\hat{\rho}_A(\lambda)| \leqslant \mathrm{e}^{-M(A)}, \quad \lambda \in \mathbb{R},$$

with $M(A) \geqslant 0$ and increasing for $A \geqslant 0$. Going back to

$$F(X+\mathrm{i}) = \int_0^\infty \mathrm{e}^{-\lambda}\mathrm{e}^{\mathrm{i}X\lambda}\hat{\mu}(\lambda)\,\mathrm{d}\lambda,$$

we see that the latter *differs* from

$$\int_0^\infty \mathrm{e}^{\mathrm{i}(X+\mathrm{i})\lambda}\hat{\mu}_A(\lambda)\,\mathrm{d}\lambda$$

by a quantity in modulus

$$\leqslant \int_0^\infty \mathrm{e}^{-\lambda}|\hat{\rho}_A(\lambda)|\,\mathrm{d}\lambda \leqslant \mathrm{e}^{-M(A)}.$$

As we have already remarked, this is very small when A is large. Showing that $|F(X+\mathrm{i})|$ gets small enough for (∗) to hold thus turns out to reduce to the careful estimation of

$$\left|\int_0^\infty \mathrm{e}^{\mathrm{i}(X+\mathrm{i})\lambda}\hat{\mu}_A(\lambda)\,\mathrm{d}\lambda\right|.$$

We use Cauchy's theorem for that.

Taking Γ as shown,

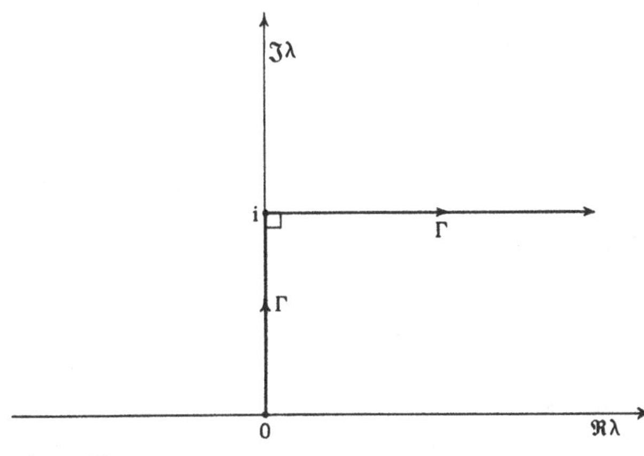

Figure 59

we have

$$\int_0^{\infty} e^{i\lambda(X+i)}\hat{\mu}_A(\lambda)\,d\lambda \;=\; \int_\Gamma e^{i\lambda(X+i)}\hat{\mu}_A(\lambda)\,d\lambda$$

because $\hat{\mu}_A(\lambda)$ is *analytic* for $\Im\lambda > 0$ and *bounded* in the strip

$$0 \leqslant \Im\lambda \leqslant 1,$$

with $|e^{i\lambda(X+i)}|$ going to zero like $e^{-\Re\lambda}$ there as $\Re\lambda \to \infty$. The integral along Γ breaks up as

$$i\int_0^1 e^{-i\tau}e^{-\tau X}\hat{\mu}_A(i\tau)\,d\tau \;+\; \int_0^{\infty} e^{i(\sigma X-1)}e^{-\sigma}e^{-X}\hat{\mu}_A(\sigma+i)\,d\sigma$$

$$= \; I \;+\; II,$$

say.

Since

$$\int_{-\infty}^{\infty} |d\mu(t)| \;\leqslant\; 1,$$

we have

$$|e^{iA\lambda}\hat{\mu}_A(\lambda)| \;=\; \left|\int_{-A}^{\infty} e^{i\lambda(t+A)}\,d\mu(t)\right| \;\leqslant\; 1$$

for $\Im\lambda \geqslant 0$. In particular, for $\sigma \in \mathbb{R}$, $|\hat{\mu}_A(\sigma+i)| \leqslant e^A$, and

$$|II| \;\leqslant\; \int_0^{\infty} e^{-(X-A)}e^{-\sigma}\,d\sigma \;=\; e^{A-X}.$$

To estimate I, we use the *theorem on two constants* given in the previous article. As we have just seen, $e^{iA\lambda}\hat{\mu}_A(\lambda)$ is in modulus $\leqslant 1$ on the closed strip $\bar{\mathscr{D}}$; it is also continuous there and analytic in $\mathscr{D}$. However, on $E \subseteq \mathbb{R}$, $\hat{\mu}(\lambda) = 0$ (by hypothesis!), so $\hat{\mu}_A(\lambda) = \hat{\mu}(\lambda) - \hat{\rho}_A(\lambda) = -\hat{\rho}_A(\lambda)$ there. Thus, for $\lambda \in E$,

$$|e^{iA\lambda}\hat{\mu}_A(\lambda)| \;=\; |\hat{\rho}_A(\lambda)| \;\leqslant\; e^{-M(A)}$$

According to the theorem on two constants we thus have

$$|e^{iA\lambda}\hat{\mu}_A(\lambda)| \;\leqslant\; e^{-M(A)\omega_{\mathscr{D}}(E,\lambda)} \cdot 1^{1-\omega_{\mathscr{D}}(E,\lambda)}$$

for $\lambda \in \mathscr{D}$, i.e.,

$$|\hat{\mu}_A(\lambda)| \;\leqslant\; e^{A\Im\lambda}e^{-M(A)\omega_{\mathscr{D}}(E,\lambda)}, \qquad \lambda \in \mathscr{D}.$$

Substituting this estimate into I, we find

$$|I| \;\leqslant\; \int_0^1 e^{A\tau - X\tau - M(A)\omega_{\mathscr{D}}(E,i\tau)}\,d\tau.$$

Recall, however, that $\omega_{\mathscr{D}}(E, \mathrm{i}\tau) \longrightarrow 1$ for $\tau \to 0$. For this reason *$\tau + \omega_{\mathscr{D}}(E, \mathrm{i}\tau)$ has a strictly positive minimum – call it θ – for $0 \leqslant \tau \leqslant 1$; θ does not depend on A or X.*

Suppose $X > 0$. *Then take* $A = X/2$. With this value of A, the previous relation becomes

$$|\mathrm{I}| \leqslant \int_0^1 \mathrm{e}^{-(X/2)\tau - M(X/2)\omega_{\mathscr{D}}(E,\mathrm{i}\tau)}\,\mathrm{d}\tau \leqslant \mathrm{e}^{-\theta M_*(X/2)},$$

where $M_*(r) = \min(r, M(r))$.

At the same time,

$$|\mathrm{II}| \leqslant \mathrm{e}^{-X/2}$$

for $A = X/2$, according to the estimate made above. Therefore, for $X > 0$,

$$\int_0^\infty \mathrm{e}^{\mathrm{i}(X+\mathrm{i})\lambda}\hat{\mu}_{X/2}(\lambda)\,\mathrm{d}\lambda = \int_\Gamma \mathrm{e}^{\mathrm{i}(X+\mathrm{i})\lambda}\hat{\mu}_{X/2}(\lambda)\,\mathrm{d}\lambda$$

is in modulus

$$\leqslant |\mathrm{I}| + |\mathrm{II}| \leqslant \mathrm{e}^{-\theta M_*(X/2)} + \mathrm{e}^{-X/2}.$$

However, the *first* of the last two integrals differs from $F(X + \mathrm{i})$ by a quantity in modulus $\leqslant \mathrm{e}^{-M(X/2)}$ as we have seen. So, for $X > 0$,

$$|F(X + \mathrm{i})| \leqslant \mathrm{e}^{-\theta M_*(X/2)} + \mathrm{e}^{-X/2} + \mathrm{e}^{-M(X/2)}.$$

There is no loss of generality in assuming $\theta \leqslant 1$. Then we get

$$|F(X + \mathrm{i})| \leqslant 3\mathrm{e}^{-\theta M_*(X/2)}, \quad X > 0.$$

Returning to (*), which we are trying to prove, we see that

$$\int_0^\infty \frac{\log|F(X + \mathrm{i})|}{1 + X^2}\,\mathrm{d}X \leqslant \int_0^\infty \frac{\log 3 - \theta M_*(X/2)}{1 + X^2}\,\mathrm{d}X,$$

and the integral on the left will diverge to $-\infty$ if

$$(\overset{*}{*}) \qquad \int_0^\infty \frac{M_*(X/2)}{1 + X^2}\,\mathrm{d}X = \infty.$$

Here,

$$M(A) = \log\left(\frac{1}{\int_{-\infty}^{-A}|\mathrm{d}\mu(t)|}\right),$$

so $\int_0^\infty (M(A)/(1 + A^2))\,\mathrm{d}A = \infty$ by the hypothesis. Therefore

$$\int_0^\infty \frac{M_*(A)}{1 + A^2}\,\mathrm{d}A = \infty$$

for $M_*(A) = \min(A, M(A))$ by the second lemma, i.e.,

$$\int_0^\infty \frac{2M_*(X/2)}{4+X^2}\,\mathrm{d}X \;=\; \infty,$$

implying ($\ast\atop\ast$), since $M_*(A) \geqslant 0$.

We conclude in this fashion that $(*)$ *holds*, whence $F(Z) \equiv 0$ for $\Im Z > 0$, i.e.,

$$\int_0^\infty \mathrm{e}^{-Y\lambda}\mathrm{e}^{\mathrm{i}X\lambda}\hat{\mu}(\lambda)\,\mathrm{d}\lambda \;\equiv\; 0$$

for $Y > 0$ and $X \in \mathbb{R}$.

One shows in like manner that $\int_{-\infty}^0 \mathrm{e}^{Y\lambda}\mathrm{e}^{\mathrm{i}X\lambda}\hat{\mu}(\lambda)\,\mathrm{d}\lambda \equiv 0$ for $Y > 0$ and $X \in \mathbb{R}$; here* one follows the above procedure to estimate

$$\left|\int_{-\infty}^0 \mathrm{e}^{\lambda}\mathrm{e}^{\mathrm{i}X\lambda}\hat{\mu}(\lambda)\,\mathrm{d}\lambda\right|$$

(*again* for $X > 0$!) using *this* contour:

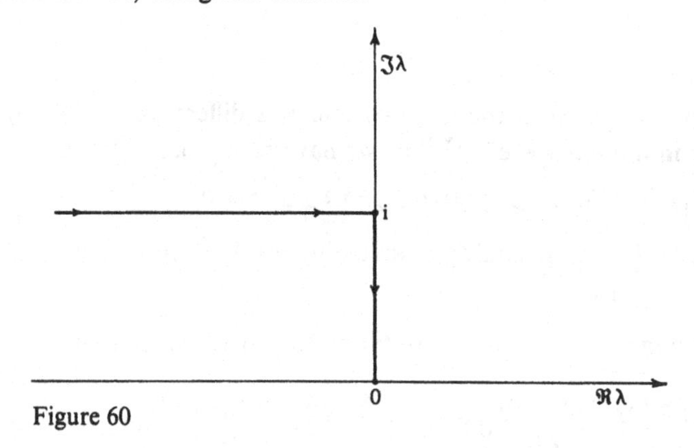

Figure 60

Aside from this change, the argument is like the one given.

The two integrals in question thus vanish identically for $Y > 0$ and $X \in \mathbb{R}$. This, as we remarked at the beginning of our proof, implies that $\mu \equiv 0$. We are done.

Remark 1. The use of the contour integral in the above argument goes back to Levinson, who assumed, however, that $\hat{\mu}(\lambda) = 0$ on an *interval* J instead of just on a set E with $|E| > 0$. In this way Levinson obtained his theorem, given in § A.5, which we now know how to prove much more easily using test functions. By bringing in harmonic measure, Beurling was able to replace the interval J by any measurable set E with $|E| > 0$, getting a qualitative improvement in Levinson's result.

* In which case the integral just written is an analytic function of $X - \mathrm{i}Y$

Remark 2. What about Beurling's gap theorem from §A.2, which says that if the measure μ *has no mass* on any of the intervals (a_n, b_n) with $0 < a_1 < b_1 < a_2 < b_2 < \cdots$ and $\sum_n((b_n - a_n)/a_n)^2 = \infty$, then $\hat{\mu}(\lambda)$ can't vanish identically *on an interval* J, $|J| > 0$, *unless* $\mu \equiv 0$? Can one improve *this* result so as to make it apply for *sets E of positive Lebesgue measure instead of just intervals J of positive length*? Contrary to what happens with Levinson's theorem, the answer *here* turns out to be *no*. This is shown by an example of P. Kargaev, to be given in §C.

3. Beurling's study of quasianalyticity

The argument used to establish the theorem of the preceding article can be applied in the investigation of a kind of quasianalyticity. Let γ be a nice Jordan arc, and look at functions $\varphi(\zeta)$ bounded and continuous on γ. A natural way of describing the *regularity* of such φ is to *measure how well* they can be *approximated* on γ by *certain analytic functions.* The regularity which we are able to specify in such fashion is not necessarily the same as differentiability; it is, however, relevant to the study of a quasianalyticity property considered by Beurling, namely, that of not being able to vanish on a subset of γ having positive (arc-length) measure without being identically zero.

A clue to the kind of regularity involved here comes from the observation that a function φ having a continuous analytic extension to a region *bordering on one side* of γ *possesses* the quasianalyticity property just described. We may thus think of such a φ as being *fully* regular. In order to make this notion of regularity quantitative, let us assume that the arc γ is part of the boundary $\partial\mathscr{D}$ of a simply connected region $\mathscr{D}$.

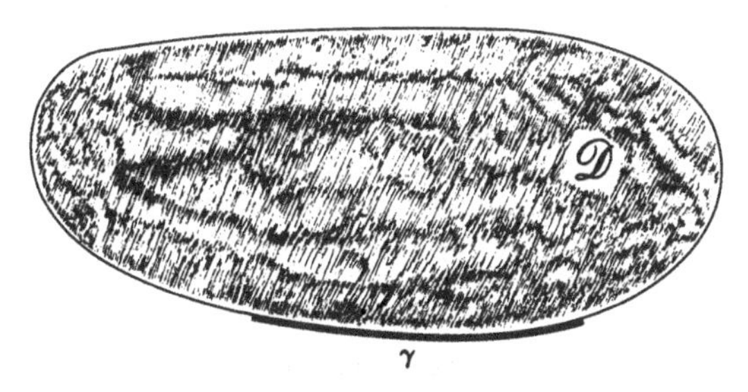

Figure 61

So as to avoid considerations foreign to the matter at hand, we take $\partial\mathscr{D}$ as 'nice' – piecewise analytic and rectifiable, for instance. Given φ bounded and continuous on γ, *define the approximation index* $M(A)$ *for* φ *by functions analytic in* $\mathscr{D}$ *as follows*:

> $e^{-M(A)}$ *is the infimum of* $\sup_{\zeta\in\gamma}|\varphi(\zeta)-f(\zeta)|$ *for* f *analytic in* $\mathscr{D}$ *and continuous on* $\bar{\mathscr{D}}$ *such that* $|f(z)|\leqslant e^A$, $z\in\mathscr{D}$.

We *should* write $M_{\mathscr{D}}(A,\varphi)$ instead of $M(A)$ in order to show the dependence of the approximation index on φ and $\mathscr{D}$; we prefer, however, to use a simpler notation.

When A is made *larger*, we have *more* competing functions f with which to try to approximate φ on γ, so $e^{-M(A)}$ gets *smaller*. In other words, $M(A)$ *increases* with A and we take *the rapidity of this increase* as a *measure* of the *regularity* of φ. Note that if φ actually *has* a bounded continuous extension to $\bar{\mathscr{D}}$ which is *analytic* in $\mathscr{D}$, we have $M(A)=\infty$ beginning with a certain value of A. Such a function φ *cannot* vanish on a set of positive (arc-length) measure on γ without being identically zero, as we have already remarked (this comes, by the way, from two well-known results of F. and M. Riesz). We see that if $M(A)$ grows *rapidly enough*, φ will surely *have* the quasi-analyticity property in question.

The *approximation index* $M(A)$ is a *conformal invariant* in the following sense. Let F map $\mathscr{D}$ conformally onto $\tilde{\mathscr{D}}$, taking the arc γ of $\partial\mathscr{D}$ onto the arc $\tilde{\gamma}\subseteq\partial\tilde{\mathscr{D}}$, and let $\tilde{\varphi}$ be the function defined on $\tilde{\gamma}$ by the relation $\tilde{\varphi}(F(\zeta))=\varphi(\zeta)$, $\zeta\in\gamma$. Then $\tilde{\varphi}$ has the *same* approximation index $M(A)$ for functions analytic in $\tilde{\mathscr{D}}$ as φ has for functions analytic in $\mathscr{D}$. *This is an evident consequence of the use of the sup-norm in defining* $M(A)$.

Our *quasianalyticity property is also* a *conformal invariant*. This follows from the famous theorem of F. and M. Riesz which says that as long as $\partial\mathscr{D}$ and $\partial\tilde{\mathscr{D}}$ are *both rectifiable*, a conformal mapping F of $\mathscr{D}$ onto $\tilde{\mathscr{D}}$ takes sets of arc-length measure zero on $\partial\mathscr{D}$ to such sets on $\partial\tilde{\mathscr{D}}$, and conversely. If $\partial\mathscr{D}$ and $\partial\tilde{\mathscr{D}}$ are really nice, that fact can also be verified directly.

Without further ado, we can now state the

Theorem (Beurling). *Suppose that, for a given bounded continuous* φ *on* $\gamma\subseteq\partial\mathscr{D}$, *the approximation index* $M(A)$ *for* φ *by functions analytic in* $\mathscr{D}$ *satisfies*

$$\int_1^\infty \frac{M(A)}{A^2}\,dA \;=\; \infty.$$

Then, if $E\subseteq\gamma$ *has positive (arc-length) measure, and* $\varphi(\zeta)\equiv 0$ *on* E, $\varphi\equiv 0$ *on* γ.

Proof. By the above statements on conformal invariance, it is enough to establish the result for the special case where $\mathscr{D}$ is the strip $0<\Im\lambda<1$ in the λ-plane and γ is the real axis. The fact that $\partial\mathscr{D}$ is *not rectifiable* here *makes no difference*. We need only show that a set of measure >0 on the rectifiable boundary of our *original* nice domain corresponds under conformal mapping to a set of measure > 0 on the boundary of the *strip*. This may be checked by first mapping the original domain onto the unit disk Δ (whose boundary *is* rectifiable) and appealing to the theorem of F. and M. Riesz mentioned above. One then maps Δ conformally onto the strip; that mapping is, however, easily obtained *explicitly* and thus seen *by inspection* to take sets of measure > 0 on $\partial\Delta$ to sets of measure > 0 on the boundary of the strip.

We have, then, a function φ bounded and continuous on the *real axis*; wlog $|\varphi| \leqslant 1$ there.

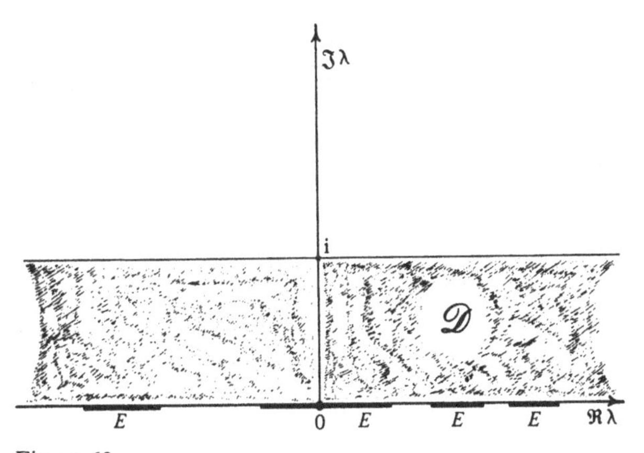

Figure 62

There is a set $E \subseteq \mathbb{R}$ (which we may wlog take to be *closed*) with $|E| > 0$* and $\varphi \equiv 0$ on E. According to the definition of $M(A)$ there is for each $A > 0$ a function $f_A(\lambda)$ analytic in $\mathscr{D}$ and continuous on $\bar{\mathscr{D}}$ with $|f_A(\lambda)| \leqslant e^A$ there and

$$(\dagger) \qquad |\varphi(\lambda) - f_A(\lambda)| \;\leqslant\; 2e^{-M(A)}, \quad \lambda \in \mathbb{R}.$$

By the discussion at the end of article 1, we certainly have

$$\omega_{\mathscr{D}}(E, \lambda) \longrightarrow 1 \quad \text{as} \quad \lambda \xrightarrow{\angle} \lambda_0$$

for some $\lambda_0 \in E$. There is no loss of generality in taking $\lambda_0 = 0$ (we may arrive at this situation by sliding $\bar{\mathscr{D}}$ along the real axis!), and this we *henceforth assume*. We have, then,

$$\omega_{\mathscr{D}}(E, i\tau) \longrightarrow 1 \quad \text{as} \quad \tau \longrightarrow 0+,$$

just as in the proof of the theorem in article 2.

* We are denoting the Lebesgue measure of $E \subseteq \mathbb{R}$ by $|E|$.

In order to show that $\varphi(\lambda) \equiv 0$ on $\mathbb{R}$ it is enough to prove that

$$\int_{-\infty}^{\infty} e^{-Y|\lambda|} e^{i\lambda X} \varphi(\lambda) \, d\lambda = 0$$

for some $Y > 0$ and all real X, for then the function $e^{-Y|\lambda|}\varphi(\lambda)$ (which belongs to $L_1(\mathbb{R})$) must *vanish* a.e. on $\mathbb{R}$ by the *uniqueness theorem* for Fourier transforms. We do this by verifying separately that

$$\int_0^{\infty} e^{i\lambda(X+iY)} \varphi(\lambda) \, d\lambda = 0 \quad \text{for} \quad Y > 0 \quad \text{and} \quad X \in \mathbb{R},$$

and that

$$\int_{-\infty}^{0} e^{i\lambda(X+iY)} \varphi(\lambda) \, d\lambda = 0 \quad \text{for} \quad Y < 0 \quad \text{and} \quad X \in \mathbb{R}.$$

Considering the first relation, write, for $Y > 0$,

$$F(X + iY) = \int_0^{\infty} e^{i\lambda(X+iY)} \varphi(\lambda) \, d\lambda;$$

the function $F(Z)$ is analytic for $\Im Z > 0$ and *bounded* in each half plane $\Im Z \geqslant h > 0$. We want to conclude that $F(Z) \equiv 0$ for $\Im Z > 0$.

Beginning here, we can practically *copy* the proof of the theorem in the previous article. In that proof, we *replace*

$$\hat{\mu}(\lambda) \ \text{by} \ \varphi(\lambda),$$
$$\hat{\mu}_A(\lambda) \ \text{by} \ f_A(\lambda)$$

and $\hat{\rho}_A(\lambda)$ by $\varphi(\lambda) - f_A(\lambda)$. *Everything* will then be *the same*, almost *word* for *word*. True, instead of the inequality $|\hat{\rho}_A(\lambda)| \leqslant e^{-M(A)}$ used above, we *here* have (†), but the extra factor of 2 makes very little difference. We also have to find an inequality for $|f_A(\lambda)|$ in the strip $\bar{\mathscr{D}}$ which will play the rôle of the relation $|\hat{\mu}_A(\lambda)| \leqslant e^{A\Im\lambda}$ used previously. Our function f_A satisfies $|f_A(\lambda)| \leqslant e^A$ on $\bar{\mathscr{D}}$ and

$$|f_A(\lambda)| \leqslant |\varphi(\lambda)| + 2e^{-M(A)} \leqslant 1 + 2e^{-M(0)} \quad \text{for} \quad \lambda \in \mathbb{R}$$

by (†), $M(A)$ being increasing. Therefore

$$|e^{iA\lambda} f_A(\lambda)| \leqslant 1 + 2e^{-M(0)}, \quad \lambda \in \partial\mathscr{D},$$

and we conclude that this inequality holds *throughout* $\bar{\mathscr{D}}$ by the extended principle of maximum (first theorem of § C, Chapter III). In other words,

$$|f_A(\lambda)| \leqslant (1 + 2e^{-M(0)}) e^{A\Im\lambda}$$

for $\lambda \in \bar{\mathscr{D}}$, and this plays the same rôle as the abovementioned inequality on

$\hat{\mu}_A(\lambda)$, the constant factor $1 + 2e^{-M(0)}$ being without real importance.

Repeating in this way the argument from the previous article, we see that the hypothesis

$$\int_1^\infty \frac{M(A)}{A^2}\,dA = \infty$$

of our present theorem implies that $F(Z) \equiv 0$ for $\Im Z > 0$. The fact that

$$\int_{-\infty}^0 e^{i\lambda(X+iY)}\varphi(\lambda)\,d\lambda \equiv 0.$$

for $Y < 0$ and $X \in \mathbb{R}$ also follows by a simple modification of that argument, as indicated at the end of the proof we have been referring to. We are done.

Corollary. *Let μ be a finite measure on $\mathbb{R}$ such that*

$$\int_{-\infty}^0 \frac{1}{1+x^2}\log\left(\frac{1}{\int_{-\infty}^x |d\mu(t)|}\right)dx = \infty,$$

and suppose that $\psi(\lambda)$ is analytic *in a* rectangle

$$\{a < \Re\lambda < b, \quad 0 < \Im\lambda < h\},$$

and continuous on the closure of that rectangle.

If $E \subseteq (a,b)$, $|E| > 0$, and $\hat{\mu}(\lambda) = \int_{-\infty}^\infty e^{i\lambda t}d\mu(t)$ coincides with $\psi(\lambda)$ on E, then $\hat{\mu}(\lambda) \equiv \psi(\lambda)$ on the whole segment $[a,b]$.

Remark. For E an *interval* $\subseteq [a,b]$, this result was proved by Levinson.

Proof of corollary. Without loss of generality, assume that $h = 1$, that $|\psi(\lambda)| \leqslant \frac{1}{2}$ on the rectangle in question, and that $\int_{-\infty}^\infty |d\mu(t)| \leqslant \frac{1}{2}$.

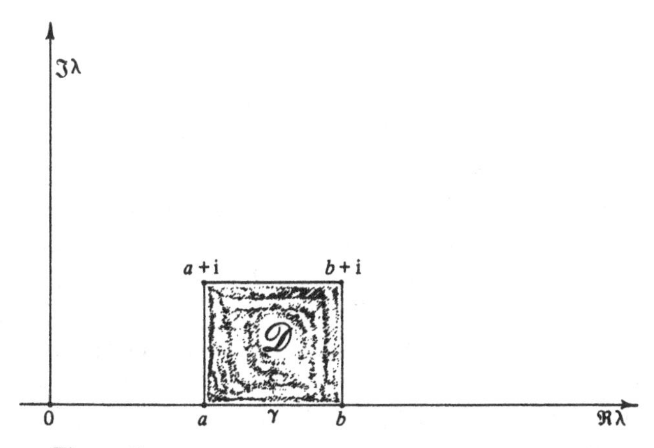

Figure 63

Take our *rectangle* as the domain $\mathscr{D}$ of the theorem, with (a,b) as the arc γ, and put

$$\varphi(\lambda) = \hat{\mu}(\lambda) - \psi(\lambda), \quad a < \lambda < b.$$

For $A > 0$, write

$$f_A(\lambda) = \int_{-A}^{\infty} e^{i\lambda t}\, d\hat{\mu}(t) - \psi(\lambda);$$

the function is analytic in $\mathscr{D}$ and continuous on $\bar{\mathscr{D}}$, and for $A > 0$,

$$|f_A(\lambda)| \leqslant \tfrac{1}{2}e^A + \tfrac{1}{2} \leqslant e^A, \qquad \lambda \in \bar{\mathscr{D}},$$

while for $a < \lambda < b$

$$|\varphi(\lambda) - f_A(\lambda)| \leqslant \int_{-\infty}^{-A} |d\mu(t)| = e^{-M(A)},$$

where $M(A) = \log(1/\int_{-\infty}^{-A} |d\mu(t)|)$.

According to our hypothesis, $\varphi(\lambda) \equiv 0$ on $E \subseteq (a,b) = \gamma$ with $|E| > 0$, and also

$$\int_1^{\infty} \frac{M(A)}{A^2}\, dA = \infty.$$

Therefore $\varphi(\lambda) \equiv 0$ on (a,b) by the theorem, and, by continuity, $\hat{\mu}(\lambda) \equiv \psi(\lambda)$ for $a \leqslant \lambda \leqslant b$. Q.E.D.

4. The spaces $\mathscr{S}_p(\mathscr{D}_0)$, especially $\mathscr{S}_1(\mathscr{D}_0)$

Suppose that $F(\vartheta) \sim \sum_{-\infty}^{\infty} a_n e^{in\vartheta}$ belongs to $L_2(-\pi,\pi)$ and that

$$\sum_{-\infty}^{-1} \frac{1}{n^2} \log\left(\frac{1}{\sum_{-\infty}^{n} |a_k|^2}\right) = \infty.$$

We would like, in analogy with the theorem of article 2, to be able to affirm that $F(\vartheta) \equiv 0$ a.e. if $F(\vartheta)$ vanishes on a set of positive measure. The trouble is that F is not necessarily bounded on $[-\pi,\pi]$, so we cannot work directly with the uniform norm used up to now in the present §. At least two ideas for getting around this difficulty come to mind; one of them is to establish L_p variants of the results in article 2 and 3. Such versions are no longer conformally invariant. Beurling gave one for *rectangular* domains; one could of course use his method to obtain similar results for other regions. In this and the next subsection we stick to rectangles.

Given a rectangle $\mathscr{D}_0$ with sides parallel to the axes, Beurling considers approximation in L_p norm by certain functions analytic in $\mathscr{D}_0$, belonging to a space $\mathscr{S}_p(\mathscr{D}_0)$ to be defined presently. We need some information about those functions which, strictly speaking, comes from the theory of H_p spaces. Although this is *not* a book about H_p spaces, we proceed to sketch that material. In various special situations (including the one mentioned at the beginning of this article), easier results would suffice, and the reader is encouraged to investigate the possibilities of such simplification.

If $\mathscr{D}_0$ is the rectangle $\{(x, y): x \in I_0, y \in J_0\}$,

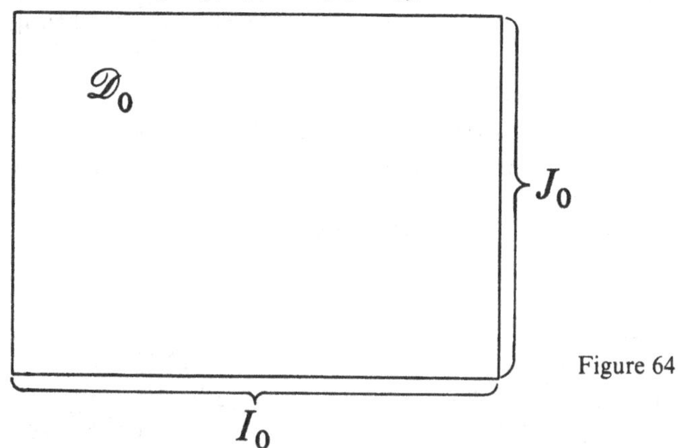

Figure 64

we denote by $\mathscr{S}_p(\mathscr{D}_0)$ the set of functions $f(z)$ analytic in $\mathscr{D}_0$ with

$$\sup_{y \in J_0} \int_{I_0} |f(x+\mathrm{i}y)|^p \,\mathrm{d}x$$

finite, and write

$$\mathscr{s}_p(f) = \sqrt[p]{\sup_{y \in J_0} \int_{I_0} |f(x+\mathrm{i}y)|^p \,\mathrm{d}x}$$

for such f. We are only interested in values of $p \geqslant 1$, and, for such p, $\mathscr{s}_p(\ )$ is a *norm.*

Note that the *compactness* of $\bar{I}_0$ makes $\mathscr{S}_p(\mathscr{D}_0) \subseteq \mathscr{S}_1(\mathscr{D}_0)$ for $p > 1$.

Lemma (Fejér and F. Riesz). *Let $f(w)$ be regular and bounded for $\Im w > 0$, continuous up to the real axis, and zero at ∞. Then*

$$\int_0^\infty |f(\mathrm{i}v)| \,\mathrm{d}v \leqslant \frac{1}{2} \int_{-\infty}^\infty |f(u)| \,\mathrm{d}u.$$

Proof. Under our assumptions on f, we have, for $v > 0$,

$$f(\mathrm{i}v) = \frac{1}{2\pi \mathrm{i}} \int_{-\infty}^\infty \frac{f(u)\,\mathrm{d}u}{u - \mathrm{i}v},$$

$$0 = \frac{1}{2\pi \mathrm{i}} \int_{-\infty}^\infty \frac{f(u)\,\mathrm{d}u}{u + \mathrm{i}v},$$

as long as $\int_{-\infty}^{\infty}|f(u)|du < \infty$, which is the only situation we need consider (see proof of lemma in § H.1, Chapter III). Adding, we get

$$f(\mathrm{i}v) = \frac{1}{\pi \mathrm{i}}\int_{-\infty}^{\infty}\frac{uf(u)\,du}{u^2+v^2},$$

whence

$$\int_0^{\infty}|f(\mathrm{i}v)|dv \leqslant \frac{1}{\pi}\int_{-\infty}^{\infty}\int_0^{\infty}\frac{|u||f(u)|}{u^2+v^2}dv\,du = \frac{1}{2}\int_{-\infty}^{\infty}|f(u)|du.$$

Q.E.D.

Lemma. *Let $F(z)$ be analytic in a rectangle $\mathscr{D}$ and continuous up to $\bar{\mathscr{D}}$. If Λ is a straight line joining the midpoints of two opposite sides of $\mathscr{D}$, we have*

$$\int_{\Lambda}|F(z)||dz| \leqslant \frac{1}{2}\int_{\partial\mathscr{D}}|F(\zeta)||d\zeta|.$$

Proof.

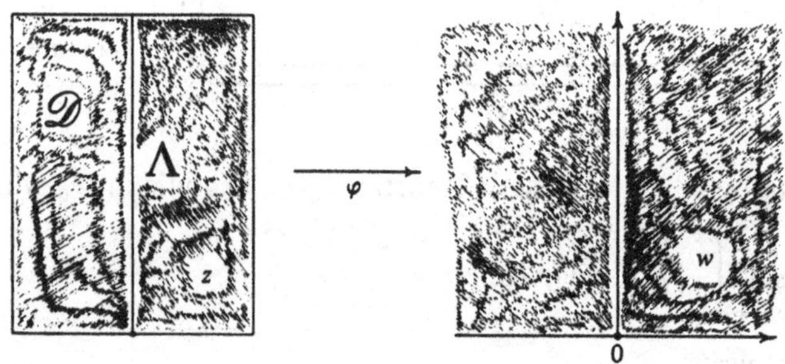

Figure 65

Let φ map $\mathscr{D}$ conformally onto $\Im w > 0$ in such a way that Λ goes onto the *positive imaginary axis*, and, for $z \in \mathscr{D}$ and $w = \varphi(z)$, put

$$f(w) = \frac{F(z)}{\varphi'(z)}.$$

When $w = \varphi(z) \longrightarrow \infty$, $\varphi'(z)$ must tend to ∞ (otherwise the upper half plane would be bounded!), so $f(w)$ must tend to *zero*, $F(z)$ being continuous on $\bar{\mathscr{D}}$. We may therefore apply the previous lemma to f. This yields

$$\int_{\Lambda}|F(z)||dz| = \int_{\Lambda}\left|\frac{F(z)}{\varphi'(z)}\right||\varphi'(z)\,dz| = \int_0^{\infty}|f(\mathrm{i}v)|dv$$

$$\leqslant \frac{1}{2}\int_{-\infty}^{\infty}|f(u)|du = \frac{1}{2}\int_{\partial\mathscr{D}}\left|\frac{F(\zeta)}{\varphi'(\zeta)}\right||\varphi'(\zeta)d\zeta| = \frac{1}{2}\int_{\partial\mathscr{D}}|F(\zeta)||d\zeta|,$$

Q.E.D.

Lemma (Beurling). *Let $\mathscr{D}_0$ be the rectangle $\{-a < \Re z < a, 0 < \Im z < h\}$, and let $f \in \mathscr{S}_1(\mathscr{D}_0)$. Then, if $-a < x < a$,*

$$\int_0^h |f(x+\mathrm{i}y)|dy \leqslant \left(1 + \frac{h}{a-|x|}\right)\sigma_1(f).$$

Proof. Wlog, let $x \geqslant 0$. Taking any small $\delta > 0$ we let $\mathscr{D}_l$, for $0 < l < a - x$, be the rectangle shown in the figure:

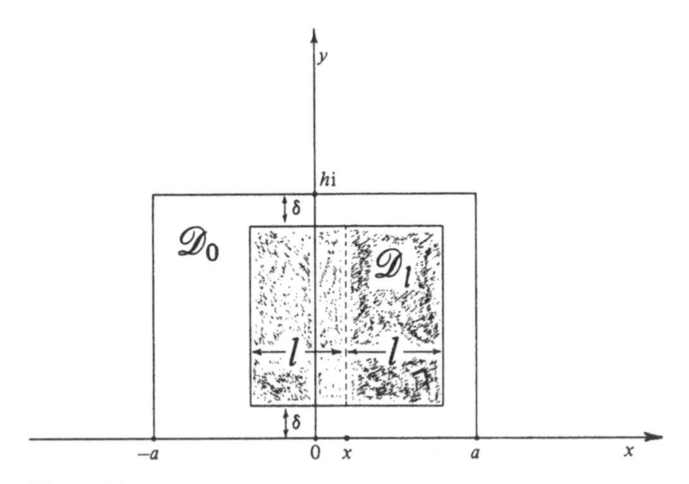

Figure 66

Applying the previous lemma to $\mathscr{D}_l$ we find that

$$\int_\delta^{h-\delta} |f(x + iy)|\,dy \;\leqslant\; \frac{1}{2}\int_{\partial\mathscr{D}_l} |f(\zeta)||d\zeta|.$$

Multiply both sides by dl and integrate l from $\frac{1}{2}(a - x)$ to $a - x$! We get

$$\frac{a-x}{2}\int_\delta^{h-\delta} |f(x + iy)|\,dy \;\leqslant\; \frac{1}{2}\int_{(a-x)/2}^{a-x}\int_{\partial\mathscr{D}_l} |f(\zeta)||d\zeta|\,dl.$$

The *lower horizontal sides* of the $\mathscr{D}_l$ contribute at most

$$\frac{a-x}{4}\int_{-(a-x)}^{a-x} |f(x + i\delta + \xi)|\,d\xi \;\leqslant\; \frac{a-x}{4}\mathfrak{s}_1(f)$$

to the expression on the right, and the *top horizontal sides* of the $\mathscr{D}_l$ contribute a similar amount. The *right vertical sides* give

$$\frac{1}{2}\int_\delta^{h-\delta}\int_{(a-x)/2}^{a-x} |f(x + iy + l)|\,dl\,dy$$

and the *left vertical sides* make a similar contribution. The sum of these last two amounts is

$$\leqslant\; \frac{1}{2}\int_\delta^{h-\delta}\int_{-(a-x)}^{(a-x)} |f(x + iy + l)|\,dl\,dy \;\leqslant\; \tfrac{1}{2}(h - 2\delta)\mathfrak{s}_1(f).$$

All told, we thus have

$$\frac{1}{2}\int_{(a-x)/2}^{a-x}\int_{\partial\mathscr{D}_l} |f(\zeta)||d\zeta|\,dl \;\leqslant\; \left(\frac{a-x}{2} + \frac{h-2\delta}{2}\right)\mathfrak{s}_1(f),$$

so by the previous relation we see that

$$\int_{\delta}^{h-\delta} |f(x+iy)|dy \leqslant \left(1+\frac{h-2\delta}{a-x}\right)\mathscr{I}_1(f).$$

Making $\delta \to 0$, we obtain the lemma for the case $x \geqslant 0$. Done.

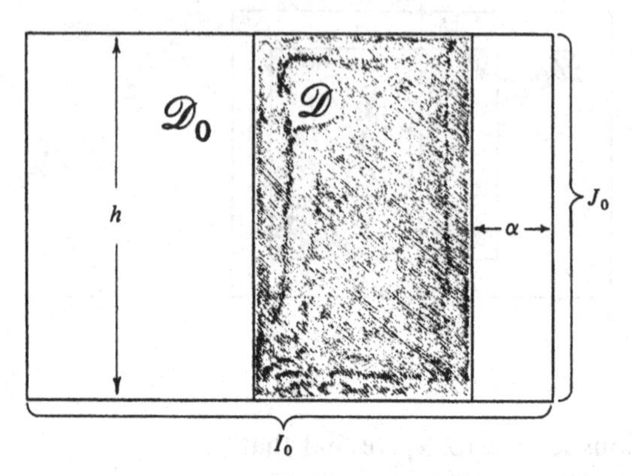

Figure 67

Let $f \in \mathscr{S}_1(\mathscr{D}_0)$, and let $\mathscr{D}$ be a rectangle lying in $\mathscr{D}_0$, in the manner shown – the vertical sides of $\mathscr{D}$ being at *positive distance*, say $\alpha > 0$, from those of $\mathscr{D}_0$. We proceed to investigate the boundary behaviour of f in $\mathscr{D}$.

In order to do this, it is convenient to *take* 0 *as the point of intersection of the diagonals of* $\mathscr{D}$. This setup makes it easy for us to imitate the discussion at the beginning of § F.1, Chapter III.

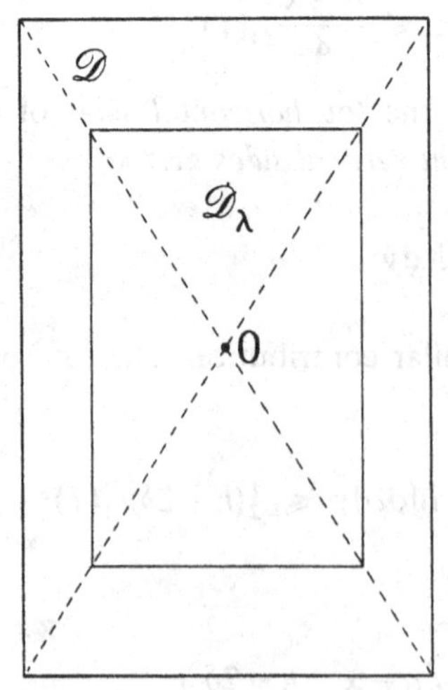

Figure 68

For $0 < \lambda < 1$ denote by $\mathscr{D}_\lambda$ the rectangle $\{\lambda z\colon z \in \mathscr{D}\}$ (see diagram). $\mathscr{D}_\lambda \subseteq \mathscr{D}$ which, in turn, has the above described disposition inside $\mathscr{D}_0$. Since $f \in \mathscr{S}_1(\mathscr{D}_0)$, we have, by the preceding lemma,

$$\int_{\partial\mathscr{D}_\lambda} |f(\zeta)||\mathrm{d}\zeta| \;\leqslant\; 2\sigma_1(f) \;+\; 2\left(1 + \frac{h}{\alpha}\right)\sigma_1(f),$$

calling h the height of $\mathscr{D}_0$. In other words,

$$(*) \qquad \int_{\mathscr{D}} |f(\lambda\zeta)||\mathrm{d}\zeta| \;\leqslant\; \frac{K}{\lambda}$$

for $0 < \lambda < 1$, where K depends on $\mathscr{D}$ and on f.

Fix any $z_0 \in \mathscr{D}$ and let $\lambda < 1$. The function $f(\lambda z)$ is certainly analytic (hence *harmonic*!) in $\mathscr{D}$ and *continuous* up to $\partial\mathscr{D}$ (when z ranges over $\bar{\mathscr{D}}$, the argument of $f(\lambda z)$ actually ranges over $\bar{\mathscr{D}}_\lambda$). Therefore, *by the discussion of article* 1,

$$f(\lambda z_0) \;=\; \int_{\partial\mathscr{D}} f(\lambda\zeta)\,\mathrm{d}\omega_{\mathscr{D}}(\zeta, z_0),$$

denoting, as usual, harmonic measure for $\mathscr{D}$ by $\omega_{\mathscr{D}}(\ \ , z)$. Since the corners of $\mathscr{D}$ makes angles (of 90°) *less than* 180° *from inside*, we know by article 1 that $\mathrm{d}\omega_{\mathscr{D}}(\zeta, z_0)/|\mathrm{d}\zeta|$ is *bounded* (and indeed *continuous*) on $\partial\mathscr{D}$, and the preceding formula can be rewritten thus:

$$f(\lambda z_0) \;=\; \int_{\partial\mathscr{D}} \frac{\mathrm{d}\omega_{\mathscr{D}}(\zeta, z_0)}{|\mathrm{d}\zeta|} \cdot f(\lambda\zeta)|\mathrm{d}\zeta|.$$

(In order to compute $\mathrm{d}\omega_{\mathscr{D}}(\zeta, z_0)/|\mathrm{d}\zeta|$ explicitly, we would have to resort to elliptic functions!)

We can now argue by $(*)$ that there is a certain complex valued measure μ on $\partial\mathscr{D}$ such that

$$f(\lambda\zeta)|\mathrm{d}\zeta| \longrightarrow \mathrm{d}\mu(\zeta) \quad \mathrm{w}^*$$

when $\lambda \to 1$ through a certain sequence of values, and thereby deduce from the previous relation that

$$(\dagger) \qquad f(z_0) = \int_{\partial\mathscr{D}} \frac{\mathrm{d}\omega_{\mathscr{D}}(\zeta, z_0)}{|\mathrm{d}\zeta|}\,\mathrm{d}\mu(\zeta).$$

(See proof of first theorem in § F.1, Chapter III.) This, of course, holds for any $z_0 \in \mathscr{D}$.

Let φ be a conformal mapping of $\mathscr{D}$ onto $\{|w| < 1\}$ and let the function F, analytic in the unit disk, be defined by the formula $F(\varphi(z)) = f(z)$, $z \in \mathscr{D}$. If ν is the complex measure on $\{|w| = 1\}$ such that $\mathrm{d}\nu(\varphi(\zeta)) = \mathrm{d}\mu(\zeta)$ for ζ varying on $\partial\mathscr{D}$, (†) becomes

$$(\dagger\dagger) \qquad F(w) = \frac{1}{2\pi}\int_{|\omega|=1}\frac{1-|w|^2}{|w-\omega|^2}\,d\nu(\omega),$$

$|w|<1$. The integral on the right therefore represents an *analytic function of* w for $|w|<1$. From *this* it follows by the celebrated *theorem of the brothers Riesz* that ν *must be absolutely continuous*, i.e.,

$$(\S) \qquad d\nu(\omega) = \psi(\omega)|d\omega|$$

with some L_1*-function* ψ *on the unit circumference.* By Chapter II, § B, and (††) we now have $F(w)\longrightarrow\psi(\omega)$ as $w \mathrel{\angle\!\!\!\rightarrow} \omega$ for almost every ω on the unit circumference. Write $g(\zeta)=\psi(\varphi(\zeta))$ for $\zeta\in\partial\mathscr{D}$. Then, going back to $\mathscr{D}$, we see by the discussion in article 1 that

$$f(z)\longrightarrow g(\zeta) \quad \text{as} \quad z \mathrel{\angle\!\!\!\rightarrow} \zeta$$

for almost every $\zeta\in\partial\mathscr{D}$.

Plugging (§) into (††) and then returning to (†), we find that

$$f(z_0) = \int_{\partial\mathscr{D}}\frac{d\omega_{\mathscr{D}}(\zeta,z_0)}{|d\zeta|}g(\zeta)|d\zeta|.$$

We have already practically obtained the

Theorem. *Let* $f\in\mathscr{S}_1(\mathscr{D}_0)$. *Then*

$$\lim_{z \angle\!\rightarrow \zeta} f(z) \text{ which we call } f(\zeta)$$

exists for almost every ζ *on the* horizontal sides *of* $\mathscr{D}_0$.

If $\mathscr{D}$ *is a rectangle in* $\mathscr{D}_0$, *disposed in the manner indicated above,*

$$\int_{\partial\mathscr{D}}|f(\zeta)||d\zeta| < \infty,$$

and, for $z\in\mathscr{D}$,

$$f(z) = \int_{\partial\mathscr{D}} f(\zeta)\,d\omega_{\mathscr{D}}(\zeta,z).$$

If B_1 *and* B_2 *denote the horizontal sides of* $\mathscr{D}_0$, *we have*

$$\int_{B_1}|f(z)|dx \leqslant \sigma_1(f),$$

$$\int_{B_2}|f(z)|dx \leqslant \sigma_1(f).$$

Proof.

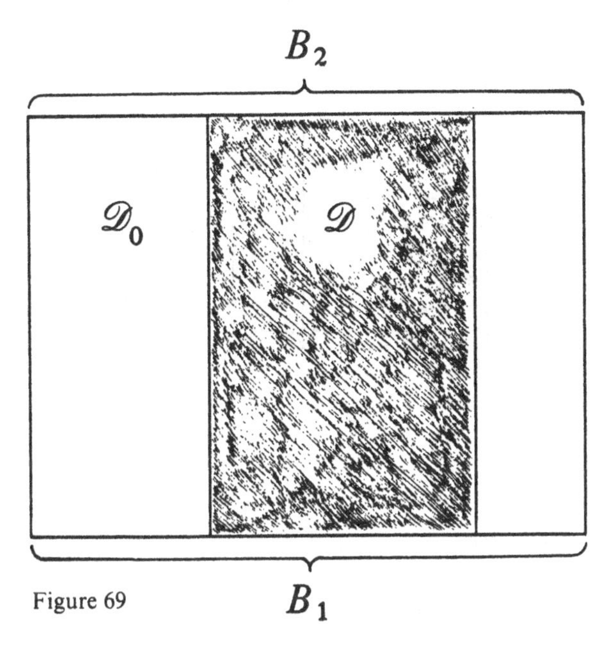

Figure 69

The *first* statement holds because $\lim_{z \not\to \zeta} f(z)$ exists for almost all ζ on the boundary of *any* rectangle $\mathscr{D}$ lying in $\mathscr{D}_0$ in the manner shown; this we have just seen. Of course, if ζ lies on the *vertical sides* of such a rectangle $\mathscr{D}$, we *know anyway* that $\lim_{z\to\zeta} f(z)$ (without the angle mark!) exists and equals $f(\zeta)$, since those vertical sides lie in $\mathscr{D}_0$, where f is given as analytic. The *second* statement therefore follows from (∗) and the *first* one, by Fatou's lemma. (In using (∗), one must take 0 as the point of intersection of the diagonals of $\mathscr{D}$.)

In view of what has just been said, the *third* statement is merely another way of expressing the formula immediately preceding this theorem. There remains the *fourth* statement. Considering, for instance, the *upper horizontal side* B_2 of $\mathscr{D}_0$, we have $f(z - i/n) \underset{n}{\longrightarrow} f(z)$ for *almost all* $z \in B_2$ (*first* statement!). Therefore, by Fatou's lemma,

$$\int_{B_2} |f(z)| \mathrm{d}z \leqslant \liminf_{n\to\infty} \int_{B_2} \left| f\left(z - \frac{i}{n}\right)\right| \mathrm{d}x.$$

The integrals on the right are all $\leqslant \mathfrak{s}_1(f)$ (by definition), at least as soon as $1/n <$ the height of $\mathscr{D}_0$. We are done.

Theorem. *Let I be any interval properly included within the base of* $\mathscr{D}_0$, *in the manner shown:*

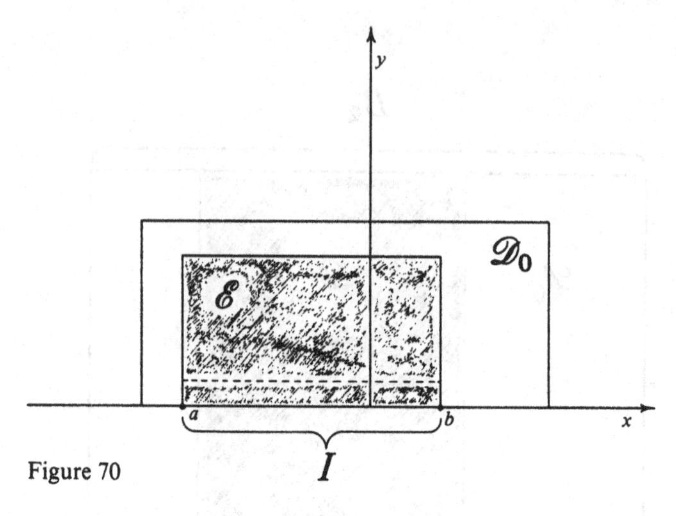

Figure 70

Then, if $f \in \mathscr{S}_1(\mathscr{D}_0)$,

$$\int_I |f(z+i\delta) - f(z)|\,dx \longrightarrow 0$$

as $\delta \to 0$.

Proof. To simplify the writing, we take the base of $\mathscr{D}_0$ to lie on the x-axis as shown in the figure.

In view of the preceding theorem, we may assume that, at the *endpoints* a and b of I, $\lim_{z \not\to a} f(z)$ and $\lim_{z \not\to b} f(z)$ *exist and are finite.* (Otherwise, just make I *a little bigger.*) Then, if we construct the rectangle $\mathscr{E} \subseteq \mathscr{D}_0$ with base on I, in the way shown in the figure, $f(z)$ *will be continuous on the top and two vertical sides of* $\mathscr{E}$, *right up to where the latter meet* I. And by *exactly the same argument* as the one used to establish the *third* statement of the preceding theorem, we can see that

$$f(z) = \int_{\partial\mathscr{E}} f(\zeta)\,d\omega_{\mathscr{E}}(\zeta, z) \quad \text{for} \quad z \in \mathscr{E}.$$

Now let $\varepsilon > 0$ be given, and take a *continuous function* $g(\zeta)$ defined on $\partial\mathscr{E}$ which *coincides with* $f(\zeta)$ *on the top and vertical sides of* $\mathscr{E}$ and is specified on I in such a way that

$$\int_I |f(\xi) - g(\xi)|\,d\xi < \varepsilon,$$

For $z \in \mathscr{E}$, put

$$g(z) = \int_{\partial\mathscr{E}} g(\zeta)\,d\omega_{\mathscr{E}}(\zeta, z);$$

$g(z)$ is at least *harmonic* in $\mathscr{E}$ (N.B. *not* necessarily *analytic* there!), and, by the discussion in article 1, *continuous up to $\partial\mathscr{E}$, where it takes the boundary values $g(\zeta)$.*

For $x \in I$ and small $\delta > 0$,

$$f(x+i\delta) - f(x) = f(x+i\delta) - g(x+i\delta) + g(x+i\delta) \\ - g(x) + g(x) - f(x).$$

We are interested in showing that $\int_I |f(x+i\delta) - f(x)|dx$ is *small* if $\delta > 0$ is small enough. We already know that $\int_I |g(x) - f(x)|dx < \varepsilon$, and, by *continuity* of g on $\bar{\mathscr{E}}$, $\int_I |g(x+i\delta) - g(x)|dx < \varepsilon$ if $\delta > 0$ is small. We will therefore be *done* if we verify that

$$\int_I |g(x+i\delta) - f(x+i\delta)|dx < \varepsilon.$$

Since $f(\zeta) = g(\zeta)$ on $\partial\mathscr{E} \sim I$,

$$f(x+i\delta) - g(x+i\delta) = \int_I (f(\xi) - g(\xi))d\omega_{\mathscr{E}}(\xi, x+i\delta).$$

However, $\mathscr{E}$ lies in the upper half-plane and I on the real axis, so, by the *principle of extension of domain* used in article 1, for $x + i\delta \in \mathscr{E}$,

$$d\omega_{\mathscr{E}}(\xi, x+i\delta) \leqslant \frac{1}{\pi}\frac{\delta d\xi}{(x-\xi)^2 + \delta^2}$$

on I, the right-hand expression being the differential of *harmonic measure* for $\{\Im z > 0\}$ as seen from $x + i\delta$. Thus, for $x \in I$,

$$|f(x+i\delta) - g(x+i\delta)| \leqslant \frac{1}{\pi}\int_I |f(\xi) - g(\xi)|\frac{\delta d\xi}{(x-\xi)^2 + \delta^2}.$$

And

$$\int_I |f(x+i\delta) - g(x+i\delta)|dx$$

$$\leqslant \frac{1}{\pi}\int_I \int_{-\infty}^{\infty} |f(\xi) - g(\xi)|\frac{\delta}{(x-\xi)^2 + \delta^2}dx\,d\xi$$

$$= \int_I |f(\xi) - g(\xi)|d\xi < \varepsilon.$$

This does it.

Corollary. *Let $f \in \mathscr{S}_1(\mathscr{D}_0)$ and let $G(z)$ be any function analytic in a region including the closure of a rectangle $\mathscr{E}$ like the one used above lying in $\mathscr{D}_0$'s* *interior. Then*

$$\int_{\partial\mathscr{E}} G(\zeta)f(\zeta)\mathrm{d}\zeta = 0.$$

Proof. Use Cauchy's theorem for the rectangles with the dotted base together with the above result:

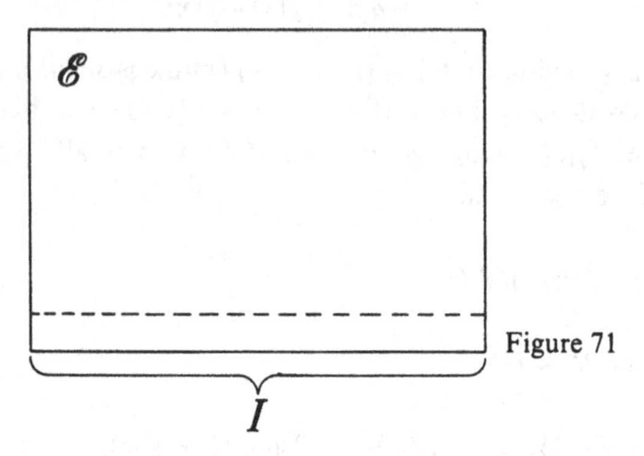

Figure 71

Note that the integrals along the *vertical sides* of $\mathscr{E}$ are absolutely convergent by the *third* lemma of this article.

We need one more result – a Jensen inequality for rectangles $\mathscr{E}$ like the one used above.

Theorem. *Let* $f \in \mathscr{S}_1(\mathscr{D}_0)$*, and let* $\mathscr{E}$ *be a rectangle like the one shown:*

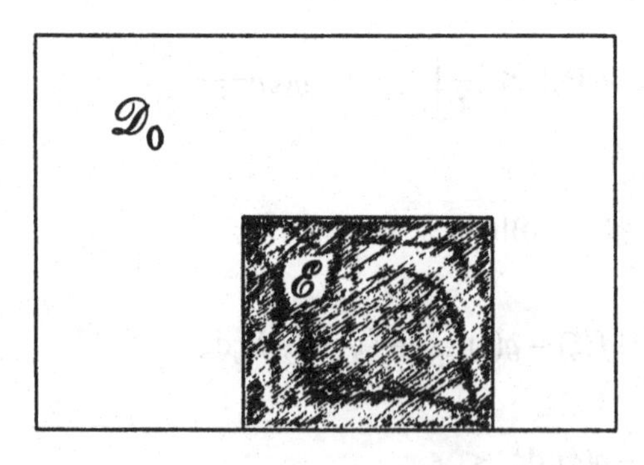

Figure 72

Then, for $z \in \mathscr{E}$,

$$\log|f(z)| \leqslant \int_{\partial\mathscr{E}} \log|f(\zeta)|\mathrm{d}\omega_{\mathscr{E}}(\zeta, z).$$

Proof. This would just be a restatement of the theorem on harmonic estimation from article 1, except that $f(z)$ is not necessarily continuous up to the base of $\mathscr{E}$. There are several ways of getting around the difficulty caused by this lack of continuity; in one such we first map $\mathscr{E}$ conformally onto the unit disk and then use properties of the space H_1. Functions in H_1 can be expressed as products of inner and outer factors, so Jensen's inequality holds for them.

In order to keep the exposition as nearly self-contained as possible, we give a different argument, based on *Szegő's theorem* (§A, Chapter II!), whose idea goes back to Helson and Lowdenslager.

Given $z_0 \in \mathscr{E}$, take a conformal mapping φ onto $\{|w|<1\}$ that sends z_0 to 0, and define a function $F(w)$ analytic in the unit disk by means of the formula

$$F(\varphi(z)) = f(z), \quad z \in \mathscr{E}.$$

The relation

$$f(z) = \int_{\partial\mathscr{E}} f(\zeta)\, d\omega_{\mathscr{E}}(\zeta, z), \quad z \in \mathscr{E},$$

used in proving the above theorem, goes over into

$$F(w) = \frac{1}{2\pi}\int_{-\pi}^{\pi} \frac{1-|w|^2}{|w-e^{i\tau}|^2} F(e^{i\tau})\, d\tau,$$

with $F(e^{i\tau}) = f(\varphi^{-1}(e^{i\tau}))$ defined almost everywhere on the unit circumference and in L_1 (see discussion preceding the first theorem of this article).

From this last relation, we have

$$\int_0^{2\pi} |F(\rho e^{i\vartheta}) - F(e^{i\vartheta})|\, d\vartheta \longrightarrow 0$$

as $\rho \to 1$. Also, for each $\rho < 1$, $\int_0^{2\pi} e^{in\vartheta} F(\rho e^{i\vartheta})\, d\vartheta = 0$ when $n = 1, 2, 3, \ldots$ by *Cauchy's theorem*. Hence

$$\int_0^{2\pi} e^{in\vartheta} F(e^{i\vartheta})\, d\vartheta = 0$$

for $n = 1, 2, 3, \ldots$, and, finally,

$$\frac{1}{2\pi}\int_0^{2\pi} \left(1 + \sum_{n>0} A_n e^{in\vartheta}\right) F(e^{i\vartheta})\, d\vartheta = F(0)$$

for *any* finite sum $\sum_{n>0} A_n e^{in\vartheta}$.

Thus,

$$|F(0)| \leqslant \frac{1}{2\pi}\int_0^{2\pi} \left|1 + \sum_{n>0} A_n e^{in\vartheta}\right| |F(e^{i\vartheta})|\, d\vartheta$$

for all such finite sums. By Szegő's theorem, the *infimum* of the expressions on the right is

$$\exp\left(\frac{1}{2\pi}\int_0^{2\pi}\log|F(e^{i\vartheta})|\,d\vartheta\right).$$

Therefore,

$$\log|F(0)| \leqslant \frac{1}{2\pi}\int_0^{2\pi}\log|F(e^{i\vartheta})|\,d\vartheta,$$

or, in terms of f and $z_0 = \varphi^{-1}(0)$:

$$\log|f(z_0)| \leqslant \int_{\partial\mathscr{E}}\log|f(\zeta)|\,d\omega_{\mathscr{E}}(\zeta, z_0).$$

That's what we wanted to prove.

5. Beurling's quasianalyticity theorem for L_p approximation by functions in $\mathscr{S}_p(\mathscr{D}_0)$.

Being now in possession of the previous article's somewhat *ad hoc* material, we are able to look at approximation by functions in $\mathscr{S}_p(\mathscr{D}_0)$ $(p \geqslant 1)$ and to prove a result about such approximation analogous to the one of article 3.

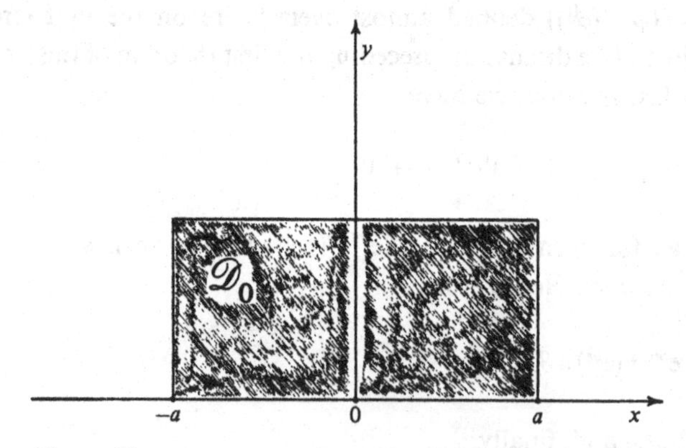

Figure 73

Throughout the following discussion, we work with a certain rectangular domain $\mathscr{D}_0$ whose base is an interval on the real axis which we take, wlog, as $[-a, a]$. If $p \geqslant 1$, $\mathscr{S}_p(\mathscr{D}_0) \subseteq \mathscr{S}_1(\mathscr{D}_0)$, so we know by the *first* theorem of the previous article that, for functions f in $\mathscr{S}_p(\mathscr{D}_0)$, the non-tangential boundary values $f(x)$ exist for almost every x on $[-a, a]$. As in the proof of that theorem we see by Fatou's lemma (there applied in the case $p = 1$) that

$$\int_{-a}^{a} |f(x)|^p \, dx \leqslant (\mathscr{s}_p(f))^p, \quad f \in \mathscr{S}_p(\mathscr{D}_0).$$

The 'restrictions' of functions $f \in \mathscr{S}_p(\mathscr{D}_0)$ to $[-a, a]$ thus belong to $L_p(-a, a)$, and we may use them to try to approximate arbitrary members of $L_p(-a, a)$ in the norm of that space.

In analogy with article 3, we define the *L_p approximation index $M_p(A)$ for any given $\varphi \in L_p(-a, a)$* (and the rectangle $\mathscr{D}_0$) as follows:

$$e^{-M_p(A)} \textit{ is the infimum of } \sqrt[p]{\int_{-a}^{a} |\varphi(x) - f(x)|^p \, dx}$$

for $f \in \mathscr{S}_p(\mathscr{D}_0)$ with $\mathscr{s}_p(f) \leqslant e^A$.

$M_p(A)$ is obviously an *increasing* function of A, and we have the following

Theorem (Beurling). *Let $\varphi \in L_p(-a, a)$, and let its L_p approximation index $M_p(A)$ (for $\mathscr{D}_0$) satisfy*

$$\int_1^{\infty} \frac{M_p(A)}{A^2} \, dA = \infty.$$

If $\varphi(x)$ vanishes on a set of positive measure *in $[-a, a]$, then $\varphi(x) \equiv 0$ a.e. on $[-a, a]$.*

Proof. We first carry out some *preliminary reductions*.

We have $\mathscr{S}_p(\mathscr{D}_0) \subseteq \mathscr{S}_1(\mathscr{D}_0)$, $L_p(-a, a) \subseteq L_1(-a, a)$, and, by Hölder's inequality, $\mathscr{s}_1(f) \leqslant a^{(p-1)/p} \mathscr{s}_p(f)$ and $\|\varphi - f\|_1 \leqslant a^{(p-1)/p} \|\varphi - f\|_p$ for $f \in \mathscr{S}_p(\mathscr{D}_0)$ and $\varphi \in L_p(-a, a)$. (We write $\| \ \|_p$ for the L_p norm on $[-a, a]$). From these facts it is clear that, if $\varphi \in L_p(-a, a)$ has L_p approximation index $M_p(A)$, *the L_1 approximation index $M_1(A)$ of $a^{p/(p-1)}\varphi$ (sic!) is $\geqslant M_p(A)$. It is therefore enough to prove the theorem for $p = 1$, for it will then follow for all values of $p > 1$.*

Suppose then that $\int_1^{\infty} (M_1(A)/A^2) \, dA = \infty$ with $M_1(A)$ the L_1 approximation index for $\varphi \in L_1(-a, a)$, and that φ vanishes on a set of positive measure in $[-a, a]$. In order to prove that $\varphi \equiv 0$ a.e. on $[-a, a]$, it is *enough* to show that it *vanishes a.e. on some interval $J \subseteq [-a, a]$* with positive length.

To see this, take any very small *fixed* $\eta > 0$ and write

$$\varphi_\eta(x) = \frac{1}{2\eta} \int_{-\eta}^{\eta} \varphi(x + t) \, dt$$

for $-a+\eta \leqslant x \leqslant a-\eta$. $\varphi_\eta(x)$ is then *continuous* on the interval $[-a+\eta,\ a-\eta]$, and *vanishes identically on an interval of positive length therein* as long as $2\eta < |J|$. Corresponding to any $f \in \mathscr{S}_1(\mathscr{D}_0)$ we also form the function

$$f_\eta(z) = \frac{1}{2\eta}\int_{-\eta}^{\eta} f(z+t)\,\mathrm{d}t;$$

let us check that $f_\eta(z)$ is *analytic* in the rectangle $\mathscr{D}_\eta$ with base $[-a+2\eta,\ a-2\eta]$ having the same height as $\mathscr{D}_0$, and is *continuous* on $\bar{\mathscr{D}}_\eta$.

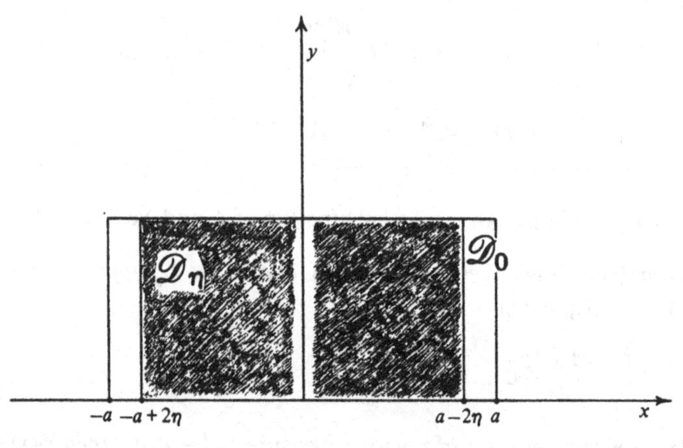

Figure 74

The *analyticity* of $f_\eta(z)$ in $\mathscr{D}_\eta$ is *clear*; so is *continuity* up to the *vertical sides* of $\mathscr{D}_\eta$. The boundary-value function $f(x)$ belongs to $L_1(-a,a)$, so $f_\eta(x)$ *is* continuous *on* $[-a+2\eta,\ a-2\eta]$. Let, then, $-a+2\eta \leqslant x_0 \leqslant a-2\eta$, and suppose that x, also on that closed interval, is *near* x_0 and that $y>0$ is *small*. We have $|f_\eta(x_0)-f_\eta(x+iy)| \leqslant |f_\eta(x_0)-f_\eta(x)| + |f_\eta(x)-f_\eta(x+iy)|$. The first term on the right is small if x is close enough to x_0. The second is

$$\leqslant \frac{1}{2\eta}\int_{x-\eta}^{x+\eta} |f(\xi)-f(\xi+iy)|\,\mathrm{d}\xi \;\leqslant\; \frac{1}{2\eta}\int_{-a+\eta}^{a-\eta} |f(\xi)-f(\xi+iy)|\,\mathrm{d}\xi$$

which, by the *second* theorem of the preceding article, tends to *zero* (independently of x!) as $y \to 0$. Thus $f_\eta(x+iy) \longrightarrow f_\eta(x_0)$ as $x+iy \longrightarrow x_0$ from within $\mathscr{D}_\eta$, and *continuity of f_η up to the lower horizontal side of $\mathscr{D}_\eta$ is established. Continuity* of f_η up to the *upper horizontal side* of $\mathscr{D}_\eta$ follows in like manner, so $f_\eta(z)$ is continuous on $\bar{\mathscr{D}}_\eta$.

The functions f_η are thus of the kind used in article 3 to *uniformly approximate continuous functions* given on $[-a+2\eta,\ a+2\eta]$. By definition of $M_1(A)$, we can find an f in $\mathscr{S}_1(\mathscr{D}_0)$ with $\mathfrak{s}_1(f) \leqslant e^A$ and $\int_{-a}^{a}|\varphi(x)-f(x)|dx \leqslant 2e^{-M_1(A)}$. With this f, $|f_\eta(z)| \leqslant (1/2\eta)e^A$ for $z \in \bar{\mathscr{D}}_\eta$ and

$$|\varphi_\eta(x) - f_\eta(x)| \;\leqslant\; \frac{1}{\eta}e^{-M_1(A)}$$

on $[-a+2\eta,\ a-2\eta]$. The *uniform approximation index* $M(A)$ *for* $\eta\varphi_\eta$ (and the domain $\mathscr{D}_\eta$) *is thus* $\geqslant M_1(A)$. Therefore, under the hypothesis of the present theorem,

$$\int_1^\infty \frac{M(A)}{A^2}\,dA \;=\; \infty,$$

so, since $\varphi_\eta(x)$ *vanishes identically* on an *interval of positive length* in $[-a+2\eta,\ a-2\eta]$ (when $\eta>0$ is small enough) *we have*

$$\varphi_\eta(x) \equiv 0, \quad -a+2\eta \leqslant x \leqslant a-2\eta$$

by the theorem of article 3.

However, as $\eta \to 0$, $\varphi_\eta(x) \longrightarrow \varphi(x)$ a.e. on $(-a,a)$. From what has just been shown we conclude, then, that $\varphi(x) \equiv 0$ a.e. on $(-a,a)$ *if it vanishes* a.e. on an *interval* J of positive length lying therein, provided that

$$\int_1^\infty \frac{M_1(A)}{A^2}\,dA \;=\; \infty.$$

Our task has thus finally boiled down to the following one. *Given* $\varphi \in L_1(-a,a)$ *with* L_1 *approximation index* $M_1(A)$ (for $\mathscr{D}_0$) *such that*

$$\int_1^\infty \frac{M_1(A)}{A^2}\,dA \;=\; \infty,$$

prove that φ *vanishes a.e. on an interval of positive length in* $(-a,a)$ *if it vanishes on a set of positive measure therein.*

Let us proceed. It is easy to see that the *increasing* function $M_1(A)$ is continuous (in the extended sense) – that's because, if $\lambda < 1$ is close to 1, λf approximates φ *almost as well* as f does in $L_1(-a,a)$. Since $\int_1^\infty (M_1(A)/A^2)\,dA = \infty$ we may therefore, starting with a suitable $A_1 > 0$, get an increasing sequence of numbers A_n tending to ∞ such that

$$M_1(A_{n+1}) = 2M_1(A_n).\text{*}$$

Assume henceforth that $\varphi(x) = 0$ on the closed set $E_0 \subseteq [-a,a]$ with

* We are allowing for the possibility that $M_1(A) \equiv \infty$ for large values of A; this happens when $\varphi(x)$ actually coincides with a function in $\mathscr{S}_p(\mathscr{D}_0)$ on $(-a,a)$, and then it is necessary to take A_1 with $M_1(A_1) = \infty$. We will, in any event, need to have A_1 *large* – see the following page.

$|E_0| > 0$*. For each $A > 0$ there is an $f \in \mathscr{S}_1(\mathscr{D}_0)$ with $\sigma_1(f) \leqslant e^A$ and

$$\int_{-a}^{a} |\varphi(x) - f(x)|\,dx \;\leqslant\; 2e^{-M_1(A)}$$

In particular,

$$\int_{E_0} |f(x)|\,dx \;\leqslant\; 2e^{-M_1(A)},$$

so, if

$$\Delta_A \;=\; \{x \in E_0 \colon |f(x)| > e^{-M_1(A)/2}\},$$

we have $|\Delta_A| \leqslant 2e^{-M_1(A)/2}$. Taking the sequence of numbers A_n just described, we thus get

$$\left|\bigcup_n \Delta_{A_n}\right| \;\leqslant\; 2\sum_n e^{-M_1(A_n)/2} \;=\; 2\sum_1^{\infty} e^{-2^{n-1}M_1(A_1)}.$$

We can choose A_1 large enough so that this sum is

$$< \frac{|E_0|}{2};$$

then the set

$$E \;=\; E_0 \sim \left(\bigcup_n \Delta_n\right)$$

has measure $> |E_0|/2$, and, by its *construction*, for each n there is an $f_n \in \mathscr{S}_1(\mathscr{D}_0)$ with $\sigma_1(f_n) \leqslant e^{A_n}$,

$$\int_{-a}^{a} |\varphi(x) - f_n(x)|\,dx \;\leqslant\; 2e^{-M_1(A_n)},$$

and

$$|f_n(x)| \;\leqslant\; e^{-M_1(A_n)/2}$$

for $x \in E$.

Take now a number b, $0 < b < a$, sufficiently close to a so that

$$|E \cap [-b, b]| \;>\; 0,$$

and construct the rectangle $\mathscr{D}$ with base on $[-b, b]$, lying within $\mathscr{D}_0$ in the manner shown:

* where $|E|$ denotes the Lebesgue measure of $E \subseteq \mathbb{R}$

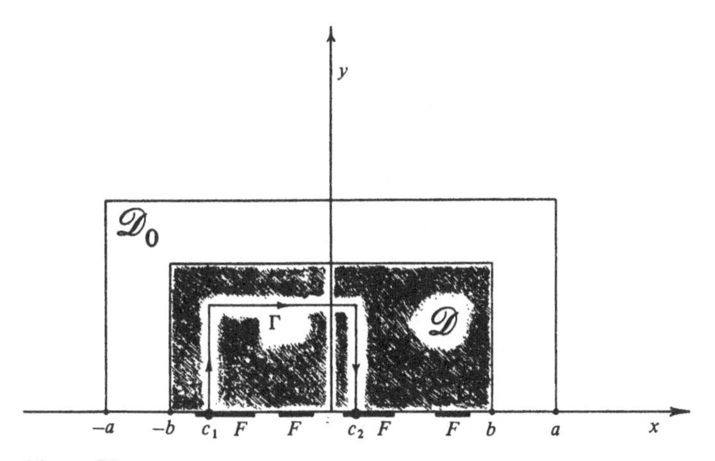

Figure 75

Take a *closed subset F of $E \cap (-b,\ b)$ having positive measure*; this set F will remain fixed during the following discussion.

As we saw at the end of article 1,

$$\omega_{\mathscr{D}}(F, x + iy) \longrightarrow 1$$

as $y \to 0+$ for almost every $x \in F$. Let c_1 and c_2, $c_1 < c_2$, be *two such x's for which this is true.* We are going to show that $\varphi(x) = 0$ a.e. for $c_1 \leqslant x \leqslant c_2$; according to what has been said above, *this is all we need to do* to finish the proof of our theorem.

The desired vanishing of φ will follow if

$$\Phi(\lambda) = \int_{c_1}^{c_2} e^{i\lambda x} \varphi(x)\,dx$$

is identically zero. Φ is, however, *an entire function of exponential type bounded on the real axis.* Hence, by §G.2 *of Chapter* III, $\Phi \equiv 0$ provided that

$$\int_1^{\infty} \frac{1}{\lambda^2} \log\left|\frac{1}{\Phi(\lambda)}\right| d\lambda = \infty.$$

We proceed to verify this relation. The reasoning here resembles that of article 2, but is more complicated.

Take one of the functions f_n (later on, n will be made to depend on λ), and write

$$\Phi(\lambda) = \int_{c_1}^{c_2} e^{i\lambda x}(\varphi(x) - f_n(x))\,dx + \int_{c_1}^{c_2} e^{i\lambda x} f_n(x)\,dx = \text{I} + \text{II, say.}$$

Here, for $\lambda > 0$,

$$|\mathrm{I}| \leqslant \int_{c_1}^{c_2} |\varphi(x) - f_n(x)| \mathrm{d}x \leqslant 2\mathrm{e}^{-M_1(A_n)},$$

and the real work is to estimate II.

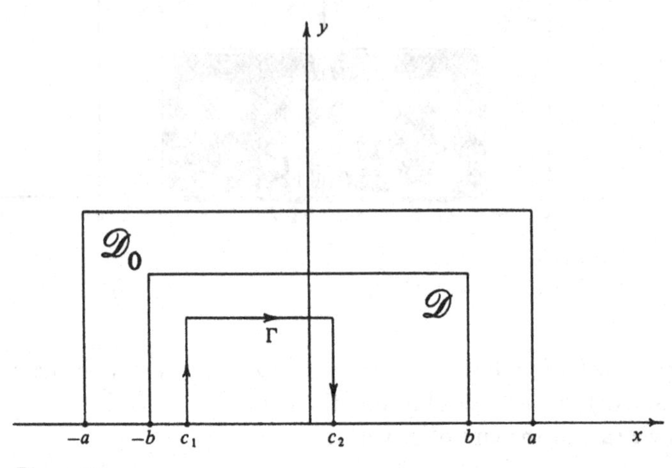

Figure 76

Let Γ be a fixed contour in $\mathscr{D}$ consisting of three sides of a rectangle with base on $[c_1, c_2]$. Because $f_n \in \mathscr{S}_1(\mathscr{D}_0)$, we have

$$\int_{c_1}^{c_2} \mathrm{e}^{\mathrm{i}\lambda x} f_n(x) \,\mathrm{d}x = \int_\Gamma \mathrm{e}^{\mathrm{i}\lambda z} f_n(z) \,\mathrm{d}z$$

by the *corollary* to the *second* theorem of the previous article. In order to estimate the integral on the right, we use the inequality

$$\log|f_n(z)| \leqslant \int_{\partial\mathscr{D}} \log|f_n(\zeta)| \,\mathrm{d}\omega_{\mathscr{D}}(\zeta, z), \qquad z \in \mathscr{D},$$

furnished by the *third* theorem in the preceding article. This we further break up so as to obtain the following for $z \in \mathscr{D}$:

$$(*) \qquad \log|f_n(z)| \leqslant \int_\Pi \log|f_n(\zeta)| \,\mathrm{d}\omega_{\mathscr{D}}(\zeta, z) + \int_F \log|f_n(\zeta)| \,\mathrm{d}\omega_{\mathscr{D}}(\zeta, z)$$
$$+ \int_{(-b,b)\sim F} \log|f_n(\zeta)| \,\mathrm{d}\omega_{\mathscr{D}}(\zeta, z).$$

Here, Π denotes $\partial\mathscr{D} \sim (-b, b)$, i.e., the *vertical* and *top horizontal sides* of $\mathscr{D}$:

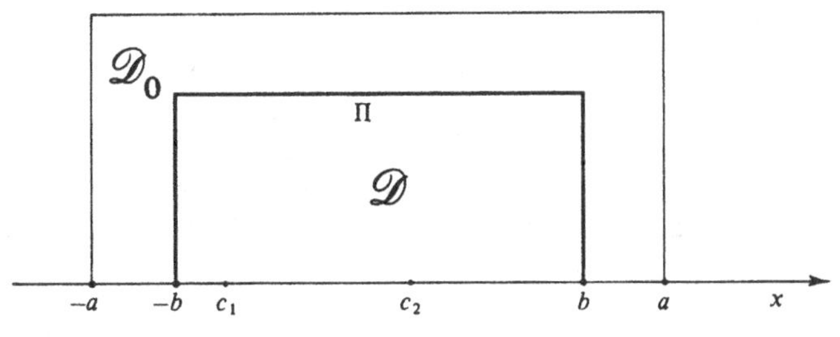

Figure 77

Consider the *first* integral on the right in (*). It equals a certain function $u(z)$ harmonic in $\mathscr{D}$. Take any *harmonic conjugate* $v(z)$ of $u(z)$ for the region $\mathscr{D}$ and put

$$g_n(z) \;=\; \mathrm{e}^{u(z)+\mathrm{i}v(z)}, \qquad z\in\mathscr{D};$$

the function $g_n(z)$ is *analytic* in $\mathscr{D}$ and we have

$$\log|g_n(z)| \;=\; \int_\Pi \log|f_n(\zeta)|\,\mathrm{d}\omega_{\mathscr{D}}(\zeta, z), \qquad z\in\mathscr{D}.$$

In the same way we get functions $h_n(z)$ and $k_n(z)$ analytic in $\mathscr{D}$ with

$$\log|h_n(z)| \;=\; \int_F \log|f_n(\zeta)|\,\mathrm{d}\omega_{\mathscr{D}}(\zeta, z), \qquad z\in\mathscr{D},$$

and

$$\log|k_n(z)| \;=\; \int_{(-b,b)\sim F} \log|f_n(\zeta)|\,\mathrm{d}\omega_{\mathscr{D}}(\zeta, z), \qquad z\in\mathscr{D}.$$

In terms of these functions, (*) becomes

$$(\dagger) \qquad |f_n(z)| \;\leqslant\; |g_n(z)|\,|h_n(z)|\,|k_n(z)|, \qquad z\in\mathscr{D}.$$

Our idea now is to estimate $\sup_{z\in\Gamma}|g_n(z)|$, $\sup_{z\in\Gamma}|h_n(z)|$ and $\int_\Gamma |k_n(z)|\,|\mathrm{d}z|$ in order to get a bound on $\int_\Gamma \mathrm{e}^{\mathrm{i}\lambda z} f_n(z)\,\mathrm{d}z$ for $\lambda > 0$. The *third* of these quantities will give us the most trouble.

We first look at $|g_n(z)|$, $z\in\Gamma$. For ζ on Π, the Poisson kernel $\mathrm{d}\omega_{\mathscr{D}}(\zeta, z)/|\mathrm{d}\zeta|$ goes to *zero* when $z\in\mathscr{D}$ tends to any point of $(-b, b)$, and does so *uniformly* for $\zeta\in\Pi$ and z tending to any point of $[c_1, c_2]$. From this we see, by reflecting the harmonic function $\mathrm{d}\omega_{\mathscr{D}}(\zeta, z)/|\mathrm{d}\zeta|$ of z across $(-b, b)$, that there is a certain constant C, depending *only* on the geometric configuration of Γ and $\mathscr{D}$, such that

$$\frac{\mathrm{d}\omega_{\mathscr{D}}(\zeta, z)}{|\mathrm{d}\zeta|} \;\leqslant\; C\,\Im z, \qquad z\in\Gamma, \quad \zeta\in\Pi.$$

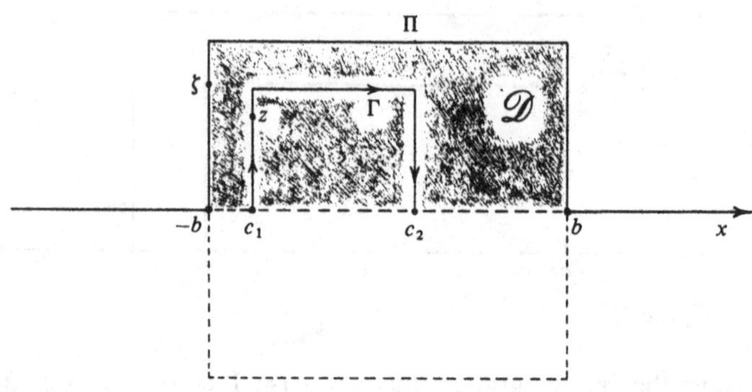

Figure 78

Substituting this into the above formula for $\log|g_n(z)|$, we get

$$\log|g_n(z)| \leqslant C\Im z \int_\Pi \log|f_n(\zeta)||d\zeta|$$

for $z\in\Gamma$, whence, by the inequality between arithmetic and geometric means,*

$$|g_n(z)| \leqslant \left(\frac{1}{|\Pi|}\right)^{|\Pi|C\Im z}\left(\int_\Pi |f_n(\zeta)||d\zeta|\right)^{|\Pi|C\Im z},$$

$z\in\Gamma$. Write now $|\Pi|C = B$. Then we have

$$|g_n(z)| \leqslant \text{const.}\left(\int_\Pi |f_n(\zeta)||d\zeta|\right)^{B\Im z}, \qquad z\in\Gamma,$$

where the constant is independent of n. Here, $f_n\in\mathscr{S}_1(\mathscr{D}_0)$ and $\sigma_1(f_n)\leqslant e^{A_n}$. Thence, by the *third* lemma of the preceding article, if h denotes the height of $\mathscr{D}_0$,

$$\int_\Pi |f_n(\zeta)||d\zeta| \leqslant \sigma_1(f_n) + \left\{1+\frac{h}{a-|c_1|}\right\}\sigma_1(f_n) + \left\{1+\frac{h}{a-|c_2|}\right\}\sigma_1(f_n) \leqslant Ke^{A_n}$$

with a constant K independent of n. Plugging this into the previous relation, we find that

$$|g_n(z)| \leqslant \text{const.}e^{BA_n\Im z}, \qquad z\in\Gamma,$$

the constant in front on the right being independent of n.

To estimate $|h_n(z)|$ on Γ we simply use the fact that

$$|f_n(\xi)| \leqslant e^{-M_1(A_n)/2} \quad \text{for } \xi \in F \subseteq E$$

* in the following relation, $|\Pi|$ is used to designate the *linear measure* (length) of Π

and get

$$|h_n(z)| = \exp\left(\int_F \log|f_n(\zeta)|\,d\omega_{\mathscr{D}}(\zeta, z)\right) \leq e^{-\omega_{\mathscr{D}}(F,z)M_1(A_n)/2}, \qquad z\in\mathscr{D}.$$

Substituting the estimates for $|g_n(z)|$ and $|h_n(z)|$ which we have already found into (†), we obtain

$$(\ast\ast) \qquad |e^{i\lambda z}f_n(z)| \leq \text{const.}e^{(BA_n-\lambda)\Im z}e^{-\omega_{\mathscr{D}}(F,z)M_1(A_n)/2}|k_n(z)|$$

for $z\in\Gamma$. Thus, in order to get a good upper bound for

$$|\text{II}| = \left|\int_\Gamma e^{i\lambda z}f_n(z)\,dz\right|,$$

it suffices to find one for $\int_\Gamma |k_n(z)||dz|$ which is independent of n.

We have

$$\int_{-a}^{a} |\varphi(x) - f_n(x)|\,dx \leq 2e^{-M_1(A_n)}.$$

Wlog,

$$\int_{-a}^{a} |\varphi(x)|\,dx \leq \frac{1}{2},$$

therefore, for all sufficiently large n,

$$(\dagger\dagger) \qquad \int_{-a}^{a} |f_n(x)|\,dx \leq 1.$$

We henceforth limit our attention to the large values of n for which this relation is true.

The formula for $\log|k_n(z)|$ can be rewritten

$$\log|k_n(z)| = \int_{\partial\mathscr{D}} \log P(\zeta)\,d\omega_{\mathscr{D}}(\zeta, z),$$

where

$$P(\zeta) = \begin{cases} |f_n(\zeta)|, & \zeta\in(-b,b)\sim F, \\ 1 \text{ elsewhere on } \partial\mathscr{D}. \end{cases}$$

From this, by the inequality between arithmetic and geometric means, we get

$$|k_n(z)| \leq \int_{\partial\mathscr{D}} P(\zeta)d\omega_{\mathscr{D}}(\zeta, z) \leq 1 + \int_{-b}^{b} |f_n(\xi)|\,d\omega_{\mathscr{D}}(\xi, z), \qquad z\in\mathscr{D}.$$

However, for $-b<\xi<b$, we can apply the principle of extension of

domain to compare $d\omega_{\mathscr{D}}(\xi, z)$ with harmonic measure for $\{\Im z > 0\}$ as we did in proving the *second* theorem of the preceding article. This gives us

$$d\omega_{\mathscr{D}}(\xi, z) \leqslant \frac{1}{\pi}\frac{\Im z \, d\xi}{|z-\xi|^2}, \quad -b < \xi < b,$$

so the previous inequality becomes

$$|k_n(z)| \leqslant 1 + \frac{1}{\pi}\int_{-b}^{b} \frac{\Im z}{|z-\xi|^2} |f_n(\xi)| d\xi, \qquad z \in \mathscr{D}.$$

Denoting by h' the *height* of $\mathscr{D}$, and using this last relation together with Fubini's theorem, we see that, for $0 < y < h'$,

$$\int_{-b}^{b} |k_n(x+iy)| dx \leqslant 2b + \int_{-b}^{b} |f_n(\xi)| d\xi \leqslant 2b + 1$$

(in view of (††)).

In other words, $k_n(z) \in \mathscr{S}_1(\mathscr{D})$ *(sic!), and the* $\mathscr{s}_1$*-norm of* k_n *for* $\mathscr{D}$ *is* $\leqslant 2b+1$ *independently of n.*

Use now the *third* lemma of the previous article *for* $\mathscr{D}$.

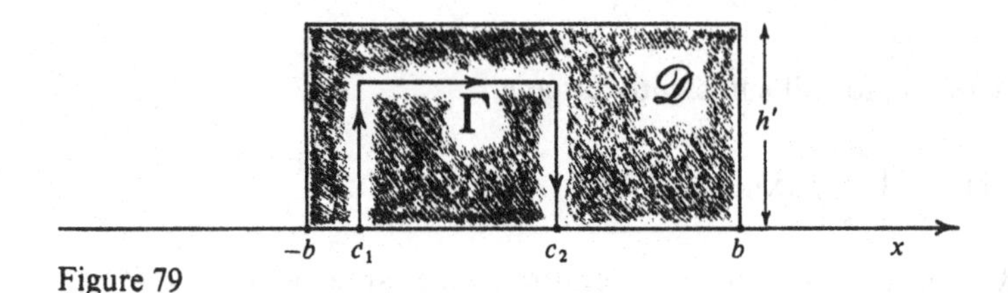

Figure 79

On account of what has just been said, we get

$$\int_{\Gamma} |k_n(z)|\,|dz| \leqslant (2b+1) + (2b+1)\left\{2 + \frac{h'}{b-|c_1|} + \frac{h'}{b-|c_2|}\right\},$$

i.e.

$$(\S) \qquad \int_{\Gamma} |k_n(z)|\,|dz| \leqslant \text{const.},$$

independently of n.

Let us return to ($*$). It is at *this point* that we *choose n according to the value of* $\lambda > 0$. We are actually only interested in *large* values of λ. For any such one, we refer to the sequence $\{A_n\}$ described above, and take n as the integer for which $2BA_n \leqslant \lambda < 2BA_{n+1}$. For *this n*, ($*$) becomes

$$|e^{i\lambda z} f_n(z)| \leqslant \text{const.} e^{-BA_n \Im z - (M_1(A_n)\omega_{\mathscr{D}}(F,z)/2)} |k_n(z)|, \qquad z \in \Gamma.$$

Recall that the two feet c_1 and c_2 of Γ were chosen so as to have

$$\lim_{y\to 0+} \omega_{\mathscr{D}}(F, c_1 + \mathrm{i}y) \;=\; \lim_{y\to 0+} \omega_{\mathscr{D}}(F, c_2 + \mathrm{i}y) \;=\; 1.$$

Therefore

$$B\Im z + \tfrac{1}{2}\omega_{\mathscr{D}}(F, z)$$

has a strictly positive minimum, say β, on Γ. β depends *only* on the geometric configuration of $\mathscr{D}$ and Γ. From the preceding relation, we have, then, when $2BA_n \leqslant \lambda < 2BA_{n+1}$, n being *large,*

$$|\mathrm{e}^{\mathrm{i}\lambda z} f_n(z)| \;\leqslant\; \text{const.}\mathrm{e}^{-\beta\min(A_n, M_1(A_n))}|k_n(z)|, \qquad z\in\Gamma.$$

Now use (§). We get

$$\left|\int_\Gamma \mathrm{e}^{\mathrm{i}\lambda z} f_n(z)\,\mathrm{d}z\right| \;\leqslant\; \text{const.}\mathrm{e}^{-\beta\min(A_n, M_1(A_n))}$$

for $2BA_n \leqslant \lambda < 2BA_{n+1}$; *this,* then, is our desired estimate for $|\mathrm{II}|$.

Now

$$|\Phi(\lambda)| \;=\; \left|\int_{c_1}^{c_2} \mathrm{e}^{\mathrm{i}\lambda x}\varphi(x)\,\mathrm{d}x\right| \;\leqslant\; |\mathrm{I}| + |\mathrm{II}|$$

where $|\mathrm{I}| \leqslant 2\mathrm{e}^{-M_1(A_n)}$, as we saw near the start of the present discussion. We may just as well take $\beta < 1$ (which is in fact *true* any way); then, by the estimate for $|\mathrm{II}|$ just found, we have, for *large n,*

$$|\Phi(\lambda)| \;\leqslant\; \text{const.}\mathrm{e}^{-\beta\min(A_n, M_1(A_n))}, \qquad 2BA_n \leqslant \lambda < 2BA_{n+1}.$$

Our aim here is to show that

$$\int_1^\infty \frac{1}{\lambda^2}\log\left|\frac{1}{\Phi(\lambda)}\right|\mathrm{d}\lambda \;=\; \infty,$$

or, what comes to the same thing, that

$$\int_{\lambda_0}^\infty \frac{1}{\lambda^2}\log\left|\frac{1}{\Phi(\lambda)}\right|\mathrm{d}\lambda \;=\; \infty$$

for some large λ_0. In view of the above inequality for $|\Phi(\lambda)|$, *this holds* if

$$\sum_n \int_{2BA_n}^{2BA_{n+1}} \frac{\min(A_n, M_1(A_n))}{\lambda^2}\,\mathrm{d}\lambda \;=\; \infty,$$

i.e., if

$$(\S\S) \qquad \sum_n \min(A_n, M_1(A_n))\left\{\frac{1}{A_n} - \frac{1}{A_{n+1}}\right\} \;=\; \infty.$$

We proceed to establish this relation. Our hypothesis says that

$$\int_1^\infty \frac{M_1(A)}{A^2}\,\mathrm{d}A \;=\; \infty.$$

The function $M_1(A)$ is increasing, so, by the *second* lemma of article 2, we also have

$$(\ddagger) \qquad \int_1^\infty \frac{\min(A, M_1(A))}{A^2}\,\mathrm{d}A \;=\; \infty.$$

Divide $\mathbb{N}$, the set of positive integers, into three disjoint subsets:

$$R = \{n \geqslant 1\colon\; A_{n+1} \leqslant 2A_n\},$$
$$S = \{n \geqslant 1\colon\; A_{n+1} > 2A_n \text{ and } A_n < M_1(A_n)\},$$
$$T = \{n \geqslant 1\colon\; A_{n+1} > 2A_n \text{ and } M_1(A_n) \leqslant A_n\}.$$

By ($\ddagger$), one of the three sums

$$\sum_{n\in R} \int_{A_n}^{A_{n+1}} \frac{\min(A, M_1(A))}{A^2}\,\mathrm{d}A,$$

$$\sum_{n\in S} \int_{A_n}^{A_{n+1}} \frac{\min(A, M_1(A))}{A^2}\,\mathrm{d}A,$$

$$\sum_{n\in T} \int_{A_n}^{A_{n+1}} \frac{\min(A, M_1(A))}{A^2}\,\mathrm{d}A$$

must be infinite.

Suppose the *first* of those sums is infinite. Recall that the A_n were chosen so as to have $M_1(A_{n+1}) = 2M_1(A_n)$. Therefore, if $n \in R$ and $A_n \leqslant A < A_{n+1}$,

$$\min(A, M_1(A)) \;\leqslant\; \min(A_{n+1}, M_1(A_{n+1})) \;\leqslant\; 2\min(A_n, M_1(A_n)),$$

i.e.,

$$\int_{A_n}^{A_{n+1}} \frac{\min(A, M_1(A))}{A^2}\,\mathrm{d}A$$
$$\leqslant 2\min(A_n, M_1(A_n))\left\{\frac{1}{A_n} - \frac{1}{A_{n+1}}\right\}, \qquad n \in R.$$

In the present case, then, we certainly have (§§).

If the *second* of the sums in question (the one over S) is infinite, the set S *cannot be finite.* However, for $n \in S$,

$$\min(A_n, M_1(A_n))\left\{\frac{1}{A_n} - \frac{1}{A_{n+1}}\right\} \;=\; \frac{A_{n+1} - A_n}{A_{n+1}} \;>\; \frac{1}{2},$$

so (§§) holds when S is infinite.

There remains the case where the *third* sum (over T) is infinite. Here, for $n \in T$ and $A_n \leqslant A < A_{n+1}$ we have

$$\min(A_n, M_1(A_n)) = M_1(A_n) = \tfrac{1}{2}M_1(A_{n+1}) \geqslant \tfrac{1}{2}M_1(A),$$

so, for such n,

$$\min(A_n, M_1(A_n))\left\{\frac{1}{A_n} - \frac{1}{A_{n+1}}\right\} \geqslant \frac{1}{2}\int_{A_n}^{A_{n+1}} \frac{M_1(A)}{A^2}\,\mathrm{d}A$$

$$\geqslant \frac{1}{2}\int_{A_n}^{A_{n+1}} \frac{\min(A, M_1(A))}{A^2}\,\mathrm{d}A.$$

Hence, if the sum of the right-hand integrals for $n \in T$ is infinite, so is that of the left-hand expressions, and (§§) holds.

The relation (§§) is thus proved. This, however, implies that

$$\int_1^\infty \frac{1}{\lambda^2}\log\left|\frac{1}{\Phi(\lambda)}\right|\mathrm{d}\lambda = \infty$$

as we have seen, which is what we needed to show. The theorem is completely proved, and we are done.

Corollary. *Let* $f(\vartheta) \sim \sum_{-\infty}^{\infty} a_n \mathrm{e}^{\mathrm{i}n\vartheta}$ *belong to* $L_2(-\pi, \pi)$*, and suppose that*

$$\sum_{-\infty}^{-1} \frac{1}{n^2}\log\left(\frac{1}{\sum_{-\infty}^{n} |a_k|^2}\right) = \infty.$$

If $f(\vartheta)$ *vanishes on a set of positive measure, then* $f \equiv 0$ *a.e.*

Let the reader deduce the corollary from the theorem. He or she is also encouraged to examine how some of the results from the previous article can be weakened (making their proofs simpler), leaving, however, enough to establish an L_2 version of the theorem which will yield the corollary.

C. Kargaev's example

In remark 2 following the proof of the Beurling gap theorem (§B.2), it was said that that result *cannot* be improved so as to apply to measure μ with $\hat{\mu}(\lambda)$ *vanishing on a set of positive measure, instead of on a whole interval.* This is shown by an example due to P. Kargaev which we give in the present §.

Kargaev's construction furnishes a measure μ with *gaps* (a_n, b_n) in its support, $0 < a_1 < b_1 < a_2 < b_2 < \cdots$, such that

$$\sum_1^\infty \left(\frac{b_n - a_n}{a_n}\right)^2 = \infty$$

while $\hat{\mu}(\lambda) = 0$ on a set E with $|E| > 0$. His method shows that *in fact the relative size, $(b_n - a_n)/a_n$, of the gaps in μ's support has no bearing on $\hat{\mu}(\lambda)$'s capability of vanishing on a set of positive measure without being identically zero.* It is possible to obtain such measures with $(b_n - a_n)/a_n \xrightarrow[n]{} \infty$ *as rapidly as we please.* In view of Beurling's gap theorem, there is thus a *qualitative difference* between requiring that $\hat{\mu}(\lambda)$ *vanish on an interval* and merely having it *vanish on a set of positive measure.*

The measures obtained are *supported on the integers*, and their construction uses absolutely convergent Fourier series. The reasoning is elementary and somewhat reminiscent of the work of Smith, Pigno and McGehee on Littlewood's conjecture.

1. Two lemmas

Let us first introduce some *notation.* $\mathscr{A}$ denotes the collection of functions

$$f(\vartheta) = \sum_{-\infty}^{\infty} a_n e^{in\vartheta}$$

with the series on the right *absolutely convergent*. For such a function $f(\vartheta)$ we put

$$\|f\| = \sum_{-\infty}^{\infty} |a_n|$$

and frequently write $\hat{f}(n)$ instead of a_n (both of these notations are customary). $\mathscr{A}$, $\|\ \ \|$ is a Banach space; in fact, a *Banach algebra* because, if f and $g \in \mathscr{A}$, then $f(\vartheta)g(\vartheta) \in \mathscr{A}$, and

$$\|fg\| \leqslant \|f\| \, \|g\|.$$

On account of this relation, $\Phi(f) \in \mathscr{A}$ for any *entire function* Φ if $f \in \mathscr{A}$.

We will be using some simple linear operators on $\mathscr{A}$.

Definition. If $f(\vartheta) = \sum_{-\infty}^{\infty} \hat{f}(n) e^{in\vartheta}$ belongs to $\mathscr{A}$,

$$(P_+ f)(\vartheta) = \sum_{n=0}^{\infty} \hat{f}(n) e^{in\vartheta}$$

and $P_- f = f - P_+ f$. We frequently write f_+ for $P_+ f$ and f_- for $P_- f$.

Observe that, for $f \in \mathscr{A}$, $\|P_+ f\| \leqslant \|f\|$ and $\|P_- f\| \leqslant \|f\|$.

Definition. For N an integer $\geqslant 1$ and $f \in \mathscr{A}$,

$$(H_N f)(\vartheta) = f(N\vartheta).$$

(The H stands for 'homothety'.)

The following relations are obvious:

$$\begin{aligned} H_N(fg) &= (H_N f)(H_N g), \qquad f, g \in \mathscr{A}, \\ \| H_N f \| &= \| f \|, \\ P_+(H_N f) &= H_N(P_+ f), \end{aligned}$$

and $H_N\Phi(f) = \Phi(H_N f)$ for $f \in \mathscr{A}$ and Φ an entire function.

Lemma. *For each integer $N \geqslant 1$ and each $\delta > 0$ there is a linear operator $T_{N,\delta}$ on $\mathscr{A}$ together with a set $E_{N,\delta} \subseteq [0, 2\pi)$ such that:*

(i) *For each $f \in \mathscr{A}$, $g = T_{N,\delta} f$ has $\hat{g}(n) = 0$ for $-N \leqslant n < N$* (sic!);
(ii) *For each $f \in \mathscr{A}$, $(T_{N,\delta} f)(\vartheta) = f(\vartheta)$ for $\vartheta \in E_{N,\delta}$;*
(iii) *$\| T_{N,\delta} f \| \leqslant C(\delta) \| f \|$ with $C(\delta)$ depending* only *on δ and* not *on N;*
(iv) *$|E_{N,\delta}| = 2\pi(1 - \delta)$.*

Proof. The idea is as follows: starting with an $f \in \mathscr{A}$, we try to cook functions $g_+(\vartheta)$ and $g_-(\vartheta)$ in $\mathscr{A}$, the first having only positive frequencies and the second only negative ones, in such a way as to get

$$g_+(\vartheta)e^{iN\vartheta} + g_-(\vartheta)e^{-iN\vartheta}$$

'almost' equal to $f(\vartheta)$.

We take a certain $\psi \in \mathscr{A}$ (to be described in a moment) and write

$$(*) \qquad q = e^{i(\psi_+ - \psi_-)}.$$

According to the observations preceding the lemma, $q \in \mathscr{A}$. Our construction of $T_{N,\delta}$ and $E_{N,\delta}$ is based on the following *identity* valid for $f \in \mathscr{A}$:

$$\boxed{f = ((fq)_+ e^{-2i\psi_+})e^{i\psi} + ((fq)_- e^{2i\psi_-})e^{-i\psi}.}$$

To check this, just observe that the right-hand side is

$$\begin{aligned} &(fq)_+ e^{-i(\psi_+ - \psi_-)} + (fq)_- e^{i(\psi_- - \psi_+)} \\ &\quad = ((fq)_+ + (fq)_-)q^{-1} = fq \cdot q^{-1} = f. \end{aligned}$$

Here is the way we choose ψ. Take any 2π-periodic $\mathscr{C}_\infty$-function $\varphi_\delta(\vartheta)$ with a graph like this on the range $0 \leqslant \vartheta \leqslant 2\pi$:

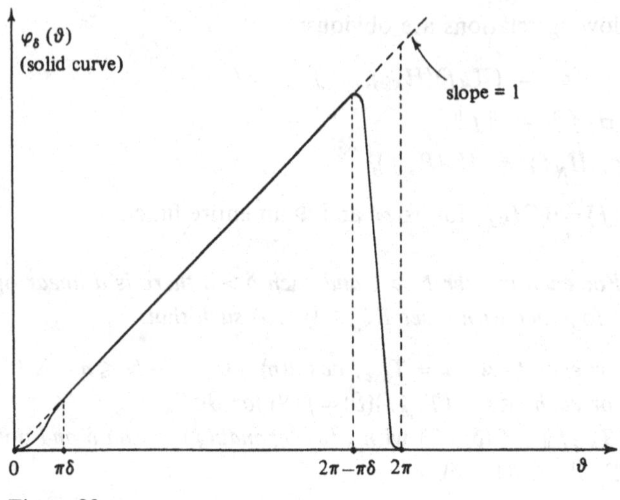

Figure 80

Then put $\psi = H_N\varphi_\delta$; ψ thus depends on N and δ. Note that $\varphi_\delta \in \mathscr{A}$ because φ_δ is infinitely differentiable ($|\hat{\varphi}_\delta(n)| \leqslant O(|n|^{-k})$ for *every* $k > 0$!). Therefore ψ belongs to $\mathscr{A}$.

With $q \in \mathscr{A}$ related by $(*)$ to the ψ just specified, put, for $f \in \mathscr{A}$,

$$T_{N,\delta}f = ((fq)_+ e^{-2i\psi_+})e^{iN\vartheta} + ((fq)_- e^{2i\psi_-})e^{-iN\vartheta}.$$

$T_{N,\delta}$ obviously takes $\mathscr{A}$ into $\mathscr{A}$; let us show that there is a set $E_{N,\delta} \subseteq [0, 2\pi)$ independent of f such that (ii) holds.

The set

$$\Delta_{N,\delta} = \{\vartheta,\ 0 \leqslant \vartheta < 2\pi\colon\ 0 < N\vartheta < \pi\delta \bmod 2\pi \text{ or } 2\pi - \pi\delta < N\vartheta < 2\pi \bmod 2\pi\}$$

consists of $2N$ disjoint intervals, each of length $\pi\delta/N$, so $|\Delta_{N,\delta}| = 2\pi\delta$. Taking into account the 2π-periodicity of the function $\varphi_\delta(\vartheta)$ we see, by looking at its graph, that

$$e^{i\varphi_\delta(N\vartheta)} = e^{iN\vartheta}, \qquad \vartheta \in [0, 2\pi) \sim \Delta_{N,\delta};$$

i.e.,

$$e^{i\psi(\vartheta)} = e^{iN\vartheta}, \qquad \vartheta \in [0, 2\pi) \sim \Delta_{N,\delta}.$$

Put, therefore, $E_{N,\delta} = [0, 2\pi) \sim \Delta_{N,\delta}$; then, by comparing the formula for $T_{N,\delta}f$ with the boxed identity following $(*)$, we see that $(T_{N,\delta}f)(\vartheta) = f(\vartheta)$ for $\vartheta \in E_{N,\delta}$, proving (ii).

We also have (iv), since

$$|E_{N,\delta}| = 2\pi - |\Delta_{N,\delta}| = 2\pi - 2\pi\delta.$$

We come to (i). The function $(fq)_+$ only has *non-negative frequencies* in its Fourier series. The same is true for $e^{-2i\psi_+}$. Indeed, the latter function equals

$$1 - 2i\psi_+ + \frac{(2i\psi_+)^2}{2!} - \frac{(2i\psi_+)^3}{3!} + \cdots$$

with the series *convergent in the norm* $\| \ \|$, and each *power* $(\psi_+)^n$ has a Fourier series involving only frequencies $\geqslant 0$. The Fourier series of the *product* $(fq)_+ e^{-2i\psi_+}$ thus only involves frequencies $\geqslant 0$, and finally, that for

$$((fq)_+ e^{-2i\psi_+})e^{iN\vartheta}$$

only has frequencies $\geqslant N$. One verifies in the same way that

$$((fq)_- e^{2i\psi_-})e^{-iN\vartheta}$$

has a Fourier series involving only the frequencies $< -N$, and (i) now follows from our definition of $T_{N,\delta}$.

There remains (iii). We have, for example,

$$\begin{aligned}\|(fq)_+ e^{-i\psi_+}\| &\leqslant \|(fq)_+\| \, \|e^{-i\psi_+}\| \\ &\leqslant \|fq\| \, \|e^{-2i\psi_+}\| \leqslant \|f\| \, \|q\| \, \|e^{-2i\psi_+}\|.\end{aligned}$$

Here,

$$e^{-2i\psi_+} = e^{-2iP_+ H_N \varphi_\delta} = e^{-2iH_N P_+ \varphi_\delta} = H_N e^{-2iP_+ \varphi_\delta},$$

according to the elementary relations preceding the lemma, so

$$\|e^{-2i\psi_+}\| = \|H_N e^{-2iP_+\varphi_\delta}\| = \|e^{-2iP_+\varphi_\delta}\|,$$

a *finite quantity, depending on δ but not on N*. In like manner,

$$\|q\| = \|e^{i(\psi_+ - \psi_-)}\| = \|H_N e^{i(P_+\varphi_\delta - P_-\varphi_\delta)}\| = \|e^{i(P_+\varphi_\delta - P_-\varphi_\delta)}\|,$$

a finite quantity depending on δ but independent of N. We thus have

$$\|(fq)_+ e^{-2i\psi_+} e^{iN\vartheta}\| = \|(fq)_+ e^{-2i\psi_+}\| \leqslant A_\delta \|f\|,$$

where A_δ depends only on δ.

The norm $\|(fq)_- e^{2i\psi_-} e^{-iN\vartheta}\|$ is handled in exactly the same way, and found to be $\leqslant B_\delta \|f\|$ with B_δ depending only on δ. Referring to the definition of $T_{N,\delta}$, we see that (iii) holds.

The lemma is thus proved.

We now take two positive integers L and N; N will usually be much larger than $2L$.

Definition.

$$\mathscr{M}(N,L) \;=\; \bigcup_{k=-2L-1}^{2L+1}{}' \;[Nk-L,\; Nk+L].$$

Here, the prime next to the union sign means that the term corresponding to the value $k=0$ is *omitted.*

For $N>2L$, $\mathscr{M}(N,L)$ is the union of $4L+2$ separate intervals, each of length $2L$:

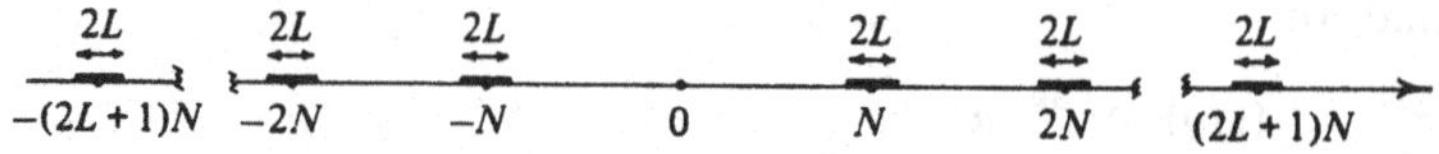

$2L$ $2L$ $2L$ $2L$ $2L$ $2L$

$-(2L+1)N$ $-2N$ $-N$ 0 N $2N$ $(2L+1)N$

Figure 81

In proving the following lemma we use another linear operator on $\mathscr{A}$.

Definition. For $f \in \mathscr{A}$, put

$$(S_L f)(\vartheta) \;=\; \sum_{n=-L}^{L} \hat{f}(n)e^{in\vartheta}.$$

Observe that $\|S_L f\| \leqslant \|f\|$ and $\|f - S_L f\| \longrightarrow 0$ as $L \to \infty$ whenever $f \in \mathscr{A}$. We also have

$$P_+ S_L f \;=\; S_L P_+ f.$$

Lemma. *For each $\delta > 0$ and pair N, L of positive integers there is a linear operator $T_{N,\delta}^{(L)}$ on $\mathscr{A}$ such that*

(1) *For any $f \in \mathscr{A}$, the Fourier coefficients $\hat{g}(n)$ of $g = T_{N,\delta}^{(L)} f$ are all* zero *when $n \notin \mathscr{M}(N,L)$;*
(2) *For $f \in \mathscr{A}$, $\|T_{N,\delta}^{(L)} f\| \leqslant C(\delta)\|f\|$ with $C(\delta)$* independent *of N and L;*
(3) *If $T_{N,\delta}$ is the operator furnished by the previous lemma, we have*

$$\|T_{N,\delta} f - T_{N,\delta}^{(L)} f\| \longrightarrow 0$$

uniformly *in N as $L \to \infty$, for each $f \in \mathscr{A}$ and $\delta > 0$.*

Remarks. Actually, the spectrum of $T_{N,\delta}^{(L)} f$ is contained in a *smaller* set than $\mathscr{M}(N,L)$ when $f \in \mathscr{A}$. It is the *uniformity with respect to N* in property 3 which will turn out to be especially important in Kargaev's construction.

Proof of lemma. Fix $\delta > 0$ and take the function φ_δ used in proving the preceding lemma – *here we just denote it by φ.* In terms of

$$q_0 \;=\; e^{i(\varphi_+ - \varphi_-)},$$

we observe that the definition of $T_{N,\delta}f$ given in the proof of the previous lemma can be rewritten thus:

$$T_{N,\delta}f = (fH_Nq_0)_+(H_N e^{-2i\varphi_+})\cdot e^{iN\vartheta} + (fH_Nq_0)_-(H_N e^{2i\varphi_-})\cdot e^{-iN\vartheta}.$$

Put

$$T_{N\delta}^{(L)}f = (S_Lf\cdot H_NS_Lq_0)_+(H_NS_Le^{-2i\varphi_+})\cdot e^{iN\vartheta} + (S_Lf\cdot H_NS_Lq_0)_-(H_NS_Le^{2i\varphi_-})\cdot e^{-iN\vartheta}.$$

Since $\|g - S_Lg\| \longrightarrow 0$ as $L \to \infty$ for every $g \in \mathscr{A}$, $T_{N,\delta}^{(L)}f$ is clearly a kind of approximation to $T_{N,\delta}f$.

We proceed to verify property (1). The Fourier coefficients of S_Lf are all zero save for those with index in the set

$$\{-L, \ -L+1, \ \ldots, \ 0, \ 1, \ \ldots, \ L\}.$$

The non-zero Fourier coefficients of $H_NS_Lq_0$ have their indices in the set

$$\{-NL, \ -N(L-1), \ \ldots, \ -N, \ 0, \ N, \ \ldots, \ NL\}.$$

Therefore the Fourier coefficients of $(S_Lf\cdot H_NS_Lq_0)_+$ with index *outside* the set

$$\{0, 1, \ldots, L\} \cup \{N-L, N-L+1, \ldots, N, N+1, \ldots, N+L\} \cup \{2N-L, 2N-L+1, \ldots, 2N+L\} \cup \cdots \cup \{NL-L, NL-L+1, \ldots, NL+L\}$$

are *surely zero.*

Again, the Fourier coefficients of $H_NS_Le^{-2i\varphi_+}$ are all zero save for those with index in the set $\{0, N, 2N, \ldots, LN\}$. So, finally, the Fourier coefficients of

$$(S_Lf\cdot H_NS_Lq_0)_+(H_NS_Le^{-2i\varphi_+})e^{iN\vartheta}$$

(the *first* of the two terms making up $T_{N,\delta}^{(L)}f$) are *all zero, save for those with index in the union of intervals*

$$[N, \ N+L] \cup \bigcup_{k=2}^{2L+1} [Nk-L, \ Nk+L].$$

Treating the *second* term of $T_{N,\delta}^{(L)}f$ in the same way, we see that property 1 holds (and that indeed *more* is true regarding the spectrum of $T_{N,\delta}^{(L)}f$).

To check property (2), we have, for the *first term* of $T_{N,\delta}^{(L)}f$,

$$\begin{aligned} &\|(S_Lf\cdot H_NS_Lq_0)_+(H_NS_Le^{-2i\varphi_+})\cdot e^{iN\vartheta}\| \\ &\quad \leqslant \|S_Lf\| \, \|H_NS_Lq_0\| \, \|H_NS_Le^{-2i\varphi_+}\| \\ &\quad \leqslant \|f\| \, \|S_Lq_0\| \, \|S_Le^{-2i\varphi_+}\| \leqslant \|f\| \, \|q_0\| \, \|e^{-2i\varphi_+}\|; \end{aligned}$$

we have used the fact that $\|H_N g\| = \|g\|$ for $g \in \mathscr{A}$. In the extreme right-hand member of the chain of inequalities just written, the factors $\|q_0\|$ and $\|e^{-2i\varphi_+}\|$ are finite and *only involve* $\varphi = \varphi_\delta$; therefore they depend *only on* δ. The *second term* of $T_{N,\delta}^{(L)}f$ is handled in exactly the same fashion, and, putting together the estimates obtained for *both* terms, we arrive at property 2.

Verification of property (3) remains. This is somewhat long-winded. It is really nothing but an elaborate version of the argument presented in good elementary calculus courses to show that the limit of a product equals the product of the limits. In order not to lose sight of the main idea, let's just compare the *first terms* of $T_{N,\delta}f$ and $T_{N,\delta}^{(L)}f$. The *difference* of these first terms has norm equal to

$$\begin{aligned}
&\|(S_L f \cdot H_N S_L q_0)_+ (H_N S_L e^{-2i\varphi_+}) \cdot e^{iN\vartheta} \;-\; (f \cdot H_N q_0)_+ (H_N e^{-2i\varphi_+}) e^{iN\vartheta}\| \\
&\leqslant \|(S_L f \cdot H_N S_L q_0)_+ - (f H_N q_0)_+\| \, \|H_N S_L e^{-2i\varphi_+}\| \\
&\qquad\qquad + \|(f H_N q_0)_+\| \, \|H_N e^{-2i\varphi_+} - H_N S_L e^{-2i\varphi_+}\| \\
&\leqslant \|e^{-2i\varphi_+}\| \, \|(S_L f - f) H_N q_0 + (S_L f)(H_N S_L q_0 - H_N q_0)\| \\
&\qquad\qquad + \|f\| \, \|q_0\| \, \|e^{-2i\varphi_+} - S_L e^{-2i\varphi_+}\| \\
&\leqslant \|e^{-2i\varphi_+}\| \{ \|S_L f - f\| \, \|q_0\| + \|f\| \, \|S_L q_0 - q_0\| \} \\
&\qquad\qquad + \|f\| \, \|q_0\| \, \|e^{-2i\varphi_+} - S_L e^{-2i\varphi_+}\|
\end{aligned}$$

This last expression *does not involve N at all*, and, for fixed $f \in \mathscr{A}$, tends to *zero* as $L \to \infty$. (It depends on δ through the functions φ and $q_0 = e^{i(\varphi_+ - \varphi_-)}$.)

The difference of the *second terms* of $T_{N,\delta}^{(L)}f$ and $T_{N,\delta}f$ is treated in the same way, and we see that property (3) holds. The lemma is proved, and we are done.

2. The example

Theorem (Kargaev). *Let* $\Lambda \subseteq \mathbb{Z}$. *Suppose that for each positive integer L there is some positive integer* N_L *with*

$$\Lambda \supseteq \mathscr{M}(N_L, L) \cap \mathbb{Z},$$

where the sets $\mathscr{M}(N, L)$ *are those defined in the previous article. Then, given* $\varepsilon > 0$ *and* $g \in \mathscr{A}$ *we can find a* $g_\varepsilon \in \mathscr{A}$ *such that*

(i) $\|g_\varepsilon\| \leqslant K_\varepsilon \|g\|$, *where* K_ε *depends* only *on* ε;
(ii) $\hat{g}_\varepsilon(n) = 0$ *for* $n \notin \Lambda$;
(iii) $g_\varepsilon(\vartheta) = g(\vartheta)$ *for* $\vartheta \in [0, 2\pi) \sim \Delta$, *where* $|\Delta| < 2\pi\varepsilon$.

Proof. Taking $\varepsilon > 0$, we put $\delta_n = \varepsilon/2^n$ and $\varepsilon_n = 1/2^n C(\delta_{n+1})$ with $C(\delta)$ from property (2) of the *second* lemma in the previous article. There is no harm in supposing that $C(\delta) > 1$; this we *do* in the following construction.

The function g_ε is obtained from a given $g \in \mathscr{A}$ by a process of successive approximations, using the operators $T_{N,\delta}$ and $T_{N,\delta}^{(L)}$ from the two lemmas of the preceding article.

According to the *second* of those lemmas, we can choose an L_1 such that

$$(*) \qquad \| T_{N,\delta_1}^{(L_1)} g - T_{N,\delta_1} g \| \leqslant \varepsilon_1 \| g \|$$

for all values of N simultaneously. If we take any positive integer N, the Fourier coefficients $\hat{h}(n)$ of $h = T_{N,\delta_1}^{(L_1)} g$ all vanish for $n \notin \mathscr{M}(N, L_1)$ by that second lemma. The hypothesis now furnishes a value of N such that

$$\mathscr{M}(N, L_1) \cap \mathbb{Z} \subseteq \Lambda.$$

Fix such a value of N, calling it N_1. Then, if we put $h_1 = T_{N_1,\delta_1}^{(L_1)} g$, we have $\hat{h}_1(n) = 0$ for $n \notin \Lambda$. Let us also write $r_1 = T_{N_1,\delta_1} g - h_1$. Then $(*)$ says that $\| r_1 \| \leqslant \varepsilon_1 \| g \|$, and, by the *first* lemma of the preceding article,

$$g(\vartheta) - h_1(\vartheta) - r_1(\vartheta) = g(\vartheta) - (T_{N_1,\delta_1} g)(\vartheta) = 0$$

for $\vartheta \in E_{N_1,\delta_1}$, a certain subset of $[0, 2\pi)$ with $|E_{N_1,\delta_1}| = 2\pi(1 - \delta_1)$.

We proceed, treating r_1 the way our given function g was just handled. First use the second lemma to get an L_2 such that

$$\| T_{N,\delta_2}^{(L_2)} r_1 - T_{N,\delta_2} r_1 \| \leqslant \varepsilon_2 \| g \|$$

for all positive N simultaneously, then choose (and fix) a value N_2 of N for which $\mathscr{M}(N_2, L_2) \cap \mathbb{Z} \subseteq \Lambda$, such choice being *possible according to the hypothesis.* Writing

$$h_2 = T_{N_2,\delta_2}^{(L_2)} r_1$$

and

$$r_2 = T_{N_2,\delta_2} r_1 - h_2,$$

we will have $\hat{h}_2(n) = 0$ for $n \notin \mathscr{M}(N_2, L_2)$ by the second lemma, hence, *a fortiori*, $\hat{h}_2(n) = 0$ for $n \notin \Lambda$. Our choice of L_2 makes

$$\| r_2 \| \leqslant \varepsilon_2 \| g \|,$$

and, by the first lemma, we have $r_1(\vartheta) = (T_{N_2,\delta_2} r_1)(\vartheta)$, i.e., $r_1(\vartheta) = h_2(\vartheta) + r_2(\vartheta)$ for $\vartheta \in E_{N_2,\delta_2}$, a subset of $[0, 2\pi)$ with $|E_{N_2,\delta_2}| = 2\pi(1 - \delta_2)$. *According to the preceding step*, we then have

$$g(\vartheta) = h_1(\vartheta) + h_2(\vartheta) + r_2(\vartheta) \qquad \text{for } \vartheta \in E_{N_1,\delta_1} \cap E_{N_2,\delta_2}.$$

And $\hat{h}_1(n) + \hat{h}_2(n) = 0$ for $n \notin \Lambda$.

Suppose that functions $h_1, h_2, \ldots, h_{k-1}$ and r_{k-1} (in $\mathscr{A}$) and positive integers $N_1, N_2, \ldots, N_{k-1}$ have been determined with $\|r_{k-1}\| \leqslant \varepsilon_{k-1}\|g\|$, $\hat{h}_j(n) = 0$ for $n \notin \Lambda$, $j = 1, 2, \ldots, k-1$, and $g = h_1 + h_2 + \cdots + h_{k-1} + r_{k-1}$ on the intersection $\bigcap_{j=1}^{k-1} E_{N_j,\delta_j}$. Then choose L_k in such a way that

$$\| T_{N,\delta_k}^{(L_k)} r_{k-1} - T_{N,\delta_k} r_{k-1} \| \leqslant \varepsilon_k \|g\|$$

simultaneously for all N (second lemma), and afterwards pick an N_k with $\mathscr{M}(N_k, L_k) \cap \mathbb{Z} \subseteq \Lambda$ (hypothesis). Putting

$$h_k = T_{N_k,\delta_k}^{(L_k)} r_{k-1}$$

and

$$r_k = T_{N_k,\delta_k} r_{k-1} - h_k,$$

we see that $\hat{h}_k(n) = 0$ for $n \notin \Lambda$, that $\|r_k\| \leqslant \varepsilon_k \|g\|$, and that $g = h_1 + h_2 + \cdots + h_{k-1} + h_k + r_k$ on $\bigcap_{j=1}^{k} E_{N_j,\delta_j}$ (first lemma).

Observe now that, by the second lemma, we *also* have

$$\|h_k\| = \|T_{N_k,\delta_k}^{(L_k)} r_{k-1}\| \leqslant C(\delta_k)\|r_{k-1}\| \leqslant C(\delta_k)\varepsilon_{k-1}\|g\| \leqslant \|g\|/2^{k-1}$$

for $k \geqslant 2$ on account of the way the numbers ε_k were rigged at the beginning of this proof. The series $h_1 + h_2 + h_3 + \cdots$ therefore *converges in the space* $\mathscr{A}$ (hence *uniformly* on $[0, 2\pi]$). Putting

$$g_\varepsilon(\vartheta) = \sum_{k=1}^{\infty} h_k(\vartheta),$$

we have

$$\|g_\varepsilon\| \leqslant (C(\delta_1) + \tfrac{1}{2} + \tfrac{1}{4} + \cdots)\|g\| = (1 + C(\varepsilon/2))\|g\|,$$

and $\hat{g}_\varepsilon(n) = 0$ for $n \notin \Lambda$ since, for such n, we have $\hat{h}_k(n) = 0$ for every k. Finally, since

$$|r_k(\vartheta)| \leqslant \|r_k\| \leqslant \varepsilon_k \|g\| \xrightarrow[k]{} 0,$$

we have

$$\sum_{j=1}^{k} h_j(\vartheta) + r_k(\vartheta) \xrightarrow[k]{} g_\varepsilon(\vartheta)$$

uniformly for $0 \leqslant \vartheta \leqslant 2\pi$, so $g_\varepsilon(\vartheta) = g(\vartheta)$ on the intersection

$$E = \bigcap_{j=1}^{\infty} E_{N_j,\delta_j}.$$

Here, since $|E_{N_j,\delta_j}| = 2\pi(1 - \delta_j)$ and the sets E_{N_j,δ_j} all lie in $[0, 2\pi)$, we have

$$|E| \geqslant 2\pi(1 - \delta_1 - \delta_2 - \delta_3 - \cdots) = 2\pi\left(1 - \frac{\varepsilon}{2} - \frac{\varepsilon}{4} - \cdots\right) = 2\pi(1 - \varepsilon).$$

The theorem is proved.

Our example is now furnished by the following

Corollary. *There exists a non-zero measure μ having gaps (a_n, b_n) in its support, with*

$$0 < a_1 < b_1 < a_2 < b_2 < a_3 < \cdots$$

and

$$\sum_1^\infty \left(\frac{b_l - a_l}{a_l}\right)^2 = \infty$$

(and the ratios $(b_l - a_l)/a_l$ even tending to ∞ as rapidly as we want!), while $\hat{\mu}(\lambda) = 0$ on a set of positive measure.

Proof. For $l = 1, 2, 3, \ldots$, take the sets

$$\mathscr{M}_l = \bigcup_{k=-2l-1}^{2l+1}{}' \ [N_l k - l,\ N_l k + l]$$

(term with $k = 0$ omitted), with the positive integers N_l so chosen that $N_l > 2l$ and that N_{l+1} is *much larger* than $(2l+1)(N_l+1)$. *There is no obstacle to our taking N_{l+1} as large as we wish in relation to* $(2l+1)(N_l+1)$ *for each l.*

Put

$$\Lambda = \bigcup_{l=1}^{\infty} (\mathscr{M}_l \cap \mathbb{Z});$$

it is clear that Λ satisfies the hypothesis of the theorem.

Choose any $g \in \mathscr{A}$ such that $g(\vartheta) > 0$ for $\pi/2 < \vartheta < 3\pi/2$ and $g(\vartheta) = 0$ for $0 \leqslant \vartheta \leqslant \pi/2$ and for $3\pi/2 \leqslant \vartheta \leqslant 2\pi$.

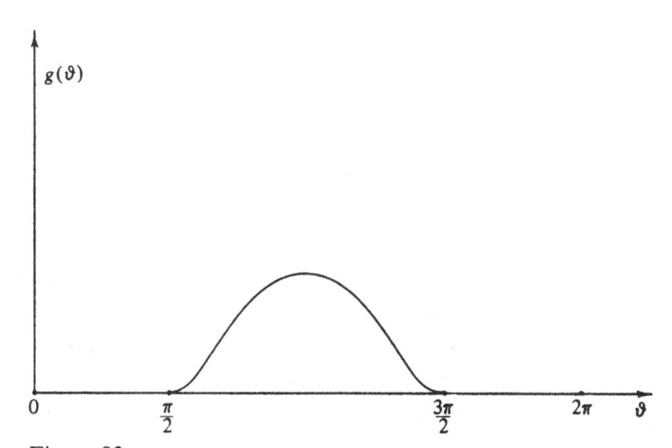

Figure 82

There are plenty of such functions g; we fix one of them.

Apply the theorem with $\varepsilon = \frac{1}{4}$, getting a function g_ε in $\mathscr{A}$ with $\hat{g}_\varepsilon(n) = 0$ for $n \notin \Lambda$ and $g_\varepsilon(\vartheta) = g(\vartheta)$ for all $\vartheta \in [0, 2\pi)$ outside a set of measure $\leqslant \pi/2$. Then *certainly* $g_\varepsilon(\vartheta)$ must be > 0 on a set in $[0, 2\pi)$ of measure $\geqslant \pi/2$ (*hence, in particular*, $g_\varepsilon \not\equiv 0$), while *at the same time* $g_\varepsilon(\vartheta) = 0$ *on a set of measure* $\geqslant \pi/2$ lying *in* $[0, 2\pi)$.

We have

$$g_\varepsilon(\vartheta) = \sum_{n \in \Lambda} \hat{g}_\varepsilon(n) e^{in\vartheta}$$

with

$$\sum_{n \in \Lambda} |\hat{g}_\varepsilon(n)| < \infty,$$

so, if we define a measure μ *supported on* $\Lambda \subseteq \mathbb{Z}$ by putting $\mu(E) = \sum_{n \in E} \hat{g}_\varepsilon(n)$, we have $\mu \not\equiv 0$, but $\hat{\mu}(\vartheta) = g_\varepsilon(\vartheta)$ *vanishes on a set of positive measure.*

The support ($\subseteq \Lambda$) of μ has the *gaps* $((2l+1)N_l + l,\ N_{l+1} - l - 1)$ *in it*. By choosing N_{l+1} *sufficiently large in relation to* $(2l+1)(N_l + 1)$ for each l, *we can make the ratios*

$$\frac{(N_{l+1} - l - 1) - ((2l+1)N_l + l)}{(2l+1)N_l + l}$$

go to ∞ *as rapidly as we please for* $l \longrightarrow \infty$.

We are done.

D. Volberg's work

Let $f(\vartheta) \in L_1(-\pi, \pi)$; say

$$f(\vartheta) \sim \sum_{-\infty}^{\infty} a_n e^{in\vartheta}.$$

Suppose that the Fourier coefficients a_n with *negative* indices n are *small enough* to satisfy the relation

$$(*) \qquad \sum_{-\infty}^{-1} \frac{1}{n^2} \log\left(\frac{1}{\sum_{-\infty}^{n} |a_k|} \right) = \infty.$$

According to a corollary to Levinson's theorem (§ A.5), $f(\vartheta)$ then *cannot vanish on an interval of positive length* unless $f \equiv 0$. If we also assume (for instance) that $\sum_k |a_k| < \infty$, Beurling's improvement of Levinson's theorem (§B.2) shows that $f(\vartheta)$ *cannot even vanish on a set of positive measure* without being identically zero when $(*)$ holds.

It is therefore natural to ask *how small* $|f(\vartheta)|$ *can actually be* for a *non-*

zero f whose Fourier coefficients a_n satisfy ($*$), or something like it. Suppose for instance, that

$$|a_n| \leqslant e^{-M(|n|)}, \qquad n < 0,$$

with a regularly increasing $M(m)$ for which

$$\sum_1^\infty \frac{M(m)}{m^2} = \infty.$$

Volberg's surprising result is that if the behaviour of $M(m)$ is *regular enough*, then we *must have*

$$\int_{-\pi}^{\pi} \log|f(\vartheta)|\,d\vartheta > -\infty$$

unless $f \equiv 0$. *Very* loosely speaking, this amounts to saying that if $f \not\equiv 0$ and

$$\sum_{-\infty}^{-1} \frac{1}{n^2} \log\left|\frac{1}{\hat{f}(n)}\right| = \infty,$$

then

$$\int_{-\pi}^{\pi} \log|f(\vartheta)|\,d\vartheta > -\infty,$$

at least when the *decrease* of $|\hat{f}(n)|$ for $n \longrightarrow -\infty$ is *sufficiently regular. If one logarithmic integral* (the *sum*) *diverges, the other must converge!*

One could improve this result *only* by finding a way to relax the regularity conditions imposed on $M(m)$.

Indeed, if $p(\vartheta) \geqslant 0$ is *any* function in $L_1(-\pi, \pi)$ with

$$\int_{-\pi}^{\pi} \log p(\vartheta)\,d\vartheta > -\infty,$$

we can *get* a function

$$f(\vartheta) \sim \sum_0^\infty a_n e^{in\vartheta}$$

such that $|f(\vartheta)| = p(\vartheta)$ a.e. by putting

$$f(\vartheta) = \lim_{\substack{z \not\longrightarrow e^{i\vartheta} \\ |z| < 1}} \exp\left\{\frac{1}{2\pi}\int_{-\pi}^{\pi} \frac{e^{it}+z}{e^{it}-z} \log p(t)\,dt\right\}$$

(see Chapter II, § A). Here, the Fourier coefficients of *negative* index *are all zero*, i.e., for $n < 0$,

$$|a_n| \leqslant e^{-M(|n|)}$$

with $M(|n|) \equiv \infty$. This means that from the condition

$$|a_n| \leq e^{-M(|n|)}, \qquad n<0,$$

with

$$\sum_1^\infty \frac{M(m)}{m^2} = \infty$$

one can *never hope to deduce a more stringent restriction* on the *smallness* of $|f(\vartheta)|$ than

$$\int_{-\pi}^{\pi} \log|f(\vartheta)|d\vartheta > -\infty.$$

Also, if $M(m)$ is *increasing*, from a *less* stringent condition than

$$\sum_1^\infty \frac{M(m)}{m^2} = \infty$$

one can *never hope to deduce any limitation on the smallness of* $|f(\vartheta)|$ for functions $f \not\equiv 0$ with $|a_n| \leq e^{-M(|n|)}$, $n<0$. That is the content of

Problem 12

Let $M(m)>0$ be increasing for $m>0$, and such that

$$\sum_1^\infty \frac{M(m)}{m^2} < \infty.$$

Given h, $0<h<\pi$, show that there is a function $f(\vartheta)$, continuous and of period 2π, with $f(\vartheta)=0$ for $h \leq |\vartheta| \leq \pi$ but $f \not\equiv 0$, such that

$$|a_n| \leq e^{-M(|n|)}, \quad n \neq 0\,(sic!),$$

for the Fourier coefficients a_n of $f(\vartheta)$. (Hint: Use the theorems of Chapter IV, § D and Chapter III, § D. Take a suitable convolution.)

It is important to note that Volberg's theorem *relates specifically to the unit circle; its analogue for the real line is false.* Take, namely, $F(x) = e^{-x^2}$, so that

$$\int_{-\infty}^{\infty} \frac{1}{1+x^2} \log F(x)\,dx = -\infty.$$

Here,

$$\hat{F}(\lambda) = \int_{-\infty}^{\infty} e^{i\lambda x} e^{-x^2}\,dx = \sqrt{\left(\frac{\pi}{2}\right)} e^{-\lambda^2/4},$$

so

$$\int_{-\infty}^{0} \frac{1}{1+\lambda^2} \log\left(\frac{1}{\hat{F}(\lambda)}\right) d\lambda \;=\; \infty.$$

and even

$$\int_{-\infty}^{0} \frac{1}{1+\lambda^2} \log\left(\frac{1}{\int_{-\infty}^{\lambda} \hat{F}(t)\,dt}\right) d\lambda \;=\; \infty.$$

This example shows that a *function and its Fourier transform can both get very small on* $\mathbb{R}$ (in terms of the logarithmic integral).

1. The planar Cauchy transform

Notation. If $G(z)$ is differentiable as a function of x and y we write

$$\frac{\partial G(z)}{\partial z} \;=\; G_z(z) \;=\; \frac{\partial G(z)}{\partial x} - i\frac{\partial G(z)}{\partial y}$$

and

$$\frac{\partial G(z)}{\partial \bar{z}} \;=\; G_{\bar{z}}(z) \;=\; \frac{\partial G(z)}{\partial x} + i\frac{\partial G(z)}{\partial y}.$$

Nota bene. Nowadays, most people take $\partial G/\partial z$ and $\partial G/\partial \bar{z}$ as *one-half* of the respective right-hand quantities.

Remark. If $G = U + iV$ with *real* functions U and V, the equation $G_{\bar{z}} = 0$ reduces to

$$\begin{cases} U_x = V_y, \\ U_y = -V_x, \end{cases}$$

i.e., the *Cauchy–Riemann equations* for U and V. The condition that $G_{\bar{z}} \equiv 0$ in a domain $\mathscr{D}$ is thus *equivalent to analyticity of* $G(z)$ *in* $\mathscr{D}$.

Theorem. *Let* $F(z)$ *be bounded and* $\mathscr{C}_1$ *in a bounded domain* $\mathscr{D}$*, and put*

$$G(z) \;=\; \frac{1}{2\pi}\iint_{\mathscr{D}} \frac{F(\zeta)\,d\xi\,d\eta}{z-\zeta},$$

where, as usual, $\zeta = \xi + i\eta$*. Then* $G(z)$ *is* $\mathscr{C}_1$ *in* $\mathscr{D}$ *and*

$$\frac{\partial G(z)}{\partial \bar{z}} \;=\; F(z), \qquad z \in \mathscr{D}.$$

Remark. The integral in question *converges absolutely* for each z, as is seen by going over to the polar coordinates (ρ, ψ) with

$$\zeta - z = \rho e^{i\psi}.$$

$G(z)$ is called the *planar Cauchy transform of* $F(z)$.

Proof of theorem. We first establish the differentiability of $G(z)$ in $\mathscr{D}$.

Let $z_0 \in \mathscr{D}$ with $\operatorname{dist}(z_0, \partial\mathscr{D}) = 3\rho$, say. Take any infinitely differentiable function $\varphi(\zeta)$ of ζ with $0 \leqslant \varphi(\zeta) \leqslant 1$ and

$$\varphi(\zeta) = \begin{cases} 1, & |\zeta - z_0| \leqslant \rho, \\ 0, & |\zeta - z_0| \geqslant 2\rho. \end{cases}$$

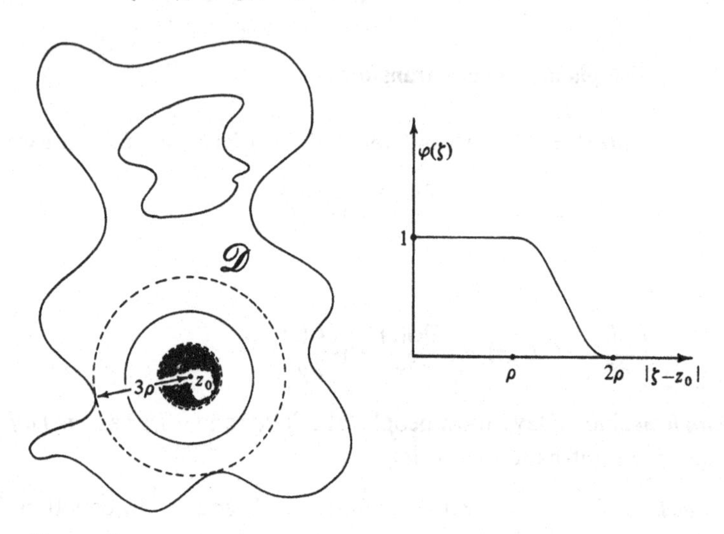

Figure 83

We can write

$$G(z) = \frac{1}{2\pi} \iint_{|\zeta - z_0| \leqslant 2\rho} \frac{\varphi(\zeta)F(\zeta)}{z - \zeta} \, d\xi \, d\eta$$

$$+ \frac{1}{2\pi} \iint_{\substack{|\zeta - z_0| > \rho \\ \zeta \in \mathscr{D}}} \frac{(1 - \varphi(\zeta))F(\zeta)}{z - \zeta} \, d\xi \, d\eta.$$

The *second* integral on the right is obviously a $\mathscr{C}_\infty$ function of z for $|z - z_0| < \rho$; it remains to consider the *first* one. After a change of variable, the *latter can be rewritten as*

$$\frac{1}{2\pi} \iint_{\mathbb{C}} \frac{F_1(z - w)}{w} \, du \, dv$$

(where $w = u + iv$, as usual) with $F_1(\zeta) = \varphi(\zeta)F(\zeta)$. Here, $F_1(\zeta)$ is *of compact support*, and has as much differentiability as $F(\zeta)$. Hence, since

$$\iint_{|w|<R} \frac{du\,dv}{|w|} < \infty$$

for any finite R, we can differentiate $(1/2\pi)\iint_{\mathbb{C}}(F_1(z-w)/w)\,du\,dv$ with respect to x and y *under the integral sign*, and thus see that that expression is $\mathscr{C}_1$ in those variables.

We have shown that $G(z)$ is $\mathscr{C}_1$ in the neighborhood of any $z_0 \in \mathscr{D}$; there remains the evaluation of $G_{\bar{z}}(z_0)$ in terms of F. This turns out to be surprisingly difficult if we try to do it directly, and we resort to the following dodge.

Let $r > 0$ be small, and $z_0 \in \mathscr{D}$. By the differentiability of $G(z)$ at z_0,

$$\begin{aligned} G(z_0 + re^{i\vartheta}) &= G(z_0) + G_x(z_0)r\cos\vartheta + G_y(z_0)r\sin\vartheta + o(r) \\ &= G(z_0) + \tfrac{1}{2}G_z(z_0)re^{i\vartheta} + \tfrac{1}{2}G_{\bar{z}}(z_0)re^{-i\vartheta} + o(r). \end{aligned}$$

Multiplying the last expression by $e^{i\vartheta}\,d\vartheta$ and integrating ϑ from 0 to 2π, we find the value $\pi r G_{\bar{z}}(z_0) + o(r)$; therefore

$$G_{\bar{z}}(z_0) = \lim_{r\to 0} \frac{1}{\pi r}\int_0^{2\pi} G(z_0 + re^{i\vartheta})e^{i\vartheta}\,d\vartheta.$$

Plugging in the expression for G in terms of F and changing the order of integration, this becomes

$$G_{\bar{z}}(z_0) = \lim_{r\to 0} \frac{1}{2\pi^2 r}\iint_{\mathscr{D}}\int_0^{2\pi} \frac{F(\zeta)e^{i\vartheta}}{z_0 + re^{i\vartheta} - \zeta}\,d\vartheta\,d\xi\,d\eta.$$

However,

$$\int_0^{2\pi} \frac{ire^{i\vartheta}\,d\vartheta}{re^{i\vartheta} - (\zeta - z_0)} = \begin{cases} 2\pi i, & |\zeta - z_0| < r, \\ 0, & |\zeta - z_0| > r. \end{cases}$$

Therefore, by the previous relation, we have

$$G_{\bar{z}}(z_0) = \lim_{r\to 0}\frac{1}{\pi r^2}\iint_{|\zeta - z_0|<r} F(\zeta)\,d\xi\,d\eta = F(z_0),$$

F having been assumed to be $\mathscr{C}_1$ in $\mathscr{D}$. We are done.

Corollary. *Let $\mathscr{D}$ be a bounded domain. Suppose that $F(z)$ is $\mathscr{C}_2$ in $\mathscr{D}$, that $|F(z)| > 0$ there, and that there is a constant C such that*

$$\left|\frac{\partial F(z)}{\partial \bar{z}}\right| \leqslant C|F(z)|, \qquad z \in \mathscr{D}.$$

Then

$$\Phi(z) \;=\; F(z)\exp\left\{\frac{1}{2\pi}\iint_{\mathscr{D}}\frac{F_{\bar{\zeta}}(\zeta)\,\mathrm{d}\xi\,\mathrm{d}\eta}{F(\zeta)(\zeta - z)}\right\}$$

is analytic in $\mathscr{D}$, *and* $|\Phi(z)|$ *lies between two constant multiples of* $|F(z)|$ *therein.*

Proof. $F_{\bar{z}}(z)/F(z)$ is $\mathscr{C}_1$ in $\mathscr{D}$ and bounded there by hypothesis, so we can apply the theorem, which tells us first of all that $\Phi(z)$ is differentiable in $\mathscr{D}$, and secondly that

$$\frac{\partial\Phi(z)}{\partial\bar{z}} \;=\; \left(F_{\bar{z}}(z) - F(z)\frac{F_{\bar{z}}(z)}{F(z)}\right)\exp\left(\frac{1}{2\pi}\iint_{\mathscr{D}}\frac{F_{\bar{\zeta}}(\zeta)\,\mathrm{d}\xi\,\mathrm{d}\eta}{F(\zeta)(\zeta - z)}\right) \;=\; 0$$

there. The *Cauchy–Riemann equations for* $\Re\Phi(z)$ *and* $\Im\Phi(z)$ *are thus satisfied* (see remark at the beginning of this article), so $\Phi(z)$ is *analytic* in $\mathscr{D}$.

If R is the *diameter* of $\mathscr{D}$, we easily check that

$$e^{-CR}|F(z)| \;\leqslant\; |\Phi(z)| \;\leqslant\; e^{CR}|F(z)|$$

for $z\in\mathscr{D}$. This does it.

The corollary has been extensively used by Lipman Bers and by Vekua in the study of partial differential equations. Volberg also uses it so as to bring analytic functions into his treatment.

Problem 13

Show that the condition that $|F(z)| > 0$ in $\mathscr{D}$ can be *dropped* from the hypothesis of the corollary, provided that we *maintain* the assumption that $|F_{\bar{z}}(z)| \leqslant C|F(z)|$, $z\in\mathscr{D}$, and define the ratio $F_{\bar{z}}(z)/F(z)$ in a satisfactory way on the set where $F(z) = 0$. Hence show that a function F satisfying the inequality $|F_{\bar{z}}(z)| \leqslant C|F(z)|$ *can have only isolated zeros in* $\mathscr{D}$, unless $F \equiv 0$ there. (Hint. On $E = \{z\in\mathscr{D}\colon F(z) = 0\}$, assign any constant value to the ratio $F_{\bar{z}}(z)/F(z)$. The function $\Phi(z)$ defined in the statement of the corollary is surely *analytic* in $\mathscr{D} \sim E$; it is also analytic in E° (if that set is non-empty) because it vanishes identically there. To check *existence* of

$$\Phi'(z_0) \;=\; \lim_{z\to z_0}\frac{\Phi(z) - \Phi(z_0)}{z - z_0}$$

at a point $z_0 \in \partial E\cap\mathscr{D}$, note that both $F(z_0)$ and $F_{\bar{z}}(z_0)$ must vanish, so, *near* z_0,

$$F(z) \;=\; \tfrac{1}{2}F_z(z_0)(z - z_0) + o(|z - z_0|).$$

If $F_z(z_0) = 0$, $\Phi'(z_0)$ *exists* and equals *zero*. If $F_z(z_0) \neq 0$, $|F(z)| > 0$ in some punctured neighborhood $0 < |z - z_0| < \eta$ of z_0, so such a punctured neighborhood is included in $\mathscr{D} \sim E$.)

2. The function $M(\nu)$ and its Legendre transform $h(\xi)$

As explained at the beginning of this chapter, Volberg's work deals with functions

$$f(\vartheta) \sim \sum_{-\infty}^{\infty} a_n e^{in\vartheta}$$

for which the a_n with negative index are very small; more precisely,

$$|a_{-n}| \leqslant e^{-M(n)}, \quad n > 0,$$

where $M(n)$ is *increasing* and such that

$$\sum_1^{\infty} \frac{M(n)}{n^2} = \infty.$$

▶ It will be convenient to assume throughout this § that $M(\nu)$ *is defined for all real values of* $\nu \geqslant 0$ *and not just the integral ones, and is increasing on* $[0, \infty)$. We do not, to begin with, exclude the possibility that $M(0) = -\infty$. Whether this happens or not will turn out to make no difference as far as our final result is concerned.

Volberg's treatment makes essential use of a *weight* $w(r) > 0$ defined for $0 < r < 1$ by means of the formula

$$\log\left(\frac{1}{w(r)}\right) = \sup_{\nu > 0}\left(M(\nu) - \nu \log\frac{1}{r}\right).$$

It is therefore necessary to make a study of the relation between $M(\nu)$ and the function

$$\boxed{h(\xi) = \sup_{\nu > 0}(M(\nu) - \nu\xi),}$$

defined for $\xi > 0$, and to find out how various properties of $M(\nu)$ are connected to others of $h(\xi)$. We take up these matters in the present article.

The formula for the function $h(\xi)$ (sometimes called the *Legendre transform* of $M(\nu)$) is reminiscent of material discussed extensively in Chapter IV, beginning with § A.2 therein. It is perhaps a good idea to start by showing how the situation now under consideration is related to that of Chapter IV, and especially how it *differs* from the latter.

Our present function $M(\nu)$ can be interpreted as $\log T(\nu)$, where $T(r)$ is the Ostrowski function used in Chapter IV. ($M(n)$ is *not*, as the similarity in letters might lead one to believe, a version of the $\{M_n\}$ – or of $\log M_n$ –

from Chapter IV!) Suppose indeed that

$$f(\vartheta) \sim \sum_{-\infty}^{\infty} a_n e^{in\vartheta}$$

is infinitely differentiable and in the class $\mathscr{C}(\{M_n\})$ considered in Chapter IV – in order to simplify matters, let us say that

$$|f^{(n)}(\vartheta)| \leqslant M_n, \quad n \geqslant 0.$$

We have

$$a_n = \frac{1}{2\pi}\int_{-\pi}^{\pi} e^{-in\vartheta} f(\vartheta)\, d\vartheta,$$

and the right side, after k integrations by parts, becomes

$$\frac{1}{2\pi}\int_{-\pi}^{\pi} (in)^{-k} f^{(k)}(\vartheta)\, d\vartheta$$

when $n \neq 0$. Using the above inequality on the derivatives of $f(\vartheta)$ in this integral, we see that

$$|a_n| \leqslant \inf_{k \geqslant 0} \frac{M_k}{|n|^k} = \frac{1}{T(|n|)}$$

where, as in Chapter IV,

$$T(r) = \sup_{k \geqslant 0} \frac{r^k}{M_k} \quad \text{for} \quad r > 0.$$

On putting $T(v) = e^{M(v)}$, we get

$$|a_n| \leqslant e^{-M(|n|)}.$$

This connection makes it possible to apply the final result of the present § to certain classes $\mathscr{C}(\{M_n\})$ of periodic functions, of period 2π. But that application does not show its real scope. The inequality for the a_n obtained by assuming that $f \in \mathscr{C}(\{M_n\})$ is a *two-sided* one; it shows that the a_n go to zero rapidly as $n \longrightarrow \pm\infty$. The *hypothesis* for the theorem on the logarithmic integral is, however, *one-sided*; it is only necessary to assume that

$$|a_{-n}| \leqslant e^{-M(n)}$$

for $n > 0$ in order to reach the desired conclusion.

There is another essential difference between our present situation and that of Chapter IV. *Here* we look at the function

$$h(\xi) = \sup_{v > 0} (M(v) - v\xi),$$

i.e., in terms of $T(\nu)$,

$$h(\xi) \;=\; \sup_{\nu>0}(\log T(\nu) - \nu\xi).$$

There we used the convex logarithmic regularisation $\{\underline{M}_n\}$ given by

$$\log \underline{M}_n = \sup_{\nu>0}((\log \nu)n - \log T(\nu)).$$

There is, first of all, a *change in sign.* Besides this, the former expression involves terms $\nu\xi$, *linear in the parameter* ν, where the latter has terms *linear in* $\log \nu$. On account of these differences it usually turns out that the function $h(\xi)$ considered here *tends to* ∞ *for* $\xi \to 0$, whereas $\log \underline{M}_n$ *usually tended to* ∞ *for* $n \to \infty$.

Let us begin our examination of $h(\xi)$ by verifying the statement just made about its behaviour for $\xi \to 0$.

Lemma. *If* $M(\nu) \to \infty$ *for* $\nu \to \infty$, $h(\xi) \to \infty$ *for* $\xi \to 0$.

Proof. Take any ν_0. Then, if $0 < \xi < \frac{1}{2}M(\nu_0)/\nu_0$,

$$h(\xi) \;\geqslant\; M(\nu_0) - \nu_0\xi \;>\; \tfrac{1}{2}M(\nu_0). \qquad \text{Q.E.D.}$$

The function

$$h(\xi) = \sup_{\nu>0}(M(\nu) - \nu\xi),$$

as the supremum of *decreasing* functions of ξ, is *decreasing.* As the *supremum of linear functions* of ξ, *it is convex.* The *upper supporting line of slope* ξ *to the graph of* $M(\nu)$ vs ν *has ordinate intercept equal to* $h(\xi)$:

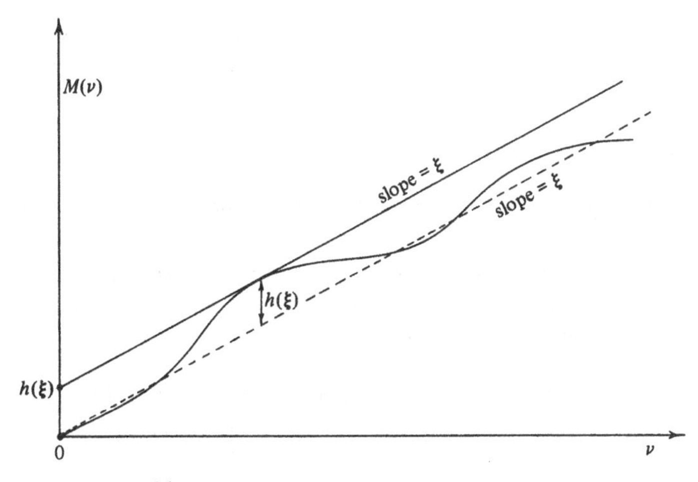

Figure 84

From this picture, we see immediately that

$$M^*(\nu) \;=\; \inf_{\xi>0}\,(h(\xi)+\xi\nu)$$

is the smallest concave increasing function which is $\geqslant M(\nu)$. Therefore, *if* $M(\nu)$ is *also concave*, $M^*(\nu) = M(\nu)$. We will come back to this relation later on.

Here is a graph dual to the one just drawn:

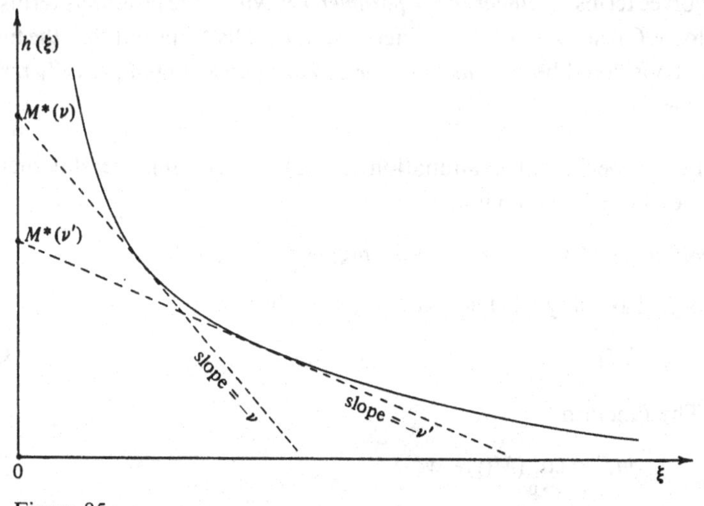

Figure 85

We see that $M^*(\nu)$ is the *ordinate intercept of the* (lower) *supporting line to the convex graph of* $h(\xi)$ having slope $-\nu$.

Volberg's construction depends in an essential way on a theorem of Dynkin, to be proved in the next article, which requires *concavity* of the function $M(\nu)$. Insofar as inequalities of the form

$$|a_{-n}| \;\leqslant\; e^{-M(n)}$$

are concerned, this concavity is pretty much *equivalent* to the cruder property that $M(\nu)/\nu$ be *decreasing*. It is, first of all, fairly evident that *the concavity of* $M(\nu)$ *makes* $M(\nu)/\nu$ *decreasing* (and even *strictly* decreasing, save in the trivial case where $M(\nu)/\nu \equiv$ const.) *for all sufficiently large* ν. We have, in the other direction, the following

Theorem. *Let* $M(\nu)$ *be* >0 *and increasing for* $\nu>0$, *and denote by* $M^*(\nu)$ *the* smallest concave majorant *of* $M(\nu)$. *If* $M(\nu)/\nu$ *is* decreasing.

$$M^*(\nu) \;<\; 2M(\nu).$$

Problem 14(a)

Prove this result. (Hint: The graph of $M^*(\nu)$ vs ν coincides with that of $M(\nu)$, save on certain open intervals (a_n, b_n) on each of which $M^*(\nu)$ is *linear*, with $M^*(a_n) = M(a_n)$ and $M^*(b_n) = M(b_n)$:

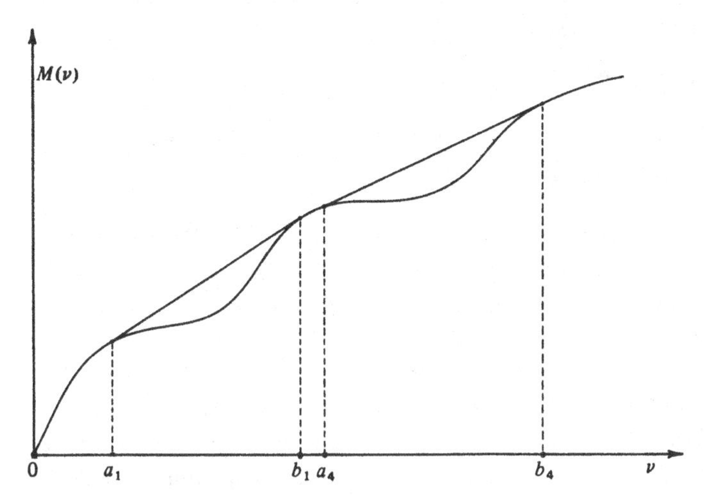

Figure 86

The (a_n, b_n) *may*, of course, be disposed like the contiguous intervals to the Cantor set, for instance. Consider any one of them, say, wlog, (a_1, b_1):

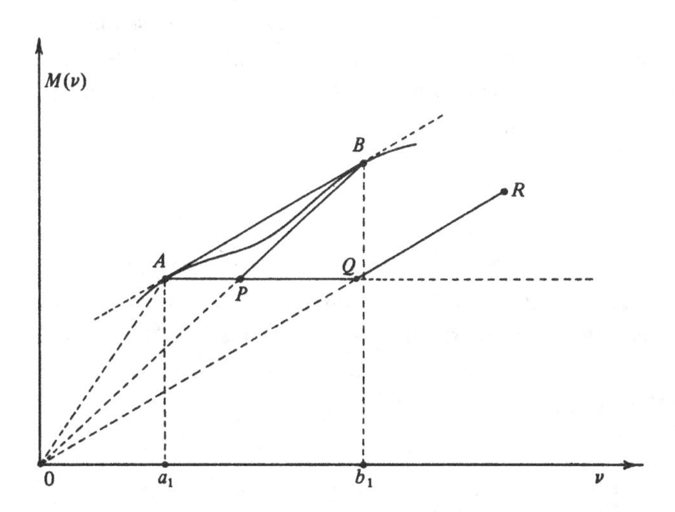

Figure 87

For $a_1 \leqslant \nu \leqslant b_1$, $(\nu, M(\nu))$ must lie *above* the broken line path APB, and $(\nu, M^*(\nu))$ lies on the segment $\overline{AB}$. Work with the broken line path $\overline{AQR}$, where OR is a line through the origin parallel to $\overline{AB}$.)

Because of this fact, the Fourier coefficients a_n of a given function which satisfy an inequality of the form

$$|a_{-n}| \leqslant e^{-M(n)}, \qquad n \geqslant 1,$$

with an *increasing* $M(\nu) > 0$ *such that* $M(\nu)/\nu$ *decreases* also satisfy

$$|a_{-n}| \leqslant e^{-M^*(n)/2}, \qquad n \geqslant 1$$

with the *concave majorant* $M^*(\nu)$ of $M(\nu)$. Clearly, $\sum_1^\infty M^*(n)/n^2 = \infty$ if $\sum_1^\infty M(n)/n^2 = \infty$. This circumstance makes it possible to simplify much of the computational work by *supposing to begin with that* $M(\nu)$ *is concave as well as increasing.*

A further (really, mainly formal) simplification results if we consider only functions $M(\nu)$ for which $M(\nu)/\nu \longrightarrow 0$ as $\nu \to \infty$ (see the next lemma). *As far as Volberg's work is concerned, this entails no restriction.* Since we will be assuming (at least) that $M(\nu)/\nu$ is decreasing, $\lim_{\nu\to\infty}(M(\nu)/\nu)$ *certainly exists. In case that limit is strictly positive,* the inequalities

$$|a_{-n}| \leqslant e^{-M(n)}, \qquad n \geqslant 1,$$

imply that

$$F(z) = \sum_{-\infty}^{\infty} a_n z^n$$

is *analytic in some annulus* $\{\rho < |z| < 1\}$, $\rho < 1$. This makes it possible for us to apply the *theorem on harmonic estimation* (§B.1), at least when $F(z)$ is continuous up to $\{|z| = 1\}$ (which will be the case in our version of Volberg's result). We find in this way that

$$\int_{-\pi}^{\pi} \log|F(e^{i\vartheta})|\,d\vartheta > -\infty$$

unless $F(z) \equiv 0$, using a simple estimate for harmonic measure in an annulus. (If the reader has any trouble working out that estimate, he or she may find it near the *very end* of the proof of Volberg's theorem in article 6 below.) The *conclusion* of Volberg's theorem is thus *verified* in the special case that $\lim_{\nu\to\infty}(M(\nu)/\nu) > 0$.

For this reason, we will mostly only consider functions $M(\nu)$ *for which* $\lim_{\nu\to\infty}(M(\nu)/\nu) = 0$ *in the present §.*

Once we decide to work with *concave* functions $M(v)$, it costs but little to *further restrict our attention* to *strictly concave infinitely differentiable* $M(v)$'s. Given any concave increasing $M(v)$, we may, first of all, add to it a *bounded strictly concave* increasing function (with second derivative < 0

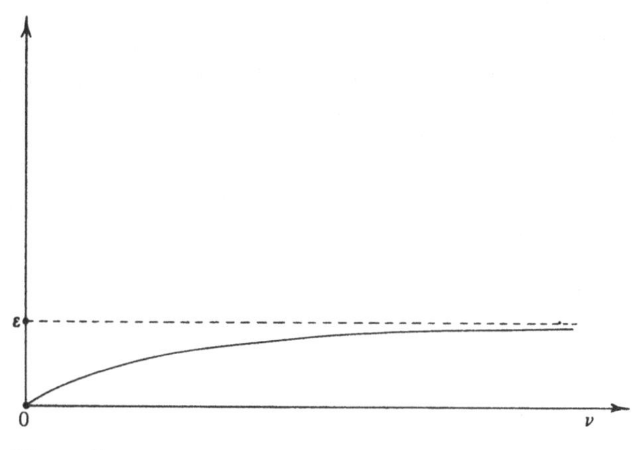

Figure 88

on $(0, \infty)$) whose graph has a *horizontal asymptote* of height ε, and thus obtain a new *strictly concave* increasing function $M_1(v)$, with $M_1''(v) < 0$, differing by at most ε from $M(v)$. We may then take an infinitely differentiable positive function φ supported on $[0, 1]$ and having $\int_0^1 \varphi(t)\,dt = 1$,

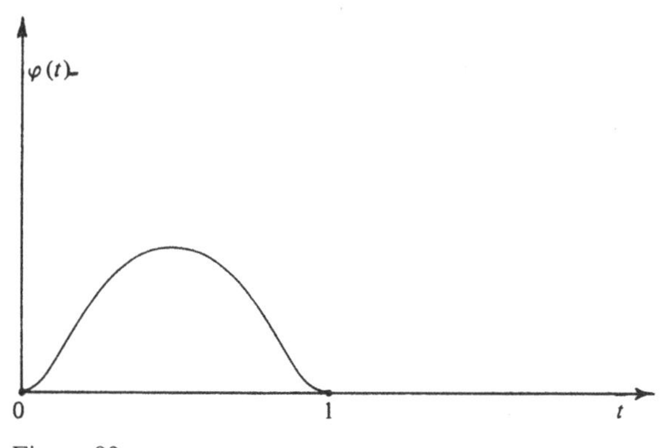

Figure 89

and form the function

$$M_2(v) = \frac{1}{h}\int_0^h M_1(v + \tau)\varphi(\tau/h)\,d\tau,$$

using a small value of $h > 0$. $M_2(v)$ will also be strictly concave with

$M_2''(\nu) < 0$ on $(0, \infty)$, and increasing, and *infinitely differentiable besides* for $0 < \nu < \infty$. It will differ by less than ε from $M_1(\nu)$ for $\nu \geqslant a$ when a is any given number > 0, if $h > 0$ is small enough (depending on a). That's because $0 \leqslant M_1'(\nu) \leqslant M_1'(a) < \infty$ for $\nu \geqslant a$.

Our function $M_2(\nu)$, infinitely differentiable, increasing, and strictly concave, thus differs by less than 2ε from $M(\nu)$ when ν is large. This, however, means that $h_2(\xi) = \sup_{\nu>0}(M_2(\nu) - \nu\xi)$ differs by less than 2ε from

$$h(\xi) \;=\; \sup_{\nu>0}(M(\nu) - \nu\xi)$$

for small values of $\xi > 0$, the suprema in question being attained for *large values* of ν if ξ is *small*:

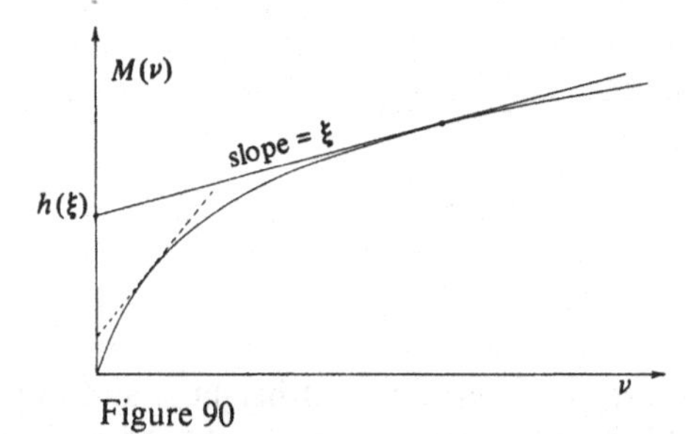

Figure 90

Hence, in studying the *order of magnitude of* $h(\xi)$ *for* ξ *near zero* (which is what we will be mainly concerned with in this §), *we may as well assume to begin with that* $M(\nu)$ is *strictly concave* and *infinitely differentiable.*

When this restriction holds, one can obtain some useful relations in connection with the duality between $M(\nu)$ and $h(\xi)$.*

Lemma. *If* $M(\nu)$ *is strictly concave and increasing with* $M(\nu)/\nu \longrightarrow 0$ *for* $\nu \to \infty$, *there is for each* $\xi > 0$ *a* unique $\nu = \nu(\xi)$ *such that*

$$h(\xi) \;=\; M(\nu) - \nu\xi.$$

$h(\xi)$ *has a* derivative *for* $\xi > 0$, *and* $h'(\xi) = -\nu(\xi)$.

Proof. Since $M(\nu)/\nu \longrightarrow 0$ as $\nu \to \infty$, the supporting line of slope ξ to the graph of $M(\nu)$ vs ν *does* touch that graph *somewhere* (see preceding diagram), say at $(\nu_1, M(\nu_1))$. Thus,

$$h(\xi) \;=\; M(\nu_1) - \nu_1\xi.$$

* In the following 3 lemmas, it is tacitly assumed that $\xi > 0$ ranges over some *small* interval with left endpoint at the origin, for they will be used only for such values of ξ. This eliminates our having to worry about the behaviour of $M(\nu)$ for small ν.

Suppose that $\nu_2 \neq \nu_1$ and also

$$h(\xi) = M(\nu_2) - \nu_2\xi;$$

wlog say that $\nu_2 > \nu_1$. Then

$$M(\nu_2) = M(\nu_1) + \xi(\nu_2 - \nu_1).$$

Therefore, for $\nu_1 < \nu < \nu_2$, by *strict* concavity of $M(\nu)$,

$$M(\nu) > M(\nu_1) + \xi(\nu - \nu_1),$$

i.e.,

$$M(\nu) - \nu\xi > M(\nu_1) - \nu_1\xi = h(\xi).$$

This, however, contradicts the definition of $h(\xi)$, so there can *be* no $\nu_2 \neq \nu_1$ with

$$h(\xi) = M(\nu_2) - \nu_2\xi.$$

Since $M(\nu)$ is already concave, it is *equal* to its smallest concave majorant, $M^*(\nu)$, i.e.,

$$M(\nu) = \inf_{\xi > 0} (h(\xi) + \xi\nu).$$

The function $h(\xi)$ is convex, so if it does *not* have a derivative at a point $\xi_0 > 0$, it has a *corner* there, with two *different* supporting lines, of slopes $-\nu_1$ and $-\nu_2$, touching the graph of $h(\xi)$ vs ξ at $(\xi_0, h(\xi_0))$:

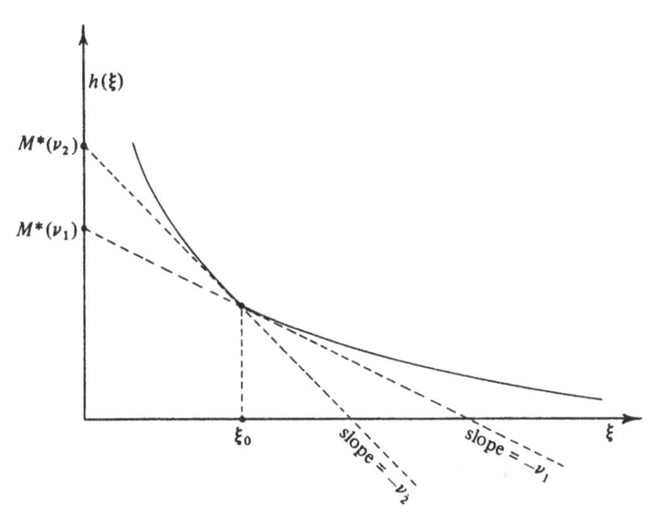

Figure 91

Those two supporting lines have ordinate intercepts equal to $M^*(\nu_1)$ and $M^*(\nu_2)$, i.e., to $M(\nu_1)$ and $M(\nu_2)$. But then $h(\xi_0) = M(\nu_1) - \nu_1\xi_0 = M(\nu_2) - \nu_2\xi_0$, which we have already seen to be impossible. $h'(\xi_0)$ must therefore *exist*, and it is now clear that derivative must have the value $-\nu(\xi_0)$, the slope of the *unique* supporting line to the graph of $h(\xi)$ vs ξ at the point $(\xi_0, h(\xi_0))$.

Lemma. *If $M(\nu)$ is differentiable and strictly concave and $M(\nu)/\nu \longrightarrow 0$ for $\nu \to \infty$,*

$$\frac{\mathrm{d}M(\nu)}{\mathrm{d}\nu} = \xi \quad \textit{for} \quad \nu = \nu(\xi).$$

Proof. $\nu(\xi)$ is the abscissa at which the supporting line of slope ξ to the graph of $M(\nu)$ vs ν touches that graph.

Recall that, for the strictly concave functions $M(\nu)$ we are dealing with here, *we actually have* $M''(\nu) < 0$ on $(0, \infty)$ – refer to the above construction of $M_1(\nu)$ and $M_2(\nu)$ from $M(\nu)$.

Lemma. *If $M(\nu)$ is twice continuously differentiable and $M''(\nu) < 0$ on $(0, \infty)$, and if $M(\nu)/\nu \longrightarrow 0$ for $\nu \to \infty$, $h''(\xi)$ exists for $\xi > 0$.*

Proof. $M(\nu)$ is certainly strictly concave, so, by the preceding two lemmas, $h'(\xi)$ exists and we have the implicit relation

$$M'(-h'(\xi)) = \xi.$$

Since $M''(\nu)$ exists, is continuous, and is < 0, we can apply the *implicit function theorem* to conclude that $h''(\xi)$ exists and equals $-1/M''(-h'(\xi))$.

Volberg's construction, besides depending (through Dynkin's theorem) on the concavity of $M(\nu)$, makes essential use of one *additional special property*, namely, that

$$\boxed{\xi^{-K} \leqslant \text{const.} h(\xi)}$$

for some $K > 1$ as $\xi \to 0$. Let us express this in terms of $M(\nu)$.

Lemma. *For concave $M(\nu)$, the preceding boxed relation holds with some $K > 1$ for $\xi \to 0$ iff*

$$M(\nu) \geqslant \text{const.}\nu^{K/(K+1)} \quad \textit{for large } \nu.$$

Proof. Since $M(\nu)$ is concave, it is equal to $\inf_{\xi>0} (h(\xi) + \nu\xi)$. If the boxed relation holds and ν is large, this expression is $\geqslant \inf_{\xi>0} (\text{const.}\xi^{-K} + \nu\xi)$

whose value is readily seen to be of the form const.$\nu^{K/(K+1)}$.

To go the other way, compute $\sup_{\nu>0}(\text{const.}\nu^{K/(K+1)} - \nu\xi)$.

Remark. One *might think* that the *concavity of* $M(\nu)$ and the fact that

$$\sum_1^\infty M(n)/n^2 \;=\; \infty$$

together imply that $M(\nu) \geqslant \nu^\rho$ with *some* positive ρ (say $\rho = \frac{1}{2}$) for large ν. That, however, is *not so.* A counter example may easily be constructed by building the graph of $M(\nu)$ vs ν out of exceedingly long straight segments chosen one after the other so as to alternately cut the graph of ν^ρ vs. ν *from below* and *from above.*

Here is one more rather trivial fact which we will have occasion to use.

Lemma. *For* increasing $M(\nu)$,

$$h(\xi) \;\geqslant\; M(0) \quad for \quad \xi > 0$$

and hence $\lim_{\xi\to\infty} h(\xi)$ *is* finite *if* $M(0) > -\infty$.

Proof. $h(\xi)$ is *decreasing*, so $\lim_{\xi\to\infty} h(\xi)$ *exists*, but is perhaps equal to $-\infty$. The rest is clear.

The principal result on the connection between $M(\nu)$ and $h(\xi)$ was published independently by Beurling and by Dynkin in 1972. It says that, if $a > 0$ is sufficiently small (so that $\log h(\xi) > 0$ for $0 < \xi \leqslant a$), the convergence of $\int_0^a \log h(\xi)\, d\xi$ is equivalent to that of $\int_1^\infty (M(\nu)/\nu^2)\, d\nu$ (compare with the material in §C of Chapter IV). More precisely:

Theorem. *If* $M(\nu)$ *is* increasing *and* concave, *and*

$$h(\xi) \;=\; \sup_{\nu>0}(M(\nu) - \nu\xi),$$

there is an $a > 0$ *such that*

$$\int_0^a \log h(\xi)\, d\xi \;<\; \infty$$

iff

$$\int_1^\infty \frac{M(\nu)}{\nu^2}\, d\nu \;<\; \infty.$$

Proof. In the first place, if $\lim_{\nu\to\infty} M(\nu)/\nu = c > 0$, the function $h(\xi) = \sup_{\nu>0}(M(\nu) - \nu\xi)$ is *infinite* for $0 < \xi < c$. In this case, the integrals involved in the theorem *both diverge*. For the remainder of the proof we may thus suppose that $M(\nu)/\nu \longrightarrow 0$ as $\nu \to \infty$.

Again, by the *first* lemma of this article, $h(\xi) \longrightarrow \infty$ for $\xi \to 0$ unless $M(\nu)$ is bounded for $\nu \to \infty$, and in that case both of the integrals in question are obviously finite. There is thus no loss of generality in supposing that $h(\xi) \longrightarrow \infty$ for $\xi \to 0$, and we may take an $a > 0$ with $h(a) \geqslant 2$, say.

These things being granted, let us, as in the previous discussion, approximate $M(\nu)$ to within ε on $[A, \infty)$, $A > 0$, by an infinitely differentiable *strictly* concave function $M_\varepsilon(\nu)$, with $M_\varepsilon''(\nu) < 0$. If $\varepsilon > 0$ and $A > 0$ are small enough, the corresponding function

$$h_\varepsilon(\xi) \;=\; \sup_{\nu > 0}(M_\varepsilon(\nu) - \nu\xi)$$

approximates $h(\xi)$ *to within* 1 *unit* (say) on $(0, a]$. But then

$$\int_0^a \log h(\xi)\,d\xi \quad \text{and} \quad \int_0^a \log h_\varepsilon(\xi)\,d\xi$$

converge simultaneously, and the *same* is true for the integrals

$$\int_1^\infty \frac{M(\nu)}{\nu^2}\,d\nu \quad \text{and} \quad \int_1^\infty \frac{M_\varepsilon(\nu)}{\nu^2}\,d\nu.$$

It is therefore enough to establish the theorem for $M_\varepsilon(\nu)$ and $h_\varepsilon(\xi)$; in other words, we may, wlog, *assume to begin with that* $M(\nu)$ *is infinitely differentiable and strictly concave, with* $M''(\nu) < 0$, *and that* $M(\nu)/\nu \longrightarrow 0$ *for* $\nu \to \infty$.

In these circumstances, we can use the relations furnished by the preceding lemmas. It is convenient to work with $\log|h'(\xi)|$ instead of $\log h(\xi)$, so for this purpose let us first show that

$$\int_0^a \log h(\xi)\,d\xi \quad \text{and} \quad \int_0^a \log|h'(\xi)|\,d\xi$$

converge simultaneously. First of all,

$$h(\xi) \;\leqslant\; h(a) + (a - \xi)|h'(\xi)| \;\leqslant\; h(a) + a|h'(\xi)| \quad \text{for} \quad 0 < \xi < a$$

by the *convexity* of $h(\xi)$, as the following diagram shows:

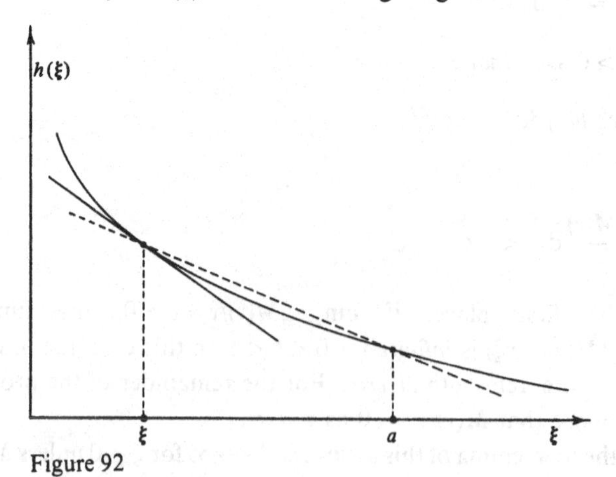

Figure 92

Therefore convergence of the *second* integral implies that of the *first*. Again, for $0 < \xi < a$, $h(\xi) \geqslant 2$, so

$$h\left(\frac{\xi}{2}\right) \geqslant 2 + \frac{\xi}{2}|h'(\xi)| \geqslant \frac{\xi}{2}|h'(\xi)|$$

for such ξ:

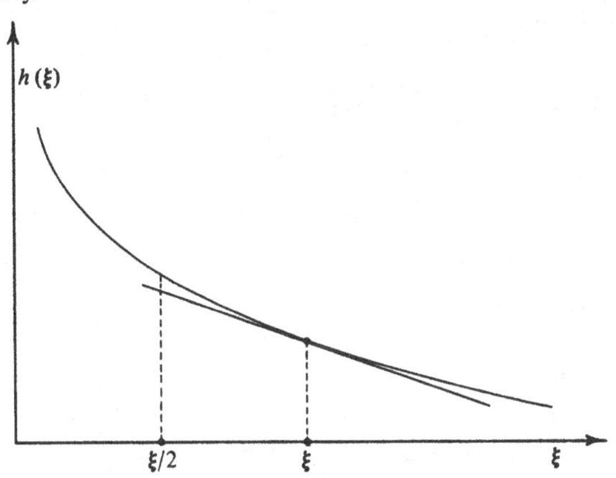

Figure 93

So, since $\int_0^a |\log \xi| \mathrm{d}\xi < \infty$, convergence of the *first* integral implies that of the *second.*

We have

$$h(\xi) = M(\nu(\xi)) - \xi\nu(\xi)$$

with $\nu(\xi) = -h'(\xi)$, and $M'(\nu(\xi)) = \xi$. Therefore

$$\xi \mathrm{d} \log |h'(\xi)| = M'(\nu(\xi)) \frac{\mathrm{d}\nu(\xi)}{\nu(\xi)}.$$

Taking a number b, $0 < b < a$, and integrating by parts, we find that

$$\begin{aligned}\int_b^a \xi \mathrm{d} \log |h'(\xi)| &= a \log |h'(a)| - b \log |h'(b)| - \int_b^a \log |h'(\xi)| \mathrm{d}\xi \\ &= \frac{M(\nu(a))}{\nu(a)} - \frac{M(\nu(b))}{\nu(b)} + \int_{\nu(b)}^{\nu(a)} \frac{M(\nu)}{\nu^2} \mathrm{d}\nu.\end{aligned}$$

Here, $\nu(\xi)$ is decreasing, so $\nu(b) \geqslant \nu(a)$. Turning things around, we thus have

$$\begin{aligned}&\int_b^a \log|h'(\xi)|\mathrm{d}\xi + b\log|h'(b)| - a\log|h'(a)| \\ &\quad = \frac{M(\nu(b))}{\nu(b)} - \frac{M(\nu(a))}{\nu(a)} + \int_{\nu(a)}^{\nu(b)} \frac{M(\nu)}{\nu^2}\mathrm{d}\nu.\end{aligned}$$

$M(v)/v$ is decreasing (concavity of $M(v)$!) and, as $b\to 0$, $v(b)\to\infty$. We see, then, that

$$\int_0^a \log|h'(\xi)|\,d\xi \;<\; \infty$$

if

$$\int_{v(a)}^{\infty} \frac{M(v)}{v^2}\,dv \;<\; \infty.$$

Also, $|h'(\xi)|$ *decreases*, so $b\log|h'(b)| \leqslant \int_0^b \log|h'(\xi)|\,d\xi$. Therefore

$$\int_{v(a)}^{v(b)} \frac{M(v)}{v^2}\,dv$$

is bounded above for $b\to 0$ if $\int_0^a \log|h'(\xi)|\,d\xi < \infty$, i.e., $\int_{v(a)}^{\infty}(M(v)/v^2)\,dv < \infty$. We are done.

Problem 14(b)

Let $H(\xi)$ be decreasing for $\xi > 0$ with $H(\xi)\to\infty$ for $\xi\to 0$, and denote by $h(\xi)$ the *largest convex minorant* of $H(\xi)$. Show that, if, for some small $a > 0$, $\int_0^a \log h(\xi)\,d\xi < \infty$, then $\int_0^a \log H(\xi)\,d\xi < \infty$. Hint: Use the following picture:

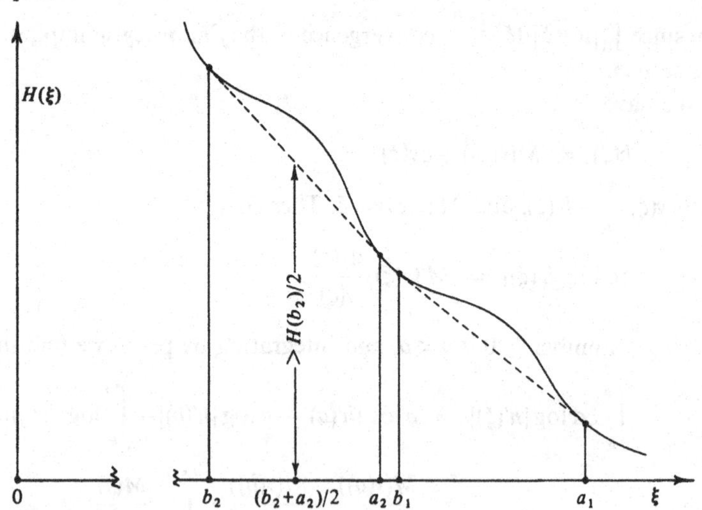

Figure 94

Problem 14(c)

If $M(v)$ is increasing, it is in general *false* that $\int_1^{\infty}(M(v)/v^2)\,dv < \infty$ makes $\int_1^{\infty}(M^*(v)/v^2)\,dv < \infty$ for the *smallest concave majorant* $M^*(v)$ of $M(v)$. (Hint: In one counter example, $M^*(v)$ has a broken line graph with vertices on the one of $v/\log v$ (v large).)

Theorem. *Let* $H(\xi)$ *be decreasing for* $\xi > 0$ *and tend to* ∞ *as* $\xi \to 0$. *For* $\nu > 0$, *put*

$$M(\nu) \;=\; \inf_{\xi>0}\,(H(\xi) + \xi\nu).$$

Then

$$\int_0^a \log H(\xi)\,\mathrm{d}\xi \;<\; \infty$$

for some (*and hence for all*) *arbitrarily small values of* $a > 0$ *iff*

$$\int_1^\infty \frac{M(\nu)}{\nu^2}\,\mathrm{d}\nu \;<\; \infty.$$

Proof. As the infimum of linear functions of ν, $M(\nu)$ *is* concave; it is obviously *increasing*. The function

$$h(\xi) \;=\; \sup_{\nu>0}\,(M(\nu) - \nu\xi)$$

is the largest *convex minorant* of $H(\xi)$ because its height at any abscissa ξ is the supremum of the heights of all the (lower) supporting lines with slopes $-\nu < 0$ to the graph of H:

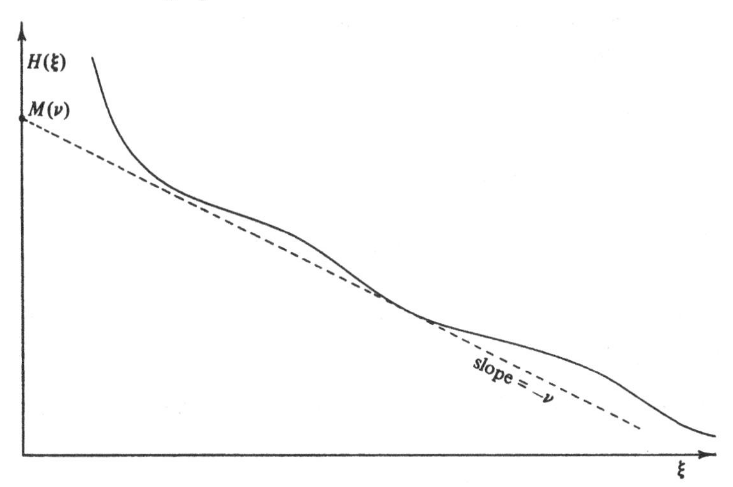

Figure 95

Therefore $\int_0^a \log H(\xi)\mathrm{d}\xi = \infty$ makes $\int_0^a \log h(\xi)\mathrm{d}\xi = \infty$ *by problem* 14(b), so in that case $\int_1^\infty (M(v)/v^2)\mathrm{d}v = \infty$ *by the preceding theorem*. If, on the other hand, $\int_1^\infty (M(v)/v^2)\mathrm{d}v$ *does diverge*, $\int_0^a \log h(\xi)\mathrm{d}\xi = \infty$ for each small enough $a > 0$ by that same theorem, so certainly $\int_0^a \log H(\xi)\mathrm{d}\xi = \infty$ for such a. This does it.

3. Dynkin's extension of $F(e^{i\vartheta})$ to $\{|z| \leqslant 1\}$ with control on $|F_{\bar{z}}(z)|$.

As stated near the beginning of the previous article, a very important role in Volberg's construction is played by a weight $w(r) > 0$ defined for $0 < r < 1$ by the formula

$$w(r) = \exp\left(-h\left(\log\frac{1}{r}\right)\right),$$

where, for $\xi > 0$,

$$h(\xi) = \sup_{v>0}(M(v) - v\xi).$$

Here $M(v)$ is an increasing (usually concave) function such that $\int_1^\infty (M(v)/v^2)\,dv = \infty$; this makes $h(\xi)$ increase to ∞ rather rapidly as ξ decreases towards 0, so that *$w(r)$ decreases very rapidly towards zero* as $r \to 1$.

A typical example of the kind of functions $M(v)$ figuring in Volberg's theorem is obtained by putting

$$M(v) = \frac{v}{\log v}$$

for $v \geqslant e^2$, say, and defining $M(v)$ in any convenient fashion for $0 \leqslant v < e^2$ so as to keep it increasing and concave on that range. Here we find without difficulty that

$$h(\xi) \sim \frac{\xi^2}{e} e^{1/\xi} \quad \text{for} \quad \xi \longrightarrow 0,$$

and $w(r)$ decreases towards zero like

$$\exp\left(-\frac{(1-r)^2}{e} e^{1/(1-r)}\right)$$

as $r \to 1$; this is *really* fast. It is good to keep this example in mind during the following development.

Lemma. *Let $M(v)$ be increasing and strictly concave for $v > 0$ with $M(v)/v \longrightarrow 0$ for $v \to \infty$, put*

$$h(\xi) = \sup_{v>0}(M(v) - v\xi),$$

and write $w(r) = \exp(-h(\log(1/r)))$ for $0 < r < 1$. Then

$$\int_0^1 r^{n+2} w(r)\,dr > \frac{\text{const.}}{n} e^{-M(n)}$$

for $n \geqslant 1$.

Proof. In terms of $\xi = \log(1/r)$, $r^n w(r) = \exp(-h(\xi) - \xi n)$. Since $M(v)$ is strictly concave, we have, by the previous article,

$$\inf_{\xi > 0} (h(\xi) + \xi n) = M(n),$$

the infimum being attained at the value $\xi = \xi_n = M'(n)$. Put $r_n = e^{-\xi_n}$. Then, $r_n^n w(r_n) = e^{-M(n)}$. Because $w(r)$ decreases, we now see that

$$\int_0^1 r^{n+2} w(r)\,dr \geqslant w(r_n) \int_0^{r_n} r^{n+2}\,dr = \frac{r_n^3}{n+3}(r_n^n w(r_n))$$
$$= \frac{r_n^3}{n+3} e^{-M(n)}.$$

Here,

$$r_n^3 = e^{-3M'(n)},$$

and this is $\geqslant e^{-3M'(1)}$ since $M'(v)$ decreases, when $n \geqslant 1$. From the previous relation, we thus find that

$$\int_0^1 r^{n+2} w(r)\,dr \geqslant \frac{e^{-3M'(1)}}{4n} e^{-M(n)}$$

for $n \geqslant 1$, Q.E.D.

Theorem (Dynkin (the younger), 1972). *Let $M(v)$ be increasing on $(0, \infty)$, as well as strictly concave and infinitely differentiable on $(0, \infty)$, with $M''(v) < 0$ there and $M(v)/v \longrightarrow 0$ for $v \to \infty$. Let $M(0) > -\infty$.*

For $0 < r < 1$, put

$$w(r) = \exp\left(-h\left(\log\frac{1}{r}\right)\right),$$

where $h(\xi)$ is related to $M(v)$ in the usual fashion.

Suppose that

$$F(e^{i\vartheta}) \sim \sum_{-\infty}^{\infty} a_n e^{in\vartheta}$$

is continuous on the unit circumference, and that

$$\sum_1^{\infty} |n^2 a_{-n}| e^{M(n)} < \infty.$$

Then F has a continuous extension $F(z)$ onto $\{|z| \leqslant 1\}$ with $F(z)$ continuously differentiable for $|z| < 1$ and $|\partial F(z)/\partial \bar{z}| \leqslant \text{const.} w(|z|)$, $|z| < 1$.

Remark. The sense of Dynkin's theorem is that *rapid growth* of $M(n)$

to ∞ for $n \to \infty$ (which corresponds to *rapid growth* of $h(\xi)$ to ∞ for ξ tending to 0) makes it possible to extend F continuously to $\{|z| \leqslant 1\}$ in such a way as to have $|\partial F(z)/\partial \bar{z}|$ *dropping off to zero very quickly for* $|z| \to 1$.

Proof of theorem. We start by taking a continuously differentiable function $\Omega(e^{it})$, to be determined presently, and putting*

$$(*) \qquad G(z) = \sum_0^\infty a_n z^n + \frac{1}{2\pi}\int_0^{2\pi}\int_0^1 \frac{\Omega(e^{it})r^2 w(r) r\,dr\,dt}{re^{it} - z}$$

for $|z| \leqslant 1$. The reason for using the factor r^2 with $w(r)$ will soon be apparent.

The idea now is to specify $\Omega(e^{it})$ in such fashion as to make $G(e^{i\vartheta})$ have the same Fourier series as $F(e^{i\vartheta})$. If we can do that, the function $G(z)$ will be a continuous extension of $F(e^{i\vartheta})$ to $\{|z| \leqslant 1\}$.

To see this, observe that our hypothesis certainly makes the trigonometric series

$$\sum_{-\infty}^{-1} a_n e^{in\vartheta}$$

absolutely convergent, so, since $F(e^{i\vartheta})$ is continuous,

$$\sum_0^\infty a_n e^{in\vartheta}$$

must also be the Fourier series of some continuous function, and hence the power series on the right in $(*)$ a continuous function of z for $|z| \leqslant 1$. According to a lemma from the previous article, the property $M(0) > -\infty$ makes $h(\xi)$ *bounded below* for $\xi > 0$ and hence $w(r)$ *bounded above* in $(0, 1)$. The right-hand integral in $(*)$ is thus of the form

$$\frac{1}{2\pi}\iint_{|\zeta|<1} \frac{b(\zeta)\,d\xi\,d\eta}{\zeta - z}$$

with a *bounded function* $b(\zeta)$. (Here, we are writing $\zeta = \xi + i\eta$ which *conflicts* with our frequent use of ξ to denote $\log(1/r)$. *No confusion should thereby result.*) It is well known that such an integral gives a *continuous* function of z; that's because it's a *convolution* on $\mathbb{R}^2$, with

$$\iint_{|\zeta|<R} \frac{d\xi\,d\eta}{|\zeta|}$$

finite for each finite R. We see in this way that the function $G(z)$ given by $(*)$ *will* be continuous for $|z| \leqslant 1$. If also the Fourier series of $G(e^{i\vartheta})$ and $F(e^{i\vartheta})$ coincide, those two functions must obviously be equal.

We wish to apply the theorem of article 1 to the right-hand integral in $(*)$. In order to stay honest, we should therefore check *continuous*

* The power series on the right in $(*)$ may not actually be *convergent* for $|z| = 1$, but *does* represent a *continuous function* for $|z| \leqslant 1$, as will be clear in a moment.

differentiability (for $|\zeta| < 1$) of the function $b(\zeta)$ figuring in that double integral, viz.,

$$b(re^{it}) = r^2 w(r)\Omega(e^{it}),$$

because continuous differentiability (at least) is required in the hypothesis of the theorem. It is for this purpose that the factor r^2 has been included; that factor ensures differentiability of $b(\zeta)$ at 0. Thanks to it, the desired property of $b(\zeta)$ follows from the continuous differentiability of $\Omega(e^{it})$ together with the continuity of $rw'(r)$ on $[0, 1)$ which we now verify.

We have $rw'(r) = h'(\log(1/r))w(r)$ for $0 < r < 1$. By a lemma in the previous article, $h''(\xi)$ exists for each $\xi > 0$ since $M''(v) < 0$. This certainly makes $h'(\log(1/r))$ continuous for $0 < r < 1$, so $rw'(r)$ is continuous for such r. When $r \to 0$, $w(r)$ *increases* and tends to a *finite* limit (since $M'(0) > -\infty$), and $h'(\log(1/r))$ *increases* (convexity of $h'(\xi)$), remaining, however, always $\leqslant 0$. Hence $rw'(r)$ *tends to a finite limit* as $r \to 0$, and (with obvious definition of $rw'(r)$ *at* the origin) is thus continuous at 0.

Having justified the application of the theorem from article 1 by this rather fussy argument, we see through its use that

$$\frac{\partial G(z)}{\partial \bar{z}} = -|z|^2 w(|z|)\Omega(e^{i\vartheta})$$

for

$$z = |z|e^{i\vartheta}, \quad |z| < 1,$$

after taking account of the fact that

$$\frac{\partial}{\partial \bar{z}}\left(\sum_0^\infty a_n z^n\right) = 0, \qquad |z| < 1.$$

This relation certainly makes

$$\left|\frac{\partial G(z)}{\partial \bar{z}}\right| \leqslant \text{const.} w(|z|)$$

for $|z| < 1$, so, if $G(z)$ *does* coincide with $F(z)$ for $|z| = 1$, *we will* have the theorem on putting $F(z) = G(z)$ for $|z| < 1$.

Everything thus depends on our being able to determine a continuously differentiable $\Omega(e^{it})$ which will make

$$(*) \qquad \frac{1}{2\pi}\int_0^{2\pi}\int_0^1 \frac{\Omega(e^{it})r^2 w(r) r\,dr\,dt}{re^{it} - e^{i\vartheta}}$$

have the Fourier series

$$\sum_{-\infty}^{-1} a_n e^{in\vartheta}.$$

Expanding the integrand from (⁑) in powers of re^{it}, we obtain for that expression the value

$$-\sum_{n=1}^{\infty}\left(\int_0^1 r^{n+2}w(r)\,dr\right)\left(\frac{1}{2\pi}\int_0^{2\pi} e^{i(n-1)t}\Omega(e^{it})\,dt\right)e^{-in\vartheta}.$$

We see that we need to have

$$\Omega(e^{it}) \sim \sum_{-\infty}^{\infty} b_n e^{int}$$

where, for $n > 1$,

$$(\dagger)\qquad b_{1-n} = -\frac{a_{-n}}{\int_0^1 r^{n+2}w(r)\,dr}.$$

We may choose the b_n with *positive* index in any manner compatible with the continuous differentiability of $\Omega(e^{it})$; let us simply *put them all equal to zero.*

By the lemma, the *right side* of (†) is in modulus

$$\leqslant \text{const.}|na_{-n}|e^{M(n)}$$

for $n \geqslant 1$. The b_m given by (†) therefore satisfy

$$\sum_{-\infty}^{0} |mb_m| < \infty$$

according to the hypothesis of our theorem. This means that there *is* a function $\Omega(e^{it})$ satisfying our requirements whose *differentiated Fourier series is absolutely convergent.* Such a function is surely continuously differentiable; that is what was needed.

The theorem is proved.

Remark 1. We are going to use the extension of F to $\{|z| < 1\}$ furnished by Dynkin's theorem in conjunction with the *corollary at the end of article* 1. That corollary involves the integral

$$\frac{1}{2\pi}\iint_{\mathscr{D}} \frac{F_{\bar{\zeta}}(\zeta)}{F(\zeta)}\frac{d\xi\,d\eta}{(\zeta - z)}$$

where, on the open set $\mathscr{D}$, $|F(\zeta)| > 0$ and $|F_{\bar{\zeta}}(\zeta)| \leqslant \text{const.}|F(\zeta)|$. The theorem of article 1 is used to take $\partial/\partial\bar{z}$ of this integral, and the *hypothesis of the corollary* requires that $F(\zeta)$ be $\mathscr{C}_2$ in $\mathscr{D}$ in order to guarantee the legitimacy of that theorem's application. *Our extension $F(z)$ furnished by Dynkin's theorem is, however, only ensured to be $\mathscr{C}_1$* for $|z| < 1$. Are we not in *trouble*?

Not to worry. All that the corollary really *uses* is continuous differentiability of the *quotient* $F_{\bar{\zeta}}(\zeta)/F(\zeta)$ in $\mathscr{D}$. Our F *is*, however, $\mathscr{C}_1$ in $\mathscr{D}$, where it is also $\neq 0$. And, *from the proof of Dynkin's theorem,* $F_{\bar\zeta}(\zeta) = -|\zeta|^2 w(|\zeta|)\Omega(e^{it})$ for $\zeta = |\zeta|e^{it}$ with $|\zeta| < 1$. *The expression on the right is, however,* $\mathscr{C}_1$ for $|\zeta| < 1$; we indeed checked that it *had* that property during the proof (as we *had* to do in order to justify using the theorem of article 1 to show that $F_{\bar\zeta}(\zeta)$ was equal to it!). We are all right.

Remark 2. Under the conditions of Volberg's theorem, there is no essential distinction between the functions $M(v)$ and $2M(v)$, and $e^{M(v)}$ goes to infinity much faster than any power of v as $v \to \infty$. (We will need to require that $M(v) \geqslant \text{const.}\, v^\alpha$ for some $\alpha > \frac{1}{2}$ as has already been remarked in article 2.) In the application of Dynkin's theorem to be made below, we will therefore be able to replace the condition

$$\sum_1^\infty |n^2 a_{-n}| e^{M(n)} \;<\; \infty$$

figuring in its hypothesis by

$$|a_{-n}| \;\leqslant\; \text{const.}e^{-2M(n)}, \qquad n \geqslant 1,$$

or even (after a suitable unessential modification in the description of $w(r)$) by

$$|a_{-n}| \;\leqslant\; \text{const.}e^{-M(n)}, \qquad n \geqslant 1.$$

4. Material about weighted planar approximation by polynomials

Lemma. *Let* $w(r) \geqslant 0$ *for* $0 \leqslant r < 1$, *with*

$$\int_0^1 w(r) r\, dr \;<\; \infty.$$

If $F(z)$ *is any function analytic in* $\{|z| < 1\}$ *such that*

$$\iint_{|z|<1} |F(z)|^2 w(|z|)\, dx\, dy \;<\; \infty,$$

there are polynomials $Q(z)$ *making*

$$\iint_{|z|<1} |F(z) - Q(z)|^2 w(|z|)\, dx\, dy$$

arbitrarily small.

Proof. The basic idea is that $w(|z|)$ depends only on the modulus of z.

Given $\varepsilon > 0$, take $\rho < 1$ so close to 1 that

$$\iint_{\rho<|z|<1} |F(z)|^2 w(|z|)\,dx\,dy \;<\; \varepsilon.$$

Note that if $0 < \lambda < 1$ and $0 < r < 1$,

$$\int_0^{2\pi} |F(\lambda r e^{i\vartheta})|^2\,d\vartheta \;\leqslant\; \int_0^{2\pi} |F(re^{i\vartheta})|^2\,d\vartheta.$$

Therefore,

$$\iint_{\rho<|z|<1} |F(\lambda z)|^2 w(|z|)\,dx\,dy \;=\; \int_\rho^1 \int_0^{2\pi} |F(\lambda r e^{i\vartheta})|^2 w(r) r\,d\vartheta\,dr$$

$$\leqslant\; \int_\rho^1 \int_0^{2\pi} |F(re^{i\vartheta})|^2 w(r) r\,d\vartheta\,dr \;<\; \varepsilon.$$

Once $\rho < 1$ has been fixed, $F(z)$ is *uniformly continuous* for $|z| \leqslant \rho$, so $\iint_{|z|<\rho} |F(z) - F(\lambda z)|^2 w(|z|)\,dx\,dy \longrightarrow 0$ as $\lambda \to 1$ (we use the integrability of $rw(r)$ on $[0, 1)$ here). In view of the preceding calculation, we can thus find (and fix) a $\lambda < 1$ such that

$$\iint_{|z|<1} |F(z) - F(\lambda z)|^2 w(|z|)\,dx\,dy \;<\; 5\varepsilon.$$

The Taylor series for $F(\lambda z)$ converges *uniformly* for $|z| \leqslant 1$. We may therefore take a suitable *partial sum* $Q(z)$ of that Taylor series so as to make

$$\iint_{|z|<1} |F(\lambda z) - Q(z)|^2 w(|z|)\,dx\,dy \;<\; \varepsilon.$$

Then

$$\iint_{|z|<1} |F(z) - Q(z)|^2 w(|z|)\,dx\,dy \;<\; 16\varepsilon.$$

That does it.

At this point, we begin to make systematic use of a corollary to the theorem of Levinson given in §A.5. Oddly enough, Beurling's stronger results from §B are never called for in Volberg's work.

Theorem on simultaneous polynominal approximation (Kriete, Volberg). *Let, for* $0 < r < 1$,

$$w(r) \;=\; \exp\left(-H\left(\log\frac{1}{r}\right)\right)$$

where $H(\xi)$ is decreasing and bounded below on $(0, \infty)$, and suppose that

$$\int_0^a \log H(\xi)\,\mathrm{d}\xi = \infty$$

for all sufficiently small $a > 0$.

Let E be any proper closed subset of the unit circumference, let $p(\mathrm{e}^{\mathrm{i}\vartheta}) \in L_2(E)$, and suppose that $f(z)$ is analytic in $\{|z| < 1\}$, and such that

$$\iint_{|z|<1} |f(z)|^2 w(|z|)\,\mathrm{d}x\,\mathrm{d}y < \infty.$$

Then there is a sequence of polynomials $P_n(z)$ with

$$\int_E |p(\mathrm{e}^{\mathrm{i}\vartheta}) - P_n(\mathrm{e}^{\mathrm{i}\vartheta})|^2\,\mathrm{d}\vartheta + \iint_{|z|<1} |f(z) - P_n(z)|^2 w(|z|)\,\mathrm{d}x\,\mathrm{d}y \underset{n}{\longrightarrow} 0.$$

Proof. We use the fact that the collection of functions $F(z)$ analytic in $\{|z| < 1\}$ with $\iint_{|z|<1} |F(z)|^2 w(|z|)\,\mathrm{d}x\,\mathrm{d}y < \infty$ forms a Hilbert space if we bring in the inner product

$$\langle F_1, F_2 \rangle_w = \iint_{|z|<1} F_1(z)\overline{F_2(z)}\,w(|z|)\,\mathrm{d}x\,\mathrm{d}y.$$

This is evident, except perhaps for the completeness property. To verify the latter, it is clearly enough to show that, for any L, the functions $F(z)$ analytic in $\{|z| < 1\}$ and satisfying

$$\iint_{|z|<1} |F(z)|^2 w(|z|)\,\mathrm{d}x\,\mathrm{d}y \leqslant L$$

form a *normal family* in the open unit disk. However, the weight $w(r)$ we are using is *strictly positive* and *decreasing* on $(0, 1)$. Hence, for any $r < 1$, the previous relation makes

$$\iint_{|z|<r} |F(z)|^2\,\mathrm{d}x\,\mathrm{d}y \leqslant \frac{L}{w(r)}.$$

It is well known that such functions $F(z)$ form a normal family in $\{|z| < r\}$. Here, $r < 1$ is arbitrary.

Let us turn to the proof of the theorem, reasoning by duality in the Hilbert space $L_2(E) \oplus \mathscr{H}$ where $\mathscr{H}$ is the Hilbert space just described.* Suppose, then, that there is a $p(\mathrm{e}^{\mathrm{i}\vartheta}) \in L_2(E)$ and an $f(z) \in \mathscr{H}$ for which the *conclusion of the theorem fails to hold.* There must then be a *non-zero* element (q, F) of $L_2(E) \oplus \mathscr{H}$ *orthogonal* in that space to *all* the elements of the form $(P(\mathrm{e}^{\mathrm{i}\vartheta}), P(z))$ with *polynomials* P. We are going to obtain a *contradiction* by showing that *in fact* $q = 0$ and $F = 0$.

* We are dealing here with the *direct sum* of $L_2(E)$ and $\mathscr{H}$.

The orthogonality in question is equivalent to the relations

$$\int_E \overline{q(e^{i\vartheta})}e^{in\vartheta}d\vartheta + \iint_{|z|<1} \overline{F(z)}z^n w(|z|)\,dx\,dy = 0, \quad n = 0,1,2,3,\ldots.$$

Define $q(e^{i\vartheta})$ for *all* of $\{|z|=1\}$ by making it *zero* for $e^{i\vartheta} \notin E$. Then $q(e^{i\vartheta}) \in L_2(-\pi,\pi)$, and if we write

$$(*) \qquad q(e^{i\vartheta}) \sim \sum_{-\infty}^{\infty} \alpha_n e^{in\vartheta}$$

we find from the previous relation that

$$\bar{\alpha}_n = -\frac{1}{2\pi}\iint_{|z|<1} \overline{F(z)}z^n w(|z|)\,dx\,dy, \qquad n \geqslant 0.$$

Since

$$\iint_{|z|<1} |F(z)|^2 w(|z|)\,dx\,dy < \infty,$$

the integral on the right is in modulus

$$\leqslant \text{const.}\sqrt{\int_0^1 r^{2n}w(r)r\,dr}$$

by Schwarz' inequality. However, in terms of $\xi = \log(1/r)$ and the function $H(\xi)$,

$$r^{2n}w(r) = e^{-(H(\xi)+2n\xi)}.$$

Denoting $\inf_{\xi>0}(H(\xi)+\xi v)$ by $M(v)$ as in the last theorem of article 2, we see that the right side is $\leqslant e^{-M(2n)}$. The preceding expression is therefore $\leqslant \text{const.}e^{-M(2n)/2}$, i.e.

$$(\ast\ast) \qquad |\alpha_n| \leqslant \text{const.}e^{-M(2n)/2}, \qquad n \geqslant 1.$$

Since E is a *proper closed subset of* $\{|z|=1\}$, *its complement on the unit circumference contains an arc* J *of positive length.* The function $q(e^{i\vartheta})$ *vanishes* outside E, hence on J, and certainly belongs to $L_1(-\pi,\pi)$. Also, *by the last theorem in article 2,*

$$\int_1^\infty \frac{M(v)}{v^2}dv = \infty$$

on account of our hypothesis on $H(\xi)$. Therefore

$$\sum_1^\infty \frac{M(2n)}{2n^2} = \infty,$$

$M(\nu)$ being increasing, so, *by virtue of the corollary at the end of* §A.5, $(*)$ and $(\ast\!\ast)$ imply that $q(e^{i\vartheta}) = 0$ a.e.

We see that $\alpha_n = 0$ for all n, which means that

$$\iint_{|z|<1} \overline{F(z)}\, z^n w(|z|)\, dx\, dy \;=\; 0$$

for $n = 0, 1, 2, \ldots$. Since $H(\xi)$ is bounded below on $(0, \infty)$, $w(r)$ is *bounded above* for $0 < r < 1$, and we can invoke the lemma, concluding that *polynomials are dense in the Hilbert space* $\mathscr{H}$. The previous relation therefore implies that $F(z) \equiv 0$.

We have thus reached a contradiction by showing that $q = 0$ and $F = 0$. The theorem is proved.

Remark. Some applications involve a weight

$$w(r) \;=\; \exp\left(-h\left(\log\frac{1}{r}\right)\right)$$

where, for $\xi > 0$,

$$h(\xi) \;=\; \sup_{\nu>0}\,(M(\nu) - \nu\xi),$$

the function $M(\nu)$ being *merely* supposed *increasing*, and such that $M(0) > -\infty$.

In this situation, we can, from the condition

$$\sum_1^\infty \frac{M(n)}{n^2} \;=\; \infty,$$

conclude that *the rest of the above theorem's statement is valid.*

This can be seen without appealing to the last theorems of article 2. We have here, with $\xi = \log(1/r)$,

$$r^{2n} w(r) \;=\; e^{-(h(\xi) + 2n\xi)},$$

and, since, for any $\xi > 0$,

$$h(\xi) \;\geqslant\; M(\nu) - \nu\xi$$

for each $\nu > 0$, $h(\xi) + 2n\xi \geqslant M(2n)$. We now arrive at $(\ast\!\ast)$ in the same way as above, so, since $M(\nu)$ is increasing,

$$\sum_1^\infty \frac{M(n)}{n^2} \;=\; \infty \quad \text{makes} \quad \sum_1^\infty \frac{M(2n)}{2n^2} \;=\; \infty$$

and we can conclude by direct application of the corollary from §A.5. (Here, boundedness of $w(r)$ is ensured by the condition $M(0) > -\infty$.)

Remark on a certain change of cariable

If the weight $w(r) = \exp(-H(\log(1/r)))$ *satisfies the hypothesis of the theorem on simultaneous polynomial approximation, so does the weight* $w(r^L) = \exp(-H(L\log(1/r)))$ *for any positive constant* L. That's simply because

$$\int_0^a H(L\xi)\,d\xi \;=\; \frac{1}{L}\int_0^{aL} H(\xi)\,d\xi \;!$$

That theorem therefore remains valid if we replace the weight $w(r)$ *figuring in its statement by* $w(r^L)$, L *being any positive constant.*

We will use this fact several times in what follows.

5. Volberg's theorem on harmonic measures

The result to be proved here plays an important role in the establishment of the main theorem of this §. It is also of interest in its own right.

Definition. Let $\mathcal{O}$ be an open subset of $\{|z| < 1\}$, and J any open *arc* of $\{|z| = 1\}$. We say that $\mathcal{O}$ *abuts* on J if, for each $\zeta \in J$, there is a neighborhood V_ζ of ζ with

$$V_\zeta \cap \{|z| < 1\} \;\subseteq\; \mathcal{O}.$$

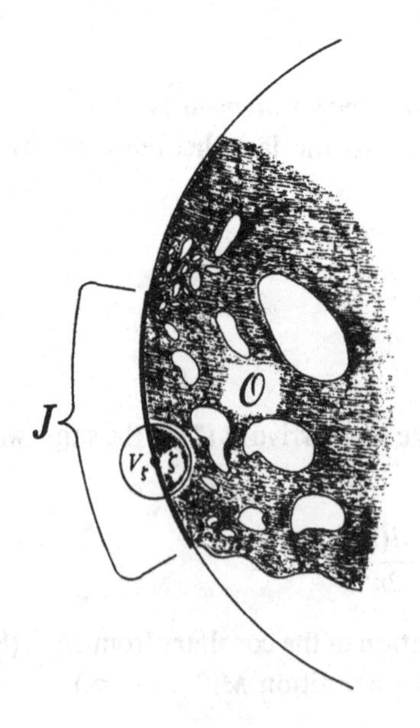

Figure 96

Now we come to the

Theorem on harmonic measures (Volberg). *Let, for* $0 < r < 1$, $w(r) = \exp(-H(\log(1/r)))$, *where* $H(\xi)$ *is* decreasing *and* bounded *below on* $(0, \infty)$, *and tends to* ∞ *sufficiently rapidly as* $\xi \to 0$ *to make* $w(r) = O((1-r)^2)$ *for* $r \to 1$. *(In the situation of Volberg's theorem, we have* $H(\xi) \geqslant \text{const}.\xi^{-c}$ *with* $c > 0$, *so this will certainly be the case.)*

Assume furthermore that

$$\int_0^a \log H(\xi)\,d\xi \;=\; \infty$$

for all sufficiently small $a > 0$.

Let $\mathcal{O}$ *be any connected open set in* $\{|z| < 1\}$ *whose boundary is regular enough to permit the solution of Dirichlet's problem for* $\mathcal{O}$. *Suppose that there are two open arcs* I *and* J *of positive length on* $\{|z| = 1\}$ *such that:*

(i) $\partial\mathcal{O} \cap J$ *is* empty;

(ii) $\mathcal{O}$ abuts *on* I.

Then, if $\omega_{\mathcal{O}}(\ ,z)$ *denotes* harmonic measure *for* $\mathcal{O}$ *(as seen from* $z \in \mathcal{O}$), *we have*

$$\int_{\{|\zeta|<1\}\cap\partial\mathcal{O}} \log\left(\frac{1}{w(|\zeta|)}\right) d\omega_{\mathcal{O}}(\zeta, z_0) \;=\; \infty$$

for each $z_0 \in \mathcal{O}$.

Remark 1. The integral is taken over the part of $\partial\mathcal{O}$ lying *inside* $\{|z| < 1\}$.

Remark 2. The assumption that $\mathcal{O}$ *abuts* on an *arc I* can be relaxed. But the proof uses the full strength of the assumption that $\partial\mathcal{O}$ *avoids J*.

Proof of theorem. We work with the weight $w_1(r) = w(r^3)$. By the *theorem on simultaneous polynomial approximation* and *remark on a change of variable* (previous article), there are polynomials $P_n(z)$ with

$$\int_{\{|\zeta|=1\}\sim J} |P_n(e^{i\vartheta})|^2\,d\vartheta \;\underset{n}{\longrightarrow}\; 0$$

and at the same time

$$\iint_{|z|<1} |P_n(z) - 1|^2 w_1(|z|)\,dx\,dy \;\underset{n}{\longrightarrow}\; 0.$$

The second relation certainly implies that

$$\iint_{|z|<1} |P_n(z)|^2 w_1(|z|)\,dx\,dy \;\leqslant\; C$$

for some $C < \infty$, and all n.

Take any z_0, $|z_0| < 1$; we use the last inequality to get a uniform upper estimate for the values $|P_n(z_0)|$. Put $\rho = \frac{1}{2}(1 - |z_0|)$.

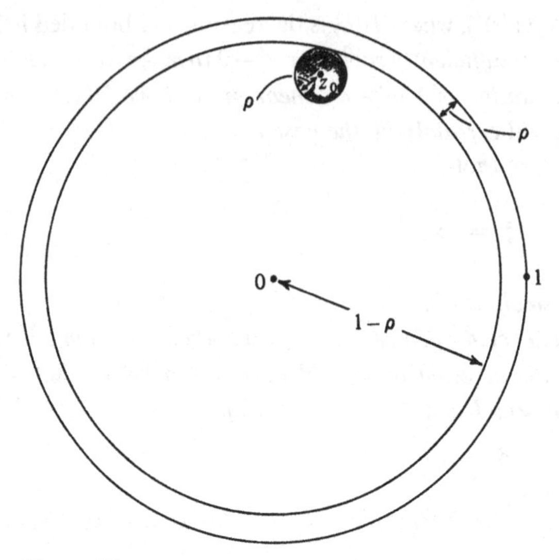

Figure 97

We have

$$|P_n(z_0)|^2 \leqslant \frac{1}{\pi\rho^2}\iint_{|z-z_0|<\rho} |P_n(z)|^2 \, dx \, dy.$$

$w_1(r)$ *decreases*, so the right side is

$$\leqslant \frac{1}{\pi\rho^2 w_1(1-\rho)}\iint_{|z-z_0|<\rho} |P_n(z)|^2 w_1(|z|) \, dx \, dy$$

which, in turn, is

$$\leqslant \frac{C}{\pi\rho^2 w_1(1-\rho)}$$

by the above inequality.

Here, $w_1(1-\rho) = w(1 - 3\rho + 3\rho^2 - \rho^3)$ is $\geqslant$ $w(1-2\rho) = w(|z_0|)$ when $3\rho^2 - \rho^3 \leqslant \rho$, i.e., for $\rho \leqslant (3-\sqrt{5})/2$. Also, $w(r)$ is bounded above ($H(\xi)$ being bounded below), so, for $(3-\sqrt{5})/2 < \rho < \frac{1}{2}$, $w_1(1-\rho) \geqslant w(\sqrt{5}-2) \geqslant \text{const.} w(|z_0|)$. The result just found therefore reduces to

$$|P_n(z_0)|^2 \leqslant \frac{\text{const.}}{(1-|z_0|)^2 w(|z_0|)},$$

with the right-hand side in turn

$$\leqslant \frac{\text{const.}}{(w(z_0))^2}$$

according to the hypothesis. Thus, since z_0 was arbitrary,

$$(*) \qquad \log|P_n(z)| \leqslant \text{const.} + \log\left(\frac{1}{w(|z|)}\right), \qquad |z|<1.$$

A similar (and simpler) argument, applied to $P_n(z)-1$, shows that

$$(\ast\!\ast) \qquad P_n(z) \underset{n}{\longrightarrow} 1, \qquad |z|<1.$$

Let us now fix our attention on $\partial\mathcal{O}\cap\{|\zeta|=1\}$, *which we henceforth denote by* S, in order to simplify the notation. The *open unit disk* Δ includes $\mathcal{O}$, therefore, by the *principle of extension of domain* (see §B.1), for $z_0\in\mathcal{O}$,

$$\mathrm{d}\omega_{\mathcal{O}}(\zeta, z_0) \leqslant \mathrm{d}\omega_{\Delta}(\zeta, z_0)$$

for ζ varying on S. In other words,

$$(\dagger) \qquad \mathrm{d}\omega_{\mathcal{O}}(\zeta, z_0) \leqslant K(z_0)|\mathrm{d}\zeta|$$

for ζ on S with a number $K(z_0)$ depending only on z_0.

Since $\mathcal{O}$ *abuts* on the arc I with $|I|>0$, *we surely have*

$$\omega_{\mathcal{O}}(S, z_0) > 0$$

for each $z_0\in\mathcal{O}$ by Harnack's theorem. Since $S\subseteq\{|\zeta|=1\}\sim J$ ((i) of the hypothesis), we have, by ($\dagger$) and the relation between arithmetic and geometric means,

$$\int_S \log|P_n(\mathrm{e}^{\mathrm{i}\vartheta})|\,\mathrm{d}\omega_{\mathcal{O}}(\mathrm{e}^{\mathrm{i}\vartheta}, z_0)$$

$$\leqslant \tfrac{1}{2}\omega_{\mathcal{O}}(S, z_0)\log\left(\frac{1}{\omega_{\mathcal{O}}(S, z_0)}\int_S |P_n(\mathrm{e}^{\mathrm{i}\vartheta})|^2\,\mathrm{d}\omega_{\mathcal{O}}(\mathrm{e}^{\mathrm{i}\vartheta}, z_0)\right)$$

$$\leqslant \tfrac{1}{2}\omega_{\mathcal{O}}(S, z_0)\log\left(\frac{K(z_0)}{\omega_{\mathcal{O}}(S, z_0)}\int_{\{|\zeta|=1\}\sim J} |P_n(\mathrm{e}^{\mathrm{i}\vartheta})|^2\,\mathrm{d}\vartheta\right)$$

for $z_0\in\mathcal{O}$. *This last expression, however, tends to* $-\infty$ *as* $n\to\infty$ because

$$\int_{\{|\zeta|=1\}\sim J} |P_n(\mathrm{e}^{\mathrm{i}\vartheta})|^2\,\mathrm{d}\vartheta \underset{n}{\longrightarrow} 0$$

At the same time, for any $z_0\in\mathcal{O}$, $\log|P_n(z_0)| \underset{n}{\longrightarrow} 0$ by ($\ast\!\ast$). Therefore, by the *theorem on harmonic estimation* in §B.1 (whose extension to possibly infinitely connected domains $\mathcal{O}$ of the kind considered here presents no difficulty, at least for *polynomials* $P_n(z)$), we see that

$$\int_S \log|P_n(e^{i\vartheta})|\, d\omega_{\mathcal{O}}(e^{i\vartheta}, z_0) + \int_{\partial\mathcal{O}\cap\Delta} \log|P_n(\zeta)|\, d\omega_{\mathcal{O}}(\zeta, z_0)$$

$$= \int_{\partial\mathcal{O}} \log|P_n(\zeta)|\, d\omega_{\mathcal{O}}(\zeta, z_0) \geqslant \log|P_n(z_0)| \underset{n}{\longrightarrow} 0.$$

As we have just shown, the *first* of the two integrals in the left-hand member *tends to* $-\infty$ as $n \to \infty$. *Hence the second must tend to* ∞ *as* $n \to \infty$ (!). However, by (*),

$$\log|P_n(\zeta)| \leqslant \text{const.} + \log\left(\frac{1}{w(|\zeta|)}\right)$$

for $\zeta \in \partial\mathcal{O} \cap \Delta$. So we must have

$$\int_{\partial\mathcal{O}\cap\Delta} \log\left(\frac{1}{w(|\zeta|)}\right) d\omega_{\mathcal{O}}(\zeta, z_0) = \infty \quad \text{for} \quad z_0 \in \mathcal{O}.$$

The theorem is proved.

Remark. The result just established holds in particular for weights $w(r) = \exp(-h(\log(1/r)))$ with $h(\xi) = \sup_{v>0}(M(v) - v\xi)$ for $\xi > 0$, $M(v)$ being increasing, provided that $M(0) > -\infty$, that $\sum_1^\infty (M(n)/n^2) = \infty$, and that $M(v) \to \infty$ as $v \to \infty$ fast enough to make $w(r) = O((1-r)^2)$ for $r \to 1$. See remark following the theorem on simultaneous polynomial approximation (previous article).

Corollary. *Let the connected open set* $\mathcal{O}$ *and the weight* $w(r)$ *be as in the theorem (or the last remark). Let* $G(z)$ *be* analytic *in* $\mathcal{O}$ *and continuous up to* $\partial\mathcal{O}$, *and suppose that, for some* ρ, $0 < \rho < 1$, *we have*

$$|G(\zeta)| \leqslant w(|\zeta|) \quad \textit{for } \zeta \in \partial\mathcal{O} \textit{ with } 1 - \rho < |\zeta| < 1 \ (\textit{sic!}).$$

Then $G(z) \equiv 0$ *in* $\mathcal{O}$.

Proof. Take any $z_0 \in \mathcal{O}$. Since $G(z)$ is continuous on $\bar{\mathcal{O}}$, it is bounded there, so, since $w(r)$ *decreases*, we surely have $G(z) \leqslant \text{const.}w(|z|)$ for $z \in \bar{\mathcal{O}}$ and $|z| \leqslant 1 - \rho$. Therefore our hypothesis in fact implies that

$$|G(\zeta)| \leqslant Cw(|\zeta|) \quad \text{for} \quad \zeta \in \partial\mathcal{O} \cap \{|\zeta| < 1\}$$

with a certain constant C.

Write $\partial\mathcal{O} \cap \{|\zeta| < 1\} = \gamma$, and denote the intersection $\partial\mathcal{O} \sim \gamma$ of $\partial\mathcal{O}$ with the unit circumference by S as in the proof of the theorem. By the *theorem on harmonic estimation* (§B.1), we have

$$\log|G(z_0)| \leqslant \int_S \log|G(\zeta)|\, d\omega_{\mathcal{O}}(\zeta, z_0) + \int_\gamma \log|G(\zeta)|\, d\omega_{\mathcal{O}}(\zeta, z_0).$$

If M is a bound for $G(z)$ on $\bar{\mathcal{O}}$, the *first* integral on the right is $\leqslant \log M$. The *second* is

$$\leqslant \log C + \int_{\gamma} \log w(|\zeta|)\, d\omega_{\mathcal{O}}(\zeta, z_0)$$

by the above inequality. The integral just written is, however, equal to $-\infty$ by the theorem on harmonic measures. Hence $G(z_0) = 0$, as required.

Scholium. L. Carleson observed that the result furnished by the theorem on harmonic measures cannot be essentially improved. By this he meant the following:

If $w(r) = \exp(-h(\log(1/r)))$ *with* $h(\xi)$ *strictly decreasing, convex, and bounded below on* $(0, \infty)$, *and if*

$$\int_0^a \log h(\xi)\, d\xi \;<\; \infty$$

for all sufficiently small $a > 0$, *there is a simply connected open set* $\mathcal{O}$ *in* $\Delta = \{|z| < 1\}$ *fulfilling the conditions of the above theorem for which*

$$\int_{\partial\mathcal{O}\cap\Delta} \log\left(\frac{1}{w(|\zeta|)}\right) d\omega_{\mathcal{O}}(\zeta, z_0) \;<\; \infty, \qquad z_0 \in \mathcal{O}.$$

To see this, observe that the convergence of $\int_0^a \log h(\xi)\, d\xi$ for all sufficiently small $a > 0$ implies that

$$(\S) \qquad \int_0^a \log|h'(\xi)|\, d\xi \;<\; \infty$$

for such a. (See the proof of the *second* theorem in article 2.) We use (§) in order to construct a domain $\mathcal{O}$ like this

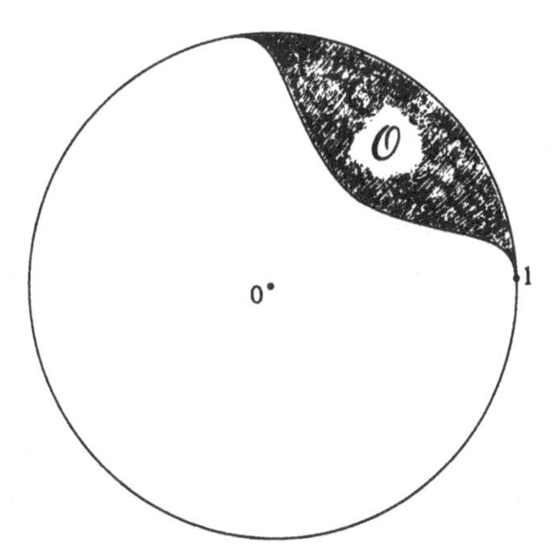

Figure 98

for which

$$\int_{\partial\mathcal{O}\cap\Delta} \log\left(\frac{1}{w(|\zeta|)}\right) d\omega_{\mathcal{O}}(\zeta, z_0) \;<\; \infty, \qquad z_0 \in \mathcal{O},$$

with $w(r) = \exp(-h(\log(1/r)))$. It is convenient to map our (as yet undetermined) region $\mathcal{O}$ *conformally* onto another one, $\mathcal{D}$, by taking $z = re^{i\vartheta}$ to $\varphi = i\log(1/z) = \vartheta + i\log(1/r) = \vartheta + i\xi$. Here ξ has its usual significance.

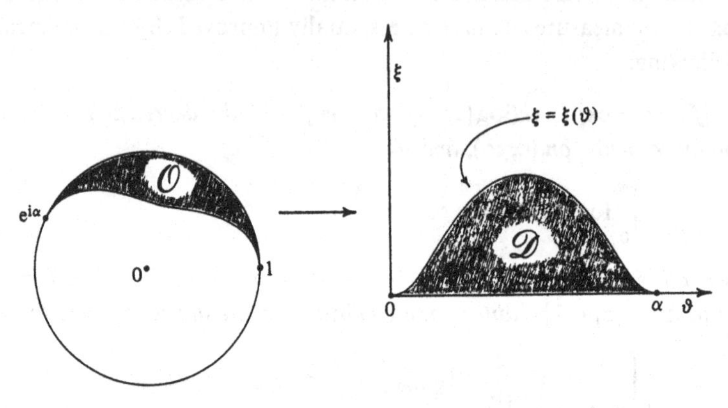

Figure 99

If, in this mapping, the point $z_0 \in \vartheta$ goes over to $p \in \mathcal{D}$, we have, clearly,

$$\int_{\partial\mathcal{O}\cap\Delta} \log\left(\frac{1}{w(|\zeta|)}\right) d\omega_{\mathcal{O}}(\zeta, z_0) \;=\; \int_{\partial\mathcal{D}\cap\{\xi>0\}} h(\xi)\, d\omega_{\mathcal{D}}(\varphi, p).$$

We see that it is enough to determine the equation $\xi = \xi(\vartheta)$ of the *upper bounding curve* of $\mathcal{D}$ (see picture) in such fashion as to have

$$\int_0^\alpha h(\xi(\vartheta))\, d\omega_{\mathcal{D}}(\vartheta + i\xi(\vartheta), p) \;<\; \infty$$

when $p \in \mathcal{D}$. The easiest way to proceed is to construct a function $\xi(\vartheta) = \xi(\alpha - \vartheta)$, making the upper bounding curve *symmetric about the vertical line through its midpoint*. Then we need only determine an increasing function $\xi(\vartheta)$ on the range $0 \leqslant \vartheta \leqslant \alpha/2$ in such a way that

$$\int_0^{\theta_0} h(\xi(\vartheta))\, d\omega_{\mathcal{D}}(\vartheta + i\xi(\vartheta), p) \;<\; \infty$$

for some $\theta_0 > 0$ and *some* $p \in \mathcal{D}$. From this, the same inequality will follow for every $p \in \mathcal{D}$ by Harnack's theorem, and we can arrive at the full result by adding two such integrals.

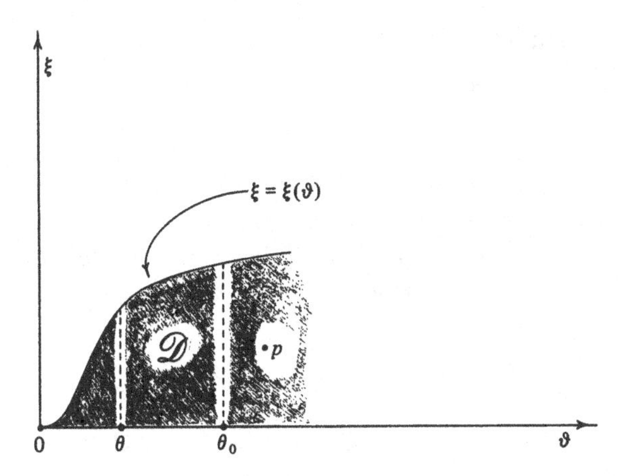

Figure 100

For $0<\theta<\theta_0$, we have, integrating by parts,

$$\int_\theta^{\theta_0} h(\xi(\vartheta))\mathrm{d}\omega_{\mathscr{D}}(\vartheta+\mathrm{i}\xi(\vartheta),p) \;=\; h(\xi(\theta_0))\omega(\theta_0)-h(\xi(\theta))\omega(\theta)$$

$$-\int_\theta^{\theta_0}\omega(\vartheta)\mathrm{d}h(\xi(\vartheta)),$$

where $\omega(\vartheta)$ denotes the *harmonic measure* (at p) of the segment of the *upper bounding curve* having abscissae between 0 and ϑ, viz.,

$$\omega(\vartheta) \;=\; \int_0^\vartheta \mathrm{d}\omega_{\mathscr{D}}(\tau+\mathrm{i}\xi(\tau),p).$$

Making $\theta\to 0$ and remembering that $h(\xi)$ *decreases*, we see that what we *want* is

$$(\S\S)\qquad \int_0^{\theta_0}\omega(\vartheta)|h'(\xi(\vartheta))|\mathrm{d}\xi(\vartheta) \;<\; \infty.$$

For $\omega(\vartheta)$ we may use the *Carleman–Ahlfors estimate for harmonic measure in curvilinear strips*, to be derived in Chapter IX.* According to that, if p is *fixed* and to the *right* of θ_0,

$$\omega(\vartheta) \;\leqslant\; \text{const.}\exp\left(-\pi\int_\vartheta^{\theta_0}\frac{\mathrm{d}\tau}{\xi(\tau)}\right)$$

for $0<\vartheta<\theta_0$. The most simple-minded way of ensuring (§§) is then to cook

* See Remark 1 following the third theorem of §E.1 in that chapter. The upper bound arrived at by the method explained there applies in fact to a harmonic measure *larger* than $\omega(\vartheta)$.

the positive increasing function $\xi(\tau)$ so as to have

$$\log|h'(\xi(\vartheta))| \;-\; \pi\int_{\vartheta}^{\theta_0}\frac{d\tau}{\xi(\tau)} \;=\; \text{const.}$$

It is the relation (§) which makes it possible for us to do this.

In order to avoid being fussy, *let us at this point make the additional* (and not really restrictive) *assumption that $h'(\xi)$ is continuously differentiable.* Then we can differentiate the previous equation with respect to ϑ, getting

$$\frac{d\log|h'(\xi)|}{d\xi}\frac{d\xi}{d\vartheta} \;=\; -\frac{\pi}{\xi}$$

for our unknown increasing function $\xi = \xi(\vartheta)$. Calling $\vartheta(\xi)$ the *inverse* to the function $\xi(\vartheta)$, we have

$$\frac{d\vartheta(\xi)}{d\xi} \;=\; -\frac{1}{\pi}\xi\frac{d\log|h'(\xi)|}{d\xi}.$$

In view of (§), this has the solution

$$\vartheta(\xi) \;=\; \frac{1}{\pi}\int_0^{\xi}(\log|h'(t)| - \log|h'(\xi)|)dt$$

with $\vartheta(0) = 0$. Since $h'(t)$ is < 0 and *increasing* (i.e., $\log|h'(t)|$ *decreases*), we see that the function $\vartheta(\xi)$ given by this formula *is* strictly increasing, and therefore *has* an increasing inverse $\xi(\vartheta)$ for which (§§) holds.

This completes our construction, and Carleson's observation is verified.

Let us remark that one can, by the same method, establish a version of the Levinson loglog theorem which we will give at the end of this § (accompanied, however, by a proof based on a different idea). V.P. Gurarii showed me this simple argument (Levinson's original proof of the loglog theorem, found in his book, is quite hard) at the 1966 International Congress in Moscow.

6. Volberg's theorem on the logarithmic integral

We are finally in a position to undertake the proof of the main result of this §. This is what we will establish:*

Theorem. *Let $M(v)$ be increasing for $v \geqslant 0$.*

Suppose that

$$M(v)/v \text{ is decreasing,}$$

that

$$M(v) \;\geqslant\; \text{const.}v^{\alpha}$$

* A refinement of the following result due to Brennan is given in the *Addendum* at the end of the present volume.

with some $\alpha > \frac{1}{2}$ for all large ν, and that

$$\sum_1^\infty \frac{M(n)}{n^2} = \infty.$$

Let

$$F(e^{i\vartheta}) \sim \sum_{-\infty}^{\infty} a_n e^{in\vartheta}$$

be continuous and not identically zero.

Then, if

$$|a_{-n}| \leqslant e^{-M(n)} \quad \text{for} \quad n \geqslant 1,$$

we have

$$\int_{-\pi}^{\pi} \log|F(e^{i\vartheta})|\,d\vartheta > -\infty.$$

Remark. Volberg states this theorem for functions $F(e^{i\vartheta}) \in L_1(-\pi, \pi)$.* He replaces our second displayed condition on $M(\nu)$ by a weaker one, requiring only that

$$\nu^{-\frac{1}{2}} M(\nu) \longrightarrow \infty$$

for $\nu \to \infty$, but includes an additional restrictive one, to the effect that

$$\nu^{1/2} M(\nu^{1/2}) \leqslant \text{const.} M(\nu)$$

for large ν. This extra requirement serves to ensure that the function

$$h(\xi) = \sup_{\nu > 0} (M(\nu) - \nu\xi)$$

satisfies the relation $h(K\xi) \leqslant (h(\xi))^{1-c}$ with some $K > 1$ and $c > 0$ for small $\xi > 0$; here we have entirely dispensed with it.

Proof of theorem (essentially Volberg's). This will be quite long.

We start by making some simple reductions. First of all, we assume that $M(\nu)/\nu \longrightarrow 0$ for $\nu \to \infty$, since, in the contrary situation, the theorem is easily verified directly (see article 2).

According to the first theorem of article 2, our condition that $M(\nu)/\nu$ decrease implies that the *smallest concave majorant* $M^*(\nu)$ of $M(\nu)$ is $\leqslant 2M(\nu)$; this means that the hypothesis of the theorem is satisfied if, in it, we replace $M(\nu)$ by the *concave increasing function* $M^*(\nu)/2$.

There is thus no loss of generality in supposing to begin with that $M(\nu)$ is *also concave*. We may also assume that $M(0) \geqslant 3$. To see this, suppose that $M(0) < 3$; in that case we may draw a straight line $\mathscr{L}$ from $(0, 3)$ *tangent to the graph* of $M(\nu)$ vs. ν:

* See the addendum for such an extension of Brennan's result.

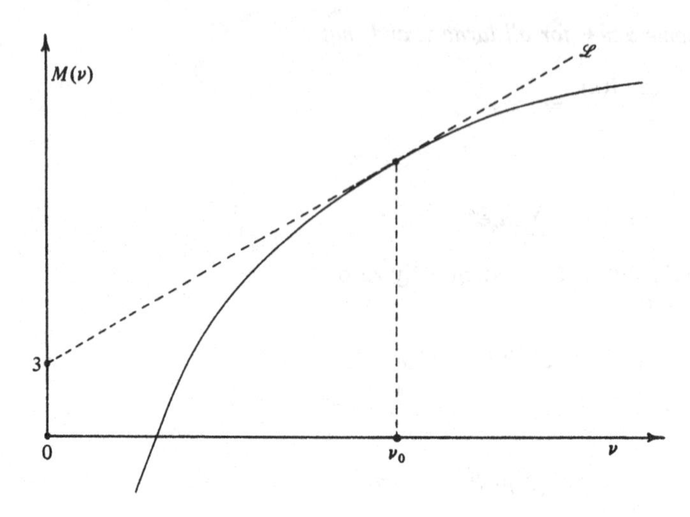

Figure 101

If the point of tangency is at $(\nu_0, M(\nu_0))$, we may then take the new increasing concave function $M_0(\nu)$ equal to $M(\nu)$ for $\nu \geqslant \nu_0$ and to the height of $\mathscr{L}$ at the abscissa ν for $0 \leqslant \nu < \nu_0$. Our Fourier coefficients a_n will satisfy

$$|a_{-n}| \leqslant \text{const.} e^{-M_0(n)}$$

for $n \geqslant 1$, and the rest of the hypothesis will hold with $M_0(\nu)$ in place of $M(\nu)$.

We may now use the simple constructions of $M_1(\nu)$ and $M_2(\nu)$ given in article 2 to obtain an infinitely differentiable, increasing and *strictly* concave function $M_2(\nu)$ (with $M_2''(\nu) < 0$) which is *uniformly close* (within $\frac{1}{2}$ unit, say) to $M_0(\nu)$ on $[0, \infty)$. (Here uniformly close on *all* of $[0, \infty)$ because our present function $M_0(\nu)$ has a *bounded* first derivative on $(0, \infty)$.) We will then still have

$$|a_{-n}| \leqslant \text{const.} e^{-M_2(n)}$$

for $n \geqslant 1$, and the rest of the hypothesis will hold with $M_2(\nu)$ in place of $M(\nu)$.

Since $M(\nu) \geqslant \text{const.}\nu^\alpha$ for large ν, where $\alpha > \frac{1}{2}$, we certainly (and by far!) have

$$n^4 \exp(-M_2(n)/2) \longrightarrow 0, \quad n \to \infty.$$

Therefore

$$\sum_{n=1}^{\infty} |n^2 a_{-n}| e^{M_2(n)/2} < \infty.$$

So, putting $\bar{M}(\nu) = M_2(\nu)/2$, we have

$$\sum_1^\infty |n^2 a_{-n}| e^{\bar{M}(n)} < \infty$$

with a function $\bar{M}(\nu)$ which is *increasing, strictly concave, and infinitely differentiable on* $(0, \infty)$, *having* $\bar{M}''(\nu) < 0$ *there.* The hypothesis of the theorem *holds* with $\bar{M}(\nu)$ standing in place of $M(\nu)$. Besides, $\bar{M}(0)$ $(= \lim_{\nu \to 0} \bar{M}(\nu))$ is $\geqslant 1$ since $M_0(0) \geqslant 3$. Later on, this property will be helpful technically.

▶ *Let us henceforth simply write* $M(\nu)$ *instead of* $\bar{M}(\nu)$. Our *new* function $M(\nu)$ thus *satisfies the hypothesis of Dynkin's theorem* (article 3). We put, as usual,

$$h(\xi) = \sup_{\nu > 0}(M(\nu) - \nu\xi) \quad \text{for} \quad \xi > 0,$$

and then form the weight

$$w(r) = \exp\left(-h\left(\log\frac{1}{r}\right)\right), \qquad 0 < r < 1.$$

Because $M(0) \geqslant 1$, we have $h(\xi) \geqslant 1$ for $\xi > 0$, so $w(r) \leqslant 1/e$. *Applying* Dynkin's theorem, we obtain a *continuous extension* $F(z)$ of our given function $F(e^{i\vartheta})$ to $\{|z| \leqslant 1\}$ with $F(z)$ *continuously differentiable in the open unit disk* and

$$\left|\frac{\partial F(z)}{\partial \bar{z}}\right| \leqslant \text{const.} w(|z|), \qquad |z| < 1.$$

We note here that the properties of $M(\nu)$ assumed in the hypothesis certainly make $w(r) \to 0$ (indeed rather rapidly) as $r \to 1$; see article 2.

Let

$$B_0 = \{z\colon\ |z| < 1 \text{ and } |F(z)| \leqslant w(|z|)\}.$$

We cover each of the closed sets

$$B_0 \cap \left\{1 - \frac{1}{n} \leqslant |z| \leqslant 1 - \frac{1}{n+1}\right\}, \qquad n = 1, 2, 3, \ldots,$$

by a finite number of open disks lying in $\{|z| < 1\}$ on which

$$|F(z)| < 2w(|z|).$$

This gives us altogether a countable collection of open disks lying in the unit circle and covering B_0; *the closure of the union of those disks is denoted by* B. We then have

$$|F(z)| \leqslant 2w(|z|), \qquad z \in B,^*$$

* *including* for $z \in B$ of modulus 1, as long as we take $w(1) = 0$! See argument for *Step 1*, p. 361

and

$$|F(z)| \;>\; w(|z|) \quad \text{for } z\notin B \text{ and } |z|<1.$$

Put

$$\mathcal{O} \;=\; \{|z|<1\}\cap(\sim B);$$

$\mathcal{O}$ is an *open* subset of the unit disk and $|F(z)|>w(|z|)$ therein, as we have just seen. We will see presently that $\mathcal{O}$ fills much of the unit disk. Let us at this point simply observe that $\mathcal{O}$ *is certainly not empty*. If, indeed, it *were* empty, B would fill the unit disk and we would have $|F(z)|\leqslant 2w(|z|)$ for $|z|<1$. The fact that $w(r)\to 0$ for $r\to 1$ would then make $F(e^{i\vartheta})\equiv 0$, *contrary to our hypothesis*, by virtue of the continuity of $F(z)$ on $\{|z|\leqslant 1\}$.

Although the open set $\mathcal{O}$ may have an exceedingly complicated structure, *the Dirichlet problem for it can be solved*. This will follow from a well-known result in elementary potential theory (for a proof of which see, for instance, pp. 35–6 of Gamelin's book, the latter part of the one by Kellog, or any other work on potential theory – I cannot, after all, prove *everything*!), if we show that the *Poincaré cone condition for $\mathcal{O}$ is satisfied at every point ζ of $\partial\mathcal{O}$*, i.e., if, for each such ζ, *there is a small triangle with vertex at ζ lying outside $\mathcal{O}$*. To check this, observe that the small disks used to build up B can accumulate only at points of the unit circumference. Hence, if $|\zeta|<1$ and $\zeta\in\partial\mathcal{O}$, ζ *is on the boundary of one of those small disks, inside* of which a triangle with vertex at ζ may be drawn. If, however, $|\zeta|=1$, we may take a triangle lying *outside the unit disk* with vertex at ζ.

Since $|F_{\bar z}(z)|\leqslant Cw(|z|)$ in $\{|z|<1\}$ while $|F(z)|>w(|z|)$ in $\mathcal{O}$, we have

$$\left|\frac{1}{F(z)}\,\frac{\partial F(z)}{\partial \bar z}\right| \;\leqslant\; C, \qquad z\in\mathcal{O}.$$

Volberg's idea is to take advantage of this relation and use the function

$$\Phi(z) \;=\; F(z)\exp\left\{\frac{1}{2\pi}\int\!\!\int_{\mathcal{O}} \frac{F_{\bar\zeta}(\zeta)}{F(\zeta)}\,\frac{d\xi\, d\eta}{(\zeta-z)}\right\}$$

on $\{|z|\leqslant 1\}$. (N.B. *Again we are writing $\zeta=\xi+i\eta$, in conflict with the notation $\xi=\log(1/r)$ used in discussing $h(\xi)$.*) According to *Remark* 1 to Dynkin's theorem (end of article 3), $F(z)$ has enough differentiability in $\{|z|<1\}$ for us to be able to use the *corollary from the end of article* 1. By that corollary, $\Phi(z)$ *is analytic in $\mathcal{O}$*, and $|\Phi(z)|$ *lies between two constant multiples of $|F(z)|$ there* (*and, actually, on* $\{|z|\leqslant 1\}$ as the last part of that corollary's proof shows). We thus certainly have $|\Phi(z)|>0$ in $\mathcal{O}$ since $|F(z)|>w(|z|)$ there.

Volberg now applies the *theorem on harmonic estimation* (§B.1) to

$\Phi(z)$ and the open set $\mathcal{O}$ in order to eventually get at

$$\int_{-\pi}^{\pi} \log|F(e^{i\vartheta})|\,d\vartheta.$$

His procedure is to show that the whole unit circumference is contained in $\partial\mathcal{O}$, and that the part of $\partial\mathcal{O}$ lying *inside* the unit disk is so unimportant as to make *harmonic measure for* $\mathcal{O}$ act like *ordinary Lebesgue measure* on the unit circumference. We will carry out this program *in 5 steps*. Before going to step 1, we should, however, acknowledge that here the theorem on harmonic estimation will be used under conditions somewhat more general than those allowed for in §B.1. The open set $\mathcal{O}$ may not even be connected, and its components may be infinitely connected! Nevertheless, extension of the theorem in question to the present situation involves no real difficulty – $\Phi(z)$ *is* continuous on $\{|z| \leqslant 1\}$ and the Dirichlet problem for $\mathcal{O}$ *is* solvable.

We proceed.

Step 1. *$B \cap \{|\zeta| = 1\}$ contains no arc of positive length.*

Assume the contrary. By construction of B, each of its points on the unit circumference is a *limit* of a sequence of z_n having modulus < 1 for which $|F(z_n)| < 2w(|z_n|)$. Since $w(r) \to 0$ for $r \to 1$ we thus have $F(e^{i\vartheta}) = 0$ for every point of the form $e^{i\vartheta}$ in B.

If now this happens for each point *belonging to an arc J of positive length* on the unit circumference, we will have $F(e^{i\vartheta}) \equiv 0$ on J. At the same time, the *Fourier coefficients a_n* of $F(e^{i\vartheta})$ satisfy $|a_{-n}| \leqslant \text{const.}e^{-M(n)}$ for $n \geqslant 1$ by hypothesis. *Therefore $F(e^{i\vartheta})$ vanishes identically by the corollary to Levinson's theorem at the end of* §A.5. This, however, is *contrary to our hypothesis.*

Before going on, we note that the properties of $M(v)$ came into play in the preceding argument *only* when we looked at the Fourier coefficients of $F(e^{i\vartheta})$ and *not* when we brought in $w(r) = \exp(-h(\log(1/r)))$, even though $h(\xi)$ *is* related to $M(v)$ in the usual way. *Any other* weight $w(r) \geqslant 0$ tending to zero as $r \to 1$ would have worked just as well.

Thanks to what we found in step 1, $\{|\zeta| = 1\} \cap (\sim B)$ is *non-empty* (and even *dense* on the unit circumference). *It is open, hence equal to a countable union of disjoint open arcs I_k on $\{|\zeta| = 1\}$.* (B, remember, is *closed.*) *$\mathcal{O} = \{|z| < 1\} \cap (\sim B)$ clearly abuts on each of the I_k.* (See definition, beginning of article 5.)

Take any ρ_0, $0 < \rho_0 < 1$, and denote by $\Omega_k(\rho_0)$ (or just by Ω_k, if it is not necessary to keep the value of ρ_0 in mind) the *connected component of $\mathcal{O} \cap \{\rho_0 < |z| < 1\}$ abutting on I_k.*

Step 2. *All the $\Omega_k(\rho_0)$ are the same. In other words, $\bigcup_k \Omega_k(\rho_0)$ is connected.*

Assume the contrary. Then we must have two different arcs I_k – call them I_1 and I_2 – for which the corresponding components Ω_1 and Ω_2 are *disjoint.*

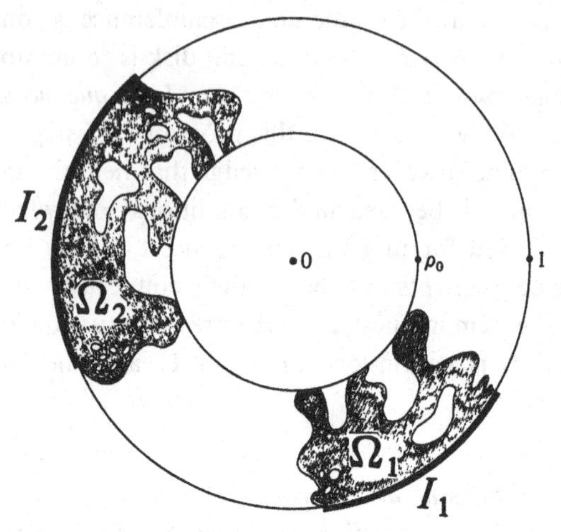

Figure 102

The function $\Phi(z)$ is analytic in Ω_1 and continuous on its closure. As stated above, it's *continuous on* $\{|z| \leqslant 1\}$ – that's because $F(z)$ is, and because the ratio $F_{\bar{\zeta}}(\zeta)/F(\zeta)$ appearing in the formula for $\Phi(z)$ is *bounded* on the region $\mathcal{O}$ over which the double integral figuring therein is taken (see beginning of the proof of the theorem in article 1).

The points ζ on $\partial\Omega_1$ with $\rho_0 < |\zeta| < 1$ (*sic!*) must belong to B, therefore $|F(\zeta)| \leqslant 2w(|\zeta|)$ for them. So, since $A|F(z)| \leqslant |\Phi(z)| \leqslant A'|F(z)|$ for $|z| \leqslant 1$, we have

$$|\Phi(\zeta)| \;\leqslant\; \text{const.}w(|\zeta|) \qquad \text{for } \zeta \in \partial\Omega_1 \text{ and } \rho_0 < |\zeta| < 1.$$

We have $w(r) = \exp(-h(\log(1/r)))$ with $h(\zeta)$ *decreasing* and *bounded below* for $\xi > 0$. In the present case, where $h(\xi) = \sup_{v>0}(M(v) - v\xi)$ and $\int_1^\infty (M(v)/v^2)\mathrm{d}v = \infty$, $\int_0^a \log h(\xi)\mathrm{d}\xi = \infty$ for all sufficiently small $a > 0$ (next to last theorem of article 2 – in the present circumstances we could even bypass that theorem as in the remark following the one of article 4). Finally, the condition (given!) that $M(v) \;\geqslant\; \text{const.}v^\alpha$ for large v, with $\alpha > \frac{1}{2}$, makes $h(\xi) \;\geqslant\; \xi^{-\lambda}$ for small $\xi > 0$, where $\lambda > 1$, by a lemma of article 2. Therefore (and by far!) $w(r) = O((1-r)^2)$ as $r \to 1$.

In our present situation, Ω_1 *abuts* on I_1 and $\partial\Omega_1$ *avoids* I_2. Here, *all the conditions* of *Volberg's theorem on harmonic measures* (previous article) *are fulfilled.* Therefore, by the corollary to that theorem, $\Phi(z) \equiv 0$ in Ω_1. *This, however, is impossible since $\Omega_1 \subseteq \mathcal{O}$ on which $|\Phi(z)| > 0$.*

As we have just seen, the union $\bigcup_k \Omega_k(\rho_0)$ is *connected. We denote that union by* $\Omega(\rho_0)$, or sometimes just by Ω. $\Omega(\rho_0)$ is an open subset of $\mathbb{C}$ lying in the ring $\rho_0 < |z| < 1$ and *abutting on each arc of* $\{|\zeta| = 1\}$ *contiguous to* $\{|\zeta| = 1\} \cap B$.

Step 3. *If* $|\zeta| = 1$, *there are values of* $r < 1$ *arbitrarily close to* 1 *with* $r\zeta \in \Omega(\rho_0)$, *and hence, in particular, with*

$$|F(r\zeta)| > w(r).$$

Take, wlog, $\zeta = 1$, and assume that for some a, $\rho_0 < a < 1$, *the whole segment* $[a, 1]$ *fails to intersect* Ω. The function $\sqrt{(z-a)/(1-az)}$ can then be defined so as to be analytic and single valued in Ω, and, if we introduce the new variable

$$s = \sqrt{\frac{z-a}{1-az}},$$

the mapping $z \to s$ takes Ω conformally onto a new domain – call it $\Omega_{\surd}$ – lying in $\{|s| < 1\}$:

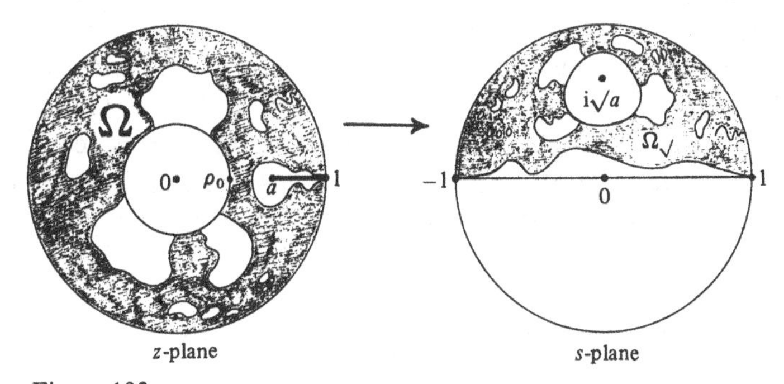

Figure 103

In terms of the variable s, write

$$\Phi(z) = \Psi(s), \qquad z \in \Omega;$$

$\Psi(s)$ is obviously analytic in $\Omega_{\surd}$ and continuous on its closure. If $s \in \Omega_{\surd}$ has $|s| > \sqrt{a}$, we have, since

$$z = \frac{s^2 + a}{1 + as^2},$$

that $1 - |z| \leqslant ((1+a)/(1-a))(1 - |s|^2)$; proof of this inequality is an elementary exercise in the geometry of linear fractional transformations which the reader should do. Hence, for $s \in \Omega_{\surd}$ with $|s| > \sqrt{a}$,

$$|z| \geqslant 1 - \frac{1+a}{1-a}(1 - |s|^2),$$

and, if $|s|$ is *close* to 1, this last expression is $\geqslant |s|^L$, where we can take for L a number $> 2(1+a)/(1-a)$ (depending on the closeness of $|s|$ to 1). The same relation between $|s|$ and $|z|$ holds for $s \in \partial\Omega_\surd$ with $|s|$ close to 1.

Suppose $|s| < 1$ is *close* to 1 and $s \in \partial\Omega_\surd$. The corresponding z then lies on $\partial\mathcal{O}$ with $|z| < 1$, so $|\Phi(z)| \leqslant \text{const.}w(|z|)$; therefore $|\Psi(s)| = |\Phi(z)| \leqslant \text{const.}w(|s|^L)$ by the relation just found, $w(r)$ being *decreasing*. The open set $\Omega_\surd$ *certainly abuts* on some arcs of $\{|s| = 1\}$ having positive length, since Ω abuts on the I_k. And $\partial\Omega_\surd$ *does not intersect the arc* $\pi < \arg s < 2\pi$ on $\{|s| = 1\}$ – that's why we *did* the conformal mapping $z \to s$! Here, the weight $w(r^L)$ is just as good (or just as bad) as $w(r)$ – see the *remark on a certain change of variable* at the end of article 4. We can therefore apply the *corollary* of the *theorem on harmonic measures* (end of article 5) to $\Psi(s)$ and the domain $\Omega_\surd$, and conclude that $\Psi(s) \equiv 0$ in $\Omega_\surd$. This, however, would make $\Phi(z) \equiv 0$ in Ω *which is impossible*, since $\Omega \subseteq \mathcal{O}$ where $|\Phi(z)| > 0$. *Step* 3's assertion must therefore hold.

The result just proved certainly implies that $\partial\Omega(\rho_0)$ *includes the unit circumference*. Since the Dirichlet problem can be solved for $\mathcal{O}$, it can be solved for Ω. It therefore makes sense to speak of the *harmonic measure* $\omega_\Omega(E, z)$ of an arbitrary closed subset E of $\{|\zeta| = 1\}$ ($\subseteq \partial\Omega$) relative to Ω, as seen from $z \in \Omega$. As was said above, our *aim* is to show that $\omega_\Omega(E, z_0) \geqslant k(z_0)|E|$ for such sets E; the analyticity of $\Phi(z)$ in Ω together with the fact that $|\Phi(z)|$ is > 0 and lies between two constant multiples of $|F(z)|$ there will *then* make

$$\int_{-\pi}^{\pi} \log|F(e^{i\vartheta})|d\vartheta > -\infty$$

by the *theorem on harmonic estimation* (§B.1), $|F(z)|$ being in any case *bounded above* in the closed unit disk. According to Harnack's theorem, the desired inequality for $\omega_\Omega(E, z_0)$ will follow from a *local version* of it, which is thus all that we need establish.

At this point the condition, assumed in the hypothesis, that $M(v) \geqslant \text{const.}v^\alpha$ with some $\alpha > \frac{1}{2}$ for large v, begins to play a more important rôle in our construction. We have already made *some* use of that property; it has not yet, however, been used *essentially*.

According to a lemma in article 2, the condition is equivalent to the property that $h(\xi) \geqslant \text{const.}\xi^{-\alpha/(1-\alpha)}$ for small $\xi > 0$. There is, in other words, an η, $0 < \eta < \frac{1}{2}$, with

$$\frac{1}{\xi} \leqslant \text{const.}(h(\xi))^{1-2\eta}$$

for small $\xi > 0$. *We fix such an* η *and put*

$$H(\xi) = (h(2\xi))^\eta$$

for $\xi > 0$.

Since $h(\xi)$ is *decreasing*, *so is* $H(\xi)$. Also, the property $h(\xi) \geqslant 1$ (due to the condition $M(0) \geqslant 1$) makes $H(\xi) \leqslant h(2\xi)$, whence, *a fortiori*,

$$H(\xi) \leqslant h(\xi) \quad \text{for} \quad \xi > 0.$$

We have

$$\int_0^a \log H(\xi)\,\mathrm{d}\xi = \frac{\eta}{2}\int_0^{2a} \log h(x)\,\mathrm{d}x = \infty$$

for all sufficiently small $a > 0$ by a theorem in article 2, since, as we are assuming, $\sum_1^\infty M(n)/n^2 = \infty$.

Using $H(\xi)$, let us form the new weight

$$w_1(r) = \exp\left(-H\left(\log\frac{1}{r}\right)\right), \quad 0 \leqslant r < 1.$$

Then

$$w_1(r) \geqslant w(r), \quad 0 \leqslant r < 1;$$

we still have, however,

$$w_1(r) = \mathrm{O}((1-r)^2) \quad \text{for} \quad r \to 1,$$

although, when r is close to 1, $w_1(r)$ is *much larger than* $w(r)$. Starting with $w_1(r)$, we proceed to construct a new open set $\mathcal{O}_1 \subseteq \mathcal{O}$ on which $|F(z)| > w_1(|z|)$ in much the same fashion as $\mathcal{O}$ was formed by use of $w(r)$.

Take first the set

$$B_0' = \{z\colon\ |z| < 1 \text{ and } |F(z)| \leqslant w_1(|z|)\}.$$

Since $w_1(r) \geqslant w(r)$, B_0' contains the set B_0 used above in the construction of B. Note that on each of the little open disks used to cover B_0 and form the set B – call those disks Δ_j – we have $|F(z)| < 2w_1(|z|)$ since $|F(z)| < 2w(|z|)$ on them. If the Δ_j also cover B_0', we take $B_1 = B$. Otherwise, we form the difference

$$B_0'' = B_0' \cap \sim \bigcup_j \Delta_j$$

and cover each closed set

$$B_0'' \cap \left\{1 - \frac{1}{n} \leqslant |z| \leqslant 1 - \frac{1}{n+1}\right\}, \qquad n = 1, 2, 3, \ldots,$$

by a *finite number* of open disks $\tilde{\Delta}_k(n)$ lying in $\{|z| < 1\}$, on which $|F(z)| < 2w_1(|z|)$. We then take B_1 as the *closure of the union*

$$\left(\bigcup_j \Delta_j\right) \cup \left(\bigcup_{n=1}^{\infty} \bigcup_k \tilde{\Delta}_k(n)\right).$$

For $z \in B_1$, $|F(z)| \leqslant 2w_1(|z|)$. B_1 contains B and has clearly the same general structure as B; B_1 includes all the points z of $\{|z| < 1\}$ for which $|F(z)| \leqslant w_1(|z|)$.

We now put

$$\mathcal{O}_1 = \{|z| < 1\} \cap \sim B_1.$$

The set $\mathcal{O}_1$ is open and contained in $\mathcal{O}$ since $B_1 \supseteq B$. For $z \in \mathcal{O}_1$, $|F(z)| > w_1(|z|)$, and on $\partial\mathcal{O}_1 \cap \{|z| < 1\}$ we have $|F(z)| \leqslant 2w_1(|z|)$ since the points of the later set must belong to B_1. The function $\Phi(z)$ introduced above is thus *analytic in* $\mathcal{O}_1$ (and continuous on its closure) and, since $A|F(z)| \leqslant |\Phi(z)| \leqslant A'|F(z)|$ for $|z| < 1$, satisfies

$$|\Phi(z)| > \text{const.} w_1(|z|), \qquad z \in \mathcal{O}_1,$$

as well as

$$|\Phi(z)| \leqslant \text{const.} w_1(|z|) \quad \text{for } z \in \partial\mathcal{O}_1 \text{ and } |z| < 1 \ (sic!).$$

Our new weight $w_1(r)$ and the function $H(\xi)$ to which it is associated *fulfill the conditions for the theorem on harmonic measures* (article 5). *Hence, in view of the above two inequalities satisfied by* $\Phi(z)$, *there is nothing to prevent our going through steps* 1, 2, *and* 3 *again, with* $\mathcal{O}_1$ *in place of* $\mathcal{O}$ *and* $w_1(r)$ *in place of* $w(r)$. *We henceforth consider this done.*

Once step 3 for $\mathcal{O}_1$ and $w_1(r)$ is carried out, we know that *for each* ζ, $|\zeta| = 1$, *there are* $r < 1$ *arbitrarily close to* 1 *for which* $|F(r\zeta)| > w_1(r)$. The open set $\mathcal{O}_1$ was brought into our discussion in order to obtain this result, which will be used to play off $w_1(r)$ against $w(r)$. Having now served its purpose, $\mathcal{O}_1$ will not appear again.

Given ζ_0, $|\zeta_0| = 1$, consider any $\rho < 1$ for which

$$(*) \qquad |F(\rho\zeta_0)| > w_1(\rho)$$

and form the domain $\Omega(\rho^2)$; this is the *connected* (step 2) component of $\mathcal{O} \cap \{\rho^2 < |z| < 1\}$ ($\mathcal{O}$ and *not* $\mathcal{O}_1$ here!) which *abuts* on *each* of the arcs I_k making up $\{|\zeta| = 1\} \cap (\sim B)$.

Step 4. *If, for given* ζ_0 *of modulus* 1, (*) *holds with* ρ *close enough to* 1, *we have* $\rho\zeta_0 \in \Omega(\rho^2)$.

Assuming the contrary, we shall obtain a contradiction. Wlog, $\zeta_0 = 1$.

When $\rho \to 1$, the ratio

$$w_1(\rho)/w(\rho) \;=\; \exp\left(h\left(\log\frac{1}{\rho}\right) - \left(h\left(2\log\frac{1}{\rho}\right)\right)^{\eta}\right)$$

tends to ∞ since $h(\log(1/\rho))$ tends to ∞ then, h is decreasing, and $0 < \eta < 1$. Hence, if $|F(\rho)| > w_1(\rho)$ and $\rho > 1$ is close enough to 1, we surely have $|F(\rho)| > 2w(\rho)$, i.e., $\rho \notin B$, so $\rho \in \mathcal{O}$. The point ρ must then belong to *some* component of $\mathcal{O} \cap \{\rho^2 < |z| < 1\}$, so, *if it is not* in the *component* $\Omega(\rho^2)$, abutting on the I_k, of that intersection, *it must be in some other one, which* we may call $\mathscr{D}$. $\mathscr{D}$, being *disjoint* from $\Omega(\rho^2)$, *can thus abut on none of the arcs I_k of $\{|\zeta| = 1\}$ contiguous to $\{|\zeta| = 1\} \cap B$.*

It is now claimed that

$$\partial\mathscr{D} \cap \{\rho^2 \leqslant |z| \leqslant 1\} \quad (sic!)$$

is contained in B. Let ζ be in that intersection; if $|\zeta| < 1$, $\zeta \in \partial\mathcal{O} \cap \{|z| < 1\} \subseteq B$, so suppose that $|\zeta| = 1$. If $\zeta \notin B$, then ζ must lie in some contiguous arc I_k, say $\zeta \in I_1$. Then, for some open disk V_ζ with centre at ζ, $V_\zeta \cap \{|z| < 1\} \subseteq \mathcal{O}$. The intersection on the left is, however, *connected*, and, since $\zeta \in \partial\mathscr{D}$, it contains some points from the (*connected*!) open set $\mathscr{D}$. Therefore, $V_\zeta \cap \{|z| < 1\}$ must lie entirely in $\mathscr{D}$, and $\mathscr{D}$ *intersects* with the component $\Omega(\rho^2)$ of $\mathcal{O} \cap \{\rho^2 < |z| < 1\}$ abutting on I_1. But it doesn't! This contradiction shows that we must have $\zeta \in B$, as claimed.

Because $\partial\mathscr{D} \cap \{\rho^2 < |z| \leqslant 1\}$ is contained in B, we have

$$|\Phi(\zeta)| \;\leqslant\; \text{const.}|F(\zeta)| \;\leqslant\; \text{const.}w(|\zeta|)$$

for $\zeta \in \partial\mathscr{D}$ and $\rho^2 < |z| \leqslant 1$ (*sic*!).* The function $\Phi(z)$ is of course *analytic* in $\mathscr{D} \subseteq \mathcal{O}$, and continuous up to $\partial\mathscr{D}$. In order to help the reader follow the argument, let us *try* to draw a picture of $\mathscr{D}$:

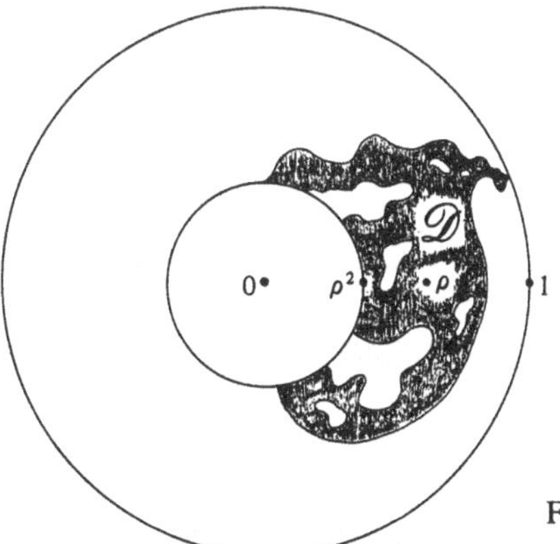

Figure 104

* We are taking $w(1) = 0$. See footnote, p. 359.

(N.B. $\mathscr{D}$ *can't really look like this* as we can see by applying an argument like the one of *step* 3 to $\Omega(\rho^2)$. One of the reasons why the present material is so hard is the *difficulty in drawing correct pictures* of what can happen.)

Call $\Gamma = \partial\mathscr{D} \cap \{\rho^2 < |z| \leqslant 1\}$, and denote harmonic measure for $\mathscr{D}$ by $\omega_{\mathscr{D}}(\ ,z)$. Since $|\Phi(z)|$ is bounded above on the unit disk and $|\Phi(\zeta)| \leqslant \text{const.}w(|\zeta|)$ on Γ, we have, by the theorem on harmonic estimation (§B.1),

$$\log|\Phi(\rho)| \ \leqslant\ \text{const.} \ +\ \omega_{\mathscr{D}}(\Gamma,\rho)\log w(\rho^2),$$

for $|\zeta| > \rho^2$ on Γ and $w(r)$ decreases. An almost trivial application of the *principle of extension of domain* shows that $\omega_{\mathscr{D}}(\Gamma,\rho)$ is *larger* than the *harmonic measure of the circle* $\{|\zeta| = 1\}$ *for the ring* $\{\rho^2 < |z| < 1\}$, *seen from* ρ. However, *that harmonic measure is* $\frac{1}{2}$! We thus see by the previous inequality that

$$\log|\Phi(\rho)| \ \leqslant\ \text{const.} \ +\ \tfrac{1}{2}\log w(\rho^2),$$

i.e., since

$$|\Phi(\rho)| \ \geqslant\ \text{const.}\,|F(\rho)| \ \geqslant\ \text{const.}w_1(\rho)$$

by (*), that

$$\frac{1}{2}h\left(\log\frac{1}{\rho^2}\right) \ \leqslant\ \text{const.} \ +\ \left(h\left(\log\frac{1}{\rho^2}\right)\right)^{\bar{\eta}}$$

in terms of the function h, with a constant independent of ρ.

Now $0 < \bar{\eta} < 1$ and $h(\xi) \longrightarrow \infty$ for $\xi \to 0$. *This means that the relation just obtained is impossible for values of* ρ *sufficiently close to* 1. Therefore, if $\rho < 1$ is close enough to 1 and $|F(\rho)| > w_1(\rho)$, we *must have* $\rho \in \Omega(\rho^2)$, the conclusion we desired to make. (Part of the idea for the preceding argument is due to Peter Jones.)

Take now any ζ_0, $|\zeta_0| = 1$, and pick a $\rho < 1$ *very close to* 1 such that

$$|F(\rho\zeta_0)| \ >\ w_1(\rho)$$

(which is possible, as we have already observed), and that therefore $\rho\zeta_0 \in \Omega(\rho^2)$ (by step 4, just completed). We are going to show that if E is a closed set on the arc of the unit circumference going from $\zeta_0 e^{i\log\rho}$ to $\zeta_0 e^{-i\log\rho}$, then

$$\omega_{\Omega(\rho^2)}(E,\rho\zeta_0) \ \geqslant\ C(\zeta_0,\rho)|E|$$

with some constant $C(\zeta_0,\rho)$ depending on ζ_0 and on ρ. Here, of course, $\omega_{\Omega(\rho^2)}(\ ,z)$ denotes harmonic measure for the domain $\Omega(\rho^2)$.

In order to do this, we write

$$\gamma_\rho = \partial\Omega(\rho^2) \cap \{\rho^2 < |z| < 1\} \quad (sic!)$$

and carry out

Step 5. *If ρ, chosen according to the above specifications, is close enough to* 1,

$$\int_{\gamma_\rho} \frac{1}{1 - |\zeta|} \mathrm{d}\omega_{\Omega(\rho^2)}(\zeta, \rho\zeta_0)$$

is as small as we please.

In order to simplify the notation, let us write

$$\omega(\ \ , z) \quad \text{for} \quad \omega_{\Omega(\rho^2)}(\ \ , z)$$

during the *remainder* of the *present discussion.* The proof of our statement uses *almost the full strength* of the property that

$$\frac{1}{\xi} \leqslant \text{const.}\, (h(\xi))^{1-2\eta}$$

for small $\xi > 0$ (with $0 < \eta < \frac{1}{2}$), equivalent to our condition that

$$M(v) \geqslant \text{const.} v^\alpha$$

(with $\alpha > \frac{1}{2}$) for large v.

Take, wlog, $\zeta_0 = 1$. Then, if $\rho \in \Omega(\rho^2)$, we have, by the theorem on harmonic estimation (§ B.1),

$$\log|\Phi(\rho)| \leqslant \int_{\partial\Omega(\rho^2)} \log|\Phi(\zeta)| \mathrm{d}\omega(\zeta, \rho),$$

$\Phi(z)$ being analytic in $\Omega(\rho^2)$ and continuous up to that set's boundary. The subset γ_ρ of $\partial\Omega(\rho^2)$ is of course contained in B, so, for $\zeta \in \gamma_\rho$,

$$|\Phi(\zeta)| < \text{const.} |F(\zeta)| \leqslant \text{const.} w(|\zeta|);$$

that is,

$$\log|\Phi(\zeta)| \leqslant \text{const.} - h\left(\log\frac{1}{|\zeta|}\right), \qquad \zeta \in \gamma_\rho.$$

The function $|\Phi(z)|$ is in any event bounded on $\{|z| \leqslant 1\}$, so this relation, together with the previous one, yields

$$\log|\Phi(\rho)| \leqslant \text{const.} - \int_{\gamma_\rho} h\left(\log\frac{1}{|\zeta|}\right) \mathrm{d}\omega(\zeta, \rho).$$

If ρ is chosen in such a way that we also have $|F(\rho)| > w_1(\rho)$, this becomes

$$(\ast) \qquad \int_{\gamma_\rho} h\left(\log\frac{1}{|\zeta|}\right) d\omega(\zeta,\rho) \;\leqslant\; \text{const.} + \left(h\left(\log\frac{1}{\rho^2}\right)\right)^\eta.$$

Since $1/\xi \leqslant \text{const.}(h(\xi))^{1-2\eta}$ for small $\xi > 0$, we have (when $\rho < 1$ is close to 1),

$$\int_{\gamma_\rho} \frac{1}{1-|\zeta|} d\omega(\zeta,\rho) \;\leqslant\; \text{const.} \int_{\gamma_\rho} \left(h\left(\log\frac{1}{|\zeta|}\right)\right)^{1-2\eta} d\omega(\zeta,\rho).$$

Rewrite (!) the right-hand integral as

$$\text{const.} \int_{\gamma_\rho} \frac{\left(h\left(\log\frac{1}{|\zeta|}\right)\right)^{1-\eta}}{\left(h\left(\log\frac{1}{|\zeta|}\right)\right)^{\eta}} d\omega(\zeta,\rho).$$

Since γ_ρ lies in the ring $\{\rho^2 < |z| < 1\}$ and $h(\xi)$ decreases, the expression just written is

$$\leqslant \frac{\text{const.}}{\left(h\left(\log\frac{1}{\rho^2}\right)\right)^\eta} \int_{\gamma_\rho} \left(h\left(\log\frac{1}{|\zeta|}\right)\right)^{1-\eta} d\omega(\zeta,\rho),$$

so, since $h(\xi) \longrightarrow \infty$ for $\xi \to 0$, we see that

$$\int_{\gamma_\rho} \frac{1}{1-|\zeta|} d\omega(\zeta,\rho) \;\leqslant\; \frac{\delta_\rho}{\left(h\left(\log\frac{1}{\rho^2}\right)\right)^\eta} \int_{\gamma_\rho} h\left(\log\frac{1}{|\zeta|}\right) d\omega(\zeta,\rho)$$

with a quantity δ_ρ going to zero for $\rho \to 1$. Plugging $(\ast)$ into the right-hand side, we obtain finally

$$\int_{\gamma_\rho} \frac{1}{1-|\zeta|} d\omega(\zeta,\rho) \;\leqslant\; \delta_\rho\left(1 + \frac{\text{const.}}{(h(\log(1/\rho^2)))^\eta}\right).$$

Here, the right side tends to *zero* as $\rho \to 1$. *Step* 5 is finished.

Now we can make the *local estimate* of $\omega(E, z)$ for closed E on the unit circumference, promised just before step 5. Taking any fixed ζ_0, $|\zeta_0| = 1$, we pick a $\rho < 1$ very close to 1, in such fashion that the conclusions reached in *steps* 4 and 5 apply. Let E be any closed set on the arc of the unit circumference going from $\zeta_0 e^{i\log\rho}$ to $\zeta_0 e^{-i\log\rho}$.

In order to keep the notation simple, consider, as before, the case where $\zeta_0 = 1$.

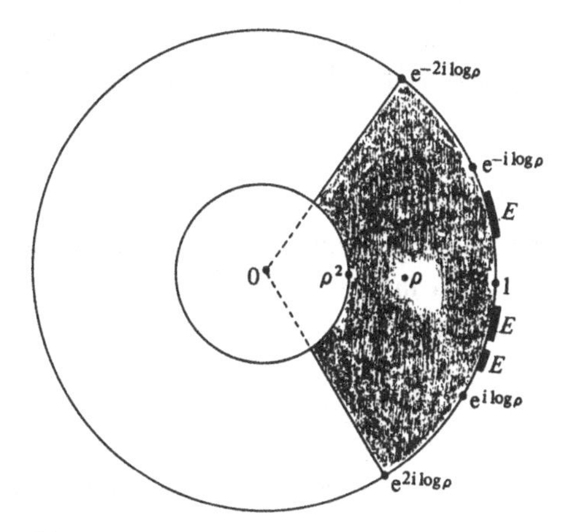

Figure 105

Denote harmonic measure for the *ring* $\{\rho^2 < |z| < 1\}$ by $\bar{\omega}(\ \ ,z)$. If we *compare* $\bar{\omega}(E,z)$ with the harmonic measure of *E for the sectorial box shown in the above diagram*, we see immediately that

$$(\dagger) \qquad \bar{\omega}(E,\rho) \;\geqslant\; C\frac{|E|}{1-\rho}$$

with a numerical constant C independent of $|E|$ and of ρ.

Put, as in *step* 5,

$$\gamma_\rho \;=\; \partial\Omega(\rho^2)\cap\{\rho^2 < |z| < 1\},$$

and continue to denote the harmonic measure for $\Omega(\rho^2)$ by $\omega(\ \ ,z)$. Since

$$\Omega(\rho^2) \;\subseteq\; \{\rho^2 < |z| < 1\}$$

while

$$\partial\Omega(\rho^2) \;\supseteq\; \{|\zeta| = 1\},$$

we can apply a formula established near the end of § B.1, getting

$$\omega(E,z) \;=\; \bar{\omega}(E,z) - \int_{\gamma_\rho} \bar{\omega}(E,\zeta)\,\mathrm{d}\omega(\zeta,z)$$

for $z\in\Omega(\rho^2)$.

Comparing $\bar{\omega}(E,z)$ with harmonic measure of E for the *whole unit disk*, we get the estimate

$$\bar{\omega}(E,\zeta) \;\leqslant\; \frac{|E|}{\pi(1-|\zeta|)}.$$

Substitute this into the *integral* in the above formula for $\omega(E,z)$, specialize to $z=\rho$, and use (†) in the first right-hand term of that formula. We find

$$\omega(E,\rho) \geqslant |E|\left(\frac{C}{1-\rho} - \frac{1}{\pi}\int_{\gamma_\rho}\frac{d\omega(\zeta,\rho)}{1-|\zeta|}\right).$$

It was, however, seen in *step* 5 that $\rho<1$ could be chosen in accordance with our requirements so as to make the *integral* in this expression *small.* For a suitable $\rho<1$ close to 1, we will thus have (and by far !)

$$\omega(E,\rho) \geqslant \frac{C}{2(1-\rho)}|E|$$

provided that the closed set E lies on the (shorter) arc from $e^{i\log\rho}$ to $e^{-i\log\rho}$ on the unit circle.

This is our local estimate. What it says is that, corresponding to any ζ, $|\zeta|=1$, we can *get* a $\rho_\zeta<1$ such that, for closed sets E lying on *the smaller arc* J_ξ *of the unit circle joining* $\zeta e^{i\log\rho_\zeta}$ *to* $\zeta e^{-i\log\rho_\zeta}$,

(††) $\qquad \omega_{\Omega(\rho_\zeta^2)}(E,\rho_\zeta\zeta) \geqslant C_\zeta|E|,$

with $C_\zeta>0$ depending (*a priori*) on ζ. Observe that, if $0<\rho\leqslant\rho_\zeta^2$, $\Omega(\rho)$ (the component of $\mathcal{O}\cap\{\rho<|z|<1\}$ abutting on the arcs I_k) must by definition contain $\Omega(\rho_\zeta^2)$. Therefore

(§) $\qquad \omega_{\Omega(\rho)}(E,\rho_\zeta\zeta) \geqslant \omega_{\Omega(\rho_\zeta^2)}(E,\rho_\zeta\zeta)$

by the *principle of extension of domain* when $E\subseteq\{|z|=1\}$, a subset of *both* boundaries $\partial\Omega(\rho)$, $\partial\Omega(\rho_\zeta^2)$.

A finite number of the arcs J_ζ serve to cover the unit circumference; denote them by $J_1, J_2,\ldots,J_n$, calling the corresponding values of ζ, $\zeta_1,\ldots,\zeta_n$ and the corresponding ρ_ζ's $\rho_1,\rho_2,\ldots,\rho_n$. Let ρ be the *least* of the quantities ρ_j^2, $j=1,2,\ldots,n$, and denote the *least* of the C_{ζ_j} by k, which is thus >0. If E is a closed subset of J_j, (††) and (§) give

$$\omega_{\Omega(\rho)}(E,\rho_j\zeta_j) \geqslant k|E|.$$

Fix any $z_0\in\Omega(\rho)$. Using Harnack's inequality in $\Omega(\rho)$ for each of the pairs of points $(z_0,\ \rho_j\zeta_j)$, $j=1,2,\ldots,n$, we obtain, from the preceding relation,

(§§) $\qquad \omega_{\Omega(\rho)}(E,z_0) \geqslant K(z_0)|E|$

for closed subsets E of *any* of the arcs $J_1, J_2,\ldots,J_n$. Here, $K(z_0)>0$ depends on z_0. Now we see finally that (§§) *in fact holds for any closed subset E of the unit circumference,* large or small. That is an obvious consequence of the additivity of the set function $\omega_{\Omega(\rho)}(\ \ ,z_0)$, the arcs J_j forming a covering of $\{|\zeta|=1\}$.

We are at long last able to conclude our proof of Volberg's theorem on the logarithmic integral. Our chosen z_0 in $\Omega(\rho)$ lies in $\mathcal{O}$, therefore $|\Phi(z_0)| > 0$. By the *theorem on harmonic estimation* applied to the function $\Phi(z)$ analytic in $\Omega(\rho)$ and continuous on $\{|z| \leqslant 1\}$,

$$-\infty \;<\; \log|\Phi(z_0)| \;\leqslant\; \int_{\partial\Omega(\rho)} \log|\Phi(\zeta)|\,d\omega_{\Omega(\rho)}(\zeta, z_0)$$

$$\leqslant\; \text{const.} + \int_{|\zeta|=1} \log|\Phi(\zeta)|\,d\omega_{\Omega(\rho)}(\zeta, z_0).$$

According to (§§), this last is in turn

$$\leqslant\; \text{const.} - K(z_0)\int_{-\pi}^{\pi} \log^-|\Phi(e^{i\vartheta})|\,d\vartheta,$$

$\log|\Phi(e^{i\vartheta})|$ being in any case *bounded above*. Thus,

$$\int_{-\pi}^{\pi} \log^-|\Phi(e^{i\vartheta})|\,d\vartheta \;<\; \infty.$$

However, $|\Phi(e^{i\vartheta})|$ lies, as we know, between *two constant multiples of* $|F(e^{i\vartheta})|$. Therefore

$$\int_{-\pi}^{\pi} \log^-|F(e^{i\vartheta})|\,d\vartheta \;<\; \infty,$$

i.e.,

$$\int_{-\pi}^{\pi} \log|F(e^{i\vartheta})|\,d\vartheta \;>\; -\infty.$$

Volberg's theorem is thus completely proved, and *we are finally done.*

Remark. Of the two regularity conditions required of $M(v)$ for this theorem, viz., that $M(v)/v$ be *decreasing* and that $M(v) \geqslant \text{const.}v^\alpha$ for large v with some $\alpha > \frac{1}{2}$, the *first* served to make possible the use of Dynkin's result (article 3) by means of which the analytic function $\Phi(z)$ was brought into the proof.

Decisive use of the *second* was not made until *step* 5, where we estimated

$$\int_{\gamma_\rho} \frac{1}{1-|\zeta|}\,d\omega(\zeta, \rho)$$

in terms of $\int_{\gamma_\rho} h(\log(1/|\zeta|))\,d\omega(\zeta, \rho)$.

Examination of the argument used there shows that *some* relaxation of the condition $M(v) \geqslant \text{const.}v^\alpha$ (with $\alpha > \frac{1}{2}$) *is* possible if one is willing to replace it by another with considerably more complicated statement. The *method* of Volberg's proof necessitates, however, that $M(v)$ be *at least* $\geqslant$ const.$v^{\frac{1}{2}}$ for large v. For, by a lemma of article 3, that relation is equivalent to the property that

$$\frac{1}{\xi} = O(h(\xi))$$

for $\xi \to 0$, and we need at least *this* in order to make the abovementioned estimate for $\rho < 1$ near 1.

We needed $\int_{\gamma_\rho}(1/(1-|\zeta|))d\omega(\zeta,\rho)$ in the computation following step 5, where we got a *lower bound* on $\omega(E,\rho)$. The integral came in there on account of the inequality

$$\tilde{\omega}(E,\zeta) \leqslant \frac{|E|}{\pi(1-|\zeta|)}$$

for harmonic measure $\tilde{\omega}(E,\zeta)$ (of sets $E \subseteq \{|\zeta| = 1\}$) in the ring $\{\rho^2 < |\zeta| < 1\}$. And, aside from a constant factor, this inequality is best possible.

7. Scholium. Levinson's log log theorem

Part of the material in articles 2 and 5 is closely related to some older work of Levinson which, because of its usefulness, should certainly be taken up before ending the present chapter.

During the proof of the *first* theorem in §F.4, Chapter VI, we came up with an entire function $L(z)$ satisfying an inequality of the form $|L(z)| \leqslant \text{const.}e^{K|z|}/|\Im z|$, and wished to conclude that $L(z)$ was of exponential type. Here there is an obvious difficulty for the points z lying near the real axis. We dealt with it by using the subharmonicity of $\sqrt{|L(z)|}$ and convergence of

$$\int_{-1}^{1} |y|^{-\frac{1}{2}} dy$$

in order to *integrate out* the denominator $|\Im z|$ from the inequality and thus strengthen the latter to an estimate $|L(z)| \leqslant \text{const.}e^{K|z|}$ for z near $\mathbb{R}$. A more elaborate version of the same procedure was applied in the proof of the *second* theorem of §F.4, Chapter VI, where subharmonicity of $\log|S(z)|$ was used to get rid of a troublesome term tending to ∞ for z approaching the real axis.

It is natural to ask *how far* such tricks can be pushed. Suppose that $f(z)$ is known to be analytic in some rectangle straddling the real axis, and we are assured that

$$|f(z)| \leqslant \text{const.}L(y)$$

in that rectangle, where, unfortunately, the majorant $L(y)$ goes to ∞ as $y \to 0$. What conditions on $L(y)$ will permit us to deduce *finite bounds* on $|f(z)|$, *uniform in the interior* of the given rectangle, from the preceding relation? One's first guess is that a condition of the form $\int_{-a}^{a} \log L(y)\,dy < \infty$ *will do*, but that *nothing much weaker than that can suffice*, because $\log|f(z)|$ *is* subharmonic while functions of $|f(z)|$ which increase *more slowly* than the logarithm are *not*, in general. This conservative appraisal turns out to be *wrong by a whole* (exponential) *order of magnitude*. Levinson found that *it is already enough to have*

$$\int_{-a}^{a} \log\log L(y)\,dy \ < \ \infty,$$

and that this condition *cannot be further weakened.*

Levinson's result is extremely useful. One application could be to *eliminate* the rough and ready but somewhat clumsy *hall of mirrors* argument from many of the places where it occurs in Chapter VI. Let us, for instance, consider again the proof of Akhiezer's first theorem from §B.1 of that chapter. If, in the circumstances of that theorem, we have $\|P\|_W \leqslant 1$ for a polynomial P, the relation

$$\log|P(z)| \ \leqslant \ \frac{1}{\pi}\int_{-\infty}^{\infty} \frac{|\Im z|}{|z-t|^2} \log|P(t)|\,dt$$

$$\leqslant \ \frac{1}{\pi}\int_{-\infty}^{\infty} \frac{|\Im z|}{|z-t|^2} \log W_+(t)\,dt$$

and the estimate of $\sup_{t \in \mathbb{R}} |(t-i)/(t-z)|$ from §A.2 (Chapter VI) tell us immediately that

$$\log|P(z)| \ \leqslant \ M\frac{(1+|z|)^2}{|\Im z|},$$

where

$$M \ = \ \frac{1}{\pi}\int_{-\infty}^{\infty} \frac{\log W_*(t)}{1+t^2}\,dt.$$

Taking any rectangle

$$\mathscr{D}_R \ = \ \{z\colon \ |\Re z| \leqslant R, \quad |\Im z| \leqslant 2\}$$

and putting

$$L(y) \ = \ \exp\left(M\frac{R^2+5}{|y|} \right),$$

we have $|P(z)| \leqslant L(|\Im z|)$ on $\mathscr{D}_R$. Here,

$$\int_{-1}^{1} \log\log L(y)\,dy \;<\; \infty,$$

so the result of Levinson gives us a control on the size of $|P(z)|$ *in the interior of* $\mathscr{D}_R$ (even *right on* the real axis). In this way we can see that the polynomials P with $\|P\|_W \leqslant 1$ form a *normal family* in any strip straddling the real axis. The relation $\log|P(z)| \leqslant M(1+|z|)^2/|\Im z|$ already shows that those polynomials form a normal family *outside* such a strip, and the main part of the proof of Akhiezer's first theorem is complete.

One can easily envision the possibility of other applications like the one just shown to situations where the *hall of mirrors* argument would not be available. There is thus no doubt about the worth of the result in question; let us, then, proceed to its precise statement and proof without further ado.

Levinson's* log log *theorem. *Consider any rectangle*

$$\mathscr{D} \;=\; \{z\colon\ a<x<a' \text{ and } -b<y<b\}.$$

Let $L(y)$ be Lebesgue measurable and $\geqslant$ e for $-b<y<b$.

Suppose that

$$\int_{-b}^{b} \log\log L(y)\,dy \;<\; \infty.$$

Then there is a decreasing function $m(\delta)$, depending only *on $L(y)$ and finite for $\delta>0$, such that, if $f(z)$ is analytic in $\mathscr{D}$ and if*

$$|f(z)| \;\leqslant\; L(\Im z)$$

there, we also have

$$|f(z)| \;\leqslant\; m(\text{dist.}(z,\partial\mathscr{D})) \quad \textit{for } z\in\mathscr{D}.$$

Remark. In this version (due to Y. Domar), *no regularity properties whatever are required of $L(y)$.* The assumption that $L(y)\geqslant$ e is of course made *merely to ensure positivity* of $\log\log L(y)$.

Proof of theorem (Y. Domar). Denote by $\mu(\lambda)$ the *distribution function* for $\log L(y)$; i.e.,

$$\mu(\lambda) \;=\; |\{y\colon\ -b<y<b \text{ and } \log L(y)>\lambda\}|.^{*}$$

Let $z_0\in\mathscr{D}$ and $R<\text{dist}(z_0,\partial\mathscr{D})$, and write $u(z)=\log|f(z)|$. Since $u(z)$ is

* We are continuing to denote by $|E|$ the Lebesgue measure of sets $E\subseteq\mathbb{R}$.

subharmonic in $\mathscr{D}$, we have

$$u(z_0) \leqslant \frac{1}{\pi R^2} \iint_{|z - z_0| < R} u(z)\,\mathrm{d}x\,\mathrm{d}y.$$

We are going to show that, if $u(z_0)$ is *large* and z_0 *far enough* from $\partial\mathscr{D}$, this inequality leads to a *contradiction* when $u(z) \leqslant \log L(\Im z)$ in $\mathscr{D}$.

Call Δ the disk

$$\{z\colon\ |z - z_0| < R\}.$$

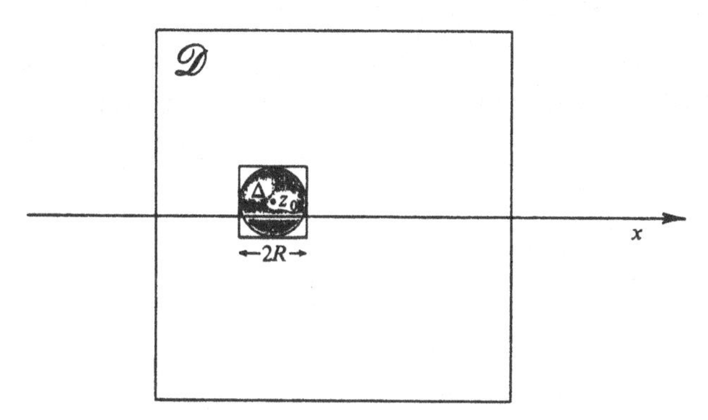

Figure 106

Use $\|E\|$ to denote the two-dimensional Lebesque measure of $E \subseteq \mathbb{C}$ (since $|\ \ |$ is being used for one-dimensional Lebesque measure of sets on $\mathbb{R}$). Then, by boxing Δ into a square of side $2R$ in the manner shown, we see from the ineqality $u(z) \leqslant \log L(\Im z)$ that

$$\|\{z \in \Delta\colon\ u(z) > M/2\}\| \ \leqslant\ 2R\mu(M/2) \quad \text{for } M > 0.$$

Suppose now that $u(z_0) \geqslant M$, but that at the same time we have $u(z) \leqslant 2M$ on Δ. From the previous subharmonicity relation we will then have

$$(*) \qquad u(z_0) \ \leqslant\ \frac{M}{2} + \frac{2M}{\|\Delta\|}\left\|\left\{z \in \Delta\colon\ u(z) \geqslant \frac{M}{2}\right\}\right\| \ <\ \frac{M}{2} + \frac{4M}{\pi R}\mu(M/2).$$

Because $\log\log L(y)$ is integrable on $[-b, b]$ we certainly have $\mu(\lambda) \to 0$ for $\lambda \to \infty$. We can therefore take M *so large* that $\mu(M/2)$ is *much smaller* than $\operatorname{dist}(z_0, \partial\mathscr{D})$. *With such a value* of M, put

$$R \ =\ R_0 \ =\ \frac{16}{\pi}\mu(M/2).$$

The *right side of* $(*)$ *will then* be $\leqslant \frac{3}{4}M$. This means that *if* $u(z_0) \geqslant M$, the *assumption under which* $(*)$ *was derived is untenable*, i.e., *that* $u(z) > 2M$ *somewhere in* Δ, say for $z = z_1$, $|z_1 - z_0| \ <\ R_0$.

Supposing, then, that $u(z_0) \geqslant M$, we have a z_1, $|z_1 - z_0| < R_0$, with $u(z_1) > 2M$. We can then *repeat* the argument just given, making z_1 play the rôle of z_0, $2M$ that of M,

$$R_1 = \frac{16}{\pi}\mu(M)$$

that of R_0, and $\{z: |z - z_1| < R_1\}$ that of Δ. As long as $R_1 > \operatorname{dist}(z_1, \partial\mathscr{D})$, hence, surely, provided that

$$R_0 + R_1 < \operatorname{dist}(z_0, \partial\mathscr{D}),$$

we will get a z_2, $|z_2 - z_1| < R_1$, with $u(z_2) > 4M$. Then we can take $R_2 = (16/\pi)\mu(2M)$, have $4M$ play the rôle held by $2M$ in the previous step, and keep on going.

If, for the numbers

$$R_k = \frac{16}{\pi}\mu(2^{k-1}M),$$

we have

$$\sum_{k=0}^{\infty} R_k < \operatorname{dist}(z_0, \partial\mathscr{D}),$$

the process never stops, and we get a sequence of points $z_k \in \mathscr{D}$, $|z_k - z_{k-1}| < R_{k-1}$, with

$$u(z_k) \geqslant 2^k M.$$

Evidently, $z_k \underset{k}{\longrightarrow}$ a point $z_\infty \in \mathscr{D}$.

The function $f(z)$ is *analytic* in $\mathscr{D}$, so $u(z) = \log|f(z)|$ is *continuous* (in the extended sense) at z_∞, where $u(z_\infty) < \infty$. This is certainly incompatible with the inequalities $u(z_k) \geqslant 2^k M$ when $z_k \underset{k}{\longrightarrow} z_\infty$. *Therefore we cannot have $u(z_0) \geqslant M$ if we can take M so large that $\sum_0^\infty R_k$* $<$ dist $(z_0, \partial\mathscr{D})$, i.e., *that*

$$(\dagger) \qquad \sum_{k=0}^{\infty} \frac{16}{\pi}\mu(2^{k-1}M) < \operatorname{dist}(z_0, \partial\mathscr{D}).$$

In order to complete the proof, it suffices, then, to show that the left-hand sum in (†) tends to *zero* as $M \to \infty$. By Abel's rearrangement,

$$\begin{aligned}\sum_{k=0}^{n} \mu(2^{k-1}M) = {} & \left\{\mu\left(\frac{M}{2}\right) - \mu(M)\right\} + 2\{\mu(M) - \mu(2M)\} \\ & + 3\{\mu(2M) - \mu(4M)\} + \cdots \\ & + n\{\mu(2^{n-2}M) - \mu(2^{n-1}M)\} + (n+1)\mu(2^{n-1}M).\end{aligned}$$

Remembering that $\mu(\lambda)$ is a *decreasing* function of μ, we see that as long as $M \geqslant 4$, the sum on the right is

$$\leqslant \sum_{k=0}^{n-1} \int_{2^{k-1}M}^{2^k M} \frac{\log\lambda - \log(M/4)}{\log 2}(-\mathrm{d}\mu(\lambda))$$

$$+ \int_{2^{n-1}M}^{\infty} \frac{\log\lambda - \log(M/4)}{\log 2}(-\mathrm{d}\mu(\lambda))$$

$$\leqslant -\frac{1}{\log 2}\int_{M/2}^{\infty} \log\lambda\,\mathrm{d}\mu(\lambda) = \frac{1}{\log 2}\int_{\substack{\log L(y) \geqslant M/2 \\ -b<y<b}} \log\log L(y)\,\mathrm{d}y.$$

Since $\int_{-b}^{b} \log\log L(y)\,\mathrm{d}y < \infty$, the previous expression, and hence the left-hand side of (†), tends to 0 as $M \to \infty$. Therefore, given $z_0 \in \mathscr{D}$, we can get an M sufficiently large for (†) to *hold, and, with that* M, $u(z_0) < M$, i.e., $|f(z_0)| < \mathrm{e}^M$. Let, then,

$$m(\delta) = \inf\left\{\mathrm{e}^M : \sum_{k=0}^{\infty} \frac{16}{\pi}\mu(2^{k-1}M) < \delta\right\}.$$

As we have just seen, $m(\delta)$ is *finite* for $\delta > 0$; it is obviously decreasing. And, if $f(z)$ is analytic in $\mathscr{D}$ with $|f(z)| \leqslant L(\Im z)$ there, we have $|f(z_0)| \leqslant m(\operatorname{dist}(z_0, \partial\mathscr{D}))$ for $z_0 \in \mathscr{D}$. We are done.

Remark. This beautiful proof is quite recent. The procedure of the scholium at the end of article 5 will yield the same result for sufficiently regular majorants $L(y)$.

We now consider the possibility of relaxing the condition

$$\int_{-b}^{b} \log\log L(y)\,\mathrm{d}y < \infty$$

required in the above theorem. If, for some majorant $L(y)$, *the conclusion of the theorem holds*, any set of *polynomials* P with $|P(z)| \leqslant L(\Im z)$ for $z \in \mathscr{D}$ *must form a normal family* in $\mathscr{D}$. This observation enables us to give a simple proof of the fact that the requirement.

$$\int_{-b}^{b} \log\log L(y)\,\mathrm{d}y < \infty$$

is essential in Levinson's result, at least for majorants $L(y)$ of sufficiently regular behaviour.

Theorem. *Let* $L(y)$ *be continuous and* $\geqslant$ e *for* $0 < |y| \leqslant b$, *with* $L(y) \to \infty$

for $y \to 0$ *and* $L(y)$ decreasing *on* $(0, b)$. *Suppose also that*

$$\int_0^b \log\log L(y)\,dy \;=\; \infty.$$

Then there is a sequence of polynomials $P_n(z)$ *with, for* $\Im z \neq 0$,

$$|P_n(z)| \;\leqslant\; \text{const.}L(\Im z)$$

on the rectangle

$$\mathscr{D} \;=\; \{z\colon\; -1 < \Re z < 1 \text{ and } -b < \Im z < b\},$$

while at the same time

$$P_n(z) \underset{n}{\longrightarrow} \begin{cases} \;\;1, & z \in \mathscr{D} \text{ and } \Im z > 0, \\ -1, & z \in \mathscr{D} \text{ and } \Im z < 0. \end{cases}$$

Thus the conclusion of Levinson's theorem does not hold *for the majorant* $L(y)$.

Remark. We only require $L(y)$ to be monotone *on one side* of the origin, on the side over which the integral of $\log\log L(y)$ *diverges.* Levinson already had this result under the assumption of more regularity for $L(y)$.

Proof of theorem (Beurling, 1972 – compare with the proof of the theorem on simultaneous polynomial approximation in article 4). The last sentence in the statement follows from the existence of a sequence of polynomials P_n having the asserted properties. For, if the conclusion of Levinson's theorem *held*, the sequence $\{P_n\}$ would form a normal family in $\mathscr{D}$ and there would hence be a function *analytic* in $\mathscr{D}$, equal to $+1$ *above* the real axis, and to -1 *below* it. This is absurd.

Put

$$\varphi(z) \;=\; \begin{cases} \;\;1, & z \in \mathscr{D} \text{ and } \Im z > 0 \\ -1, & z \in \mathscr{D} \text{ and } \Im z < 0, \end{cases}$$

and let us argue by *duality* to obtain a sequence of polynomials $P_n(z)$ for which

$$\sup_{z \in \mathscr{D}} \left(\frac{|\varphi(z) - P_n(z)|}{L(\Im z)} \right) \;\underset{n}{\longrightarrow}\; 0.$$

Such P_n will clearly satisfy the conclusion of our theorem.

Note that, since $L(y) \to \infty$ for $y \to 0$, the ratio $\varphi(z)/L(\Im z)$ is *continuous on* $\bar{\mathscr{D}}$ if we *define* $L(0)$ *to be* ∞, which we *do, for the rest of this proof.* Therefore, if a sequence of polynomials P_n fulfilling the above condition *does not exist*, we can, by the *Hahn–Banach theorem*, find a finite complex-valued measure μ on $\bar{\mathscr{D}}$ with

$$(\S) \qquad \iint_{\mathscr{D}} \frac{z^n}{L(\Im z)}\,\mathrm{d}\mu(z) \;=\; 0 \quad \text{for} \quad n = 0, 1, 2, \ldots,$$

whilst

$$\iint_{\mathscr{D}} \frac{\varphi(z)}{L(\Im z)}\,\mathrm{d}\mu(z) \;\neq\; 0.$$

The proof will be completed by showing that *in fact* (§) implies

$$\iint_{\mathscr{D}} \frac{\varphi(z)}{L(\Im z)}\,\mathrm{d}\mu(z) \;=\; 0.$$

Given any measure μ satisfying (§), write

$$\mathrm{d}\nu(z) \;=\; \frac{1}{L(\Im z)}\,\mathrm{d}\mu(z).$$

The measure ν has *very little mass near the real axis, and none at all on it*. For each complex λ, the power series for $\mathrm{e}^{\mathrm{i}\lambda z}$ converges uniformly for $z \in \bar{\mathscr{D}}$, so from (§) we get

$$\iint_{\mathscr{D}} \mathrm{e}^{\mathrm{i}\lambda z}\,\mathrm{d}\nu(z) \;=\; 0.$$

Write now

$$\mathscr{D}_+ \;=\; \bar{\mathscr{D}} \cap \{\Im z > 0\} \quad (\textit{sic!})$$

and

$$\mathscr{D}_- \;=\; \bar{\mathscr{D}} \cap \{\Im z < 0\} \quad (\textit{sic!}).$$

Then put

$$\Phi_+(\lambda) \;=\; \iint_{\mathscr{D}_+} \mathrm{e}^{\mathrm{i}\lambda z}\,\mathrm{d}\nu(z),$$

$$\Phi_-(\lambda) \;=\; \iint_{\mathscr{D}_-} \mathrm{e}^{\mathrm{i}\lambda z}\,\mathrm{d}\nu(z);$$

since ν has no mass on $\mathbb{R}$, the previous relation becomes

$$(\dagger\dagger) \qquad \Phi_+(\lambda) \;+\; \Phi_-(\lambda) \;\equiv\; 0.$$

We are going to show that in fact each of the left-hand terms in (††) vanishes identically.

Consider $\Phi_+(\lambda)$. Since $\mathscr{D}_+$ is a *bounded* domain; $\Phi_+(\lambda)$ is *entire and of exponential type*. It is also *bounded* on the *real axis*. Indeed, for $\lambda > 0$,

$$|\Phi_+(\lambda)| \leqslant \iint_{\mathscr{D}_+} e^{-\lambda \Im z}|d\nu(z)| \leqslant \iint_{\mathscr{D}_+} |d\nu(z)|,$$

and for $\lambda < 0$ we can, *on account of* (††), use the relation $\Phi_+(\lambda) = -\Phi_-(\lambda)$ and make a similar estimate involving $\mathscr{D}_-$. These properties of $\Phi_+(\lambda)$ and the theorem of Chapter III, §G.2, imply that $\Phi_+(\lambda) \equiv 0$ provided that

$$(\ast) \qquad \int_{-\infty}^{\infty} \frac{\log|\Phi_+(\lambda)|}{1+\lambda^2} d\lambda = -\infty.$$

We proceed to establish this relation.

Write

$$H(y) = \begin{cases} \log L(y), & 0 < y \leqslant b, \\ \log L(b), & y > b. \end{cases}$$

Then $H(y)$ is *decreasing* for $y > 0$ by hypothesis. For $\lambda > 0$,

$$|\Phi_+(\lambda)| = \left| \iint_{\mathscr{D}_+} \frac{e^{i\lambda z}}{L(\Im z)} d\mu(z) \right| \leqslant \iint_{\mathscr{D}_+} e^{-H(\Im z) - \lambda \Im z}|d\mu(z)|.$$

If, as in article 5, we put

$$M(\lambda) = \inf_{y>0} (H(y) + y\lambda),$$

we see by the previous relation that

$$(\S\S) \qquad |\Phi_+(\lambda)| \leqslant \text{const.}e^{-M(\lambda)}, \qquad \lambda > 0.$$

Since $H(y)$ is decreasing for $y > 0$ and $\geqslant 1$ there, and

$$\int_0^b \log H(y)\, dy = \int_0^b \log\log L(y)\, dy = \infty$$

by hypothesis, we have

$$\int_1^{\infty} \frac{M(\lambda)}{\lambda^2} d\lambda = \infty$$

according to the *last* theorem of article 2. This, together with (§§), gives us ($\ast$), and hence $\Phi_+(\lambda) \equiv 0$.

Referring again to (††), we see that also $\Phi_-(\lambda) \equiv 0$. Specializing to $\lambda = 0$ (!), we obtain the two relations

$$\iint_{\mathscr{D}_+} \frac{1}{L(\Im z)} d\mu(z) = 0, \qquad \iint_{\mathscr{D}_-} \frac{1}{L(\Im z)} d\mu(z) = 0,$$

from which, by subtraction,

$$\iint_{\mathfrak{D}} \frac{\varphi(z)}{L(\Im z)} \, d\mu(z) \;=\; 0,$$

what we had set out to show. The proof of our theorem is thus finished, and we are done.

And thus ends this long (aye, too long!) seventh chapter of the present book.

VIII

Persistence of the form $dx/(1+x^2)$

Up to now, integrals like

$$\int_{-\infty}^{\infty} \frac{\log|F(x)|}{1+x^2}\,dx$$

have appeared so frequently in this book mainly on account of the specific form of the Poisson kernel for a half plane. If $\omega(S,z)$ denotes the harmonic measure (at z) of $S \subseteq \mathbb{R}$ for the half plane $\{\Im z > 0\}$, we simply have

$$\omega(S,\mathrm{i}) \;=\; \frac{1}{\pi}\int_S \frac{dt}{1+t^2}.$$

Suppose now that we *remove* certain finite open intervals – perhaps infinitely many – from $\mathbb{R}$, leaving a certain residual set E, and that E *looks* something like $\mathbb{R}$ when seen from far enough away. E should, in particular, have infinite extent on both sides of the origin and not be too sparse. Denote by $\mathscr{D}$ the (multiply connected – perhaps even infinitely connected) domain $\mathbb{C} \sim E$, and by $\omega_{\mathscr{D}}(\ ,z)$ the harmonic measure for $\mathscr{D}$.

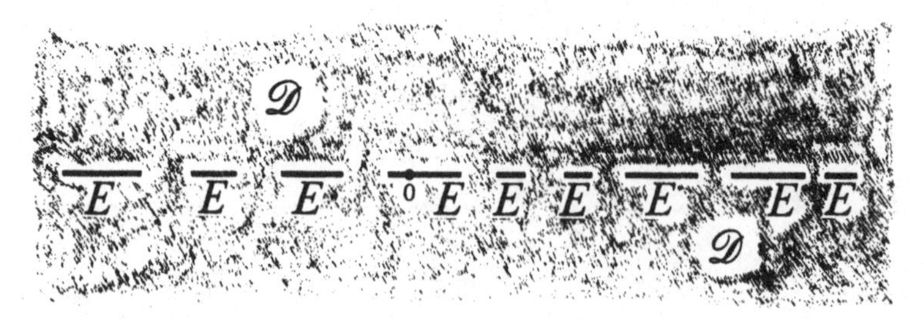

Figure 107

It is a remarkable fact that a formula like the above one for $\omega(S,\mathrm{i})$ *subsists*, to a certain extent, for $\omega_{\mathscr{D}}(\ ,\mathrm{i})$, provided that the degradation suffered by $\mathbb{R}$

in its reduction to E is not too great. We have, for instance,

$$\omega_{\mathscr{D}}(J \cap E, \mathrm{i}) \;\leqslant\; C_E(\alpha) \int_{J\cap E} \frac{\mathrm{d}t}{1+t^2}$$

for *intervals* J with $|J \cap E| \geqslant \alpha > 0$, where $C_E(\alpha)$ depends on α as well as on the set E. In other words, $\mathrm{d}\omega_{\mathscr{D}}(t, \mathrm{i})$ *still acts* (crudely) *like the restriction of* $\mathrm{d}x/(1+x^2)$ *to* E. It is this *tendency of the form* $\mathrm{d}x/(1+x^2)$ *to persist* when we reduce $\mathbb{R}$ to certain smaller sets E (and enlarge the upper half plane to $\mathscr{D} = \mathbb{C} \sim E$) that constitutes the theme of the present chapter.

The persistence is well illustrated in the situation of *weighted approximation* (whether by polynomials or by functions of exponential type) on the sets E. If a function $W(x) \geqslant 1$ is given on E, with $W(x) \to \infty$ as $x \to \pm\infty$ in E (for weighted *polynomial* approximation on E this must of course take place faster than any power of x), we can look at *approximation on* E (by polynomials or by functions of exponential type bounded on $\mathbb{R}$) *using the weight* W. It turns out that *precise formal analogues* of many of the results established for weighted approximation *on* $\mathbb{R}$ in §§A, B and E of Chapter VI *are valid here*; the only change consists in the *replacement* of the integrals of the form

$$\int_{-\infty}^{\infty} \frac{\log M(t)}{1+t^2}\,\mathrm{d}t$$

occurring in Chapter VI by the corresponding expressions

$$\int_E \frac{\log M(t)}{1+t^2}\,\mathrm{d}t.$$

The integrand, involving $\mathrm{d}t/(1+t^2)$, remains unchanged.

This chapter has three sections. The *first* is mainly devoted to the case where E has *positive lower uniform density* on $\mathbb{R}$ – a typical example is furnished by the set

$$E \;=\; \bigcup_{n=-\infty}^{\infty} [n-\rho,\; n+\rho]$$

where $0 < \rho < \frac{1}{2}$.

In §B, we study the limiting case of the example just mentioned which arises when $\rho = 0$, i.e., when $E = \mathbb{Z}$. There is of course no longer any harmonic measure for $\mathscr{D} = \mathbb{C} \sim \mathbb{Z}$. It is therefore remarkable that *something nevertheless remains true of the results established in* §A. If $P(z)$ is a *polynomial* such that

$$\sum_{n=-\infty}^{\infty} \frac{1}{1+n^2} \log^+ |P(n)| \;\leqslant\; \eta$$

with $\eta > 0$ *sufficiently small* (this restriction turns out to be crucial!), we *still* have, for $z \in \mathbb{C}$,

$$|P(z)| \leqslant K(z, \eta),$$

where $K(z, \eta)$ depends *only* on z and η, and not on P. The proof of this fact is very long, and hard to grasp as a whole. It uses specific properties of polynomials. Since $\mathscr{D}$ has no harmonic measure, a corresponding statement with $\log^+|P(z)|$ replaced by a *general* continuous subharmonic function of at most logarithmic growth is *false.*

We return in §C to the study of harmonic estimation in $\mathscr{D}$ when its boundary, E, *does not* reduce to a discrete set. *Here,* we assume that E contains all $x \in \mathbb{R}$ of sufficiently large absolute value, that situation being general enough for applications. The purpose of §C is to connect up the behavior of a *Phragmén–Lindelöf function* for $\mathscr{D}$ (i.e., one harmonic in $\mathscr{D}$ and acting like $|\Im z|$ there, with boundary value *zero* on E) to that of *harmonic measure* for $\mathscr{D}$. There is a quantitative relation between the former and the latter. Harmonic measure still acts (very crudely!) like the restriction of $\mathrm{d}t/(1+t^2)$ to E. This § is independent of §B to a large extent, but does use a fair amount of material from §A. Results obtained in it are needed for Chapter XI.

A. The set E has positive lower uniform density

During most of this §, we consider sets E of the special form

$$\bigcup_{n=-\infty}^{\infty} [a_n - \delta_n,\ a_n + \delta_n],$$

the intervals $[a_n - \delta_n,\ a_n + \delta_n]$ being *disjoint. We will assume that there are four constants, A, B, δ and Δ, with*

$$0 < A < a_n - a_{n-1} < B, \qquad 0 < \delta < \delta_n < \Delta,$$

for all n.

The following *notation* will be used throughout:

$$\begin{aligned} E_n &= [a_n - \delta_n,\ a_n + \delta_n], \\ \mathcal{O}_n &= (a_n + \delta_n,\ a_{n+1} - \delta_{n+1}), \\ \mathscr{D} &= \mathbb{C} \sim E. \end{aligned}$$

Here is a picture of the setup we are studying:

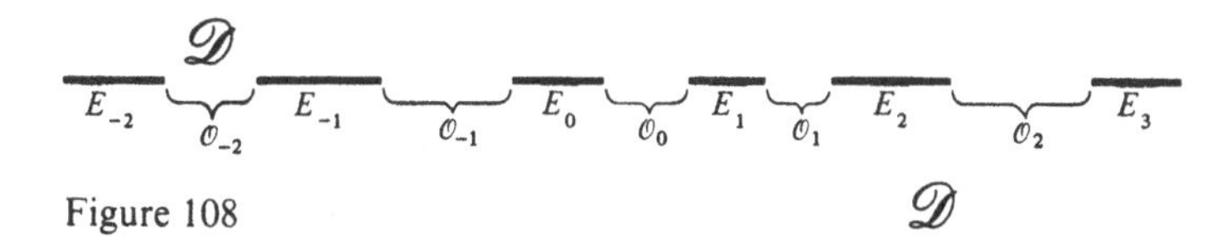

Figure 108

The above boxed conditions on the a_n and δ_n *clearly* imply the existence of two constants C_1 and $C_2 > 0$ (*each* depending on the four numbers A, B, δ and Δ) such that:

(i) if $k \neq l$ and $x \in \mathcal{O}_k$, $x' \in \mathcal{O}_l$, we have $C_1|k-l| \leqslant |x-x'| \leqslant C_2|k-l|$;
(ii) if $k \neq l-1$, l, or $l+1$ and $x \in E_k$, $x' \in E_l$, we have
$C_1|k-l| \leqslant |x-x'| \leqslant C_2|k-l|$.

The restriction on the pair (k, l) in (ii) is due to the fact that the *lengths* of the $\mathcal{O}_k$ are *not assumed to be bounded away from zero*; their lengths are only bounded above. It is *the lengths of the E_k* that are bounded above *and* away from zero.

Heavy use will be made of properties (i) and (ii) during the following development. Clearly, if E is any set for which the above boxed condition holds (with given A, B, δ and Δ), so is each of its translates $E + h$ (with the same constants A, B, δ and Δ). The properties (i) and (ii) are thus valid for each of those translates, with the same constants C_1 and C_2 as for E. For this reason there is no real loss of generality in supposing that $0 \in \mathcal{O}_0$, and we will frequently do so when that is convenient.

1. Harmonic measure for $\mathscr{D}$

The Dirichlet problem can be solved for the kind of domains $\mathscr{D}$ we are considering and (*at least*, certainly!) for continuous boundary data on E given by functions tending to 0 as $x \to \pm\infty$ in E. Let us, without going into too much detail, indicate how this fact can be verified.

Take large values of R, and put

$$\mathscr{D}_R = \mathscr{D} \cap \sim \{(-\infty, -R] \cup [R, \infty)\}:$$

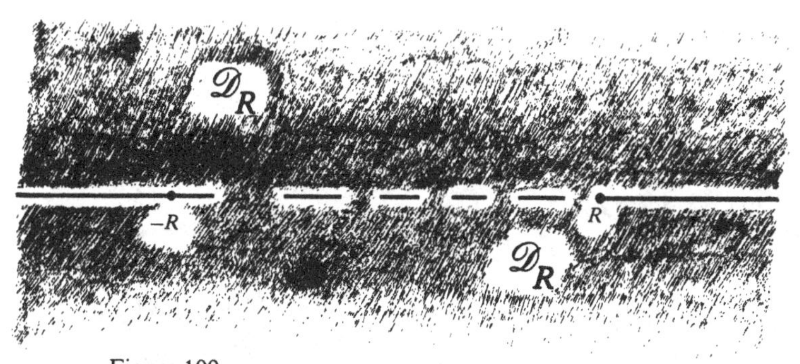

Figure 109

Each of the regions $\mathscr{D}_R$ is *finitely connected*, and the Dirichlet problem can be solved in it. This is *known*; it is true because the straight segment boundary components ('slits') of $\mathscr{D}_R$ are practically as nice as Jordan curve boundary components. One can indeed map $\mathscr{D}_R$ conformally onto a region *bounded* by Jordan curves by using a succession of Joukowski transformations, one for each slit (including the infinite one through ∞):

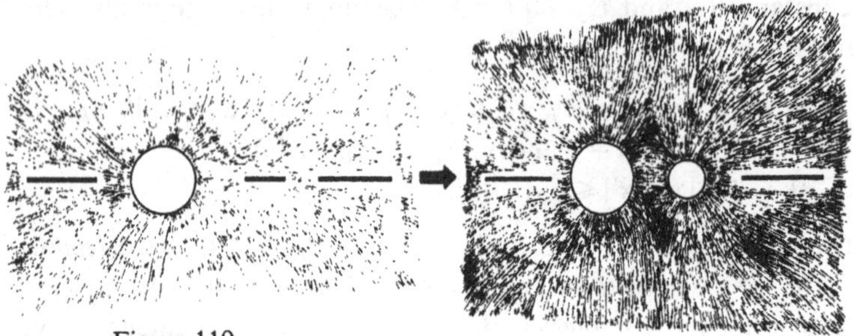

Figure 110

The inverse to the conformal mapping thus obtained *does* take the Jordan curve boundary components back *continuously* onto the original slits, so the Dirichlet problem *can* be solved for $\mathscr{D}_R$ if it can be solved for regions bounded by a finite number of Jordan curves. (This same idea will be used again at the end of article 2, in proving the symmetry of Green's function.)

Once we are sure that the Dirichlet problem can be solved in each $\mathscr{D}_R$ we can, by examining how certain solutions behave for $R \to \infty$, convince ourselves that the Dirichlet problem for $\mathscr{D}$ is also solvable, at least for boundary data of the abovementioned kind. Details of this examination are left to the reader.

Since $\mathscr{D}$ is regular for the Dirichlet problem, harmonic measure is available for it. We know from the rudiments of conformal mapping theory *that a slit should be considered as having two sides, or edges.* Given (say) an interval $J \subseteq$ one of the boundary components E_n of $\mathscr{D}$, we should *distinguish* between *two* intervals coinciding with J: J_+ (lying on the *upper* side of E_n), and J_- (lying on the *lower* side of E_n):

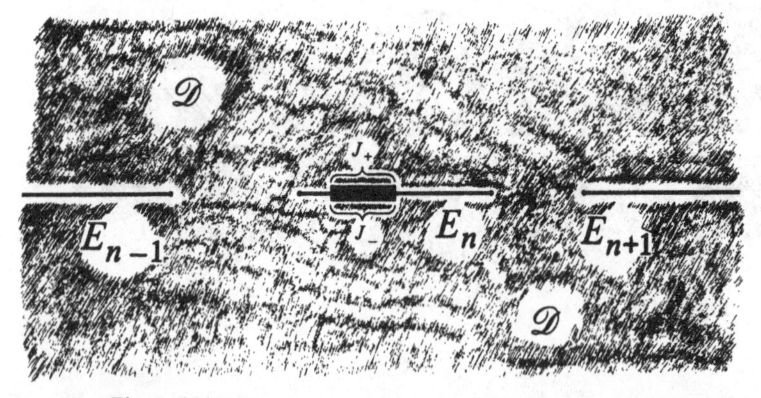

Figure 111

It makes sense, then, to talk about the *two* harmonic measures:

$$\omega_{\mathscr{D}}(J_+, z)$$

(which tends to *zero* when z tends to an interior point of J_-), and

$$\omega_{\mathscr{D}}(J_-, z).$$

In most of our work, however, *separation of J into J_+ and J_- will serve no purpose.* It will in fact be sufficient to work with the *sum*

$$\omega_{\mathscr{D}}(J_+, z) + \omega_{\mathscr{D}}(J_-, z).$$

▶ *This harmonic function tends to* 1 *when z tends from either side of the real axis to an interior point of J, and it is what we take as the harmonic measure*

$$\omega_{\mathscr{D}}(J, z)$$

of J. The harmonic measure $\omega_{\mathscr{D}}(S, z)$ *of any $S \subseteq E$ is defined in the same way.*

Consider now any of the boundary components E_k of $E = \partial\mathscr{D}$, and write

$$\omega_k(z) = \omega_{\mathscr{D}}(E_k, z)$$

for the harmonic measure of E_k, as seen from $z \in \mathscr{D}$. We are going to show that there is a constant C, depending on the four numbers A, B, δ and Δ associated with E, such that

$$\omega_k(x) \leqslant \frac{C}{(l-k)^2 + 1} \quad \text{for } x \in \mathscr{O}_l.$$

($\mathscr{O}_l$, recall, is the part of $\mathscr{D} \cap \mathbb{R}$ lying between E_l and E_{l+1}.) The proof is due to Carleson; one or two of its ideas go back to earlier work. We need two lemmas, the first of which could almost be given as an exercise.

Lemma. *Denote by $\Omega_k(\ \ , z)$ harmonic measure for the domain*

$$\mathscr{D}_k = \{\Im z > 0\} \cup \mathscr{O}_k \cup \{\Im z < 0\}.$$

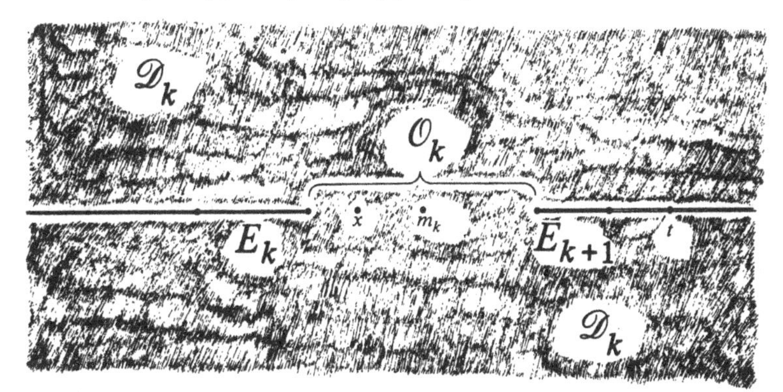

Figure 112

There is a constant K', depending only on the numbers B and δ associated with the set E, such that for $x \in \mathcal{O}_k$ and $t \in \partial\mathcal{D}_k$ lying outside both *of the segments E_k and E_{k+1} we have*

$$d\Omega_k(t, x) \leqslant \frac{K'}{(x-t)^2}\, dt.$$

Proof. By conformal mapping of $\mathcal{D}_k$ onto the unit disk. Calling m_k the midpoint of $\mathcal{O}_k$, we apply to $z \in \mathcal{D}_k$ the chain of mappings

$$z \longrightarrow \zeta = 2\frac{z - m_k}{|\mathcal{O}_k|} \longrightarrow w = \frac{1}{\zeta} - \sqrt{\left(\frac{1}{\zeta^2} - 1\right)}.$$

We write $w = \varphi(z)$ and, if t is on $\partial\mathcal{D}_k$, denote $\varphi(t)$ by ω. (In the latter case we must of course distinguish between points t lying on the *upper side* of $\partial\mathcal{D}_k$ and those on its *lower side* – see the preceding remarks. On this distinction depends the choice of the branch of $\sqrt{\ }$ to be used in computing $\omega = \varphi(t)$.)

For t on $\partial\mathcal{D}_k$ outside both E_k and E_{k+1} we have

$$\omega = \varphi(t) = \frac{1}{\tau} - \sqrt{\left(\frac{1}{\tau^2} - 1\right)},$$

where $\tau = 2(t - m_k)/|\mathcal{O}_k|$ satisfies the inequality

$$|\tau| - 1 \geqslant 2\frac{\min(|E_k|, |E_{k+1}|)}{|\mathcal{O}_k|} > \frac{4\delta}{B},$$

in view of the relations $|E_l| = 2\delta_l > 2\delta$, $|\mathcal{O}_k| < a_{k+1} - a_k < B$. In terms of τ,

$$d\omega = -\frac{d\tau}{\tau^2}\left(1 \pm \frac{i}{\sqrt{(\tau^2 - 1)}}\right),$$

i.e.,

$$|d\omega| = \left(\frac{\tau^2}{\tau^2 - 1}\right)^{\frac{1}{2}} \frac{|d\tau|}{\tau^2}.$$

For t outside both E_k and E_{k+1}, the expression on the right is

$$< \left(\frac{1 + 4\delta/B}{4\delta/B}\right)^{\frac{1}{2}} \frac{|d\tau|}{\tau^2} = \left(\frac{B}{4\delta} + 1\right)^{\frac{1}{2}} \frac{|d\tau|}{\tau^2}$$

by the inequality for $|\tau| - 1$.

Let $x \in \mathcal{O}_k$. Then, remembering that $d\Omega_k(t, x)$ is the harmonic measure of *two infinitesimal intervals* $[t, t + dt]$ lying on $\partial\mathcal{D}_k$ – one on *the upper edge* and one on *the lower*, we see that

$$d\Omega_k(t, x) = \frac{1}{\pi} \frac{1 - |\varphi(x)|^2}{|\varphi(x) - \omega|^2} |d\omega|,$$

with $|d\omega|$ being given by the above formula. Since, for $t \notin E_k \cup E_{k+1}$,

$|\tau| - 1 > 4\delta/B$, the image, ω, of t on the unit circumference must lie *outside* two arcs thereof entered at 1 and at -1, and having *lengths* that depend on the ratio $4\delta/B$. We do not need to know the exact form of this dependence.

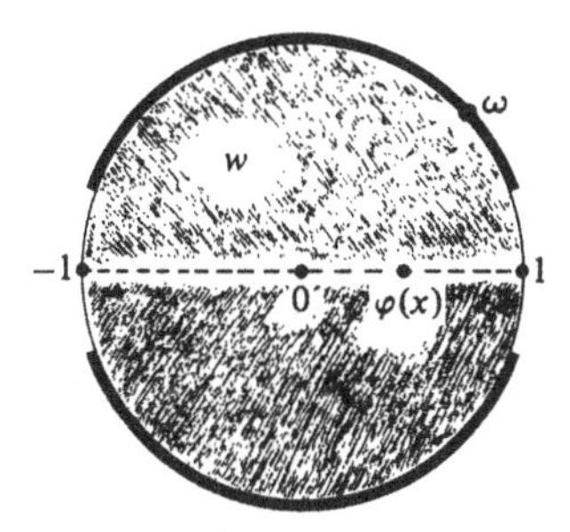

Figure 113

Hence, since $-1 < \varphi(x) < 1$, it is clear from the picture that $|\varphi(x) - \omega|$ is $\geqslant$ a positive quantity depending on the lengths of the excluded arcs about 1 and -1, and thence on the ratio $4\delta/B$. The factor $(1 - |\varphi(x)|^2)/|\varphi(x) - \omega|^2$ in the above formula for $d\Omega_k(t, x)$ is thus bounded above by a number depending on $4\delta/B$ when $t \notin E_k \cup E_{k+1}$. Substituting into that formula the inequality for $|d\omega|$ already found, we get

$$d\Omega_k(t, x) \leqslant C \frac{|d\tau|}{\tau^2}$$

for $t \notin E_k \cup E_{k+1}$, with a constant C depending on $4\delta/B$. In terms of $t = m_k + \frac{1}{2}|\mathcal{O}_k|\tau$, the right side is

$$\leqslant \frac{C|\mathcal{O}_k|}{2} \frac{dt}{(t - m_k)^2} \leqslant \frac{CB}{2} \frac{dt}{(t - m_k)^2}.$$

Since $x \in \mathcal{O}_k$ and $t \notin \mathcal{O}_k$, $\quad |t - x| \leqslant 2|t - m_k|$:

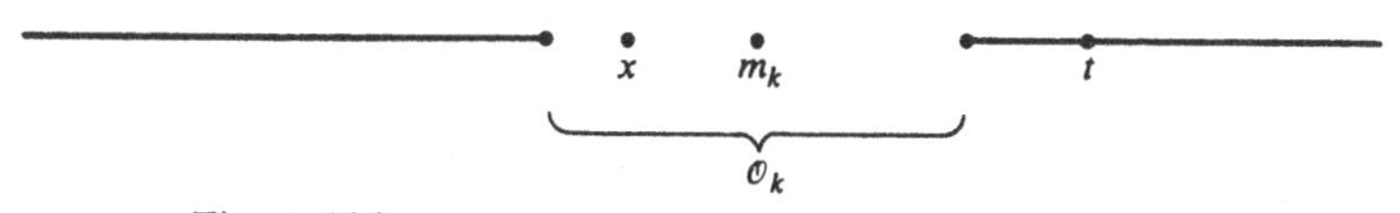

Figure 114

So finally,

$$d\Omega_k(t, x) \leqslant 2BC \frac{dt}{(t - x)^2},$$

Q.E.D.

Computational lemma (Carleson, 1982; see also Benedicks' 1980 *Arkiv* paper). *Let $A_{k,l} \geqslant 0$ for $k, l \in \mathbb{Z}$. Suppose there are constants K and λ, with $0 < \lambda < 1$, such that*

$$(*) \qquad A_{k,l} \leqslant \frac{K}{(l-k)^2+1},$$

and

$$(\overset{*}{*}) \qquad \sum_{l=-\infty}^{\infty} A_{k,l} \leqslant \lambda \qquad \text{for all} \quad k.$$

Then there is a number $\eta > 0$ depending on K and λ such that, for any sequence $\{y_l\}$ with $0 \leqslant y_l \leqslant \eta$ and $0 \leqslant y_l \leqslant 1/(l^2+1)$, we have.

$$\sum_{l=-\infty}^{\infty} A_{k,l} y_l \leqslant \lambda \sup_l y_l \qquad \text{for all} \quad k,$$

and

$$\sum_{l=-\infty}^{\infty} A_{k,l} y_l \leqslant \frac{1+\lambda}{2} \cdot \frac{1}{k^2+1}.$$

Remark. The *first* of the asserted inequalities is manifest; it is the *second* that is non-trivial. If the constant K is small enough, the second inequality is also clear; it is when K is *not small* that the latter is difficult to verify.

Proof of lemma. As we have just remarked, the *first* inequality is obvious (by $(\overset{*}{*})$); let us therefore see to the *second*, endeavoring first of all to prove it for *large* values of $|k|$, say $|k| \geqslant$ some k_0.

Assume, wlog, that $k > 0$, and take some *small* number μ, $0 < \mu < \frac{1}{2}$, about whose precise value we will decide later on. Write the sum

$$\sum_{l=-\infty}^{\infty} A_{k,l} y_l$$

as

$$\sum_{|l| < \mu k} + \sum_{\substack{|l| \geqslant \mu k \\ |l-k| \geqslant \mu k}} + \sum_{|l-k| < \mu k} = \mathrm{I} + \mathrm{II} + \mathrm{III},$$

say. Use of $(*)$ together with the inequality $0 \leqslant y_l \leqslant \eta$ (where η is *as yet* unspecified) gives us first of all

$$\mathrm{I} \leqslant \frac{K}{(1-\mu)^2 k^2 + 1} \cdot 2k\mu \cdot \eta$$

We get III out of the way by combining $(\overset{*}{*})$ and the inequality

$0 \leqslant y_l \leqslant 1/(l^2 + 1)$, with the result that

$$\text{III} \leqslant \frac{\lambda}{(1-\mu)^2 k^2 + 1}.$$

It is the middle sum, II, that gives us trouble. We break II up further as

$$\sum_{l \leqslant -\mu k} + \sum_{\mu k \leqslant l < k/2} + \sum_{k/2 \leqslant l \leqslant (1-\mu)k} + \sum_{l \geqslant (1+\mu)k}.$$

The *first* of these sums is

$$\leqslant \sum_{l \leqslant -\mu k} \frac{1}{l^2+1} \cdot \frac{K}{(k-l)^2+1} \leqslant \frac{K}{k^2+1} \sum_{m \geqslant \mu k} \frac{1}{m^2} \leqslant \frac{2K}{\mu k(k^2+1)}.$$

The *second* is similarly

$$\leqslant \sum_{\mu k \leqslant l < k/2} \frac{1}{l^2+1} \cdot \frac{K}{(k-l)^2+1} \leqslant \frac{K}{(k/2)^2+1} \sum_{m \geqslant \mu k} \frac{1}{m^2} \leqslant \frac{8K}{\mu k(k^2+4)}$$

The *third* sum is

$$\leqslant \sum_{l \leqslant (1-\mu)k} \frac{1}{(k/2)^2+1} \cdot \frac{K}{(k-l)^2+1} \leqslant \frac{8K}{\mu k(k^2+4)},$$

and the *fourth*

$$\leqslant \sum_{l \geqslant (1+\mu)k} \frac{1}{l^2+1} \cdot \frac{K}{(l-k)^2+1} \leqslant \frac{2K}{\mu k(k^2+1)}.$$

All told, then,

$$\text{II} \leqslant \frac{20K}{\mu k(k^2+1)}.$$

Adding this last estimate to those already obtained for I and III, we get finally

$$\sum_{l=-\infty}^{\infty} A_{k,l} y_l \leqslant \frac{2k\mu\eta K}{(1-\mu)^2(k^2+1)} + \frac{20K}{\mu k(k^2+1)} + \frac{\lambda}{(1-\mu)^2(k^2+1)}.$$

The idea now is to *first* put η equal to a *very small quantity* η_0, and *then*, assuming k is *large*, put $\mu = 1/\eta_0^{1/2} k$; this will *also be small* for large enough k. For such large k, the previous inequality will reduce to

$$\sum_{l=-\infty}^{\infty} A_{k,l} y_l \leqslant \frac{2\eta_0^{1/2} K + 20\eta_0^{1/2} K + \lambda}{(1-\mu)^2(k^2+1)}.$$

Choosing first $\eta_0^{1/2}$ *small* enough and then taking k_0 so *large* that $1/\eta_0^{1/2} k_0$

is *also* small, we will make the right-hand side of this inequality

$$\leqslant \frac{1+\lambda}{2} \cdot \frac{1}{k^2+1}$$

for $|k| \geqslant k_0$ by putting $\mu = 1/\eta_0^{1/2}|k|$. When $\eta < \eta_0$ and $0 \leqslant y_l \leqslant \eta$ we then have, *a fortiori*,

$$\sum_{l=-\infty}^{\infty} A_{k,l} y_l \leqslant \frac{1+\lambda}{2} \cdot \frac{1}{k^2+1} \qquad \text{for} \quad |k| \geqslant k_0.$$

With such y_l, however, the sum on the left is also $\leqslant \lambda\eta$. So, taking finally

$$\eta = \min\left(\eta_0, \frac{1}{k_0^2+1}\right)$$

makes the left side $\leqslant ((1+\lambda)/2)(1/(k^2+1))$ for $|k| < k_0$ *as well*, i.e.,

$$\sum_{l=-\infty}^{\infty} A_{k,l} y_l \leqslant \frac{1+\lambda}{2} \cdot \frac{1}{k^2+1}$$

for all k, provided that $0 \leqslant y_l \leqslant \eta$ and $0 \leqslant y_l \leqslant 1/(l^2+1)$.

The lemma is proved.

Theorem (Carleson, 1982; see also Benedicks' 1980 *Arkiv* paper). *In the domain $\mathscr{D}$, the harmonic measure $\omega_0(z)$ of the component E_0 of $\partial\mathscr{D}$ satisfies*

$$\omega_0(x) \leqslant \frac{C}{k^2+1} \qquad \textit{for } x \in \mathcal{O}_k,$$

with a constant C depending only on the four numbers A, B, δ and Δ associated with $E = \partial\mathscr{D}$.

Proof (Carleson). Call u_k the *maximum value* of $\omega_0(x)$ on $\mathcal{O}_k$; we are to show that

$$u_k \leqslant \frac{C}{k^2+1}.$$

For $k = -1$ and $k = 0$ this is certainly true if we put $C = 2$; we may therefore restrict our discussion to the *values of k different from* -1 and 0.

As in the first of the above lemmas, denote by $\mathscr{D}_k$ the domain

$$\{\Im z > 0\} \cup \mathcal{O}_k \cup \{\Im z < 0\}$$

and by $\Omega_k(\ ,z)$ the harmonic measure for $\mathscr{D}_k$. $\mathscr{D}_k$ is of course *contained* in $\mathscr{D}$:

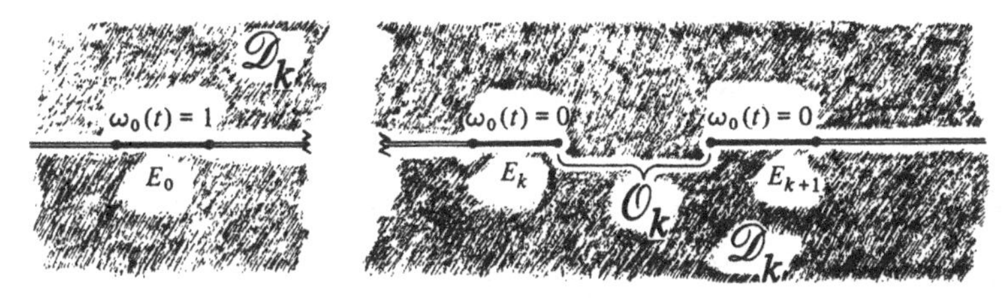

Figure 115

$\omega_0(z)$ is thus harmonic in $\mathscr{D}_k$; since it is clearly bounded there and continuous up to $\partial\mathscr{D}_k$, we may recover it from its values on $\partial\mathscr{D}_k$ by the Poisson formula

$$\omega_0(z) \;=\; \int_{\partial\mathscr{D}_k} \omega_0(t)\mathrm{d}\Omega_k(t, z), \qquad z \in \mathscr{D}_k.$$

Since we are assuming that $k \neq -1, 0$, we have (by *definition* of harmonic measure!) $\omega_0(t) = 0$ for $t \in E_k \cup E_{k+1}$. In fact, $\omega_0(t)$ is identically zero on *all* the E_l save E_0, where $\omega_0(t) = 1$. The above formula hence becomes

$$\omega_0(z) \;=\; \int_{E_0} \mathrm{d}\Omega_k(t, z) \;+\; \sum_{l \neq k} \int_{\mathscr{O}_l} \omega_0(t)\mathrm{d}\Omega_k(t, z),$$

$\mathbb{R}$ being the disjoint union of the intervals $\mathscr{O}_l$ and E_l.

Let x_l be the point in $\mathscr{O}_l$ where $\omega_0(x)$ *assumes its maximum* u_l *therein*, and write

$$A_{k,l} \;=\; \int_{\mathscr{O}_l} \mathrm{d}\Omega_k(t, x_k).$$

Then, by the previous relation,

$$u_k \;\leqslant\; \Omega_k(E_0, x_k) \;+\; \sum_{l \neq k} A_{k,l} u_l.$$

Here, the integrals

$$\int_{E_0} \mathrm{d}\Omega_k(t, x_k) \;=\; \Omega_k(E_0, x_k)$$

and

$$\int_{\mathscr{O}_l} \mathrm{d}\Omega_k(t, x_k) \;=\; A_{k,l}, \qquad l \neq k,$$

are taken over sets *disjoint from* E_k and E_{k+1}, whereas $x_k \in \mathscr{O}_k$. We may therefore apply the *first* lemma to estimate $\mathrm{d}\Omega_k(t, x_k)$ in these integrals, getting

$$\Omega_k(E_0, x_k) \;\leqslant\; K' \int_{E_0} \frac{\mathrm{d}t}{(t - x_k)^2}$$

and

$$A_{k,l} \leqslant K' \int_{\mathcal{O}_l} \frac{dt}{(t-x_k)^2}, \qquad l \neq k,$$

where K' is a constant depending on the numbers B and δ associated with E. By properties (i) and (ii), given at the beginning of this §, we have

$$(t-x_k)^2 \geqslant C_1^2 k^2 \quad \text{for} \quad t \in E_0$$

and

$$(t-x_k)^2 \geqslant C_1^2 (k-l)^2 \quad \text{for} \quad t \in \mathcal{O}_l.$$

So, since $|E_0| = 2\delta_0 < 2\Delta$ and $|\mathcal{O}_l| < B$, the preceding relations become

$$\Omega_k(E_0, x_k) \leqslant \frac{K}{k^2+1}$$

and

$$(*) \qquad A_{k,l} \leqslant \frac{K}{(k-l)^2+1}, \qquad l \neq k;$$

here, K is a constant depending on the four numbers A, B, δ and Δ.

The numbers $A_{k,l}$ also satisfy the inequality

$$(\ast\ast) \qquad \sum_{l \neq k} A_{k,l} \leqslant \lambda < 1$$

with λ depending *only on the ratio* δ/B. Indeed,

$$\sum_{l \neq k} A_{k,l} = \sum_{l \neq k} \Omega_k(\mathcal{O}_l, x_k)$$

is $\leqslant \Omega_k(\partial\mathscr{D}_k \sim E_k \sim E_{k+1}, x_k)$. Since $|\mathcal{O}_k| < B$ and $|E_k| > 2\delta$, $|E_{k+1}| > 2\delta$, a simple change of variable shows that the latter quantity is *less* than the harmonic measure of the set $1 + 4\delta/B \leqslant |t| < \infty$ on the boundary of the domain $\mathbb{C} \sim (-\infty, -1] \sim [1, \infty)$, seen from some point in $(-1, 1)$:

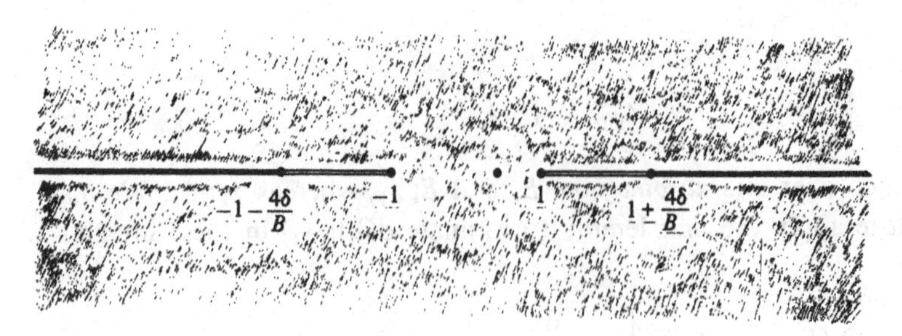

Figure 116

And *that* harmonic measure is clearly *at most* equal to some number $\lambda < 1$ depending on $4\delta/B$.

Let us return to the inequality

$$u_k \leqslant \Omega_k(E_0, x_k) + \sum_{l \neq k} A_{k,l} u_l$$

established above. By plugging into it the estimates just found, we get

$$(\dagger) \qquad u_k - \sum_{l \neq k} A_{k,l} u_l \leqslant \frac{K}{k^2 + 1},$$

where the $A_{k,l}$ are $\geqslant 0$, *and satisfy* $(*)$ *and* $(\ast\!\!\ast)$. This has been proved for $k \neq -1$ and 0, but it *also holds (a fortiori!)* for *those* values of k, provided that we take $K \geqslant 2$. *Then our* (unknown) *maxima* $u_k \geqslant 0$ *will satisfy* $(\dagger)$ *for all* k; *this we henceforth assume.*

The idea now is to invert the relations $(\dagger)$ in order to obtain bounds on the u_k. *It is convenient to define* $A_{k,l}$ *for* $l = k$ *by putting* $A_{k,k} = 0$. Then, calling

$$(\dagger\dagger) \qquad v_k = u_k - \sum_l A_{k,l} u_l,$$

we can recover the u_k *from the* v_k by virtue of $(\ast\!\!\ast)$. Write $A_{k,l}^{(1)} = A_{k,l}$; then put

$$A_{k,l}^{(2)} = \sum_{j=-\infty}^{\infty} A_{k,j} A_{j,l},$$

and in general

$$A_{k,l}^{(n+1)} = \sum_{j=-\infty}^{\infty} A_{k,j} A_{j,l}^{(n)}.$$

The numbers $A_{k,l}^{(n)}$ are $\geqslant 0$ (since the $A_{k,l}$ are), and from $(\ast\!\!\ast)$, we have

$$(\S) \qquad \sum_{l=-\infty}^{\infty} A_{k,l}^{(n)} \leqslant \lambda^n.$$

This makes it possible for us to invert $(\dagger\dagger)$, getting

$$u_k = v_k + \sum_l A_{k,l}^{(1)} v_l + \sum_l A_{k,l}^{(2)} v_l + \cdots + \sum_l A_{k,l}^{(n)} v_l + \cdots,$$

the Neumann series on the right being absolutely convergent. Since the $A_{k,l}^{(n)}$ are $\geqslant 0$ and

$$v_l \leqslant \frac{K}{l^2 + 1}$$

by (†) and (††), the previous relation gives

$$u_k \leqslant K\left\{\frac{1}{k^2+1} + \sum_{n=1}^{\infty}\sum_{l=-\infty}^{\infty}\frac{A_{k,l}^{(n)}}{l^2+1}\right\}.$$

We proceed to examine the right-hand side of this inequality.

We have

$$\sum_{l=-\infty}^{\infty}\frac{1}{(k-l)^2+1}\cdot\frac{1}{l^2+1} \leqslant \frac{\text{const.}}{k^2+1}$$

(look at the reproduction property of the Poisson kernel $y/((x-t)^2+y^2)$ on which the *hall of mirrors* argument used in Chapter 6 is based!). Hence, by (∗), there is a constant L with

$$0 \leqslant A_{k,l}^{(n)} \leqslant \frac{L^n}{(k-l)^2+1},$$

and the summand

$$\sum_l A_{k,l}^{(n)}\cdot\frac{1}{l^2+1}$$

on the right side of the above estimate for u_k is

$$\leqslant \frac{\text{const.}\, L^n}{k^2+1}.$$

We have, however, to add up infinitely many of these summands. *It is here that we must resort to the computational lemma.*

Call

$$v_k^{(n)} = \sum_{l=-\infty}^{\infty}\frac{A_{k,l}^{(n)}}{l^2+1};$$

we certainly have $v_k^{(n)} \geqslant 0$. According to the computational lemma, there is an $\eta > 0$ depending on λ and the K in (∗) such that

$$\sum_{l=-\infty}^{\infty} A_{k,l}y_l \leqslant \frac{1+\lambda}{2}\cdot\frac{1}{k^2+1}$$

if $0 \leqslant y_l \leqslant \eta$ and $y_l \leqslant 1/(l^2+1)$. *Fix* such an η. By (§) we can certainly find an m such that $0 \leqslant v_k^{(m)} \leqslant \eta$ for all k. *Fix* such an m. As we have just seen, there is an M depending on m such that

$$v_k^{(m)} = \sum_l \frac{A_{k,l}^{(m)}}{l^2+1} \leqslant \frac{M}{k^2+1},$$

and we may of course suppose that $M \geqslant 1$. Apply now the computational lemma with

$$y_l = v_l^{(m)}/M;$$

we get (after multiplying by M again – this trick works because $\eta/M \leqslant \eta$!),

$$v_k^{(m+1)} = \sum_l A_{k,l} v_l^{(m)} \leqslant \frac{1+\lambda}{2} \cdot \frac{M}{l^2+1}.$$

We also have, of course,

$$0 \leqslant v_k^{(m+1)} \leqslant \lambda\eta$$

by (⁑).

We may now use the computational lemma again with

$$y_l = \frac{2}{(1+\lambda)M} v_l^{(m+1)};$$

note that *here*

$$0 \leqslant y_l \leqslant \frac{2\eta\lambda}{(1+\lambda)M} < \eta.$$

After multiplying back by $(1+\lambda)M/2$, we find

$$v_k^{(m+2)} = \sum_l A_{k,l} v_l^{(m+1)} \leqslant \left(\frac{1+\lambda}{2}\right)^2 \frac{M}{k^2+1}.$$

In this fashion, we can continue indefinitely and prove that

$$v_k^{(m+p)} \leqslant \left(\frac{1+\lambda}{2}\right)^p \cdot \frac{M}{k^2+1}$$

for $p = 1, 2, 3, \ldots$. Therefore, since $\lambda < 1$,

$$\sum_{n=m}^{\infty} v_k^{(n)} \leqslant \frac{2M}{(1-\lambda)(k^2+1)}.$$

This, however, implies that

$$u_k \leqslant K\left\{\frac{1}{k^2+1} + \sum_{n=1}^{\infty} \sum_l \frac{A_{k,l}^{(n)}}{l^2+1}\right\}$$

$$= K\left\{\frac{1}{k^2+1} + v_k^{(1)} + \cdots + v_k^{(m-1)} + \sum_{n=m}^{\infty} v_k^{(n)}\right\} \leqslant \frac{C}{k^2+1}$$

with a certain constant C, since

$$v_k^{(n)} \leqslant \frac{\text{const. } L^n}{k^2+1}$$

for each n. We have proved that $\omega_0(x)$ (which is *at most* u_k on $\mathcal{O}_k$) is

$$\leqslant \frac{C}{k^2+1} \qquad \text{for } x \in \mathcal{O}_k.$$

Q.E.D

Problem 15

In this problem, the set E is as described at the beginning of the present §, with the boxed condition given there.

(a) Let $U_R(z)$ be the harmonic measure (for $\mathscr{D}$) of the subset $E \cap [-R/2, R/2]$ of $\partial\mathscr{D}$, seen from $z \in \mathscr{D}$. Show that there is a number $\alpha > 0$ depending *only* on the four quantities A, B, δ and Δ associated with E, such that $U_R(z) \leqslant \frac{1}{2}$ for $|z| = R$ and $|\Im z| \leqslant \alpha R$. (Hint: First look at $U_R(z)$ for $|z| = R$ and $|\Im z| \leqslant 1$; then use Harnack.)

(b) Let $V_R(z)$ be the harmonic measure (for $\mathscr{D}$) of

$$E \cap \{(-\infty, -R/2] \cup [R/2, \infty)\},$$

seen from $z \in \mathscr{D}$. Show that there is a number $\beta > 0$ depending only on A, B, δ and Δ such that $V_R(z) \geqslant \beta$ for $|z| = R$. (Hint: Use (a) and Harnack.)

(c) For $R > 0$, call $\mathscr{D}_R = \mathscr{D} \cap \{|z| < R\}$ and let $\omega_R(z)$ be the harmonic measure of $\{|z| = R\}$ *for* $\mathscr{D}_R$, as seen from $z \in \mathscr{D}_R$. Prove *Benedicks' lemma*, which says that

$$\omega_R(0) \leqslant \frac{C}{R}$$

with a constant C depending *only* on the four quantities A, B, δ and Δ. (Hint: Compare the $V_R(z)$ of (b) with $\omega_R(z)$ in $\mathscr{D}_R$.)

2. Green's function and a Phragmén–Lindelöf function for $\mathscr{D}$

A Green's function is available for domains $\mathscr{D}$ of the kind considered here. Let us remind the reader who may not remember that, for given $w \in \mathscr{D}$, the Green's function $G(z, w)$ is a positive function of z, *harmonic in* $\mathscr{D}$ *save at* w *where it acts like*

$$\log \frac{1}{|z-w|},$$

which is *bounded* in $\mathscr{D}$ *outside of disks of positive radius centered at* w, and *tends to zero when* z *tends to any point on the boundary* E of $\mathscr{D}$. Existence of $G(z, w)$ for our domains $\mathscr{D}$ follows by standard general arguments, found in many books on complex variable theory. For the sake of completeness, we will show that $G(z, w)$ is *symmetric* in z and w at the end of this article. The *last part* of the argument given there may easily be adapted so as to furnish an existence proof for $G(z, w)$.

Theorem. *Let $\mathscr{D} = \mathbb{C} \sim E$, where $E \subseteq \mathbb{R}$ has the properties given at the beginning of this §. Assume that $0 \in \mathcal{O}_0$. Then there is a constant C, depending only on the four numbers A, B, δ and Δ associated with E, such that $G(x, 0) \leqslant C/(x^2 + 1)$ for $x \in \mathbb{R} \sim \mathcal{O}_0$, $G(z, w)$ being the Green's function for $\mathscr{D}$.*

Proof. Draw a circle Γ with diameter running from the *left endpoint* of E_0 to the *right endpoint* of E_1:

Figure 117

Let us show first of all that there is a number α, depending only on δ, Δ and B, such that

$$G(z, 0) \leqslant \alpha, \quad z \in \Gamma.$$

To see this, observe first of all that $G(z, 0) \leqslant g(z, 0)$, the Green's function for $(\mathbb{C} \sim E_1) \cup \{\infty\}$. This follows by looking at the *difference* $g(z, 0) - G(z, 0)$ on E. The latter is *harmonic* and *bounded* in $\mathscr{D}$, since the logarithmic poles of $g(z, 0)$ and $G(z, 0)$ at 0 cancel each other out. It is thus enough to get an upper bound for $g(z, 0)$ on Γ, and that bound will also serve for $G(z, 0)$ there.

Translation along $\mathbb{R}$ to the midpoint of E_1 followed by scaling down, using the factor $2/|E_1|$, takes $(\mathbb{C} \sim E_1) \cup \{\infty\}$ conformally onto the standard domain $\mathscr{E} = (\mathbb{C} \sim [-1, 1]) \cup \{\infty\}$:

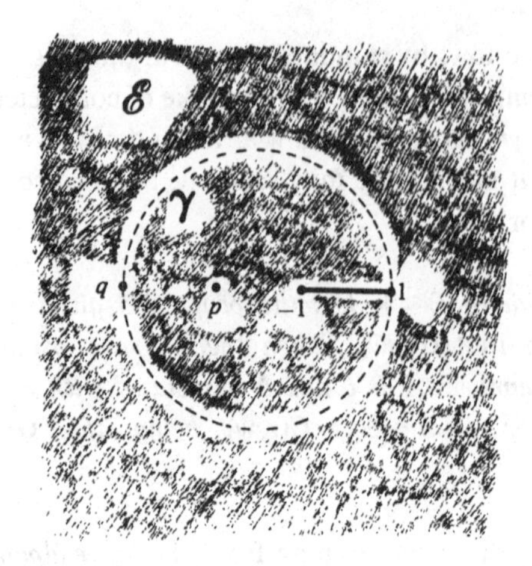

Figure 118

In this reduction, 0 goes to a point p on the real axis, $p < -1$, and the circle Γ goes to another, γ, having $[q, 1]$ as its diameter, where $q < p$. $g(z, 0)$ is of course *equal* to the Green's function for $\mathscr{E}$ with pole at p.

We have

$$|p| - 1 \leqslant \frac{|\mathcal{O}_0|}{|E_1|/2} \leqslant \frac{B}{\delta}$$

and

$$|q| - |p| \geqslant \frac{2|E_0|}{|E_1|} \geqslant \frac{2\delta}{\Delta}.$$

Therefore the Green's function for $\mathscr{E}$ with pole at p is bounded above on γ by some number α depending on B/δ and $2\delta/\Delta$. (The nature of this dependence could be worked out by mapping $\mathscr{E}$ conformally onto $\{1 < |w| \leqslant \infty\}$; we, however, do not need to know it.) This means that $g(z, 0) \leqslant \alpha$ on Γ and finally $G(z, 0) \leqslant \alpha$, $z \in \Gamma$.

This being verified, we take the centre m of Γ and, with R equal to that circle's radius, map the exterior of Γ conformally onto the domain $\mathscr{E}$ just considered by taking z to $w = \frac{1}{2}\{(z-m)/R + R/(z-m)\}$. That mapping takes Γ to the slit $E_1' = [-1, 1]$ *and each of the components* E_n *of* $\partial\mathscr{D}$, $n \neq 0, 1$, *onto segments* E_n' *on the real axis*. The function $\varphi(z) = \frac{1}{2}\{(z-m)/R + R/(z-m)\}$ thus takes

$$\mathscr{D}_\Gamma = \mathscr{D} \cap \{|z - m| > R\}$$

conformally onto a domain

$$\mathscr{D}' = \mathbb{C} \sim \bigcup_{-\infty}^{\infty}{}' E_n':$$

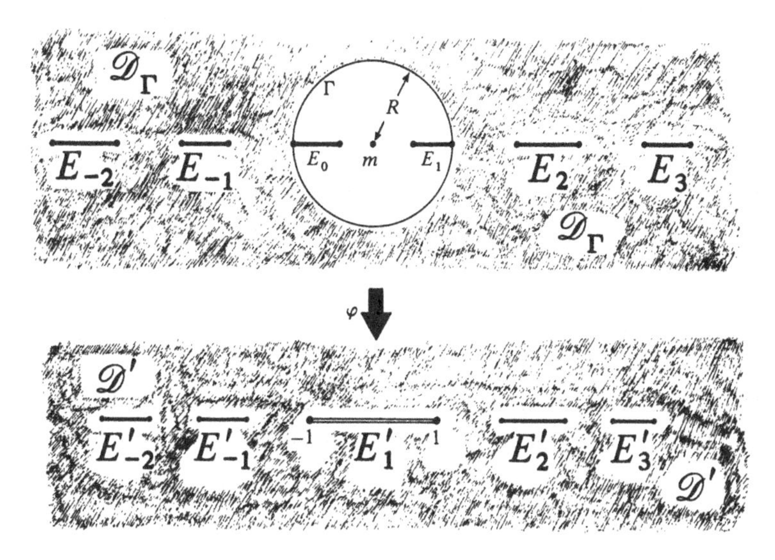

Figure 119

Define a harmonic function $\Omega(w)$ for $w \in \mathscr{D}'$ by putting

$$\Omega(\varphi(z)) = G(z, 0)$$

for $z \in \mathscr{D}_\Gamma$. $\Omega(w)$ is evidently bounded in $\mathscr{D}'$, and has boundary value *zero* on *each* of the components $E'_n (= \varphi(E_n))$ of $\partial\mathscr{D}'$, *save on* E'_1. $\Omega(w)$ *is*, however, continuous up to the latter one, and *on* it

$$\Omega(w) \leqslant \alpha,$$

since $E'_1 = \varphi(\Gamma)$ and $G(z, 0) \leqslant \alpha$ on Γ. We therefore have $\Omega(w) \leqslant \alpha\omega_{\mathscr{D}'}(E'_1, w)$ in $\mathscr{D}'$, where $\omega_{\mathscr{D}'}(\ , w)$ denotes *harmonic measure* for $\mathscr{D}'$.

The set

$$E' = \bigcup_{-\infty}^{\infty}{}' E'_n$$

has, however, the properties specified for our sets E at the beginning of this §. Indeed, for real x,

$$\varphi(x) = \frac{1}{2R}x - \frac{m}{2R} + O\left(\frac{1}{x}\right)$$

when $|x|$ is large, with R lying between the two numbers 2δ and $B/2 + 2\Delta$. Hence, each of the intervals $E'_n (n \neq 0)$ is of the form $[a'_n - \delta'_n,\ a'_n + \delta'_n]$, where, for certain numbers A', B', γ' and Δ' depending on the original A, B, δ and Δ,

$$0 < A' < a'_{n+1} - a'_n < B'$$

and

$$0 < \delta' < \delta'_n < \Delta'.$$

(Again, the exact *form* of the dependence does not concern us here.) *We can therefore apply Carleson's theorem from the previous article to the domain* $\mathscr{D}'$, and find that

$$\omega_{\mathscr{D}'}(E'_1, u) \leqslant \frac{C'}{1+u^2}, \qquad u \in \mathbb{R},$$

with a constant C' depending on A', B', δ' and Δ' and hence, finally, on A, B, δ and Δ. Thence, in view of the previous relation,

$$\Omega(u) \leqslant \frac{\alpha C'}{1+u^2} \quad \text{for} \quad u \in \mathbb{R},$$

i.e.

$$G(x, 0) \leqslant \frac{\alpha C'}{1+(\varphi(x))^2}$$

for real x lying *outside* the circle Γ. Using the fact that $0 \in \mathcal{O}_0$ (whence $|m| \leqslant B + 2\Delta$), the bounds on R given above, and the asymptotic formula for $\varphi(x)$, we see that the right side of the preceding inequality is in turn

$$\leqslant \frac{C}{1+x^2}$$

with a constant C depending only on A, B, δ and Δ. Thus

$$G(x, 0) \leqslant \frac{C}{1+x^2}$$

for real x outside of E_0, $\mathcal{O}_0$ and E_1. But this also holds on E_0 and E_1 since $G(x, 0) = 0$ on those sets! So it holds for real x outside of $\mathcal{O}_0$, which is what we had to prove. We are done.

Problem 16

Let $E \subseteq \mathbb{R}$ fulfill the conditions set forth at the beginning of this §, and assume that $0 \in \mathcal{O}_0$. Let $\omega_{\mathscr{D}}(\ , z)$ be *harmonic measure* for $\mathscr{D} = \mathbb{C} \sim E$. *Prove Benedicks' theorem*, which says that there is a constant C depending only on the four numbers A, B, δ and Δ associated with E, such that, for t in any component

$$E_n = [a_n - \delta_n,\ a_n + \delta_n]$$

of E, and $n \neq 0, 1$,

$$d\omega_{\mathscr{D}}(t, 0) \;\leqslant\; \frac{C}{1+t^2} \cdot \frac{dt}{\sqrt{(\delta_n^2 - (t - a_n)^2)}}.$$

(This is a most beautiful result, by the way!) (Hint: Let G be the Green's function for $\mathscr{D}$. According to a classical elementary formula, if, for instance, we consider points t_+ lying on the *upper edge* of E_n, we have

$$\frac{d\omega_{\mathscr{D}}(t_+, 0)}{dt} \;=\; \frac{1}{2\pi} G_y(t_+, 0) \;=\; \lim_{\Delta y \to 0+} \frac{G(t + i\Delta y, 0)}{2\pi \Delta y},$$

since $G(t, 0) = 0$. (Green *introduced* the functions bearing his name for this very reason!) Take the ellipse Γ given by the equation

$$\frac{(x - a_n)^2}{2\delta_n^2} + \frac{y^2}{\delta_n^2} \;=\; 1$$

and compare $G(z, 0)$ with

$$U(z) \;=\; \log\left| \frac{z - a_n}{\delta_n} + \sqrt{\left(\left(\frac{z - a_n}{\delta_n}\right)^2 - 1\right)} \right|$$

on Γ. Note that $G(x, 0)$ and $U(x)$ *both vanish* on E_n, that $U(z)$ is *harmonic* in the region $\mathscr{E}$ between E_n and Γ, and that $G(z, 0)$ is at least *subharmonic* there (not *necessarily* harmonic because some of the E_k with $k \neq n$ may intrude into $\mathscr{E}$).)

The work in Chapter VI frequently involved entire functions of exponential type bounded on the real axis. If $f(z)$ is such a function, of exponential type A say, and we know that

$$|f(x)| \;\leqslant\; 1, \quad x \in \mathbb{R},$$

we can deduce that

$$|f(z)| \;\leqslant\; e^{A|\Im z|}$$

for all z. This follows by the *third* Phragmén–Lindelöf theorem of Chapter III, §C, whose *proof* depends on the availability of the function $|\Im z|$, *harmonic and* $\geqslant 0$ *in each of the half planes* $\{\Im z > 0\}$, $\{\Im z < 0\}$, *and zero on the real axis.*

Suppose now that we are presented with such a function $f(z)$, known to be *bounded* (with, however, an *unknown bound*) on $\mathbb{R}$, such that, for some closed $E \subseteq \mathbb{R}$,

$$|f(x)| \leqslant 1, \quad x \in E.$$

If there is a function $Y(z)$, *harmonic in* $\mathscr{D} = \mathbb{C} \sim E$, *having boundary value zero on E and such that* $Y(z) \geqslant |\Im z|$, $z \in \mathscr{D}$, we can argue as in the proof of the

Phragmén–Lindelöf theorem just mentioned, and conclude that

$$|f(z)| \leqslant e^{AY(z)}$$

for $z \in \mathscr{D}$ if $f(z)$ is of exponential type A.* We are therefore interested in the *existence* of such functions $Y(z)$ for given closed sets $E \subseteq \mathbb{R}$.

In order to avoid situations involving irregular boundary points for the Dirichlet problem, whose investigation has nothing to do with the material of the present book, we *limit* the following discussion to closed sets E which can be expressed as *disjoint unions of segments on* $\mathbb{R}$ *not accumulating at any finite point*. We do *not*, however, assume in that discussion that the sets E have the form specified at the beginning of this §.

Definition. A Phragmén–Lindelöf function $Y(z)$ for $\mathscr{D} = \mathbb{C} \sim E$ is one *harmonic* in $\mathscr{D}$ and *continuous* up to E, such that

(i) $Y(x) = 0, \qquad x \in E$

(ii) $Y(z) \geqslant |\Im z|, \qquad z \in \mathscr{D}$,

(iii) $Y(iy) = |y| + o(|y|)$ for $y \to \pm\infty$.

It turns out that for given closed $E \subseteq \mathbb{R}$ of the form just described, the existence of $Y(z)$ is governed by the *behaviour of the Green's function* $G(z, w)$ for $\mathscr{D} = \mathbb{C} \sim E$. Before going into this matter, let us mention a simple example (not without its own usefulness) which the reader should keep in mind.

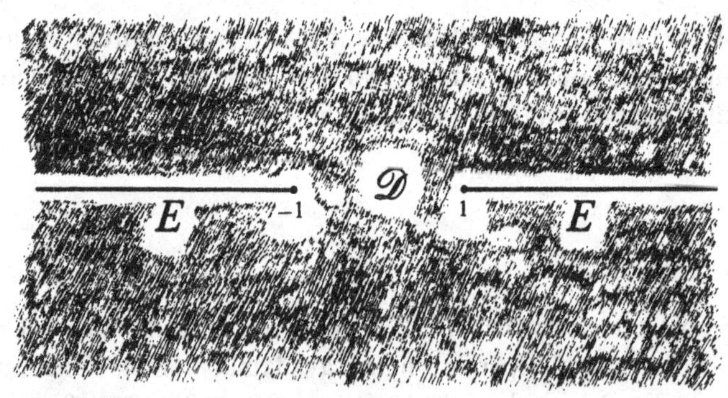

Figure 120

* In fact, *boundedness of* f on $\mathbb{R}$ is *not necessary* here. If $|f(x)| \leqslant 1$, $x \in E$, and $|f(z)| \leqslant C\exp(A|z|)$, the function $v(z) = \log|f(z)| - (A\sec\delta)\,Y(z)$ is subharmonic in $\mathscr{D}$ and bounded above on each of the lines $x = \pm y\tan\delta$ – here, $0 < \delta < \pi/2$. Since $v(z) \leqslant \text{const.}|z|$ in $\mathscr{D}$, the *second* Phragmén–Lindelöf theorem of Chapter III §C shows that v is bounded above in the vertical sectors $|x| < \pm y\tan\delta$. Because $v(x) \leqslant 0$ on E, the *proof* of that same theorem can be adapted without change to show that v is *also* bounded above in $\mathscr{D} \cap \{x > |y|\tan\delta\}$ and $\mathscr{D} \cap \{x < -|y|\tan\delta\}$, even though the latter domains are not full sectors. Therefore v is bounded above in $\mathscr{D}$, so by the *first* theorem of §C, Chapter III, $v(z) \leqslant 0$ in $\mathscr{D}$. Hence $|f(z)| \leqslant \exp(A\sec\delta \cdot Y(z))$, $z \in \mathscr{D}$, and, making $\delta \to 0$, we get $|f(z)| \leqslant \exp(AY(z))$.

Here, $E = (-\infty, -1] \cup [1, \infty)$. In $\mathscr{D} = \mathbb{C} \sim E$, we can put

$$Y(z) = \Im(\sqrt{(z^2 - 1)}),$$

using the branch of the square root which is *positive imaginary* for $z \in (-1, 1)$. It is easy to check that this $Y(z)$ is a Phragmén–Lindelöf function for $\mathscr{D}$.

The Green's function $G(z, w)$ for one of our domains $\mathscr{D}$ enjoys a *symmetry property*:

$$G(z, w) = G(w, z), \qquad z, w \in \mathscr{D}.$$

The reader who does not remember how this is proved may find a proof, general enough to cover our situation, at the end of this article. It is convenient to define $G(z, w)$ for all z and w in $\bar{\mathscr{D}}$ (which *here* is just $\mathbb{C}$!) by taking $G(z, w) = 0$ if *either* z or w belongs to $\partial\mathscr{D}$. Then we have

$$G(z, w) = G(w, z) \qquad \text{for} \quad z, w \in \bar{\mathscr{D}}.$$

(N.B. $G(z, w)$ as thus defined is not *quite* continuous from $\bar{\mathscr{D}} \times \bar{\mathscr{D}}$ to $[0, \infty]$ (*sic*!). We can take sequences $\{z_n\}$ and $\{w_n\}$ of points in $\mathscr{D}$, both tending to limits on $E = \partial\mathscr{D}$, but with $|z_n - w_n| \xrightarrow[n]{} 0$ sufficiently rapidly to make $G(z_n, w_n) \xrightarrow[n]{} \infty$.)

The connection between $G(z, w)$ and $Y(z)$ (when the latter exists) can be made to depend on the elementary formula

$$\lim_{R \to \infty} \int_{-R}^{R} \log\left|1 - \frac{z}{t}\right| dt = \pi |\Im z|,$$

which may be derived by contour integration. The reader is invited (nay, urged!) to do the computation. Here is the result.

Theorem. *A Phragmén–Lindelöf function $Y(z)$ exists for $\mathscr{D}$, a domain of the kind considered here, iff*

$$\int_{-\infty}^{\infty} G(z, t)\,dt < \infty$$

for some $z \in \mathscr{D}$, *G being the Green's function for that domain. If the integral just written converges for* any *such z, it converges for* all $z \in \mathscr{D}$, *and then*

$$Y(z) = |\Im z| + \frac{1}{\pi} \int_{-\infty}^{\infty} G(z, t)\,dt.$$

Remark. In his 1980 *Arkiv* paper, Benedicks has versions of this result for $\mathbb{R}^{n+1}$

Proof of theorem. The idea is very simple, and is expressed by the identity

$$|\Im z| + \frac{1}{\pi}\int_{-\infty}^{\infty} G(z,t)\,dt$$

$$= \frac{1}{\pi}\int_{-A}^{A}\left(\log\frac{1}{|t|} + \log|z-t| + G(z,t)\right)dt$$

$$+ \frac{1}{\pi}\int_{A}^{\infty}\log\left|1-\frac{z^2}{t^2}\right|dt + \frac{1}{\pi}\int_{|t|\geqslant A} G(z,t)\,dt,$$

an obvious consequence of the formula just mentioned. Here, $A > 0$ is arbitrary.

Suppose, indeed, that the left-hand integral is convergent. That integral then equals a positive harmonic function in each of the half planes $\{\Im z > 0\}$, $\{\Im z < 0\}$, and we can use Harnack to show that it is $o(|y|)$ for $z = iy$ and $y \to \pm\infty$. Denoting the left side of our identity by $Y(z)$, we thus see that $Y(z)$ is harmonic in the upper and lower half planes and has property (iii). It is clear that $Y(z)$ has the properties (i) and (ii). Only the harmonicity of $Y(z)$ at points of $\mathscr{D}\cap\mathbb{R}$ remains to be verified. This, however, can be checked in the neighborhood of any such point by taking $A > 0$ sufficiently large and looking at the *right side* of our identity. The *first* right-hand term will be harmonic in $\mathscr{D}\cap(-A, A)$, because, for each t therein, the logarithmic pole of $G(z,t)$ at t is cancelled by the term $\log|z-t|$. The *second* term on the right is *clearly* harmonic for $|z| < A$, and the *third* harmonic in $\mathscr{D}\cap(-A, A)$.

This explanation will probably satisfy the experienced analyst. The general mathematical reader may, however, well desire more justification, based if possible on general principles, so that he or she may avoid having to search through specialized books on potential theory. We proceed to furnish this justification. Its details make the following development somewhat long.

Let us begin with a preliminary remark. Suppose we have any open subset $\mathcal{O}$ of $\mathscr{D}$, and a compact $F \subseteq \bar{\mathscr{D}}$ *disjoint* from $\mathcal{O}$. By the symmetry of G, $G(z, w) = G(w, z)$ is, for each *fixed* $z\in\mathcal{O}$, *continuous* (as a function of w) on F, so, if μ is any finite positive measure on F, the integral

$$\int_F G(z,w)\,d\mu(w)$$

is obtainable as a limit of Riemann sums in the usual way. As a *function* of z, *any one* of those sums is *positive and harmonic in* $\mathcal{O}$. So the *integral*, being a *pointwise limit* of such functions (of z), *is itself* a *positive and harmonic function of* z in $\mathcal{O}$. We will make repeated use of this observation.

Suppose, now, that $\int_{-\infty}^{\infty} G(z,t)\mathrm{d}t < \infty$ for some *non-real* z, say wlog that

$$\int_{-\infty}^{\infty} G(\mathrm{i},t)\mathrm{d}t \;<\; \infty.$$

For each N, the function

$$H_N(z) \;=\; \int_{-N}^{N} G(z,t)\mathrm{d}t$$

is positive and harmonic in both $\{\Im z > 0\}$ and $\{\Im z < 0\}$ according to the remark just made. Therefore, since $G(z,t) \geqslant 0$, $H_{N+1}(z) \geqslant H_N(z)$, and

$$H(z) \;=\; \lim_{N\to\infty} H_N(z)$$

is either *harmonic* (and finite!) in $\{\Im z > 0\}$ or else *everywhere infinite* there. Because $H(\mathrm{i}) < \infty$, the first alternative holds, and $H(z)$ is then *also* finite (and harmonic) in $\Im z < 0$, since obviously

$$G(z,t) \;=\; G(\bar{z},t)$$

for real t, $E \;=\; \partial\mathscr{D}$ being on $\mathbb{R}$.

Consider now some real $x_0 \notin E$. Take $A > \max(|x_0|, 1)$. The integrals $\int_{-A}^{A} G(x_0,t)\mathrm{d}t$ and $\int_{-A}^{A} G(\mathrm{i},t)\mathrm{d}t$ are *both finite*, so we can show that $\int_{-\infty}^{\infty} G(x_0,t)\mathrm{d}t$ and $\int_{-\infty}^{\infty} G(\mathrm{i},t)\mathrm{d}t$ are either both finite or else both infinite by comparing

$$\int_{|t|\geqslant A} G(x_0,t)\,\mathrm{d}t$$

and

$$\int_{|t|\geqslant A} G(\mathrm{i},t)\,\mathrm{d}t.$$

In $\mathscr{D}_A = \mathscr{D} \cap \{|z| < A\}$, the function $\int_{|t|\geqslant A} G(z,t)\mathrm{d}t$ is the limit of the increasing sequence of functions

$$\int_{A\leqslant|t|\leqslant N} G(z,t)\mathrm{d}t,$$

each of which is *positive and harmonic in* $\mathscr{D}_A$. So $\int_{|t|\geqslant A} G(z,t)\mathrm{d}t$ is either *harmonic* (and finite) in $\mathscr{D}_A$, or else everywhere infinite there. It is thus *finite* for $z = \mathrm{i}$ *if and only if* it is finite for $z = x_0$. We see that $\int_{-\infty}^{\infty} G(x_0,t)\mathrm{d}t < \infty$ iff $\int_{-\infty}^{\infty} G(\mathrm{i},t)\mathrm{d}t < \infty$, and, if this inequality holds,

$$H(z) \;=\; \int_{-\infty}^{\infty} G(x,t)\mathrm{d}t$$

is finite for every $z \in \mathscr{D}$.

If $H(z)$ is finite, let us show that

$$H(\mathrm{i}y) = \mathrm{o}(|y|) \quad \text{for} \quad y \to \pm\infty.$$

Pick any large N, and write

$$(*) \qquad H(\mathrm{i}y) = \int_{|t|\leqslant N} G(\mathrm{i}y, t)\,dt + \int_{|t|>N} G(\mathrm{i}y, t)\,dt.$$

Since $H(\mathrm{i}) < \infty$ we can, given any $\varepsilon > 0$, take N so large that $\int_{|t|\geqslant N} G(\mathrm{i}, t)\,dt < \varepsilon$. The function $\int_{|t|\geqslant N} G(z, t)\,dt$ is, by the previous discussion, *positive and harmonic* in $\{\Im z > 0\}$. Therefore, by Harnack's theorem,*

$$\int_{|t|\geqslant N} G(\mathrm{i}y, t)\,dt \leqslant y\int_{|t|\geqslant N} G(\mathrm{i}, t)\,dt < \varepsilon y$$

for $y > 1$. This takes care of the *second* term on the right in $(*)$.

The *first* term from the right side of $(*)$ remains; our claim is that it is *bounded*. This (and more) follows from a simple estimate which will be used several times in the proof.

Take any component E_0 of $E = \partial\mathscr{D}$, and put $\mathscr{D}_0 = (\mathbb{C} \sim E_0) \cup \{\infty\}$. If E_0 is of *infinite length*, replace it *by any segment of length* 2 *thereon* in this last expression. We have $\mathscr{D} \subseteq \mathscr{D}_0$, so, if $g(z, w)$ is the Green's function for $\mathscr{D}_0$,

$$G(z, w) \leqslant g(z, w), \qquad z, w \in \mathscr{D}$$

(cf. beginning of the proof of *first* theorem in this article). We compute $g(z, w)$ by first mapping $\mathscr{D}_0$ conformally onto the unit disk $\{|\zeta| < 1\}$, thinking of ζ as a new *coordinate variable* for $\mathscr{D}_0$:

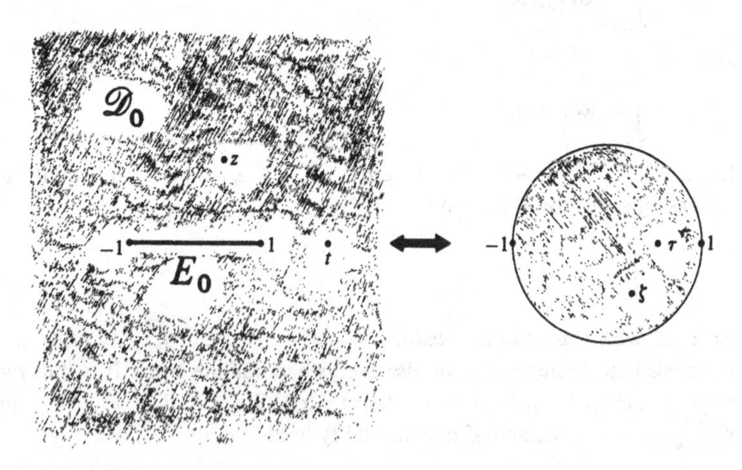

Figure 121

Say, for instance, that E_0 is $[-1, 1]$ so that we can use $z = \frac{1}{2}(\zeta + 1/\zeta)$. Then, if $t \in \mathscr{D}_0$ is

* Actually, by the Poisson representation for $\{\Im z > 0\}$ of functions positive and harmonic there. Using the ordinary form of Harnack's inequality gives us a factor of $2y$ instead of y on the right. That, of course, makes no difference in this discussion.

real, we can put $t = \frac{1}{2}(\tau + 1/\tau)$, where $-1 < \tau < 1$, and, in terms of ζ and τ,

$$g(z, t) = \log\left|\frac{1 - \tau\zeta}{\zeta - \tau}\right|,$$

the expression on the right being simply the Green's function for the unit disk.

If $N > 1$, we have

$$\int_{|t| \leqslant N} G(z, t)\,\mathrm{d}t \leqslant \int_{1 \leqslant |t| \leqslant N} g(z, t)\,\mathrm{d}t,$$

since $G(z, t)$ and $g(z, t)$ *vanish* for $t \in E_0 = [-1, 1]$. For $1 \leqslant |t| \leqslant N$, the parameter τ satisfies $C_N \leqslant |\tau| \leqslant 1$, $C_N > 0$ being a number depending on N which we need not calculate. Also, for such t,

$$\mathrm{d}t = -\frac{1}{2}\left(\frac{1 - \tau^2}{\tau^2}\right)\mathrm{d}\tau.$$

Therefore,

$$(\dagger) \qquad \int_{|t| \leqslant N} G(z, t)\,\mathrm{d}t \leqslant \frac{1}{2}\int_{C_N \leqslant |\tau| \leqslant 1} \log\left|\frac{1 - \tau\zeta}{\zeta - \tau}\right|\left(\frac{1 - \tau^2}{\tau^2}\right)\mathrm{d}\tau.$$

Since $C_N > 0$, the right side is clearly bounded for $|\zeta| < 1$; we see already that the *first right-hand term of* $(*)$ *is bounded*, verifying our claim.

As we have already shown, the *second* term on the right in $(*)$ will be $\leqslant \varepsilon y$ for $y \geqslant 1$ if N is large enough. Combining this result with the preceding, we have, from $(*)$,

$$H(\mathrm{i}y) \leqslant O(1) + \varepsilon y, \quad y > 1,$$

so, since $\varepsilon > 0$ is arbitrary, $H(\mathrm{i}y) = o(|y|)$, $y \to \infty$. Because $H(\bar{z}) = H(z)$, the same holds good for $y \to -\infty$.

Having established this fact, let us return for a moment to $(\dagger)$. For each τ, $C_N \leqslant |\tau| < 1$,

$$\log\left|\frac{1 - \tau\zeta}{\zeta - \tau}\right| \longrightarrow 0 \quad \text{as} \quad |\zeta| \longrightarrow 1.$$

Starting from this relation, one can, by a straightforward argument, check that

$$\int_{C_N \leqslant |\tau| \leqslant 1} \log\left|\frac{1 - \tau\zeta}{\zeta - \tau}\right|\left(\frac{1 - \tau^2}{\tau^2}\right)\mathrm{d}\tau \longrightarrow 0$$

as $|\zeta| \to 1$. (One may, for instance, break up the integral into two pieces.)

Problem 17 (a)

Carry out this verification

This means, by $(\dagger)$, that

$$\int_{|t| \leqslant N} G(z, t)\,\mathrm{d}t \longrightarrow 0$$

when $z \in \mathscr{D}$ *tends to any point of* E_0. We could, however, have taken E_0 to be *any of the components of* E with finite length, or any segment of length 2 on one of the unbounded ones (if there are any); that would not have essentially changed the above argument.* *Hence*

$$\int_{|t| \leqslant N} G(z,t)\,dt$$

tends to zero whenever z *tends to any point on* $E = \partial\mathscr{D}$ (besides being *bounded* in $\mathscr{D}$).

We can now prove that

$$H(z) \;=\; \int_{-\infty}^{\infty} G(z,t)\,dt \;\longrightarrow\; 0$$

whenever z tends to any point x_0 of E. *Given* such an x_0, take a circle γ about x_0 so small that *precisely one* of the components of E (the one containing x_0) cuts γ, passing into its inside:

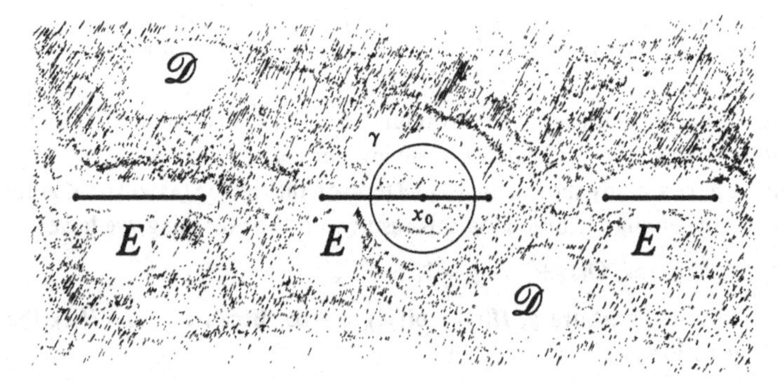

Figure 122

If x_0 is an endpoint of one of the components of E, our picture looks like this:

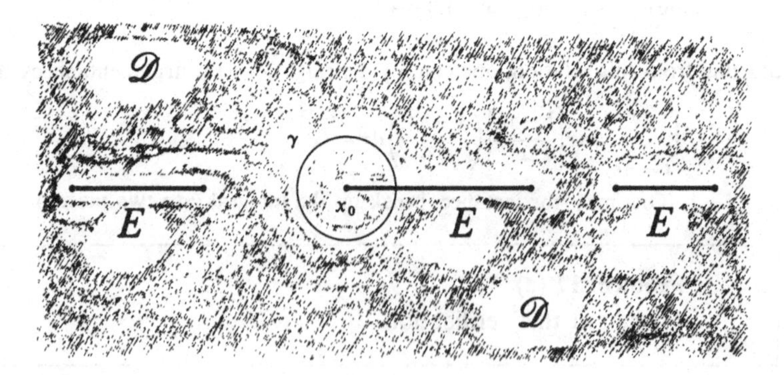

Figure 123

Call $\mathscr{D}_\gamma$ the part of $\mathscr{D}$ lying inside γ, and E_γ the part of E therein.

* As long as $E_0 \subseteq \{|t| < N\}$. If this is not so, we can *increase* N until the argument in the text applies. Since that only makes the integral in question *larger*, the one corresponding to the original value of N must (*a fortiori*!) have the asserted

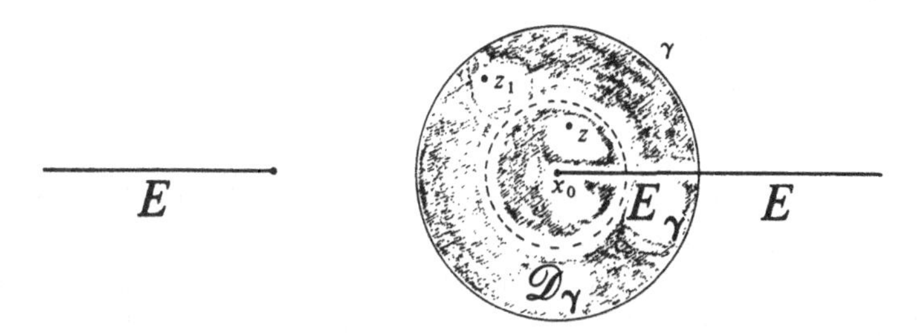

Figure 124

Fix any $z_1 \in \mathscr{D}_\gamma$. Then, there is a constant K, depending on z_1, such that, for any function $V(z)$, *positive* and *harmonic* in $\mathscr{D}_\gamma$ and continuous on its closure, *with* $V(x) = 0$ *on* E_γ, we have $V(z) \leqslant KV(z_1)$ for $|z - x_0| < \frac{1}{2}$ radius of γ.

Problem 17 (b)
Prove the statement just made.

This being granted, choose N so large that

$$\int_{|t| \geqslant N} G(z_1, t)\,\mathrm{d}t \;<\; \varepsilon/K,$$

ε being any number > 0. For each $M > N$, the function

$$V_M(z) \;=\; \int_{N \leqslant |t| \leqslant M} G(z, t)\,\mathrm{d}t$$

is positive and harmonic in $\mathscr{D}_\gamma$, and certainly continuous up to $\gamma \cap \mathscr{D}$. Also, $V_M(z) \leqslant \int_{|t| \leqslant M} G(z, t)$ which, *by the previous discussion*, tends to *zero* whenever z tends to any point of E. $V_M(z)$ is therefore continuous up to E_γ, *where it equals zero*. By the above statement, we thus have

$$V_M(z) \;\leqslant\; KV_M(z_1) \;\leqslant\; K\int_{|t| \geqslant N} G(z_1, t)\,\mathrm{d}t \;<\; \varepsilon$$

for $|z - x_0| < \frac{1}{2}$ radius of γ. This holds for all $M > N$, so making $M \to \infty$, we get $\int_{|t| \geqslant N} G(z, t)\,\mathrm{d}t \leqslant \varepsilon$ for $|z - x_0| < \frac{1}{2}$ radius of γ. Hence, since

$$H(z) \;=\; \int_{|t| \leqslant N} G(z, t)\,\mathrm{d}t \;+\; \int_{|t| \geqslant N} G(z, t)\,\mathrm{d}t,$$

and, as we already know, the first integral on the right tends to zero when $z \to x_0$, we must have $H(z) < 2\varepsilon$ for z close enough to x_0. This shows that $H(z) \to 0$ whenever z tends to any point of E.

We now see by the preceding arguments *that*

$$Y(z) \;=\; |\Im z| \;+\; \frac{1}{\pi}H(z)$$

enjoys the properties (i), (ii) *and* (iii) *required of Phragmén–Lindelöf functions, and is also harmonic in both the lower and upper half planes, and continuous everywhere.* Therefore, to complete the proof of the fact that $Y(z)$ *is* a Phragmén–Lindelöf function for $\mathscr{D}$, *we need only verify that it is harmonic at the points of* $\mathscr{D}\cap\mathbb{R}$.

For this purpose, we bring in the formula

$$|\Im z| \;=\; \lim_{A\to\infty}\frac{1}{\pi}\int_{-A}^{A}\log\left|1 \;-\; \frac{z}{t}\right|\mathrm{d}t$$

mentioned earlier. From it, and the definition of $H(z)$, we get

$$(\ast\!\ast)\qquad Y(z) \;=\; |\Im z| \;+\; \frac{1}{\pi}\int_{-\infty}^{\infty}G(z,t)\,\mathrm{d}t$$

$$= \frac{1}{\pi}\int_{-A}^{A}\left(\log\frac{1}{|t|} + \log|z-t| + G(z,t)\right)\mathrm{d}t$$

$$+\; \frac{1}{\pi}\int_{A}^{\infty}\log\left|1 - \frac{z^2}{t^2}\right|\mathrm{d}t \;+\; \frac{1}{\pi}\int_{|t|\geqslant A}G(z,t)\,\mathrm{d}t.$$

The number $A > 0$ may be chosen at pleasure.

Let $x_0\in\mathscr{D}\cap\mathbb{R}$; pick A larger than $|x_0|$. The function $\int_A^\infty\log|1-z^2/t^2|\mathrm{d}t$ is certainly harmonic near x_0; we have also seen previously that $\int_{|t|\geqslant A}G(z,t)\,\mathrm{d}t$ is harmonic in $\mathscr{D}\cap\{|z|<A\}$, so, in particular, at x_0. Again, $\int_{-A}^{A}\log(1/|t|)\mathrm{d}t$ is *finite.* Our task thus boils down to showing harmonicity of

$$\int_{-A}^{A}(\log|z-t| + G(z,t))\,\mathrm{d}t$$

at x_0.

Take a $\delta > 0$ such that $(x_0 - 5\delta,\; x_0 + 5\delta)\subseteq\mathscr{D}$. According to observations already made,

$$\int_{\substack{|t-x_0|>\delta\\ |t|<A}} G(z,t)\,\mathrm{d}t$$

is harmonic for $|z - x_0| < \delta$; so is (clearly)

$$\int_{\substack{|t-x_0|>\delta\\ |t|<A}} \log|z-t|\,\mathrm{d}t.$$

We therefore need only check the harmonicity of

$$\int_{x_0-\delta}^{x_0+\delta}(\log|z-t| + G(z,t))\,\mathrm{d}t.$$

Here, we must use the symmetry of $G(z, w)$. In order not to get bogged down in notation, *let us assume that* $x_0 = \alpha > 0$ *and that the segment* $[-2\alpha, 6\alpha]$ *lies entirely in* $\mathscr{D}$. The general situation can always be *reduced* to this one by *suitable translation*. It will be enough to show that

$$\int_0^{2\alpha} (\log|z-t| + G(z,t))\,dt$$

is harmonic for $|z-\alpha| < \alpha$.

For each *fixed* z,

$$\log|z-w| + G(z,w) \;=\; \log|w-z| + G(w,z)$$

is a certain *harmonic function*, $h_z(w)$, of $w \in \mathscr{D}$; this is where the symmetry of G comes in. ($h_z(w)$ is harmonic in w *even at the point* z, for addition of the term $\log|w-z|$ *removes* the logarithmic singularity of $G(w,z)$ there.) Hence, if $\rho < \operatorname{dist}(w, E)$,

$$h_z(w) \;=\; \frac{1}{2\pi}\int_0^{2\pi} h_z(w + \rho e^{i\vartheta})\,d\vartheta.$$

This relation makes a *trick* available. In it, put $w = t$ where $0 < t < 2\alpha$, and use $\rho = t + 2\alpha$.

We get

$$h_z(t) \;=\; \frac{1}{2\pi}\int_0^{2\pi} h_z(t + (t+2\alpha)e^{i\vartheta})\,d\vartheta,$$

whence,

$$\int_0^{2\alpha} (\log|z-t| + G(z,t))\,dt \;=\; \frac{1}{2\pi}\int_0^{2\alpha}\int_0^{2\pi} h_z(t + (t+2\alpha)e^{i\vartheta})\,d\vartheta\,dt.$$

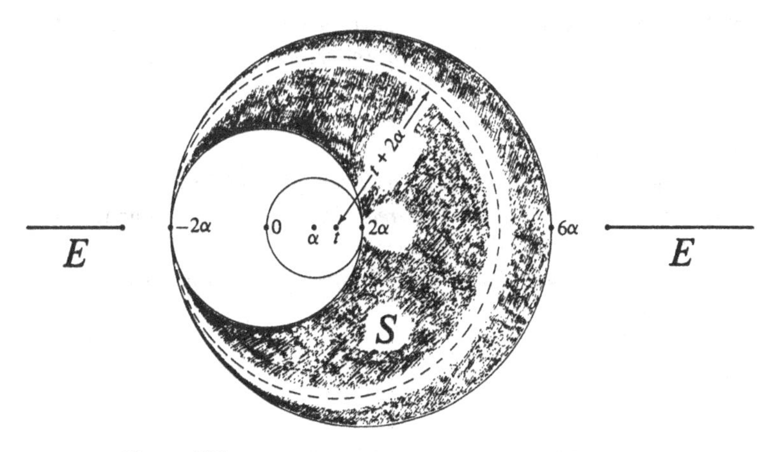

Figure 125

The double integral on the right can be expressed as one over the region

$$S = \{|\zeta - 2\alpha| < 4\alpha\} \cap \{|\zeta| > 2\alpha\}$$

shown in the above picture. Indeed, the mapping

$$(t, \vartheta) \longrightarrow (\xi, \eta)$$

given by $\xi + i\eta = \zeta = t + (t+2\alpha)e^{i\vartheta}$ takes $\{0 < t < 2\alpha\} \times \{0 < \vartheta < 2\pi\}$ in one–one fashion onto S, and the Jacobian

$$\frac{\partial(\xi, \eta)}{\partial(t, \vartheta)}$$

works out to $(t + 2\alpha)(1 + \cos\vartheta) = \xi + 2\alpha$.
Hence,

$$\frac{1}{2\pi}\int_0^{2\alpha}\int_0^{2\pi} h_z(t + (t+2\alpha)e^{i\vartheta})\,d\vartheta\,dt \;=\; \frac{1}{2\pi}\iint_S h_z(\zeta)\frac{d\xi\,d\eta}{\xi + 2\alpha},$$

so, by the previous relation,

$$\int_0^{2\alpha} (\log|z - t| + G(z, t))\,dt$$

$$= \frac{1}{2\pi}\iint_S \frac{\log|z - \zeta|}{\xi + 2\alpha}\,d\xi\,d\eta \;+\; \frac{1}{2\pi}\iint_S \frac{G(z, \zeta)}{\xi + 2\alpha}\,d\xi\,d\eta.$$

Here, we have

$$\iint_S \frac{d\xi\,d\eta}{\xi + 2\alpha} \;=\; \int_0^{2\alpha}\int_0^{2\pi} d\vartheta\,dt \;<\; \infty,$$

so both of the above double integrals must equal *harmonic functions of z in the disk* $\{|z - \alpha| < \alpha\}$, *disjoint from* $\bar{S}$. (This follows for the *second* of those integrals by the remark at the very beginning of this proof.) We see that the left-hand expression is *harmonic in z for z near* $x_0 = \alpha$. According, then, to (*) and the observations immediately following, *the same is true for* $Y(z)$.

We have finished proving that *$Y(z)$ is a Phragmén–Lindelöf function for $\mathscr{D}$ if the integral $\int_{-\infty}^{\infty} G(z, t)\,dt$ is finite for any z therein.* The *second half* of our theorem thus remains to be established. That is easier.

In the second half, *we assume that $\mathscr{D}$ has a Phragmén–Lindelöf function $Y(z)$*, and set out to show that

$$\int_{-\infty}^{\infty} G(z, t)\,dt \;<\; \infty$$

for *each* $z \in \mathscr{D}$, G being that domain's Green's function.

Given any $A > 0$, consider the expression

$$Y_A(z) = |\Im z| + \frac{1}{\pi}\int_{-A}^{A} G(z,t)\,dt$$
$$= \frac{1}{\pi}\int_{-A}^{A}\left(\log\frac{1}{|t|} + \log|z-t| + G(z,t)\right)dt$$
$$+ \frac{1}{\pi}\int_{A}^{\infty}\log\left|1-\frac{z^2}{t^2}\right|dt.$$

From the preceding arguments, we know that the *first integral on the right* is *harmonic* for $z\in\mathscr{D}$ – proof of this fact *did not depend on the convergence of*

$$\int_{-\infty}^{\infty} G(z,t)\,dt.$$

What we have already done also tells us that $Y_A(z)$ *tends to zero when z tends to any point of E* (again, *whether* $\int_{-\infty}^{\infty} G(z,t)\,dt$ *converges or not*) and that, for any fixed A,

$$\int_{-A}^{A} G(z,t)\,dt$$

is bounded in the complex plane. The expression

$$\int_{A}^{\infty}\log\left|1-\frac{z^2}{t^2}\right|dt$$

is evidently *subharmonic* in the complex plane.

The function $Y_A(z)$ given by the above formula is thus *subharmonic*, and *zero on E*, and moreover,

$$Y_A(z) = |\Im z| + O(1), \quad z\in\mathscr{D}.$$

Our Phragmén–Lindelöf function $Y(z)$ (presumed to exist!) is, *however, harmonic and* $\geqslant|\Im z|$ *in* $\mathscr{D}$, *and zero on E*. The difference $Y_A(z) - Y(z)$ is therefore *subharmonic and bounded above in* $\mathscr{D}$, *and zero on E*. We can conclude by the *extended maximum principle* (subharmonic version of *first* theorem in § C, Chapter III) that $Y_A(z) - Y(z) \leqslant 0$ for $z\in\mathscr{D}$. In other words,

$$|\Im z| + \frac{1}{\pi}\int_{-A}^{A} G(z,t)\,dt \leqslant Y(z).$$

Fixing $z\in\mathscr{D}$ and then making $A\to\infty$, we see that

$$\frac{1}{\pi}\int_{-\infty}^{\infty} G(z,t)\,dt \leqslant Y(z) - |\Im z| < \infty.$$

This is what we wanted. The *second half of the theorem is proved.*

We are done.

We apply the result just proved to domains $\mathscr{D}$ of the special form described at the beginning of the present §, using the *first* theorem of this article. In that way we obtain the important

Theorem (Benedicks). *If E is a union of segments on $\mathbb{R}$ fulfilling the conditions given at the beginning of this § (involving the four constants A, B, δ and Δ), there is a Phragmén–Lindelöf function for the domain $\mathscr{D} = \mathbb{C} \sim E$.*

Proof. Assume wlog that $0 \in \mathscr{D}$, and call $\mathscr{O}_0$ the component of $\mathbb{R} \sim E$ containing 0. By the first theorem of the present article,

$$G(t, 0) \leqslant \frac{C}{1+t^2}$$

for $t \in \mathscr{O}_0$, and clearly

$$G(t, 0) \leqslant \log^+ \frac{1}{|t|} + O(1), \qquad t \in \mathscr{O}_0.$$

Therefore (symmetry again!)

$$\int_{-\infty}^{\infty} G(0, t)\,dt = \int_{-\infty}^{\infty} G(t, 0)\,dt < \infty.$$

Now refer to the preceding theorem.

We are done.

This result will be applied to the study of weighted approximation on sets E in the next article. We cannot, however, end *this* one without keeping our promise about proving symmetry of the Green's function. So, here we go:

Theorem. *In $\mathscr{D} = \mathbb{C} \sim E$,*

$$G(z, w) = G(w, z).$$

Proof. Let us first treat the case where *E consists of a finite number of intervals, of finite or infinite length.* (If E contains *two* semi-infinite intervals at opposite ends of $\mathbb{R}$, we consider them as forming *one* interval passing through ∞.)

We first proceed as at the beginning of article 1, *and map $\mathscr{D}$* (or $\mathscr{D} \cup \{\infty\}$, if $\infty \notin E$) *conformally onto a bounded domain, bounded by a finite number of analytic Jordan curves.* This useful trick simplifies a lot of work; let us describe (in somewhat more detail than at the beginning of article 1) how it is done.

Suppose that $E_1, E_2, \ldots, E_N$ are the components of E. First map $(\mathbb{C} \cup \{\infty\}) \sim E_1$ conformally onto the disk $\{|z| < 1\}$; in this mapping, E_1 (which gets split down its middle, with its two edges spread apart) goes onto $\{|z| = 1\}$, and $E_2, \ldots, E_N$ are taken onto *analytic* Jordan arcs, $A_2, \ldots, A_N$ respectively, lying inside the unit disk. (Actually, in our situation, where the E_k lie on $\mathbb{R}$, we can choose the mapping of $(\mathbb{C} \cup \{\infty\}) \sim E_1$ onto $\{|z| < 1\}$ so that $\mathbb{R} \sim E_1$ is taken onto $(-1, 1)$. Then $A_2, \ldots, A_N$ will be *segments* on $(-1, 1)$.) In this fashion, $\mathscr{D}$ is mapped conformally onto

$$\{|z| < 1\} \sim A_2 \sim A_3 \sim \cdots \sim A_N.$$

Now map $(\mathbb{C} \cup \{\infty\}) \sim A_2$ conformally onto $\{|w| < 1\}$. In this transformation, $\{|z| = 1\}$ goes onto a certain analytic Jordan curve $\mathscr{C}_1$ lying inside the unit disk, A_2 (after having its two sides spread apart) goes onto $\{|w| = 1\}$, and, if $N > 2$, the arcs $A_3, \ldots, A_N$ go onto other analytic arcs $A'_3, \ldots, A'_N$, lying inside $\{|w| < 1\}$. ($A'_3, \ldots, A'_N$ are indeed *segments* on $(-1, 1)$ in our present situation, if this second conformal mapping is properly chosen.) So far, composition of our two mappings yields a conformal transformation of $\mathscr{D}$ onto the region lying in $\{|w| < 1\}$, bounded by the unit circumference, the analytic Jordan curve $\mathscr{C}_1$, and the analytic Jordan arcs $A'_3, \ldots, A'_N$ (in the case where $N > 2$).

It is evident how one may continue this process when $N > 2$. Do the same thing with A'_3 that was done with A_2, and so forth, until all the boundary components are used up. The final result is a conformal mapping of $\mathscr{D}$ onto a region bounded by the *unit circumference* and $N - 1$ *analytic Jordan curves situated within it.*

Under conformal mapping, Green's functions correspond to Green's functions. Therefore, in order to prove that $G(z, w) = G(w, z)$, we *may as well assume that G is the Green's function for a bounded domain* Ω like the one arrived at by the process just described, i.e., with $\partial\Omega$ *consisting of a finite number of analytic Jordan curves.* For such domains Ω we can establish symmetry using methods going back to Green himself. (Green's original proof – the result *is* due to him, by the way – is a little different from the one we are about to give. Adapted to two dimensions, it amounts to the observation that

$$\begin{aligned} G(z, w) &= \log\frac{1}{|z - w|} + \int_{\partial\Omega} \log|\zeta - w| \, d\omega_\Omega(\zeta, z) \\ &= \log\frac{1}{|z - w|} + \int_{\partial\Omega}\int_{\partial\Omega} \log|\zeta - \sigma| \, d\omega_\Omega(\sigma, w) \, d\omega_\Omega(\zeta, z), \end{aligned}$$

where $\omega_\Omega(\ \ , z)$ is the harmonic measure for Ω. This argument can easily be made rigorous for our domains Ω. The interested reader may want to consult Green's collected papers, reprinted by Chelsea in 1970.)

If $\zeta \in \partial\Omega$ and the function F is $\mathscr{C}_1$ in a *neighborhood* of ζ, we denote by

$$\frac{\partial F(\zeta)}{\partial n_\zeta}$$

the *directional derivative of F in the direction of the unit outward normal* $\mathbf{n}_\zeta$ to $\partial\Omega$ at ζ:

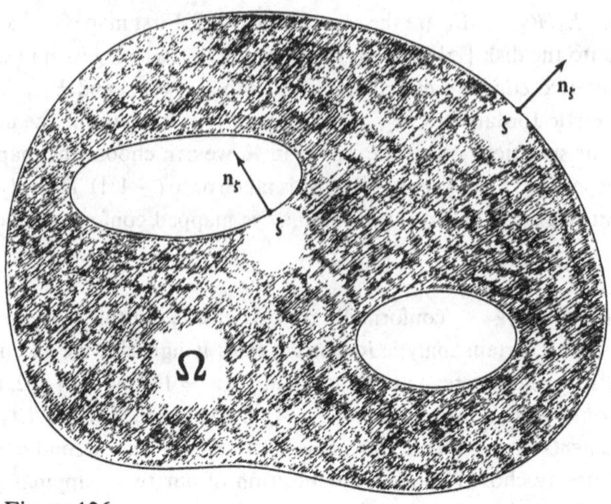

Figure 126

If $w\in\Omega$ is *fixed*, $G(z, w)$ is harmonic as a function of $z\in\Omega$ (for z away from w) and continuous up to $\partial\Omega$, where it equals *zero*. *Analyticity* of the components of $\partial\Omega$ means that, given any $\zeta_0 \in \partial\Omega$, we can find a *conformal mapping* of a small disk centered at ζ_0 which takes the *part of* $\partial\Omega$ *lying in that disk* to a *segment* σ *on the real axis.* If we compose $G(z, w)$ with this conformal mapping for $z\in\Omega$ near ζ_0, we see, by *Schwarz' reflection principle*, that the *composed* function is actually *harmonic in a neighborhood of* σ, and thence that $G(z, w)$ is *harmonic* (in z) *in a neighborhood of* ζ_0. $G(z, w)$ is, in particular, *a* $\mathscr{C}_\infty$ *function of z in the neighborhood of every point on* $\partial\Omega$.

This regularity, together with the smoothness of the components of $\partial\Omega$, makes it possible for us to apply *Green's theorem.* Given z and $w\in\Omega$ with $z \neq w$, take two small non-intersecting circles γ_z and γ_w lying in Ω, about z and w respectively. Call Ω' the domain obtained from Ω by *removing* therefrom the small disks bounded by γ_z and γ_w:

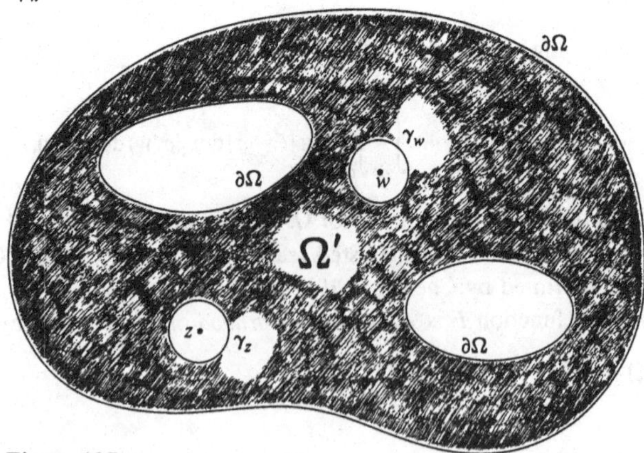

Figure 127

Denote by **grad** the *vector gradient* with respect to (ξ, η), where $\zeta = \xi + i\eta$, and by '$\cdot$' the *dot product* in $\mathbb{R}^2$. We have

$$\int_{\partial\Omega'} \left(G(\zeta, w) \frac{\partial G(\zeta, z)}{\partial n_\zeta} - G(\zeta, z) \frac{\partial G(\zeta, w)}{\partial n_\zeta} \right) |d\zeta|$$

$$= \int_{\partial\Omega'} (G(\zeta, w)\, \mathbf{grad}\, G(\zeta, z) - G(\zeta, z)\, \mathbf{grad}\, G(\zeta, w)) \cdot \mathbf{n}_\zeta |d\zeta|.$$

Since the vector-valued function

$$G(\zeta, w)\, \mathbf{grad}\, G(\zeta, z) - G(\zeta, z)\, \mathbf{grad}\, G(\zeta, w)$$

of ζ is $\mathscr{C}_\infty$ in and on $\bar{\Omega}'$, ($\mathscr{C}_\infty$ on $\partial\Omega$ by what was said above), we can apply *Green's theorem* to the second of these integrals, and find that it equals

$$\iint_{\Omega'} \operatorname{div} (G(\zeta, w)\, \mathbf{grad}\, G(\zeta, z) - G(\zeta, z)\, \mathbf{grad}\, G(\zeta, w))\, d\xi\, d\eta,$$

where div denotes *divergence* with respect to (ξ, η). However, by *Green's identity*,

$$\operatorname{div} (G(\zeta, w)\, \mathbf{grad}\, G(\zeta, z) - G(\zeta, z)\, \mathbf{grad}\, G(\zeta, w))$$
$$= G(\zeta, w)\nabla^2 G(\zeta, z) - G(\zeta, z)\nabla^2 G(\zeta, w),$$

where $\nabla^2 = \partial^2/\partial\xi^2 + \partial^2/\partial\eta^2$. Because $z \notin \Omega'$ and $w \notin \Omega'$, $G(\zeta, z)$ and $G(\zeta, w)$ are *harmonic* in ζ, $\zeta \in \Omega'$. Hence

$$\nabla^2 G(\zeta, z) = \nabla^2 G(\zeta, w) = 0, \qquad \zeta \in \Omega',$$

and the above double integral vanishes identically. Therefore the *first* of the above *line integrals* around $\partial\Omega'$ must be *zero*.

Now $\partial\Omega' = \partial\Omega \cup \gamma_z \cup \gamma_w$, and $G(\zeta, w) = G(\zeta, z) = 0$ for $\zeta \in \partial\Omega$. That line integral therefore reduces to

$$\left\{ \int_{\gamma_z} + \int_{\gamma_w} \right\} \left(G(\zeta, w) \frac{\partial G(\zeta, z)}{\partial n_\zeta} - G(\zeta, z) \frac{\partial G(\zeta, w)}{\partial n_\zeta} \right) |d\zeta|,$$

which must thus *vanish*. Near z, $G(\zeta, z)$ equals $\log(1/|\zeta - z|)$ plus a harmonic function of ζ; with this in mind we see that the integral around γ_z is *very nearly* $2\pi G(z, w)$ if the radius of γ_z is small. The integral around γ_w is seen in the same way to be very nearly equal to $-2\pi G(w, z)$ when that circle has small radius, so, making the radii of both γ_z and γ_w tend to *zero*, we find in the limit that

$$2\pi G(z, w) - 2\pi G(w, z) = 0,$$

i.e., $G(z, w) = G(w, z)$ for $z, w \in \Omega$. This same symmetry must then hold for the Green's functions belonging to *finitely connected domains* $\mathscr{D}$ of the kind we are considering. **How much must we admire George Green, self taught, who did such beautiful work isolated in provincial England at the beginning of the nineteenth century. One wonders what he might have done had he lived longer than he did.**

AN ESSAY

ON THE

APPLICATION

OF

MATHEMATICAL ANALYSIS TO THE THEORIES OF ELECTRICITY AND MAGNETISM.

BY

GEORGE GREEN.

Nottingham:

PRINTED FOR THE AUTHOR, BY T. WHEELHOUSE.

SOLD BY HAMILTON, ADAMS & Co. 33, PATERNOSTER ROW; LONGMAN & Co.; AND W. JOY, LONDON; J. DEIGHTON, CAMBRIDGE;

AND S. BENNETT, H. BARNETT, AND W. DEARDEN, NOTTINGHAM.

1828.

Once the symmetry of Green's function for *finitely connected domains* $\mathscr{D}$ *is known*, we can establish that property *in the general case* by a limiting argument. By a slight modification of the following procedure, one can actually *prove existence* of the Green's function for infinitely connected domains $\mathscr{D} = \mathbb{C} \sim E$ of the kind being considered here, and the reader is invited to see how such a proof would go. Let us, however, content ourselves with what we set out to do.

Put $E_R = E \cap [-R, R]$ and take $\mathscr{D}_R = (\mathbb{C} \cup \{\infty\}) \sim E_R$. With our sets E, E_R consists of a finite number of intervals, so $\mathscr{D}_R$ is finitely connected, and, by what we have just shown, $G_R(z, w) = G_R(w, z)$ for the Green's function G_R belonging to $\mathscr{D}_R$. (Provided, of course, that R is large enough to make $|E_R| > 0$, so that $\mathscr{D}_R$ *has* a Green's function! *This we henceforth assume.*) We have $\mathscr{D}_R \supseteq \mathscr{D}$, whence, for $z, w \in \mathscr{D}$,

$$G(z, w) \leqslant G_R(z, w).$$

If we can show that

$$G_R(z, w) \longrightarrow G(z, w)$$

for z, $w \in \mathscr{D}$ as $R \to \infty$, we will obviously have $G(z, w) = G(w, z)$.

To verify this convergence, observe that

$$G_{R'}(z, w) \leqslant G_R(z, w)$$

for z and w in $\mathscr{D}_{R'}$ (hence *certainly* for $z, w \in \mathscr{D}$!) when $R' \geqslant R$, because then $\mathscr{D}_{R'} \subseteq \mathscr{D}_R$. The limit

$$\tilde{G}(z, w) = \lim_{R \to \infty} G_R(z, w)$$

thus *certainly exists* for $z, w \in \mathscr{D}$, and is $\geqslant 0$. If we can prove that $\tilde{G}(z, w) = G(z, w)$, we will be done.

Fix any $w \in \mathscr{D}$. Outside any small circle about w lying in $\mathscr{D}$, $\tilde{G}(z, w)$ is the limit of a decreasing sequence of positive harmonic functions of z, and is therefore itself harmonic in that variable. Let $x_0 \in E$. Take $R > |x_0|$; then, since $0 \leqslant \tilde{G}(z, w) \leqslant G_R(z, w)$ for $z \in \mathscr{D}$ and $G_R(z, w) \longrightarrow 0$ as $z \to x_0$, we have

$$\tilde{G}(z, w) \longrightarrow 0 \qquad \text{for } z \longrightarrow x_0.$$

If we fix any large R, we have, for $z \in \mathscr{D}$,

$$0 \leqslant \tilde{G}(z, w) \leqslant G_R(z, w) = \log \frac{1}{|z - w|} + O(1).$$

Therefore, since

$$G(z, w) = \log \frac{1}{|z - w|} + O(1),$$

we have

$$\tilde{G}(z, w) \leqslant G(z, w) + O(1), \qquad z \in \mathscr{D}.$$

However, this last inequality can be *turned around*. Indeed, for $z \in \mathscr{D}$ and *every* sufficiently large R,

$$G_R(z, w) \geqslant G(z, w),$$

from which we get

$$\tilde{G}(z, w) \geqslant G(z, w), \qquad z \in \mathscr{D}$$

on making $R \to \infty$.

We see finally that $0 \leqslant \tilde{G}(z, w) - G(z, w) \leqslant O(1)$ for $z \in \mathscr{D}$ (at least when $z \neq w$); the difference in question is, moreover, *harmonic in* z (for $z \in \mathscr{D}$, $z \neq w$) and *tends*, according to what we have shown above, to *zero when* z *tends to any point of* $E = \partial\mathscr{D}$. Hence

$$\tilde{G}(z, w) - G(z, w) = 0, \qquad z \in \mathscr{D},$$

i.e.,

$$G_R(z, w) \longrightarrow G(z, w)$$

for $z \in \mathscr{D}$ when $R \to \infty$, which is what we needed to establish the symmetry of $G(z, w)$.
We are done.

3. Weighted approximation on the sets E

Let E be a closed set on $\mathbb{R}$, having infinite extent in both directions and consisting of (at most) countably many closed intervals not accumulating at any finite point. Suppose that we are given a function $W(x) \geqslant 1$, defined and continuous on E, such that $W(x) \to \infty$ for $x \to \pm\infty$ in E. Then, in analogy with Chapter VI, we make the

Definition. $\mathscr{C}_W(E)$ is the set of functions φ defined and continuous on E, such that

$$\frac{\varphi(x)}{W(x)} \longrightarrow 0 \quad \text{for} \quad x \longrightarrow \pm\infty \textit{ in } E.$$

And we put

$$\|\varphi\|_{W,E} = \sup_{x\in E}\left|\frac{\varphi(x)}{W(x)}\right|$$

for $\varphi \in \mathscr{C}_W(E)$.

For $A > 0$, we denote by $\mathscr{C}_W(A, E)$ the $\|\ \ \|_{W,E}$-closure in $\mathscr{C}_W(E)$ of the collection of finite sums of the form

$$\sum_{-A\leqslant\lambda\leqslant A} C_\lambda e^{i\lambda x}.$$

Also, if, for every $n > 0$,

$$\frac{x^n}{W(x)} \longrightarrow 0 \quad \text{as} \quad x \longrightarrow \pm\infty \text{ in } E,$$

we denote by $\mathscr{C}_W(0, E)$ the $\|\ \ \|_{W,E}$-closure in $\mathscr{C}_W(E)$ of the set of *polynomials*.

We are interested in obtaining criteria for equality of the $\mathscr{C}_W(A, E)$, $A > 0$, (and of $\mathscr{C}_W(0, E)$) with $\mathscr{C}_W(E)$. One *can*, of course, reduce our present situation to the one considered in Chapter VI by putting $W(x) \equiv \infty$ on $\mathbb{R} \sim E$ and working with the space $\mathscr{C}_W(\mathbb{R})$. The equality in question is then governed by Akhiezer's theorems found in §§B and E of Chapter VI, according to the remark in §B.1 of that chapter (see also the corollary at the end of §E.2 therein). In this way, one arrives at results in which the set E does not figure explicitly. Our aim, however, already mentioned at the beginning of the present chapter, is to show how the *form*

$$\int_{-\infty}^{\infty} \frac{\log W_*(x)}{1+x^2}\,dx,$$

occurring in Akhiezer's first theorem, can, in the present situation, be replaced by

$$\int_E \frac{\log W_*(x)}{1+x^2}\,dx$$

when dealing with certain kinds of sets E. That is the subject of the following discussion. Our results will depend strongly on those of the preceding two articles.

Lemma. *Let $A > 0$, and suppose that there is a finite M such that*

$$(*) \qquad \int_E \frac{\log|S(x)|}{1+x^2}\,dx \;\leqslant\; M$$

for all finite sums $S(x)$ of the form

$$\sum_{-A \leqslant \lambda \leqslant A} a_\lambda e^{i\lambda x}$$

with $\|S\|_{W,E} \leqslant 1$. Then there is a finite M' such that

$$(\S) \qquad \int_E \frac{\log^+|S(x)|}{1+x^2}\,dx \;\leqslant\; M'$$

for such S with $\|S\|_{W,E} \leqslant 1$.

Proof. Given a sum $S(x)$ of the specified form with $\|S\|_{W,E} \leqslant 1$, we wish to show that (§) holds for some M' independent of S. Let us assume, to begin with, that the *exponents λ figuring in the sum $S(x)$ are in arithmetic progression*, more precisely, that

$$S(x) \;=\; \sum_{n=-N}^{N} C_n e^{inhx}$$

where $h = A/N$, N being some large integer. There is then another sum

$$T(x) \;=\; \sum_{n=-N}^{N} a_n e^{inhx}$$

(which is thus *also* of the form $\sum_{-A \leqslant \lambda \leqslant A} C_\lambda e^{i\lambda x}$) such that

$$1 + S(x)\overline{S(x)} \;=\; T(x)\overline{T(x)} \qquad \text{for } x \in \mathbb{R}.$$

This we can see by an elementary argument, going back to Fejér and

Riesz. For $x \in \mathbb{R}$, we have

$$1 + S(x)\overline{S(x)} \;=\; \sum_{n=-2N}^{2N} \gamma_n e^{inhx}$$

with certain coefficients γ_n. Write, for the moment,

$$e^{ihx} = \zeta;$$

then

$$1 + S(x)\overline{S(x)} \;=\; R(\zeta),$$

where

$$R(\zeta) \;=\; \sum_{n=-2N}^{2N} \gamma_n \zeta^n$$

is a certain *rational* function of ζ. We have $R(\zeta) \geqslant 1$ for $|\zeta| = 1$, so, by the Schwarz reflection principle,

$$R(\overline{1/\zeta}) \;=\; \overline{R(\zeta)}.$$

Therefore, if α, $0 < |\alpha| < 1$, is a zero of $R(\zeta)$, so is $1/\bar{\alpha}$, and the latter has the same multiplicity as α. Also, if $-m$ denotes the *least* integer n for which $\gamma_n \neq 0$, we must have $\gamma_n = 0$ for $n > m$ (*sic*!), as follows on comparing the orders of magnitude of $R(\zeta)$ for $\zeta \to 0$ and for $\zeta \to \infty$.

The *polynomial* $\zeta^m R(\zeta)$ is thus of degree $2m$, and of the form

$$\text{const.} \prod_{k=1}^{m} (\zeta - \alpha_k)\left(\zeta - \frac{1}{\bar{\alpha}_k}\right).$$

Thence,

$$R(\zeta) \;=\; C \prod_{k=1}^{m} (\zeta - \alpha_k)\left(\frac{1}{\zeta} - \bar{\alpha}_k\right),$$

and $C > 0$ since $R(\zeta) \geqslant 1$ for $|\zeta| = 1$. Going back to the real variable x, we see that

$$1 + S(x)\overline{S(x)} \;=\; C \prod_{k=1}^{m} (e^{ihx} - \alpha_k)(e^{-ihx} - \bar{\alpha}_k) \;=\; T(x)\overline{T(x)},$$

where

$$T(x) \;=\; C^{\frac{1}{2}} e^{-iNhx} \prod_{k=1}^{m} (e^{ihx} - \alpha_k)$$

is of the form

$$\sum_{n=-N}^{N} a_n e^{inhx},$$

since $m \leqslant 2N$.

Once this is known, it is easy to deduce (§) for sums $S(x)$ of the special form just considered with $\|S\|_{W,E} \leqslant 1$. Take any such S; we have another sum $T(x)$ of the same kind with $1 + |S(x)|^2 = |T(x)|^2$ on $\mathbb{R}$. Since $W(x) \geqslant 1$ on E, the condition $\|S\|_{W,E} \leqslant 1$ implies that $\|T\|_{W,E} \leqslant \sqrt{2}$, i.e.,

$$\|T/\sqrt{2}\|_{W,E} \;\leqslant\; 1.$$

For this reason, $T(x)/\sqrt{2}$ satisfies (*), by hypothesis. Hence

$$\int_E \frac{\log|T(x)|}{1+x^2}\,dx \;\leqslant\; M + \int_E \frac{\log\sqrt{2}}{1+x^2}\,dx.$$

But $\log|T(x)| = \log\sqrt{(1+|S(x)|^2)} \geqslant \log^+|S(x)|$. Therefore

$$\int_E \frac{\log^+|S(x)|}{1+x^2}\,dx \;\leqslant\; M + \pi\log\sqrt{2},$$

and we have obtained (§).

We must still consider the case where the exponents λ in the *finite* sum

$$\sum_{-A\leqslant\lambda\leqslant A} a_\lambda e^{i\lambda x} \;=\; S(x)$$

are *not* in arithmetic progression, the condition $\|S\|_{W,E} \leqslant 1$ being, however, satisfied. Here, we may associate to each λ figuring in the expression just written a *rational multiple* λ' *of* A, with $|\lambda' - \lambda|$ exceedingly small. Since $W(x) \to \infty$ for $x \to \pm\infty$ in E, the sum

$$S'(x) \;=\; \sum_{-A\leqslant\lambda\leqslant A} a_\lambda e^{i\lambda' x}$$

will then be *as close as we like* in $\|\quad\|_{W,E}$-norm to $S(x)$ (depending on the closeness of the individual λ' to their corresponding λ). In this way, we can get a sequence of sums $S_n(x)$ of the form in question, *each one having its exponents in arithmetic progression*, such that $\|S_n\|_{W,E} \leqslant 1$ and $S_n(x) \xrightarrow[n]{} S(x)$ u.c.c. on $\mathbb{R}$. By what we have already shown, $\int_E(\log^+|S_n(x)|/(1+x^2))\,dx \leqslant M + \pi\log\sqrt{2}$ for each n. Therefore

$$\int_E \frac{\log^+|S(x)|}{1+x^2}\,dx \;\leqslant\; M + \pi\log\sqrt{2}$$

by Fatou's lemma. We are done.

For the sets E described at the beginning of the present § we can establish analogues, involving integrals over E, of the Akhiezer and Pollard theorems given in Chapter VI, §§E.2 and E.4. These are included in the following theorem which, in one direction, assumes as little as possible and concludes that $\mathscr{C}_W(A, E) \subset \mathscr{C}_W(E)$ properly. In the other, it assumes the proper inclusion and asserts as much as possible.

For $z \in \mathbb{C}$, *denote by* $W_{A,E}(z)$ *the supremum of* $|S(z)|$ *for the finite sums*

$$S(z) \;=\; \sum_{-A \leqslant \lambda \leqslant A} a_\lambda e^{i\lambda z}$$

with $\|S\|_{W,E} \leqslant 1$. (If we agree that $W(x) \equiv \infty$ on $\mathbb{R} \sim E$, $W_{A,E}(z)$ and $\mathscr{C}_W(A, E)$ reduce respectively to the function $W_A(z)$ and the space $\mathscr{C}_W(A)$ already considered in Chapter VI, §§E.2ff.)

Theorem. *Let E be one of the sets described at the beginning of this §, the conditions involving the four numbers* A, B, δ *and* Δ *being fulfilled. If, for some* $C > 0$, *the supremum of*

$$\int_E \frac{\log|S(x)|}{1+x^2}\,dx$$

for all finite sums

$$S(x) \;=\; \sum_{-C \leqslant \lambda \leqslant C} a_\lambda e^{i\lambda x}$$

with $\|S\|_{W,E} \leqslant 1$ *is* finite, *then* $\mathscr{C}_W(C, E) \subset \mathscr{C}_W(E)$ properly.

If, conversely, that proper inclusion holds, then

$$\int_E \frac{\log W_{C,E}(x)}{1+x^2}\,dx \;<\; \infty.$$

Proof. All the work here is in the establishment of the *first* part of the statement.

Define $W(x)$ on all of $\mathbb{R}$ by putting it equal to ∞ on $\mathbb{R} \sim E$. This makes it possible for us to apply results about $\mathscr{C}_W(C)$ from Chapter VI, §E, in the present situation. According to the Pollard theorem of Chapter VI, §E.4, and the remark thereto, we will have $\mathscr{C}_W(C) \neq \mathscr{C}_W(\mathbb{R})$ as soon as $W_C(i) < \infty$; in other words, $\mathscr{C}_W(C, E) \neq \mathscr{C}_W(E)$ provided that

$$W_{C,E}(i) \;<\; \infty.$$

We proceed to show this inequality, using the results of articles 1 and 2.

According to the lemma, our hypothesis for the first part of the theorem implies that

$$(*) \qquad \int_E \frac{\log^+|S(x)|}{1+x^2}\,dx \;\leqslant\; M' \;<\; \infty$$

for all sums $S(x)$ of the stipulated form with $\|S\|_{W,E} \leqslant 1$.

We have

$$E = \bigcup_{n=-\infty}^{\infty} [a_n - \delta_n,\ a_n + \delta_n],$$

where

$$0 < A < a_{n+1} - a_n < B,$$
$$0 < \delta < \delta_n < \Delta.$$

Given any finite sum

$$S(z) = \sum_{-C \leqslant \lambda \leqslant C} a_\lambda e^{i\lambda z}$$

with $\|S\|_{W,E} \leqslant 1$, let us put

$$v_S(z) = \frac{1}{\delta}\int_{-\delta/2}^{\delta/2} \log^+ |S(z+t)|\,dt$$

(using the δ associated to E by the above inequalities). The function $v_S(z)$ is then *continuous* and *subharmonic* in the complex plane. Obviously,

$$|S(z)| \leqslant \text{const.}e^{C|\Im z|}$$

(the constant may be *enormous*, but we don't care!), so

$$(\dagger) \qquad v_S(z) \leqslant O(1) + C|\Im z|, \quad z \in \mathbb{C}.$$

Put now

$$E' = \bigcup_{n=-\infty}^{\infty} \left[a_n - \frac{\delta_n}{2},\ a_n + \frac{\delta_n}{2}\right].$$

On the component

$$E'_n = \left[a_n - \frac{\delta_n}{2},\ a_n + \frac{\delta_n}{2}\right]$$

of E' we have

$$v_S(x) \leqslant \frac{1}{\delta}\int_{E_n} \log^+ |S(t)|\,dt,$$

where (as usual)

$$E_n = [a_n - \delta_n,\ a_n + \delta_n].$$

Denoting the right side of the previous relation by v_n, we have

$$(\S\S) \qquad v_S(x) \leqslant v_n, \quad x \in E'_n.$$

The set E' (like E) is one of the kind specified at the beginning of the present §; the numbers A, B, $\delta/2$ and $\Delta/2$ are associated to it. *The*

▶ *results of the previous two articles are therefore valid for E' and the domain* $\mathscr{D}' = \mathbb{C} \sim E'$. We can, in particular, apply Carleson's theorem from article 1. Assume, wlog, that

$$0 \in \mathcal{O}_0' = \left(a_0 + \frac{\delta_0}{2},\ a_1 - \frac{\delta_1}{2}\right).$$

Then, if we denote by $\omega_n'(z)$ the *harmonic measure of E_n'* in $\mathscr{D}'$ (as seen from z), that theorem tells us that

$$\omega_n'(0) \leqslant \frac{K}{1+n^2}$$

with a constant K depending on A, B, $\delta/2$ and $\Delta/2$. By Harnack's theorem, there is thus a function $K(z)$, *continuous in $\mathscr{D}'$*, such that

$$\omega_n'(z) \leqslant \frac{K(z)}{1+n^2}, \qquad z \in \mathscr{D}'$$

(see discussion near the beginning of §B.1, Chapter VII).

Using properties (i) and (ii) from the beginning of this § we see that the quantities v_n introduced above satisfy

$$\frac{v_n}{1+n^2} \leqslant \alpha \int_{E_n} \frac{\log^+ |S(t)|}{1+t^2} dt,$$

α being a certain constant depending only on the set E. Combined with the previous estimate, this yields

$$v_n \omega_n'(z) \leqslant \alpha K(z) \int_{E_n} \frac{\log^+ |S(t)|}{1+t^2} dt, \qquad z \in \mathscr{D}',$$

so, for the sum

$$P(z) = \sum_{n=-\infty}^{\infty} v_n \omega_n'(z),$$

we have

$$(\dagger\dagger) \qquad P(z) \leqslant \alpha M' K(z), \qquad z \in \mathscr{D}',$$

by virtue of ($\ast\atop\ast$).

Because $|S(x)|$ is bounded on $\mathbb{R}$, the v_n (which are $\geqslant 0$, by the way) are bounded. The series used to define $P(z)$ is therefore u.c.c. convergent, so that function is *continuous up to* $E' = \partial\mathscr{D}'$ as well as being *positive* and *harmonic* in $\mathscr{D}'$. On the component E_n' of E', $P(x)$ takes the constant value v_n. Hence the function

$$v_S(z) - P(z)$$

is *subharmonic in* $\mathscr{D}'$ and continuous up to E', where it is $\leqslant 0$ by (§§). It is, moreover, $\leqslant O(1) + C|\Im z|$ by (†).

Now according to *Benedicks' theorem* (article 2), a *Phragmén–Lindelöf function* $Y(z)$ is available for $\mathscr{D}'$. The function

$$v_S(z) - P(z) - CY(z)$$

is *subharmonic and bounded above* in $\mathscr{D}'$ and continuous up to $E' = \partial\mathscr{D}'$ where it is $\leqslant 0$. It is thence $\leqslant 0$ *throughout* $\mathscr{D}'$ by the extended maximum principle (Chapter III, §C). Referring to (††), we see that

$$(\ddagger) \qquad v_S(z) \;\leqslant\; \alpha M'K(z) + CY(z), \qquad z \in \mathscr{D}'.$$

Let $\rho = \min(\frac{1}{2}, \delta/2)$. Since $\log^+|S(z)|$ is *subharmonic*, we have

$$\log^+|S(\mathrm{i})| \;\leqslant\; \frac{1}{\pi\rho^2}\iint_{|z-\mathrm{i}|<\rho} \log^+|S(z)|\,\mathrm{d}x\,\mathrm{d}y.$$

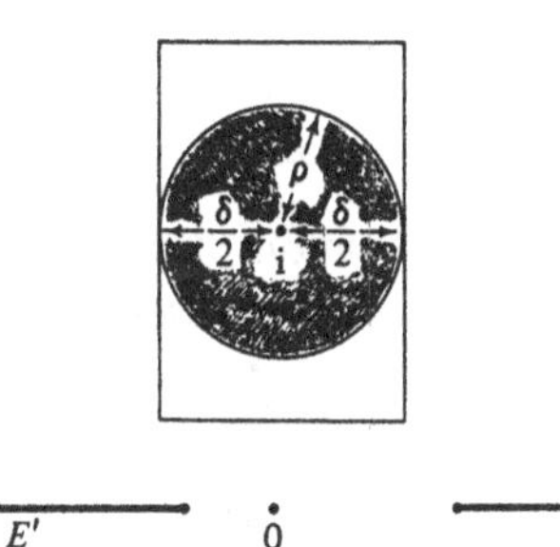

Figure 128

The integral on the right is

$$\leqslant \frac{1}{\pi\rho^2}\int_{1-\rho}^{1+\rho}\int_{-\delta/2}^{\delta/2} \log^+|S(x+\mathrm{i}y)|\,\mathrm{d}x\,\mathrm{d}y \;=\; \frac{\delta}{\pi\rho^2}\int_{1-\rho}^{1+\rho} v_S(\mathrm{i}y)\,\mathrm{d}y.$$

Plugging (‡) into the last expression, we obtain

$$\log^+|S(\mathrm{i})| \;\leqslant\; \frac{\delta}{\pi\rho^2}\int_{1-\rho}^{1+\rho} (\alpha M'K(\mathrm{i}y) + CY(\mathrm{i}y))\,\mathrm{d}y.$$

This, then, is valid for *any* finite sum $S(z)$ of the form

$$\sum_{-C\leqslant\lambda\leqslant C} a_\lambda \mathrm{e}^{\mathrm{i}\lambda z}$$

with $\|S\|_{W,E} \leqslant 1$.

The *right side* of the inequality just found is a *finite quantity*, dependent on M' and C, and on the set E (through α, $K(\mathrm{i}y)$ and $Y(\mathrm{i}y)$); it is, however, *completely independent* of the *particular* sum $S(z)$ under consideration. Therefore $W_{C,E}(\mathrm{i})$, the supremum of $|S(\mathrm{i})|$ for such sums S with $\|S\|_{W,E} \leqslant 1$, is *finite*. This is what we needed to show in order to infer the proper inclusion of $\mathscr{C}_W(C, E)$ in $\mathscr{C}_W(E)$. The *first part* of our theorem is thus *proved.*

There remains the *second part*. *That*, however, is not new! Putting, as before, $W(x)$ equal to ∞ on $\mathbb{R} \sim E$, which makes $\mathscr{C}_W(C, E)$ coincide with the subspace $\mathscr{C}_W(C)$ of $\mathscr{C}_W(\mathbb{R})$ considered in Chapter VI, §E, we have, in the notation of that §,

$$\int_{-\infty}^{\infty} \frac{\log W_C(x)}{1 + x^2} dx \; < \; \infty$$

if $\mathscr{C}_W(C) \neq \mathscr{C}_W(\mathbb{R})$, according to *Akhiezer's theorem* (Chapter VI, §E.2). Our function $W_{C,E}(x)$ is simply $W_C(x)$. Hence, since $E \subseteq \mathbb{R}$ (!),

$$\int_E \frac{\log W_{C,E}(x)}{1 + x^2} dx \; < \; \infty$$

when $\mathscr{C}_W(C, E) \neq \mathscr{C}_W(E)$.

The theorem is completely proved. We are done.

Remark. If we do not assume anything about the continuity of a weight $W(x) \geqslant 1$ defined on E, it is still possible to characterize the equality of $\mathscr{C}_W(C, E)$ with $\mathscr{C}_W(E)$ by an analogue of *Mergelian's second theorem* involving an integral over E. The establishment of such a result proceeds very much along the lines of the proof just finished, and is left to the reader.

Problem 18

Let E be a closed set on $\mathbb{R}$ of the kind specified at the beginning of this §. Show that there are two constants a, b, depending on E, such that, for any entire function $f(z)$ of exponential type $\leqslant C$, bounded on $\mathbb{R}$, we have

$$\int_{-\infty}^{\infty} \frac{\log(1 + |f(x)|^2)}{1 + x^2} dx \; \leqslant \; aC \; + \; b \int_E \frac{\log(1 + |f(x)|^2)}{1 + x^2} dx.$$

(Hint: One may apply the *third* and *fourth* theorems from Chapter III, §G.3, and reason as in the above proof. Another procedure is to use the proof of the lemma in §E.1 of Chapter VI so as to first approximate $f(z)$ by finite sums $S(z)$ of the form considered above, having exponents in arithmetic progression.)

If $W(x)$, continuous and $\geqslant 1$ on E, is such that

$$\frac{x^n}{W(x)} \longrightarrow 0 \quad \text{for} \quad x \longrightarrow \pm\infty \quad \text{in} \quad E$$

and $n = 1, 2, 3, \ldots$, we denote by $W_{0,E}(z)$ the *supremum* of $|P(z)|$ for all *polynomials* P with $\|P\|_{W,E} \leqslant 1$.

Theorem. *Let $E \subseteq \mathbb{R}$ be a set of the kind specified at the beginning of this §, and let $W(x)$, continuous and $\geqslant 1$ on E, tend to ∞ faster than any power of x as $x \to \pm\infty$ in E.*

If, for polynomials $P(z)$ *with* $\|P\|_{W,E} \leqslant 1$, *the integrals*

$$\int_E \frac{\log|P(x)|}{1+x^2}\,dx$$

are bounded above, *then* $\mathscr{C}_W(0, E)$ *is* properly contained *in* $\mathscr{C}_W(E)$.

If $\mathscr{C}_W(0, E)$ is *properly contained in* $\mathscr{C}_W(E)$, *then*

$$\int_E \frac{\log W_{0,E}(x)}{1+x^2}\,dx \;<\; \infty.$$

Proof. The *second* part reduces (as at the end of the preceding demonstration) to a known result of Akhiezer (in this case from §B.1, Chapter VI) on putting $W(x) = \infty$ on $\mathbb{R} \sim E$. Hence only the *first* part requires discussion here.

According to *Pollard's theorem* (Chapter VI, §B.3), proper inclusion of $\mathscr{C}_W(0, E)$ in $\mathscr{C}_W(E)$ will certainly follow if the integrals

$$\int_{-\infty}^{\infty} \frac{\log(1+|P(x)|^2)}{1+x^2}\,dx$$

are *bounded above* for P ranging over the *polynomials* with $\|P\|_{W,E} \leqslant 1$. It is therefore enough to *show* this, under the assumption that

$$\int_E \frac{\log|P(x)|}{1+x^2}\,dx \;\leqslant\; M, \text{ say,}$$

for *any* polynomial P with $\|P\|_{W,E} \leqslant 1$.

We may, first of all, argue as in the proof of the above lemma to conclude that our assumption implies a seemingly stronger property: we have

$$\int_E \frac{\log(1+|P(x)|^2)}{1+x^2}\,dx \;\leqslant\; 2M + \pi\log 2$$

for the polynomials P with $\|P\|_{W,E} \leqslant 1$. The proof will therefore be complete if we can verify that

$$\int_{-\infty}^{\infty} \frac{\log(1+|P(x)|^2)}{1+x^2}\,dx \;\leqslant\; b\int_E \frac{\log(1+|P(x)|^2)}{1+x^2}\,dx$$

for polynomials P, b being a certain constant depending on the set E. This we do, using the result of *problem* 18.

Take any polynomial P, of degree N, say. With an arbitrary $\eta > 0$, put

$$f_\eta(z) = \left(\frac{\sin \eta z}{\eta z}\right)^N P(z);$$

$f_\eta(z)$ is then entire, of exponential type $N\eta$, and *bounded* on $\mathbb{R}$. By *problem* 18, we thus have

$$\int_{-\infty}^{\infty} \frac{\log(1+|f_\eta(x)|^2)}{1+x^2}\,dx \;\leqslant\; aN\eta \;+\; b\int_E \frac{\log(1+|f_\eta(x)|^2)}{1+x^2}\,dx.$$

Here, $|f_\eta(x)| \leqslant |P(x)|$ on $\mathbb{R}$ and $f_\eta(x) \longrightarrow P(x)$ as $\eta \to 0$, so the desired inequality follows on making $\eta \to 0$. We are done.

4. What happens when the set E is sparse

The sets E described at the beginning of this § have the property that

$$|E \cap I|/|I| \;\geqslant\; c \;>\; 0$$

for all intervals I on $\mathbb{R}$ of length exceeding some L. In other words, their *lower uniform density* is *positive*. One suspects that the continual occurrence of the form $dx/(1+x^2)$ in the integrals over E figuring in the preceding article is somehow connected with this positivity. As a first step towards finding out whether our hunch has any basis in fact, let us try to see what happens to the form $dx/(1+x^2)$ when E becomes *sparse*. We do this in the special case where

$$E \;=\; \bigcup_{n=-\infty}^{\infty} [a_n - \delta,\; a_n + \delta]$$

with $a_n = |n|^p \operatorname{sgn} n$ and $p > 1$. This example was worked out by Benedicks (see his preprint), and all the material in the present article is due to him.

► In order that there may be no doubt, we point out that *the sets E now under consideration are no longer of the sort described at the beginning of the present* §.

Lemma. *Let S be the square*

$$\{(x,y)\colon\; -a < x < a \text{ and } -a < y < a\},$$

and denote by H the union *of its* two horizontal sides, *and by V the* union *of its* two vertical sides. *Then, if* $-a < x < a$,

$$\omega_S(H,x) \;\leqslant\; \omega_S(H,0)$$

and

$$\omega_S(V,x) \ \geqslant \ \omega_S(V,0),$$

where, as usual, $\omega_S(\ ,z)$ *denotes harmonic measure for S.*

Proof (Benedicks). Let, wlog, $0 < x_0 < a$ and consider the harmonic function

$$\Delta(z) \ = \ \omega_S(H,z) - \omega_S(H,z+x_0)$$

defined in the rectangle

$$T \ = \ \{(x,y)\colon \ -a < x < a - x_0 \text{ and } -a < y < a\}.$$

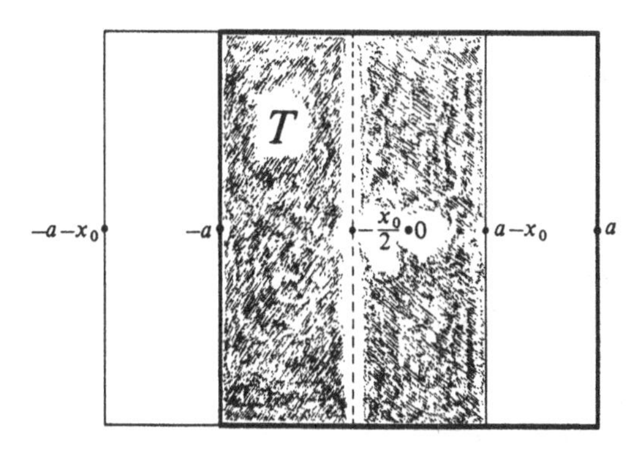

Figure 129

It is clear by symmetry that, for $z \in S$, $\omega_S(H,z) = \omega_S(H,\bar{z})$, and also $\omega_S(H,x+\mathrm{i}y) = \omega_S(H,-x+\mathrm{i}y)$. Therefore $\Delta(-\frac{1}{2}x_0 + \mathrm{i}y) = 0$ on the vertical bisector of T (see figure). Again, on T's *right vertical side,*

$$\begin{aligned}\Delta(a - x_0 + \mathrm{i}y) \ &= \ \omega_S(H, a - x_0 + \mathrm{i}y) - \omega_S(H, a + \mathrm{i}y) \\ &= \ \omega_S(H, a - x_0 + \mathrm{i}y) \ \geqslant \ 0\end{aligned}$$

(and similarly, on the opposite side of T,

$$\Delta(-a+\mathrm{i}y) \ = \ -\omega_S(H, -a + x_0 + \mathrm{i}y) \ \leqslant \ 0).$$

It is clear on the other hand that $\Delta(z) = 0$ on the *top* and *bottom sides* of T $(1-1=0)$. By the principle of maximum we thus have $\Delta(z) \geqslant 0$ in the *right half* of T; in particular,

$$\Delta(0) \ = \ \omega_S(H,0) - \omega_S(H,x_0) \ \geqslant \ 0,$$

and $\omega_S(H,x_0) \ \leqslant \ \omega_S(H,0)$, proving the *first* inequality asserted by the lemma.

The *second* inequality follows from the first one because

$$\omega_S(H,z) + \omega_S(V,z) \equiv 1$$

in S and clearly $\omega_S(V,0) = \omega_S(H,0)$. We are done.

Lemma (Benedicks). *Let $E \subseteq \mathbb{R}$ be any 'reasonable' closed set (for instance, a finite union of closed intervals), let S be the square of the preceding lemma, and put*

$$\Omega = S \cap \sim E.$$

If H denotes the union of the two horizontal sides of S and V that of the vertical ones, we have

$$\omega_\Omega(V,0) \leqslant \omega_\Omega(H,0)$$

for the harmonic measure $\omega_\Omega(\ ,z)$ associated with the domain Ω.

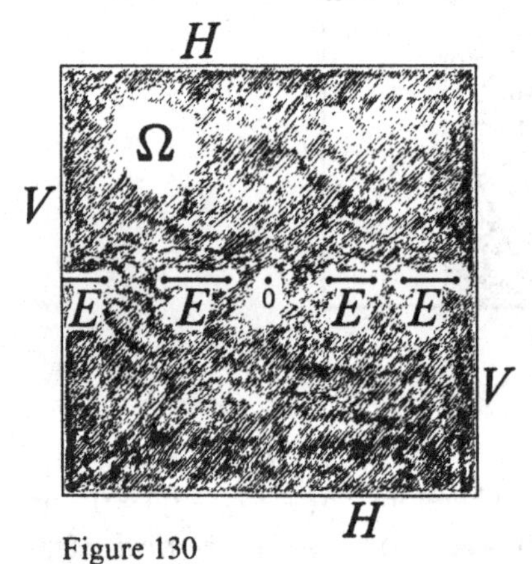

Figure 130

Proof. By a formula derived near the end of §B.1, Chapter VII, for $z \in \Omega$,

$$\omega_\Omega(H,z) = \omega_S(H,z) - \int_E \omega_S(H,\xi)\,d\omega_\Omega(\xi,z)$$

$$\omega_\Omega(V,z) = \omega_S(V,z) - \int_E \omega_S(V,\xi)\,d\omega_\Omega(\xi,z).$$

From the previous lemma,

$$\omega_S(H,\xi) \leqslant \omega_S(H,0) = \omega_S(V,0) \leqslant \omega_S(V,\xi)$$

for real ξ lying in S; in particular, for $\xi \in E$. Substituting this relation into the preceding ones and then making $z = 0$, we get $\omega_\Omega(V,0) \leqslant \omega_\Omega(H,0)$.

Q.E.D.

Corollary. *In the above configuration,*

$$\omega_\Omega(\partial S, 0) \;\leqslant\; 2\omega_\Omega(H, 0).$$

Proof. Clear.

Lemma. *Let $p > 1$ and put*

$$E \;=\; \bigcup_{n=-\infty}^{\infty} \; [|n|^p \operatorname{sgn} n - \delta,\; |n|^p \operatorname{sgn} n + \delta],$$

$\delta > 0$ being taken small enough so that the intervals figuring in the union do not intersect. With $x_0 > 0$, let S_{x_0} be the square

$$\left\{ \frac{x_0}{2} < \Re z < \frac{3x_0}{2}, \quad -\frac{x_0}{2} < \Im z < \frac{x_0}{2} \right\},$$

and Ω_{x_0} the domain

$$S_{x_0} \cap \sim E.$$

For large x_0, the harmonic measure $\omega_{\Omega_{x_0}}(\quad, z)$ associated with Ω_{x_0} satisfies

$$\omega_{\Omega_{x_0}}(\partial S_{x_0},\; x_0) \;\leqslant\; \text{const.} \frac{\log x_0}{x_0^{1/p}}.$$

Proof (Benedicks). By use of a test function and application of the preceding corollary.

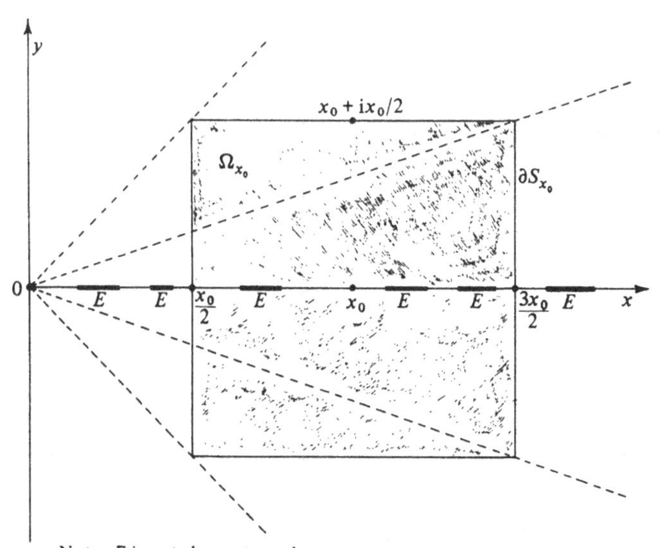

Figure 131

The function $z^{1/p}$ (taken as positive on the positive real axis) is analytic for $\Re z > 0$; so, therefore, is

$$\sin \pi z^{1/p}.$$

In $\Re z > 0$, this function vanishes only at the midpoints n^p of the intervals making up E, and, at $x = n^p$,

$$\frac{\mathrm{d}(\sin \pi x^{1/p})}{\mathrm{d}x} = (-1)^n \frac{\pi}{pn^{p-1}}.$$

This means that, if we take $x_0 > 0$ *large* and put $C_0 = 1/x_0^{(p-1)/p}$, we have $|\sin \pi x^{1/p}| \geqslant kC_0\delta$ for x *outside* E on the interval $(x_0/2,\ 3x_0/2)$, $k > 0$ being a constant depending on p, but independent of x_0 and δ. Recalling the behaviour of the Joukowski transformation

$$w \longrightarrow w + \sqrt{(w^2 - 1)},$$

we see that for a suitable definition of $\sqrt{\ }$, the function

$$v(z) = \log\left|\frac{\sin \pi z^{1/p}}{kC_0\delta} + \sqrt{\left(\frac{\sin^2 \pi z^{1/p}}{(kC_0\delta)^2} - 1\right)}\right|$$

is *positive and harmonic* in Ω_{x_0}.

For this reason, when $x \in \mathbb{R} \cap \Omega_{x_0}$,

$$v(x) \geqslant \inf_{\zeta \in H} v(\zeta) \cdot \omega_{\Omega_{x_0}}(H, x),$$

H denoting the union of the two *horizontal* sides of ∂S_{x_0}. However,

$$v(\zeta) \geqslant \text{const.} x_0^{1/p} \qquad \text{for } \zeta \in H$$

as is easily seen (almost without computation, if one refers to the above diagram). Also,

$$v(x) \leqslant \log \frac{2}{kC_0\delta} = (1 - 1/p)\log x_0 + O(1), \qquad x \in \mathbb{R} \cap \Omega_{x_0}.$$

Therefore

$$\omega_{\Omega_{x_0}}(H, x) \leqslant \text{const.} \frac{\log x_0}{x_0^{1/p}}, \qquad x \in \mathbb{R} \cap \Omega_{x_0}.$$

Since x_0 lies at the centre of the square S_{x_0}, the corollary to the previous lemma gives

$$\omega_{\Omega_{x_0}}(\partial S_{x_0}, x_0) \leqslant 2\omega_{\Omega_{x_0}}(H, x_0).$$

Combining this and the preceding relations, we obtain the desired result.

Theorem (Benedicks). *Let G be the Green's function for the domain*

$$\mathscr{D} = \mathbb{C} \sim E = \mathbb{C} \sim \bigcup_{n=-\infty}^{\infty} [|n|^p \operatorname{sgn} n - \delta,\ |n|^p \operatorname{sgn} n + \delta],$$

where $p > 1$ and $\delta > 0$ is small enough so that the intervals in the union do not intersect.

Then, for real x of large modulus,

$$G(x, \mathrm{i}) \leqslant C \frac{\log|x|}{|x|^{(p+1)/p}},$$

with a constant C depending on p and δ.

Proof. $G(z, \mathrm{i})$ is certainly bounded above – by M say – in the sector $\{0 \leqslant |\Im z| < \Re z\}$. Given $x_0 > 0$, the square S_{x_0} considered in the previous lemma lies in that sector, so $G(\zeta, \mathrm{i}) \leqslant M$ on ∂S_{x_0}. $G(z, \mathrm{i})$ is, moreover, harmonic in $\Omega_{x_0} \subseteq \mathscr{D}$ and *zero on E*, whence $G(x_0, \mathrm{i}) \leqslant M \cdot \omega_{\Omega_{x_0}}(\partial S_{x_0}, x_0)$. By the last lemma we therefore have

$$(*) \qquad G(x_0, \mathrm{i}) \leqslant \text{const.} \frac{\log|x_0|}{|x_0|^{1/p}}$$

for large $x_0 > 0$.

Benedicks' idea is to now use Poisson's formula for the half plane, so as to improve $(*)$ by iteration. Take any fixed α with $0 < \alpha < 1/p$. Then $(*)$ certainly implies (by symmetry of E) that

$$G(x, \mathrm{i}) \leqslant \frac{\text{const.}}{|x|^\alpha + 1}, \qquad x \in \mathbb{R},$$

$G(x, \mathrm{i})$ being at any rate bounded on the real axis. The function $G(z, \mathrm{i})$ is in fact *bounded and harmonic in* $\Im z < 0$, so

$$G(z, \mathrm{i}) = \frac{1}{\pi} \int_{-\infty}^{\infty} \frac{|\Im z|}{|z - t|^2} G(t, \mathrm{i}) \, \mathrm{d}t, \qquad \Im z < 0.$$

Plugging in the previous relation, we get

$$G(z, \mathrm{i}) \leqslant \text{const.} \int_{-\infty}^{\infty} \frac{|y|}{(x - t)^2 + y^2} \cdot \frac{\mathrm{d}t}{|t|^\alpha + 1}, \qquad y < 0.$$

Let, wlog, $x > 0$. Then, the integral just written can be broken up as $\int_{-x/2}^{x/2} + \int_{|t| \geqslant x/2}$. Since $\int_{-\infty}^{\infty} (|y|/((x - t)^2 + y^2)) \mathrm{d}t = \pi$, the *second* of these terms is obviously $O(x^{-\alpha})$ for large x. The *first*, on the other hand, is

$$\leqslant \text{const.} \frac{4|y|}{x^2 + 4y^2} \cdot x^{1-\alpha},$$

so, all in all,

$$G(z, \mathrm{i}) \leqslant \text{const.}|x|^{-\alpha}$$

for $0 > y > -|x|$, $|x|$ being large.

The inequality just found *remains true*, however, for $0 < y < |x|$, in spite of the logarithmic singularity that $G(z,\mathrm{i})$ has at i. This follows from the fact that $0 < \alpha < 1/p < 1$ and the relation

$$G(z, \mathrm{i}) - G(\bar{z}, \mathrm{i}) = \log\left|\frac{z+\mathrm{i}}{z-\mathrm{i}}\right|.$$

To verify the latter, just subtract the right side from the left. The difference is *harmonic* in $\Im z > 0$ and *bounded* there (the logarithmic poles at i cancel each other out). It is also clearly zero on $\mathbb{R}$, so hence zero for $\Im z > 0$. For large $|z|$, $\log|(z+i)/(z-i)| = O(1/|z|)$, and we see that

$$G(z, \mathrm{i}) \leqslant \text{const.}|x|^{-\alpha} \qquad \text{for } 0 \leqslant |y| < x$$

since this inequality is true for $0 > y > -|x|$.

Suppose that $x_0 > 0$ is large; we can use the previous lemma again. By what has just been shown,

$$G(\zeta, \mathrm{i}) \leqslant \text{const.}|x_0|^{-\alpha}, \qquad \zeta \in \partial S_{x_0}.$$

Arguing as at the beginning of this proof, we get

$$G(x_0, \mathrm{i}) \leqslant \text{const.}|x_0|^{-\alpha}\omega_{\Omega_{x_0}}(\partial S_{x_0}, x_0) \leqslant \text{const.}\frac{\log|x_0|}{|x_0|^{\alpha}|x_0|^{1/p}}.$$

Hence, since $0 < \alpha < 1/p$, we have

$$G(x, \mathrm{i}) \leqslant \frac{\text{const.}}{|x|^{2\alpha}+1}, \qquad x \in \mathbb{R}.$$

The exponent α in the inequality we started with has been improved to 2α.

If now $2\alpha < 1$, we may start from the inequality just obtained and *repeat* the above argument, ending with the relation

$$G(x, \mathrm{i}) \leqslant \frac{\text{const.}}{|x|^{3\alpha}+1}, \qquad x \in \mathbb{R}.$$

The process may evidently be continued so as to yield successively the estimates

$$G(x, \mathrm{i}) \leqslant \frac{\text{const.}}{|x|^{n\alpha}+1}, \qquad x \in \mathbb{R},$$

with $n = 3, 4, \ldots,$ *as long as* $(n-1)\alpha < 1$. Choosing α, $0 < \alpha < 1/p$, *to not be of the form* $1/m$, $m = 1, 2, 3, \ldots$, we arrive at an estimate

$$(\ddagger) \qquad G(x, \mathrm{i}) \leqslant \frac{\text{const.}}{|x|^{n\alpha} + 1}, \qquad x \in \mathbb{R},$$

where $n\alpha$ *is the first integral multiple of* α *strictly* > 1.

Because the exponent $n\alpha$ in ($\ddagger$) is > 1, we have

$$\int_{-\infty}^{\infty} G(t, \mathrm{i})\,\mathrm{d}t < \infty.$$

As before, for $y < 0$, we can write

$$G(z, \mathrm{i}) = \frac{1}{\pi}\int_{-|x|/2}^{|x|/2} \frac{|y|}{|z - t|^2} G(t, \mathrm{i})\,\mathrm{d}t + \frac{1}{\pi}\int_{|t| \geqslant |x|/2} \frac{|y|}{|z - t|^2} G(t, \mathrm{i})\,\mathrm{d}t.$$

For $|t| \leqslant |x|/2$, $|y|/|z - t|^2 \leqslant 1/|x|$, so the *first* term on the right is $\leqslant \text{const.}/|x|$ in view of the preceding relation. The *second* is $\leqslant \text{const.}/|x|^{n\alpha} = o(1/|x|)$ by ($\ddagger$). Thence, for $|x|$ large,

$$G(z, \mathrm{i}) \leqslant \frac{\text{const.}}{|x|}, \qquad y < 0.$$

Using the relation

$$G(z, \mathrm{i}) - G(\bar{z}, \mathrm{i}) = \log\left|\frac{\mathrm{i} + z}{\mathrm{i} - z}\right|$$

as above, we find that in fact

$$G(z, \mathrm{i}) \leqslant \frac{\text{const.}}{|x|} \qquad \text{for } |z| \text{ large.}$$

Take this relation and apply the preceding lemma *one more time*. For large x_0, we have

$$G(\zeta, \mathrm{i}) \leqslant \text{const.}/x_0 \qquad \text{on } \partial S_{x_0}.$$

Therefore

$$G(x_0, \mathrm{i}) \leqslant \frac{\text{const.}}{x_0}\,\omega_{\Omega_{x_0}}(\partial S_{x_0}, x_0) \leqslant \frac{\text{const.}\log x_0}{x_0^{1 + (1/p)}}.$$

This is what we wanted to prove.

We are done.

Corollary. *A Phragmén–Lindelöf function* $Y(z)$ *exists for the domain*

$$\mathbb{C} \sim \bigcup_{n=-\infty}^{\infty} [|n|^p \operatorname{sgn} n - \delta,\ |n|^p \operatorname{sgn} n + \delta].$$

Proof. By the theorem, we certainly have

$$\int_{-\infty}^{\infty} G(x,\mathrm{i})\,\mathrm{d}x \;<\; \infty.$$

The result then follows by the second theorem of article 2.

Remark. Although the theorem tells us that, *on the real axis,*

$$G(x,\mathrm{i}) \;\leqslant\; \text{const.}\frac{\log|x|}{|x|^{1+1/p}}$$

when $|x|$ is large, the inequality

$$G(z,\mathrm{i}) \;\leqslant\; \frac{\text{const.}}{|x|},$$

valid for $|z|$ large, obtained near the end of the theorem's proof, *cannot be improved* in the sector $0 \leqslant |y| \leqslant |x|$.

Indeed, since $G(t,\mathrm{i}) \geqslant 0$ we have, for large $|x|$,

$$G(x - \mathrm{i}|x|,\ \mathrm{i}) \;=\; \frac{1}{\pi}\int_{-\infty}^{\infty} \frac{|x|}{(x-t)^2 + x^2} G(t,\mathrm{i})\,\mathrm{d}t$$

$$\geqslant\; \frac{4}{13\pi|x|}\int_{-|x|/2}^{|x|/2} G(t,\mathrm{i})\,\mathrm{d}t \;\sim\; \frac{4}{13\pi|x|}\int_{-\infty}^{\infty} G(t,\mathrm{i})\,\mathrm{d}t.$$

A better bound on $G(z,\mathrm{i})$ *can* be obtained if $|y|$ is *much smaller* than $|x|$. The following result is used in the next exercise.

Lemma. *For large* $|x|$,

$$G(z,\mathrm{i}) \;\leqslant\; \text{const.}\frac{\log|x|}{|x|^{1+1/p}}, \qquad 0 \leqslant |y| \leqslant |x|^{1-1/p}.$$

Proof. Taking wlog $x > 0$, consider first the case where $y < 0$. By the theorem,

$$G(z,\mathrm{i}) \;=\; \frac{1}{\pi}\int_{-\infty}^{\infty} \frac{|y|}{(x-t)^2 + y^2} G(t,\mathrm{i})\,\mathrm{d}t$$

$$\leqslant\; \text{const.}\int_{-\infty}^{\infty} \frac{|y|}{(x-t)^2 + y^2}\cdot\frac{\log^+|t| + 1}{|t|^{1+1/p} + 1}\,\mathrm{d}t.$$

As usual, we break up the right-hand integral into

$$\int_{-x/2}^{x/2} + \int_{|t|\geqslant x/2}.$$

The first term is $\leqslant$ const.$|y|/x^2$ (because $1 + 1/p > 1$!), and this is

$\leqslant$ const.$/x^{1+1/p}$ for $|y| \leqslant x^{1-1/p}$. The second term is clearly

$$\leqslant \frac{\text{const.}\log x}{x^{1+1/p}}.$$

This handles the case of negative y.

For $0 < y < x^{1-1/p}$, use the relation

$$G(z, \mathrm{i}) - G(\bar{z}, \mathrm{i}) = \log\left|\frac{z+\mathrm{i}}{z-\mathrm{i}}\right|$$

already applied in the proof of the theorem. Note that the *right hand* side is

$$\Re \log\left(\frac{1+(\mathrm{i}/z)}{1-(\mathrm{i}/z)}\right) = 2\frac{\Im z}{|z|^2} + O\left(\frac{1}{|z|^3}\right)$$

for large $|z|$. For $0 \leqslant \Im z \leqslant |x|^{1-1/p}$, this is

$$\leqslant \text{const.}\frac{1}{|x|^{1+1/p}}.$$

The lemma thus follows because it is true for negative y.

In the following problem the reader is asked to work out the analogue, for our present sets E, of Benedicks' beautiful result about the ones with positive lower uniform density (Problem 16).

Problem 19

If t is on the component $[n^p - \delta,\ n^p + \delta]$ of

$$\mathscr{D} = \mathbb{C} \sim \bigcup_{-\infty}^{\infty} [|k|^p \operatorname{sgn} k - \delta,\quad |k|^p \operatorname{sgn} k + \delta],$$

show that

$$\frac{\mathrm{d}\omega_{\mathscr{D}}(t, \mathrm{i})}{\mathrm{d}t} \leqslant \frac{\text{const.}}{t^{1+1/p} + 1} \cdot \frac{1}{\sqrt{(\delta^2 - (t - n^p)^2)}},$$

where $\omega_{\mathscr{D}}(\ , z)$ denotes harmonic measure for $\mathscr{D}$. Here, the constant depends only on $p > 1$ and $\delta > 0$.

Remark. The result is due to Benedicks. We see that the factor $\log|t|$ in the estimate for $G(t, \mathrm{i})$ furnished by the above theorem *disappears* when we evaluate *harmonic measure*.

Hint for the problem: One proceeds as in the solution of Problem 16, here comparing $G(z, \mathrm{i})$ with

$$U(z) = \log\left|\frac{z - n^p}{\delta} + \sqrt{\left(\left(\frac{z - n^p}{\delta}\right)^2 - 1\right)}\right|$$

on the ellipse Γ_n with foci at $n^p \pm \delta$ and semi-minor-axis equal to $n^{p-1}\delta$:

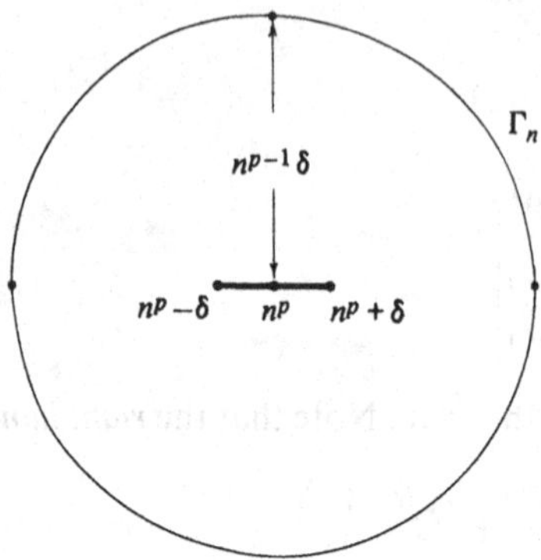

Figure 132

By Problem 19, we have, for the harmonic measure of the component

$$E_n \;=\; [|n|^p \operatorname{sgn} n - \delta,\; |n|^p \operatorname{sgn} n + \delta]$$

of $E = \partial\mathscr{D}$,

$$\omega_{\mathscr{D}}(E_n, \mathrm{i}) \;\leqslant\; \frac{\text{const.}}{|n|^{p+1}+1}.$$

Using this estimate, one can establish a result corresponding to the *first part* of the first theorem in article 3.

Theorem. *Let* $W(x) \geqslant 1$ *be continuous on*

$$E \;=\; \bigcup_{-\infty}^{\infty} [|n|^p \operatorname{sgn} n - \delta,\; |n|^p \operatorname{sgn} n + \delta],$$

and suppose that $W(x) \longrightarrow \infty$ *for* $x \to \pm\infty$ *in E. If, for some* $C > 0$, *the supremum of*

$$\int_E \frac{\log|S(t)|}{1+|t|^{1+1/p}}\,dt$$

for S ranging over all finite sums of the form

$$S(t) \;=\; \sum_{-C \leqslant \lambda \leqslant C} a_\lambda e^{i\lambda t}$$

with $\|S\|_{W,E} \leqslant 1$ *is* finite, *then* $\mathscr{C}_W(E, C) \neq \mathscr{C}_W(E)$.

Proof. We have the above boxed estimate for the harmonic measure (in $\mathscr{D} = \mathbb{C} \sim E$) of the components of E, and a previous corollary gives us a Phragmén–Lindelöf function $Y(z)$ for $\mathscr{D}$. Using these facts, one proceeds exactly as in the proof of the first theorem of article 3.

Remark. The sparsity of the set E involved here has caused the form $dt/(1+t^2)$ occurring in the result of article 3 to be replaced by $dt/(1+|t|^{1+1/p})$.

Remark. The statement of the above theorem goes in only one direction, unlike that of the corresponding one in article 3. There, since we were dealing with the restriction of the form $dt/(1+t^2)$ to E, we were able to obtain a *converse* by simply appealing to Akhiezer's theorem from §E.2 of Chapter VI. In the present situation we can't do that, because we are dealing with $dt/(1+|t|^{1+1/p})$ instead of $dt/(1+t^2)$, and $1/p < 1$. *It would be interesting to see whether* (as seems likely) *the converse is true here.*

In case $W(x) \to \infty$ faster than any power of $|x|$ as $x \to \pm\infty$ in E, we can formulate a result like the above one for *polynomial approximation* on E in the weight W. The statement of it is exactly like that of the *first part* of the *second* theorem in article 3, save that the integrals

$$\int_E \frac{\log|P(x)|}{1+x^2}\,dx$$

figuring there are here replaced by

$$\int_E \frac{\log|P(x)|}{1+|x|^{1+1/p}}\,dx.$$

The proof runs much like that of the result in article 3. Details are left to the reader.

B. The set E reduces to the integers

Consider the set

$$E_\rho \;=\; \bigcup_{n=-\infty}^{\infty} \;[n-\rho,\; n+\rho],$$

where $0 \leqslant \rho < \frac{1}{2}$. If $\rho > 0$, the results of §§A.1–A.3 apply to E_ρ, and there is, in particular, a constant b_ρ such that the inequality

$$\int_{-\infty}^{\infty} \frac{\log(1+|P(x)|^2)}{1+x^2}\,dx \;\leqslant\; b_\rho \int_{E_\rho} \frac{\log(1+|P(x)|^2)}{1+x^2}\,dx,$$

used in proving the second theorem of §A.3, holds for polynomials P.

For this reason, given any M, the set of polynomials P such that

$$\int_{E_\rho} \frac{\log^+|P(x)|}{1+x^2}\,dx \leqslant M$$

forms a normal family in the complex plane.

Suppose now that $\rho = 0$. Then $E_\rho = \mathbb{Z}$, and the proof in §A of the above inequality involving b_ρ, available when $\rho > 0$, cannot be made to work so as to yield a relation of the form

$$\int_{-\infty}^{\infty} \frac{\log(1+|P(x)|^2)}{1+x^2}\,dx \leqslant C\sum_{-\infty}^{\infty} \frac{\log(1+|P(n)|^2)}{1+n^2}$$

That proof depends on the properties of harmonic measure for $\mathscr{D}_\rho = \mathbb{C} \sim E_\rho$ worked out in §A.1 (for $\rho > 0$); there is, however, *no harmonic measure* for $\mathscr{D} = \mathbb{C} \sim \mathbb{Z}$. This makes it seem very unlikely that the set of polynomials P satisfying

$$\sum_{-\infty}^{\infty} \frac{\log^+|P(n)|}{1+n^2} \leqslant M$$

for arbitrary given M would form a normal family in the complex plane, and it is in fact easy to construct a counter example to such a claim.

Take simply

$$P_N(x) = (1-x^2)^{[N/\log N]} \prod_{k=1}^{N} \left(1-\frac{x^2}{k^2}\right)$$

for $N \geqslant 2$, with $[p]$ denoting the greatest integer $\leqslant p$ as usual. Then it is not hard to verify that

$$(*) \qquad \sum_{-\infty}^{\infty} \frac{\log^+|P_N(n)|}{1+n^2} \leqslant 20 \qquad \text{for } N \geqslant 8.$$

At the same time,

$$P_N(\mathrm{i}) \geqslant 2^{[N/\log N]} \underset{N}{\longrightarrow} \infty.$$

Problem 20

Prove $(*)$.

(Hint: $\displaystyle\sum_1^{\infty} \frac{1}{n^2}\log^+|P_N(n)| \leqslant \left[\frac{N}{\log N}\right] \sum_{n=N+1}^{\infty} \frac{\log(n^2-1)}{n^2}$

$$+ \sum_{n=N+1}^{\infty} \frac{1}{n^2}\left[\sum_{k=1}^{N} \log\left|\frac{n^2}{k^2}-1\right|\right]^+.$$

After replacing the sums on the right by suitable integrals and doing some calculation, one obtains the upper bound

$$2 + \frac{2}{\log N} + 2 + 2\int_1^\infty \log\left(\frac{\xi+1}{\xi-1}\right)\frac{d\xi}{\xi}.$$

Here, the integral can be worked out by contour integration.)

This example, however, does not invalidate the analogue (with obvious *statement*) of Akhiezer's theorem for weighted polynomial approximation on $\mathbb{Z}$. In order to disprove such a conjecture, one would (at least) need similar examples with the number 20 standing on the right side of (*) replaced by *arbitrarily small quantities* > 0. No matter how one tries to construct such examples, something always seems to go wrong. It seems impossible to diminish the number in (*) to less than a certain strictly positive quantity without forcing boundedness of the $|P_N(i)|$. One comes in such fashion to believe in the existence of a number $C > 0$ such that the set of polynomials P with

$$\sum_{-\infty}^{\infty} \frac{\log^+ |P(n)|}{1+n^2} \leqslant C$$

does form a normal family in the complex plane.

This partial extension of the result from §A.3 to the limiting case $E_p = \mathbb{Z}$ turns out to be *valid*. With its help one can establish the *complete analogue* of Akhiezer's theorem for weighted polynomial approximation on $\mathbb{Z}$; its interest is not, however, limited to that application. The extension is easily reduced to a special version of it for polynomials P of the particular form

$$P(x) = \prod_k \left(1 - \frac{x^2}{x_k^2}\right)$$

with *real* roots $x_k > 0$, and most of the *real work* is involved in the treatment of this case, taking up all but the last two of the following articles. The investigation is straightforward but very laborious; although I have tried hard to simplify it, I have not succeeded too well.

The difficulties are what they are, and there is no point in stewing over them. It is better to just take hold of the traces and forge ahead.

1. Using certain sums as upper bounds for integrals corresponding to them

Our situation from now up to almost the end of the present § is as follows: we have a polynomial $P(z)$ of the special form

$$P(z) = \prod_k \left(1 - \frac{z^2}{x_k^2}\right),$$

where the x_k are >0 (in other words, $P(z)$ is *even*, with all of its zeros *real*, and $P(0) = 1$), and we are given an *upper bound* for the sum

$$\sum_{-\infty}^{\infty} \frac{\log^+|P(m)|}{1+m^2},$$

or, what amounts to the same thing *here*, for

$$\sum_1^{\infty} \frac{1}{m^2} \log^+|P(m)|.$$

From this information we desire to *obtain* a bound on $|P(z)|$ for each complex z.

The first idea that comes to mind is to try to use our knowledge about the preceding *sum* in order to control *the integral*

$$\int_{-\infty}^{\infty} \frac{\log^+|P(x)|}{1+x^2} dx;$$

we have indeed seen in Chapter VI, §B.1, how to deduce an upper bound on $|P(z)|$ from one for this integral. This plan, although probably too simple to be carried out as it stands, does suggest a start on the study of our problem. For *certain intervals* $I \subset (0, \infty)$,

$$\int_I \frac{\log|P(x)|}{x^2} dx$$

is comparable with

$$\sum_{m \in I \cap \mathbb{Z}} \frac{\log^+|P(m)|}{m^2}.$$

We have

$$\frac{d^2 \log|P(x)|}{dx^2} = -2\sum_k \frac{x^2 + x_k^2}{(x^2 - x_k^2)^2} < 0,$$

so $\log|P(x)|$ is *concave* (downward) on any real interval *free of the zeros* $\pm x_k$ *of* P. This means that, if $a < b$ and P has no *zeros* on $[a, b]$,

$$\int_a^b \log|P(x)| dx \leqslant (b-a)\log|P(m)|$$

for the *midpoint* m of $[a, b]$:

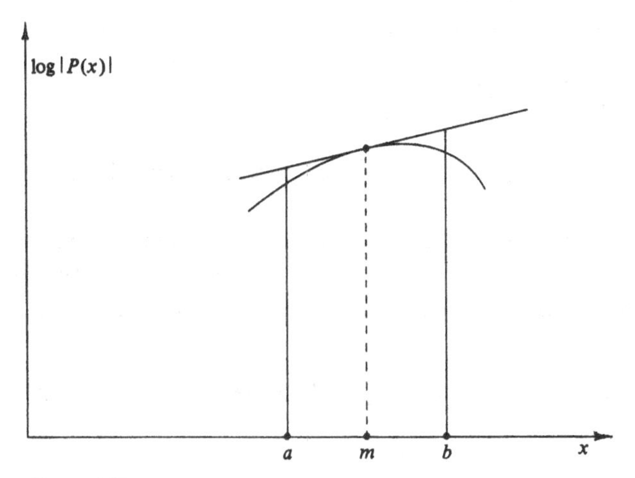

Figure 133

Of course $\int_a^b \log|P(x)|\mathrm{d}x$ is not the integral we are dealing with here. If, however, $a>0$ is large and $b-a$ not too big, the presence of the factor $1/x^2$ in front of $\log|P(x)|$ does not make much difference. A similar formula still holds, except that m is no longer *exactly the midpoint* of $[a, b]$.

Lemma. *Let* $0<a<m<b$, *and suppose that* P *has no zeros on* $[a, b]$. *Then*

$$\int_a^b \frac{\log|P(x)|}{x^2}\mathrm{d}x \;\leqslant\; (\log|P(m)|)\int_a^b \frac{\mathrm{d}x}{x^2},$$

provided that

$$\boxed{\log\frac{b}{a} \;=\; \frac{m}{a} \;-\; \frac{m}{b}.}$$

Proof. Let M denote the *slope* of the graph of $\log|P(x)|$ vs. x at $x=m$. Then, since $\log|P(x)|$ is *concave* on $[a, b]$, we have there

$$\log|P(x)| \;\leqslant\; \log|P(m)| + M(x-m)$$

(see the previous figure). Hence

$$\int_a^b \frac{\log|P(x)|}{x^2}\mathrm{d}x \;\leqslant\; (\log|P(m)|)\int_a^b \frac{\mathrm{d}x}{x^2} \;+\; M\int_a^b\left(\frac{1}{x}-\frac{m}{x^2}\right)\mathrm{d}x.$$

The second term on the right is

$$M\log\frac{b}{a} - M\left(\frac{m}{a}-\frac{m}{b}\right),$$

and this is zero if the boxed condition on m is satisfied. Done.

We will be interested in situations where the number m figuring in the above boxed relation is a *positive integer*, and where *one* of the two numbers a, b $(a \leqslant m \leqslant b)$ is to be *found*, the other being *given*. Regarding these, we have two estimates.

Lemma. *If* $m \geqslant 7$ *and* $m-1 \leqslant a \leqslant m$, *the number* $b \geqslant m$ *such that*

$$\log\frac{b}{a} = \frac{m}{a} - \frac{m}{b}$$

is $\leqslant m+2$.

Proof. Write $\rho = a/b$; then $0 < \rho \leqslant 1$, and the relation to be satisfied becomes $\log(1/\rho) = (m/a)(1-\rho)$. If $a = m$, this is obviously satisfied for $\rho = 1$, i.e., $m = b$; otherwise $0 < \rho < 1$, and we have

$$\frac{m}{a} = \frac{\log\frac{1}{\rho}}{1-\rho}.$$

Now

$$\log\frac{1}{\rho} = 1-\rho+\tfrac{1}{2}(1-\rho)^2+\tfrac{1}{3}(1-\rho)^3+\cdots,$$

so the preceding relation implies that

$$\frac{m}{a} \geqslant 1+\tfrac{1}{2}(1-\rho),$$

i.e.,

$$1-\rho \leqslant 2\frac{m-a}{a},$$

and

$$\rho \geqslant \frac{3a-2m}{a}.$$

Therefore

$$b = \frac{a}{\rho} \leqslant \frac{a^2}{3a-2m},$$

and

$$b - m \leqslant \frac{(2m-a)(m-a)}{3a-2m} \leqslant \frac{m+1}{m-3}.$$

Here the right-hand side is $\leqslant 2$ for $m \geqslant 7$. We are done.

Lemma. *If $m \geqslant 2$ and $m \leqslant b \leqslant m+1$, the number $a \leqslant m$ such that*

$$\log\frac{b}{a} = \frac{m}{a} - \frac{m}{b}$$

is $> m-2$.

Proof. Put $\rho = a/b$ as in proving the preceding lemma; here, it is also convenient to write

$$y = \frac{m}{b}.$$

Then $0 < \rho \leqslant 1$ and $0 < y \leqslant 1$. In terms of y and ρ, our equation becomes

$$\log\frac{1}{\rho} = \frac{y}{\rho} - y.$$

When $y < 1$, we must also have $\rho < 1$, and then

$$y = \frac{\rho\log(1/\rho)}{1-\rho}.$$

This yields, for $0 < \rho < 1$,

$$\frac{\mathrm{d}y}{\mathrm{d}\rho} = \frac{\log(1/\rho) - (1-\rho)}{(1-\rho)^2} = \tfrac{1}{2} + \tfrac{1}{3}(1-\rho) + \tfrac{1}{4}(1-\rho)^2 + \cdots \geqslant \tfrac{1}{2}.$$

Hence, since the value $y = 1$ corresponds to $\rho = 1$, we have, for $0 < y < 1$,

$$\tfrac{1}{2}(1-\rho) \leqslant 1-y,$$

i.e.,

$$\rho \geqslant 1 - 2(1-y).$$

It was given that $m \leqslant b \leqslant m+1$, so

$$1 - y = \frac{b-m}{b} \leqslant \frac{1}{m+1}$$

(the middle term here is a monotone function of b). Therefore, by the previous relation,

$$\rho \geqslant 1 - \frac{2}{m+1},$$

and finally,

$$a = \rho b \geqslant \rho m \geqslant m - \frac{2m}{m+1} > m - 2.$$

We are done.

Theorem. *Let $6 \leqslant a < b$. There is a number b^*, $b \leqslant b^* < b+3$, such that*

$$\int_a^{b^*} \frac{\log|P(x)|}{x^2}\,dx \leqslant 5 \sum_{a<m<b^*} \frac{\log^+|P(m)|}{m^2},$$

provided that P has no zeros on $[a, b^*]$. The sum on the right is taken over the integers *m with $a < m < b^*$.*

> ***Definition.*** During the rest of this §, we will say that b^* is *well disposed* with respect to a.

Proof. By repeated application of the first two of the above lemmas.

Let the integer m_1 be such that $m_1 - 1 \leqslant a < m_1$; then $m_1 \geqslant 7$, so, by the *second* lemma, we can find a number a_1, $m_1 < a_1 \leqslant m_1 + 2$, with

$$\log\frac{a_1}{a} = \frac{m_1}{a} - \frac{m_1}{a_1}.$$

We have $a_1 \leqslant a + 3$, so, since $b > a$, $a_1 < b + 3$.

By the *first* lemma, if $P(x)$ is free of zeros on $[a, a_1]$,

$$\int_a^{a_1} \frac{\log|P(x)}{x^2}\,dx \leqslant \log|P(m_1)| \int_a^{a_1} \frac{dx}{x^2}.$$

Here, $a_1 - a \leqslant 3$ and $m_1/a \leqslant \frac{7}{6}$, so

$$\int_a^{a_1} \frac{dx}{x^2} \leqslant \frac{5}{m_1^2}.$$

Therefore,

$$\int_a^{a_1} \frac{\log|P(x)|}{x^2}\,dx \leqslant \frac{5\log^+|P(m_1)|}{m_1^2}.$$

If now $a_1 \geqslant b$, we simply put $b^* = a_1$ and the theorem is proved. Otherwise, $a_1 < b$ and we take the integer m_2 such that $m_2 - 1 \leqslant a_1 < m_2$. Since $a_1 > m_1$, $m_2 > m_1$, and we can find an a_2, $m_2 < a_2 \leqslant m_2 + 2$, with

$$\log\frac{a_2}{a_1} = \frac{m_2}{a_1} - \frac{m_2}{a_2}.$$

We have $a_2 \leqslant a_1 + 3 < b + 3$, and, by the first lemma,

$$\int_{a_1}^{a_2} \frac{\log|P(x)|}{x^2}\mathrm{d}x \leqslant \log|P(m_2)| \int_{a_1}^{a_2} \frac{\mathrm{d}x}{x^2} \leqslant \frac{5\log^+|P(m_2)|}{m_2^2}$$

just as in the preceding step, provided that P has no zeros on $[a_1, a_2]$.

If $a_2 \geqslant b$, we put $b^* = a_2$. If not, we continue as above, getting numbers $a_3 > a_2$, $a_4 > a_3$, and so forth, $a_{k+1} \leqslant a_k + 3$, until we first reach an a_l with $a_l \geqslant b$. We will then have $a_l < b + 3$, and we put $b^* = a_l$. There are integers m_k, $m_2 < m_3 < \cdots < m_l$, with $a_{k-1} < m_k < a_k$, $k = 3,\ldots,l$, and, as in the previous steps,

$$\int_{a_{k-1}}^{a_k} \frac{\log|P(x)|}{x^2}\mathrm{d}x \leqslant \frac{5\log^+|P(m_k)|}{m_k^2}$$

for $k = 3,\ldots,l$, as long as P has no zeros on $[a_{k-1}, a_k]$.

Write $a_0 = a$. Then, if P has no zeros on $[a, b^*] = [a_0, a_l]$,

$$\int_a^{b^*} \frac{\log|P(x)|}{x^2}\mathrm{d}x = \sum_{k=1}^{l} \int_{a_{k-1}}^{a_k} \frac{\log|P(x)|}{x^2}\mathrm{d}x$$

$$\leqslant \sum_{k=1}^{l} \frac{5\log^+|P(m_k)|}{m_k^2} \leqslant \sum_{\substack{a<m<b^* \\ m\in\mathbb{Z}}} \frac{5\log^+|P(m)|}{m^2}.$$

We are done.

In the result just proved, a is kept fixed and we move from b to a point b^* well disposed with respect to a, lying between b and $b + 3$. One can obtain the same effect keeping b fixed and moving downward from a.

Theorem. *Let $10 \leqslant a < b$. There is an a^*, $a - 3 < a^* \leqslant a$, such that b is well disposed with respect to a^*, i.e.,*

$$\int_{a^*}^{b} \frac{\log|P(x)|}{x^2}\mathrm{d}x \leqslant 5 \sum_{a^*<m<b} \frac{\log^+|P(m)|}{m^2},$$

provided that $P(x)$ has no zeros on $[a^*, b]$.

The proof uses the *first* and *third* of the above lemmas, and is otherwise very much like the one of the previous theorem. Its details are left to the reader.

2. Construction of certain intervals containing the zeros of $P(x)$

We have seen in the preceding article how certain intervals $I \subseteq (0, \infty)$ can be obtained for which

$$\int_I \frac{\log|P(x)|}{x^2} dx \leq 5 \sum_{m \in I} \frac{\log^+|P(m)|}{m^2}$$

as long as they are free of zeros of P. Our next step is to split up $(0, \infty)$ into two kinds of intervals: zero-free ones of the sort just mentioned and then some residual ones which, together, contain all the positive zeros of $P(x)$. The latter are closely related to some intervals used earlier by Vladimir Bernstein (*not* the S. Bernstein after whom the weighted polynomial approximation problem is named) in his study of Dirichlet series, and it is to their construction we now turn.

▶ As is customary, we denote by $n(t)$ *the number of zeros* x_k *of* $P(x)$ *in the interval* $[0, t]$ for $t \geq 0$ (*counting multiplicities as in Chapter* III). When $t < 0$, we take $n(t) = 0$. The function $n(t)$ is thus *integer-valued and increasing.* It is *zero* for all $t > 0$ *sufficiently close to* 0 (because the $x_k > 0$), and *constant* for sufficiently large t (P being a polynomial).

The graph of $n(t)$ vs. t consists of some *horizontal portions* separated by *jumps.* At each jump, $n(t)$ increases by an integral multiple of unity. In this quantization must lie the essential difference between the behaviour of subharmonic functions of the *special form* $\log|P(x)|$ with P a polynomial, and that of *general* ones having at most logarithmic growth at ∞, for which there holds no valid analogue of the theorem to be established in this §. (Just look at the subharmonic functions $\eta \log|P_N(z)|$, where $\eta > 0$ is arbitrarily small and the P_N are the polynomials considered in Problem 20.) During the present article we will see precisely how the quantization affects matters.

For the following work we *fill in the vertical portions* of the *graph* of $n(t)$ vs. t. In other words, if $n(t)$ has a jump discontinuity at t_0, we consider the *vertical segment joining* $(t_0, n(t_0 -))$ *to* $(t_0, n(t_0 +))$ *as forming part of that graph.*

Our constructions are arranged in three stages.

First stage. Construction of the Bernstein intervals

We begin by taking an arbitrary *small* number $p > 0$ (requiring, say, that $p < 1/20$). Once chosen, p is kept *fixed* during most of the discussion of this and the following articles.

Denote by $\mathcal{O}$ the set of points $t_0 \in \mathbb{R}$ with the property that *a straight line of slope* p through $(t_0, n(t_0))$ *cuts or touches the graph of* $n(t)$ *vs.* t *only once.* $\mathcal{O}$ is open and its *complement* in $\mathbb{R}$ consists of a *finite number* of closed intervals B_0, B_1, B_2,... called the Bernstein intervals for slope p associated with the polynomial $P(x)$. (Together, the B_k make up what V. Bernstein called a *neighborhood set* for the positive zeros of P – see page 259 of his book on Dirichlet series. His construction of the B_k is different from the one given here.) It is best to show the formation of the B_k by a diagram:

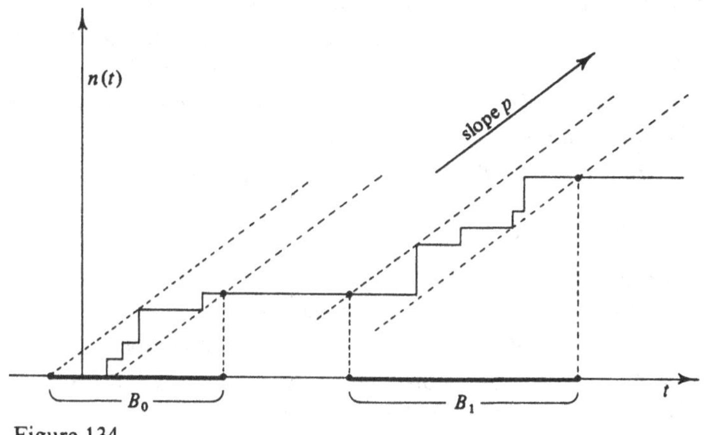

Figure 134

We see that all the positive zeros of P (points of discontinuity of $n(t)$) are contained in the union of the B_k. Also, *taking any B_k and denoting it by $[a, b]$*:

The part of the graph of $n(t)$ vs. t corresponding to the values of t in B_k lies between the two parallel lines of slope p through the points $(a, n(a))$ and $(b, n(b))$.

For a closed interval $I = [\alpha, \beta]$, say, let us write $n(I)$ for $n(\beta+) - n(\alpha-)$. The statement just made then implies that

$$n(B_k)/p|B_k| \leqslant 1$$

for each Bernstein interval. An inequality in the opposite sense is less apparent.

Lemma (Bernstein). *For each of the B_k*,

$$n(B_k)/p|B_k| \geqslant 1/2.$$

Proof. It is geometrically evident that a line of slope p which cuts (or touches) the graph of $n(t)$ vs. t *more than once* must come into contact with some *vertical portion* of it – let the reader make a diagram.

Take any interval B_k, denote it by $[a,b]$, and denote the portion of the graph of $n(t)$ vs. t corresponding to the values $a \leqslant t \leqslant b$ by G. We indicate by L and M the lines of slope p through the points $(a,\ n(a))$ and $(b,\ n(b))$ respectively. According to our definition, any line N of slope p *between* L and M must cut (or touch) the graph of $n(t)$ vs. t *at least twice*, and hence come into contact with some *vertical portion* of that graph. Otherwise such a line N, which surely cuts G, would intersect the graph *only once*, at some point with abscissa $t_0 \in (a,b)$; t_0 would then belong to $\mathcal{O}$ and thus *not* to B_k. The line N must in fact come into contact with a vertical portion of G, for, as a glance at the preceding figure shows, it can never touch any part of the graph that does not lie over $[a,b]$.

In order to prove the lemma, it is therefore enough to show that *if*

$$n(B_k)/p|B_k| \;<\; 1/2.$$

there must be some line N of slope p, lying between L and M, that does not come into contact with any vertical portion of G.

Let V be the *union* of the *vertical portions* of G, and for $X \in V$, denote by $\Pi(X)$ *the downward projection, along a line with slope p,* of the point X onto the horizontal line through $(a,\ n(a))$.

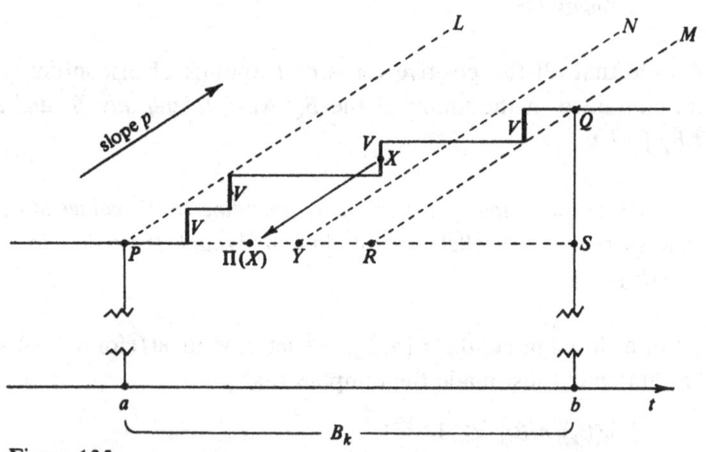

Figure 135

In this figure, $|B_k| = \overline{PS}$ and $n(B_k) = \overline{QS}$. The result, $\Pi(V)$, of applying Π to all the points of V is a certain closed subset of the segment PR, and, if we use $|\;\;|$ to denote linear Lebesgue measure, it is clear that

$$|\Pi(V)| \;\leqslant\; |V|/p.$$

We have $p\cdot\overline{RS} = \overline{QS}$, so, if $n(B_k) \;=\; \overline{QS} \;<\; \tfrac{1}{2}p|B_k| \;=\; \tfrac{1}{2}p\cdot\overline{PS}$, $p\cdot\overline{RS} \;<\; \tfrac{1}{2}p\cdot\overline{PS}$, and therefore $\overline{PR} \;>\; \tfrac{1}{2}\overline{PS} \;>\; \overline{QS}/p \;=\; |V|/p$. With the preceding relation, this yields

$$|\Pi(V)| < \overline{PR}.$$

There is thus a point Y on PR *not belonging* to the projection $\Pi(V)$. If, then, N is the line of slope p through Y, N cannot come into contact with V. This line N lies between L and M, so we are done.

Second stage. Modification of the Bernstein intervals

The Bernstein intervals B_k just constructed include all the positive zeros of $P(x)$, and

$$\frac{1}{2} \leqslant \frac{n(B_k)}{p|B_k|} \leqslant 1.$$

We are going to modify them so as to obtain new closed intervals $I_k \subseteq (0, \infty)$ containing all the positive zeros of $P(x)$, positioned so as to make

$$\int_I \frac{\log|P(x)|}{x^2}dx \leqslant 5\sum_{m\in I} \frac{\log^+|P(m)|}{m^2}$$

for each of the interval components I of

$$(0, \infty) \sim \bigcup_k I_k.$$

(Note that B_0 need not even be contained in $[0, \infty)$.) For the calculations which come later on, it is also very useful to have *all the ratios* $n(I_k)/|I_k|$ *the same*, and we carry out the construction so as to ensure this.

Specifically, the intervals I_k, which we will write as $[\alpha_k, \beta_k]$ with $k = 0, 1, 2, \ldots$ and $0 < \alpha_0 < \beta_0 < \alpha_1 < \beta_1 < \cdots$, are to have the following properties:

(i) *All the positive zeros of* $P(x)$ *are contained in the union of the* I_k,

(ii) $n(I_k)/p|I_k| = \frac{1}{2}, \quad k = 0, 1, 2, \ldots,$

(iii) *For* $\alpha_0 \leqslant t \leqslant \beta_0$,

$$n(\beta_0) - n(t) \leqslant \frac{p}{1-3p}(\beta_0 - t),$$

and, for $\alpha_k \leqslant t \leqslant \beta_k$ *with* $k \geqslant 1$,

$$n(t) - n(\alpha_k) \leqslant \frac{p}{1-3p}(t - \alpha_k),$$

$$n(\beta_k) - n(t) \leqslant \frac{p}{1-3p}(\beta_k - t),$$

(recall that we are assuming $0 < p < \frac{1}{20}$),

(iv) *For* $k \geqslant 1$, α_k *is well disposed with respect to* β_{k-1} (see the preceding article).

Denote the Bernstein intervals B_k, $k = 0, 1, 2, \ldots$, by $[a_k, b_k]$, arranging the indices so as to have $b_{k-1} < a_k$. We begin by constructing I_0. Take α_0 as the *smallest* positive zero of $P(x)$; α_0 is the first point of discontinuity of $n(t)$ and $a_0 < \alpha_0 < b_0$. We have

$$\frac{n([\alpha_0, b_0])}{p(b_0 - \alpha_0)} = \frac{n(B_0)}{p(b_0 - \alpha_0)} > \frac{n(B_0)}{p|B_0|} \geqslant \frac{1}{2}$$

by the lemma from the preceding (first) stage. For $\tau \geqslant b_0$, let J_τ be the interval $[\alpha_0, \tau]$. As we have just seen,

$$n(J_\tau)/p|J_\tau| > 1/2$$

for $\tau = b_0$. When τ increases from b_0 to a_1 (assuming that there *is* a Bernstein interval B_1; there need not be!) the numerator of the left-hand ratio remains equal to $n(B_0)$, while the denominator increases. The ratio itself therefore decreases when τ goes from b_0 to a_1, and either gets down to $\frac{1}{2}$ in (b_0, a_1), or else remains $> \frac{1}{2}$ there. (In case there is no Bernstein interval B_1 we may take $a_1 = \infty$, and then the first possibility is realized.)

Suppose that we *do* have $n(J_\tau)/p|J_\tau| = \frac{1}{2}$ for some τ, $b_0 < \tau < a_1$. Then we put β_0 *equal to that value of* τ, and property (ii) certainly holds for $I_0 = [\alpha_0, \beta_0]$. Property (iii) does also. Indeed, by construction of the B_k, the line of slope p through $(\beta_0, n(\beta_0))$ cuts the graph of $n(t)$ vs. t *only once*, so the portion of the graph corresponding to values of $t < \beta_0$ *lies entirely to the left* of that line (look at the *first* of the diagrams in this article). That is,

$$n(\beta_0) - n(t) \leqslant p(\beta_0 - t), \qquad t \leqslant \beta_0,$$

whence, *a fortiori*,

$$n(\beta_0) - n(t) \leqslant \frac{p}{1 - 3p}(\beta_0 - t), \qquad t \leqslant \beta_0$$

(since $0 < p < 1/20$, $0 < 1 - 3p < 1$).

It may happen, however, that $n(J_\tau)/p|J_\tau|$ remains $> \frac{1}{2}$ for $b_0 < \tau < a_1$. Then that ratio is *still* $\geqslant \frac{1}{2}$ for $\tau = b_1$. This is true because $n(B_1)/p|B_1| \geqslant \frac{1}{2}$ (lemma from the preceding stage), and

$$n([\alpha_0, b_1]) = n(a_1 -) - n(\alpha_0 -) + n(B_1),$$

while

$$b_1 - \alpha_0 = a_1 - \alpha_0 + |B_1|.$$

Thus, in our present case, $n(J_\tau)/p|J_\tau|$ is $\geqslant \frac{1}{2}$ for $\tau = b_1$ and again decreases as τ moves from b_1 towards $a_2 > b_1$. (If there *is* no interval B_2 we may take $a_2 = \infty$.) If, for some $\tau \in [b_1, a_2)$, we have $n(J_\tau)/p|J_\tau| = \frac{1}{2}$, we take β_0 equal to that value of τ, and property (ii) holds for $I_0 = [\alpha_0, \beta_0]$. Also, for $\beta_0 \in [b_1, a_2)$, the part of the graph of $n(t)$ vs. t corresponding to the values $t \leqslant \beta_0$ lies *on or entirely to the left* of the line of slope p through $(\beta_0, n(\beta_0))$, as in the situation already discussed. Therefore, $n(\beta_0) - n(t) \leqslant (p/(1-3p))(\beta_0 - t)$ for $t \leqslant \beta_0$ as before, and property (iii) holds for I_0.

In case $n(J_\tau)/p|J_\tau|$ still remains $> \frac{1}{2}$ for $b_1 \leqslant \tau < a_2$, we will have $n(J_\tau)/p|J_\tau| \geqslant \frac{1}{2}$ for $\tau = b_2$ by an argument like the one used above, and we look for β_0 in the interval $[b_2, a_3)$. The process continues in this way, and we either get a β_0 lying between two successive intervals B_k, B_{k+1} (perhaps coinciding with the right endpoint of B_k), or else pass through the half open interval separating the *last two* of the B_k without ever bringing the ratio $n(J_\tau)/p|J_\tau|$ down to $\frac{1}{2}$. If this second eventuality occurs, suppose that $B_l = [a_l, b_l]$ is the *last* B_k; then $n(J_\tau)/p|J_\tau| \geqslant \frac{1}{2}$ for $\tau = b_l$ by the reasoning already used. Here, $n(J_\tau)$ remains equal to $n([0, b_l])$ for $\tau \geqslant b_l$ while $|J_\tau|$ increases without limit, so a value β_0 of $\tau \geqslant b_l$ will make $n(J_\tau)/p|J_\tau| = \frac{1}{2}$. There is then only one interval I_k, namely, $I_0 = [\alpha_0, \beta_0]$, and our construction is finished, because properties (i) and (ii) obviously hold, while (iii) does by the above reasoning and (iv) is vacuously true.

In the event that the process gives us a β_0 *lying between two successive Bernstein intervals*, we have to construct $I_1 = [\alpha_1, \beta_1]$. In these circumstances we must first choose α_1 so as to have it *well disposed* with respect to β_0, ensuring property (iv) for $k = 1$.

▶ ***It is here that we make crucial use of the property that each jump in $n(t)$ has height $\geqslant$ 1.***

Assume that $b_k \leqslant \beta_0 < a_{k+1}$. We have $p(\beta_0 - \alpha_0) = 2n(I_0) \geqslant 2$ with $0 < p < \frac{1}{20}$; therefore $\beta_0 > 40$ and there is by the *first* theorem of the preceding article a number α_1, $a_{k+1} \leqslant \alpha_1 < a_{k+1} + 3$, which is *well disposed with respect to* β_0.

Now α_1 may well lie to the right of a_{k+1}. It is nevertheless true that $n(\alpha_1) = n(a_{k+1} -)$, and moreover

$$n(t) - n(\alpha_1) \leqslant \frac{p}{1-3p}(t - \alpha_1) \qquad \text{for } t \geqslant \alpha_1.$$

The following diagram shows how these properties follow from two facts:

that $n(t)$ increases by at least 1 at each jump, and that $1/p > 3$:

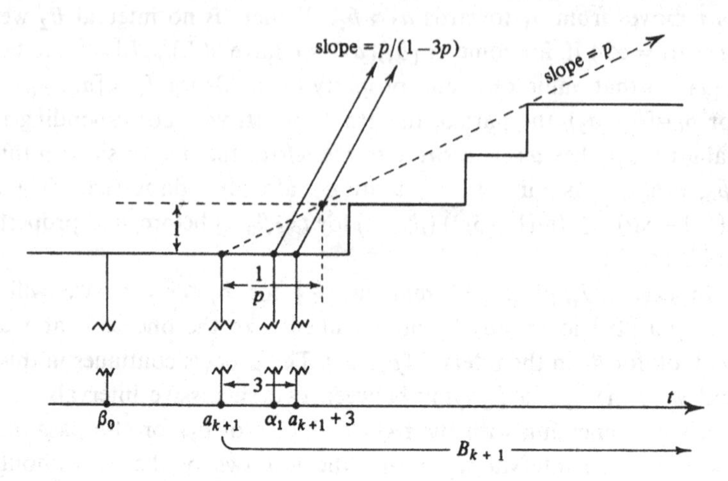

Figure 136

For this choice of α_1, properties (i)–(iv) will hold, provided that β_1, α_2 and so forth are correctly determined.

We go on to specify β_1. This is very much like the determination of β_0. Since

$$n(b_{k+1}) - n(\alpha_1) \;=\; n(b_{k+1}) - n(a_{k+1}-) \;=\; n(B_{k+1}),$$

we certainly have

$$\frac{n([\alpha_1, b_{k+1}])}{p(b_{k+1}-\alpha_1)} \;=\; \frac{n(B_{k+1})}{p(b_{k+1}-\alpha_1)} \;\geqslant\; \frac{n(B_{k+1})}{p|B_{k+1}|} \;\geqslant\; \frac{1}{2}$$

by the lemma from the preceding stage. For $\tau \geqslant b_{k+1}$, denote by J'_τ the interval $[\alpha_1, \tau]$; then $n(J'_\tau)/p|J'_\tau|$ is $\geqslant \frac{1}{2}$ for $\tau = b_{k+1}$ and *diminishes* as τ increases along $[b_{k+1}, a_{k+2})$. (If there *is* no B_{k+2} we take $a_{k+2} = \infty$.) We may evidently proceed just as above to get a $\tau \geqslant b_{k+1}$, lying either in a half open interval separating two successive Bernstein intervals or else beyond all of the latter, such that $n(J'_\tau)/p|J'_\tau| = \frac{1}{2}$. *That value* of τ is taken as β_1. The part of the graph of $n(t)$ vs. t corresponding to values of $t \leqslant \beta_1$ lies, as before, *on or to the left* of the line through $(\beta_1, n(\beta_1))$ with slope p. Hence, *a fortiori*,

$$n(\beta_1) - n(t) \;\leqslant\; \frac{p}{1-3p}(\beta_1 - t) \qquad \text{for } t \leqslant \beta_1.$$

We see that properties (ii) and (iii) hold for I_0 and $I_1 = [\alpha_1, \beta_1]$.

If $I_0 \cup I_1$ does not already include all of the B_k, β_1 must lie between two of them, and we may proceed to find an α_2 in the way that α_1 was found above. Then we can construct an I_2. Since there are only a finite number of B_k, the process will eventually stop, and we will end with a finite number of intervals $I_k = [\alpha_k, \beta_k]$ having properties (ii)–(iv). Property (i) will then also hold, since, when we finish, the union of the I_k includes that of the B_k.

Here is a picture showing the relation of the intervals I_k to the graph of $n(t)$ vs. t:

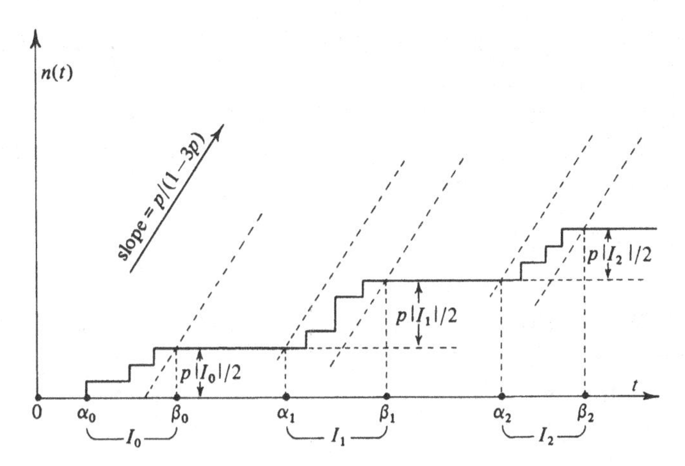

Figure 137

Let us check the statement made before starting the construction of the I_k, to the effect that

$$\int_I \frac{\log|P(x)|}{x^2}\,dx \;\leqslant\; 5\sum_{m\in I}\frac{\log^+|P(m)|}{m^2}$$

for each of the interval components I of the *complement*

$$(0,\infty) \sim \bigcup_k I_k.$$

Since, for $k > 0$, α_k is well disposed with respect to β_{k-1}, this is certainly true for the components I of the form (β_{k-1}, α_k), $k \geqslant 1$ (if there *are* any!), by the first theorem of the preceding article. This is also true, and trivially so, for $I = (0, \alpha_0)$, because

$$|P(x)| \;=\; \prod_k \left|1 - \frac{x^2}{x_k^2}\right| \;<\; 1$$

for $0 < x < \alpha_0$, all the positive zeros x_k of $P(x)$ being $\geqslant \alpha_0$. Finally, if I_l is the *last* of the I_k, our relation is true for $I = (\beta_l, \infty)$. This follows because we can obviously get *arbitrarily large* numbers $A > \beta_l$ which are well disposed with respect to β_l. We then have

$$\int_{\beta_l}^{A} \frac{\log|P(x)|}{x^2}\,dx \;\leqslant\; 5 \sum_{\beta_l < m < A} \frac{\log^+|P(m)|}{m^2}$$

for each such A by the first theorem of the preceding article, and need only make A tend to ∞.

Third stage. Replacement of the first few intervals I_k by a single one if $n(t)/t$ is not always $\leqslant p/(1-3p)$

Recall that the problem we are studying is as follows: we are presented with an unknown even polynomial $P(z)$ having only real roots and such that $P(0) = 1$, and *told* that

$$\sum_1^\infty \frac{\log^+|P(m)|}{m^2}$$

is small. We are *asked to obtain,* for $z \in \mathbb{C}$, a *bound on* $|P(z)|$ depending on that sum, but independent of P.

As a control on the size of $|P(z)|$ we will use the quantity

$$\sup_{t>0} \frac{n(t)}{t}.$$

A computation like the one at the end of §B, Chapter III, shows indeed that

$$\log|P(z)| \;\leqslant\; \pi|z| \sup_{t>0} \frac{n(t)}{t}.$$

We are therefore interested in obtaining an upper bound on $\sup_{t>0}(n(t)/t)$ from a suitable (small) one for

$$\sum_1^\infty \frac{\log^+|P(m)|}{m^2}.$$

Our procedure is to *work backwards,* assuming that $\sup_{t>0}(n(t)/t)$ is *not small* and thence deriving a *strictly positive lower bound for the sum.* We begin with the following simple

Lemma. *If* $\sup_{t>0}(n(t)/t) > p/(1-3p)$, *we have* $|I_0|/\beta_0 \geqslant \frac{2}{3}$ *for the interval* $I_0 = [\alpha_0, \beta_0]$ *arrived at in the previous stage of our construction.*

Proof. Let us examine carefully the *initial portion* of the last diagram given above:

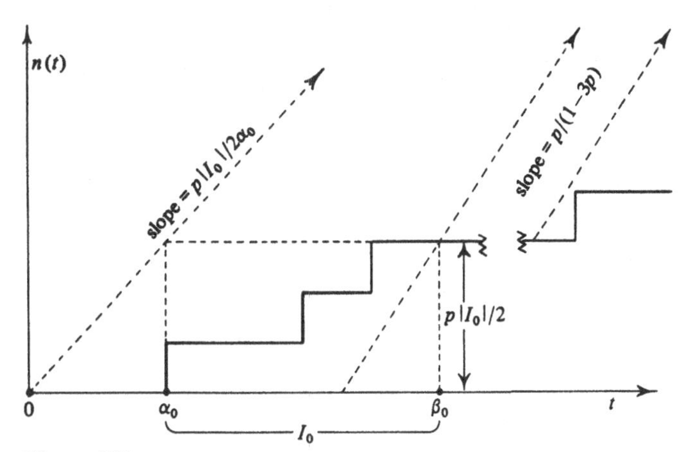

Figure 138

We see that, for $t > 0$,

$$\frac{n(t)}{t} \leqslant \max\left\{\frac{p}{1-3p},\ \frac{p|I_0|}{2\alpha_0}\right\},$$

whether or not the first term in curly brackets is less than the second.

Here,

$$\frac{p|I_0|}{2\alpha_0} = \frac{p}{2}\cdot\frac{|I_0|/\beta_0}{1-|I_0|/\beta_0},$$

and this is $< p < p/(1-3p)$ (making the above maximum *equal* to $p/(1-3p)$) if $|I_0|/\beta_0 < \frac{2}{3}$. Done.

Our construction of the intervals I_k involved the parameter p. We now bring in another quantity, η, which will continue to intervene during most of the articles of this §. For the time being, we require only that $0 < \eta < \frac{2}{3}$ and take the value of η as fixed during the work that follows. From time to time we will obtain various intermediate results whose validity will depend on η's having been chosen sufficiently small to begin with. A final decision about η's size will be made when we put together those results.

In accordance with the above indication of our procedure, we assume henceforth that

$$\sup_{t>0}\frac{n(t)}{t} > \frac{p}{1-3p}.$$

By the lemma we then certainly have

$$|I_0|/\beta_0 > \eta,$$

since we are taking $0 < \eta < \frac{2}{3}$. This being the case, we replace the first few intervals I_k by a single one, according to the following construction.

Let $\omega(x)$ be the continuous and piecewise linear function defined on $[0, \infty)$ which has slope 1 on each of the intervals I_k and slope *zero* elsewhere, and vanishes at the origin:

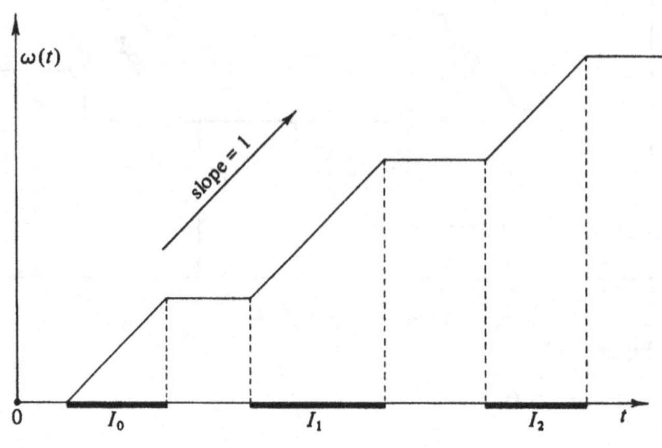

Figure 139

The ratio $\omega(t)/t$ is continuous and tends to zero as $t \to \infty$ since there are only a finite number of I_k. Clearly, $\omega(t)/t < 1$, so, if t belongs to the *interior* of an I_k,

$$\frac{\mathrm{d}}{\mathrm{d}t}\left(\frac{\omega(t)}{t}\right) = \frac{1}{t} - \frac{\omega(t)}{t^2} > 0;$$

i.e., *$\omega(t)/t$ is strictly increasing on each I_k.*

We have

$$\omega(\beta_0)/\beta_0 = |I_0|/\beta_0 > \eta,$$

so, in view of what has just been said, there must be a *largest* value of t $(> \beta_0)$ for which

$$\omega(t)/t = \eta$$

and that value *cannot lie in the interior, or be a left endpoint,* of any of the intervals I_k. Denote by d that value of t. Then, since $d > \beta_0$, there must be a *last interval I_k* – call it I_m – *lying entirely to the left* of d. If I_m is also the last of the intervals I_k we write

$$d_0 = d,$$
$$c_0 = (1 - \eta)d,$$

and denote the interval $[c_0, d_0]$ by J_0. In this case all the positive zeros of $P(x)$ (discontinuities of $n(t)$) lie to the *left* of d_0.

It may be, however, that I_m is *not* the last of the I_k; then there is an interval

$$I_{m+1} = [\alpha_{m+1}, \beta_{m+1}],$$

and we must have $d < \alpha_{m+1}$ according to the above observation. Since $d > \beta_0 > 2/p > 40$ (remember that we are taking $0 < p < \frac{1}{20}$), we can apply the *second* theorem of article 1 to conclude that there is a d_0,

$$d - 3 \; < \; d_0 \; \leqslant \; d,$$

such that α_{m+1} is *well disposed with respect to* d_0. We then put

$$c_0 \;=\; d_0 - \eta d$$

and denote by J_0 the interval $[c_0,\, d_0]$. The intervals $I_{m+1}, I_{m+2}, \ldots$ are also relabeled as follows:

$$I_{m+1} \;=\; J_1,$$
$$I_{m+2} \;=\; J_2,$$

and we write $\alpha_{m+1} = c_1$, $\beta_{m+1} = d_1$, $\alpha_{m+2} = c_2$, $\beta_{m+2} = d_2$, and so forth, so as to have the uniform notation

$$J_k \;=\; [c_k,\, d_k], \qquad k = 0, 1, 2, \ldots$$

In the present case, $\beta_m \leqslant d < \alpha_{m+1}$ (*sic*!) so, referring to the previous (second) stage of our construction, we see that the part of the graph of $n(t)$ vs. t corresponding to values of $t \leqslant d$ lies entirely to the left of, or on, the line of slope p (*sic*!) through $(d,\; n(d))$. By an argument very much like the one near the end of the second stage, based on the fact that $n(t)$ increases by *at least* 1 at each of its jumps, this implies that d_0, *although it may lie to the left of* d, *still lies to the right of all the zeros of* $P(x)$ *in* $I_0, \ldots, I_m$, *and that*

$$n(d_0) - n(t) \;\leqslant\; \frac{p}{1-3p}(d_0 - t) \qquad \text{for } \; t \leqslant d_0.$$

(The diagram used here is obtained by *rotating through 180°* the one from the argument just referred to.)

We have, in the first place, $c_0 \;\geqslant\; (1-\eta)d - 3 \;>\; 0$, because $\eta < \frac{2}{3}$ and $d > \beta_0 > 40$.

In the second place,

$$|J_0|/d_0 \;\geqslant\; |J_0|/d \;=\; \eta,$$

by choice of d. Also,

$$\frac{|J_0|}{d_0} \;<\; \frac{|J_0|}{d-3} \;=\; \frac{\eta d}{d-3} \;<\; \frac{40\eta}{37},$$

since $d > 40$.

Finally,

$$\frac{n(d_0)}{p|J_0|} = \frac{1}{2}.$$

Indeed, both d_0 and d lie strictly between all the discontinuities of $n(t)$ in $I_0, I_1, \ldots, I_m$ and those in $I_{m+1}, I_{m+2}, \ldots$ (or to the *right* of the last I_k if our construction yields *only one* interval J_0), so

$$n(d_0) = \sum_{k=0}^{m} n(I_k) = \frac{1}{2}\sum_{k=0}^{m} p|I_k|$$

by property (ii) of the I_k. And

$$\sum_{k=0}^{m} |I_k| = \omega(d) = \eta d = |J_0|$$

by the choice of d and the definition of J_0. Thus $n(d_0) = \frac{1}{2}p|J_0|$, as claimed.

Denote by J the union of the J_k, and put for the moment

$$\tilde{\omega}(t) = |[0, t] \cap J|.$$

The function $\tilde{\omega}(t)$ is similar to $\omega(t)$, considered above, and differs from the latter only in that it increases (with constant slope 1) on each of the J_k instead of doing so on the I_k. The ratio $\tilde{\omega}(t)/t$ is therefore *increasing* on each J_k (see above), so in particular

$$\frac{\tilde{\omega}(t)}{t} \leqslant \frac{\tilde{\omega}(d_0)}{d_0} = \frac{|J_0|}{d_0} < \frac{40\eta}{37}$$

for $t \in J_0 = [c_0, d_0]$. This inequality remains (trivially) true for $0 \leqslant t < c_0$, since $\tilde{\omega}(t) = 0$ there. It also remains true for $d_0 \leqslant t \leqslant d$, for $\tilde{\omega}(t)$ is constant on that interval. And finally,

$$\tilde{\omega}(t) = \omega(t) \qquad \text{for } t > d,$$

so $\tilde{\omega}(t)/t = \omega(t)/t < \eta$ for such t by choice of d. Thus, we surely have

$$\frac{\tilde{\omega}(t)}{t} < 2\eta \qquad \text{for } t \geqslant 0.$$

The quantity on the left is, however, equal to $|J_0|/d_0 \geqslant \eta$ for $t = d_0$.

The purpose of the constructions in this article has been to arrive at the intervals J_k, and the remaining work of this § concerned with even polynomials having real zeros *deals exclusively with them.* The preceding discussions amount to a proof of the following

Theorem. *Let p, $0<p<\frac{1}{20}$, and η, $0<\eta<\frac{2}{3}$, be given, and suppose that*

$$\sup_t \frac{n(t)}{t} \; > \; \frac{p}{1-3p}.$$

Then there is a finite collection of intervals $J_k=[c_k,\, d_k]$, $k\geqslant 0$, lying in $(0,\infty)$, such that

(i) *all the discontinuities of $n(t)$ lie in $(0,d_0)\cup\bigcup_{k\geqslant 1}J_k$;*

(ii) $$\frac{n(d_0)}{p|J_0|} \; = \; \frac{n(J_k)}{p|J_k|} \; = \; \frac{1}{2} \qquad \textit{for } k\geqslant 1$$

(if there are intervals J_k with $k\geqslant 1$);

(iii) *for $0\leqslant t\leqslant d_0$,*

$$n(d_0)-n(t) \; \leqslant \; \frac{p}{1-3p}(d_0-t),$$

whilst, for $c_k\leqslant t\leqslant d_k$ when $k\geqslant 1$,

$$n(t)-n(c_k) \; \leqslant \; \frac{p}{1-3p}(t-c_k)$$

and

$$n(d_k)-n(t) \; \leqslant \; \frac{p}{1-3p}(d_k-t);$$

(iv) *for $k\geqslant 1$, c_k is well disposed with respect to d_{k-1} (if there are J_k with $k\geqslant 1$);*

(v) *for $t\geqslant 0$,*

$$\frac{1}{t}\left|[0,t]\cap\bigcup_{k\geqslant 0}J_k\right| \; < \; 2\eta,$$

and the quantity on the left is $\geqslant\eta$ for $t=d_0$.

Remark 1. The J_k with $k\geqslant 1$ (if there are any) are just certain of the I_r from the *second stage*. So, for $k\geqslant 1$, the above property (ii) is just property (ii) for the I_r.

Remark 2. By property (iv) and the theorems of article 1, we have

$$\int_{d_{k-1}}^{c_k} \frac{\log|P(x)|}{x^2}\,dx \; \leqslant \; 5 \sum_{d_{k-1}<m<c_k} \frac{\log^+|P(m)|}{m^2}$$

for each of the intervals $(d_{k-1},\, c_k)$ with $k\geqslant 1$ (if there are any). And, if J_1

is the *last* of the J_k,

$$\int_{d_l}^{\infty} \frac{\log|P(x)|}{x^2}\,dx \;\leqslant\; 5 \sum_{d_l<m<\infty} \frac{\log^+|P(m)|}{m^2}.$$

See the end of the *second* stage of the preceding construction.

Here is a picture of the graph of $n(t)$ vs. t, showing the intervals J_k:

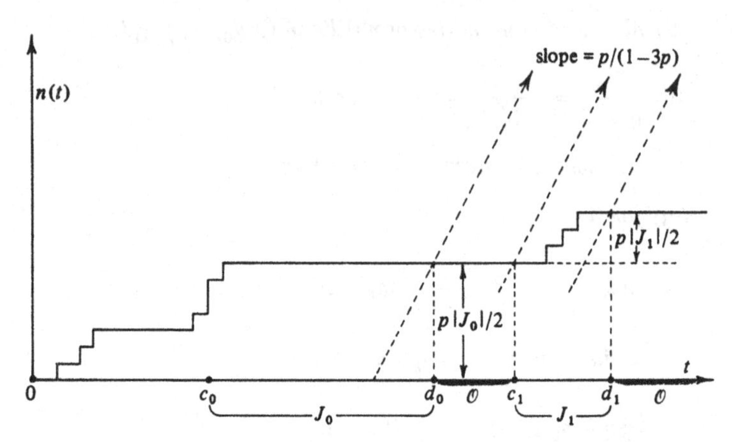

Figure 140

3. Replacement of the distribution $n(t)$ by a continuous one

Having chosen p, $0<p<1/20$, and η, $0<\eta<\frac{2}{3}$, we continue with our program, assuming that

$$\sup_{t>0} \frac{n(t)}{t} > \frac{p}{1-3p},$$

our aim being to obtain a *lower bound* for

$$\sum_{1}^{\infty} \frac{\log^+|P(m)|}{m^2}.$$

Our assumption makes it possible, by the work of the preceding article, to get the intervals

$$J_k = [c_k, d_k] \subset (0,\infty), \quad k=0,1,\ldots,$$

related to the (unknown) increasing function $n(t)$ in the manner described by the theorem at the end of that article.

Let J_l be the *last* of those J_k; *during this article we will denote the union*

$$(d_0, c_1)\cup(d_1, c_2)\cup\cdots\cup(d_{l-1}, c_l)\cup(d_l, \infty)$$

by $\mathcal{O}$ – see the preceding diagram. (Note that this is *not* the same set $\mathcal{O}$ as the one used at the beginning of article 2!) Our idea is to estimate

$$\sum_{m\in\mathcal{O}} \frac{\log^+|P(m)|}{m^2}$$

from below, this quantity being certainly *smaller* than the one we are interested in. According to Remark 2 following the theorem about the J_k, we have

$$\sum_{m\in\mathcal{O}} \frac{\log^+|P(m)|}{m^2} \geqslant \frac{1}{5}\int_{\mathcal{O}} \frac{\log|P(x)|}{x^2}\,dx.$$

What we want, then, is a *lower bound* for *the integral* on the right. This is the form that our initial simplistic plan of 'replacing' sums by integrals finally assumes.

In terms of $n(t)$,

$$\log|P(x)| = \sum_k \log\left|1 - \frac{x^2}{x_k^2}\right| = \int_0^\infty \log\left|1 - \frac{x^2}{t^2}\right| dn(t),$$

so the object of our interest is the expression

$$\frac{1}{5}\int_{\mathcal{O}}\int_0^\infty \log\left|1 - \frac{x^2}{t^2}\right| dn(t)\frac{dx}{x^2}.$$

Here, $n(t)$ is constant on each component of $\mathcal{O}$, and increases only on that set's complement.

We are now able to render our problem more tractable by replacing $n(t)$ with another increasing function $\mu(t)$ of much more simple and regular behaviour, continuous and piecewise linear on $\mathbb{R}$ and constant on each of the intervals complementary to the J_k. The slope $\mu'(t)$ will take only *two* values, 0 and $p/(1-3p)$, and, on each J_k, $\mu(t)$ will increase by $p|J_k|/2$. What we have to do is find such a $\mu(t)$ which makes

$$\frac{1}{5}\int_{\mathcal{O}}\int_0^\infty \log\left|1 - \frac{x^2}{t^2}\right| d\mu(t)\frac{dx}{x^2}$$

smaller than the expression written above, yet *still* (we hope) *strictly positive.*

Part of our requirement on $\mu(t)$ is that $\mu(t) = n(t)$ for $t \in \mathcal{O}$, so we will have

$$\int_0^\infty \log\left|1 - \frac{x^2}{t^2}\right| d\mu(t) - \int_0^\infty \log\left|1 - \frac{x^2}{t^2}\right| dn(t)$$

$$= \int_0^{d_0} \log\left|1 - \frac{x^2}{t^2}\right| d(\mu(t) - n(t))$$

$$+ \sum_{k\geqslant 1} \int_{c_k}^{d_k} \log\left|1 - \frac{x^2}{t^2}\right| d(\mu(t) - n(t)).$$

We are interested in values of x in $\mathcal{O}$, and for them, each of the above terms can be integrated by parts. Since $\mu(t) = n(t) = 0$ for t near 0 and $\mu(d_0) = n(d_0)$, $\mu(c_k) = n(c_k)$ and $\mu(d_k) = n(d_k)$ for $k \geqslant 1$, we obtain in this way the expression

$$\int_0^{d_0} \frac{2x^2}{x^2 - t^2} \frac{\mu(t) - n(t)}{t} \mathrm{d}t + \sum_{k \geqslant 1} \int_{c_k}^{d_k} \frac{2x^2}{x^2 - t^2} \frac{\mu(t) - n(t)}{t} \mathrm{d}t.$$

Therefore

$$\int_{\mathcal{O}} \int_0^{\infty} \log\left|1 - \frac{x^2}{t^2}\right| \mathrm{d}(\mu(t) - n(t)) \frac{\mathrm{d}x}{x^2}$$

$$= \int_0^{d_0} \int_{\mathcal{O}} \frac{2\mathrm{d}x}{x^2 - t^2} \cdot \frac{\mu(t) - n(t)}{t} \mathrm{d}t$$

$$+ \sum_{k \geqslant 1} \int_{c_k}^{d_k} \int_{\mathcal{O}} \frac{2\mathrm{d}x}{x^2 - t^2} \cdot \frac{\mu(t) - n(t)}{t} \mathrm{d}t,$$

and we desire to find a function $\mu(t)$ fitting our requirements, for which each of the terms on the right comes out *negative*.

Put

$$F(t) = 2\int_{\mathcal{O}} \frac{\mathrm{d}x}{x^2 - t^2}$$

for $t \notin \mathcal{O}$. We certainly have $F(t) > 0$ for $0 < t < d_0$, so the first right-hand term, which equals

$$\int_0^{d_0} F(t) \frac{\mu(t) - n(t)}{t} \mathrm{d}t$$

is $\leqslant 0$ if $\mu(t) \leqslant n(t)$ on $[0, d_0]$. Referring to the diagram at the end of the previous article, we see that this will happen if, for $0 \leqslant t \leqslant d_0$, $\mu(t)$ has the form shown here:

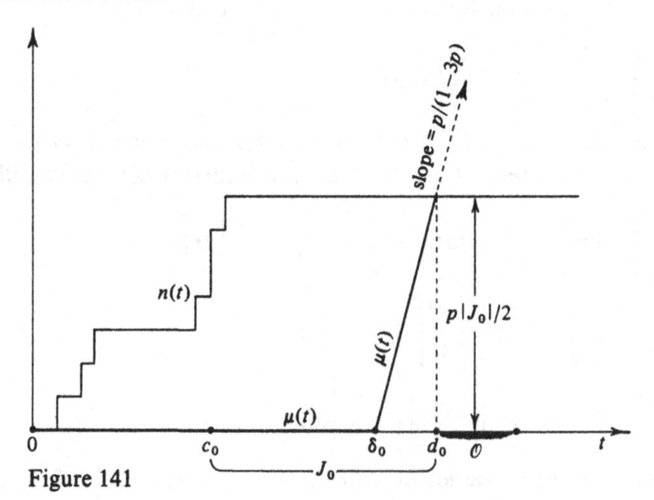

Figure 141

For $k \geqslant 1$, we need to define $\mu(t)$ on $[c_k, d_k]$ in a manner compatible with our requirements, so as to make

$$\int_{c_k}^{d_k} F(t)\frac{\mu(t)-n(t)}{t}\,dt \;\leqslant\; 0.$$

Here, $\mathcal{O}$ includes intervals of the form

$$(c_k - \delta,\; c_k)$$

and

$$(d_k,\; d_k + \delta)$$

where $\delta > 0$, so, when $t \in (c_k, d_k)$, $F(t) \to -\infty$ for $t \to c_k$ and $F(t) \to \infty$ for $t \to d_k$. Moreover, for such t,

$$F'(t) \;=\; 4t\int_{\mathcal{O}} \frac{dx}{(x^2-t^2)^2} \;>\; 0,$$

so there is *precisely one point* $t_k \in (c_k, d_k)$ *where* $F(t)$ *vanishes*, and $F(t) < 0$ for $c_k < t < t_k$, while $F(t) > 0$ for $t_k < t < d_k$. We see that in order to make

$$\int_{c_k}^{d_k} F(t)\frac{\mu(t)-n(t)}{t}\,dt \;\leqslant\; 0,$$

it is enough to define $\mu(t)$ so as to make

$$\mu(t) \;\geqslant\; n(t) \qquad \text{for } c_k \leqslant t < t_k$$

and

$$\mu(t) \;\leqslant\; n(t) \qquad \text{for } t_k \leqslant t \leqslant d_k.$$

The following diagram shows how to do this:

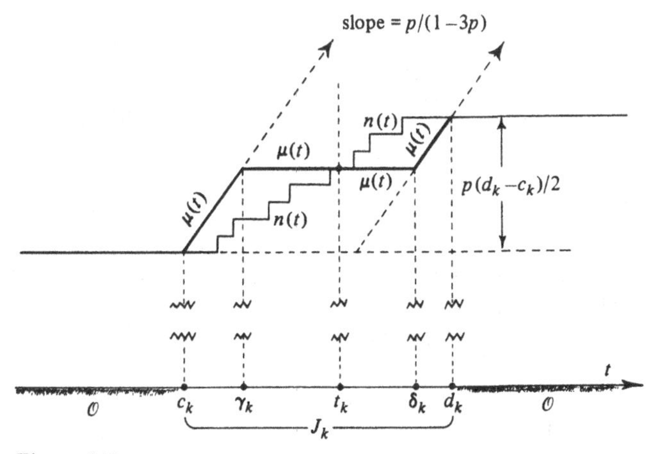

Figure 142

We carry out this construction on each of the J_k. When we are done we will have a function $\mu(t)$, defined for $t \geqslant 0$, with the following properties:

(i) $\mu(t)$ is piecewise linear and increasing, and constant on each interval component of

$$(0,\infty) \sim \bigcup_{k\geqslant 0} J_k;$$

(ii) on each of the intervals J_k, $\mu(t)$ increases by $p|J_k|/2$;

(iii) on J_0, $\mu(t)$ has *slope zero* for $c_0 < t < \delta_0$ and *slope* $p/(1-3p)$ for $\delta_0 < t < d_0$, where $(d_0 - \delta_0)/(d_0 - c_0) = (1-3p)/2$;

(iv) on each J_k, $k \geqslant 1$, $\mu(t)$ has *slope zero* for $\gamma_k < t < \delta_k$ and *slope* $p/(1-3p)$ in the intervals (c_k, γ_k) and (δ_k, d_k), where $c_k < \gamma_k < \delta_k < d_k$ and

$$\frac{\gamma_k - c_k + d_k - \delta_k}{d_k - c_k} = \frac{1-3p}{2};$$

(v) $$\int_{\mathcal{O}}\int_0^{\infty} \log\left|1-\frac{x^2}{t^2}\right| d\mu(t)\frac{dx}{x^2} \leqslant \int_{\mathcal{O}}\int_0^{\infty} \log\left|1-\frac{x^2}{t^2}\right| dn(t)\frac{dx}{x^2}.$$

Here is a drawing of the graph of $\mu(t)$ vs. t which the reader will do well to look at from time to time while reading the following articles:

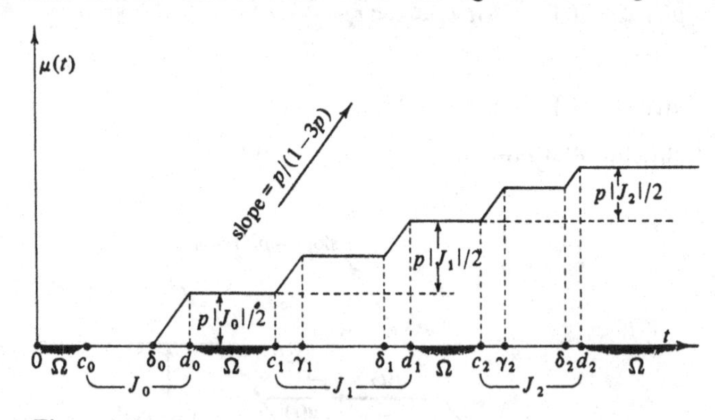

Figure 143

In what follows, we will in fact be working with integrals not over $\mathcal{O}$, but over the set $\Omega = (0, c_0) \cup \mathcal{O} = (0,\infty) \sim \bigcup_{k>0} J_k$ (see the diagram). Since our function $\mu(t)$ is *zero* for $t \leqslant c_0$, we certainly have

$$\int_0^{\infty} \log\left|1-\frac{x^2}{t^2}\right| d\mu(t) \leqslant 0$$

for $0 < t < c_0$. Hence, by property (v),

$$\int_\Omega \int_0^\infty \log\left|1 - \frac{x^2}{t^2}\right| d\mu(t) \frac{dx}{x^2} \leqslant \int_{\mathcal{O}} \frac{\log|P(x)|}{x^2} dx$$

for our polynomial P. And, as we have seen at the beginning of this article, the right-hand integral is in turn

$$\leqslant 5\sum_1^\infty \frac{\log^+|P(m)|}{m^2}.$$

What we have here is a

Theorem. *Let $0 < p < 1/20$ and $0 < \eta < \frac{2}{3}$, and suppose that*

$$\sup_{t>0} \frac{n(t)}{t} > \frac{p}{1-3p}.$$

Then there are intervals $J_k \subset (0, \infty)$, $k \geqslant 0$, fulfilling the conditions enumerated in the theorem of the preceding article, and a piecewise linear increasing function $\mu(t)$, related to those J_k in the manner just described, such that

$$\int_\Omega \int_0^\infty \log\left|1 - \frac{x^2}{t^2}\right| d\mu(t) \frac{dx}{x^2} \leqslant 5\sum_1^\infty \frac{\log^+|P(m)|}{m^2}$$

for the polynomial $P(x)$.

Here,

$$\Omega = (0, \infty) \sim \bigcup_{k \geqslant 0} J_k.$$

Our problem has thus boiled down to the *purely analytical one* of finding a positive lower bound for

$$\int_\Omega \int_0^\infty \log\left|1 - \frac{x^2}{t^2}\right| d\mu(t) \frac{dx}{x^2}$$

when $\mu(t)$ has the very special form shown in the above diagram. Note that here $|J_0|/d_0 \geqslant \eta$ according to the theorem of the preceding article.

4. Some formulas

The problem, formulated at the end of the last article, to which we have succeeded in reducing our original one seems at first glance to be rather easy – one feels that one can just sit down and *compute*

$$\int_\Omega \int_0^\infty \log\left|1 - \frac{x^2}{t^2}\right| d\mu(t) \frac{dx}{x^2}.$$

This, however, is far from being the case, and quite formidable difficulties still stand in our way. The trouble is that the intervals J_k to which μ is related may be exceedingly numerous, and we have no control over their positions relative to each other, nor on their relative lengths. To handle our task, we are going to need all the formulas we can muster.

Lemma. *Let $\nu(t)$ be increasing on $[0, \infty)$, with $\nu(0) = 0$ and $\nu(t) = O(t)$ for $t \to 0$ and for $t \to \infty$. Then, for $x \in \mathbb{R}$,*

$$\int_0^\infty \log\left|1 - \frac{x^2}{t^2}\right| d\nu(t) \;=\; -x\int_0^\infty \log\left|\frac{x+t}{x-t}\right| d\left(\frac{\nu(t)}{t}\right).$$

Proof. Both sides are even functions of x and zero for $x = 0$, so we may as well assume that $x > 0$. If $\nu(t)$ has a (jump) discontinuity at x, both sides are clearly equal to $-\infty$, so we may suppose $\nu(t)$ continuous at x.

We have

$$\int_0^x \log\left|\frac{x+t}{x-t}\right| d\left(\frac{\nu(t)}{t}\right) \;=\; \int_0^x \frac{1}{t}\log\left|\frac{x+t}{x-t}\right| d\nu(t)$$

$$-\;\int_0^x \frac{1}{t^2}\log\left|\frac{x+t}{x-t}\right| \nu(t)\,dt.$$

Using the identity

$$\int \frac{1}{t^2}\log\left|\frac{x+t}{x-t}\right| dt \;=\; -\frac{1}{t}\log\left|\frac{x+t}{x-t}\right| - \frac{1}{x}\log\left|1 - \frac{x^2}{t^2}\right|,$$

we integrate the second term on the right by parts, obtaining for it the value

$$-\;\frac{2\nu(x)\log 2}{x} \;+\; \int_0^x\left(\frac{1}{t}\log\left|\frac{x+t}{x-t}\right| + \frac{1}{x}\log\left|1 - \frac{x^2}{t^2}\right|\right) d\nu(t),$$

taking into account the given behaviour of $\nu(t)$ near 0. Hence

$$\int_0^x \log\left|\frac{x+t}{x-t}\right| d\left(\frac{\nu(t)}{t}\right) \;=\; \frac{2\nu(x)\log 2}{x} - \frac{1}{x}\int_0^x \log\left|1 - \frac{x^2}{t^2}\right| d\nu(t).$$

In the same way, we get

$$\int_x^\infty \log\left|\frac{x+t}{x-t}\right| d\left(\frac{\nu(t)}{t}\right)$$

$$= \;-\left(\frac{2\nu(x)\log 2}{x}\right) - \frac{1}{x}\int_x^\infty \log\left|1 - \frac{x^2}{t^2}\right| d\nu(t).$$

Adding these last two relations gives us the lemma.

Corollary. *Let* $\nu(t)$ *be* increasing and bounded *on* $[0, \infty)$, *and* zero *for all* t *sufficiently close to* 0. *Let* $\omega(x)$ *be increasing on* $[0, \infty)$, constant *for all sufficiently large* x, *and continuous at* 0. *Then*

$$\int_0^\infty \int_0^\infty \log\left|1 - \frac{x^2}{t^2}\right| \mathrm{d}\nu(t) \frac{\mathrm{d}x - \mathrm{d}\omega(x)}{x^2}$$

$$= \int_0^\infty \int_0^\infty \log\left|\frac{x+t}{x-t}\right| \mathrm{d}\left(\frac{\nu(t)}{t}\right) \frac{\mathrm{d}\omega(x)}{x}.$$

Proof. By the lemma, the left-hand side equals

$$\int_0^\infty \int_0^\infty \log\left|\frac{x+t}{x-t}\right| \mathrm{d}\left(\frac{\nu(t)}{t}\right) \frac{\mathrm{d}\omega(x) - \mathrm{d}x}{x}.$$

Our condition on ν makes

$$\int_0^\infty \int_0^\infty \log\left|\frac{x+t}{x-t}\right| \mathrm{d}\left(\frac{\nu(t)}{t}\right) \frac{\mathrm{d}x}{x}$$

absolutely convergent, so we can change the order of integration. For $t > 0$,

$$\int_0^\infty \log\left|\frac{x+t}{x-t}\right| \frac{\mathrm{d}x}{x}$$

assumes a constant value (equal to $\pi^2/2$ as shown by contour integration – see Problem 20), so, since in our present circumstances

$$\int_0^\infty \mathrm{d}\left(\frac{\nu(t)}{t}\right) = 0,$$

the previous double integral vanishes, and the corollary follows.*

In our application of these results we will take

$$\nu(t) = \frac{1-3p}{p} \mu(t),$$

$\mu(t)$ being the function constructed in the previous article. This function $\nu(t)$ *increases with constant slope* 1 on each of the intervals $[\delta_k, d_k]$, $k \geqslant 0$, and $[c_k, \gamma_k]$, $k \geqslant 1$, and is *constant* on each of the intervals *complementary* to those. Therefore, if

$$\tilde{\Omega} = (0, \infty) \sim \bigcup_{k \geqslant 1} [c_k, \gamma_k] \sim \bigcup_{k \geqslant 0} [\delta_k, d_k]$$

* The two sides of the relation established may both be infinite, e.g., when $\nu(t)$ and $\omega(t)$ have some coinciding jumps. But the meaning of the two iterated integrals in question is always unambiguous; in the second one, for instance, the outer integral of the *negative part* of the inner one converges.

(note that this set $\tilde{\Omega}$ *includes* our Ω), we have

$$\int_{\tilde{\Omega}}\int_0^\infty \log\left|1-\frac{x^2}{t^2}\right| d\mu(t)\frac{dx}{x^2}$$

$$= \frac{p}{1-3p}\int_0^\infty\int_0^\infty \log\left|1-\frac{x^2}{t^2}\right| d\nu(t)\frac{dx - d\nu(x)}{x^2}.$$

The corollary shows that this expression (which we can think of as a first approximation to

$$\int_{\Omega}\int_0^\infty \log\left|1-\frac{x^2}{t^2}\right| d\mu(t)\frac{dx}{x^2}\ )$$

is equal to

$$\frac{p}{1-3p}\int_0^\infty\int_0^\infty \log\left|\frac{x+t}{x-t}\right| d\left(\frac{\nu(t)}{t}\right)\frac{d\nu(x)}{x}.$$

This double integral can be given a symmetric form thanks to the

Lemma. *Let $\nu(t)$ be continuous, increasing, and piecewise continuously differentiable on $[0,\infty]$. Suppose, moreover, that $\nu(0) = 0$, that $\nu(t)$ is constant for t sufficiently large, and, finally*, that $(d/dt)(\nu(t)/t)$ remains bounded when $t \to 0+$. Then,*

$$\int_0^\infty\int_0^\infty \log\left|\frac{x+t}{x-t}\right| d\left(\frac{\nu(t)}{t}\right)\frac{\nu(x)}{x^2}dx = -\frac{\pi^2}{4}(\nu'(0))^2.$$

Proof. Our assumptions on ν make reversal of the order of integrations in the left-hand expression legitimate, so it is equal to

$$\int_0^\infty\int_0^\infty \log\left|\frac{x+t}{x-t}\right| \frac{\nu(x)}{x^2}dx\, d\left(\frac{\nu(t)}{t}\right)$$

$$= \int_0^\infty\left(\int_0^\infty \log\left|\frac{\xi+1}{\xi-1}\right| \frac{\nu(t\xi)}{t\xi}\frac{d\xi}{\xi}\right) d\left(\frac{\nu(t)}{t}\right).$$

Since

$$\int_0^\infty \log\left|\frac{\xi+1}{\xi-1}\right|\frac{d\xi}{\xi} = \frac{\pi^2}{2}$$

(which may be verified by contour integration), we have

$$\int_0^\infty \log\left|\frac{\xi+1}{\xi-1}\right| \frac{\nu(t\xi)}{t\xi}\frac{d\xi}{\xi} \longrightarrow \frac{\pi^2}{2}\nu'(0)$$

for $t \to 0$, and integration by parts of the outer integral in the previous

* This last condition can be relaxed. See problem 28(b), p. 569.

expression yields the value

$$-\frac{\pi^2}{2}(v'(0))^2 - \int_0^\infty \frac{v(t)}{t}\frac{d}{dt}\left(\int_0^\infty \log\left|\frac{\xi+1}{\xi-1}\right|\frac{v(t\xi)}{t\xi}\frac{d\xi}{\xi}\right)dt.$$

Under the conditions of our hypothesis, the differentiation with respect to t can be carried out under the inner integral sign. The last expression thus becomes

$$-\frac{\pi^2}{2}(v'(0))^2 - \int_0^\infty \frac{v(t)}{t}\int_0^\infty \log\left|\frac{\xi+1}{\xi-1}\right|\frac{d}{dt}\left(\frac{v(t\xi)}{t\xi}\right)\frac{d\xi}{\xi}\,dt$$

$$= -\frac{\pi^2}{2}(v'(0))^2 - \int_0^\infty \frac{v(t)}{t}\int_0^\infty \log\left|\frac{x+t}{x-t}\right|\frac{x}{t}\frac{d}{dx}\left(\frac{v(x)}{x}\right)\frac{dx}{x}\,dt.$$

In other words

$$\int_0^\infty\int_0^\infty \log\left|\frac{x+t}{x-t}\right| d\left(\frac{v(t)}{t}\right)\frac{v(x)}{x^2}dx$$

$$= -\frac{\pi^2}{2}(v'(0))^2 - \int_0^\infty\int_0^\infty \log\left|\frac{t+x}{t-x}\right| d\left(\frac{v(x)}{x}\right)\frac{v(t)}{t^2}dt.$$

The second term on the right obviously equals the left-hand side, so the lemma follows.

Corollary. *Let $v(t)$ be increasing, continuous, and piecewise linear on* $[0,\infty)$, constant *for all sufficiently large t and* zero *for t near* 0. *Then*

$$\int_0^\infty\int_0^\infty \log\left|1-\frac{x^2}{t^2}\right| dv(t)\frac{dx - dv(x)}{x^2}$$

$$= \int_0^\infty\int_0^\infty \log\left|\frac{x+t}{x-t}\right| d\left(\frac{v(t)}{t}\right)d\left(\frac{v(x)}{x}\right).$$

Proof. By the previous corollary, the left-hand expression equals

$$\int_0^\infty\int_0^\infty \log\left|\frac{x+t}{x-t}\right| d\left(\frac{v(t)}{t}\right)\frac{dv(x)}{x}.$$

In the present circumstances, $v'(0)$ exists and equals *zero*. Therefore by the lemma

$$\int_0^\infty\int_0^\infty \log\left|\frac{x+t}{x-t}\right| d\left(\frac{v(t)}{t}\right)\frac{v(x)}{x^2}dx = 0,$$

and the previous expression is equal to

$$\int_0^\infty\int_0^\infty \log\left|\frac{x+t}{x-t}\right| d\left(\frac{v(t)}{t}\right)d\left(\frac{v(x)}{x}\right).$$

Problem 21

Prove the last lemma using contour integration. (Hint: For $\Im z > 0$, consider the analytic function

$$F(z) = \frac{1}{\pi}\int_0^\infty \log\left(\frac{z+t}{z-t}\right) d\left(\frac{v(t)}{t}\right),$$

and examine the boundary values of $\Re F(z)$ and $\Im F(z)$ on the real axis. Then look at $\int_\Gamma ((F(z))^2/z)\,dz$ for a suitable contour Γ.)

5. The energy integral

The expression, quadratic in $d(v(t)/t)$, arrived at near the end of the previous article, namely,

$$\int_0^\infty \int_0^\infty \log\left|\frac{x+t}{x-t}\right| d\left(\frac{v(t)}{t}\right) d\left(\frac{v(x)}{x}\right),$$

has a simple physical interpretation. Let us assume that a flat metal plate of infinite extent, perpendicular to the z-plane, intersects the latter along the y-axis. This plate we suppose grounded. Let electric charge be continuously distributed on a very large thin sheet, made of non-conducting material, and intersecting the z-plane perpendicularly along the positive x-axis. Suppose the charge density on that sheet to be constant along lines perpendicular to the z-plane, and that the total charge contained in any rectangle of height 2 thereon, bounded by two such lines intersecting the x-axis at x and at $x + \Delta x$, is equal to the *net change* of $v(t)/t$ along $[x, x + \Delta x]$. This set-up will produce an *electric field* in the region lying to the right of the grounded metal plate; *near* the z-plane, the *potential function* for that field is equal, very nearly, to

$$u(z) = \int_0^\infty \log\left|\frac{z+t}{z-t}\right| d\left(\frac{v(t)}{t}\right).$$

The quantity

$$\int_0^\infty u(x) d\left(\frac{v(x)}{x}\right) = \int_0^\infty \int_0^\infty \log\left|\frac{x+t}{x-t}\right| d\left(\frac{v(t)}{t}\right) d\left(\frac{v(x)}{x}\right)$$

is then proportional to the *total energy* of the electric field generated by our distribution of electric charge (and inversely proportional to the height of the charged sheet). We therefore expect it to be *positive*, even though *charges of both sign* be present at different places on the non-conducting sheet, i.e., when $d(v(t)/t)/dt$ is *not of constant sign.*

Under quite general circumstances, the *positivity* of the quadratic form in question turns out to be *valid*, and plays a crucial rôle in the computations of the succeeding articles. In the present one, we derive two formulas, either of which makes that property evident.

The first formula is familiar from physics, and goes back to Gauss. It is convenient to write

$$\rho(t) = \frac{\nu(t)}{t}.$$

Lemma. *Let $\rho(t)$ be continuous on $[0,\infty)$, piecewise $\mathscr{C}_3$ there (say), and differentiable at 0. Suppose furthermore that $\rho(t)$ is uniformly* Lip 1 *on $[0,\infty)$ and $t\rho(t)$* constant *for sufficiently large t.*

If we write

$$u(z) = \int_0^\infty \log\left|\frac{z+t}{z-t}\right| \mathrm{d}\rho(t),$$

we have

$$\int_0^\infty\int_0^\infty \log\left|\frac{x+t}{x-t}\right| \mathrm{d}\rho(t)\,\mathrm{d}\rho(x) = \frac{1}{\pi}\int_0^\infty\int_0^\infty \{(u_x(z))^2 + (u_y(z))^2\}\,\mathrm{d}x\,\mathrm{d}y.$$

Remark 1. Note that we *do not require* that $\rho(t)$ *vanish* for t *near zero*, although $\rho(t) = \nu(t)/t$ has this property when $\nu(t)$ is the function introduced in the previous article.

Remark 2. The factor $1/\pi$ occurs on the right, and not $1/2\pi$ which one might expect from physics, because the right-hand integral is taken over the *first quadrant instead* of over the *whole right half plane* (where the 'electric field' is present). The right-hand expression is of course the *Dirichlet integral* of u over the first quadrant.

Remark 3. The function $u(z)$ is *harmonic* in each separate quadrant of the z-plane. Since

$$\log\left|\frac{z+\bar{w}}{z-w}\right|$$

is the Green's function for the right half plane, $u(z)$ is frequently referred to as the *Green potential of the charge distribution* $\mathrm{d}\rho(t)$ (for that half plane).

Proof of lemma. For $y > 0$, we have

$$u_y(z) = \int_0^\infty \left(\frac{y}{(x+t)^2+y^2} - \frac{y}{(x-t)^2+y^2}\right)\mathrm{d}\rho(t),$$

and, when $x > 0$ is not a point of discontinuity for $\rho'(t)$, the right side tends to $-\pi\rho'(x)$ as $y \to 0+$ by the usual (elementary) approximate identity property of the Poisson kernel. Thus,

$$u_y(x + \mathrm{i}0) \;=\; -\pi\rho'(x),$$

and

$$\int_0^\infty \int_0^\infty \log\left|\frac{x+t}{x-t}\right| \mathrm{d}\rho(t)\,\mathrm{d}\rho(x) \;=\; -\frac{1}{\pi}\int_0^\infty u(x)u_y(x+\mathrm{i}0)\,\mathrm{d}x.$$

At the same time, $u(\mathrm{i}y) = 0$ for $y > 0$, so the left-hand double integral from the previous relation is equal to

$$-\frac{1}{\pi}\int_0^\infty u(x)u_y(x+\mathrm{i}0)\,\mathrm{d}x \;-\; \frac{1}{\pi}\int_0^\infty u(\mathrm{i}y)u_x(\mathrm{i}y)\,\mathrm{d}y.$$

We have here a line integral around the boundary of the first quadrant. Applying Green's theorem to it in cook-book fashion, we get the value

$$\frac{1}{\pi}\int_0^\infty \int_0^\infty \left(\frac{\partial}{\partial y}(u(z)u_y(z)) + \frac{\partial}{\partial x}(u(z)u_x(z))\right)\mathrm{d}x\,\mathrm{d}y,$$

which reduces immediately to

$$\frac{1}{\pi}\int_0^\infty \int_0^\infty ((u_y(z))^2 + (u_x(z))^2)\,\mathrm{d}x\,\mathrm{d}y$$

(proving the lemma), since u is *harmonic* in the first quadrant, making $u\nabla^2 u = 0$ there.

We have, however, to justify our use of Green's theorem. The way to do that here is to adapt to our present situation the common 'non-rigorous' derivation of the theorem (using squares) found in books on engineering mathematics. Letting $\mathscr{D}_A$ denote the square with vertices at 0, A, $A + \mathrm{i}A$ and $\mathrm{i}A$, we verify in that way without difficulty (and without any being created by the discontinuities of $\rho'(x) = -u_y(x+\mathrm{i}0)/\pi$), that

$$\int_{\partial\mathscr{D}_A} (uu_x\mathrm{d}y - uu_y\mathrm{d}x) \;=\; \iint_{\mathscr{D}_A} (u_x^2 + u_y^2)\,\mathrm{d}x\,\mathrm{d}y.*$$

The line integral on the left equals

$$-\int_0^A u(x)u_y(x+\mathrm{i}0)\,\mathrm{d}x \;+\; \int_{\Gamma_A} (uu_x\mathrm{d}y - uu_y\mathrm{d}x),$$

where Γ_A denotes the *right side* and *top* of $\mathscr{D}_A$:

* The simplest procedure is to take $h > 0$ and write the corresponding relation involving $u(z + \mathrm{i}h)$ in place of $u(z)$, whose truth is certain here. Then one can make $h \to 0$. Cf the discussion on pp. 506–7.

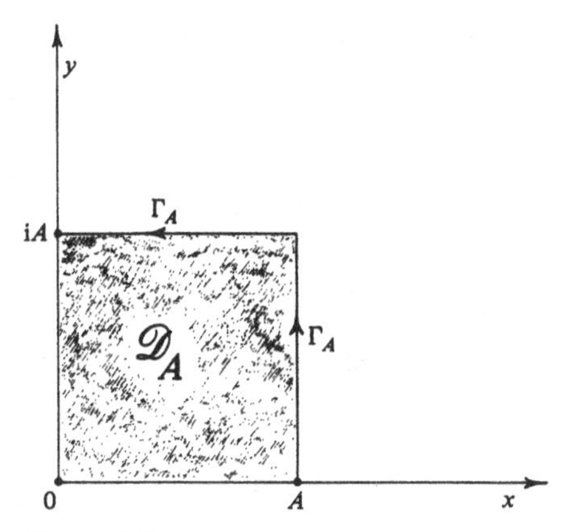

Figure 144

We will be done if we show that

$$\int_{\Gamma_A} (uu_x\mathrm{d}y - uu_y\mathrm{d}x) \longrightarrow 0 \qquad \text{for } A \longrightarrow \infty.$$

For this purpose, one may break up $u(z)$ as

$$\int_0^M \log\left|\frac{z+t}{z-t}\right| \mathrm{d}\rho(t) \; + \; \int_M^\infty \log\left|\frac{z+t}{z-t}\right| \mathrm{d}\rho(t),$$

M being chosen large enough so as to have $\rho(t) = C/t$ on $[M, \infty)$. Calling the *first* of these integrals $u_1(z)$, we easily find, for $|z| > M$ (by expanding the logarithm in powers of t/z), that

$$|u_1(z)| \;\leqslant\; \frac{\text{const.}}{|z|}$$

and that the first partial derivatives of $u_1(z)$ are $\mathrm{O}(1/|z|^2)$.

Denote by $u_2(z)$ the *second* of the above integrals, which, by choice of M, is actually equal to

$$-C\int_M^\infty \log\left|\frac{z+t}{z-t}\right| \frac{\mathrm{d}t}{t^2}.$$

The substitution $t = |z|\tau$ enables us to see after very little calculation that this expression is in modulus

$$\leqslant \; \text{const.}\frac{\log|z|}{|z|}$$

for large $|z|$.

To investigate the partial derivatives of $u_2(z)$ in the open first quadrant, we take the function

$$F(z) \;=\; \int_M^\infty \log\left(\frac{z+t}{z-t}\right)\frac{\mathrm{d}t}{t^2},$$

analytic in that region, and note that by the Cauchy–Riemann equations,

$$\frac{\partial u_2(z)}{\partial x} - i\frac{\partial u_2(z)}{\partial y} = -CF'(z)$$

there. Here,

$$F'(z) = \int_M^\infty \frac{dt}{t^2(z+t)} - \int_M^\infty \frac{dt}{t^2(z-t)}.$$

The first term on the right is obviously $O(1/|z|)$ in modulus when $\Re z$ and $\Im z > 0$. The second works out to

$$\int_M^\infty \left(\frac{1}{zt^2} + \frac{1}{z^2 t} + \frac{1}{z^2(z-t)} \right) dt = \frac{1}{zM} + \frac{1}{z^2}\log\left(\frac{z-M}{M}\right),$$

using a suitable determination of the logarithm. This is evidently $O(1/|z|)$ for large $|z|$, so $|F'(z)| = O(1/|z|)$ for z with large modulus in the first quadrant. The same is thus true for the first partial derivatives of $u_2(z)$.

Combining the estimates just made on $u_1(z)$ and $u_2(z)$, we find for $u = u_1 + u_2$ that

$$|u(z)| \leqslant \text{const.}\frac{\log|z|}{|z|}$$

$$|u_x(z)| \leqslant \text{const.}\frac{1}{|z|}$$

$$|u_y(z)| \leqslant \text{const.}\frac{1}{|z|}$$

when $\Re z > 0$, $\Im z > 0$, $|z|$ being large. Therefore

$$\int_{\Gamma_A} (uu_x dy - uu_y dx) = O\left(\frac{\log A}{A}\right)$$

for large A, and the line integral tends to zero as $A \to \infty$. This is what was needed to finish the proof of the lemma. We are done.

Corollary. *If $\rho(t)$ is* real *and satisfies the hypothesis of the lemma,*

$$\int_0^\infty \int_0^\infty \log\left|\frac{x+t}{x-t}\right| d\rho(t)\, d\rho(x) \geqslant 0.$$

Proof. Clear.

Notation. We write

$$E(d\rho(t), d\sigma(t)) = \int_0^\infty \int_0^\infty \log\left|\frac{x+t}{x-t}\right| d\rho(t)\, d\sigma(x)$$

for *real* measures ρ and σ on $[0,\infty)$ without point mass at the origin making both of the integrals

$$\int_0^\infty\int_0^\infty \log\left|\frac{x+t}{x-t}\right| \mathrm{d}\rho(t)\,\mathrm{d}\rho(x), \qquad \int_0^\infty\int_0^\infty \log\left|\frac{x+t}{x-t}\right| \mathrm{d}\sigma(t)\,\mathrm{d}\sigma(x)$$

absolutely convergent. (Vanishing of $\rho(\{0\})$ and $\sigma(\{0\})$ is required because $\log|(x+t)/(x-t)|$ cannot be defined at $(0,0)$ so as to be continuous there.)

Note that, in the case of functions $\rho(t)$ and $\sigma(t)$ satisfying the hypothesis of the above lemma, the integrals just written *do* converge absolutely. In terms of $E(\mathrm{d}\rho(t), \mathrm{d}\sigma(t))$, we can state the very important

Corollary. *If $\rho(t)$ and $\sigma(t)$, defined and real valued on $[0,\infty)$, both satisfy the hypothesis of the lemma,*

$$|E(\mathrm{d}\rho(t), \mathrm{d}\sigma(t))| \leqslant \sqrt{(E(\mathrm{d}\rho(t), \mathrm{d}\rho(t)))}\cdot\sqrt{(E(\mathrm{d}\sigma(t), \mathrm{d}\sigma(t)))}.$$

Proof. Use the preceding corollary and proceed as in the usual derivation of Schwarz' inequality.

Remark. The result remains valid as long as ρ and σ, with $\rho(\{0\}) = \sigma(\{0\}) = 0$, are such that the abovementioned absolute convergence holds. We will see that at the end of the present article.

Scholium and warning. The results just given should not mislead the reader into believing that the energy integral corresponding to the *ordinary logarithmic potential* is necessarily positive. Example:

$$\int_0^{2\pi}\int_0^{2\pi} \log\frac{1}{|2e^{i\vartheta} - 2e^{i\varphi}|}\,\mathrm{d}\vartheta\,\mathrm{d}\varphi \;=\; \int_0^{2\pi} 2\pi\log\tfrac{1}{2}\,\mathrm{d}\varphi \;=\; -4\pi^2\log 2 \text{ !}$$

It is strongly recommended that the reader find out exactly where the argument used in the proof of the lemma *goes wrong*, when one attempts to adapt it to the potential

$$u(z) \;=\; \int_0^{2\pi} \log\frac{1}{|2e^{i\vartheta} - z|}\,\mathrm{d}\vartheta.$$

For 'nice' real measures μ of compact support, *it is true that*

$$\int_{\mathbb{C}}\int_{\mathbb{C}} \log\frac{1}{|z-w|}\,\mathrm{d}\mu(z)\,\mathrm{d}\mu(w) \;\geqslant\; 0$$

provided that $\int_{\mathbb{C}}\mathrm{d}\mu(z) = 0$. *The reader should verify this fact* by applying a suitable version of Green's theorem to the potential $\int_{\mathbb{C}}\log(1/|z-w|)\,\mathrm{d}\mu(w)$.

The formula for $E(\mathrm{d}\rho(t), \mathrm{d}\rho(t))$ furnished by the above lemma exhibits that quantity's positivity. The same service is rendered by an analogous relation involving the values of $\rho(t)$ on $[0, \infty)$. Such representations go back to Jesse Douglas; we are going to use one based on a beautiful identity of Beurling. In order to encourage the reader's participation, we set as a problem the derivation of Beurling's result.

Problem 22

(a) Let m be a real measure on $\mathbb{R}$. Suppose that $h > 0$ and that $\iint_{|\xi-\eta|\leqslant h} dm(\xi)\,dm(\eta)$ converges absolutely. Show that

$$\int_{-\infty}^{\infty} (m(x+h) - m(x))^2\,dx = \int_{-\infty}^{\infty}\int_{-\infty}^{\infty} (h - |\xi - \eta|)^+\,dm(\xi)dm(\eta).$$

(Hint: Trick:

$$(m(x+h) - m(x))^2 = \int_x^{x+h}\int_x^{x+h} dm(\xi)dm(\eta).\ )$$

(b) Let $K(x)$ be even and positive, $\mathscr{C}_2$ and *convex* for $x > 0$, and such that $K(x) \to 0$ for $x \to \infty$. Show that, for $x \neq 0$,

$$K(x) = \int_0^{\infty} (h - |x|)^+ K''(h)\,dh.$$

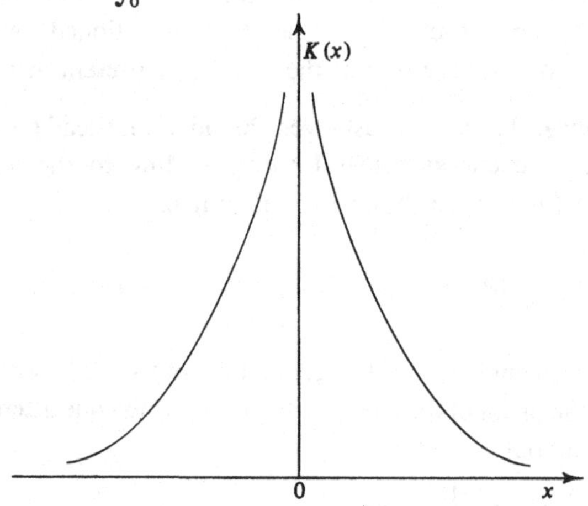

Figure 145

(Hint: First observe that $K'(x)$ must also $\to 0$ for $x \to \infty$.)

(c) If $K(x)$ is as in (b) and m is a real measure on $\mathbb{R}$ with $\int_{-\infty}^{\infty}\int_{-\infty}^{\infty} K(|\xi - \eta|)\,dm(\xi)\,dm(\eta)$ absolutely convergent, that integral is equal to

$$\int_{-\infty}^{\infty}\int_0^{\infty} [m(x+h) - m(x)]^2 K''(h)\,dh\,dx$$

$$= \frac{1}{2}\int_{-\infty}^{\infty}\int_{-\infty}^{\infty} [m(y) - m(x)]^2 K''(|x-y|)\,dy\,dx.$$

(Hint: The assumed absolute convergence guarantees that m fulfills, for each $h > 0$, the condition required in part (a). The order of integration in

$$\int_{-\infty}^{\infty}\int_{0}^{\infty} K''(h)(m(x+h)-m(x))^2 \, dh \, dx$$

may be reversed, yielding, by part (a), an iterated triple integral. Here, that triple integral is absolutely convergent and we may conclude by the help of part (b).)

Lemma. *Let the real measure ρ on $[0, \infty)$, without point mass at the origin, be such that*

$$\int_{0}^{\infty}\int_{0}^{\infty} \log\left|\frac{x+t}{x-t}\right| d\rho(t)\, d\rho(x)$$

is absolutely convergent. Then

$$\int_{0}^{\infty}\int_{0}^{\infty} \log\left|\frac{x+t}{x-t}\right| d\rho(t)\, d\rho(x) = \int_{0}^{\infty}\int_{0}^{\infty} \left(\frac{\rho(x)-\rho(y)}{x-y}\right)^2 \frac{x^2+y^2}{(x+y)^2} \, dx\, dy.$$

Proof. The left-hand double integral is of the form

$$\int_{0}^{\infty}\int_{0}^{\infty} k\left(\frac{x}{t}\right) d\rho(x)\, d\rho(t),$$

where

$$k(\tau) = \log\left|\frac{1+\tau}{1-\tau}\right| = k\left(\frac{1}{\tau}\right),$$

so we can reduce that integral to one figuring in Problem 22(c) by making the substitutions $x = e^{\xi}$, $t = e^{\eta}$, $\rho(x) = m(\xi)$, $\rho(t) = m(\eta)$, and

$$k\left(\frac{x}{t}\right) = K(\xi-\eta) = \log\left|\coth\left(\frac{\xi-\eta}{2}\right)\right|.$$

$K(h)$, besides being obviously even and positive, tends to zero for $h \to \infty$. Also

$$K'(h) = \frac{1}{2}\tanh\frac{h}{2} - \frac{1}{2}\coth\frac{h}{2},$$

and

$$K''(h) = \frac{1}{4}\operatorname{sech}^2\frac{h}{2} + \frac{1}{4}\operatorname{cosech}^2\frac{h}{2} > 0,$$

so $K(h)$ is convex for $h > 0$. The application of Beurling's formula from problem 22(c) is therefore legitimate, and yields

$$\int_0^\infty\int_0^\infty \log\left|\frac{x+t}{x-t}\right| d\rho(t)\,d\rho(x)$$

$$= \int_{-\infty}^\infty\int_{-\infty}^\infty K(|\xi-\eta|)\,dm(\eta)\,dm(\xi)$$

$$= \frac{1}{2}\int_{-\infty}^\infty\int_{-\infty}^\infty K''(|\xi-\eta|)[m(\xi)-m(\eta)]^2\,d\xi\,d\eta$$

(note that the first of these integrals, *and hence the second*, is absolutely convergent by hypothesis).

Here,

$$K''(|\xi-\eta|) = \frac{1}{4}\frac{\sinh^2\left(\frac{\xi-\eta}{2}\right) + \cosh^2\left(\frac{\xi-\eta}{2}\right)}{\sinh^2\left(\frac{\xi-\eta}{2}\right)\cosh^2\left(\frac{\xi-\eta}{2}\right)}$$

$$= \frac{\cosh(\xi-\eta)}{\sinh^2(\xi-\eta)} = 2e^\xi e^\eta\frac{e^{2\xi}+e^{2\eta}}{(e^{2\xi}-e^{2\eta})^2},$$

so the third of the above expressions reduces to

$$\int_{-\infty}^\infty\int_{-\infty}^\infty \frac{e^{2\xi}+e^{2\eta}}{(e^\xi+e^\eta)^2}\left(\frac{m(\xi)-m(\eta)}{e^\xi-e^\eta}\right)^2 e^\xi e^\eta\,d\xi\,d\eta$$

$$= \int_0^\infty\int_0^\infty \frac{x^2+t^2}{(x+t)^2}\left(\frac{\rho(x)-\rho(t)}{x-t}\right)^2 dx\,dt.$$

We are done.

Remark. This certainly implies that the *first* of the above corollaries is true for *any real measure* ρ *with* $\rho(\{0\}) = 0$ *rendering absolutely convergent* the double integral used to define $E(d\rho(t), d\rho(t))$. The *second* corollary is then *also true* for such real measures ρ and σ.

The formula provided by this second lemma is one of the main ingredients in our treatment of the question discussed in the present §. It is the basis for the important calculation carried out in the next article.

6. A lower estimate for $\displaystyle\int_{\tilde{\Omega}}\int_0^\infty \log\left|1-\frac{x^2}{t^2}\right|\mathrm{d}\mu(t)\,\frac{\mathrm{d}x}{x^2}$

We return to where we left off near the end of article 4, focusing our attention on the quantity

$$\int_{\tilde{\Omega}}\int_0^\infty \log\left|1-\frac{x^2}{t^2}\right|\mathrm{d}\mu(t)\frac{\mathrm{d}x}{x^2},$$

where $\mu(t)$ is the function constructed in article 3 and

$$\tilde{\Omega} = (0,\ \infty) \sim \{x\colon\ \mu'(x)>0\}.$$

Before going any further, the reader should refer to the graph of $\mu(t)$ found near the end of article 3. As explained in article 4, we prefer to work not with $\mu(t)$, but with

$$\nu(t) = \frac{1-3p}{p}\,\mu(t);$$

the graph of $\nu(t)$ looks just like that of $\mu(t)$, save that its slanting portions all have slope 1, and not $p/(1-3p)$. Those slanting portions lie over certain intervals $[c_k,\ \gamma_k]$, $k\geqslant 1$, $[\delta_k,\ d_k]$, $k\geqslant 0$, contained in the $J_k=[c_k,\ d_k]$, and

$$\tilde{\Omega} = (0,\infty)\sim\bigcup_{k\geqslant 0}[\delta_k,d_k]\sim\bigcup_{k\geqslant 1}[c_k,\gamma_k].$$

This set $\tilde{\Omega}$ is obtained from the one Ω shown on the graph of $\mu(t)$ by adjoining to the latter the intervals $(c_0,\ \delta_0)\subseteq J_0$ and $(\gamma_k,\ \delta_k)\subseteq J_k$, $k\geqslant 1$.

By the corollary at the end of article 4,

$$\begin{aligned}
&\int_{\tilde{\Omega}}\int_0^\infty \log\left|1-\frac{x^2}{t^2}\right|\mathrm{d}\mu(t)\frac{\mathrm{d}x}{x^2}\\
&\quad= \frac{p}{1-3p}\int_0^\infty\int_0^\infty \log\left|1-\frac{x^2}{t^2}\right|\mathrm{d}\nu(t)\frac{\mathrm{d}x-\mathrm{d}\nu(x)}{x^2}\\
&\quad= \frac{p}{1-3p}\int_0^\infty\int_0^\infty \log\left|\frac{x+t}{x-t}\right|\mathrm{d}\left(\frac{\nu(t)}{t}\right)\mathrm{d}\left(\frac{\nu(x)}{x}\right),
\end{aligned}$$

and this is just

$$\frac{p}{1-3p}\,E\left(\mathrm{d}\left(\frac{\nu(t)}{t}\right),\mathrm{d}\left(\frac{\nu(x)}{x}\right)\right),$$

$E(\ \ ,\ \ )$ being the bilinear form defined and studied in the previous article.

This identification is a key step in our work. It, and the results of article

5, enable us to see that

$$\int_{\tilde{\Omega}}\int_0^\infty \log\left|1-\frac{x^2}{t^2}\right| d\mu(t)\frac{dx}{x^2}$$

is at least *positive* (until now, we were not even sure of this). The second lemma of article 5 actually makes it possible for us to estimate that integral from below in terms of a sum,

$$\sum_{k\geq 1}\left(\frac{\gamma_k - c_k}{\gamma_k}\right)^2 + \sum_{k\geq 0}\left(\frac{d_k-\delta_k}{d_k}\right)^2,$$

like one which occurred previously in Chapter VII, §A.2. In our estimate, that sum is affected with a certain *coefficient*.

On account of the theorem of article 3, we are *really* interested in

$$\int_{\Omega}\int_0^\infty \log\left|1-\frac{x^2}{t^2}\right| d\mu(t)\frac{dx}{x^2}$$

rather than the quantity considered here. It will turn out later on that the passage from integration over $\tilde{\Omega}$ to that over Ω involves a serious *loss*, in whose evaluation the sum just written again figures. For this reason we have to take care to get *a large enough numerical value* for the coefficient mentioned above. That circumstance requires us to be somewhat fussy in the computation made to derive the following result. *From now on, in order to make the notation more uniform, we will write*

$$\gamma_0 = c_0.$$

Theorem. *If* $\nu(t) = ((1-3p)/p)\mu(t)$ *with the function* $\mu(t)$ *from article* 3, *and the parameter* $\eta > 0$ *used in the construction of the* J_k *(see the theorem, end of article* 2) *is* sufficiently small, *we have*

$$E\left(d\left(\frac{\nu(t)}{t}\right), d\left(\frac{\nu(t)}{t}\right)\right)$$

$$\geq \left(\tfrac{3}{2}-\log 2 - K\eta\right)\sum_{k\geq 0}\left\{\left(\frac{\gamma_k-c_k}{\gamma_k}\right)^2+\left(\frac{d_k-\delta_k}{d_k}\right)^2\right\}.$$

Here, K *is a purely numerical constant, independent of* p *or the configuration of the* J_k.

Remark. Later on, we will need the numerical value

$$\tfrac{3}{2}-\log 2 = 0.80685\ldots.$$

Proof of theorem. By the second lemma of article 5 and brute force. The lemma gives

$$E\left(\mathrm{d}\left(\frac{v(t)}{t}\right), \mathrm{d}\left(\frac{v(t)}{t}\right)\right)$$

$$= \int_0^\infty \int_0^\infty \left(\frac{\dfrac{v(x)}{x} - \dfrac{v(y)}{y}}{x-y}\right)^2 \frac{x^2+y^2}{(x+y)^2}\,\mathrm{d}x\,\mathrm{d}y$$

$$\geqslant \frac{1}{2}\sum_{k\geqslant 0} \int_{J_k}\int_{J_k} \left(\frac{\dfrac{v(x)}{x} - \dfrac{v(y)}{y}}{x-y}\right)^2 \mathrm{d}x\,\mathrm{d}y.$$

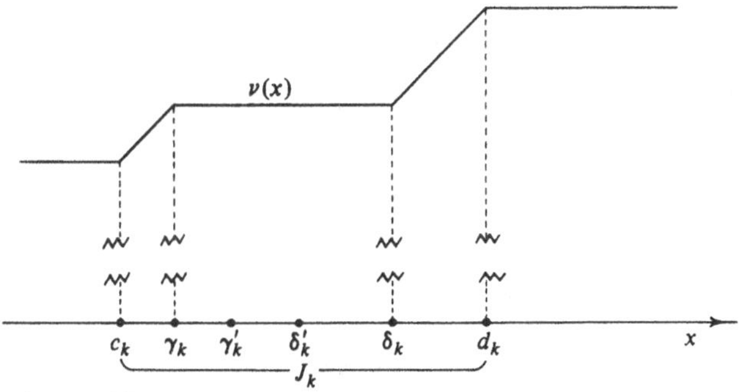

Figure 146

On each interval $J_k = [c_k, d_k]$ we take

$$\begin{aligned} \gamma_k' &= c_k + 2(\gamma_k - c_k) \\ \delta_k' &= d_k - 2(d_k - \delta_k) \end{aligned}$$

(see figure). Since

$$\frac{\gamma_k - c_k + d_k - \delta_k}{d_k - c_k} = \frac{1-3p}{2} < \frac{1}{2}$$

(properties (iii), (iv) of the description near the end of article 3) we have $\gamma_k' < \delta_k'$. Therefore, for each k,

$$\int_{J_k}\int_{J_k} \left(\frac{\dfrac{v(x)}{x} - \dfrac{v(y)}{x}}{x-y}\right)^2 \mathrm{d}x\,\mathrm{d}y$$

$$\geqslant \left\{\int_{c_k}^{\gamma_k'}\int_{c_k}^{\gamma_k'} + \int_{\delta_k'}^{d_k}\int_{\delta_k'}^{d_k}\right\}\left(\frac{\dfrac{v(x)}{x} - \dfrac{v(y)}{y}}{x-y}\right)^2 \mathrm{d}x\,\mathrm{d}y.$$

We estimate the *second* of the integrals on the right – the other one is handled similarly.

We begin by writing

$$\int_{\delta_k'}^{d_k}\int_{\delta_k'}^{d_k}\left(\frac{\frac{v(x)}{x}-\frac{v(x)}{y}}{x-y}\right)^2 dx\,dy$$

$$\geqslant \left\{\int_{\delta_k}^{d_k}\int_{\delta_k}^{d_k}+\int_{\delta_k'}^{\delta_k}\int_{\delta_k}^{d_k}+\int_{\delta_k}^{d_k}\int_{\delta_k'}^{\delta_k}\right\}\left(\frac{\frac{v(x)}{x}-\frac{v(y)}{y}}{x-y}\right)^2 dx\,dy.$$

Of the three double integrals on the right, the *first* is easiest to evaluate. Things being bad enough as they are, *let us lighten the notation by dropping, for the moment, the subscript k, putting*

δ' for δ_k',

δ for δ_k

and

d for d_k.

Since $v'(x) = 1$ for $\delta_k = \delta < x < d = d_k$,

$$\frac{v(x)}{x} = 1+\frac{v(\delta)-\delta}{x}, \qquad \delta \leqslant x \leqslant d.$$

Using this, we easily find that

$$\int_{\delta}^{d}\int_{\delta}^{d}\left(\frac{\frac{v(x)}{x}-\frac{v(y)}{y}}{x-y}\right)^2 dx\,dy = \left(1-\frac{v(\delta)}{\delta}\right)^2\left(\frac{d-\delta}{d}\right)^2.$$

In terms of

$$\mathscr{J} = \bigcup_{k\geqslant 0}((c_k, \gamma_k)\cup(\delta_k, d_k))$$

and

$$J = \bigcup_{k\geqslant 0} J_k,$$

we have clearly

$$v(t) = |[0, t]\cap\mathscr{J}| \leqslant |[0, t]\cap J|, \qquad t > 0.$$

The right-hand quantity is, however, $\leqslant 2\eta t$ by construction of the J_k (property (v) in the theorem at the end of article 2). Therefore

$\nu(\delta)/\delta \;=\; \nu(\delta_k)/\delta_k \;\leqslant\; 2\eta$, and the integral just evaluated is

$$\geqslant\; (1-2\eta)^2\left(\frac{d-\delta}{d}\right)^2.$$

We pass now to the *second* of the three double integrals in question, continuing to omit the subscript k. To simplify the work, we make the changes of variable

$$x = \delta + s, \qquad y = \delta - t,$$

and denote $d-\delta = \delta-\delta'$ by Δ. Then

$$\int_{\delta'}^{\delta}\int_{\delta}^{d}\left(\frac{\dfrac{\nu(x)}{x}-\dfrac{\nu(y)}{y}}{x-y}\right)^2 dx\,dy \;=\; \int_0^{\Delta}\int_0^{\Delta}\left(\frac{\dfrac{\nu(\delta)+s}{\delta+s}-\dfrac{\nu(\delta)}{\delta-t}}{s+t}\right)^2 ds\,dt,$$

since $\nu(y) = \nu(\delta)$ for $\delta' \leqslant y \leqslant \delta$ (see the above figure). The expression on the right simplifies to

$$\int_0^{\Delta}\int_0^{\Delta}\left(\frac{s}{(\delta+s)(t+s)}-\frac{\nu(\delta)}{(\delta-t)(\delta+s)}\right)^2 ds\,dt$$

which in turn is

$$\geqslant\; \frac{1}{d^2}\int_0^{\Delta}\int_0^{\Delta}\left(\frac{s}{t+s}\right)^2 dt\,ds \;-\; 2\frac{\nu(\delta)}{\delta}\cdot\frac{1}{\delta'\delta}\int_0^{\Delta}\int_0^{\Delta}\frac{s}{t+s}\,ds\,dt$$

$$\geqslant\; \frac{\Delta^2}{d^2}(1-\log 2)-\frac{4\eta\Delta^2}{\delta'\delta}$$

(we have again used the fact that $\nu(\delta)/\delta \leqslant 2\eta$). We have $\nu(d) \geqslant \nu(d)-\nu(\delta) = d-\delta = \Delta$, so, since $\nu(d)/d \leqslant 2\eta$,

$$\delta \;=\; d-\Delta \;\geqslant\; (1-2\eta)d$$

and

$$\delta' \;=\; d-2\Delta \;\geqslant\; (1-4\eta)d.$$

By the computation just made we thus have

$$\int_{\delta'}^{\delta}\int_{\delta}^{d}\left(\frac{\dfrac{\nu(x)}{x}-\dfrac{\nu(y)}{y}}{x-y}\right)^2 dx\,dy$$

$$\geqslant\; \left(1-\log 2-\frac{4\eta}{(1-2\eta)(1-4\eta)}\right)\left(\frac{d-\delta}{d}\right)^2.$$

For the *third* of our three double integrals we have exactly the same estimate. Hence, *restoring now the subscript k,*

$$\int_{\delta_k'}^{d_k}\int_{\delta_k'}^{d_k}\left(\frac{\dfrac{v(x)}{x}-\dfrac{v(y)}{y}}{x-y}\right)^2 \mathrm{d}x\,\mathrm{d}y \;\geqslant\; (3-2\log 2-15\eta)\left(\frac{d_k-\delta_k}{d_k}\right)^2,$$

as long as $\eta > 0$ is sufficiently small.

In the same way, one finds that

$$\int_{c_k}^{\gamma_k'}\int_{c_k}^{\gamma_k'}\left(\frac{\dfrac{v(x)}{x}-\dfrac{v(y)}{y}}{x-y}\right)^2 \mathrm{d}x\,\mathrm{d}y \;\geqslant\; (3-2\log 2-K\eta)\left(\frac{\gamma_k-c_k}{\gamma_k}\right)^2$$

for small enough $\eta > 0$, K being a certain numerical constant. Adding this to the previous relation gives us a lower estimate for

$$\int_{J_k}\int_{J_k}\left(\frac{\dfrac{v(x)}{x}-\dfrac{v(y)}{y}}{x-y}\right)^2 \mathrm{d}x\,\mathrm{d}y;$$

adding these estimates and referring again to the relation at the beginning of this proof, we obtain the theorem.

Q.E.D.

From the initial discussion of this article, we see that the theorem has the following

Corollary. *Let $\mu(t)$ be the function constructed in article 3 and $\tilde{\Omega}$ be the complement, in $(0,\infty)$, of the set on which $\mu(t)$ is increasing. Then, if the parameter $\eta > 0$ used in constructing the J_k is sufficiently small,*

$$\int_{\tilde{\Omega}}\int_0^\infty \log\left|1-\frac{x^2}{t^2}\right|\mathrm{d}\mu(t)\frac{\mathrm{d}x}{x^2}$$

$$\geqslant\; \frac{p}{1-3p}\left(\tfrac{3}{2}-\log 2-K\eta\right)\sum_{k\geqslant 0}\left(\left(\frac{\gamma_k-c_k}{\gamma_k}\right)^2+\left(\frac{d_k-\delta_k}{d_k}\right)^2\right).$$

Here K is a numerical constant, independent of p or of the particular configuration of the J_k.

In the following work, our guiding idea will be to show that $\int_\Omega\int_0^\infty \log|1-x^2/t^2|\,\mathrm{d}\mu(t)(\mathrm{d}x/x^2)$ is *not too much less* than the left-hand integral in the above relation, in terms of the sum on the right.

7. Effect of taking x to be constant on each of the intervals J_k

We continue to write

$$\Omega = (0,\infty)\sim J,$$

where $J = \bigcup_{k\geqslant 0} J_k$ with $J_k = [c_k, d_k]$, and

$$\tilde{\Omega} = (0, \infty) \sim \tilde{J},$$

with

$$\tilde{J} = \bigcup_{k\geqslant 0} ((c_k, \gamma_k) \cup (\delta_k, d_k))$$

being the set on which $\mu(t)$ is increasing. The comparison of $\int_\Omega \int_0^\infty \log|1 - x^2/t^2| \mathrm{d}\mu(t)(\mathrm{d}x/x^2)$, object of our interest, with $\int_{\tilde{\Omega}} \int_0^\infty \log|1 - x^2/t^2| \mathrm{d}\mu(t)(\mathrm{d}x/x^2)$ is simplified by using two approximations to those quantities.

As in the previous article, we work in terms of

$$v(t) = \frac{1-3p}{p}\mu(t)$$

instead of $\mu(t)$. Put

$$\boxed{u(z) = \int_0^\infty \log\left|\frac{z+t}{z-t}\right| \mathrm{d}\left(\frac{v(t)}{t}\right).}$$

Then, by the corollary to the first lemma in article 4,

$$\int_\Omega \int_0^\infty \log\left|1 - \frac{x^2}{t^2}\right| \mathrm{d}\mu(t)\frac{\mathrm{d}x}{x^2} = \frac{p}{1-3p}\int_J u(x)\frac{\mathrm{d}x}{x}$$

and

$$\int_{\tilde{\Omega}} \int_0^\infty \log\left|1 - \frac{x^2}{t^2}\right| \mathrm{d}\mu(t)\frac{\mathrm{d}x}{x^2} = \frac{p}{1-3p}\int_{\tilde{J}} u(x)\frac{\mathrm{d}x}{x}.$$

Our approximation consists in the replacement of

$$\int_J u(x)\frac{\mathrm{d}x}{x} \quad \text{by} \quad \sum_{k\geqslant 0}\frac{1}{d_k}\int_{J_k} u(x)\,\mathrm{d}x$$

and of

$$\int_{\tilde{J}} u(x)\frac{\mathrm{d}x}{x} \quad \text{by} \quad \sum_{k\geqslant 0}\frac{1}{d_k}\left(\int_{c_k}^{\gamma_k} + \int_{\delta_k}^{d_k}\right)u(x)\,\mathrm{d}x.$$

To estimate the difference between the left-hand and right-hand quantities we use the positivity of the bilinear form $E(\ ,\ )$, proved in article 5.

Theorem. *If the parameter $\eta > 0$ used in the construction of the J_k is sufficiently small.*

$$\left| \int_J u(x)\frac{dx}{x} - \sum_{k\geq 0} \frac{1}{d_k} \int_{J_k} u(x)\,dx \right|$$

and

$$\left| \int_{\tilde{J}} u(x)\frac{dx}{x} - \sum_{k\geq 0} \frac{1}{d_k} \left(\int_{c_k}^{\gamma_k} + \int_{\delta_k}^{d_k} \right) u(x)\,dx \right|$$

are both

$$\leqslant C\eta^{\frac{1}{4}} E\left(d\left(\frac{v(t)}{t}\right), d\left(\frac{v(t)}{t}\right) \right),$$

where C is a purely numerical constant, independent of $p < \frac{1}{20}$ or the configuration of the J_k.

Remark. Here,

$$E\left(d\left(\frac{v(t)}{t}\right), d\left(\frac{v(t)}{t}\right) \right) = \int_{\tilde{J}} u(x)\frac{dx}{x}$$

according to the corollary at the end of article 4.

Proof. Let us treat the *second* difference; the first is handled similarly. Take

$$\varphi(x) = \begin{cases} \dfrac{1}{x} - \dfrac{1}{d_k}, & c_k < x < \gamma_k, \quad k \geq 1; \\[2mm] \dfrac{1}{x} - \dfrac{1}{d_k}, & \delta_k < x < d_k, \quad k \geq 0; \\[2mm] 0 & \text{elsewhere.} \end{cases}$$

(Recall that $\gamma_0 = c_0$, so (c_0, γ_0) is empty.) The second of the expressions in question is then just the absolute value of

$$\int_0^\infty u(x)\varphi(x)\,dx = \int_0^\infty \int_0^\infty \log\left|\frac{x+t}{x-t}\right| d\left(\frac{v(t)}{t}\right) \varphi(x)\,dx,$$

i.e., of $E(d(v(t)/t), \varphi(t)dt)$, in the notation of article 5. By the second corollary in that article and the remark at the end of it,

$$\left| E\left(d\left(\frac{v(t)}{t}\right), \varphi(t)\,dt \right) \right| \leqslant \sqrt{\left(E\left(d\left(\frac{v(t)}{t}\right), d\left(\frac{v(t)}{t}\right) \right) \right)}$$
$$\times \sqrt{(E(\varphi(t)\,dt, \varphi(t)\,dt))}.$$

The function $\varphi(x)$ is surely zero outside of the J_k, and, on J_k,

$$0 \leqslant \varphi(x) \leqslant \frac{d_k - x}{xd_k} \leqslant \frac{|J_k|}{xd_k}$$

with $|J_k|/d_k \leqslant 2\eta$ as in the proof of the theorem of article 6. Therefore,

$$0 < \int_0^\infty \log\left|\frac{x+t}{x-t}\right| \varphi(t)\,dt \leqslant 2\eta \int_0^\infty \log\left|\frac{x+t}{x-t}\right| \frac{dt}{t} = \pi^2\eta,$$

and

$$E(\varphi(t)dt, \varphi(t)dt) = \int_0^\infty \int_0^\infty \log\left|\frac{x+t}{x-t}\right| \varphi(t)\,dt\, \varphi(x)\,dx$$

$$\leqslant \pi^2\eta \sum_{k\geqslant 0} \int_{J_k} \frac{d_k - x}{xd_k}\,dx \leqslant \tfrac{1}{2}\pi^2\eta \sum_{k \geqslant 0} \frac{|J_k|^2}{c_k d_k}.$$

We have $c_k = d_k - |J_k| \geqslant (1-2\eta)d_k$ (see above), and, according to property (iv) from the list near the end of article 3,

$$|J_k| = d_k - c_k = \frac{2}{1-3p}\{(\gamma_k - c_k) + (d_k - \delta_k)\}.$$

Since we are assuming (throughout this §) that $p < \frac{1}{20}$, this makes

$$|J_k|^2 < 2\left(\frac{40}{17}\right)^2 \{(\gamma_k - c_k)^2 + (d_k - \delta_k)^2\},$$

yielding, by the preceding relation,

$$\frac{|J_k|^2}{c_k d_k} \leqslant \frac{12}{1-2\eta}\left\{\left(\frac{\gamma_k - c_k}{\gamma_k}\right)^2 + \left(\frac{d_k - \delta_k}{d_k}\right)^2\right\}.$$

Substitute this inequality into the previous estimate and then apply the theorem from the preceding article. One obtains

$$E(\varphi(t)dt, \varphi(t)dt)$$

$$\leqslant \frac{6\pi^2\eta}{(1-2\eta)(\frac{3}{2} - \log 2 - K\eta)} E\left(d\left(\frac{v(t)}{t}\right), d\left(\frac{v(t)}{t}\right)\right).$$

Using this in the above inequality for $|E(d(v(t)/t), \varphi(t)\,dt)|$, we immediately arrive at the desired bound on the difference in question. We are done.

8. An auxiliary harmonic function

We desire to use the lower bound furnished by the theorem of article 6 for

$$\int_{\mathscr{J}} u(x)\frac{dx}{x} = E\left(d\left(\frac{v(t)}{t}\right), d\left(\frac{v(t)}{t}\right)\right)$$

in order to obtain one for $\int_J u(x)(dx/x)$, the quantity of interest to us. Our plan is to pass from

$$\int_{\tilde{J}} u(x)\frac{dx}{x} \quad \text{to} \quad \sum_{k\geq 0}\frac{1}{d_k}\left(\int_{c_k}^{\gamma_k} + \int_{\delta_k}^{d_k}\right)u(x)\,dx$$

and from

$$\sum_{k\geq 0}\frac{1}{d_k}\int_{c_k}^{d_k} u(x)\,dx \quad \text{to} \quad \int_J u(x)\frac{dx}{x};$$

according to the result of the preceding article (whose notation we maintain here), this will entail only small losses (relative to $\int_{\tilde{J}} u(x)(dx/x)$), if $\eta > 0$ is small. This procedure still requires us, however, to get from the *first* sum to the *second.*

The simplest idea that comes to mind is to just compare corresponding terms of the two sums. That, however, would not be quite right, for in $\int_{c_k}^{d_k} u(x)\,dx$, the integration takes place over a *set with larger Lebesgue measure* than in $(\int_{c_k}^{\gamma_k} + \int_{\delta_k}^{d_k})u(x)\,dx$. In order to correct for this discrepancy, one should take an *appropriate multiple* of the second integral and then match the result against the first. The factor to be used here is obviously

$$\frac{2}{1-3p},$$

since (article 3),

$$\frac{\gamma_k - c_k + d_k - \delta_k}{d_k - c_k} = \frac{1-3p}{2}.$$

We are looking, then, at

$$\int_{c_k}^{d_k} u(x)dx - \frac{2}{1-3p}\left(\int_{c_k}^{\gamma_k} + \int_{\delta_k}^{d_k}\right)u(x)dx$$
$$= \int_{\gamma_k}^{\delta_k} u(x)dx - \frac{1+3p}{1-3p}\left(\int_{c_k}^{\gamma_k} + \int_{\delta_k}^{d_k}\right)u(x)dx.$$

From now on, it will be convenient to write

▶
$$\lambda = \frac{1+3p}{1-3p};$$

λ is >1 and *very close to* 1 *if* $p > 0$ *is small.* It is also useful to split up each interval (γ_k, δ_k) into two pieces, associating the left-hand one with (c_k, γ_k) and the other with (δ_k, d_k), and doing this in such a way that each piece has λ *times the length* of the interval to which it is associated. This is of course possible because

$$\frac{\delta_k - \gamma_k}{\gamma_k - c_k + d_k - \delta_k} = \frac{1+3p}{1-3p} = \lambda;$$

we thus take $g_k \in (\gamma_k, \delta_k)$ with

$$g_k = \gamma_k + \lambda(\gamma_k - c_k)$$

(and hence also $g_k = \delta_k - \lambda(d_k - \delta_k)$), and look at each of the two differences

$$\int_{\gamma_k}^{g_k} u(x)\mathrm{d}x - \lambda\int_{c_k}^{\gamma_k} u(x)\mathrm{d}x, \qquad \int_{g_k}^{\delta_k} u(x)\mathrm{d}x - \lambda\int_{\delta_k}^{d_k} u(x)\mathrm{d}x$$

separately; what we want to show is that *neither comes out too negative*, for we are trying to obtain a *positive lower bound* on $\int_J u(x)(\mathrm{d}x/x)$.

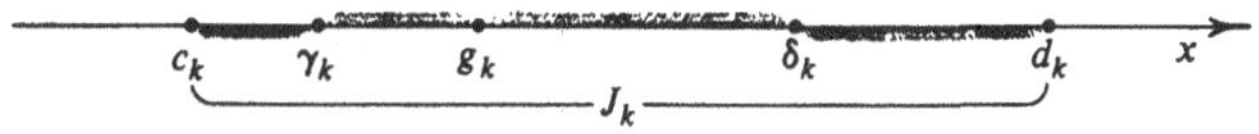

Figure 147

It is a fact that the two differences just written *can* be estimated in terms of $E(\mathrm{d}(\nu(t)/t), \mathrm{d}(\nu(t)/t))$.

Problem 23

(a) Show that for our function

$$u(z) = \int_0^\infty \log\left|\frac{z+t}{z-t}\right| \mathrm{d}\left(\frac{\nu(t)}{t}\right),$$

one has

$$E\left(\mathrm{d}\left(\frac{\nu(t)}{t}\right), \mathrm{d}\left(\frac{\nu(t)}{t}\right)\right) = \frac{1}{4\pi^2}\int_{-\infty}^{\infty}\int_{-\infty}^{\infty}\left(\frac{u(x)-u(y)}{x-y}\right)^2 \mathrm{d}x\,\mathrm{d}y.$$

This is Jesse Douglas' formula – I hope the coefficient on the right is correct. (Hint: Here, $u(x) = -(1/x)\int_0^\infty \log|1 - x^2/t^2|\,\mathrm{d}\nu(t)$ belongs to $L_2(-\infty, \infty)$ (it is *odd* on $\mathbb{R}$), so we can use Fourier–Plancherel transforms. In terms of

$$\hat{u}(\lambda) = \int_{-\infty}^{\infty} \mathrm{e}^{\mathrm{i}\lambda t} u(t)\,\mathrm{d}t$$

we have

$$u(x+\mathrm{i}y) = \frac{1}{2\pi}\int_{-\infty}^{\infty} \mathrm{e}^{-|\lambda|y}\mathrm{e}^{-\mathrm{i}\lambda x}\hat{u}(\lambda)\,\mathrm{d}\lambda$$

for $y > 0$ (the left side being just the Poisson harmonic extension of the function $u(x)$ to $\Im z > 0$), and

$$\frac{u(x+h)-u(x)}{h} = \frac{1}{2\pi}\int_{-\infty}^{\infty} e^{-i\lambda x}\frac{e^{-i\lambda h}-1}{h}\hat{u}(\lambda)\,d\lambda.$$

(All the right-hand integrals are to be understood in the l.i.m. sense.) Use Plancherel's theorem to express

$$\int_{-\infty}^{\infty}\left(\frac{u(x+h)-u(x)}{h}\right)^2 dx \qquad \text{and} \qquad \int_{-\infty}^{\infty}[(u_x(z))^2+(u_y(z))^2]\,dx$$

in terms of integrals involving $|\hat{u}(\lambda)|^2$, then integrate h from $-\infty$ to ∞ and y from 0 to ∞, and compare the results. Refer finally to the first lemma of article 5.)

(b) Show that

$$\left|\int_{\gamma_k}^{g_k} u(x)dx - \lambda\int_{c_k}^{\gamma_k} u(x)dx\right|$$

$$\leqslant \sqrt{\left(\frac{(1+\lambda)^4-1-\lambda^4}{12}\right)\cdot(\gamma_k-c_k)}\cdot\sqrt{\left(\int_{c_k}^{\gamma_k}\int_{\gamma_k}^{g_k}\left(\frac{u(x)-u(y)}{x-y}\right)^2 dy\,dx\right)},$$

and obtain a similar estimate for

$$\int_{g_k}^{\delta_k} u(x)dx - \lambda\int_{\delta_k}^{d_k} u(x)dx.$$

(Hint: *Trick*:

$$\int_{\gamma_k}^{g_k} u(x)dx - \lambda\int_{c_k}^{\gamma_k} u(x)dx = \frac{1}{\gamma_k-c_k}\int_{c_k}^{\gamma_k}\int_{\gamma_k}^{g_k}[u(y)-u(x)]\,dy\,dx.\ )$$

(c) Use the result of article 6 with those of (a) and (b) to estimate

$$\left|\sum_{k\geqslant 0}\frac{1}{d_k}\left(\int_{\gamma_k}^{\delta_k} u(x)dx - \lambda\left(\int_{c_k}^{\gamma_k}+\int_{\delta_k}^{d_k}\right)u(x)dx\right)\right|$$

in terms of $E(d(\nu(t)/t), d(\nu(t)/t))$.

By working the problem, one finds that the difference considered in part (c) is in absolute value $\leqslant C\int_J u(x)(dx/x)$ for a certain numerical constant C. The trouble is, however, that the value of C obtained in this way comes out quite a bit *larger* than 1, so that the result cannot be used to yield a *positive* lower bound on $\int_J u(x)(dx/x)$, λ being near 1. Too much is lost in following the simple reasoning of part (b); we need a more refined argument that will bring the value of C down below 1.

Any such refinement that works seems to involve bringing in (by use of Green's theorem, for instance) certain double integrals taken over portions of the first quadrant, in which the partial derivatives of u occur. Let us see how this comes about, considering the difference

$$\int_{\gamma_k}^{g_k} u(x)\,dx \;-\; \lambda \int_{c_k}^{\gamma_k} u(x)\,dx.$$

The latter can be rewritten as

$$\int_0^{(1+\lambda)\Delta_k} u(c_k + x)s_k(x)\,dx,$$

where $\Delta_k = \gamma_k - c_k$, and

$$s_k(x) \;=\; \begin{cases} -\lambda, & 0 < x < \Delta_k, \\ \;\;1, & \Delta_k < x < (1+\lambda)\Delta_k. \end{cases}$$

Suppose that we can find a function $V(z) = V_k(z)$, *harmonic* in the half-strip

$$S_k \;=\; \{z\colon\ 0 < \Re z < (1+\lambda)\Delta_k \ \text{ and } \ \Im z > 0\}$$

and having the following boundary behaviour:

$$V_y(x + i0) \;=\; -s_k(x), \qquad 0 < x < (1+\lambda)\Delta_k$$

($V_y(x + i0)$ will be discontinuous at $x = \Delta_k$),

$$V_x(iy) \;=\; 0, \qquad y > 0,$$

$$V_x(iy + (1+\lambda)\Delta_k) \;=\; 0, \qquad y > 0.$$

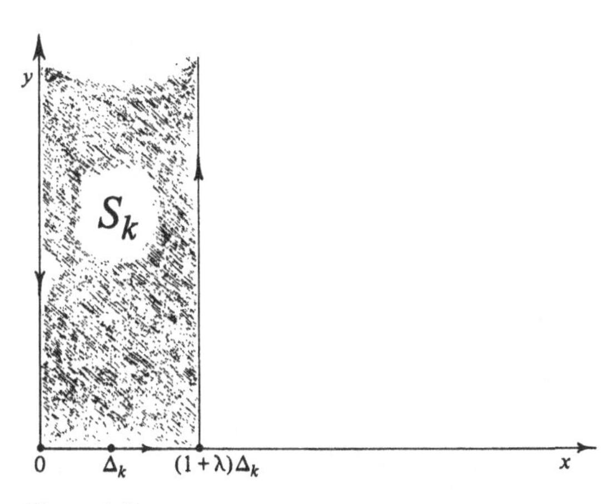

Figure 148

Then the previous integral becomes

$$\int_{\partial S_k} (-u(c_k + z)V_y(z)\,dx \;+\; u(c_k + z)V_x(z)\,dy),$$

∂S_k being oriented in the usual counterclockwise sense. Application of Green's theorem, if legitimate (which is easily shown to be the case here, as we shall see in due time), converts the line integral to

$$\iint_{S_k} (u_y(c_k + z)V_y(z) \;+\; u_x(c_k + z)V_x(z))\,dx\,dy$$

$$+ \iint_{S_k} u(c_k + z)[V_{yy}(z) + V_{xx}(z)]\,dx\,dy.$$

The harmonicity of V in S_k will make the *second* integral vanish, and finally the *difference under consideration* will be equal to the *first* one. Referring to the first lemma of article 5, we see that the successful use of this procedure in order to get what we want necessitates our actually *obtaining* such a harmonic function $V = V_k$ and then *computing* (at least) its *Dirichlet integral*

$$\iint_{S_k} (V_x^2 + V_y^2)\,dx\,dy.$$

We will in fact need to know a little more than that. Let us proceed with the necessary calculations.

Our harmonic function $V_k(z)$ (assuming, of course, that there *is* one) will depend on *two* parameters, Δ_k and $\lambda = (1 + 3p)/(1 - 3p)$. The dependence on the *first* of these is nothing but a kind of homogeneity. Let $v(z, \lambda)$ be the function $V(z)$ corresponding to the special value $\pi/(1 + \lambda)$ of Δ_k, using the value of λ figuring in $V_k(z)$; $v(z, \lambda)$ is, in other words, to be harmonic in the half-strip

$$S \;=\; \{z\colon\ 0 < \Re z < \pi \text{ and } \Im z > 0\}$$

with $v_x(z, \lambda) = 0$ on the *vertical sides* of S and

$$v_y(x + i0,\ \lambda) \;=\; \begin{cases} \lambda, & 0 < x < \dfrac{\pi}{1 + \lambda}, \\[2ex] -1, & \dfrac{\pi}{1 + \lambda} < x < \pi. \end{cases}$$

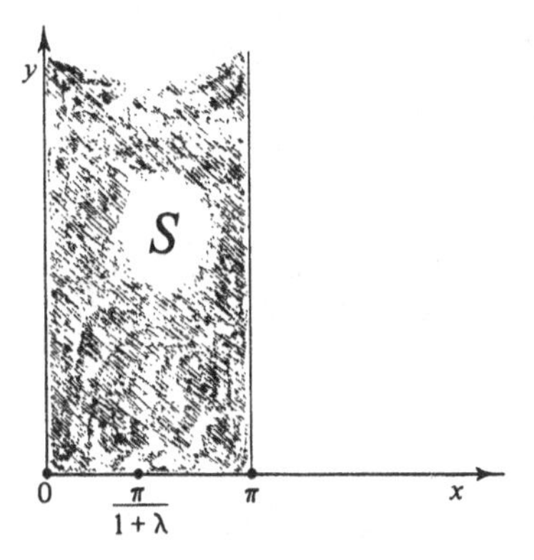

Figure 149

On the half-strip S_k of width $(1 + \lambda)\Delta_k$ shown previously, the function

$$\frac{1}{\pi}(1 + \lambda)\Delta_k v(\pi z/(1 + \lambda)\Delta_k,\ \lambda)$$

is harmonic, and its partial derivatives clearly satisfy the boundary conditions on those of $V_k(z)$ stipulated above. We may therefore take

$$V_k(z) \;=\; \frac{1}{\pi}(1 + \lambda)\Delta_k v(\pi z/(1 + \lambda)\Delta_k,\ \lambda)\,;$$

this permits us to do all our calculations with the standard function v. Note that we will have, by simple change of variables,

$$\iint_{S_k}\left(\frac{\partial V_k}{\partial x}\right)^2 \mathrm{d}x\,\mathrm{d}y \;=\; \left(\frac{(1+\lambda)\Delta_k}{\pi}\right)^2 \iint_S (v_x(z,\lambda))^2\,\mathrm{d}x\,\mathrm{d}y$$

and

$$\iint_{S_k}\left[\left(\frac{\partial V_k}{\partial y}\right)^+\right]^2 \mathrm{d}x\,\mathrm{d}y \;=\; \left(\frac{(1+\lambda)\Delta_k}{\pi}\right)^2 \iint_S [(v_y(z,\lambda))^+]^2\,\mathrm{d}x\,\mathrm{d}y,$$

while

$$\iint_{S_k}\left|\frac{\partial V_k}{\partial y}\right| \mathrm{d}x\,\mathrm{d}y \;=\; \left(\frac{(1+\lambda)\Delta_k}{\pi}\right)^2 \iint_S |v_y(z,\lambda)|\,\mathrm{d}x\,\mathrm{d}y.$$

Lemma. *Given $\lambda \geqslant 1$, we can find a function $v(z,\lambda)$ harmonic in S whose partial derivatives satisfy the boundary conditions specified above. If $\varepsilon > 0$ is* *given, we have, for all $\lambda \geqslant 1$ sufficiently close to* 1,

$$\pi \iint_S (v_x(z,\lambda))^2 \, dx\,dy \;<\; 4\left(1 + \frac{1}{3^3} + \frac{1}{5^3} + \cdots + \varepsilon\right),$$

$$\pi \iint_S [(v_y(z,\lambda))^+]^2 \, dx\,dy \;<\; 2\left(1 + \frac{1}{3^3} + \frac{1}{5^3} + \cdots + \varepsilon\right),$$

and

$$\iint_S |v_y(z,\lambda)| \, dx\,dy \;\leqslant\; C,$$

C being a numerical constant, whose value we do not bother to calculate.

Remark. In the next article we will need the *numerical approximation*

$$\boxed{\frac{4}{\pi^2}\left(1 + \frac{1}{3^3} + \frac{1}{5^3} + \cdots\right) \;<\; 0.4268.}$$

Proof of lemma. The method followed here (plain old 'separation of variables' from engineering mathematics) was suggested to me by Cedric Schubert. We look for a function v represented in the form

$$v(z,\lambda) \;=\; \sum_1^\infty A_n(\lambda) e^{-ny} \cos nx$$

The series on the right, if convergent, *will* represent a function harmonic in S (each of its *terms* is harmonic!), and, for $y > 0$,

$$v_x(z,\lambda) \;=\; -\sum_1^\infty n A_n(\lambda) e^{-ny} \sin nx$$

will vanish for $x = 0$ and $x = \pi$, for the exponentially decreasing factors e^{-ny} will make the series absolutely convergent.

For $y = 0$, by Abel's theorem,

$$v_y(x + i0,\ \lambda) \;=\; -\sum_1^\infty n A_n(\lambda) \cos nx$$

at each x for which the series on the right is convergent. Let us choose the $A_n(\lambda)$ so as to make the right side the *Fourier cosine series* of the function

$$s(x,\lambda) \;=\; \begin{cases} \lambda, & 0 < x < \dfrac{\pi}{1+\lambda}, \\[2ex] -1, & \dfrac{\pi}{1+\lambda} < x < \pi. \end{cases}$$

We know from the very rudiments of Fourier series theory that this is accomplished by taking

$$-nA_n(\lambda) = \frac{2}{\pi}\int_0^\pi s(x,\lambda)\cos nx\,dx,$$

and that the resulting cosine series *does* converge to $s(x,\lambda)$ for $0<x<\pi/(1+\lambda)$ and for $\pi/(1+\lambda)<x<\pi$. We can therefore *get* in this way a function $v(z\,\lambda)$ meeting all of our requirements.

Let us continue as long as we can without resorting to explicit computations. For fixed $y>0$, Parseval's formula yields

$$\int_0^\pi (v_y(z,\lambda))^2\,dx = \frac{\pi}{2}\sum_1^\infty n^2(A_n(\lambda))^2 e^{-2ny},$$

and, in like manner,

$$\int_0^\pi [v_y(z,\lambda)-v_y(z,1)]^2\,dx = \frac{\pi}{2}\sum_1^\infty n^2(A_n(\lambda)-A_n(1))^2 e^{-2ny}.$$

Integrating both sides of this last relation with respect to y, we find that

$$\int_0^\infty\int_0^\pi (v_y(z,\lambda)-v_y(z,1))^2\,dx\,dy = \frac{\pi}{4}\sum_1^\infty n[A_n(\lambda)-A_n(1)]^2.$$

By Parseval's formula, we have, however,

$$\sum_1^\infty n^2[A_n(\lambda)-A_n(1)]^2 = \frac{2}{\pi}\int_0^\pi [s(x,\lambda)-s(x,1)]^2\,dx,$$

and it is evident that the right-hand integral tends to *zero* as $\lambda\to 1$. Hence, by the preceding relation,

$$\int_0^\infty\int_0^\pi [v_y(z,\lambda)-v_y(z,1)]^2\,dx\,dy \longrightarrow 0$$

for $\lambda\to 1$.

Now clearly

$$|(v_y(z,\lambda))^+ - (v_y(z,1))^+| \leqslant |v_y(z,\lambda)-v_y(z,1)|;$$

the result just obtained therefore implies that

$$\int_0^\infty\int_0^\pi [(v_y(z,\lambda))^+]^2\,dx\,dy \longrightarrow \int_0^\infty\int_0^\pi [(v_y(z,1))^+]^2\,dx\,dy$$

as $\lambda\to 1$. We see in the same fashion that

$$\int_0^\infty\int_0^\pi (v_x(z,\lambda))^2\,dx\,dy = \frac{\pi}{4}\sum_1^\infty n(A_n(\lambda))^2,$$

which $\longrightarrow$ $(\pi/4)\sum_1^\infty n(A_n(1))^2$ as $\lambda\to 1$.

For our purpose, it thus suffices to make the calculations for the limiting case $\lambda = 1$. Here,

$$-nA_n(1) = \frac{2}{\pi}\left(\int_0^{\pi/2} - \int_{\pi/2}^{\pi}\right)\cos nx\,dx = \frac{4\sin\frac{\pi}{2}n}{\pi n},$$

so

$$\frac{\pi}{4}\sum_1^\infty n(A_n(1))^2 = \frac{4}{\pi}\left(1 + \frac{1}{3^3} + \frac{1}{5^3} + \cdots\right),$$

whence, if $\lambda \geqslant 1$ is sufficiently close to 1,

$$\pi\int_0^\infty\int_0^\pi (v_x(z,\lambda))^2\,dx\,dy < 4\left(1 + \frac{1}{3^3} + \frac{1}{5^3} + \cdots + \varepsilon\right).$$

Again

$$v_y(x + i0, 1) = \begin{cases} 1, & 0 < x < \frac{\pi}{2}, \\ -1, & \frac{\pi}{2} < x < \pi, \end{cases}$$

so by symmetry, for $y > 0$,

$$v_y\left(\frac{\pi}{2} - h + iy, 1\right) = -v_y\left(\frac{\pi}{2} + h + iy, 1\right) > 0, \qquad 0 < h < \frac{\pi}{2}.$$

Hence,

$$\begin{aligned}\int_0^\infty\int_0^\pi [(v_y(z,1))^+]^2\,dx\,dy &= \int_0^\infty\int_0^{\pi/2} (v_y(z,1))^2\,dx\,dy \\ = \frac{1}{2}\int_0^\infty\int_0^\pi (v_y(z,1))^2\,dx\,dy &= \frac{\pi}{8}\sum_1^\infty n(A_n(1))^2 \\ &= \frac{2}{\pi}\left(1 + \frac{1}{3^3} + \frac{1}{5^3} + \cdots\right).\end{aligned}$$

Therefore, by the above observation,

$$\pi\int_0^\infty\int_0^\pi [(v_y(z,\lambda))^+]^2\,dx\,dy < 2\left(1 + \frac{1}{3^3} + \frac{1}{5^3} + \cdots + \varepsilon\right)$$

for $\lambda \geqslant 1$ close enough to 1.

We are left with the integral $\int_0^\infty\int_0^\pi |v_y(z,\lambda)|\,dx\,dy$. This, by Schwarz' inequality, is

$$\leqslant \int_0^\infty \left(\pi \int_0^\pi (v_y(z,\lambda))^2 \, dx \right)^{\frac{1}{2}} dy$$

$$= \int_0^\infty \left(\frac{\pi^2}{2} \sum_1^\infty n^2 (A_n(\lambda))^2 e^{-2ny} \right)^{\frac{1}{2}} dy$$

$$= \int_0^\infty e^{-y/2} \left(\frac{\pi^2}{2} \sum_1^\infty n^2 (A_n(\lambda))^2 e^{-(2n-1)y} \right)^{\frac{1}{2}} dy$$

$$\leqslant \sqrt{\left(\int_0^\infty e^{-y} dy \cdot \int_0^\infty \frac{\pi^2}{2} \sum_1^\infty n^2 (A_n(\lambda))^2 e^{-(2n-1)y} dy \right)}$$

$$= \sqrt{\left(\frac{\pi^2}{2} \sum_1^\infty \frac{n^2}{2n-1} (A_n(\lambda))^2 \right)} \leqslant \frac{\pi}{\sqrt{2}} \sqrt{\left(\sum_1^\infty n (A_n(\lambda))^2 \right)}.$$

We have already seen that the sum inside the radical in the last of these terms tends to a definite (finite) limit as $\lambda \to 1$. So

$$\int_0^\infty \int_0^\pi |v_y(z,\lambda)| \, dx \, dy$$

is certainly *bounded* for $\lambda \geqslant 1$ near 1. The lemma is proved.

Referring to the remarks made just before the lemma and to the boxed numerical estimate immediately following its statement, we obtain, regarding our original functions V_k, the following

Corollary. *Given $\lambda \geqslant 1$ there is, for each k, a function $V_k(z)$ (depending on λ), harmonic in $S_k = \{z\colon\ 0 < \Re z < (1+\lambda)\Delta_k$ and $\Im z > 0\}$, with $\partial V_k(z)/\partial x = 0$ on the* vertical sides *of S_k and, on the latter's* base, *$\partial V_k/\partial y$ taking the boundary values λ and -1 along $(0,\ \Delta_k)$ and $(\Delta_k,\ (1+\lambda)\Delta_k)$ respectively.*

If $\lambda \geqslant 1$ is close enough *to 1, we have.*

$$\pi \iint_{S_k} \left(\frac{\partial V_k}{\partial x} \right)^2 dx\,dy \leqslant 0.44(1+\lambda)^2 \Delta_k^2,$$

$$\pi \iint_{S_k} \left[\left(\frac{\partial V_k}{\partial y} \right)^+ \right]^2 dx\,dy \leqslant 0.22(1+\lambda)^2 \Delta_k^2,$$

and

$$\iint_{S_k} \left| \frac{\partial V_k}{\partial y} \right| dx\,dy \leqslant \alpha(1+\lambda)^2 \Delta_k^2,$$

α being a certain numerical constant.

9. Lower estimate for $\int_\Omega \int_0^\infty \log|1 - x^2/t^2| \mathrm{d}\mu(t)(\mathrm{d}x/x^2)$

We return to the termwise comparison of

$$\sum_{k\geq 0} \frac{1}{d_k} \int_{c_k}^{d_k} u(x)\,\mathrm{d}x \quad \text{and} \quad \frac{2}{1-3p} \sum_{k \geq 0} \frac{1}{d_k} \left(\int_{c_k}^{\gamma_k} + \int_{\delta_k}^{d_k} \right) u(x)\,\mathrm{d}x,$$

which, as we saw in the first half of the preceding article, leads to the task of estimating

$$\int_{\gamma_k}^{g_k} u(x)\,\mathrm{d}x \; - \; \lambda \int_{c_k}^{\gamma_k} u(x)\,\mathrm{d}x$$

and

$$\int_{g_k}^{\delta_k} u(x)\,\mathrm{d}x \; - \; \lambda \int_{\delta_k}^{d_k} u(x)\,\mathrm{d}x$$

from below. The notation of the previous two articles is maintained here.

Following the idea of the last article, we use the harmonic function $V_k(z)$ described there to express the *first* of the above differences as a *line integral*

$$\int_{\partial S_k} \left(-u(c_k + z) \frac{\partial V_k(z)}{\partial y} \mathrm{d}x \; + \; u(c_k + z) \frac{\partial V_k(z)}{\partial x} \mathrm{d}y \right)$$

around the vertical half-strip S_k whose base is the segment $[0, (1+\lambda)\Delta_k] = [0, (1+\lambda)(\gamma_k - c_k)]$ of the real axis. By use of Green's theorem, this line integral is converted to

$$\iint_{S_k} \left(u_x(c_k + z) \frac{\partial V_k(z)}{\partial x} \; + \; u_y(c_k + z) \frac{\partial V_k(z)}{\partial y} \right) \mathrm{d}x\,\mathrm{d}y,$$

thanks to the harmonicity of $V_k(z)$ in S_k. The justification of the present application of Green's theorem proceeds as follows.

We have

$$\int_0^{(1+\lambda)(\gamma_k - c_k)} u(c_k + x)(V_k)_y(x + \mathrm{i}0)\,\mathrm{d}x$$

$$= \lim_{h \to 0} \int_0^{(1+\lambda)(\gamma_k - c_k)} u(c_k + x + \mathrm{i}h)(V_k)_y(x + \mathrm{i}h)\,\mathrm{d}x,$$

because $u(z)$ is continuous up to the real axis, and, as one verifies by referring to the computations with v and v_y near the end of the previous article,

$$\int_0^{(1+\lambda)(\gamma_k - c_k)} [(V_k)_y(x + \mathrm{i}h) \; - \; (V_k)_y(x + \mathrm{i}0)]^2\,\mathrm{d}x \; \longrightarrow \; 0$$

for $h \to 0$.

However, for $h>0$ and $0<x<(1+\lambda)(\gamma_k-c_k)$,

$$\int_h^\infty \frac{\partial}{\partial y}\left(u(c_k+z)\frac{\partial V_k(z)}{\partial y}\right)\mathrm{d}y \;=\; -u(c_k+x+ih)(V_k)_y(x+ih),$$

since

$$|u(c_k+z)| \;\leqslant\; \text{const.}\frac{\log|z|}{|z|}, \qquad z\in S_k,$$

by an estimate used in proving the first lemma of article 5, and $V_k(z)$, together with its partial derivatives, tends (exponentially) to zero as $z\to\infty$ in S_k (see the calculations at end of the previous article).

Again, since $\partial V_k(z)/\partial x = 0$ on the vertical sides of S_k,

$$\int_0^{(1+\lambda)(\gamma_k-c_k)} \frac{\partial}{\partial x}\left(u(c_k+z)\frac{\partial V_k(z)}{\partial x}\right)\mathrm{d}x \;=\; 0, \qquad y>0.$$

By integrating y in this formula from h to ∞ and x in the previous one from 0 to $(1+\lambda)(\gamma_k-c_k)$, and then adding the results, we express

$$-\int_0^{(1+\lambda)(\gamma_k-c_k)} u(c_k+x+ih)(V_k)_y(x+ih)\,\mathrm{d}x$$

as the sum of two iterated integrals. For $h>0$, both of the latter are *absolutely convergent*, and the order of integration in *one of them* may be *reversed*. Doing this and remembering that $\nabla^2 V_k = 0$ in S_k, we see that the sum in question boils down to

$$\int_h^\infty\int_0^{(1+\lambda)(\gamma_k-c_k)}\left(u_x(c_k+z)\frac{\partial V_k(z)}{\partial x} \;+\; u_y(c_k+z)\frac{\partial V_k(z)}{\partial y}\right)\mathrm{d}x\,\mathrm{d}y.$$

Making $h\to 0$ in this expression finally gives us the corresponding double integral over S_k (whose absolute convergence readily follows from the first lemma in article 5 and the work at the end of the previous one by Schwarz' inequality).

This, together with our initial observation, shows that the double integral over S_k is equal to

$$-\int_0^{(1+\lambda)(\gamma_k-c_k)} u(c_k+x)(V_k)_y(x+i0)\,\mathrm{d}x,$$

a quantity clearly identical with the above line integral around ∂S_k.* In this way, we see that our use of Green's theorem is legitimate.

The line integral is, as we recall (and as we see by glancing at the preceding expression), the same as

$$\int_{\gamma_k}^{g_k} u(x)\,\mathrm{d}x \;-\; \lambda\int_{c_k}^{\gamma_k} u(x)\,\mathrm{d}x.$$

* and actually coinciding with the original expression on p. 499 (the second one displayed there) from which the line integral was elaborated

That difference is therefore equal to

$$\iint_{S_k} \left(u_x(c_k + z) \frac{\partial V_k(z)}{\partial x} + u_y(c_k + z) \frac{\partial V_k(z)}{\partial y} \right) \mathrm{d}x\,\mathrm{d}y.$$

What we want is a *lower bound* for the difference, and that means we have to *find one for this double integral.*

Our intention is to express such a lower bound as a certain portion of $E(\mathrm{d}(v(t)/t), \mathrm{d}(v(t)/t))$, the hope being that when all these portions are added (and also all the ones corresponding to the differences

$$\int_{g_k}^{\delta_k} u(x)\,\mathrm{d}x \; - \; \lambda \int_{\delta_k}^{d_k} u(x)\,\mathrm{d}x \;),$$

we will end with a multiple of $E(\mathrm{d}(v(t)/t), \mathrm{d}(v(t)/t))$ that is not too large. In view, then, of the first lemma of article 5, we are interested in getting a lower bound in terms of

$$\frac{1}{\pi} \iint_{S_k} [(u_x(c_k + z))^2 + (u_y(c_k + z))^2]\,\mathrm{d}x\,\mathrm{d}y.$$

The present situation allows for very little leeway, and we have to be quite careful.

We start by writing

$$\iint_{S_k} u_x(c_k + z) \frac{\partial V_k(z)}{\partial x} \mathrm{d}x\,\mathrm{d}y$$

$$\geqslant - \sqrt{\left(\pi \iint_{S_k} \left(\frac{\partial V_k(x)}{\partial x} \right)^2 \mathrm{d}x\,\mathrm{d}y \right)} \times \sqrt{\left(\frac{1}{\pi} \iint_{S_k} (u_x(c_k + z))^2\,\mathrm{d}x\,\mathrm{d}y \right)}.$$

According to the corollary at the end of the last article, the right side is in turn

$$\geqslant \; -(0.44)^{\frac{1}{2}}(1 + \lambda)(\gamma_k - c_k) \sqrt{\left(\frac{1}{\pi} \iint_{S_k} (u_x(c_k + z))^2\,\mathrm{d}x\,\mathrm{d}y \right)},$$

provided that $\lambda = (1 + 3p)/(1 - 3p)$ is close enough to 1 (recall that the Δ_k of the previous article equals $\gamma_k - c_k$).

For the estimation of

$$\iint_{S_k} u_y(c_k + z) \frac{\partial V_k(z)}{\partial y} \mathrm{d}x\,\mathrm{d}y,$$

we split up S_k into *two pieces*,

$$S_k^+ \;=\; \left\{z\in S_k:\;\; \frac{\partial V_k(z)}{\partial y} > 0\right\}$$

and

$$S_k^- \;=\; S_k \sim S_k^+ .$$

We have

$$\iint_{S_k^+} u_y(c_k+z)\frac{\partial V_k(z)}{\partial y}\,\mathrm{d}x\,\mathrm{d}y$$

$$\geqslant -\sqrt{\left(\pi\iint_{S_k^+}\left(\frac{\partial V_k(z)}{\partial y}\right)^2\mathrm{d}x\,\mathrm{d}y\right)}$$

$$\times\sqrt{\left(\frac{1}{\pi}\iint_{S_k^+}(u_y(c_k+z))^2\,\mathrm{d}x\,\mathrm{d}y\right)},$$

which, by the corollary of the preceding article, is

$$\geqslant\; -(0.22)^{\frac{1}{2}}(1+\lambda)(\gamma_k-c_k)\sqrt{\left(\frac{1}{\pi}\iint_{S_k^+}(u_y(c_k+z))^2\,\mathrm{d}x\,\mathrm{d}y\right)}$$

for λ close enough to 1. In this last expression, the integral involving u_y may, if we wish, be replaced by one over S_k, yielding a worse result.

We are left with

$$\iint_{S_k^-} u_y(c_k+z)\frac{\partial V_k(z)}{\partial y}\,\mathrm{d}x\,\mathrm{d}y,$$

in which $\partial V_k(z)/\partial y \leqslant 0$. To handle this integral, we recall that

$$u(z) \;=\; \int_0^\infty \log\left|\frac{z+t}{z-t}\right|\mathrm{d}\left(\frac{\nu(t)}{t}\right),$$

which makes

$$u_y(z) \;=\; \int_0^\infty\left[\frac{y}{(x+t)^2+y^2}-\frac{y}{(x-t)^2+y^2}\right]\mathrm{d}\left(\frac{\nu(t)}{t}\right),$$

with the quantity in brackets obviously *negative* for x, y and $t>0$. Since $\nu(t)/t\leqslant 2\eta$ by our construction of the intervals J_k, we have

$$\mathrm{d}\left(\frac{\nu(t)}{t}\right) \;=\; \frac{\mathrm{d}\nu(t)}{t} \;-\; \frac{\nu(t)\,\mathrm{d}t}{t^2} \;\geqslant\; -2\eta\frac{\mathrm{d}t}{t},$$

and therefore, for x and $y > 0$,

$$u_y(z) \leqslant 2\eta \int_0^\infty \left(\frac{y}{(x-t)^2+y^2} - \frac{y}{(x+t)^2+y^2} \right) \frac{dt}{t}$$

$$= 2\eta \lim_{\delta\to 0} \int_{-\infty}^\infty \frac{t}{t^2+\delta^2} \frac{y}{(x-t)^2+y^2} dt$$

$$= 2\eta\pi \lim_{\delta\to 0} \frac{x}{x^2+(y+\delta)^2} = 2\pi\eta \frac{x}{x^2+y^2}.$$

(We have simply used the Poisson representation for the function $\Re(1/(z+i\delta))$, harmonic in $\Im z > 0$.) Thus,

$$u_y(c_k + z) \leqslant \frac{2\pi\eta}{c_k}, \qquad z \in S_k,$$

whence

$$\iint_{S_k^-} u_y(c_k+z) \frac{\partial V_k(z)}{\partial y} dx\,dy \geqslant -\frac{2\pi\eta}{c_k} \iint_{S_k^-} \left| \frac{\partial V_k(z)}{\partial y} \right| dx\,dy.$$

For λ close to 1, the right side is

$$\geqslant -2\pi\alpha\eta \frac{(1+\lambda)^2(\gamma_k - c_k)^2}{c_k}$$

by the corollary from the previous article, α being a numerical constant.

Combining the three estimates just obtained, we find with the help of Schwarz' inequality that

$$\iint_{S_k} \left(u_x(c_k+z) \frac{\partial V_k(z)}{\partial x} + u_y(c_k+z) \frac{\partial V_k(z)}{\partial y} \right) dx\,dy$$

$$\geqslant -(0.44)^{\frac{1}{2}}(1+\lambda)(\gamma_k - c_k) \sqrt{\left(\frac{1}{\pi} \iint_{S_k} (u_x(c_k+z))^2\,dx\,dy \right)}$$

$$- (0.22)^{\frac{1}{2}}(1+\lambda)(\gamma_k - c_k) \sqrt{\left(\frac{1}{\pi} \iint_{S_k} (u_y(c_k+z))^2\,dx\,dy \right)}$$

$$- 2\pi\alpha\eta \frac{(1+\lambda)^2(\gamma_k - c_k)^2}{c_k}$$

$$\geqslant -(0.66)^{\frac{1}{2}}(1+\lambda)(\gamma_k - c_k)$$

$$\times \sqrt{\left(\frac{1}{\pi} \iint_{S_k} ((u_x(c_k+z))^2 + (u_y(c_k+z))^2)\,dx\,dy \right)}$$

$$- 2\pi\alpha\eta \frac{(1+\lambda)^2(\gamma_k - c_k)^2}{c_k},$$

provided that λ is close enough to 1. The double integral on the left is nothing but a complicated expression for the first of the two differences with which we are concerned – that was, indeed, our reason for bringing the function $V_k(z)$ into this work. Hence the relation just proved can be rewritten

$$\int_{\gamma_k}^{g_k} u(x)\,dx \;-\; \lambda\int_{c_k}^{\gamma_k} u(x)\,dx$$
$$\geqslant\; -\;(0.66)^{\frac{1}{2}}(1+\lambda)(\gamma_k - c_k)\sqrt{\left(\frac{1}{\pi}\int_0^\infty\int_{c_k}^{g_k}((u_x(z))^2+(u_y(z))^2)\,dx\,dy\right)}$$
$$-\;2\pi\alpha\eta\,\frac{(1+\lambda)^2(\gamma_k - c_k)^2}{c_k}.$$

The difference $\int_{g_k}^{\delta_k} u(x)\,dx - \lambda\int_{\delta_k}^{d_k} u(x)\,dx$ can also be estimated by the method of this and the preceding articles. One finds in exactly the same way as above that

$$\int_{g_k}^{\delta_k} u(x)\,dx \;-\; \lambda\int_{\delta_k}^{d_k} u(x)\,dx$$
$$\geqslant\; -\;(0.66)^{\frac{1}{2}}(1+\lambda)(d_k - \delta_k)\sqrt{\left(\frac{1}{\pi}\int_0^\infty\int_{g_k}^{d_k}((u_x(z))^2+(u_y(z))^2)\,dx\,dy\right)}$$
$$-\;2\pi\alpha\eta\,\frac{(1+\lambda)^2(d_k - \delta_k)^2}{g_k}$$

for λ close enough to 1. The following diagram shows the regions over which the double integrals involved in this and the previous inequalities are taken:

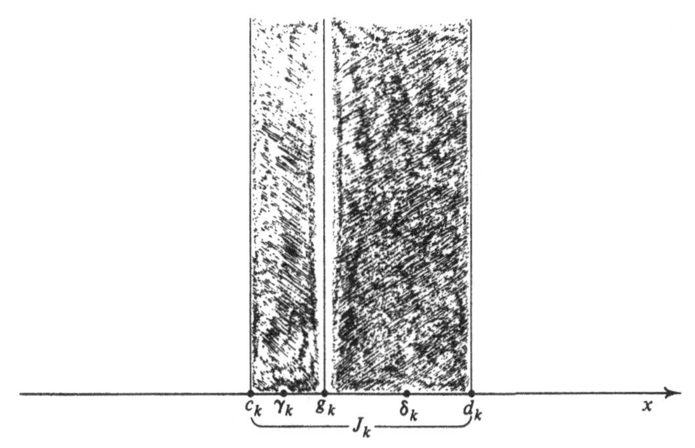

Figure 150

We now add the two relations just obtained. After dividing by d_k and using Schwarz' inequality again together with the fact that

$$c_k \leqslant \gamma_k \leqslant g_k < \delta_k < d_k < (1+2\eta)c_k,$$

we get, recalling that $\lambda = (1+3p)/(1-3p)$,

$$\frac{1}{d_k}\int_{c_k}^{d_k} u(x)\,dx \;-\; \frac{2}{1-3p}\cdot\frac{1}{d_k}\left(\int_{c_k}^{\gamma_k} + \int_{\delta_k}^{d_k}\right)u(x)\,dx$$

$$\geqslant \; - \; (0.66)^{\frac{1}{2}}\frac{2}{1-3p}(1+2\eta)\sqrt{\left(\left(\frac{\gamma_k - c_k}{\gamma_k}\right)^2 + \left(\frac{d_k-\delta_k}{d_k}\right)^2\right)}$$

$$\times\sqrt{\left(\frac{1}{\pi}\int_0^\infty\int_{c_k}^{d_k}(u_x^2+u_y^2)\,dx\,dy\right)}$$

$$-\;\frac{8\pi\alpha(1+2\eta)}{(1-3p)^2}\eta\left[\left(\frac{\gamma_k-c_k}{\gamma_k}\right)^2 + \left(\frac{d_k-\delta_k}{d_k}\right)^2\right]$$

for λ close enough to 1, *in other words, for $p > 0$ close enough to zero.*

We have now carried out the program explained in the first half of article 8 and at the beginning of the present one. Summing the preceding relation over k and using Schwarz' inequality once more, we obtain, for small $p > 0$,

$$\sum_{k\geqslant 0}\frac{1}{d_k}\int_{c_k}^{d_k} u(x)\,dx \;-\; \frac{2}{1-3p}\sum_{k\geqslant 0}\frac{1}{d_k}\left(\int_{c_k}^{\gamma_k} + \int_{\delta_k}^{d_k}\right)u(x)\,dx$$

$$\geqslant \; - \; (0.66)^{\frac{1}{2}}(1+2\eta)\frac{2}{1-3p}\sqrt{\sum_{k\geqslant 0}\left(\left(\frac{\gamma_k - c_k}{\gamma_k}\right)^2 + \left(\frac{d_k-\delta_k}{d_k}\right)^2\right)}$$

$$\times\sqrt{\left(\frac{1}{\pi}\sum_{k\geqslant 0}\int_0^\infty\int_{J_k}(u_x^2+u_y^2)\,dx\,dy\right)}$$

$$-\;\frac{8\pi\alpha(1+2\eta)}{(1-3p)^2}\eta\sum_{k\geqslant 0}\left(\left(\frac{\gamma_k-c_k}{\gamma_k}\right)^2 + \left(\frac{d_k-\delta_k}{d_k}\right)^2\right).$$

To the right-hand expression we apply the theorem of article 6 together with its remark and the first lemma of article 5. In this way, we find that the right side is

$$\geqslant \; -\frac{2}{1-3p}\sqrt{\left(\frac{0.66}{0.80-K\eta}\right)}\cdot(1+2\eta)E\left(d\left(\frac{v(t)}{t}\right), d\left(\frac{v(t)}{t}\right)\right)$$

$$-\;K'\eta E\left(d\left(\frac{v(t)}{t}\right), d\left(\frac{v(t)}{t}\right)\right)$$

for small enough positive values of η and p, K and K' being certain numerical constants independent of p and of the configuration of the J_k.

According to the theorem of article 7, the left-hand difference in the above relation is within

$$\frac{3-3p}{1-3p}C\eta^{\frac{1}{2}}E\left(\mathrm{d}\left(\frac{\nu(t)}{t}\right),\ \mathrm{d}\left(\frac{\nu(t)}{t}\right)\right)$$

of

$$\int_J u(x)\frac{\mathrm{d}x}{x} \ - \ \frac{2}{1-3p}\int_{\tilde{J}} u(x)\frac{\mathrm{d}x}{x}$$

for small enough $\eta > 0$, where C is a numerical constant independent of p or the configuration of the J_k. So, since

$$\int_{\tilde{J}} u(x)\frac{\mathrm{d}x}{x} \ = \ E\left(\mathrm{d}\left(\frac{\nu(t)}{t}\right), \mathrm{d}\left(\frac{\nu(t)}{t}\right)\right)$$

(see remark to the theorem of article 7), what we have boils down, for small enough p and $\eta > 0$, to

$$\int_J u(x)\frac{\mathrm{d}x}{x} \ \geqslant \ \frac{2}{1-3p}\left(1 - \sqrt{\left(\frac{0.66}{0.80}\right)} - A\eta - B\sqrt{\eta}\right) \times E\left(\mathrm{d}\left(\frac{\nu(t)}{t}\right), \mathrm{d}\left(\frac{\nu(t)}{t}\right)\right)$$

with numerical constants A and B independent of p and the configuration of the J_k. Here,

$$\sqrt{\left(\frac{0.66}{0.80}\right)} \ = \ 0.9083^-,$$

so, the coefficient on the right is

$$\geqslant \ \frac{2}{1-3p}(0.0917 - A\eta - B\sqrt{\eta}).$$

Not much at all, but still enough!

We have finally arrived at the point where a *value* for the parameter η must be chosen. This quantity, independent of p, was introduced during the *third stage* of the long construction in article 2, where it was necessary to take $0 < \eta < \frac{2}{3}$. Aside from that requirement, we were free to assign any value we liked to it. *Let us now choose, once and for all, a numerical value* > 0 *for* η, *small enough to ensure that all the estimates of articles* 6, 7 *and the present one hold good, and that besides*

$$0.0917 - A\eta - B\sqrt{\eta} \ > \ 1/20.$$

That value is henceforth fixed. This matter having been settled, the relation finally obtained above reduces to

$$\int_J u(x)\frac{\mathrm{d}x}{x} \geqslant \frac{1}{10(1-3p)} E\left(\mathrm{d}\left(\frac{\nu(t)}{l}\right), \mathrm{d}\left(\frac{\nu(t)}{t}\right)\right).$$

To get a lower bound on the right-hand member, we use again the inequality

$$E\left(\mathrm{d}\left(\frac{\nu(t)}{t}\right), \mathrm{d}\left(\frac{\nu(t)}{t}\right)\right)$$
$$\geqslant (0.80 - K\eta) \sum_{k \geqslant 0} \left(\left(\frac{\gamma_k - c_k}{\gamma_k}\right)^2 + \left(\frac{d_k - \delta_k}{d_k}\right)^2 \right)$$

(valid for *our* fixed value of η!), furnished by the theorem of article 6. In article 2, the intervals J_k were constructed so as to make $d_0 - c_0 = |J_0| \geqslant \eta d_0$ (see property (v) in the description near the end of that article), and in the construction of the function $\mu(t)$ we had

$$\frac{d_0 - \delta_0}{d_0 - c_0} = \frac{1-3p}{2}$$

(property (iii) of the specification near the end of article 3). Therefore

$$\frac{d_0 - \delta_0}{d_0} \geqslant \frac{1-3p}{2}\eta,$$

which, substituted into the previous inequality, yields

$$E\left(\mathrm{d}\left(\frac{\nu(t)}{t}\right), \mathrm{d}\left(\frac{\nu(t)}{t}\right)\right) \geqslant (0.80 - K\eta)\left(\frac{1-3p}{2}\right)^2 \eta^2.$$

We substitute this into the relation written above, and get

$$\boxed{\int_J u(x)\frac{\mathrm{d}x}{x} \geqslant (1-3p)c}$$

with a certain purely numerical constant c. (We see that it is finally just the ratio $|J_0|/d_0$ associated with the *first* of the intervals J_k that enters into these last calculations. If only we had been able to avoid consideration of the other J_k in the above work!) In terms of the function $\mu(t) = (p/(1-3p))\nu(t)$ constructed in article 3, we have, as at the beginning of article 7,

$$\int_\Omega\int_0^\infty \log\left|1-\frac{x^2}{t^2}\right|\mathrm{d}\mu(t)\frac{\mathrm{d}x}{x^2} = \frac{p}{1-3p}\int_J u(x)\frac{\mathrm{d}x}{x}.$$

By the preceding boxed formula and the work of article 3 we therefore have the

Theorem. *If $p \geqslant 0$ is small enough and if, for our original polynomial $P(x)$, the zero counting function $n(t)$ satisfies*

$$\sup_{t>0}\frac{n(t)}{t} > \frac{p}{1-3p},$$

then, for the function $\mu(t)$ constructed in article 3, we have

$$\int_\Omega\int_0^\infty \log\left|1-\frac{x^2}{t^2}\right|\mathrm{d}\mu(t)\frac{\mathrm{d}x}{x^2} \geqslant pc,$$

c being a numerical constant independent of $P(x)$. Here,

$$\Omega = (0,\infty)\sim\bigcup_{k\geqslant 0} J_k,$$

where the J_k are the intervals constructed in article 2.

In this way the task described at the very end of article 3 has been carried out, and the main work of the present § completed.

Remark. One reason why the present article's estimations have had to be so delicate is the *smallness* of the lower bound on

$$E\left(\mathrm{d}\left(\frac{\nu(t)}{t}\right),\mathrm{d}\left(\frac{\nu(t)}{t}\right)\right)$$

obtained in article 6. If we could be sure that this quantity was considerably *larger*, a much simpler procedure could be used to get from $\int_{\tilde{J}}u(x)(\mathrm{d}x/x)$ to $\int_J u(x)(\mathrm{d}x/x)$; the one of problem 23 (article 8) for instance.

It is possible that $E(\mathrm{d}(\nu(t)/t), \mathrm{d}(\nu(t)/t))$ *is* quite a bit larger than the lower bound we have found for it. One can write

$$E\left(\mathrm{d}\left(\frac{\nu(t)}{t}\right),\mathrm{d}\left(\frac{\nu(t)}{t}\right)\right) = \int_{\tilde{J}}\int_{\tilde{J}}\frac{1}{x^2}\log\left|\frac{1}{1-x^2/t^2}\right|\mathrm{d}t\,\mathrm{d}x.$$

If the intervals J_k are *very far apart from each other* (so that the cross terms

$$\int_{J_k\cap\tilde{J}}\int_{J_l\cap\tilde{J}}\frac{1}{x^2}\log\left|\frac{1}{1-x^2/t^2}\right|\mathrm{d}t\,\mathrm{d}x, \qquad k\neq l,$$

are all *very small*), the right-hand integral behaves like a constant multiple of

$$\sum_{k\geqslant 0}\left(\frac{|J_k|}{d_k}\right)^2\log\left(\frac{d_k}{|J_k|}\right).$$

When $\eta>0$ is taken to be *small*, this, on account of the inequality $|J_k|/d_k\leqslant 2\eta$, is much *larger* than the bound furnished by the theorem of article 6, which is essentially a *fixed* constant multiple of

$$\sum_{k\geqslant 0}\left(\frac{|J_k|}{d_k}\right)^2.$$

I have not been able to verify that the first of the above sums can be used to give a lower bound for $E(d(\nu(t)/t),\ d(\nu(t)/t))$ when the J_k are *not* far apart. That, however, is perhaps still worth trying.

10. Return to polynomials

Let us now combine the theorem from the end of article 3 with the one finally arrived at above. We obtain, without further ado, the

Theorem. *If $p>0$ is sufficiently small and $P(x)$ is any polynomial of the form*

$$\prod_k\left(1-\frac{x^2}{x_k^2}\right)$$

with the $x_k>0$, the condition

$$\sup_{t>0}\frac{n(t)}{t} > \frac{p}{1-3p}$$

for $n(t)=$ number of x_k (counting multiplicities) in $[0,t]$ implies that

$$\sum_1^\infty\frac{\log^+|P(m)|}{m^2} \geqslant \frac{cp}{5}.$$

Here, $c>0$ is a numerical constant independent *of p and of $P(x)$.*

Corollary. *Let $Q(z)$ be any even polynomial (with, in general, complex zeros) such that $Q(0)=1$. There is an absolute constant k, independent of Q, such that, for all z,*

$$\frac{\log|Q(z)|}{|z|} \leqslant k\sum_1^\infty\frac{\log^+|Q(m)|}{m^2},$$

provided that the sum on the right is less than some number $\gamma>0$, also independent of Q.

Proof. We can write

$$Q(z) \;=\; \prod_k \left(1 - \frac{z^2}{\zeta_k^2}\right).$$

Put $x_k = |\zeta_k|$ and then let

$$P(z) \;=\; \prod_k \left(1 - \frac{z^2}{x_k^2}\right);$$

we have $|P(x)| \leqslant |Q(x)|$ on $\mathbb{R}$, so

$$\sum_1^\infty \frac{\log^+ |P(m)|}{m^2} \;\leqslant\; \sum_1^\infty \frac{\log^+ |Q(m)|}{m^2}.$$

To $P(x)$ we apply the theorem, which clearly implies that

$$\sup_{t>0} \frac{n(t)}{t} \;\leqslant\; \frac{10}{c} \sum_1^\infty \frac{\log^+ |P(m)|}{m^2}$$

for $n(t)$, the number of x_k in $[0, t]$, whenever the sum on the right is small enough. For $z \in \mathbb{C}$,

$$\log|Q(z)| \;\leqslant\; \sum_k \log\left(1 + \frac{|z|^2}{|\zeta_k|^2}\right) \;=\; \int_0^\infty \log\left(1 + \frac{|z|^2}{t^2}\right) \mathrm{d}n(t),$$

and partial integration converts the last expression to

$$\int_0^\infty \frac{n(t)}{t} \frac{2|z|^2}{|z|^2 + t^2} \mathrm{d}t \;\leqslant\; \pi|z| \sup_{t>0} \frac{n(t)}{t}.$$

In view of our initial relation, we therefore have

$$\frac{\log|Q(z)|}{|z|} \;\leqslant\; \frac{10\pi}{c} \sum_1^\infty \frac{\log^+ |Q(m)|}{m^2}$$

whenever the right-hand sum is small enough. Done.

Remark 1. These results hold for objects more general than polynomials. Instead of $|Q(z)|$, we can consider any *finite product of the form*

$$\prod_k \left|1 - \frac{z^2}{\zeta_k^2}\right|^{\lambda_k}$$

where the exponents λ_k are all $\geqslant$ some fixed $\alpha > 0$. Taking $|P(x)|$ as

$$\prod_k \left|1 - \frac{x^2}{x_k^2}\right|^{\lambda_k}$$

with $x_k = |\zeta_k|$, and writing

$$n(t) = \sum_{x_k \in [0,t]} \lambda_k$$

(so that each 'zero' x_k is counted with 'multiplicity' λ_k), we easily convince ourselves that the arguments and constructions of articles 1 and 2 go through for *these* functions $|P(x)|$ and $n(t)$ without essential change. What was important there is the property, valid here, that $n(t)$ *increase by at least some fixed amount* $\alpha > 0$ *at each of its jumps*, crucial use having been made of this during the second and third stages of the construction in article 2. The work of articles 3–8 can thereafter be taken over *as is*, and we end with analogues of the above results for our present functions $|P(x)|$ and $|Q(z)|$.

Thus, in the case of *polynomials* $P(z)$, *it is not so much the single-valuedness* of the *analytic function* with *modulus* $|P(z)|$ as the *quantization of the point masses* associated with the *subharmonic function* $\log|P(z)|$ that is essential in the preceding development.

Remark 2. The specific arithmetic character of $\mathbb{Z}$ plays no rôle in the above work. Analogous results hold if we replace the sums

$$\sum_1^\infty \frac{\log^+|P(m)|}{m^2}, \qquad \sum_1^\infty \frac{\log^+|Q(m)|}{m^2},$$

by others of the form

$$\sum_{\lambda\in\Lambda} \frac{\log^+|P(\lambda)|}{\lambda^2}, \qquad \sum_{\lambda\in\Lambda} \frac{\log^+|Q(\lambda)|}{\lambda^2},$$

Λ being any fixed set of points in $(0, \infty)$ having at least one element in each interval of length $\geqslant h$ with $h > 0$ and fixed. This generalization requires some rather self-evident modification of the work in article 1. The reasoning in articles 2–8 then applies with hardly any change.

Problem 24

Consider entire functions $F(z)$ *of very small exponential type* α having the special form

$$F(z) = \prod_k \left(1 - \frac{z^2}{x_k^2}\right)$$

where the x_k are > 0, and such that

$$\int_{-\infty}^{\infty} \frac{\log^+|F(x)|}{1+x^2}\,dx < \infty.$$

Investigate the possibility of adapting the development of this § to such functions $F(z)$ (instead of *polynomials* $P(z)$).

Here, if the small numbers 2η and p are both several times larger than α, the constructions of article 2 can be made to work (by problem 1(a), Chapter I!), yielding an infinite number of intervals J_k. The statement of the *second lemma* from article 4 has to be modified.

I have not worked through this problem.

We now come to the *principal result* of this whole §, an extension of the above corollary to *general polynomials.* To establish it, we need a simple

Lemma. *Let* $\alpha > 0$ *be given. There is a number* M_α *depending on* α *such that, for* any *real valued function* f *on* $\mathbb{Z}$ *satisfying*

$$\sum_{-\infty}^{\infty} \frac{\log^+ |f(n)|}{1+n^2} \leqslant \alpha,$$

we have

$$\sum_1^{\infty} \frac{1}{n^2} \log\left(1 + \frac{n^2(f(n)+f(-n))^2}{M_\alpha^2}\right) \leqslant 6\alpha$$

and

$$\sum_1^{\infty} \frac{1}{n^2} \log\left(1 + \frac{(f(n)-f(-n))^2}{M_\alpha^2}\right) \leqslant 6\alpha$$

Proof. When $q \geqslant 0$, the function $\log(1+q) - \log^+ q$ assumes its maximum for $q = 1$. Hence

$$\log(1+q) \leqslant \log 2 + \log^+ q, \qquad q \geqslant 0.$$

Also,

$$\log^+(qq') \leqslant \log^+ q + \log^+ q', \qquad q,\ q' \geqslant 0.$$

Therefore, if $M \geqslant 1$,

$$\log\left(1 + \frac{n^2(f(n)+f(-n))^2}{M^2}\right)$$
$$\leqslant \log 2 + 2\log^+ n + 2\log^+(|f(n)| + |f(-n)|)$$
$$\leqslant 3\log 2 + 2\log n + 2\max(\log^+|f(n)|,\ \log^+|f(-n)|)$$

for $n \geqslant 1$.

Given $\alpha > 0$, choose (and then *fix*) an N sufficiently large to make

$$\sum_{n>N} \frac{3\log 2 + 2\log n}{n^2} < \alpha$$

Then, if f is any real valued function with

$$\sum_{-\infty}^{\infty} \frac{\log^+|f(n)|}{1+n^2} \leqslant \alpha,$$

we will surely have

$$\sum_{n>N} \frac{1}{n^2} \log\left(1 + \frac{n^2(f(n)+f(-n))^2}{M^2}\right) < 5\alpha$$

by the previous relation, as long as $M \geqslant 1$. Similarly,

$$\sum_{n>N} \frac{1}{n^2} \log\left(1 + \frac{(f(n)-f(-n))^2}{M^2}\right) < 5\alpha$$

for such f, if $M \geqslant 1$.

Our condition on f certainly implies that

$$\log^+|f(n)| \leqslant \alpha(1+n^2),$$

so

$$|f(n)| + |f(-n)| \leqslant 2e^{(1+N^2)\alpha}$$

for $1 \leqslant n \leqslant N$. Choosing $M_\alpha \geqslant 1$ sufficiently large so as to have

$$\sum_{1}^{N} \frac{1}{n^2} \log\left(1 + \frac{4n^2 e^{2\alpha(1+N^2)}}{M_\alpha^2}\right) < \alpha$$

will thus ensure that

$$\sum_{1}^{N} \frac{1}{n^2} \log\left(1 + \frac{n^2(f(n)+f(-n))^2}{M_\alpha^2}\right) < \alpha$$

and

$$\sum_{1}^{N} \frac{1}{n^2} \log\left(1 + \frac{(f(n)-f(-n))^2}{M_\alpha^2}\right) < \alpha.$$

Adding each of these relations to the corresponding one obtained above, we have the lemma.

Theorem. *There are numerical constants $\alpha_0 > 0$ and k such that, for any polynomial $p(z)$ with*

$$\sum_{-\infty}^{\infty} \frac{\log^+|p(n)|}{1+n^2} = \alpha \leqslant \alpha_0,$$

we have, for all z,

$$|p(z)| \leqslant K_\alpha e^{3k\alpha|z|},$$

where K_α is a constant depending only on α (and not on p).

Proof. Given a polynomial p, we may as well assume to begin with that $p(x)$ is *real for real* x – otherwise we just work separately with the polynomials $(p(z)+\overline{p(\bar{z})})/2$ and $(p(z)-\overline{p(\bar{z})})/2\mathrm{i}$ which both have that property.

Considering, then, p to be real on $\mathbb{R}$ and assuming that it satisfies the condition in the hypothesis, we take the number M_α furnished by the lemma and form each of the *polynomials*

$$Q_1(z) = 1 + \frac{z^2(p(z)+p(-z))^2}{M_\alpha^2},$$

$$Q_2(z) = 1 + \frac{(p(z)-p(-z))^2}{M_\alpha^2}.$$

The polynomials Q_1 and Q_2 are both *even*, and

$$Q_1(0) = Q_2(0) = 1.$$

By the lemma,

$$\sum_1^\infty \frac{1}{n^2}\log^+|Q_1(n)| \leqslant 6\alpha$$

and

$$\sum_1^\infty \frac{1}{n^2}\log^+|Q_2(n)| \leqslant 6\alpha,$$

since (here) $Q_1(x) \geqslant 1$ and $Q_2(x) \geqslant 1$ on $\mathbb{R}$.

If $\alpha > 0$ is *small enough*, these inequalities imply *by the above corollary* that $(\log|Q_1(z)|)/|z|$ and $(\log|Q_2(z)|)/|z|$ are both $\leqslant 6k\alpha$, k being a certain numerical constant. Therefore

$$\left|1 + \frac{z^2(p(z)+p(-z))^2}{M_\alpha^2}\right| \leqslant \mathrm{e}^{6k\alpha|z|},$$

and

$$\left|1 + \frac{(p(z)-p(-z))^2}{M_\alpha^2}\right| \leqslant \mathrm{e}^{6k\alpha|z|}.$$

From these relations we get

$$|z^2(p(z)+p(-z))^2| \leqslant M_\alpha^2(1+\mathrm{e}^{6k\alpha|z|}) \leqslant 2M_\alpha^2\mathrm{e}^{6k\alpha|z|},$$

whence

$$|p(z)+p(-z)| \leqslant \sqrt{2}M_\alpha\mathrm{e}^{3k\alpha|z|} \qquad \text{for } |z| \geqslant 1,$$

and similarly

$$|p(z) - p(-z)| \leqslant \sqrt{2M_\alpha} e^{3k\alpha|z|}$$

Hence

$$|p(z)| \leqslant 2\sqrt{2M_\alpha} e^{3k\alpha|z|}$$

for $|z| \geqslant 1$, and from this, by the principle of maximum,

$$|p(z)| \leqslant 2\sqrt{2M_\alpha} e^{3k\alpha} \qquad \text{for } |z| < 1.$$

The theorem therefore holds with

$$K_\alpha = 2\sqrt{2M_\alpha} e^{3k\alpha}.$$

Q.E.D.

Corollary. *If* $\alpha > 0$ *is* small enough, *the polynomials* $p(z)$ *satisfying*

$$\sum_{-\infty}^{\infty} \frac{\log^+ |p(n)|}{1+n^2} \leqslant \alpha$$

form a normal family *in the complex plane, and the* limit *of any convergent sequence of such polynomials is an* entire function of exponential type $\leqslant 3k\alpha$, *k being an absolute constant.*

It is thus somewhat as if harmonic measure were available for the domain $\mathbb{C} \sim \mathbb{Z}$, *even though that is not the case.*

11. Weighted polynomial approximation on $\mathbb{Z}$

Given a weight $W(n) \geqslant 1$ defined on $\mathbb{Z}$, we consider the Banach space $\mathscr{C}_W(\mathbb{Z})$ of functions $\varphi(n)$ defined on $\mathbb{Z}$ for which

$$\frac{\varphi(n)}{W(n)} \longrightarrow 0 \quad \text{as} \quad n \to \pm\infty,$$

and write

$$\|\varphi\|_{W,\mathbb{Z}} = \sup_{n\in\mathbb{Z}} \frac{|\varphi(n)|}{W(n)}$$

for such φ. (This is the notation of §A.3.)

Provided that

$$\frac{n^k}{W(n)} \longrightarrow 0 \quad \text{as} \quad n \to \pm\infty$$

for each $k = 0, 1, 2, 3, \ldots$, we can form the $\| \ \|_{W,\mathbb{Z}}$ *closure*, $\mathscr{C}_W(0, \mathbb{Z})$, of the set of *polynomials in n*, in $\mathscr{C}_W(\mathbb{Z})$. The *Bernstein approximation problem for* $\mathbb{Z}$ requires us to find necessary and sufficient conditions on weights $W(n)$ having the property just stated in order that $\mathscr{C}_W(0,\mathbb{Z})$ and $\mathscr{C}_W(\mathbb{Z})$ be the same.

The preceding work enables us to give a complete solution in terms of the Akhiezer function

$$W_*(n) = \sup\{|p(n)|: p \text{ a polynomial and } \|p\|_{W,\mathbb{Z}} \leqslant 1\}$$

introduced in §B.1 of Chapter VI.

Theorem. *Let* $W(n)$, *defined and* $\geqslant 1$ *on* $\mathbb{Z}$, *tend to* ∞ faster than any power of n as $n \to \pm\infty$. *Then* $\mathscr{C}_W(0,\mathbb{Z}) = \mathscr{C}_W(\mathbb{Z})$ *if and only if*

$$\sum_{-\infty}^{\infty} \frac{\log W_*(n)}{1+n^2} = \infty.$$

Proof. Let us get the easier *if* part out of the way first – this is not really new, and depends only on the work of Chapter VI, §B.1.

As in §A.3, we take $W(x)$ to be specified on *all* of $\mathbb{R}$ by putting $W(x) = \infty$ for $x \notin \mathbb{Z}$, and define $W_*(z)$ for all $z \in \mathbb{C}$ using the formula

$$W_*(z) = \sup\{|p(z)|: p \text{ a polynomial and } \|p\|_{W,\mathbb{Z}} \leqslant 1\}.$$

Then $\mathscr{C}_W(\mathbb{Z})$ can be identified in obvious fashion with the space $\mathscr{C}_W(\mathbb{R})$ constructed from the (discontinuous) weight $W(x)$, and $\mathscr{C}_W(0,\mathbb{Z})$ identified with $\mathscr{C}_W(0)$, the closure of the set of polynomials in $\mathscr{C}_W(\mathbb{R})$. *Proper* inclusion of $\mathscr{C}_W(0,\mathbb{Z})$ in $\mathscr{C}_W(\mathbb{Z})$ is thus *the same* as that of $\mathscr{C}_W(0)$ in $\mathscr{C}_W(\mathbb{R})$, and we can apply the *if* part of Akhiezer's theorem from §B.1 of Chapter III (whose validity *does not* depend on the continuity of $W(x)$!) to conclude that

$$\int_{-\infty}^{\infty} \frac{\log W_*(t)}{t^2+1}\,dt < \infty$$

when that proper inclusion holds.

If p is any polynomial with $\|p\|_{W,\mathbb{Z}} \leqslant 1$, the hall of mirrors argument at the beginning of the proof of Akhiezer's theorem's *only if* part shows that

$$\log|p(x)| \leqslant \frac{1}{\pi}\int_{-\infty}^{\infty} \frac{2\log W_*(t)}{(x-t)^2+4}\,dt$$

for $x \in \mathbb{R}$. Taking the supremum over such polynomials p gives us

$$\log W_*(n) \leqslant \frac{1}{\pi}\int_{-\infty}^{\infty} \frac{2\log W_*(t)}{(n-t)^2+4}\,dt, \qquad n \in \mathbb{Z}.$$

Therefore, since $\log W_*(t) \geqslant 0$ (1 being a polynomial!), we have

$$\sum_{-\infty}^{\infty} \frac{\log W_*(n)}{1+n^2} \leqslant \frac{1}{\pi}\int_{-\infty}^{\infty} \sum_{n=-\infty}^{\infty} \frac{1}{n^2+1}\cdot\frac{2\log W_*(t)}{(n-t)^2+1}\,dt.$$

The inner sum over n may easily be compared with an integral, and we find in this way that the last expression is

$$\leqslant \text{const.}\int_{-\infty}^{\infty} \frac{\log W_*(t)}{1+t^2}\,dt.$$

This, however, is *finite* when $\mathscr{C}_W(0, \mathbb{Z}) \neq \mathscr{C}_W(\mathbb{Z})$, as we have just seen. The *if* part of our theorem is proved.

For the *only if* part, we assume that

$$\sum_{-\infty}^{\infty} \frac{\log W_*(n)}{1+n^2} < \infty,$$

and show that the function

$$\varphi_0(n) = \begin{cases} 1, & n=0 \\ 0, & n\neq 0, \end{cases}$$

cannot belong to $\mathscr{C}_W(0,\mathbb{Z})$. We do this using the *corollary* to the *first* theorem of the preceding article. It is not necessary to resort to the *second* theorem given there.

Suppose, then, that we *have* a sequence of polynomials $p_l(z)$ with

$$\|\varphi_0 - p_l\|_{W,\mathbb{Z}} \underset{l}{\longrightarrow} 0.$$

This implies in particular that

$$p_l(0) \underset{l}{\longrightarrow} \varphi_0(0) = 1,$$

so there is no loss of generality in assuming that $p_l(0) = 1$ for each l, which we *do*. The polynomials

$$Q_l(z) = \tfrac{1}{2}(p_l(z) + p_l(-z))$$

then satisfy the hypothesis of the corollary in question.

We evidently have $\|p_l\|_{W,\mathbb{Z}} \leqslant C$ for some C, so, by definition of W_*, $|p_l(n)| \leqslant CW_*(n)$ for $n\in\mathbb{Z}$ and therefore

$$|Q_l(n)| \leqslant \tfrac{1}{2}C(W_*(n) + W_*(-n)), \qquad n\in\mathbb{Z}.$$

Also, $p_l(n) \underset{l}{\longrightarrow} \varphi_0(n) = 0$ for *each non-zero* $n\in\mathbb{Z}$, so, given any N, we will have

$$|Q_l(n)| < 1 \qquad \text{for } 0<|n|<N$$

when l is sufficiently large.

Taking any $\alpha > 0$, we choose and fix an N large enough to make

$$\sum_N^\infty \frac{1}{n^2} \log^+ \left(\tfrac{1}{2}C(W_*(n) + W_*(-n))\right) \;<\; \alpha,$$

this being possible in view of our assumption on W_*. By the preceding two relations we will then have

$$\sum_1^\infty \frac{1}{n^2} \log^+ |Q_l(n)| \;=\; \sum_N^\infty \frac{1}{n^2} \log^+ |Q_l(n)| \;<\; \alpha$$

for sufficiently large values of l.

If $\alpha > 0$ is sufficiently small, the last condition implies that

$$|Q_l(z)| \;\leqslant\; e^{k\alpha|z|}$$

by the corollary to the first theorem of the preceding article, with k an absolute constant. *This must therefore hold for all sufficiently large values of l.*

A *subsequence* of the polynomials $Q_l(z)$ therefore converges u.c.c. to a certain *entire function $F(z)$ of exponential type $\leqslant k\alpha$.* We evidently have $F(0) = 1$ (so $F \not\equiv 0$!), while $F(n) = 0$ *for each non-zero* $n \in \mathbb{Z}$.

However, by problem 1(a) in Chapter I (!), *such an entire function F cannot exist*, if $\alpha > 0$ is chosen *sufficiently small to begin with.* We have thus reached a contradiction, showing that φ_0 *cannot belong* to $\mathscr{C}_W(0, \mathbb{Z})$. The latter space is thus *properly contained* in $\mathscr{C}_W(\mathbb{Z})$, and the *only if* part of our theorem is proved.

We are done.

C. Harmonic estimation in slit regions

We return to domains $\mathscr{D}$ for which the Dirichlet problem *is* solvable, having boundaries formed by removing certain finite open intervals from $\mathbb{R}$. Our interest in the present § is to see whether, from the *existence* of a *Phragmén–Lindelöf function* $Y_{\mathscr{D}}(z)$ for $\mathscr{D}$ (the reader should perhaps look at §A.2 again before continuing), one can *deduce* any *estimates* on the *harmonic measure* for $\mathscr{D}$. We would like in fact to be able to *compare* harmonic measure for $\mathscr{D}$ with $Y_{\mathscr{D}}(z)$. The reason for this desire is the following. Given $A > 0$ and $M(t) \geqslant 0$ on $\partial\mathscr{D}$, suppose that we have a function $v(z)$, *subharmonic* in $\mathscr{D}$ and *continuous* up to $\partial\mathscr{D}$, with

$$v(z) \;\leqslant\; \text{const.} - A|\Im z|, \qquad z \in \mathscr{D},$$

and

$$v(t) \;\leqslant\; M(t), \qquad t \in \partial\mathscr{D}.$$

Then, by harmonic estimation

$$v(z) \leqslant \int_{\partial\mathscr{D}} M(t)\mathrm{d}\omega_{\mathscr{D}}(t,z) - AY_{\mathscr{D}}(z), \qquad z\in\mathscr{D},$$

where (as usual) $\omega_{\mathscr{D}}(\ ,z)$ denotes (*two-sided*) harmonic measure for $\mathscr{D}$ (see §A.1). *It would be very good* if, in this relation, we had some way of *comparing* the *first term* on the right with the *second.*

As we shall see below, such comparison is indeed possible. In order to avoid fastidious justification arguments like the one occurring in the proof of the second theorem from §A.2, *we will assume throughout that* $\partial\mathscr{D}$ *consists of* $\mathbb{R}$ *minus a finite number of* (bounded) *open intervals.* The results obtained for this situation can usually be extended by means of a simple limiting procedure to cover various more general cases that may arise in practice. The domains $\mathscr{D}$ considered here thus look like this:

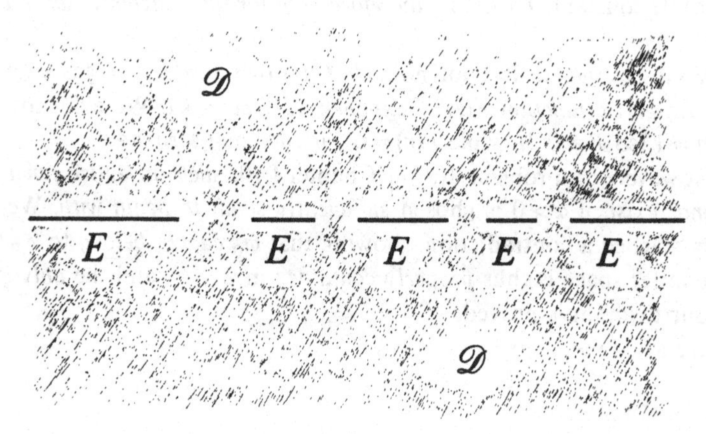

Figure 151

As in §A, we shall frequently denote $\partial\mathscr{D}$ by E. E is a closed subset of $\mathbb{R}$ which, *in this* §, *will contain all real* x *of sufficiently large absolute value.*

1. Some relations between Green's function and harmonic measure for our domains $\mathscr{D}$

During the present §, we will usually denote the Green's function for one of the domains $\mathscr{D}$ by $G_{\mathscr{D}}(z,w)$, instead of just writing $G(z,w)$ as in §A.2ff. We similarly write $Y_{\mathscr{D}}(z)$ instead of $Y(z)$ for $\mathscr{D}$'s Phragmén–Lindelöf function.

Our domains $\mathscr{D}$ *have* Phragmén–Lindelöf functions. Indeed, for *fixed* $z\in\mathscr{D}$ and *real* t, $G_{\mathscr{D}}(z,t) = G_{\mathscr{D}}(t,z)$ *vanishes* for t outside the *bounded* set $\mathbb{R} \sim E$. (We are using symmetry of the Green's function, established at the end of §A.2.) If we take $z \notin \mathbb{R}$, $G_{\mathscr{D}}(t, z)$ is also a continuous function of $t \in \mathbb{R}$. The integral

$$\int_{-\infty}^{\infty} G_{\mathscr{D}}(z, t)\, dt$$

is then certainly *finite*, and the *existence* of the *function* $Y_{\mathscr{D}}$ hence assured by the second theorem of §A.2.

According to that same theorem,

$$Y_{\mathscr{D}}(z) \;=\; |\Im z| + \frac{1}{\pi}\int_{-\infty}^{\infty} G_{\mathscr{D}}(z, t)\, dt.$$

This formula suggests that we first establish some relations between $G_{\mathscr{D}}(z,t)$ and $\omega_{\mathscr{D}}(\;, z)$ before trying to find out whether the latter is in any way governed by $Y_{\mathscr{D}}(z)$.

We prove *three* such relations here. The first of them is very well known.

Theorem. *For* $w \in \mathscr{D}$,

$$G_{\mathscr{D}}(z, w) \;=\; \log\frac{1}{|z-w|} \;+\; \int_E \log|t-w|\, d\omega_{\mathscr{D}}(t, z).$$

Proof. The right side of the asserted formula is identical with

$$\log\frac{1}{|z-w|} \;+\; \int_{\partial\mathscr{D}} \log|t-w|\, d\omega_{\mathscr{D}}(t, z),$$

and, for *bounded* domains $\mathscr{D}$, this expression clearly coincides with $G_{\mathscr{D}}(z, w)$ – just fix $w \in \mathscr{D}$ and check boundary values for z on $\partial\mathscr{D}$! (This argument, and the formula, are due to George Green himself, by the way.)

In our situation, however, $\mathscr{D}$ *is not bounded*, and the *result is not true, in general, for unbounded domains.* (Not even for those with 'nice' boundaries; example:

$$\mathscr{D} \;=\; \{|z|>1\} \cup \{\infty\}. \;)$$

What is needed then in order for it to hold is the presence of 'enough' $\partial\mathscr{D}$ near ∞. That is what we must verify in the present case.

Fixing $w \in \mathscr{D}$, we proceed to find upper and lower bounds on the integral

$$\int_E \log|t-w|\, d\omega_{\mathscr{D}}(t, z).$$

In order to get an *upper* bound, we take a function $h(z)$, positive and harmonic in $\mathscr{D}$ and continuous up to $\partial\mathscr{D}$, such that

$$h(z) \;=\; \log^+|z| + O(1).$$

In the case where E includes the interval $[-1, 1]$ (at which we can always arrive by translation), one may put

$$h(z) = \log|z + \sqrt{(z^2 - 1)}|$$

using, outside $[-1, 1]$, the determination of $\sqrt{}$ that is *positive* for $z = x > 1$. For large $A > 0$, let us write

$$h_A(z) = \min(h(z), A).$$

The function $h_A(t)$ is then *bounded and continuous* on E, so, by the elementary properties of harmonic measure (Chapter VII, §B.1), the function of z equal to

$$\int_E h_A(t)\mathrm{d}\omega_{\mathscr{D}}(t, z)$$

is *harmonic* and *bounded above* in $\mathscr{D}$, and takes the boundary value $h_A(z)$ for z on $\partial\mathscr{D}$. The difference $\int_E h_A(t)\mathrm{d}\omega_{\mathscr{D}}(t, z) - h(z)$ is thus bounded above in $\mathscr{D}$ and $\leqslant 0$ on $\partial\mathscr{D}$. Therefore, by the *extended principle of maximum* (Chapter III, §C), it is $\leqslant 0$ in $\mathscr{D}$, and we have

$$\int_E h_A(t)\,\mathrm{d}\omega_{\mathscr{D}}(t, z) \leqslant h(z), \qquad z \in \mathscr{D}.$$

For $A' \geqslant A$, $h_{A'}(t) \geqslant h_A(t)$. Hence, by the preceding relation and Lebesque's monotone convergence theorem,

$$\int_E h(t)\,\mathrm{d}\omega_{\mathscr{D}}(t, z) \leqslant h(z), \qquad z \in \mathscr{D};$$

that is,

$$\int_E \log^+|t|\mathrm{d}\omega_{\mathscr{D}}(t, z) \leqslant \log^+|z| + O(1)$$

for $z \in \mathscr{D}$. When $w \in \mathscr{D}$ is fixed, we thus have the upper bound

$$\int_E \log|t - w|\mathrm{d}\omega_{\mathscr{D}}(t, z) \leqslant \log^+|z| + O(1)$$

for z ranging over $\mathscr{D}$.

We can get some additional information with the help of the function $h(z)$. Indeed, for *each* A,

$$\int_E h_A(t)\,\mathrm{d}\omega_{\mathscr{D}}(t, z) \leqslant \int_E h(t)\,\mathrm{d}\omega_{\mathscr{D}}(t, z) \leqslant h(z)$$

when $z \in \mathscr{D}$. As we remarked above, the *left-hand* expression tends to $h_A(x_0)$

whenever $z \longrightarrow x_0 \in \partial\mathscr{D}$; at the same time, the *right-hand* member evidently tends to $h(x_0)$. Taking $A > h(x_0)$, we see that

$$\int_E h(t)\,\mathrm{d}\omega_{\mathscr{D}}(t, z) \quad\longrightarrow\quad h(x_0)$$

for $z \longrightarrow x_0 \in \partial\mathscr{D}$. On the other hand, for fixed $w \in \mathscr{D}$,

$$\log|t - w| \;-\; h(t)$$

is *continuous* and *bounded* on $\partial\mathscr{D}$. Therefore

$$\int_E (\log|t - w| - h(t))\,\mathrm{d}\omega_{\mathscr{D}}(t, z) \longrightarrow \log|x_0 - w| \;-\; h(x_0)$$

when $z \longrightarrow x_0 \in \partial\mathscr{D}$, so, on account of the previous relation, we have

$$\int_E \log|t - w|\,\mathrm{d}\omega_{\mathscr{D}}(t, z) \longrightarrow \log|x_0 - w|$$

for $z \longrightarrow x_0 \in \partial\mathscr{D}$.

To get a *lower* bound on the left-hand integral, let us, wlog, assume that $\Re z > 0$, and take an $R > 0$ sufficiently large to have $(-\infty, -R] \cup [R, \infty) \subseteq E$. Since $\mathscr{D} \supseteq \{\Im z > 0\}$, we have, for $|t| > R$,

$$\mathrm{d}\omega_{\mathscr{D}}(t, z) \;\geqslant\; \frac{1}{\pi}\frac{\Im z}{|z - t|^2}\,\mathrm{d}t$$

by the principle of *extension of domain* (Chapter VII, §B.1), the right side being just the differential of harmonic measure for the upper half plane. Hence,

$$\begin{aligned}\int_E \log|t + \mathrm{i}|\,\mathrm{d}\omega_{\mathscr{D}}(t, z) \;&\geqslant\; \int_{\{|t| \geqslant R\}} \log|t + \mathrm{i}|\,\mathrm{d}\omega_{\mathscr{D}}(t, z)\\ &\geqslant\; \frac{1}{\pi}\int_{|t| \geqslant R} \frac{\Im z \log|t + \mathrm{i}|}{|z - t|^2}\,\mathrm{d}t\\ &=\; \frac{1}{\pi}\int_{-\infty}^{\infty} \frac{\Im z \log|t + \mathrm{i}|}{|z - t|^2}\,\mathrm{d}t \;-\; O(1).\end{aligned}$$

The last integral on the right has, however, the value $\log|z + \mathrm{i}|$, as an elementary computation shows (contour integration). Thus,

$$\int_E \log|t + \mathrm{i}|\,\mathrm{d}\omega_{\mathscr{D}}(t, z) \;\geqslant\; \log|z + \mathrm{i}| - O(1)$$

for $\Im z > 0$, so, for fixed $w \in \mathscr{D}$,

$$\int_E \log|t-w|\,d\omega_{\mathscr{D}}(t,z) \;\geqslant\; \log^+|z| - O(1), \qquad z\in\mathscr{D}.$$

Taking any $w\in\mathscr{D}$, we see by the above that the function of z equal to

$$\log\frac{1}{|z-w|} + \int_E \log|t-w|\,d\omega_{\mathscr{D}}(t,z)$$

is *harmonic* in $\mathscr{D}$ save at w, differs in $\mathscr{D}$ by $O(1)$ from $\log(1/|z-w|)+\log^+|z|$, and assumes the *boundary value zero* on $\partial\mathscr{D}$. It is in particular *bounded above and below* outside of a neighborhood of w (point where it becomes infinite), and hence $\geqslant 0$ in $\mathscr{D}$ by the extended maximum principle. The expression just written thus has all the properties required of a Green's function for $\mathscr{D}$, and must coincide with $G_{\mathscr{D}}(z,w)$. We are done.

► *It will be convenient during the remainder of this § to take* $d\omega_{\mathscr{D}}(t,z)$ *as defined on all of* $\mathbb{R}$, *simply putting it equal to zero outside of* E. *This enables us to simplify our notation by writing* $\omega_{\mathscr{D}}(S,z)$ *for* $\omega_{\mathscr{D}}(S\cap E,z)$ *when* $S\subseteq\mathbb{R}$.

Lemma. *Let* $0\in\mathscr{D}$, *and write*

$$\omega_{\mathscr{D}}(x) \;=\; \begin{cases}\omega_{\mathscr{D}}([x,\infty),\,0), & x>0,\\ \omega_{\mathscr{D}}((-\infty,x],\,0), & x<0\end{cases}$$

(*note that* $\omega_{\mathscr{D}}(x)$ need not be continuous *at* 0). *Then, for* $\Im z\neq 0$,

$$G_{\mathscr{D}}(z,0) \;=\; -\int_{-\infty}^{\infty}\frac{x-t}{(x-t)^2+y^2}\,\omega_{\mathscr{D}}(t)\,\mathrm{sgn}\,t\,dt.$$

Proof. By the preceding theorem and symmetry of the Green's function (proved at the end of §A.2), we have

$$G_{\mathscr{D}}(z,0) \;=\; G_{\mathscr{D}}(0,z) \;=\; \log\frac{1}{|z|} + \int_E \log|t-z|\,d\omega_{\mathscr{D}}(t,0).$$

Thanks to our convention, we can rewrite the right-hand integral as

$$\left(\int_{-\infty}^{0} + \int_{0}^{\infty}\right)\log|t-z|\,d\omega_{\mathscr{D}}(t,0).$$

Let us accept for the moment the inequality

$$\omega_{\mathscr{D}}(t) \;\leqslant\; \frac{\text{const.}}{|t|+1},$$

postponing its verification to the end of this proof. Then partial integration yields

$$\int_0^\infty \log|t-z|\mathrm{d}\omega_{\mathscr{D}}(t,0) \;=\; \omega_{\mathscr{D}}(0+)\log|z| \;+\; \int_0^\infty \frac{t-x}{|t-z|^2}\omega_{\mathscr{D}}(t)\mathrm{d}t,$$

and

$$\int_{-\infty}^0 \log|t-z|\mathrm{d}\omega_{\mathscr{D}}(t,0) \;=\; \omega_{\mathscr{D}}(0-)\log|z| \;-\; \int_{-\infty}^0 \frac{t-x}{|t-z|^2}\omega_{\mathscr{D}}(t)\mathrm{d}t.$$

Here,

$$\omega_{\mathscr{D}}(0+)+\omega_{\mathscr{D}}(0-) \;=\; \omega_{\mathscr{D}}((-\infty,\infty),\ 0) \;=\; \omega_{\mathscr{D}}(E,0) \;=\; 1,$$

so, adding, we get

$$\begin{aligned} G_{\mathscr{D}}(z,0) \;&=\; \log\frac{1}{|z|}+\int_{-\infty}^{\infty}\log|t-z|\mathrm{d}\omega_{\mathscr{D}}(t,0) \\ &= \log\frac{1}{|z|}+\log|z|+\int_{-\infty}^{\infty}\frac{t-x}{|t-z|^2}\omega_{\mathscr{D}}(t)\,\mathrm{sgn}\,t\,\mathrm{d}t \\ &= -\int_{-\infty}^{\infty}\frac{x-t}{|z-t|^2}\omega_{\mathscr{D}}(t)\,\mathrm{sgn}\,t\,\mathrm{d}t, \end{aligned}$$

as claimed.

We still have to check the above inequality for $\omega_{\mathscr{D}}(t)$. To do this, pick an $R>0$ large enough to have

$$(-\infty,-R]\cup[R,\infty) \;\subseteq\; E,$$

and take a domain $\mathscr{E}$ equal to the *complement* of

$$(-\infty,-R]\cup[R,\infty)$$

in $\mathbb{C}$. Then $\mathscr{D}\subseteq\mathscr{E}$, so, by the *principle of extension of domain* (Chapter VII, §B.1), $\omega_{\mathscr{D}}(t)+\omega_{\mathscr{D}}(-t) \leqslant \omega_{\mathscr{E}}((-\infty,-t]\cup[t,\infty),\ 0)$ for $t>R$. The quantity on the right can, however, be worked out explicitly by mapping $\mathscr{E}$ conformally onto the unit disk so as to take $-R$ to -1, 0 to 0 and R to 1. In this way, one finds it to be $\leqslant CR/t$ (with a constant C independent of R), verifying the inequality in question. Details are left to the reader – he or she is referred to the proof of the *first* lemma from §A.1, where most of the computation involved here has already been done.

The integral figuring in the lemma just proved, viz.,

$$-\int_{-\infty}^{\infty}\frac{x-t}{|z-t|^2}\omega_{\mathscr{D}}(t)\,\mathrm{sgn}\,t\,\mathrm{d}t$$

is like one used in the scholium of §H.1, Chapter III, to express a certain *harmonic conjugate*. It differs from the latter by its sign, by the absence of the constant $1/\pi$ in front, and because its integrand involves the factor $(x-t)/|z-t|^2$ instead of the sum

$$\frac{x-t}{|z-t|^2} + \frac{t}{t^2+1}.$$

In §H of Chapter III, the main purpose of the term $t/(t^2+1)$ was really *to ensure convergence; here*, since $\omega_{\mathscr{D}}(t)$ is $O(1/(|t|+1))$, we already have convergence *without* it, and our *omission* of the term $t/(t^2+1)$ amounts merely to the *subtraction of a constant* from the value of the integral. Since harmonic conjugates are *only determined to within additive constants* anyway, *we may just as well take*

$$\frac{1}{\pi}\int_{-\infty}^{\infty} \frac{x-t}{|z-t|^2}\,\omega_{\mathscr{D}}(t)\,\mathrm{sgn}\,t\,\mathrm{d}t$$

as the *harmonic conjugate* of

$$\frac{1}{\pi}\int_{-\infty}^{\infty} \frac{\Im z}{|z-t|^2}\,\omega_{\mathscr{D}}(t)\,\mathrm{sgn}\,t\,\mathrm{d}t$$

in $\{\Im z > 0\}$. This brings the investigation of the former integral's boundary behavior on the real axis very close to the study of the *Hilbert transform* already touched on in Chapter III, §§F.2 and H.1.

In our present situation, we already know that, for real $x \neq 0$,

$$\lim_{y\to 0} G_{\mathscr{D}}(x+iy,\ 0) \;=\; G_{\mathscr{D}}(x,0)$$

exists. The identity furnished by the lemma hence shows, *independently of the general considerations in the articles just mentioned*, that

$$\lim_{y\to 0}\int_{-\infty}^{\infty} \frac{x-t}{|z-t|^2}\,\omega_{\mathscr{D}}(t)\,\mathrm{sgn}\,t\,\mathrm{d}t$$

exists (and equals $-G_{\mathscr{D}}(x,0)$) for real $x \neq 0$. According to an observation in the scholium of §H.1, Chapter III, we can express the preceding limit as an *integral*, namely

$$\int_0^{\infty} \frac{\omega_{\mathscr{D}}(x-\tau)\,\mathrm{sgn}\,(x-\tau) - \omega_{\mathscr{D}}(x+\tau)\,\mathrm{sgn}\,(x+\tau)}{\tau}\,\mathrm{d}\tau.$$

That's because this expression *converges absolutely* for $x \neq 0$, on account of the above inequality for $\omega_{\mathscr{D}}(t)$ and also of the

Lemma. *Let* $0 \in \mathscr{D}$. *Then* $\omega_{\mathscr{D}}(t)$ *is* $\mathrm{Lip}\,\tfrac{1}{2}$ *for* $t > 0$ *and for* $t < 0$.

Proof. The statement amounts to the claim that

$$\omega_{\mathscr{D}}(I,0) \;\leqslant\; \text{const.}\sqrt{|I|}$$

for any small interval $I \subseteq E$. To show this, take any *interval* $J_0 \subseteq E$ and consider small intervals $I \subseteq J_0$. Letting $\mathscr{E}$ be the region $(\mathbb{C} \cup \{\infty\}) \sim J_0$, the usual application of the *principle of extension of domain* gives us

$$\omega_{\mathscr{D}}(I,0) \;\leqslant\; \omega_{\mathscr{E}}(I,0),$$

with, in turn,

$$\omega_{\mathscr{E}}(I,0) \;\leqslant\; \text{const.}\,\omega_{\mathscr{E}}(I,\infty)$$

by Harnack's theorem.

To simplify the estimate of the right side of the *last* inequality, we may take J_0 to be $[-1,1]$; this just amounts to making a preliminary translation and change of scale – *never mind* here that $0 \in \mathscr{D}$! Then one can map $\mathscr{E}$ onto the unit disk by the Joukowski transformation

$$z \;\longrightarrow\; z - \sqrt{(z^2-1)}$$

which takes ∞ to 0, -1 to -1, and 1 to 1. In this way one easily finds that

$$\omega_{\mathscr{E}}(I,\infty) \;\leqslant\; \text{const.}\sqrt{|I|},$$

proving the lemma.

Remark. The square root is *only* necessary when I is *near one of the endpoints* of J_0. For small intervals I near the *middle* of J_0, $\omega_{\mathscr{E}}(I,\infty)$ acts like a multiple of $|I|$.

By the above two lemmas and related discussion, we have the formula

$$G_{\mathscr{D}}(x,0) \;=\; -\int_0^\infty \frac{\omega_{\mathscr{D}}(x-\tau)\operatorname{sgn}(x-\tau) - \omega_{\mathscr{D}}(x+\tau)\operatorname{sgn}(x+\tau)}{\tau}\,d\tau,$$

valid for $x \neq 0$ if 0 belongs to $\mathscr{D}$. It is customary to write the right-hand member in a different way. That expression is identical with

$$-\lim_{\delta\to 0}\int_{|t-x|\geqslant\delta} \frac{\omega_{\mathscr{D}}(t)\operatorname{sgn} t}{x-t}\,dt.$$

If a function $f(t)$, having a possible singularity at $a \in \mathbb{R}$, is integrable over each set of the form $\{|t-a| \geqslant \delta\}$, $\delta > 0$, and *if*

$$\lim_{\delta\to 0}\int_{|t-a|\geqslant\delta} f(t)\,dt$$

exists, that limit is called a Cauchy principal value, and denoted by

$$⨍_{-\infty}^{\infty} f(t)\,dt \qquad \textit{or by} \qquad \text{v.p.}\int_{-\infty}^{\infty} f(t)\,dt.$$

It is important to realize that $⨍_{-\infty}^{\infty} f(t)\,dt$ *is frequently not an integral in the ordinary sense.*

In terms of this notation, the formula for $G_{\mathscr{D}}(x, 0)$ just obtained can be expressed as in the following

Theorem. *Let* $0 \in \mathscr{D}$. *Then, for real* $x \neq 0$,

$$G_{\mathscr{D}}(x,0) = -⨍_{-\infty}^{\infty} \frac{\omega_{\mathscr{D}}(t)\operatorname{sgn} t}{x - t}\,dt,$$

where $\omega_{\mathscr{D}}(t)$ *is the function defined in the first of the above two lemmas.*

This result will be used in article 3 below. Now, however, we wish to use it to *solve for* $\omega_{\mathscr{D}}(t)\operatorname{sgn} t$ *in terms of* $G_{\mathscr{D}}(x, 0)$, obtaining the relation

$$\omega_{\mathscr{D}}(t)\operatorname{sgn} t = \frac{1}{\pi^2}⨍_{-\infty}^{\infty} \frac{G_{\mathscr{D}}(x,0)}{t - x}\,dx.$$

By the inversion theorem for the L_2 Hilbert transform, the latter formula is indeed a consequence of the boxed one above. Here, a direct proof is not very difficult, and we give one for the reader who does not know the inversion theorem.

Lemma. $\int_{-\infty}^{\infty} |G_{\mathscr{D}}(x + iy,\ 0) - G_{\mathscr{D}}(x, 0)|\,dx \longrightarrow 0$ *for* $y \to 0$.

Proof. The result follows immediately from the representation

$$G_{\mathscr{D}}(x + iy,\ 0) = \frac{1}{\pi}\int_{-\infty}^{\infty} \frac{yG_{\mathscr{D}}(t, 0)}{(x - t)^2 + y^2}\,dt, \qquad y > 0,$$

by elementary properties of the Poisson kernel, in the usual way.

The representation itself is practically obvious; here is one derivation. From the first theorem of this article,

$$G_{\mathscr{D}}(t,0) = \log\frac{1}{|t|} + \int_E \log|s - t|\,d\omega_{\mathscr{D}}(s, 0)$$

and

$$G_{\mathscr{D}}(z,0) = \log\frac{1}{|z|} + \int_E \log|s - z|\,d\omega_{\mathscr{D}}(s, 0).$$

For $\Im z > 0$, we have the elementary formula

$$\log|s - z| = \frac{1}{\pi}\int_{-\infty}^{\infty} \frac{\Im z \log|s - t|}{|z - t|^2}\,dt, \qquad s \in \mathbb{R}.$$

Use this in the right side of the preceding relation (in *both* right-hand terms !), change the order of integration (which is easily justified here), and then refer to the formula for $G_{\mathscr{D}}(t,0)$ just written. One ends with the relation in question.

Lemma. *Let* $0\in\mathscr{D}$. *Then* $G_{\mathscr{D}}(x,0)$ *is* $\operatorname{Lip}\frac{1}{2}$ *for* $x>0$ *and for* $x<0$.

Proof. The open intervals of $\mathbb{R}\sim E$ belong to $\mathscr{D}$, where $G_{\mathscr{D}}(z,0)$ is *harmonic* (save at 0), and hence $\mathscr{C}_{\infty}$. So $G_{\mathscr{D}}(x,0)$ is certainly $\mathscr{C}_1$ (hence Lip 1) in the *interior* of each of those open segments (although *not uniformly* so!) for x outside any neighborhood of 0. Also, $G_{\mathscr{D}}(x,0)=0$ on each of the *closed segments* making up E; it is thus surely Lip 1 on the interior of each of *those*.

Our claim therefore boils down to the statement that

$$|G_{\mathscr{D}}(x,0)-G_{\mathscr{D}}(a,0)| \;\leqslant\; \text{const.}\sqrt{|x-a|}$$

near any of the endpoints a of any of the segments making up E. Since $G_{\mathscr{D}}(a,0)=0$, we have to show that

$$G_{\mathscr{D}}(x,0) \;\leqslant\; \text{const.}\sqrt{|x-a|}$$

for $x\in\mathbb{R}\sim E$ *near* such an endpoint a.

Assume, wlog, that a is a *right* endpoint of a component of E and that $x>a$. Pick $b<a$ such that

$$[b,a] \;\subseteq\; E$$

and denote the domain $(\mathbb{C}\cup\{\infty\})\sim[b,a]$ by $\mathscr{E}$. We have $\mathscr{D}\subseteq\mathscr{E}$, so

$$G_{\mathscr{D}}(x,0) \;\leqslant\; G_{\mathscr{E}}(x,0)$$

by the principle of extension of domain. Here, one may compute $G_{\mathscr{E}}(x,0)$ by mapping $\mathscr{E}$ onto the unit disk conformally with the help of a Joukowski transformation. In this way one finds without much difficulty that

$$G_{\mathscr{E}}(x,0) \;\leqslant\; \text{const.}\sqrt{(x-a)}$$

for $x>a$, proving the lemma. (Cf. proof of the lemma immediately preceding the previous theorem.)

Theorem. *Let* $0\in\mathscr{D}$. *Then, for* $x\neq 0$,

$$\omega_{\mathscr{D}}(x)\operatorname{sgn}x \;=\; \frac{1}{\pi^2}\int_{-\infty}^{\infty}\frac{G_{\mathscr{D}}(t,0)}{x-t}\,dt,$$

where $\omega_{\mathscr{D}}(x)$ *is the function defined in the* first *lemma of this article.*

Proof. By the first of the preceding lemmas, for $t\in\mathbb{R}$ and $h>0$,

$$G_{\mathscr{D}}(t+ih,\;0) \;=\; -\int_{-\infty}^{\infty}\frac{t-\xi}{(t-\xi)^2+h^2}\,\omega_{\mathscr{D}}(\xi)\operatorname{sgn}\xi\,d\xi.$$

Multiply both sides by

$$\frac{x-t}{(x-t)^2+y^2}$$

and integrate the variable t. We get

$$\int_{-\infty}^{\infty} \frac{x-t}{(x-t)^2+y^2} G_{\mathscr{D}}(t+ih,\ 0)\,dt$$

$$= -\int_{-\infty}^{\infty}\int_{-\infty}^{\infty} \frac{x-t}{(x-t)^2+y^2}\cdot\frac{t-\xi}{(t-\xi)^2+h^2}\,\omega_{\mathscr{D}}(\xi)\operatorname{sgn}\xi\,d\xi\,dt.$$

Suppose for the moment that *absolute convergence* of the double integral has been established. Then we can change the order of integration therein. We have, however, for $y > 0$,

$$\int_{-\infty}^{\infty} \frac{(x-t)}{(x-t)^2+y^2}\cdot\frac{t-\xi}{(t-\xi)^2+h^2}\,dt \;=\; -\pi\frac{y+h}{(x-\xi)^2+(y+h)^2},$$

as follows from the identity

$$\int_{-\infty}^{\infty} \frac{1}{x+iy-t}\cdot\frac{1}{\xi+ih-t}\,dt \;=\; 0,$$

verifiable by contour integration (h and y are >0 here), and the semigroup convolution property of the Poisson kernel. The previous relation thus becomes

$$\int_{-\infty}^{\infty} \frac{x-t}{(x-t)^2+y^2} G_{\mathscr{D}}(t+ih,\ 0)\,dt$$

$$= \pi\int_{-\infty}^{\infty} \frac{y+h}{(x-\xi)^2+(y+h)^2}\,\omega_{\mathscr{D}}(\xi)\operatorname{sgn}\xi\,d\xi.$$

Fixing $y > 0$ for the moment, make $h \to 0$. According to the *third* of the above lemmas, the last formula then becomes

$$\int_{-\infty}^{\infty} \frac{x-t}{(x-t)^2+y^2} G_{\mathscr{D}}(t,0)\,dt \;=\; \pi\int_{-\infty}^{\infty} \frac{y}{(x-\xi)^2+y^2}\,\omega_{\mathscr{D}}(\xi)\operatorname{sgn}\xi\,d\xi.$$

Now make $y \to 0$, assuming that $x \neq 0$. Since $\omega_{\mathscr{D}}(\xi)$ is continuous at x, the *right side* tends to

$$\pi^2\omega_{\mathscr{D}}(x)\operatorname{sgn}x.$$

Also, by the *fourth* lemma, $G_{\mathscr{D}}(t,0)$ is Lip$\frac{1}{2}$ at x. The left-hand integral therefore tends to the Cauchy principal value

$$⨍_{-\infty}^{\infty} \frac{G_{\mathscr{D}}(t,0)}{x-t}\,dt$$

(which exists !), according to an observation in §H.1 of Chapter III and the discussion preceding the last theorem above. We thus have

$$\omega_{\mathscr{D}}(x)\,\mathrm{sgn}\,x \;=\; \frac{1}{\pi^2}⨍_{-\infty}^{\infty} \frac{G_{\mathscr{D}}(t,0)}{x-t}\,dt$$

for $x \neq 0$, as asserted.

The legitimacy of the above reasoning required absolute convergence of

$$\int_{-\infty}^{\infty}\int_{-\infty}^{\infty} \frac{x-t}{(x-t)^2+y^2}\cdot\frac{t-\xi}{(t-\xi)^2+h^2}\,\omega_{\mathscr{D}}(\xi)\,\mathrm{sgn}\,\xi\,dt\,d\xi$$

which we must now establish. Fixing y and $h>0$ and $x\in\mathbb{R}$, we have

$$\left|\frac{x-t}{(x-t)^2+y^2}\cdot\frac{t-\xi}{(t-\xi)^2+h^2}\right| \;\leqslant\; \frac{\text{const.}}{(|t|+1)(|\xi-t|+1)}.$$

Wlog, let $\xi>0$. Then

$$\int_{-\infty}^{\infty}\frac{dt}{(|t|+1)(|\xi-t|+1)} \;\leqslant\; 2\int_0^{\infty}\frac{dt}{(t+1)(|\xi-t|+1)},$$

which we break up in turn as

$$2\int_0^{\xi/2} \;+\; 2\int_{\xi/2}^{3\xi/2} \;+\; 2\int_{3\xi/2}^{\infty}.$$

In the *first* of these integrals we use the inequality

$$|\xi-t| \;\geqslant\; \xi/2,$$

and, in the *second*,

$$t \;\geqslant\; \xi/2,$$

taking in the latter a new variable $s=t-\xi$. Both are thus easily seen to have values

$$\leqslant\; \text{const.}\frac{\log^+\xi+1}{\xi+1}$$

In the *third* integral, use the relation

$$t-\xi \;\geqslant\; t/3.$$

This shows that expression to be $\leqslant$ const.$1/(\xi+1)$.

In fine, then,

$$\int_{-\infty}^{\infty}\left|\frac{x-t}{(x-t)^2+y^2}\cdot\frac{t-\xi}{(t-\xi)^2+h^2}\right|dt \;\leqslant\; \text{const.}\frac{\log^+|\xi|+1}{|\xi|+1}$$

for *fixed* $x \in \mathbb{R}$ and $y, h > 0$. From the proof of the *first* lemma in this article, we know, however, that

$$|\omega_{\mathscr{D}}(\xi)\operatorname{sgn}\xi| \;=\; \omega_{\mathscr{D}}(\xi) \;\leqslant\; \frac{\text{const.}}{|\xi|+1}.$$

Absolute convergence of our double integral thus depends on the convergence of

$$\int_{-\infty}^{\infty} \frac{1+\log^+|\xi|}{(|\xi|+1)^2}\,d\xi$$

which evidently *holds*. Our proof is complete.

Notation. *If* $\mathscr{D}$ is one of our domains with $0 \in \mathscr{D}$, we write, for $x > 0$,

$$\Omega_{\mathscr{D}}(x) \;=\; \omega_{\mathscr{D}}((-\infty, -x] \cup [x, \infty),\ 0).$$

Further work in this § will be based on the function $\Omega_{\mathscr{D}}$. For it, the theorem just proved has the

Corollary. *If* $0 \in \mathscr{D}$,

$$\Omega_{\mathscr{D}}(x) \;=\; \frac{2}{\pi^2}\int_{-\infty}^{\infty} \frac{xG_{\mathscr{D}}(t,0)}{x^2-t^2}\,dt \qquad \text{for } x > 0.$$

Proof. When $x > 0$,

$$\Omega_{\mathscr{D}}(x) \;=\; \omega_{\mathscr{D}}(x) + \omega_{\mathscr{D}}(-x).$$

Plug the formula furnished by the theorem into the right side.

Scholium. The preceding arguments practically suffice to work up a complete treatment of the L_2 theory of Hilbert transforms. The reader who has never studied that theory thus has an opportunity to learn it now.

If $f \in L_2(-\infty, \infty)$, let us write

$$u(z) \;=\; \frac{1}{\pi}\int_{-\infty}^{\infty} \frac{\Im z}{|z-t|^2}\,f(t)\,dt$$

and

$$\tilde{u}(z) \;=\; \frac{1}{\pi}\int_{-\infty}^{\infty} \frac{x-t}{|z-t|^2}\,f(t)\,dt$$

for $\Im z > 0$; $\tilde{u}(z)$ is a *harmonic conjugate* of $u(z)$ in the upper half plane. By taking Fourier transforms and using Plancherel's theorem, one easily checks that

$$\int_{-\infty}^{\infty} |\tilde{u}(x+\mathrm{i}y)|^2 \,\mathrm{d}x \;\leqslant\; \|f\|_2^2$$

for each $y > 0$. Following a previous discussion in this article and those of §§F.2 and H.1, Chapter III, we also see that

$$\tilde{f}(x) \;=\; \lim_{y\to 0} \tilde{u}(x+\mathrm{i}y)$$

exists a.e. Fatou's lemma then yields

$$\|\tilde{f}\|_2 \;\leqslant\; \|f\|_2$$

in view of the previous inequality.

It is in fact true that

$$\int_{-\infty}^{\infty} |\tilde{f}(x) - \tilde{u}(x+\mathrm{i}y)|^2 \,\mathrm{d}x \;\longrightarrow\; 0$$

for $y \to 0$. This may be seen by noting that

$$\int_{-\infty}^{\infty} |\tilde{u}(x+\mathrm{i}y) - \tilde{u}(x+\mathrm{i}y')|^2 \,\mathrm{d}x \;=\; \int_{-\infty}^{\infty} |u(x+\mathrm{i}y) - u(x+\mathrm{i}y')|^2 \,\mathrm{d}x$$

for y and $y' > 0$, which may be verified using Fourier transforms and Plancherel's theorem. According to elementary properties of the Poisson kernel, the right-hand integral is *small* when $y > 0$ and $y' > 0$ are, as long as $f \in L_2$. Fixing a small $y > 0$ and then making $y' \to 0$ in the *left-hand* integral, we find that

$$\int_{-\infty}^{\infty} |\tilde{f}(x) - \tilde{u}(x+\mathrm{i}y)|^2 \,\mathrm{d}x$$

is small by applying Fatou's lemma.

Once this is known, it is easy to prove that

$$\tilde{\tilde{f}}(x) \;=\; -f(x) \;\text{ a.e.}$$

by following almost exactly the argument used in proving the last theorem above. (Note that $(\log^+|\xi| + 1)/(|\xi| + 1) \in L_2(-\infty, \infty)$.) This must then imply that

$$\|f\|_2 \;\leqslant\; \|\tilde{f}\|_2,$$

so that finally

$$\|f\|_2 = \|\tilde{f}\|_2.$$

To complete this development, we need the result that the Cauchy principal value

$$\frac{1}{\pi}⨍_{-\infty}^{\infty} \frac{f(t)}{x-t}\,dt$$

exists and equals $\tilde{f}(x)$ a.e. That is the content of

Problem 25

Let $f \in L_p(-\infty, \infty)$, $p \geqslant 1$. Show that

$$\frac{1}{\pi}\int_{-\infty}^{\infty} \frac{x-t}{(x-t)^2+y^2} f(t)\,dt \;-\; \frac{1}{\pi}\int_{|t-x|\geqslant y} \frac{f(t)}{x-t}\,dt$$

tends to zero as $y \to 0$ if

$$\frac{1}{y}\int_{x-y}^{x+y} |f(t)-f(x)|\,dt \longrightarrow 0$$

for $y \to 0$, and hence for *almost every* real x. (The set of x for which the last condition holds is called the *Lebesgue set* of f.) (Hint. One may wlog take f to be of compact support, making $\|f\|_1 < \infty$. Choosing a small $\delta > 0$, one considers values of y *between* 0 and δ, for which the difference in question can be written as

$$\frac{1}{\pi}\int_0^y \frac{\tau(f(x-\tau)-f(x+\tau))}{\tau^2+y^2}\,d\tau$$

$$+\ \frac{1}{\pi}\left(\int_y^\delta + \int_\delta^\infty\right)\left(\frac{\tau}{\tau^2+y^2} - \frac{1}{\tau}\right)(f(x-\tau)-f(x+\tau))\,d\tau.$$

If the stipulated condition holds at x, the *first* of these integrals clearly $\to 0$ as $y \to 0$. For *fixed* $\delta > 0$, the integral from δ to ∞ is $\leqslant 2y^2\|f\|_1/\delta^3$ and this $\to 0$ as $y \to 0$. The integral from y to δ is in absolute value

$$\leqslant y^2 \int_y^\delta \frac{|f(x-\tau)-f(x+\tau)|}{\tau^3}\,d\tau.$$

Integrate this by parts.)

2. An estimate for harmonic measure

Given one of our domains $\mathscr{D}$ with $0 \in \mathscr{D}$, the function $\Omega_{\mathscr{D}}(x) = \omega_{\mathscr{D}}((-\infty, -x] \cup [x, \infty), 0)$ is equal to

$$\frac{2}{\pi^2}⨍_{-\infty}^{\infty} \frac{xG_{\mathscr{D}}(t,0)}{x^2-t^2}\,dt$$

by the corollary near the end of the preceding article. The Green's function $G_{\mathscr{D}}(t, 0)$ of course vanishes on $\partial\mathscr{D} = \mathbb{R} \cap (\sim \mathscr{D})$, and our attention is restricted to domains $\mathscr{D}$ having *bounded* intersection with $\mathbb{R}$. The above Cauchy principal value thus reduces to an ordinary integral for large x, and we have

$$\Omega_{\mathscr{D}}(x) \sim \frac{2}{\pi^2 x} \int_{-\infty}^{\infty} G_{\mathscr{D}}(t, 0)\, dt \qquad \text{for } x \to \infty,$$

i.e., *in terms of the Phragmén–Lindelöf function* $Y_{\mathscr{D}}(z)$ *for* $\mathscr{D}$, defined in §A.2,

$$\Omega_{\mathscr{D}}(x) \sim \frac{2 Y_{\mathscr{D}}(0)}{\pi x}, \qquad x \to \infty.$$

It is remarkable that an *inequality* resembling this asymptotic relation holds for *all* positive x; this means that the kind of comparison spoken of at the beginning of the present § is available.

Theorem. *If* $0 \in \mathscr{D}$,

$$\Omega_{\mathscr{D}}(x) \leqslant \frac{Y_{\mathscr{D}}(0)}{x} \qquad \textit{for } x > 0.$$

Proof. By comparison of harmonic measure for $\mathscr{D}$ with that for another smaller domain that depends on x.

Given $x > 0$, we let $E_x = E \cup (-\infty, -x] \cup [x, \infty)$ and then put $\mathscr{D}_x = \mathbb{C} \sim E_x$:

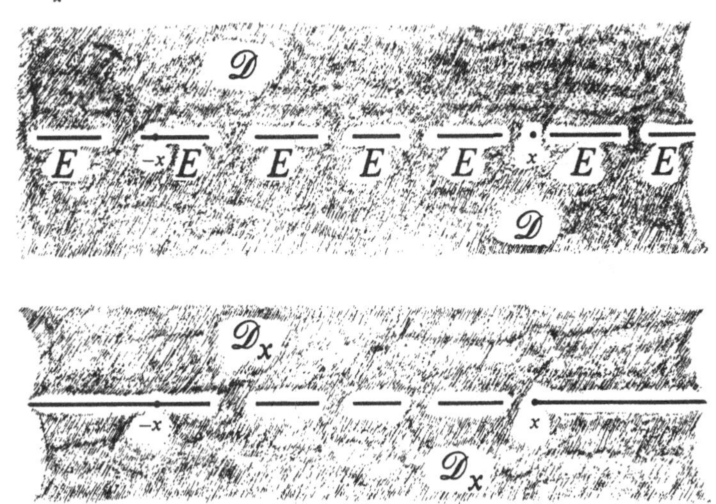

Figure 152

We have $\mathscr{D}_x \subseteq \mathscr{D}$. On comparing $\omega_{\mathscr{D}_x}((-\infty, -x] \cup [x, \infty), \zeta)$ with $\omega_{\mathscr{D}}((-\infty, -x] \cup [x, \infty), \zeta)$ on E_x, we see that the *former* is *larger* than the *latter* for $\zeta \in \mathscr{D}_x$. Hence, putting $\zeta = 0$, we get

$$\Omega_{\mathscr{D}}(x) \leqslant \Omega_{\mathscr{D}_x}(x).$$

Take any number $\rho > 1$. Applying the corollary near the end of the previous article and noting that $G_{\mathscr{D}_x}(t, 0)$ *vanishes* for $t \in E_x \supseteq (-\infty, -x] \cup [x, \infty)$, we have

$$\Omega_{\mathscr{D}_x}(\rho x) = \frac{2}{\pi^2} \int_{-x}^{x} \frac{\rho x G_{\mathscr{D}_x}(t, 0)}{\rho^2 x^2 - t^2} \, dt.$$

Since $\mathscr{D}_x \subseteq \mathscr{D}$, $G_{\mathscr{D}_x}(t, 0) \leqslant G_{\mathscr{D}}(t, 0)$, so the right-hand integral is

$$\leqslant \frac{2\rho}{\pi^2(\rho^2 - 1)x} \int_{-x}^{x} G_{\mathscr{D}}(t, 0) \, dt \leqslant \frac{2\rho}{\pi^2(\rho^2 - 1)x} \int_{-\infty}^{\infty} G_{\mathscr{D}}(t, 0) \, dt.$$

By the formula for $Y_{\mathscr{D}}(z)$ furnished by the *second* theorem of §A2, we thus get

$$\Omega_{\mathscr{D}_x}(\rho x) \leqslant \frac{2\rho}{\pi(\rho^2 - 1)} \frac{Y_{\mathscr{D}}(0)}{x}.$$

In order to complete the proof, we show that $\Omega_{\mathscr{D}_x}(\rho x)/\Omega_{\mathscr{D}_x}(x)$ is *bounded below* by a quantity *depending only on* ρ, and then use the inequality just established together with the previous one.

To compare $\Omega_{\mathscr{D}_x}(\rho x)$ with $\Omega_{\mathscr{D}_x}(x)$, take a *third* domain

$$\mathscr{E} = \mathbb{C} \sim ((-\infty, -x] \cup [x, \infty)):$$

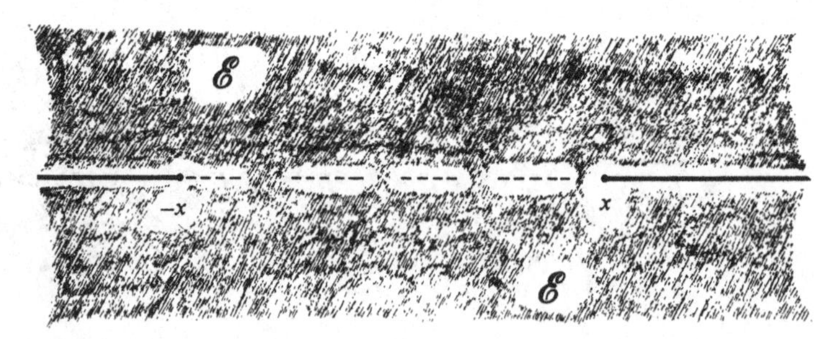

Figure 153

Note that $\mathscr{D}_x \subseteq \mathscr{E}$ and $\partial \mathscr{D}_x = E_x$ consists of $\partial \mathscr{E}$ together with the part of E lying in the segment $[-x, x]$. For $\zeta \in \mathscr{D}_x$ (and $\rho > 1$), a formula from §B.1 of Chapter VII tells us that

$$\omega_{\mathscr{D}_x}((-\infty, -\rho x] \cup [\rho x, \infty), \zeta)$$

$$= \omega_{\mathscr{E}}((-\infty, -\rho x] \cup [\rho x, \infty), \zeta)$$

$$- \int_{E \cap \mathscr{E}} \omega_{\mathscr{E}}((-\infty, -\rho x] \cup [\rho x, \infty), t) \, d\omega_{\mathscr{D}_x}(t, \zeta),$$

whence, taking $\zeta = 0$,

$$\Omega_{\mathscr{D}_x}(\rho x) = \omega_{\mathscr{E}}((-\infty, -\rho x] \cup [\rho x, \infty), 0)$$
$$- \int_{E \cap \mathscr{E}} \omega_{\mathscr{E}}((-\infty, -\rho x] \cup [\rho x, \infty), t) \mathrm{d}\omega_{\mathscr{D}_x}(t, 0).$$

Also,

$$\Omega_{\mathscr{D}_x}(x) = 1 - \int_{E \cap \mathscr{E}} \mathrm{d}\omega_{\mathscr{D}_x}(t, 0).$$

The harmonic measure $\omega_{\mathscr{E}}((-\infty, -\rho x] \cup [\rho x, \infty), t)$ can be computed explicitly by making the Joukowski mapping

$$\zeta \longrightarrow w = \frac{x}{\zeta} - \sqrt{\left(\frac{x^2}{\zeta^2} - 1\right)}$$

of $\mathscr{E}$ onto $\Delta = \{|w| < 1\}$:

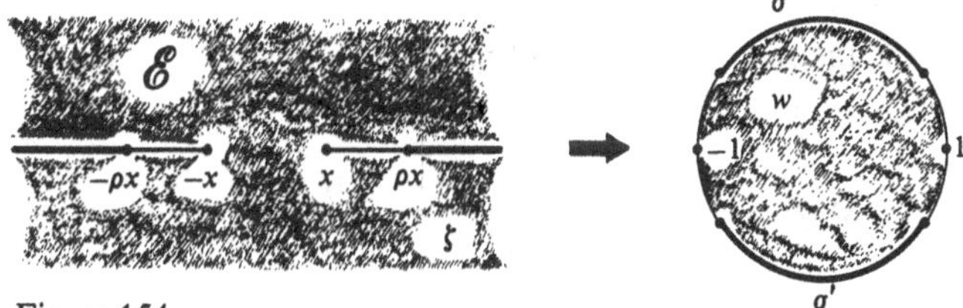

Figure 154

This conformal map takes $[-x, x]$ to the diameter $[-1, 1]$, and 0 to 0. The union of the (two-sided!) intervals $(-\infty, -\rho x]$ and $[\rho x, \infty)$ on $\partial\mathscr{E}$ is taken onto that of two arcs, σ and σ', on $\{|w| = 1\}$, the first symmetric about i and the second symmetric about $-$i. For $\zeta \in \mathscr{E}$, $\omega_{\mathscr{E}}((-\infty, -\rho x] \cup [\rho x, \infty), \zeta)$ is the *sum* of the harmonic measures of these two arcs in Δ, seen from the point w therein corresponding to ζ. When $\zeta = t$ is *real*, this sum is just $2\omega_\Delta(\sigma, u)$, u being the point of $(-1, 1)$ corresponding to t. However, from the rudiments of complex variable theory, the level lines of $\omega_\Delta(\sigma, w)$ are just the *circles* through the endpoints of σ. From a glance at the following diagram, it is hence obvious that $\omega_\Delta(\sigma, u)$ has its *maximum* for $-1 \leqslant u \leqslant 1$ when $u = 0$:

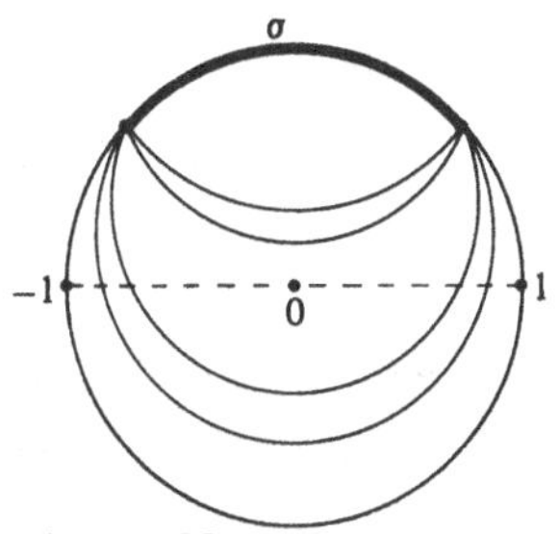

Figure 155

Going back to $\mathscr{E}$, we see that

$$\omega_{\mathscr{E}}((-\infty, -\rho x]\cup[\rho x, \infty),\ t) \ \leqslant\ \omega_{\mathscr{E}}((-\infty, -\rho x]\cup[\rho x, \infty),\ 0)$$

when $-x \leqslant t \leqslant x$. Plugging this into the above formula for $\Omega_{\mathscr{D}_x}(\rho x)$, we find that

$$\Omega_{\mathscr{D}_x}(\rho x) \ \geqslant\ \omega_{\mathscr{E}}((-\infty, -\rho x]\cup[\rho x, \infty),\ 0) \times\left\{1 - \int_{E\cap\mathscr{E}} \mathrm{d}\omega_{\mathscr{D}_x}(t, 0)\right\}.$$

The quantity in curly brackets is just $\Omega_{\mathscr{D}_x}(x)$, so we have

$$\frac{\Omega_{\mathscr{D}_x}(\rho x)}{\Omega_{\mathscr{D}_x}(x)} \ \geqslant\ \omega_{\mathscr{E}}((-\infty, -\rho x]\cup[\rho x, \infty),\ 0).$$

Here, the right side clearly depends only on ρ; this is the relation we set out to obtain.

From the inequality just found together with the two others established at the beginning of this proof, we now get

$$\Omega_{\mathscr{D}}(x) \ \leqslant\ \Omega_{\mathscr{D}_x}(x) \ \leqslant\ \frac{\Omega_{\mathscr{D}_x}(\rho x)}{\omega_{\mathscr{E}}((-\infty, -\rho x]\cup[\rho x, \infty),\ 0)}$$

$$\leqslant\ \frac{2\rho}{\pi(\rho^2 - 1)\omega_{\mathscr{E}}((-\infty, -\rho x]\cup[\rho x, \infty),\ 0)}\cdot\frac{Y_{\mathscr{D}}(0)}{x}.$$

The front factor in the right-hand member depends only on the parameter ρ; let us compute its value. The two arcs σ and σ' both subtend angles $2\arcsin(\Gamma/\rho)$ at 0. Therefore

$$\omega_{\mathscr{E}}((-\infty, -\rho x]\cup[\rho x, \infty),\ 0) \ =\ 2\omega_{\Delta}(\sigma, 0) \ =\ \frac{2}{\pi}\arcsin\frac{1}{\rho},$$

and the factor in question equals

$$\frac{\rho}{(\rho^2 - 1)\arcsin\dfrac{1}{\rho}}.$$

It is readily ascertained (put $1/\rho = \sin\alpha$!) that the expression just written *decreases* for $\rho > 1$. Making $\rho \to \infty$, we get the limit 1, whence

$$\Omega_{\mathscr{D}}(x) \ \leqslant\ Y_{\mathscr{D}}(0)/x, \qquad \text{Q.E.D.}$$

Remark. An inequality almost as good as the one just established can be obtained with considerably less effort. By the first theorem of the preceding article, we have, for $y > 0$,

$$G_{\mathscr{D}}(\mathrm{i}y, 0) = \log\frac{1}{y} + \int_{-\infty}^{\infty} \log|\mathrm{i}y - t|\,\mathrm{d}\omega_{\mathscr{D}}(t, 0)$$

$$= \int_{-\infty}^{\infty} \log\sqrt{\left(1 + \frac{t^2}{y^2}\right)}\mathrm{d}\omega_{\mathscr{D}}(t, 0),$$

a quantity clearly $\geqslant \Omega_{\mathscr{D}}(y)\log\sqrt{2}$. On the other hand,

$$G_{\mathscr{D}}(\mathrm{i}y, 0) = \frac{1}{\pi}\int_{-\infty}^{\infty} \frac{yG_{\mathscr{D}}(t, 0)}{y^2 + t^2}\mathrm{d}t$$

as in the proof of the third lemma from that article. Here, the right side is

$$\leqslant \frac{1}{\pi y}\int_{-\infty}^{\infty} G_{\mathscr{D}}(t, 0)\,\mathrm{d}t = \frac{Y_{\mathscr{D}}(0)}{y},$$

so the previous relation yields

$$\Omega_{\mathscr{D}}(y) \leqslant \frac{2}{\log 2}\frac{Y_{\mathscr{D}}(0)}{y}.$$

Problem 26

For $0 < \rho < \frac{1}{2}$, let E_ρ be the union of the segments

$$\left[\frac{2n-1}{2} - \rho,\ \frac{2n-1}{2} + \rho\right], \qquad n \in \mathbb{Z};$$

these are just the intervals of length 2ρ centered at the *half odd* integers. Denote the component $[(2n-1)/2 - \rho,\ (2n-1)/2 + \rho]$ of E_ρ by J_n (it would be more logical to write $J_n(\rho)$). $\mathscr{D}_\rho = \mathbb{C} \sim E_\rho$ is a domain of the kind considered in §A, and, by *Carleson's theorem* from §A.1,

$$\omega_{\mathscr{D}_\rho}(J_n, 0) \leqslant \frac{K_\rho}{n^2 + 1}.$$

The purpose of this problem is to obtain quantitative information about the asymptotic behaviour of the best value for K_ρ as $\rho \to 0$.

(a) Show that $Y_{\mathscr{D}_\rho}(0) \sim (1/\pi)\log(1/\rho)$ as $\rho \to 0$. (Hint. In $\mathscr{D}_\rho$, consider the harmonic function

$$\log\left|\frac{\cos \pi z}{\sin \pi\rho} + \sqrt{\left(\frac{\cos^2 \pi z}{\sin^2 \pi\rho} - 1\right)}\right|. \text{)}$$

(b) By making an appropriate limiting argument, adapt the theorem just proved to the domain $\mathscr{D}_\rho$ and hence show that

$$\Omega_{\mathscr{D}_\rho}(x) \leqslant Y_{\mathscr{D}_\rho}(0)/x \qquad \text{for } x > 0.$$

(c) For $n \geqslant 1$, show that

$$\omega_{\mathscr{D}_\rho}(J_{n+1}, 0) \leqslant \omega_{\mathscr{D}_\rho}(J_n, 0).$$

(Hint:

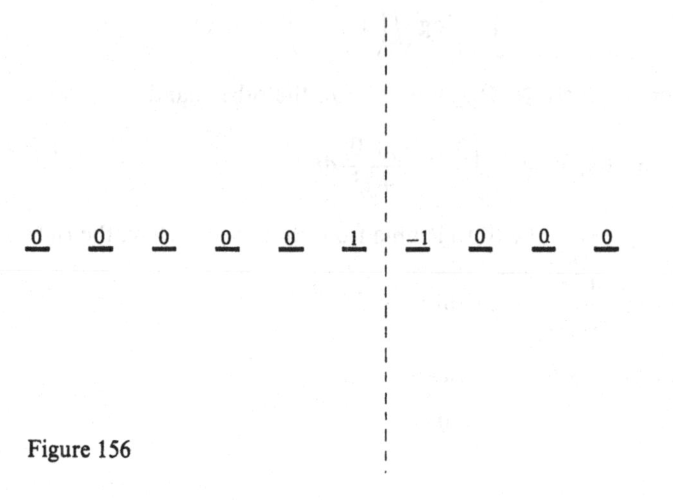

Figure 156 .)

(d) Hence show that, for $n \geqslant 3$,

$$\omega_{\mathscr{D}_\rho}(J_n, 0) \leqslant \left(C \log \frac{1}{\rho}\right) \Big/ n^2.)$$

with a numerical constant C independent of ρ.
(Hint: $\Omega_{\mathscr{D}_\rho}(n) \geqslant 2\sum_{k=n}^{2n} \omega_{\mathscr{D}_\rho}(J_{k+1}, 0)$.)

(e) Show that the *smallest* constant K_ρ such that $\omega_{\mathscr{D}_\rho}(J_n, 0) \leqslant K_\rho/(n^2+1)$ for all n satisfies

$$K_\rho \geqslant C' \log \frac{1}{\rho}$$

with a constant C' independent of ρ.
(Hint. This is harder than parts (a)–(d). Fixing any $\rho > 0$, write, for large R, $E_R = E_\rho \cup (-\infty, -R] \cup [R, \infty)$, and then put $\mathscr{D}_R = \mathbb{C} \sim E_R$. As $R \to \infty$, $G_{\mathscr{D}_R}(t, 0)$ increases to $G_{\mathscr{D}_\rho}(t, 0)$, so $Y_{\mathscr{D}_R}(0)$ increases to $Y_{\mathscr{D}_\rho}(0)$. For each R, by the *first* theorem of the previous article,

$$G_{\mathscr{D}_R}(z, w) = \log \frac{1}{|z - w|} + \int_{E_R} \log|w - s| \, d\omega_{\mathscr{D}_R}(s, z),$$

whence

$$G_{\mathscr{D}_R}(t, 0) + G_{\mathscr{D}_R}(-t, 0) = \int_{E_R} \log\left|1 - \frac{s^2}{t^2}\right| d\omega_{\mathscr{D}_R}(s, 0).$$

Fix any integer $A > 0$. Then $\int_{-A}^{A} G_{\mathscr{D}_\rho}(t,0)\mathrm{d}t$ is the limit, as $R \to \infty$, of $\int_{-\infty}^{\infty}\int_0^A \log|1-(s^2/t^2)|\mathrm{d}t\,\mathrm{d}\omega_{\mathscr{D}_R}(s,0)$. Taking an arbitrary large M, which for the moment we *fix*, we break up this double integral as

$$\int_{-M}^{M}\int_0^A + \int_{|s|>M}\int_0^A.$$

To study the two terms of this sum, first evaluate

$$\int_0^A \log\left|1-\frac{s^2}{t^2}\right|\mathrm{d}t;$$

for $|s| > A$ this can be done by direct computation, and, for $|s| < A$, by using the identity

$$\int_0^A \log\left|1-\frac{s^2}{t^2}\right|\mathrm{d}t \;=\; -\int_A^\infty \log\left|1-\frac{s^2}{t^2}\right|\mathrm{d}t.$$

Regarding $\int_{-M}^{M}\int_0^A \log|1-(s^2/t^2)|\mathrm{d}t\,\mathrm{d}\omega_{\mathscr{D}_R}(s,0)$, we may use the fact that $\omega_{\mathscr{D}_R}(S,0) \longrightarrow \omega_{\mathscr{D}_\rho}(S,0)$ as $R \to \infty$ for *bounded* $S \subseteq \mathbb{R}$, and then plug in the inequality

$$\omega_{\mathscr{D}_\rho}(J_n, 0) \;\leqslant\; K_\rho/(n^2+1)$$

together with the result of the computation just indicated. In this way we easily see that $\lim_{R\to\infty}\int_{-M}^{M}\int_0^A \;\leqslant\; CK_\rho$ with a constant C independent of A, M, and ρ.

In order to estimate

$$\int_{|s|>M}\int_0^A \log\left|1-\frac{s^2}{t^2}\right|\mathrm{d}t\,\mathrm{d}\omega_{\mathscr{D}_R}(s,0),$$

use the fact that

$$\Omega_{\mathscr{D}_R}(s) \;\leqslant\; \frac{Y_{\mathscr{D}_R}(0)}{s} \;\leqslant\; \frac{Y_{\mathscr{D}_\rho}(0)}{s}$$

(where $Y_{\mathscr{D}_\rho}(0)$, as we already know, is *finite*) together with the value of the inner integral, already computed, and integrate by parts. In this way one finds an estimate *independent* of R which, *for fixed A*, is *very small* if M is *large enough*. Combining this result with the previous one and then making $M \to \infty$, one sees that

$$\int_{-A}^{A} G_{\mathscr{D}_\rho}(t,0)\mathrm{d}t \;\leqslant\; CK_\rho$$

with C independent of A and of ρ.)

Remark. In the circumstances of the preceding problem $G_{\mathscr{D}_\rho}(z,0)$ must, when $\rho \to 0$, tend to ∞ for each z not equal to a half odd integer, and it is interesting to see how fast that happens. Fix any such $z \neq 0$. Then, given $\rho > 0$ we have, working with the domains $\mathscr{D}_R$ used in part (e) of the problem,

$$G_{\mathscr{D}_\rho}(z,0) = \lim_{R\to\infty} G_{\mathscr{D}_R}(z,0).$$

Here,

$$G_{\mathscr{D}_R}(z,0) = \log\frac{1}{|z|} + \int_{-\infty}^{\infty} \log|z-t|\,d\omega_{\mathscr{D}_R}(t,0)$$

$$= O(1) + \int_{-\infty}^{\infty} \log^+|t|\,d\omega_{\mathscr{D}_R}(t,0),$$

where the $O(1)$ term depends on z but is independent of R, and of ρ, when the latter is small enough.

Taking an $M > 1$, we rewrite the last integral on the right as

$$\int_{|t|<M} + \int_{|t|\geqslant M},$$

and thus find it to be

$$\leqslant \log M - \int_M^{\infty} \log t\, d\Omega_{\mathscr{D}_R}(t)$$

$$= \log M + \Omega_{\mathscr{D}_R}(M)\log M + \int_M^{\infty} \frac{\Omega_{\mathscr{D}_R}(t)}{t}\,dt.$$

Plug the inequalities $\Omega_{\mathscr{D}_R}(t) \leqslant Y_{\mathscr{D}_R}(0)/t$ and $Y_{\mathscr{D}_R}(0) \leqslant Y_{\mathscr{D}_\rho}(0)$ into the expression on the right. Then, referring to the previous relation and making $R \to \infty$, we see that

$$G_{\mathscr{D}_\rho}(z,0) \leqslant O(1) + \log M + Y_{\mathscr{D}_\rho}(0)\frac{\log M + 1}{M}.$$

By part (a) of the problem, $Y_{\mathscr{D}_\rho}(0) = O(1) + (1/\pi)\log(1/\rho)$. Hence, choosing $M = (1/\pi)\log(1/\rho)\log\log(1/\rho)$ in the last relation, we get

$$G_{\mathscr{D}_\rho}(z,0) \leqslant O(1) + \log\log\frac{1}{\rho} + \log\log\log\frac{1}{\rho}.$$

This order of growth seems rather slow. One would have expected $G_{\mathscr{D}_\rho}(z,0)$ to behave like $\log(1/\rho)$ for small ρ when z is fixed.

3. The energy integral again

The result of the preceding article already has some applications to the project described at the beginning of this §. Suppose that the majorant $M(t) \geqslant 0$ is defined and *even* on $\mathbb{R}$. Taking $M(t)$ to be identically zero in a neighborhood of 0 involves no real loss of generality. If $M(t)$ is also *increasing* on $[0, \infty)$, the Poisson integral

$$\int_{\partial\mathscr{D}} v(t)\mathrm{d}\omega_{\mathscr{D}}(t, 0)$$

for a function $v(z)$ subharmonic in one of our domains $\mathscr{D}$ with $0 \in \mathscr{D}$ and satisfying

$$v(t) \leqslant M(t), \qquad t \in \partial\mathscr{D},$$

has the simple majorant

$$Y_{\mathscr{D}}(0) \cdot \int_0^\infty \frac{M(t)}{t^2}\mathrm{d}t.$$

The entire dependence of the Poisson integral on the domain $\mathscr{D}$ is thus expressed by means of the single factor $Y_{\mathscr{D}}(0)$ occurring in this second expression.

To see this, recall that $\omega_{\mathscr{D}}((-\infty, -t] \cup [t, \infty),\ 0) = \Omega_{\mathscr{D}}(t)$ for $t > 0$; the given majoration on $v(t)$ therefore makes the Poisson integral $\leqslant -\int_0^\infty M(t)\mathrm{d}\Omega_{\mathscr{D}}(t)$, which here is equal to

$$\int_0^\infty \Omega_{\mathscr{D}}(t)\mathrm{d}M(t).$$

Since $M(t)$ is increasing on $[0, \infty)$, we may substitute the relation $\Omega_{\mathscr{D}}(t) \leqslant Y_{\mathscr{D}}(0)/t$ proved in the preceding article into the last expression, showing it to be

$$\leqslant Y_{\mathscr{D}}(0)\int_0^\infty \frac{\mathrm{d}M(t)}{t} = Y_{\mathscr{D}}(0)\int_0^\infty \frac{M(t)}{t^2}\mathrm{d}t.$$

This argument cannot be applied to *general* even majorants $M(t) \geqslant 0$, because the relation $\Omega_{\mathscr{D}}(t) \leqslant Y_{\mathscr{D}}(0)/t$ cannot be differentiated to yield $\mathrm{d}\omega_{\mathscr{D}}(t, 0) \leqslant (Y_{\mathscr{D}}(0)/t^2)\mathrm{d}t$. Indeed, when $x \in \partial\mathscr{D} = E$ gets near any of the *endpoints* a of the intervals making up that set, $\mathrm{d}\omega_{\mathscr{D}}(x, 0)/\mathrm{d}x$ gets *large* like a multiple of $|x - a|^{-1/2}$ (see the second lemma of article 1 and the remark following it). We are not supposing *anything* about the *disposition* of these intervals except that they be *finite in number*; there may *otherwise* be *arbitrarily many* of them. It is therefore not possible to bound $\int_{-\infty}^\infty M(t)\mathrm{d}\omega_{\mathscr{D}}(t, 0)$ by an expression involving *only* $\int_0^\infty (M(t)/t^2)\mathrm{d}t$ for *general* even majorants $M(t) \geqslant 0$; some *additional regularity properties* of $M(t)$ are required and must be taken into account. A very useful instrument for this purpose turns out to be the *energy* introduced in §B.5 which has already played such an important rôle in §B. Application of that notion to matters like the one now under discussion goes back to the 1962 paper of Beurling and Malliavin. The material of that paper will be taken up in Chapter XI, where we will use the results established in the present §.

Appearance of the energy here is due to the following

Lemma. *Let* $0 \in \mathscr{D}$. For $x \neq 0$,

$$G_{\mathscr{D}}(x,0) + G_{\mathscr{D}}(-x,0) \;=\; \frac{1}{x}\int_0^\infty \log\left|\frac{x+t}{x-t}\right| \mathrm{d}(t\Omega_{\mathscr{D}}(t)).$$

Proof. By the *second* theorem of article 1,

$$G_{\mathscr{D}}(x,0) \;=\; -\!\!\!\!\!\!\int_{-\infty}^{\infty} \frac{\omega_{\mathscr{D}}(t)\,\mathrm{sgn}\,t}{x-t}\,\mathrm{d}t \qquad \text{for } x \neq 0,$$

where

$$\omega_{\mathscr{D}}(t) \;=\; \begin{cases} \omega_{\mathscr{D}}((-\infty,t],\ 0), & t<0, \\ \omega_{\mathscr{D}}([t,\infty),\ 0), & t>0. \end{cases}$$

Thence,

$$G_{\mathscr{D}}(x,0) + G_{\mathscr{D}}(-x,0) \;=\; -\!\!\!\!\!\!\int_{-\infty}^{\infty} \frac{2t\,\mathrm{sgn}\,t\,\omega_{\mathscr{D}}(t)}{t^2-x^2}\,\mathrm{d}t$$

$$=\; -\!\!\!\!\!\!\int_0^\infty \frac{2t}{t^2-x^2}\,\Omega_{\mathscr{D}}(t)\,\mathrm{d}t,$$

since $\omega_{\mathscr{D}}(t) + \omega_{\mathscr{D}}(-t) = \Omega_{\mathscr{D}}(t)$ for $t>0$.

Assuming wlog that $x>0$, we take a small $\varepsilon>0$ and apply partial integration to the two integrals in

$$\left(\int_0^{x-\varepsilon} + \int_{x+\varepsilon}^\infty\right)\frac{2}{t^2-x^2}\,t\Omega_{\mathscr{D}}(t)\,\mathrm{d}t,$$

getting

$$\left(\frac{t\Omega_{\mathscr{D}}(t)}{x}\log\left|\frac{t-x}{t+x}\right|\right)\left(\Big]_0^{x-\varepsilon} + \Big]_{x+\varepsilon}^{\infty}\right)$$

$$+ \left(\int_0^{x-\varepsilon} + \int_{x+\varepsilon}^\infty\right)\frac{1}{x}\log\left|\frac{x+t}{x-t}\right| \mathrm{d}(t\Omega_{\mathscr{D}}(t)).$$

The function $\Omega_{\mathscr{D}}(t)$ is 1 for $t>0$ near 0 and $\mathrm{O}(1/t)$ for large t; it is moreover Lip $\frac{1}{2}$ at each $x>0$ by the second lemma of article 1. The sum of the integrated terms therefore tends to 0 as $\varepsilon \to 0$, and we see that

$$\fint_0^\infty \frac{2t}{t^2 - x^2}\Omega_{\mathscr{D}}(t)\,dt \;=\; \frac{1}{x}\int_0^\infty \log\left|\frac{x+t}{x-t}\right| d(t\Omega_{\mathscr{D}}(t)).$$

Since the left side equals $G_{\mathscr{D}}(x,0) + G_{\mathscr{D}}(-x,0)$, the lemma is proved.

In the language of §B.5, $x(G_{\mathscr{D}}(x,0) + G_{\mathscr{D}}(-x,0))$ is the *Green potential* of $d(t\Omega_{\mathscr{D}}(t))$. Here, since we are assuming $\mathscr{D} \cap \mathbb{R} = \mathbb{R} \sim E$ to be bounded,

$$\Omega_{\mathscr{D}}(x) \;=\; \frac{1}{\pi^2}\fint_0^\infty \frac{2x}{x^2 - t^2}G_{\mathscr{D}}(t,0)\,dt$$

has, for large x, a convergent expansion of the form

$$\frac{a_1}{x} + \frac{a_3}{x^3} + \frac{a_5}{x^5} + \cdots,$$

so that

$$d(t\Omega_{\mathscr{D}}(t)) \;=\; -\left(\frac{2a_3}{t^3} + \frac{4a_5}{t^5} + \cdots\right)dt$$

for large t. Using this fact it is easy to verify that

$$\int_0^\infty\int_0^\infty \log\left|\frac{x+t}{x-t}\right| d(t\Omega_{\mathscr{D}}(t))\,d(x\Omega_{\mathscr{D}}(x))$$

is *absolutely convergent; this double integral thus coincides with the energy*

$$E(d(t\Omega_{\mathscr{D}}(t)),\; d(t\Omega_{\mathscr{D}}(t)))$$

defined in §B.5.

Theorem. *If* $0 \in \mathscr{D}$,

$$\boxed{E(d(t\Omega_{\mathscr{D}}(t)),\; d(t\Omega_{\mathscr{D}}(t))) \;\leqslant\; \pi(Y_{\mathscr{D}}(0))^2.}$$

Proof. By the lemma, the left side, equal to the above double integral, can be rewritten as

$$\int_0^\infty x[G_{\mathscr{D}}(x,0) + G_{\mathscr{D}}(-x,0)]\,d(x\Omega_{\mathscr{D}}(x)).$$

Here, $G_{\mathscr{D}}(x,0) + G_{\mathscr{D}}(-x,0) \geqslant 0$ and $\Omega_{\mathscr{D}}(x)$ is *decreasing*, so the last expression is

$$\leqslant \int_0^\infty [G_{\mathscr{D}}(x,0) + G_{\mathscr{D}}(-x,0)]\,x\Omega_{\mathscr{D}}(x)\,dx.$$

From the theorem of the preceding article we have $x\Omega_{\mathscr{D}}(x) \leqslant Y_{\mathscr{D}}(0)$, so this is in turn

$$\leqslant Y_{\mathscr{D}}(0) \int_{-\infty}^{\infty} G_{\mathscr{D}}(x,0)\,\mathrm{d}x,$$

which, however equals $\pi(Y_{\mathscr{D}}(0))^2$ by the second theorem of §A.2.

We are done.

This theorem will be used in establishing the remaining results of the present §. For that work it will be convenient to have at hand an *alternative notation for the energy*

$$E(\mathrm{d}\rho(t), \mathrm{d}\rho(t)).$$

Suppose that we have a real Green potential

$$u(x) = \int_0^{\infty} \log\left|\frac{x+t}{x-t}\right| \mathrm{d}\rho(t).$$

If the double integral

$$\int_0^{\infty}\int_0^{\infty} \log\left|\frac{x+t}{x-t}\right| \mathrm{d}\rho(t)\,\mathrm{d}\rho(x)$$

used to define $E(\mathrm{d}\rho(t), \mathrm{d}\rho(t))$ *is absolutely convergent, we write*

$$\|u\|_E^2 = E(\mathrm{d}\rho(t), \mathrm{d}\rho(t)).$$

If we have another such Green potential

$$v(x) = \int_0^{\infty} \log\left|\frac{x+t}{x-t}\right| \mathrm{d}\sigma(t),$$

we similarly write

$$\langle u, v\rangle_E = E(\mathrm{d}\rho(t), \mathrm{d}\sigma(t)).$$

$\langle\ ,\ \rangle_E$ is a bilinear form on the collection of real Green potentials of this kind; according to the remark at the end of §B.5 it is *positive definite*. The reader may wonder whether our use of the symbol $\|u\|_E$ to denote $\sqrt{(E(\mathrm{d}\rho(t), \mathrm{d}\rho(t)))}$ is *legitimate*; could not *the same* function $u(x)$ be the Green potential of *two different measures*? That this cannot occur is easily seen, and boils down to showing that if $\rho(x)$ is *not constant*, the Green potential

$$u(x) = \int_0^\infty \log\left|\frac{x+t}{x-t}\right| d\rho(t)$$

cannot be $\equiv 0$ on $[0, \infty)$ (provided, of course, that the double integral used to define $E(d\rho(t), d\rho(t))$ is absolutely convergent). Here, we have $E(d\rho(t), d\rho(t)) = \int_0^\infty u(x) d\rho(x)$. Hence, if $u(x) \equiv 0$, the left-hand side is also zero. Then, however, $\rho(x)$ is constant by the second lemma of §B.5.

4. Harmonic estimation in $\mathscr{D}$

We are now able to give a fairly general result of the kind envisioned at the beginning of this §. Suppose we have an *even* majorant $M(t) \geqslant 0$ with $M(0) = 0$. In the case where $M(x)/x$ is a *Green potential*

$$\int_0^\infty \log\left|\frac{x+t}{x-t}\right| d\rho(t)$$

with the double integral defining $E(d\rho(t), d\rho(t)) = \|M(x)/x\|_E^2$ *absolutely convergent*, the following is true:

Theorem. *Let $M(t)$ be a majorant of the kind just described. Given one of our domains $\mathscr{D}$ containing* 0, *suppose we have a function $v(z)$, subharmonic in $\mathscr{D}$ and continuous up to $\partial\mathscr{D}$, with*

$$v(z) \leqslant A|\Im z| + O(1)$$

for some real (sic!) *A, and*

$$v(t) \leqslant M(t) \qquad \textit{for } t \in \partial\mathscr{D}.$$

Then

$$v(0) \leqslant Y_{\mathscr{D}}(0)\left\{A + \int_0^\infty \frac{M(t)}{t^2}dt + \sqrt{\pi}\left\|\frac{M(x)}{x}\right\|_E\right\}.$$

Remark. The assumptions on $v(z)$'s behaviour can be lightened by means of standard Phragmén–Lindelöf arguments (see footnote near beginning of §A.2, after problem 16). Such extensions are left to the reader; what we have here is general enough for the applications in this book.

Proof of theorem. The difference

$$v(z) - AY_{\mathscr{D}}(z)$$

is (by the definition of $Y_{\mathscr{D}}(z)$ in §A.2) *subharmonic* and *bounded above* in $\mathscr{D}$, and continuous up to $\partial\mathscr{D}$, where it coincides with $v(z)$. Hence, by harmonic majoration (Chapter VII, §B.1),

$$v(z) - AY_{\mathscr{D}}(z) \leqslant \int_{\partial\mathscr{D}} v(t)\mathrm{d}\omega_{\mathscr{D}}(t,z) \leqslant \int_{\partial\mathscr{D}} M(t)\mathrm{d}\omega_{\mathscr{D}}(t,z) \qquad \text{for} \quad z\in\mathscr{D}.$$

Taking $z = 0$, we see that we have to estimate $\int_{\partial\mathscr{D}} M(t)\mathrm{d}\omega_{\mathscr{D}}(t,0)$, which, in view of the definition of $\Omega_{\mathscr{D}}(t)$, equals $-\int_0^\infty M(t)\mathrm{d}\Omega_{\mathscr{D}}(t)$, $M(t)$ being *even*.

The *trick* here is to write

$$-\int_0^\infty M(t)\,\mathrm{d}\Omega_{\mathscr{D}}(t) = \int_0^\infty \frac{M(t)}{t}\Omega_{\mathscr{D}}(t)\,\mathrm{d}t - \int_0^\infty \frac{M(t)}{t}\mathrm{d}(t\Omega_{\mathscr{D}}(t)).$$

Since $M(t) \geqslant 0$, the *first* integral on the right is

$$\leqslant Y_{\mathscr{D}}(0)\int_0^\infty \frac{M(t)}{t^2}\mathrm{d}t$$

by the theorem of article 2. In view of our assumption on $M(t)$, the *second* right-hand integral can be rewritten

$$-\int_0^\infty\int_0^\infty \log\left|\frac{t+x}{t-x}\right|\mathrm{d}\rho(x)\,\mathrm{d}(t\Omega_{\mathscr{D}}(t)) = -E(\mathrm{d}\rho(t),\ \mathrm{d}(t\Omega_{\mathscr{D}}(t))).$$

Using Schwarz' inequality on the *positive definite* bilinear form $E(\ ,\ )$ (see remark, end of §B.5), we see that the last expression is in modulus

$$\leqslant \sqrt{(E(\mathrm{d}\rho(t),\ \mathrm{d}\rho(t)))}\cdot\sqrt{(E(\mathrm{d}(t\Omega_{\mathscr{D}}(t)),\ \mathrm{d}(t\Omega_{\mathscr{D}}(t))))}$$

which, by the result of the preceding article, is $\leqslant \|M(x)/x\|_E\sqrt{\pi}\,Y_{\mathscr{D}}(0)$. Putting our two estimates together, we get

$$\boxed{\int_{\partial\mathscr{D}} M(t)\mathrm{d}\omega_{\mathscr{D}}(t,0) \leqslant Y_{\mathscr{D}}(0)\left\{\int_0^\infty \frac{M(t)}{t^2}\mathrm{d}t + \sqrt{\pi}\left\|\frac{M(x)}{x}\right\|_E\right\}.}$$

As we have seen $v(0) - AY_{\mathscr{D}}(0)$ is $\leqslant$ the left-hand integral. The theorem is thus proved.

Remark. This result shows that for special majorants $M(t)$ of the kind described, the *entire dependence* of our bound for $v(0)$ *on the domain* $\mathscr{D}$ is *expressed through the quantity* $Y_{\mathscr{D}}(0)$, $Y_{\mathscr{D}}$ being the Phragmén–Lindelöf function for $\mathscr{D}$.

5. When majorant is the logarithm of an entire function of exponential type

The result in the preceding article can be extended so as to apply to certain even majorants $M(x)$ of the form

$$x\int_0^\infty \log\left|\frac{x+t}{x-t}\right| \mathrm{d}\rho(t)$$

for which the iterated integral

$$\int_0^\infty \int_0^\infty \log\left|\frac{x+t}{x-t}\right| \mathrm{d}\rho(t)\,\mathrm{d}\rho(x)$$

is *not* absolutely convergent. This can, in particular, be done in the important special case where

$$M(x) \;=\; \log|G(x)|$$

with an entire function G of exponential type, 1 at 0, having *even* modulus $\geqslant 1$ on $\mathbb{R}$, and such that

$$\int_{-\infty}^\infty \frac{\log|G(x)|}{1+x^2}\mathrm{d}x \;<\; \infty.$$

Then the right side of the boxed formula at the end of the previous article can be simplified so as to involve only $Y_{\mathscr{D}}(0)$, $\int_0^\infty (M(t)/t^2)\mathrm{d}t$, and the type of G.

The treatment of *any* majorant $M(x)$, *even or not*, of the form $\log^+|F(x)|$ with F entire, of exponential type, and such that

$$\int_{-\infty}^\infty \frac{\log^+|F(x)|}{1+x^2}\mathrm{d}x \;<\; \infty,$$

can be reduced to that of one of the kind just described. Indeed, to any such $M(x)$ corresponds another, $M_1(x) = \log|G(x)|$ with G entire and of exponential type, such that

$$M_1(x) \;\geqslant\; M(x) \qquad \text{for } |x| \geqslant 1,$$

$$M_1(x) \;=\; M_1(-x) \;\geqslant\; 0,$$

$$M_1(0) \;=\; 0,$$

and

$$\int_0^\infty \frac{M_1(x)}{x^2}\mathrm{d}x \;<\; \infty.$$

To see this, put first of all

$$\Phi(z) \;=\; 1+z^2[F(z)\overline{F(\bar{z})}+F(-z)\overline{F(-\bar{z})}];$$

$\Phi(z)$ is then entire and of exponential type, even, and $\geqslant 1$ on $\mathbb{R}$ with $\Phi(0)=1$. Clearly

$$\int_{-\infty}^{\infty}\frac{\log\Phi(x)}{1+x^2}dx \;<\; \infty$$

in view of the similar property of F, and

$$\Phi(x) \;\geqslant\; |F(x)|^2 \qquad \text{for } |x|\geqslant 1.$$

By the Riesz–Fejer theorem (the *third* one in §G.3 of Chapter III), there is an entire function $G(z)$ of exponential type, *having all its zeros in* $\Im z<0$ (since here $\Phi(x)\geqslant 1$), such that

$$\Phi(z) \;=\; G(z)\overline{G(\bar{z})}.$$

The majorant $M_1(x)=\log|G(x)|$ then has the required properties.

The result to be obtained in this article regarding even majorants $\log|G(x)|$ of the abovementioned kind can thus be used in studying problems involving the more general ones of the form $\log^+|F(x)|$.

For entire functions $G(z)$ of exponential type with $G(0)=1$, $|G(x)|=|G(-x)|\geqslant 1$, and

$$\int_{-\infty}^{\infty}\frac{\log|G(x)|}{1+x^2}dx \;<\; \infty,$$

$\log|G(x)|$ has a simple representation as a Stieltjes integral. When dealing only with the *modulus* of G on $\mathbb{R}$, we may, by the *second* theorem of §G.3, Chapter III, *assume that* $G(z)$ *has all its zeros in the lower half plane.* Forming, for the moment, the entire function $\Phi(z)=G(z)\overline{G(\bar{z})}$, we see that $\Phi(x)=\Phi(-x)$ on $\mathbb{R}$ so that $\Phi(z)=\Phi(-z)$, and every zero of $\Phi(z)$ is also one of $\Phi(-z)$. The zeros of $\Phi(z)$ are just those of $G(z)$ together with their *complex conjugates*, so, since all the former lie in $\Im z<0$, we have $G(-\bar{\lambda})=0$ whenever $G(\lambda)=0$. The zeros of $G(z)$ thus fall into three groups: those on the *negative imaginary axis*, those in the *open fourth quadrant*, and the *reflections* of *these latter ones* in the *imaginary axis.* The Hadamard factorization (Chapter III, §A) of $G(z)$ can therefore be written

$$G(z)=e^{\alpha z}\prod_k\left(1+\frac{z}{i\mu_k}\right)e^{iz/\mu_k}\cdot\prod_n\left(1-\frac{z}{\bar{\lambda}_n}\right)e^{z/\bar{\lambda}_n}\left(1+\frac{z}{\lambda_n}\right)e^{-z/\lambda_n},$$

where the $\mu_k > 0$, $\Re\lambda_n > 0$ and $\Im\lambda_n > 0$. One (or even both!) of the two products occurring on the right may of course be empty.

Since $|G(x)| = |G(-x)|$, α is pure imaginary. We also know, by the *first* theorem of §G.3, Chapter III, that

$$\sum_k \frac{1}{\mu_k} \quad \text{and} \quad \sum_n \frac{\Im\lambda_n}{|\lambda_n|^2}$$

both converge. The exponential factors figuring in the above product may therefore be grouped together and multiplied out separately, after which the expression takes the form

$$e^{ibz}\prod_k\left(1+\frac{z}{i\mu_k}\right)\cdot\prod_n\left(1-\frac{z}{\bar{\lambda}_n}\right)\left(1+\frac{z}{\lambda_n}\right),$$

with b real. Here, we are only concerned with the *modulus* $|G(x)|$, $x \in \mathbb{R}$; *we may hence take* $b = 0$. *This we do throughout the remainder of this article, working exclusively with entire functions of exponential type of the form*

$$G(z) = \prod_k\left(1+\frac{z}{i\mu_k}\right)\cdot\prod_n\left(1-\frac{z}{\bar{\lambda}_n}\right)\left(1+\frac{z}{\lambda_n}\right),$$

where the $\mu_k > 0$, $\Re\lambda_n > 0$ *and* $\Im\lambda_n > 0$. The products on the right are of course assumed to be *convergent*. Our Stieltjes integral representation for such functions is provided by the

Lemma. *Let* $G(z)$, *of exponential type, be of the form just described. Then, for* $\Im z > 0$,

$$\log|G(z)| = \int_0^\infty \log\left|1-\frac{z^2}{t^2}\right| d\nu(t)$$

with an increasing function $\nu(t)$ *given by*

$$\frac{d\nu(t)}{dt} = \frac{1}{\pi}\left(\sum_k \frac{\mu_k}{\mu_k^2+t^2} + \sum_n\left(\frac{\Im\lambda_n}{|\lambda_n-t|^2}+\frac{\Im\lambda_n}{|\lambda_n+t|^2}\right)\right).$$

Proof. *Fix* z, $\Im z > 0$. Then $\log|1+z/\lambda|$ is a *harmonic function of* λ in $\{\Im\lambda > 0\}$, *bounded* therein for λ away from 0, and *continuous* up to $\mathbb{R}$ save at $\lambda = 0$ where it has a logarithmic singularity. We can therefore apply Poisson's formula, getting

$$\log\left|1+\frac{z}{\lambda}\right| = \frac{1}{\pi}\int_{-\infty}^{\infty}\log\left|1-\frac{z}{t}\right|\cdot\frac{\Im\lambda}{|\lambda+t|^2}\,dt$$

for $\Im\lambda > 0$, from which

$$\log\left|1+\frac{z}{\lambda}\right| + \log\left|1-\frac{z}{\bar{\lambda}}\right|$$
$$= \frac{1}{\pi}\int_0^\infty \log\left|1-\frac{z^2}{t^2}\right|\left(\frac{\Im\lambda}{|\lambda+t|^2}+\frac{\Im\lambda}{|\lambda-t|^2}\right)\mathrm{d}t.$$

Similarly, for $\mu > 0$,

$$\log\left|1+\frac{z}{\mathrm{i}\mu}\right| = \frac{1}{\pi}\int_0^\infty \log\left|1-\frac{z^2}{t^2}\right|\frac{\mu}{\mu^2+t^2}\mathrm{d}t.$$

We have

$$\log|G(z)| = \sum_k \log\left|1+\frac{z}{\mathrm{i}\mu_k}\right| + \sum_n\left(\log\left|1+\frac{z}{\lambda_n}\right| + \log\left|1-\frac{z}{\bar{\lambda}_n}\right|\right).$$

When $\Im z > 0$, we can rewrite each of the terms on the right using the formulas just given, obtaining a certain *sum of integrals*. If $|\Re z| < \Im z$, the *order* of summation and integration in that sum can be *reversed*, for then

$$\log\left|1-\frac{z^2}{t^2}\right| \geqslant 0, \qquad t\in\mathbb{R}.$$

This gives

$$\log|G(z)| = \int_0^\infty \log\left|1-\frac{z^2}{t^2}\right|\mathrm{d}\nu(t),$$

at least for $|\Re z| < \Im z$, with $\nu'(t)$ as in the statement of the lemma.

Both sides of the relation just found are, however, *harmonic in z* for $\Im z > 0$; the *left* one by our assumption on $G(z)$ and the *right* one because $\int_0^\infty \log|1+y^2/t^2|\mathrm{d}\nu(t)$, being just equal to $\log|G(\mathrm{i}y)|$ for $y>0$, is *convergent* for every such y. (To show that this implies u.c.c. convergence, and hence harmonicity, of the integral involving z for $\Im z > 0$, one may argue as at the beginning of the proof of the second theorem in §A, Chapter III.) The two sides of our relation, equal for $|\Re z| < \Im z$, must therefore coincide for $\Im z > 0$ and finally for $\Im z \geqslant 0$ by a continuity argument.

Remark. Since $G(z)$ has no zeros for $\Im z \geqslant 0$, a branch of $\log G(z)$, and hence of $\arg G(z)$, is defined there. By logarithmic differentiation of the above boxed product formula for $G(z)$, it is easy to check that

$$\frac{\mathrm{d}\arg G(t)}{\mathrm{d}t} = -\pi\nu'(t)$$

with the ν of the lemma. From this it is clear that $\nu'(t)$ is certainly *continuous* (and even $\mathscr{C}_\infty$) on $\mathbb{R}$.

In what follows, we will take $\nu(0) = 0$, $\nu(t)$ being the increasing function in the lemma. Since $\nu'(t)$ is *clearly even*, $\nu(t)$ *is then odd*. With $\nu(t)$ thus specified, we have the easy

Lemma. *If* $G(z)$, *given by the above boxed formula, is of exponential type, the function* $\nu(t)$ *corresponding to it is* $\leqslant$ const.t *for* $t \geqslant 0$.

Proof. By the preceding lemma,

$$\int_0^\infty \log\left|1 - \frac{z^2}{t^2}\right| \mathrm{d}\nu(t) \;=\; \log|G(z)|$$

for $\Im z \geqslant 0$, the right side being $\leqslant K|z|$ by hypothesis, since $G(0) = 1$. Calling the left-hand integral $U(z)$, we have, however, $U(z) = U(\bar{z})$, so

$$U(z) \;\leqslant\; K|z|$$

for all z.

Reasoning as in the proof of Jensen's formula, Chapter I (what we are dealing with here is indeed nothing but a version of that formula for the subharmonic function $U(z)$), we see, for $t \neq 0$, that

$$\frac{1}{2\pi}\int_{-\pi}^{\pi} \log\left|1 - \frac{re^{i\vartheta}}{t}\right| \mathrm{d}\vartheta \;=\; \begin{cases} \log\dfrac{r}{|t|}, & |t| < r, \\ 0, & |t| \geqslant r. \end{cases}$$

Thence, by Fubini's theorem,

$$\frac{1}{2\pi}\int_{-\pi}^{\pi} U(re^{i\vartheta})\,\mathrm{d}\vartheta \;=\; \int_{-r}^{r} \log\frac{r}{|t|}\,\mathrm{d}\nu(t).$$

Integrating the right side by parts, we get the value $2\int_0^r (\nu(t)/t)\,\mathrm{d}t$, $\nu(t)$ being odd and $\nu'(0)$ finite. In view of the above inequality on $U(z)$, we thus have

$$\int_0^r \frac{\nu(t)}{t}\,\mathrm{d}t \;\leqslant\; \tfrac{1}{2}Kr.$$

From this relation we easily deduce that $\nu(r) \;\leqslant\; \frac{1}{2}\mathrm{e}Kr$ as in problem 1, Chapter I. Done.

Using the two results just proved in conjunction with the *first* lemma of §B.4, we now obtain, without further ado, the

Theorem. *Let the entire function* $G(z)$ *of exponential type be given by the above boxed formula, and let* $\nu(t)$ *be the increasing function associated to* G *in the way described above. Then, for* $x > 0$,

$$\log|G(x)| \;=\; -x\int_0^\infty \log\left|\frac{x+t}{x-t}\right| \mathrm{d}\left(\frac{\nu(t)}{t}\right).$$

For our functions $G(z)$, $(\log|G(x)|)/x$ is thus a *Green potential* on $(0, \infty)$. *This makes it possible for us to apply the result of the preceding article to majorants*

$$M(t) = \log|G(t)|.$$

With that in mind, let us give a more quantitative version of the second of the above lemmas.

Lemma. *If* $G(z)$, *given by the above boxed formula, is* $\geqslant 1$ *in modulus on* $\mathbb{R}$ *and of exponential type* a, *the increasing function* $\nu(t)$ *associated to it satisfies*

$$\frac{\nu(t)}{t} \leqslant \frac{e}{2}a + \frac{e}{\pi}\int_{-\infty}^{\infty} \frac{\log|G(x)|}{x^2}dx. \qquad t \geqslant 0.$$

Remark. We are not striving for a best possible inequality here.

Proof of lemma. The function $U(z)$ used in proving the previous lemma is *subharmonic* and $\leqslant K|z|$. Assuming that

$$\int_{-\infty}^{\infty} \frac{\log|G(t)|}{t^2}dt < \infty$$

(the only situation we need consider), let us find an explicit estimate for K.

Under our assumption, we have, for $\Im z > 0$,

$$\log|G(z)| \leqslant a\Im z + \frac{1}{\pi}\int_{-\infty}^{\infty} \frac{\Im z}{|z-t|^2}\log|G(t)|dt$$

by §E of Chapter III. When $-y \leqslant x \leqslant y$, we have, however, for $z = x + iy$,

$$|z-t|^2 = t^2 - 2xt + x^2 + y^2 \geqslant \frac{t^2}{2} + \frac{t^2}{2} - 2xt + 2x^2$$

$$\geqslant \frac{t^2}{2}, \qquad t \in \mathbb{R},$$

whence, $\log|G(t)|$ being $\geqslant 0$,

$$\log|G(z)| \leqslant ay + \frac{2y}{\pi}\int_{-\infty}^{\infty} \frac{\log|G(t)|}{t^2}dt.$$

Thus, since $U(z) = U(\bar{z}) = \log|G(z)|$ for $\Im z \geqslant 0$,

$$U(z) \leqslant \left(a + \frac{2}{\pi}\int_{-\infty}^{\infty} \frac{\log|G(t)|}{t^2}dt\right)|\Im z|$$

in both of the sectors $|\Re z| \leqslant |\Im z|$.

Because $U(z) \leqslant \text{const.}|z|$ we can apply the second Phragmén–Lindelöf theorem of §C, Chapter III, to the difference

$$U(z) \quad - \quad \left(a + \frac{2}{\pi} \int_{-\infty}^{\infty} \frac{\log|G(t)|}{t^2} \mathrm{d}t \right) \Re z$$

in the 90° sector $|\Im z| \leqslant \Re z$, and find that it is $\leqslant 0$ in that sector. One proceeds similarly in $\Re z \leqslant -|\Im z|$, and we have

$$U(z) \quad \leqslant \quad \left(a + \frac{2}{\pi} \int_{-\infty}^{\infty} \frac{\log|G(t)|}{t^2} \mathrm{d}t \right) |\Re z|$$

for $|\Im z| \leqslant |\Re z|$.

Combining the two estimates for $U(z)$ just found, we get

$$U(z) \quad \leqslant \quad K|z|$$

with

$$K \quad = \quad a + \frac{2}{\pi} \int_{-\infty}^{\infty} \frac{\log|G(t)|}{t^2} \mathrm{d}t.$$

This value of K may now be plugged into the *proof* of the previous lemma. That yields the desired result.

Problem 27

Let $\Phi(z)$ be entire and of exponential type, with $\Phi(0) = 1$. *Suppose that* $\Phi(z)$ *has all its zeros in* $\Im z < 0$ *and that* $|\Phi(x)| \geqslant 1$ *on* $\mathbb{R}$. Show that *then*

$$\int_{-\infty}^{\infty} \frac{\log|\Phi(x)|}{x^2} \mathrm{d}x \quad < \quad \infty.$$

(Hint: First use Lindelöf's theorem from Chapter III, §B, to show that the Hadamard factorization for $\Phi(z)$ can be cast in the form

$$\Phi(z) \quad = \quad \mathrm{e}^{cz} \prod_n \left(1 - \frac{z}{\lambda_n} \right) \mathrm{e}^{z\Re(1/\lambda_n)},$$

where the $\Im\lambda_n < 0$. Taking $\Psi(z) = \Phi(z)\exp(-\mathrm{i}z\Im c)$, show that $\partial \log|\Psi(z)|/\partial y \geqslant 0$ for $y \geqslant 0$, and then look at $1/\Psi(z)$.)

Suppose now that we have an entire function $G(z)$ given by the above boxed representation, of exponential type a and $\geqslant 1$ in modulus on $\mathbb{R}$. *If the double integral*

$$\int_0^{\infty} \int_0^{\infty} \log\left|\frac{x+t}{x-t}\right| \mathrm{d}\left(\frac{\nu(t)}{t}\right) \mathrm{d}\left(\frac{\nu(x)}{x}\right)$$

is absolutely convergent, we may, as in the previous two articles, speak of the *energy*

$$E\left(\mathrm{d}\left(\frac{\nu(t)}{t}\right),\ \mathrm{d}\left(\frac{\nu(t)}{t}\right)\right);$$

in terms of the Green potential

$$\frac{\log|G(x)|}{x} = -\int_0^\infty \log\left|\frac{x+t}{x-t}\right| \mathrm{d}\left(\frac{\nu(t)}{t}\right),$$

this is just $\|(\log|G(x)|)/x\|_E^2$ according to the notation introduced at the end of article 3.

To Beurling and Malliavin is due the important observation that $\|(\log|G(x)|)/x\|_E$ can be expressed in terms of a and $\int_0^\infty (\log|G(x)|/x^2)\mathrm{d}x$ under the present circumstances. Since $\log|G(t)| \geqslant 0$ and $\nu(t)$ increases, we have indeed

$$\begin{aligned}\|(\log|G(x)|)/x\|_E^2 &= -\int_0^\infty \frac{\log|G(x)|}{x}\mathrm{d}\left(\frac{\nu(x)}{x}\right) \\ &= \int_0^\infty \frac{\log|G(x)|}{x^2}\left(\frac{\nu(x)}{x}\mathrm{d}x - \mathrm{d}\nu(x)\right) \\ &\leqslant \left(\sup_{x>0}\frac{\nu(x)}{x}\right)\cdot\int_0^\infty \frac{\log|G(x)|}{x^2}\mathrm{d}x.\end{aligned}$$

Using the preceding lemma and remembering that $|G(x)|$ is even, we find that

$$\boxed{\left\|\frac{\log|G(x)|}{x}\right\|_E^2 \leqslant \left(\frac{ea}{2} + \frac{2e}{\pi}\int_0^\infty \frac{\log|G(x)|}{x^2}\mathrm{d}x\right)\cdot\int_0^\infty \frac{\log|G(x)|}{x^2}\mathrm{d}x.}$$

Take now an even majorant $M(t) \geqslant 0$ equal to $\log|G(t)|$, and consider one of our domains $\mathscr{D}$ with $0 \in \mathscr{D}$. From the result just obtained and the boxed formula near the end of the previous article, we get

$$\int_{\partial\mathscr{D}} M(t)\,\mathrm{d}\omega_{\mathscr{D}}(t,0) \leqslant Y_{\mathscr{D}}(0)\left\{J + \sqrt{\left(2eJ\left(J + \frac{\pi a}{4}\right)\right)}\right\},$$

with

$$J = \int_0^\infty \frac{\log|G(t)|}{t^2}\mathrm{d}t = \int_0^\infty \frac{M(t)}{t^2}\mathrm{d}t,$$

at least in the case where

$$\int_0^\infty \int_0^\infty \log\left|\frac{x+t}{x-t}\right| d\left(\frac{\nu(t)}{t}\right) d\left(\frac{\nu(x)}{x}\right)$$

is absolutely convergent. On the right side of this relation, the coefficient $Y_{\mathscr{D}}(0)$ is multiplied by a factor *involving only a, the type of G, and the integral* $\int_0^\infty (M(t)/t^2)\,dt$ (essentially, the one this book is about!).

It is very important that the requirement of absolute convergence on the above double integral can be lifted, and the preceding relation still remains true. This will be shown by bringing in the *completion*, for the norm $\| \ \|_E$, of the collection of real Green potentials associated with absolutely convergent energy integrals – that completion is a real Hilbert space, since $\| \ \|_E$ comes from a positive definite bilinear form. The details of the argument take up the remainder of this article.

Starting with our entire function $G(z)$ of exponential type and the increasing function $\nu(t)$ associated to it, put

$$Q(x) = \frac{\log|G(x)|}{x} = -\int_0^\infty \log\left|\frac{x+t}{x-t}\right| d\left(\frac{\nu(t)}{t}\right),$$

and, for $n = 1, 2, 3, \ldots,$

$$Q_n(x) = \frac{1}{x}\int_0^n \log\left|1 - \frac{x^2}{t^2}\right| d\nu(t).$$

In terms of

$$\nu_n(t) = \begin{cases} \nu(t), & 0 \leqslant t \leqslant n, \\ \nu(n), & t > n, \end{cases}$$

we have

$$Q_n(x) = -\int_0^\infty \log\left|\frac{x+t}{x-t}\right| d\left(\frac{\nu_n(t)}{t}\right)$$

by the first lemma of §B.4; evidently, $Q_n(x) \longrightarrow Q(x)$ u.c.c. in $[0, \infty)$ as $n \to \infty$.

Each of the integrals

$$\int_0^\infty \int_0^\infty \log\left|\frac{x+t}{x-t}\right| d\left(\frac{\nu_n(t)}{t}\right) d\left(\frac{\nu_n(x)}{x}\right)$$

is absolutely convergent. This is easily verified using the facts that

$$d\left(\frac{\nu_n(t)}{t}\right) \sim \tfrac{1}{2}\nu''(0)\,dt$$

near 0 ($\nu(t)$ being $\mathscr{C}_\infty$ by a previous remark), and that

$$d\left(\frac{\nu_n(t)}{t}\right) = -\frac{\nu(n)}{t^2}dt \qquad \text{for } t > n.$$

Lemma. *If* $|G(x)| \geqslant 1$ *on* $\mathbb{R}$, *the functions* $Q_n(x)$ *are* $\geqslant 0$ *for* $x > 0$, *and*

$$\|Q_n\|_E \leqslant \frac{\pi}{2}\nu'(0).$$

Proof. For $t > 0$, $\log|1 - x^2/t^2| \geqslant 0$ when $x \geqslant \sqrt{2}t$, so

$$xQ_n(x) = \int_0^n \log\left|1 - \frac{x^2}{t^2}\right| d\nu(t)$$

is $\geqslant 0$ for $x \geqslant \sqrt{2}n$. Again, for $0 \leqslant x \leqslant \sqrt{2}t$, $\log|1 - x^2/t^2| \leqslant 0$, so, for $0 \leqslant x \leqslant \sqrt{2}n$,

$$\int_n^\infty \log\left|1 - \frac{x^2}{t^2}\right| d\nu(t) \leqslant 0,$$

and finally $xQ_n(x)$, equal to $\log|G(x)|$ *minus* this integral, is $\geqslant 0$ since $|G(x)| \geqslant 1$.

The second lemma of §B.4 is applicable to the functions $\nu_n(t)$. Using it and the positivity of $Q_n(x)$, already established, we get

$$\begin{aligned} \|Q_n\|_E^2 &= \int_0^\infty \int_0^\infty \log\left|\frac{x+t}{x-t}\right| d\left(\frac{\nu_n(t)}{t}\right) d\left(\frac{\nu_n(x)}{x}\right) \\ &= \int_0^\infty Q_n(x)\left\{\frac{\nu_n(x)}{x^2}dx - \frac{d\nu_n(x)}{x}\right\} \\ &\leqslant \int_0^\infty Q_n(x)\frac{\nu_n(x)}{x^2}dx = \frac{\pi^2}{4}(\nu_n'(0))^2 = \frac{\pi^2}{4}(\nu'(0))^2. \end{aligned}$$

We are done.

Theorem. *Let* $G(z)$ *be an entire function of exponential type* a, 1 *at* 0, *with* $|G(x)|$ even *and* $\geqslant 1$ *on* $\mathbb{R}$, *and such that*

$$\int_{-\infty}^\infty \frac{\log|G(x)|}{1+x^2}dx < \infty.$$

If $\mathscr{D}$ *is one of our domains containing* 0, *we have*

$$\int_{\partial\mathscr{D}} \log|G(t)|\,d\omega_{\mathscr{D}}(t,0) \;\leqslant\; Y_{\mathscr{D}}(0)\left\{J + \sqrt{\left(2eJ\left(J+\frac{\pi a}{4}\right)\right)}\right\}$$

where

$$J \;=\; \int_0^\infty \frac{\log|G(t)|}{t^2}\,dt.$$

Proof. According to the discussion at the beginning of this article we may, without loss of generality,* assume that $G(z)$ has the above boxed product representation.

Beginning as in the proof of the theorem from the preceding article, we have

$$\int_{\partial\mathscr{D}} \log|G(x)|\,d\omega_{\mathscr{D}}(x,0) \;=\; \int_0^\infty \frac{\log|G(x)|}{x}\Omega_{\mathscr{D}}(x)\,dx$$

$$-\int_0^\infty \frac{\log|G(x)|}{x}\,d(x\Omega_{\mathscr{D}}(x)).$$

The *first* term on the right is of course

$$\leqslant\; Y_{\mathscr{D}}(0)\int_0^\infty \frac{\log|G(x)|}{x^2}\,dx$$

by the theorem of article 2, $\log|G(x)|$ being positive. The *second*, equal to

$$-\int_0^\infty Q(x)\,d(x\Omega_{\mathscr{D}}(x)),$$

can be looked at in two different ways.

In the first place, for $x>0$,

$$Q(x) \;=\; \lim_{n\to\infty} Q_n(x)$$

with the functions $Q_n(x)$ introduced above. Also, for each n,

* Dropping the factor $\exp(ibz)$ from the second displayed expression on p. 557 can only *diminish* the overall exponential type, for, if $G(z)$ is given by the *boxed formula* on that page, the limsups of $\log|G(iy)|/|y|$ for y tending to ∞ and to $-\infty$ are *equal*. To see that, observe that the limsup for $y\to\infty$ is actually a *limit* (see remark, p. 49), and that $\overline{G(\bar{z})}/G(z) = B(z)$ is a Blaschke product like the one figuring in the remark on p. 58. The argument of pp. 57–8 shows, however, that then the limsup of $\log|B(iy)|/y$ for $y\to\infty$ is *zero*.

$$Q_n(x) \leqslant \frac{1}{x}\int_0^n \log\left|1+\frac{x^2}{t^2}\right| \mathrm{d}\nu(t)$$

$$\leqslant \frac{1}{x}\int_0^\infty \log\left|1+\frac{x^2}{t^2}\right| \mathrm{d}\nu(t), \qquad x>0.$$

Since $\nu(t) \leqslant Kt$, the right-hand member comes out $\leqslant \pi K$ on integrating by parts. This, together with the preceding lemma, shows that

$$0 \leqslant Q_n(x) \leqslant \pi K \qquad \text{for } x>0.$$

However, for *large* x,

$$\mathrm{d}(x\Omega_{\mathscr{D}}(x)) = \left(\frac{\text{const.}}{x^3} + O\left(\frac{1}{x^5}\right)\right)\mathrm{d}x$$

(see just before the theorem of article 3). Therefore

$$\int_0^\infty Q(x)\mathrm{d}(x\Omega_{\mathscr{D}}(x)) = \lim_{n\to\infty}\int_0^\infty Q_n(x)\mathrm{d}(x\Omega_{\mathscr{D}}(x))$$

by dominated convergence.

The right-hand limit can also be expressed as an inner product in a certain real Hilbert space. The latter – call it $\mathfrak{H}$ – is the *completion with respect to the norm* $\| \ \|_E$ of the collection of real Green potentials

$$u(x) = \int_0^\infty \log\left|\frac{x+t}{x-t}\right| \mathrm{d}\rho(t)$$

such that

$$\int_0^\infty\int_0^\infty \log\left|\frac{x+t}{x-t}\right| |\mathrm{d}\rho(t)||\mathrm{d}\rho(x)| < \infty;$$

the positive definite bilinear form $\langle \ , \ \rangle_E$ extends by continuity to $\mathfrak{H}$ for which it serves as inner product. For each n, we have

$$\int_0^\infty Q_n(x)\mathrm{d}(x\Omega_{\mathscr{D}}(x)) = E\left(\mathrm{d}\left(\frac{\nu_n(t)}{t}\right), \mathrm{d}(t\Omega_{\mathscr{D}}(t))\right) = \langle Q_n, P\rangle_E,$$

where

$$P(x) = x(G_{\mathscr{D}}(x,0) + G_{\mathscr{D}}(-x,0)) ;$$

here only Green potentials associated with *absolutely convergent* energy integrals are involved. By the lemma, however,

$$\|Q_n\|_E \leqslant \frac{\pi}{2}\nu'(0),$$

so a *subsequence* of $\{Q_n\}$, which we may as well *also* denote by $\{Q_n\}$, *converges weakly in* $\mathfrak{H}$ *to some element* q *of that space.* (Here, we do *not need* to 'identify' q with the function $Q(x)$, although that can easily be done.) In view of the previous limit relation, we see that

$$\int_0^\infty Q(x)\mathrm{d}(x\Omega_{\mathscr{D}}(x)) = \lim_{n\to\infty} \langle Q_n, P\rangle_E = \langle q, P\rangle_E.$$

Thence, by Schwarz' inequality and the result of article 3,

$$\left|\int_0^\infty Q(x)\mathrm{d}(x\Omega_{\mathscr{D}}(x))\right| \leqslant \|q\|_E\|P\|_E$$
$$= \|q\|_E\sqrt{(E(\mathrm{d}(t\Omega_{\mathscr{D}}(t)), \mathrm{d}(t\Omega_{\mathscr{D}}(t))))} \leqslant \sqrt{\pi Y_{\mathscr{D}}(0)}\|q\|_E.$$

Returning to the beginning of this proof, we see that

$$\int_{\partial\mathscr{D}} \log|G(x)|\mathrm{d}\omega_{\mathscr{D}}(x,0) \leqslant Y_{\mathscr{D}}(0)\int_0^\infty \frac{\log|G(x)|}{x^2}\mathrm{d}x$$
$$+ \sqrt{\pi Y_{\mathscr{D}}(0)}\|q\|_E,$$

and thus need an estimate for $\|q\|_E$. The obvious one, $\|q\|_E \leqslant \liminf_{n\to\infty}\|Q_n\|_E \leqslant \pi\nu'(0)/2$, is not good enough to give us what we want here, so we argue as follows.

The weak convergence of Q_n to q in $\mathfrak{H}$ implies first of all that

$$\|q\|_E^2 = \lim_{n\to\infty} \langle q, Q_n\rangle_E.$$

Fix any n; then, by weak convergence again,

$$\langle q, Q_n\rangle_E = \lim_{k\to\infty}\langle Q_k, Q_n\rangle_E = -\lim_{k\to\infty}\int_0^\infty Q_k(x)\,\mathrm{d}\left(\frac{\nu_n(x)}{x}\right).$$

Here, $\mathrm{d}(\nu_n(x)/x)$ is just $-(\nu(n)/x^2)\mathrm{d}x$ for $x > n$, so, since $0 \leqslant Q_k(x) \leqslant \pi K$, we have, by dominated convergence,

$$-\lim_{k\to\infty}\int_0^\infty Q_k(x)\,\mathrm{d}\left(\frac{\nu_n(x)}{x}\right) = -\int_0^\infty Q(x)\,\mathrm{d}\left(\frac{\nu_n(x)}{x}\right)$$

which, $Q(x)$ being positive, is

$$\leqslant \int_0^\infty Q(x)\frac{\nu_n(x)}{x^2}\mathrm{d}x.$$

Again, $\nu_n(x) \leqslant \nu(x)$ for $x \geqslant 0$, so finally

$$\langle q, Q_n\rangle_E \leqslant \int_0^\infty Q(x)\frac{\nu(x)}{x^2}\mathrm{d}x = \int_0^\infty \frac{\log|G(x)|}{x^2}\frac{\nu(x)}{x}\mathrm{d}x$$

for each fixed n. The right-hand integral was already estimated above, before the preceding lemma, and found to be

$$\leqslant \frac{2e}{\pi}\left(\frac{\pi a}{4} + \int_0^\infty \frac{\log|G(x)|}{x^2}dx\right)\int_0^\infty \frac{\log|G(x)|}{x^2}dx.$$

This quantity is thus $\geqslant \lim_{n\to\infty}\langle q, Q_n\rangle_E = \|q\|_E^2$, giving us an upper bound on $\|q\|_E$.

Substituting the estimate just obtained into the above inequality for $\int_{\partial\mathscr{D}}\log|G(x)|d\omega_{\mathscr{D}}(x,0)$, we have the theorem. The proof is complete.

Corollary. *Let $G(z)$ and the domain $\mathscr{D}$ be as in the hypothesis of the theorem. If $v(z)$, subharmonic in $\mathscr{D}$ and continuous up to $\partial\mathscr{D}$, satisfies*

$$v(t) \leqslant \log|G(t)|, \qquad t\in\partial\mathscr{D},$$

and

$$v(z) \leqslant A|\Im z| + O(1)$$

with some real A, we have

$$v(0) \leqslant Y_{\mathscr{D}}(0)\left\{A + J + \sqrt{\left(2eJ\left(J+\frac{\pi a}{4}\right)\right)}\right\},$$

where

$$J = \int_0^\infty \frac{\log|G(x)|}{x^2}dx$$

and a is the type of G.

This result will be used in proving the Beurling–Malliavin multiplier theorem in Chapter XI.

Problem 28

Let $G(z)$, entire and of exponential type, be given by the above boxed product formula and satisfy the hypothesis of the preceding theorem. Suppose also that

$$\frac{\log|G(iy)|}{|y|} \longrightarrow a \quad \text{for} \quad y \longrightarrow \pm\infty.$$

The purpose of this problem is to improve the estimate of $\|(\log|G(x)|)/x\|_E$ obtained above.

(a) Show that $v'(0) = a/\pi + 2J/\pi^2$ and that $v(t)/t \longrightarrow a/\pi$ as $t\to\infty$. Here, J has the same meaning as in the statement of the theorem.
(Hint. For the second relation, one may just indicate how to adapt the argument from §H.2 of Chapter III.)

(b) Show that

$$\int_0^\infty \frac{\log|G(x)|}{x}\frac{\nu(x)}{x^2}\,dx = \frac{\pi^2}{4}\left((\nu'(0))^2 - \left(\lim_{t\to\infty}\frac{\nu(t)}{t}\right)^2\right).$$

(Hint. Integral on left is the *negative* of

$$\int_0^\infty\int_0^\infty \log\left|\frac{x+t}{x-t}\right| d\left(\frac{\nu(t)}{t}\right)\frac{\nu(x)}{x^2}\,dx.$$

Here, direct application of the method used to prove the second lemma of §B.4 is hampered by $(d/dt)(\nu(t)/t)$'s lack of regularity for large t; however, the following procedure works and is quite general.

For small $\delta > 0$ and large L one can get ε, $0 < \varepsilon < \delta$, and $R > L$ making

$$\int_\delta^L\int_\varepsilon^R \log\left|\frac{x+t}{x-t}\right| d\left(\frac{\nu(t)}{t}\right)\frac{\nu(x)}{x^2}\,dx$$

nearly equal to the above iterated integral. The order of integration can now be reversed and then the second mean value theorem applied to show that $\int_\varepsilon^\delta\int_\delta^L$ and $\int_L^R\int_\delta^L$ are *both small in magnitude* when $\delta > 0$ is small and L large. Our initial expression is thus closely approximated by

$$\int_\delta^L\int_\delta^L \log\left|\frac{x+t}{x-t}\right|\frac{\nu(x)}{x^2}\,dx\,d\left(\frac{\nu(t)}{t}\right).$$

Apply to this a suitable modification of the reasoning in the proof of the aforementioned lemma, and then make $\delta \to 0$, $L \to \infty$.)

(c) Hence show that

$$\int_0^\infty \frac{\log|G(x)|}{x^2}\frac{\nu(x)}{x}\,dx = \frac{1}{\pi^2}J(J+\pi a)$$

so that

$$\int_0^\infty\left(\int_0^\infty \log\left|\frac{x+t}{x-t}\right| d\left(\frac{\nu(t)}{t}\right)\right) d\left(\frac{\nu(x)}{x}\right) \leqslant \frac{1}{\pi^2}J(J+\pi a).$$

Addendum

Improvement of Volberg's Theorem on the Logarithmic Integral. Work of Brennan, Borichev, Jöricke and Volberg.

Writing of §D in Chapter VII was completed early in 1984, and some copies of the MS were circulated that spring. At the beginning of 1987 I learned, first from V.P. Havin and then from N.K. Nikolskii, that the persons named in the title had extended the theorem of §D.6. Expositions of their work did not come into my hands until April and May of 1987, when I had finished going through the second proof sheets for this volume.

In these circumstances, time and space cannot allow for inclusion of a thorough presentation of the recent work here. It nevertheless seems important to describe *some* of it because the strengthened version of Volberg's theorem first obtained by Brennan is very likely close to being best possible. I am thankful to Nikolskii, Volberg and Borichev for having made sure that the material got to me in time for me to be able to include the following account.

The development given below is based on the methods worked out in §D of Chapter VII, and familiarity with that § on the part of the reader is assumed. In order to save space and avoid repetition, we will refer to §D frequently and use the symbols employed there whenever possible.

1. Brennan's improvement, for $M(\nu)/\nu^{1/2}$ monotone increasing

Let us return to the proof of the theorem in §D.6 of Chapter VII, starting from the place on p. 359 where $h(\xi)$ and the weight $w(r) = \exp(-h(\log(1/r)))$ were brought into play. We take over the notation used in that discussion without explaining it anew.

What is shown by the reasoning of pp. 359–73 is that *unless* $F(e^{i\vartheta})$

vanishes identically,

$$\int_{-\pi}^{\pi} \log|F(e^{i\vartheta})|\, d\vartheta \;>\; -\infty$$

provided that

$$h(\xi) \;\geqslant\; \text{const.}\ \xi^{-(1+\delta)}$$

with some $\delta > 0$ as $\xi \longrightarrow 0$, and that

$$\int_0^a \log h(\xi)\, d\xi \;=\; \infty$$

for small $a > 0$. *Brennan's result is that the first condition on h can be replaced by the requirement that* $\xi h(\xi)$ *be decreasing for small* $\xi > 0$. (The second condition then obviously implies that $\xi h(\xi) \longrightarrow \infty$ as $\xi \longrightarrow 0$.)

Borichev and Volberg made the important observation that Brennan's result is yielded by Volberg's original argument. To see how this comes about, we begin by noting that in §D.6 of Chapter VII, *no real use of the property* $h(\xi) \geqslant$ const. $\xi^{-(1+\delta)}$ *is made until one comes to step* 5 on p. 369. Up to then, it is more than enough to have $h(\xi) \geqslant$ const. ξ^{-c} with *some* $c > 0$ together with the integral condition on $\log h(\xi)$. *Step 5* itself, however, is carried out in rather clumsy fashion (see p. 370). The reader was probably aware of this, and especially of the wasteful manner of using that step's conclusion in the subsequent local estimate of $\omega(E, z)$ (pp. 370–2). At the top of p. 372, the smallness of $\int_{\gamma_\rho}(1/(1-|\zeta|))\, d\omega(\zeta, \rho)$ was used where its *smallness in relation to* $1/(1-\rho)$ would have sufficed!

Instead of verifying the conclusion of *step* 5, let us show that *the quantity*

$$(1-\rho)\int_{\gamma_\rho} \frac{d\omega(\zeta, \rho\zeta_0)}{1-|\zeta|}$$

can be made as small as we please for ρ sufficiently close to 1 chosen according to the specifications at the bottom of p. 368, *under the assumption* that $\xi h(\xi)$ *decreases*, with the integral of $\log h(\xi)$ *divergent.*

The original argument for *step 5* is *unchanged* up to the point where the relation

$$(*) \qquad \int_{\gamma_\rho} h\left(\log\frac{1}{|\zeta|}\right) d\omega(\zeta, \rho) \;\leqslant\; \text{const.} \;+\; (h(\log(1/\rho^2)))^{\eta}$$

is obtained at the top of p. 370; here η can be chosen *at pleasure* in the interval (0, 1), the construction following *step 3* (pp. 365–6) and subsequent carrying out of *step 4* being in no way hindered. Write now

$$P(\xi) \;=\; \xi h(\xi);$$

under the present circumstances $P(\xi)$ is *decreasing* for small $\xi > 0$. Since γ_ρ, recall, lies in the ring $\{\rho^2 \leqslant |\zeta| < 1\}$, we then have, for ρ near 1,

$$\int_{\gamma_\rho} \frac{d\omega(\zeta,\rho)}{1-|\zeta|} \;\leqslant\; 2\int_{\gamma_\rho} \frac{d\omega(\zeta,\rho)}{\log(1/|\zeta|)} \;=\; 2\int_{\gamma_\rho} \frac{h(\log(1/|\zeta|))}{P(\log(1/|\zeta|))}\, d\omega(\zeta,\rho)$$

$$\leqslant\; \frac{2}{P(2\log(1/\rho))}\int_{\gamma_\rho} h(\log(1/|\zeta|))\,d\omega(\zeta,\rho).$$

Referring to (*), we see that the last expression is

$$\leqslant\; \frac{2}{P(2\log(1/\rho))}\{\text{const.} \;+\; (h(2\log(1/\rho)))^\eta\}.$$

Here, the monotoneity of $P(\xi)$ makes it tend to ∞ for $\xi \longrightarrow 0$; otherwise $\int_0^a \log h(\xi)\,d\xi$ would be *finite* for small $a > 0$ as already remarked. The function $h(\xi)$ also tends to ∞ for $\xi \longrightarrow 0$, so, for ρ close to 1 the preceding quantity is

$$\leqslant\; 3\left\{\frac{h(2\log(1/\rho))}{P(2\log(1/\rho))}\right\}^\eta \;=\; \frac{3}{(\log(1/\rho^2))^\eta} \;\leqslant\; \frac{3}{(1-\rho)^\eta}.$$

We thus have

$$\int_{\gamma_\rho} \frac{d\omega(\zeta,\rho)}{1-|\zeta|} \;\leqslant\; 3(1-\rho)^{-\eta} \;=\; o(1/(1-\rho))$$

for values of ρ tending to 1 chosen in the way mentioned above, and our substitute for *step 5* is established.

This, as already noted, is all we need for the reasoning at the top of p. 372. The local estimate for $\omega(E,\rho)$ obtained on pp. 370–2 is therefore valid, and proof of the relation

$$\int_{-\pi}^{\pi} \log|F(e^{i\vartheta})|\,d\vartheta \;>\; -\infty$$

is completed as on pp. 372–3.

It may well appear that the argument just made did not make full use of the monotoneity of $\xi h(\xi)$. However that may be, this requirement does not seem capable of further significant relaxation, as we shall see in the next two articles. At present, let us translate our conclusion into a result involving the majorant $M(\nu)$ figuring in Volberg's theorem (p. 356).

In the statement of that theorem, two regularity properties are required of the increasing function $M(\nu)$ in addition to the divergence of $\sum_1^\infty M(n)/n^2$, namely, *that $M(\nu)/\nu$ be decreasing* and that

$$M(\nu) \geqslant \text{const.}\, \nu^\alpha$$

for large ν, where $\alpha > 1/2$. The *first* of these properties is (for us) practically equivalent to *concavity* of $M(\nu)$ by the theorem on p. 326. The concavity is needed for Dynkin's theorem (p. 339) and is not at issue here. Our interest is in replacing the *second* property by a *weaker* one. That being the object, there is no point in trying to gild the lily, and we may as well phrase our result for *concave majorants* $M(\nu)$. Indeed, nothing is really lost by sticking to *infinitely differentiable ones* with $M''(\nu) < 0$ and $M'(\nu) \longrightarrow 0$ for $\nu \longrightarrow \infty$, as long as that simplifies matters. See the theorem, p. 326 and the subsequent discussion on pp. 328–30; see also the beginning of the proof of the theorem in the next article.

With this simplification granted, passage from the result just arrived at to one stated in terms of $M(\nu)$ is provided by the easy

Lemma. Let $M(\nu)$ be infinitely differentiable for $\nu > 0$ with $M''(\nu) < 0$ and $M'(\nu) \longrightarrow 0$ for $\nu \longrightarrow \infty$, and put (as usual)

$$h(\xi) = \sup_{\nu > 0} (M(\nu) - \nu\xi).$$

Then $\xi h(\xi)$ is decreasing for small $\xi > 0$ if and only if $M(\nu)/\nu^{1/2}$ is increasing for large ν.

Proof. Under the given conditions, when $\xi > 0$ is sufficiently small, $h(\xi) = M(\nu) - \nu\xi$ for the *unique* ν with $M'(\nu) = \xi$ by the lemmas on pp. 330 and 332. Thus,

$$M'(\nu)h(M'(\nu)) = M(\nu)M'(\nu) - \nu(M'(\nu))^2,$$

so, since $M'(\nu)$ tends monotonically to zero as $\nu \longrightarrow \infty$, $\xi h(\xi)$ is *decreasing* for small $\xi > 0$ if and only if the right side of the last relation is *increasing* for large ν. But

$$\frac{\mathrm{d}}{\mathrm{d}\nu}(M(\nu)M'(\nu) - \nu(M'(\nu))^2) = M''(\nu)M(\nu) - 2\nu M''(\nu)M'(\nu)$$

$$= -2\nu^{3/2}M''(\nu)\frac{\mathrm{d}}{\mathrm{d}\nu}\left(\frac{M(\nu)}{\nu^{1/2}}\right).$$

Since $M''(\nu) < 0$, the lemma is clear.

Referring now to the above result, we get, almost without further ado, the

Theorem (Brennan). Let $M(v)$ be infinitely differentiable for $v > 0$, with $M''(v) < 0$,

$$\frac{M(v)}{v^{1/2}} \text{ increasing for large } v,$$

and

$$\sum_1^\infty M(n)/n^2 = \infty.$$

Suppose that

$$F(e^{i\vartheta}) \sim \sum_{-\infty}^{\infty} a_n e^{in\vartheta}$$

is continuous, with

$$|a_n| \leqslant \text{const.}\, e^{-M(|n|)} \qquad \text{for } n < 0.$$

Then, unless $F(e^{i\vartheta})$ vanishes identically,

$$\int_{-\pi}^{\pi} \log|F(e^{i\vartheta})|\,d\vartheta > -\infty.$$

Indeed, this follows directly by the lemma unless $\lim_{v\to\infty} M'(v) > 0$. Then, however, the theorem is true anyway – see p. 328.

2. Discussion

Brennan's result *really is* more general than the theorem on p. 356. That's because the hypothesis of the former one is fulfilled for any function $F(e^{i\vartheta})$ satisfying the hypothesis of the latter, thanks to the following

Theorem. Let $M(v)$, increasing and with $M(v)/v$ decreasing, satisfy the condition $\sum_1^\infty M(n)/n^2 = \infty$ and have $M(v) \geqslant \text{const.}\, v^{\frac{1}{2}+\delta}$ for large v, where $\delta > 0$. Then there is an infinitely differentiable function $M_0(v)$, with $M_0''(v) < 0$,

$$M_0(v) \leqslant M(v) \text{ for large } v,$$

$M_0(v)/v^{1/2}$ increasing, and $\sum_1^\infty M_0(n)/n^2 = \infty$.

Proof. By the theorem on p. 326 we can, wlog, take $M(v)$ to be *actually concave*. It is then sufficient to obtain *any concave minorant* $M_*(v)$ of $M(v)$

with $M_*(\nu)/\nu^{1/2}$ *increasing* and $\int_1^\infty (M_*(\nu)/\nu^2)d\nu$ *divergent*, for from such a minorant one easily obtains an $M_0(\nu)$ with the additional regularity affirmed by the theorem.

The procedure for doing this is like the one of pp. 229–30. Starting with an $M_*(\nu)$, one first puts $M_1(\nu) = M_*(\nu) + \nu^{1/2}$ and then, using a $\mathscr{C}_\infty$ function $\varphi(\tau)$ having the graph shown on p. 329, takes

$$M_0(\nu) = c\int_0^1 M_1(\nu - \tau)\varphi(\tau)\,d\tau$$

for $\nu > 1$ with a suitable small constant c. This function $M_0(\nu)$ (defined in any convenient fashion for $0 < \nu \leqslant 1$) is readily seen to do the job.

Our main task is thus the construction of an $M_*(\nu)$. For that it is helpful to make a further reduction, arranging for $M(\nu)$ to have a *piecewise linear graph starting out from the origin.* That poses no problem; we simply replace our *given* concave function $M(\nu)$ by *another*, with graph consisting of a straight segment going from the origin to a point on the graph of the original function followed by suitably chosen *successive chords* of that graph. This having been attended to, we let $R(\nu)$ be the *largest increasing minorant* of $M(\nu)/\nu^{1/2}$ and then put

$$M_*(\nu) = \nu^{1/2}R(\nu);$$

this of course makes $M_*(\nu)/\nu^{1/2}$ automatically increasing and $M_*(\nu) \leqslant M(\nu)$.

Thanks to our initial adjustment to the graph of $M(\nu)$, we have $M(\nu)/\nu^{1/2} \longrightarrow 0$ for $\nu \longrightarrow 0$. Hence, since $M(\nu) \geqslant$ const. $\nu^{\frac{1}{2}+\delta}$ for large ν, $R(\nu)$ must tend to ∞ for $\nu \longrightarrow \infty$, and *coincides* with $M(\nu)/\nu^{1/2}$ *save on certain disjoint intervals* $(\alpha_k, \beta_k) \subset (0, \infty)$ for which

$$\frac{M(\alpha_k)}{\alpha_k^{1/2}} = R(\nu) = \frac{M(\beta_k)}{\beta_k^{1/2}}, \qquad \alpha_k \leqslant \nu \leqslant \beta_k.$$

Concavity of $M_*(\nu)$ follows from that of $M(\nu)$. The graph of $M_*(\nu)$ coincides with that of $M(\nu)$, save over the intervals (α_k, β_k), where it has *concave* arcs (along which $M_*(\nu)$ is proportional to $\nu^{1/2}$), lying *below* the corresponding arcs for $M(\nu)$ and *meeting those* at their *endpoints*. The *former* graph is thus clearly concave if the other one is.

Proving that $\sum_1^\infty M_*(n)/n^2 = \infty$ is trickier. There would be no trouble at all here if we could be sure that the ratios β_k/α_k were *bounded*, but we cannot assume that and our argument makes strong use of the fact that $\delta > 0$ in the condition $M(\nu) >$ const. $\nu^{\frac{1}{2}+\delta}$.

We again appeal to the special structure of $M(\nu)$'s graph to argue that the *local maxima* of $M(\nu)/\nu^{1/2}$, and hence the *intervals* (α_k, β_k), *cannot accumulate* at any finite point. Those intervals can therefore be indexed from left to right, and in the event that two adjacent ones should touch at their endpoints, we can consolidate them to form a single larger interval and then relabel. In this fashion, we arrive at a set-up where

$$0 < \alpha_1 < \beta_1 < \alpha_2 < \beta_2 < \cdots,$$

with $M_*(v) = M(v)$ *outside* the union of the (perhaps new) (α_k, β_k), and

$$M_*(v) = \left(\frac{v}{\alpha_k}\right)^{1/2} M(\alpha_k) = \left(\frac{\beta_k}{v}\right)^{1/2} M(\beta_k) \qquad \text{for } \alpha_k \leqslant v \leqslant \beta_k.$$

It is convenient to fix a β_0 with $0 < \beta_0 < \alpha_1$. Then, since $M(v)/v$ *decreases*, $M(\alpha_1) \leqslant (\alpha_1/\beta_0)M(\beta_0)$, so, by the preceding relation,

$$M(\beta_1) = \left(\frac{\beta_1}{\alpha_1}\right)^{1/2} M(\alpha_1) \leqslant \left(\frac{\beta_1}{\alpha_1}\right)^{1/2} \frac{\alpha_1}{\beta_0} M(\beta_0).$$

In like manner we find first that $M(\alpha_2) \leqslant (\alpha_2/\beta_1)M(\beta_1)$ and thence that $M(\beta_2) \leqslant (\beta_2/\alpha_2)^{1/2}(\alpha_2/\beta_1)M(\beta_1)$ which, substituted into the previous, yields

$$M(\beta_2) \leqslant \left(\frac{\beta_2}{\alpha_2}\right)^{1/2} \frac{\alpha_2}{\beta_1} \left(\frac{\beta_1}{\alpha_1}\right)^{1/2} \frac{\alpha_1}{\beta_0} M(\beta_0).$$

Continuing in this fashion, we see that

$$M(\beta_n) \leqslant \left(\frac{\beta_n}{\alpha_n}\right)^{1/2} \frac{\alpha_n}{\beta_{n-1}} \left(\frac{\beta_{n-1}}{\alpha_{n-1}}\right)^{1/2} \frac{\alpha_{n-1}}{\beta_{n-2}} \cdots \left(\frac{\beta_1}{\alpha_1}\right)^{1/2} \frac{\alpha_1}{\beta_0} M(\beta_0).$$

Now by hypothesis, $M(\beta_n) \geqslant C\beta_n^{\frac{1}{2}+\delta}$ where, wlog, $C = 1$. Use this with the relation just found and then divide the resulting inequality by $\alpha_n^{\frac{1}{2}+\delta}$, noting that

$$\alpha_n = \frac{\alpha_n}{\beta_{n-1}} \frac{\beta_{n-1}}{\alpha_{n-1}} \frac{\alpha_{n-1}}{\beta_{n-2}} \cdots \frac{\beta_1}{\alpha_1} \frac{\alpha_1}{\beta_0} \beta_0.$$

One gets

$$\left(\frac{\beta_n}{\alpha_n}\right)^{\frac{1}{2}+\delta} \leqslant \left(\frac{\beta_n}{\alpha_n}\right)^{\frac{1}{2}} \frac{\dfrac{\alpha_n}{\beta_{n-1}} \left(\dfrac{\beta_{n-1}}{\alpha_{n-1}}\right)^{1/2} \cdots \left(\dfrac{\beta_1}{\alpha_1}\right)^{1/2} \dfrac{\alpha_1}{\beta_0} M(\beta_0)}{\left\{\dfrac{\alpha_n}{\beta_{n-1}} \dfrac{\beta_{n-1}}{\alpha_{n-1}} \cdots \dfrac{\beta_1}{\alpha_1} \dfrac{\alpha_1}{\beta_0} \beta_0\right\}^{1/2+\delta}}.$$

After cancelling $(\beta_n/\alpha_n)^{1/2}$ from both sides and rearranging, this becomes

$$\left(\frac{\beta_n}{\alpha_n} \frac{\beta_{n-1}}{\alpha_{n-1}} \cdots \frac{\beta_1}{\alpha_1}\right)^{\delta} \leqslant \left(\frac{\alpha_n}{\beta_{n-1}} \frac{\alpha_{n-1}}{\beta_{n-2}} \cdots \frac{\alpha_1}{\beta_0}\right)^{\frac{1}{2}-\delta} \frac{M(\beta_0)}{\beta_0^{\frac{1}{2}+\delta}}.$$

There is of course no loss of generality here in assuming $\delta < 1/2$. The last formula can be rewritten

$$\sum_{k=1}^{n}\log\left(\frac{\beta_k}{\alpha_k}\right) \leqslant c + \frac{1-2\delta}{2\delta}\sum_{k=1}^{n}\log\left(\frac{\alpha_k}{\beta_{k-1}}\right)$$

where $c = (1/\delta)\log(M(\beta_0)/\beta_0^{1/2+\delta})$ is independent of n, and this estimate makes it possible for us to compare some integrals of $M(v)/v^2$ over complementary sets.

Since $M(v)/v$ is decreasing, we have

$$\int_{\beta_{n-1}}^{\alpha_n}\frac{M(v)}{v^2}\mathrm{d}v \geqslant \frac{M(\alpha_n)}{\alpha_n}\int_{\beta_{n-1}}^{\alpha_n}\frac{\mathrm{d}v}{v} = \frac{M(\alpha_n)}{\alpha_n}\log\frac{\alpha_n}{\beta_{n-1}},$$

and at the same time,

$$\int_{\alpha_n}^{\beta_n}\frac{M(v)}{v^2}\mathrm{d}v \leqslant \frac{M(\alpha_n)}{\alpha_n}\int_{\alpha_n}^{\beta_n}\frac{\mathrm{d}v}{v} = \frac{M(\alpha_n)}{\alpha_n}\log\frac{\beta_n}{\alpha_n}.$$

From the *second* inequality,

$$\sum_{n=1}^{N}\int_{\alpha_n}^{\beta_n}\frac{M(v)}{v^2}\mathrm{d}v \leqslant \sum_{n=1}^{N}\frac{M(\alpha_n)}{\alpha_n}\log\frac{\beta_n}{\alpha_n},$$

and partial summation converts the right side to

$$\sum_{n=1}^{N-1}\left\{\frac{M(\alpha_n)}{\alpha_n} - \frac{M(\alpha_{n+1})}{\alpha_{n+1}}\right\}\sum_{k=1}^{n}\log\frac{\beta_k}{\alpha_k} + \frac{M(\alpha_N)}{\alpha_N}\sum_{k=1}^{N}\log\frac{\beta_k}{\alpha_k}.$$

The ratios $M(\alpha_n)/\alpha_n$ are, however, decreasing, so we may apply the estimate obtained above to see that the last expression is

$$\leqslant \sum_{n=1}^{N-1}\left\{\frac{M(\alpha_n)}{\alpha_n} - \frac{M(\alpha_{n+1})}{\alpha_{n+1}}\right\}\left\{\frac{1-2\delta}{2\delta}\sum_{k=1}^{n}\log\frac{\alpha_k}{\beta_{k-1}} + c\right\}$$
$$+ \frac{M(\alpha_N)}{\alpha_N}\left\{\frac{1-2\delta}{2\delta}\sum_{k=1}^{N}\log\frac{\alpha_k}{\beta_{k-1}} + c\right\},$$

which, by reverse summation by parts, boils down to

$$\frac{1-2\delta}{2\delta}\sum_{n=1}^{N}\frac{M(\alpha_n)}{\alpha_n}\log\frac{\alpha_n}{\beta_{n-1}} + c\frac{M(\alpha_1)}{\alpha_1}.$$

This in turn is

$$\leqslant \frac{1-2\delta}{2\delta}\sum_{n=1}^{N}\int_{\beta_{n-1}}^{\alpha_n}\frac{M(v)}{v^2}\mathrm{d}v + c\frac{M(\alpha_1)}{\alpha_1}$$

by the *first* of the above inequalities, so, since $M(v) = M_*(v)$ on each of the intervals $[\beta_{n-1}, \alpha_n]$, we have finally

$$\sum_{n=1}^{N} \int_{\alpha_n}^{\beta_n} \frac{M(v)}{v^2} dv \leqslant \frac{1-2\delta}{2\delta} \sum_{n=1}^{N} \int_{\beta_{n-1}}^{\alpha_n} \frac{M_*(v)}{v^2} dv + c\frac{M(\alpha_1)}{\alpha_1}.$$

Adding $\sum_{n=1}^{N} \int_{\beta_{n-1}}^{\alpha_n} (M(v)/v^2)\,dv = \sum_{n=1}^{N} \int_{\beta_{n-1}}^{\alpha_n} (M_*(v)/v^2)\,dv$ to both sides of this relation one gets (*a fortiori!*)

$$\int_{\beta_0}^{\beta_N} \frac{M(v)}{v^2} dv < c\frac{M(\alpha_1)}{\alpha_1} + \frac{1}{2\delta} \int_{\beta_0}^{\alpha_N} \frac{M_*(v)}{v^2} dv,$$

and thence

$$\int_{\beta_0}^{\infty} \frac{M(v)}{v^2} dv \leqslant c\frac{M(\alpha_1)}{\alpha_1} + \frac{1}{2\delta} \int_{\beta_0}^{\infty} \frac{M_*(v)}{v^2} dv.$$

In the present circumstances, however, divergence of $\sum_1^\infty M(n)/n^2$ is equivalent to that of the left-hand integral and divergence of $\sum_1^\infty M_*(n)/n^2$ equivalent to that of the integral on the right. Our assumptions on $M(v)$ thus make $\sum_1^\infty M_*(n)/n^2 = \infty$, and the proof of the theorem is complete.

The second observation to be made about Brennan's theorem is that its *monotoneity requirement* on $M(v)/v^{1/2}$ is probably *incapable of much further relaxation.* That depends on an example mentioned at the end of Borichev and Volberg's preprint. Unfortunately, they do not describe the construction of the example, so I cannot give it here. *Let us, in the present addendum, assume that their construction is right and show how to deduce from this supposition that Brennan's result is close to being best possible in a sense to be soon made precise.*

The example of Borichev and Volberg, if correct, furnishes a decreasing function $h(\xi)$ with $\xi h(\xi) \geqslant 1$ and $\int_0^1 \log h(\xi)\,d\xi = \infty$ together with $F(z)$, bounded and $\mathscr{C}_\infty$ in $\{|z|<1\}$ and having the non-tangential boundary value $F(e^{i\vartheta})$ a.e. on $\{|z|=1\}$, such that

$$\left|\frac{\partial F(z)}{\partial \bar{z}}\right| \leqslant \exp\left(-h\left(\log\frac{1}{|z|}\right)\right) \qquad \text{for } |z|<1,$$

while

$$\int_{-\pi}^{\pi} \log|F(e^{i\vartheta})|\,d\vartheta = -\infty$$

although $F(e^{i\vartheta})$ *is not* a.e. *zero.*

The procedure we are about to follow comes from the paper of Jöricke and Volberg, and will be used again to investigate the more complicated situation taken up in the next article. In order that the reader may first see its main idea unencumbered by detail, let us *for now* make *an additional assumption that the function $F(z)$ supplied by the Borichev–Volberg construction is continuous up to* $|z| = 1$. At the end of the next article we will see that a counter-example to further extension of the L_1 version of Brennan's result given there can be obtained *without this continuity*. Assuming it here enables us to just *take over* the constructions of §D.6, Chapter VII.

The present function $F(z)$ is to be subjected to the treatment applied to the one thus denoted in §D.6, beginning on p. 359. We also employ the symbols

$$w(r) = \exp\left(-h\left(\log\frac{1}{r}\right)\right),$$

$\mathcal{O}$, B, Φ, Ω, &c with the meanings adopted there.

Starting with $F(z)$, we construct a *continuous* function $g(e^{i\vartheta})$ on $\{|z| = 1\}$ and a *concave* increasing majorant $M(v)$ having the following properties:

(i) $\quad g(e^{i\vartheta}) \not\equiv 0,$

(ii) $\quad \displaystyle\int_{-\pi}^{\pi} \log|g(e^{i\vartheta})|\,d\vartheta = -\infty,$

(iii) $\quad \displaystyle\sum_{1}^{\infty} M(n)/n^2 = \infty,$

(iv) $\quad M(v)/v^{1/2} \geqslant 2,$

(v) $\quad g(e^{i\vartheta}) \sim \displaystyle\sum_{-\infty}^{\infty} a_n e^{in\vartheta}$ with $|a_n| \leqslant \text{const.}\,e^{-M(|n|)}$ for $n < 0$.

It is clear from this *how close* Brennan's result comes to being *best possible* provided that the above assumptions are granted.

The weight $w(r)$ we are now using is decreasing, and, since $\xi h(\xi) \geqslant 1$, goes to zero rapidly enough for the reasoning followed in *steps 2* and *3* of §D.6, Chapter VII to carry over without change.

But the argument made for *step 1* on p. 361 requires modification. Here, since $F(z)$ is continuous on the closed unit disk and $\neq 0$ on its circumference, there is a non-empty open arc I of that circumference on which $|F(e^{i\vartheta})|$ is *bounded away from zero*. Then, because $w(r) \longrightarrow 0$ for $r \longrightarrow 1$, the open set $\mathcal{O}$ must have a *component* – call it $\mathcal{O}'$ – *abutting* on I. If, at the same time, B contained a non-void open arc J of the unit circumference, we would have $\partial\mathcal{O}' \cap J = \varnothing$. In that event one could reason with the analytic function $\Phi(z)$ as at the bottom of p. 362, because $|\Phi(\zeta)| \leqslant \text{const.}\, w(|\zeta|)$ on $\partial\mathcal{O}' \cap \{|\zeta| < 1\}$. In that way, one would find that $\Phi(z) \equiv 0$ in $\mathcal{O}'$, making $F(e^{i\vartheta}) \equiv 0$ on I, a contradiction. Hence no such arc as J can exist.

Once *steps 1, 2* and *3* are carried out, we fix any ρ, $0 < \rho < 1$ and take the connected set $\Omega = \Omega(\rho) \subseteq \{\rho < |z| < 1\}$ described at the top of p. 363. As pointed out on p. 364, $\partial\Omega$ *includes the whole unit circumference.*

Fix now a $z_0 \in \Omega$. Given an integer $n \geqslant 0$, let us apply *Poisson's formula* to the function $z^n\Phi(z)$, *harmonic* (since analytic!) in Ω and *continuous* up to $\partial\Omega$. We get

$$z_0^n\Phi(z_0) \;=\; \int_{\partial\Omega} \zeta^n\Phi(\zeta)\,d\omega_\Omega(\zeta, z_0)$$

where, as usual, $\omega_\Omega(\;,\;)$ is harmonic measure for Ω. The boundary $\partial\Omega$ consists of the unit circumference together with

$$\gamma \;=\; \partial\Omega \cap \{|z| < 1\},$$

so the last relation can be rewritten

$$\int_{-\pi}^{\pi} e^{in\vartheta}\Phi(e^{i\vartheta})\,d\omega_\Omega(e^{i\vartheta}, z_0) \;=\; z_0^n\Phi(z_0) \;-\; \int_\gamma \zeta^n\Phi(\zeta)\,d\omega_\Omega(\zeta, z_0).$$

Let us first examine the *right side* of this formula.

With $\log\dfrac{1}{|z_0|} = \xi_0 > 0$, the *first* term on the right has modulus $|\Phi(z_0)|e^{-n\xi_0}$. Concerning the *second* term, we recall that by the construction of $\mathcal{O}$, $|\Phi(\zeta)| \leqslant \text{const.}\, w(|\zeta|)$ on γ, including on any arcs thereof lying on $\{|\zeta| = \rho\}$ and in $\mathcal{O}$, as long as the constant is chosen large enough. Therefore, writing

$$M(v) \;=\; \inf_{\xi > 0}\,(h(\xi) + \xi v)$$

we have, since $w(|\zeta|) \;=\; \exp(-h(\xi))$ with $\xi = \log(1/|\zeta|)$,

$$|\zeta^n\Phi(\zeta)| \leqslant \text{const.}\, e^{-M(n)}, \qquad \zeta \in \gamma.$$

Harmonic measure of course has total mass 1. Our second term is hence $\leqslant \text{const.}\, e^{-M(n)}$ in magnitude, and we find that altogether, for $n \geqslant 0$,

$$\left|\int_{-\pi}^{\pi} e^{in\vartheta}\Phi(e^{i\vartheta})\,d\omega_\Omega(e^{i\vartheta}, z_0)\right| \;\leqslant\; \text{const.}\,(e^{-n\xi_0} + e^{-M(n)}).$$

It will be seen presently that $e^{-M(n)}$ dominates $e^{-n\xi_0}$ for large n, so that the latter term can be dropped from this last relation. On account of that, we next turn our attention to $M(v)$. This function is *concave* by its definition, and, since $h(\xi) \geqslant 1/\xi$, easily seen to be $\geqslant 2v^{1/2}$ and thus enjoy property (iv) of the above list. Because $h(\xi)$ is decreasing and $\int_0^1 \log h(\xi)\,d\xi = \infty$, we have $\int_1^\infty (M(v)/v^2)\,dv = \infty$ by the theorem on p. 337. That, however, implies that $\sum_1^\infty M(n)/n^2 = \infty$, which is property (iii).

We look now at the measure $\Phi(e^{i\vartheta})\,d\omega_\Omega(e^{i\vartheta}, z_0)$ appearing *on the left* in the preceding relation. In the first place, $d\omega_\Omega(e^{i\vartheta}, z_0)$ is *absolutely continuous* with respect to $d\vartheta$ on $\{|\zeta| = 1\}$, and indeed $\leqslant C\,d\vartheta$ there, the constant C depending on z_0. This follows immediately by comparison of $d\omega_\Omega(e^{i\vartheta}, z_0)$ with harmonic measure for the whole unit disk. We can therefore write

$$\Phi(e^{i\vartheta})\,d\omega_\Omega(e^{i\vartheta}, z_0) = g(e^{i\vartheta})\,d\vartheta$$

with a *bounded* function g, and have just the *moduli of* $2\pi g(e^{i\vartheta})$'s *Fourier coefficients* (of negative index) standing on the left in the above relation.

In fact, $d\omega_\Omega(e^{i\vartheta}, z_0)$ has *more regularity* than we have just noted. The *derivative* $d\omega_\Omega(e^{i\vartheta}, z_0)/d\vartheta$ is, for instance, *strictly positive* in the interior of each arc I_k of the unit circumference contiguous to B's intersection therewith. To see this one may, given I_k, construct a very shallow sectorial box $\mathscr{S}$ in the unit disk with base on I_k and *slightly shorter* than the latter. A shallow enough $\mathscr{S}$ will have none of $\partial\Omega$ in its interior since Ω *abuts* on I_k. One may therefore compare $d\omega_\Omega(e^{i\vartheta}, z)$ with harmonic measure for $\mathscr{S}$ when $z \in \mathscr{S}$ and $e^{i\vartheta}$ is on that box's base, and an application of Harnack then leads to the desired conclusion.

From this we can already see that $|g(e^{i\vartheta})|$ is *bounded away from zero* inside some of the arcs I_k, for instance, on the arc I used at the beginning of this discussion. But there is more – $g(e^{i\vartheta})$ is *continuous* on the unit circumference. That follows immediately from *four* properties: the *continuity* of $\Phi(e^{i\vartheta})$, its *vanishing* for $e^{i\vartheta} \in B$, the *boundedness* of $d\omega_\Omega(e^{i\vartheta}, z_0)/d\vartheta$, and, finally, the *continuity* of this derivative in the interior of each arc I_k contiguous to $B \cap \{|\zeta| = 1\}$. The first three of these we are sure of, so it suffices to verify the fourth.

For that purpose, it is easiest to use the formula

$$\frac{d\omega_\Omega(e^{i\vartheta}, z_0)}{d\vartheta} = \frac{d\omega_\Delta(e^{i\vartheta}, z_0)}{d\vartheta} - \int_\gamma \frac{d\omega_\Delta(e^{i\vartheta}, \zeta)}{d\vartheta}\,d\omega_\Omega(\zeta, z_0),$$

where $\omega_\Delta(\ \ , z_0)$ is ordinary harmonic measure for the unit disk Δ (cf. p. 371). For $e^{i\vartheta}$ moving along an arc I_k,

$$d\omega_\Delta(e^{i\vartheta}, \zeta)/d\vartheta = (1 - |\zeta|^2)/2\pi|\zeta - e^{i\vartheta}|^2$$

varies *continuously*, and *uniformly so*, for ζ ranging over any subset of Δ

staying away from $e^{i\vartheta}$. Continuity of $d\omega_\Omega(e^{i\vartheta}, z_0)/d\vartheta$ can then be read off from the formula since γ has no accumulation points inside the I_k.

The function $g(e^{i\vartheta})$ is thus continuous, in addition to enjoying property (i) of our list. Verification of properties (ii) and (v) thereof remains.

Because $d\omega_\Omega(e^{i\vartheta}, z_0)/d\vartheta \leqslant C$ and $|\Phi(e^{i\vartheta})|$ lies between two constant multiples of $|F(e^{i\vartheta})|$, property (ii) holds on account of the analogous condition satisfied by F and the relation of $g(e^{i\vartheta})$ to $\Phi(e^{i\vartheta})$. Passing to property (v), we note that an earlier relation can be rewritten

$$\left|\int_{-\pi}^{\pi} e^{in\vartheta} g(e^{i\vartheta}) d\vartheta\right| \leqslant \text{const.}(e^{-n\xi_0} + e^{-M(n)}), \qquad n \geqslant 0.$$

By *concavity* of $M(v)$, $M(v)/v$ eventually decreases and tends to a limit $l \geqslant 0$ as $v \longrightarrow \infty$. Were $l > 0$, the right side of the inequality just written would be $\leqslant$ const.e^{-nl_0} with $l_0 = \min(\xi_0, l) > 0$. Such a bound on the left-hand integral would, with property (ii), force $g(e^{i\vartheta})$ to vanish identically – see the bottom of p. 328. Our $g(e^{i\vartheta})$, however, does not do that, so we must have $l = 0$, making $M(n) < n\xi_0$ for large n. The right side of our inequality can therefore be replaced by const.$e^{-M(n)}$, and property (v) holds. The construction is now complete.

It is to be noted that the only objects we actually *used* were the function $h(\xi)$ with its specified properties and $\Phi(z)$, analytic in a certain domain $\mathcal{O} \subseteq \{|z| < 1\}$ and continuous up to $\partial\mathcal{O}$, satisfying $|\Phi(\zeta)| \leqslant \text{const.}\exp\left(-h\left(\log\frac{1}{|\zeta|}\right)\right)$ on $\partial\mathcal{O} \cap \{|\zeta| < 1\}$ and $|\Phi(\zeta)| > 0$ on some arc of $\{|\zeta| = 1\}$ included in $\partial\mathcal{O}$. I have a persistent nagging feeling that such functions $h(\xi)$ and $\Phi(z)$, if there really *are* any, must be lying around somewhere or at least be closely related to others whose constructions are already available. One thinks of various kinds of functions meromorphic in the unit disk but not of bounded characteristic there; especially do the ones described by Beurling at the eighth Scandinavian mathematicians' congress come back continually to mind.

This addendum, however, is already being written at the very last moment. The imminence of press time leaves me no opportunity for pursuing the matter.

3. Extension to functions $F(e^{i\vartheta})$ in $L_1(-\pi, \pi)$.

The theorem of p. 356 holds for L_1 functions $F(e^{i\vartheta})$ not a.e. zero, as does Brennan's refinement of it given in article 1 above. A procedure for handling this more general situation (absence of continuity) is worked out in the beautiful *Mat. Sbornik* paper by Jöricke and Volberg. Here we adapt their method so as to make it go with the development already familiar from §D.6, Chapter VII, hewing as closely as possible to the latter.

Our aim is to show that

$$\int_{-\pi}^{\pi} \log|F(e^{i\vartheta})|\,d\vartheta \;>\; -\infty$$

for any function $F(e^{i\vartheta}) \in L_1(-\pi,\pi)$ not a.e. zero and satisfying the hypothesis of Brennan's theorem. Let us begin by observing that the treatment of this case can be reduced to that of a *bounded* function F.

Suppose, indeed, that

$$F(e^{i\vartheta}) \;\sim\; \sum_{-\infty}^{\infty} a_n e^{in\vartheta}$$

belongs to L_1, with $|a_n| \leqslant \text{const.}\,e^{-M(|n|)}$ for $n<0$. The series $\sum_{n<0} a_n e^{in\vartheta}$ is then surely *absolutely convergent*, so

$$\sum_{0}^{\infty} a_n e^{in\vartheta}$$

is also the Fourier series of an L_1 function, which we denote by $F_+(e^{i\vartheta})$ (this belongs in fact to the space H_1). For $|z|<1$, put

$$F_+(z) \;=\; \sum_{0}^{\infty} a_n z^n;$$

for this function, analytic in $\{|z|<1\}$, we have (Chapter II, §B!),

$$F_+(z) \longrightarrow F_+(e^{i\vartheta}) \quad \text{a.e. as } z \not\angle\!\!\longrightarrow e^{i\vartheta}.$$

Using the integrable function $\log^+|F_+(e^{i\vartheta})| \geqslant 0$, we now form

$$b(z) \;=\; \frac{1}{2\pi}\int_{-\pi}^{\pi} \frac{e^{i\vartheta}+z}{e^{i\vartheta}-z}\log^+|F_+(e^{i\vartheta})|\,d\vartheta,$$

analytic and with positive real part for $|z|<1$. According to the third theorem and scholium of §F.2, Chapter III, $b(z)$ tends for almost every ϑ to a limit $b(e^{i\vartheta})$ as $z \not\angle\!\!\longrightarrow e^{i\vartheta}$, with

$$\Re b(e^{i\vartheta}) \;=\; \log^+|F_+(e^{i\vartheta})| \quad \text{a.e.}$$

A standard extension of Jensen's inequality to H_1 also tells us that

$$\log|F_+(z)| \;\leqslant\; \Re b(z), \qquad |z|<1$$

(cf. pp. 291–2 where this was proved and used for $z=0$).

We next perform the Dynkin extension (described on pp. 339–40) on the continuous function

$$F_-(e^{i\vartheta}) \;=\; \sum_{-\infty}^{-1} a_n e^{in\vartheta}.$$

This gives us $F_-(z)$, $\mathscr{C}_\infty$ in the unit disk and continuous (hence *bounded*!) up to its boundary, with

$$\left|\frac{\partial F_-(z)}{\partial \bar{z}}\right| \leqslant \text{const.}\exp\left(-h\left(\log\frac{1}{|z|}\right)\right), \qquad |z|<1,$$

where, in the present circumstances,

$$h(\xi) = \sup_{v>0}\,(M(v)/2 - v\xi)$$

(see *remark* 2, p. 343). As usual, we write

$$w(r) = \exp\left(-h\left(\log\frac{1}{r}\right)\right);$$

then, putting

$$F(z) = F_-(z) + F_+(z)$$

for $|z|<1$, we have

$$\left|\frac{\partial F(z)}{\partial \bar{z}}\right| \leqslant \text{const.}\,w(|z|)$$

there, and

$$F(z) \longrightarrow F(e^{i\vartheta}) \quad \text{a.e. for } z \not\longrightarrow e^{i\vartheta}.$$

The bounded function spoken of earlier is simply

$$F_0(z) = e^{-b(z)}F(z).$$

It *is* bounded in the unit disk by one of the previous relations; another tells us that $F_0(z)$ has a non-tangential boundary value $F_0(e^{i\vartheta}) = F(e^{i\vartheta})\exp(-b(e^{i\vartheta}))$ equal in modulus to $|F(e^{i\vartheta})|/\max(|F_+(e^{i\vartheta})|, 1)$ at almost every point of the unit circumference. Then, since $F(e^{i\vartheta})\in L_1$ is not a.e. zero, neither is $F_0(e^{i\vartheta})$. We note finally that by *analyticity* of $e^{-b(z)}$, $\partial F_0(z)/\partial\bar{z} = e^{-b(z)}\partial F(z)/\partial\bar{z}$, making

$$\left|\frac{\partial F_0(z)}{\partial \bar{z}}\right| \leqslant \text{const.}\,w(|z|), \qquad |z|<1.$$

Given that $M(v)$ satisfies the hypothesis of Brennan's theorem, our function $h(\xi)$ enjoys the two properties used in the first part of article 1, namely, that $\xi h(\xi)$ *decreases* and that $\int_0^a \log h(\xi)\,d\xi = \infty$ for small $a>0$. If, now, we can *deduce* from these together with the *preceding relation* that the *bounded* function $F_0(z)$, not a.e. zero for $|z|=1$, satisfies

$$\int_{-\pi}^{\pi} \log|F_0(e^{i\vartheta})|d\vartheta \;>\; -\infty,$$

we will certainly have *the same conclusion* for

$$\log|F(e^{i\vartheta})| \;=\; \log|F_0(e^{i\vartheta})| \;+\; \log^+|F_+(e^{i\vartheta})|.$$

The rest of our work deals exclusively with $F_0(z)$.

In order to stay as close as possible to the notation of §D.6, Chapter VII, *we denote the bounded function* $F_0(z)$ *by* $F(z)$ *from now on.* Using this new $F(z)$, we first form the sets $B \subseteq \{|z|\leqslant 1\}$ and $\mathcal{O} \subseteq \{|z|<1\}$ as on pp. 359–60, and then the function $\Phi(z)$ introduced on p. 360. The latter, *analytic in* $\mathcal{O}$, is actually defined on the whole unit disk, and has there at least as much continuity as $F(z)$ besides lying in modulus between two constant multiples of $|F(z)|$. It has, in particular, a non-tangential boundary value $\Phi(e^{i\vartheta})$ a.e. on the unit circumference, and this *does not vanish* a.e. The construction of B ensures that

$$|\Phi(\zeta)| \;\leqslant\; \text{const.}\, w(|\zeta|) \qquad \text{on } \partial\mathcal{O}\cap\{|\zeta|<1\}$$

(indeed, on B), and our task amounts to showing that

$$\int_{-\pi}^{\pi} \log|\Phi(e^{i\vartheta})|d\vartheta \;>\; -\infty$$

on account of these properties.

What makes the present situation more complicated than the one studied in §D.6 of Chapter VII is that $\Phi(z)$ need no longer be continuous up to the whole unit circumference. This causes the notion of *abutment* introduced on p. 348 to be less useful here for the examination of our set $\mathcal{O}$ than it was in §D.6, and we have to supplement it with another, that of *fatness.* The latter, based on the famous sawtooth construction of Lusin and Privalov, helps us to take account of $\Phi(z)$'s non-tangential boundary behaviour.

To describe what is meant by fatness, we need to bring in a special kind of domain together with some notation; both will also be used further on. Corresponding to each point $e^{i\alpha}$ on the unit circumference, we have an open set S_α consisting of the z with $1/2 \;<\; |z| \;<\; 1$ lying in the open 60° sector having vertex at $e^{i\alpha}$ and symmetric about the radius from 0 out to that point. Given any subset E of $\{|\zeta|=1\}$ we then write

$$S_E \;=\; \bigcup_{e^{i\alpha}\in E} S_\alpha.$$

It is evident that if we take any S_E and a ρ, $1/2 < \rho < 1$, the intersection

$$S_E \cap \{\rho < |z| < 1\}$$

breaks up into (at most) a countable number of open *connected* components, each of the form

$$S_{E_k} \cap \{\rho < |z| < 1\},$$

with the E_k making up a (disjoint) partition of the set E.

Definition. A *connected* open set of the form

$$S_E \cap \{\rho < |z| < 1\}$$

(with $1/2 < \rho < 1$) is called a *sawblade* of *depth* $1 - \rho$. We say that such a sawblade *bites on* the set E.

Now we can state the

Definition. An open subset $\mathscr{U}$ of the unit disk is called *fat* if it contains a sawblade biting on a closed $E \subseteq \{|\zeta| = 1\}$ with $|E| > 0$. In that circumstance we also say that $\mathscr{U}$ is *fat at E*.

Equipped with these tools, we endeavour to investigate the set $\mathcal{O}$ according to the procedure of §D.6, Chapter VII. In this, some modifications are necessary; we have, in the first place, to *skip over step 1* (p. 361). Then, taking ρ, $1/2 < \rho < 1$, we construct a set $\Omega(\rho)$, proceeding differently, however, than as we did on pp. 361–3.

There is, by the properties of $\Phi(z)$, a closed subset E_0 of the unit circumference, $|E_0| > 0$, such that, for the *non-tangential* boundary values $\Phi(\zeta)$, we have, wlog,

$$|\Phi(\zeta)| > 1, \qquad \zeta \in E_0.$$

Egorov's theorem enables us to in fact pick E_0 so as to have $|\Phi(z)| > 1$ for $z \in S_{E_0}$ with $\rho' < |z| < 1$ when $\rho' > \rho$ is sufficiently close to 1. But the construction of B and $\mathcal{O}$ makes $|\Phi(z)| \leqslant \text{const.}\, w(|z|)$ on B, hence on $\{|z| < 1\} \sim \mathcal{O}$. Therefore, since $w(r) \longrightarrow 0$ for $r \longrightarrow 1$, we must have

$$S_{E_0} \cap \{\rho' < |z| < 1\} \subseteq \mathcal{O}$$

if ρ', $\rho < \rho' < 1$, is near enough to 1. One of the components of the intersection on the left is a sawblade of depth $1 - \rho'$ biting on a (Borel) subset E' of E_0 with $|E'| > 0$; a suitable *closed* subset E of E' then has

$|E|>0$, and *there is a sawblade of depth* $1-\rho'$ *biting on E and contained in* $\mathcal{O}$. We now take $\Omega(\rho)$ as *the component of* $\mathcal{O}\cap\{\rho<|z|<1\}$ *including that sawblade*; $\Omega(\rho)$ is *fat at E.*

For the present set $\Omega(\rho)$ there is *a substitute* for *step 2* of p. 362:

Step 2'. *$\partial\Omega(\rho)$ includes the whole unit circumference.*

This we establish by *reductio ad absurdum.* Let us write Ω for $\Omega(\rho)$, and put

$$\gamma = \partial\Omega\cap\{|z|<1\},$$

$$\Gamma = \partial\Omega \sim \gamma;$$

Γ is thus the part of $\partial\Omega$ lying on the unit circumference. *Assume* that there is on the latter a non-empty open arc J with $J\cap\Gamma=\emptyset$; we will then deduce a contradiction.

For that it is quicker to fall back on the device used in the second half of article 2 than to adapt Volberg's theorem on harmonic measures (p. 349) to the present situation. *Fixing* $z_0\in\Omega$, we can say that

$$z_0^n\Phi(z_0) = \int_{\partial\Omega}\zeta^n\Phi(\zeta)\,d\omega_\Omega(\zeta,z_0) \qquad \text{for } n\geqslant 0,$$

whence

$$\int_\Gamma e^{in\vartheta}\Phi(e^{i\vartheta})\,d\omega_\Omega(e^{i\vartheta},z_0) = z_0^n\Phi(z_0) - \int_\gamma \zeta^n\Phi(\zeta)\,d\omega_\Omega(\zeta,z_0), \qquad n\geqslant 0.$$

Here we are using Poisson's formula for the bounded function $\zeta^n\Phi(\zeta)$ harmonic (even analytic) in Ω and continuous up to γ, but not necessarily up to Γ, where it is only known to have non-tangential boundary values a.e. Such use is legitimate; we postpone verification of that, and of a corresponding version of Jensen's inequality, to the next article, so as not to interrupt the argument now under way.

As in article 2, $d\omega_\Omega(e^{i\vartheta},z_0)$ is absolutely continuous and $\leqslant C\,d\vartheta$ on Γ, and we obtain a bounded measurable function $g(e^{i\vartheta})$ by putting

$$g(e^{i\vartheta}) = \Phi(e^{i\vartheta})\frac{d\omega_\Omega(e^{i\vartheta},z_0)}{d\vartheta} \qquad \text{for } e^{i\vartheta}\in\Gamma$$

and (here!) taking $g(e^{i\vartheta})$ to be *zero outside* Γ. From the preceding relation we then see, as in article 2, that

$$\left|\int_{-\pi}^{\pi} e^{in\vartheta} g(e^{i\vartheta})\, d\vartheta\right| \leqslant \text{const.}(e^{-n\xi_0} + e^{-M_1(n)})$$

for $n \geqslant 0$, where $\xi_0 > 0$ and

$$M_1(\nu) = \inf_{\xi > 0}\,(h(\xi) + \xi\nu).$$

This function is increasing and concave, so the right side of the last inequality can be replaced by const. $e^{-M_2(n)}$ for large n, with $M_2(n)$ equal either to $\xi_0 n$ (in case $\lim_{\nu\to\infty}(M_1(\nu)/\nu) \geqslant \xi_0$) or else to $M_1(n)$. In either event, $M_2(n)$ increases and $\sum_1^\infty M_2(n)/n^2 = \infty$ on account of the properties of $h(\xi)$. (See the theorem of p. 337 – $M_1(n)$ is actually equal to $M(n)/2$ in the present set-up.) Now we can apply Levinson's theorem, since $g(e^{i\vartheta})$ *vanishes on the arc* J. The conclusion is that $g(e^{i\vartheta}) \equiv 0$ a.e.

But $g(e^{i\vartheta})$ *does not vanish* a.e. Indeed, Ω contains a *sawblade* $\mathscr{E}$ *biting on* a closed set E, $|E| > 0$, where $|\Phi(e^{i\vartheta})| \geqslant 1$. Thence,

$$\int_E |g(e^{i\vartheta})|\, d\vartheta = \int_E |\Phi(e^{i\vartheta})|\, d\omega_\Omega(e^{i\vartheta}, z_0) \geqslant \omega_\Omega(E, z_0).$$

Harnack's theorem assures us that the quantity on the right is > 0 if, for some $z_1 \in \mathscr{E}$, $\omega_\Omega(E, z_1) > 0$. However, by the principle of extension of domain, $\omega_\Omega(E, z_1) \geqslant \omega_{\mathscr{E}}(E, z_1)$. At the same time, $\partial\mathscr{E}$ is *rectifiable*, so a conformal mapping of $\mathscr{E}$ onto the unit disk must take the *subset* E of $\partial\mathscr{E}$, *having linear measure* > 0, to a *set of measure* > 0 *on the unit circumference*. (This follows by the celebrated F. and M. Riesz theorem; a proof can be found in Zygmund or in any of the books about H_p spaces.) We therefore have $\omega_{\mathscr{E}}(E, z_1) > 0$, making $\omega_\Omega(E, z_0) > 0$ and hence, as we have seen, $\int_E |g(e^{i\vartheta})|\, d\vartheta > 0$.

Our contradiction is thus established. By it we see that the arc J cannot exist, i.e., that Γ *is the whole unit circumference*, as was to be shown.

With *step 2'* accomplished, we are ready for *step 3*. One starts out as on p. 363, using the square root mapping employed there. That gives us a domain $\Omega_\surd$, certainly *fat* at a closed subset E'', of $E_\surd$ (the image of E under our mapping), with $|E''| > 0$ (recall the earlier use of Egorov's theorem). Thereafter, one applies to $\Omega_\surd$ the argument just made for Ω in doing *step 2'*.

The weight $w_1(r)$ is next introduced as on p. 365, and the sets B_1 and $\mathcal{O}_1$ constructed (pp. 365–6). After doing *steps 2'* and *3* again with these objects, we come to *step 4*.

Jöricke and Volberg are in fact able to circumvent this step, thanks to a clever rearrangement of *step 5*. Here, however, let us continue according to the plan of §D.6, Chapter VII, for the work done there carries over practically without change to the present situation.

What is important for *step 4* is that a ζ, $|\zeta| = 1$, *not* in B must, even here, lie on an arc of the unit circumference *abutting* on $\mathscr{O}$. Such a $\zeta \notin B$ must thus, as on p. 367, have a neighborhood V_ζ with

$$V_\zeta \cap \{|z| < 1\} \subseteq \mathscr{O} \cap \{\rho^2 < |z| < 1\}.$$

The left-hand intersection therefore lies in some *connected component* of the one on the right, which, however, *can only be* $\Omega(\rho^2)$, since $\zeta \in \partial\Omega(\rho^2)$ by *step 2'*. The rest of the argument goes as on pp. 367–8.

Now we can do *step 5*, or rather the *substitute* for it carried out at the beginning of article 1. For this it is necessary to have the Jensen inequality

$$\log|\Phi(\rho)| \leqslant \int_{\partial\Omega(\rho^2)} \log|\Phi(\zeta)|\,d\omega(\zeta, \rho)$$

(notation of p. 369) available in the present circumstances, where continuity of $\Phi(z)$ up to $\{|\zeta| = 1\}$ may fail. The legitimacy of this will be established in the next article; *granting* it for now, we may proceed exactly as at the beginning of article 1.

From here on, one continues as on pp. 370–2, and reaches the desired conclusion that $\int_{-\pi}^{\pi} \log|\Phi(e^{i\vartheta})|\,d\vartheta > -\infty$ as on p. 373, after one more application of our extended Jensen inequality.

We thus arrive at the

Theorem. Let $F(e^{i\vartheta}) \in L_1(-\pi, \pi)$ not be zero a.e., and suppose that

$$F(e^{i\vartheta}) \sim \sum_{-\infty}^{\infty} a_n e^{in\vartheta}$$

with

$$|a_n| \leqslant \text{const.}\, e^{-M(|n|)}, \qquad n \leqslant 0.$$

Suppose that $M(v)$ is *concave*, that $M(v)/v^{1/2}$ is *increasing* for large v, and that

$$\sum_1^{\infty} M(n)/n^2 = \infty.$$

Then

$$\int_{-\pi}^{\pi} \log|F(e^{i\vartheta})|\,d\vartheta > -\infty.$$

Remark. In their preprint, Borichev and Volberg consider formal trigonometric series

$$\sum_{-\infty}^{\infty} a_n e^{in\vartheta}$$

in which the a_n with *negative* n satisfy the requirement of the theorem, but the a_n with $n > 0$ are allowed to *grow* like $e^{M(n)}$ as $n \longrightarrow \infty$. Assuming *more* regularity for $M(v)$ ($M(v) \geqslant$ const. v^α with an $\alpha < 1$ *close to* 1 is enough), they are able to show that under the remaining conditions of the theorem, all the a_n must vanish if

$$\liminf_{r \to 1} \int_{-\pi}^{\pi} \log \left| \sum_{-\infty}^{0} a_n e^{in\vartheta} + \sum_{1}^{\infty} a_n r^n e^{in\vartheta} \right| d\vartheta = -\infty.$$

Before ending this article let us, as promised in the *last* one, see how the example of Borichev and Volberg shows that the *monotoneity requirement* on $M(v)/v^{1/2}$ *cannot, in the above theorem at least*, be *relaxed* to $M(v)/v^{1/2} \geqslant C > 0$, *even though continuity* up to $\{|\zeta| = 1\}$ *should fail* for the function $F(z)$ supplied by their construction.

The reader should refer back to the second part of article 2. Corresponding to the bounded function $F(z)$ used there, *no longer assumed continuous* up to $\{|\zeta| = 1\}$ but having at least non-tangential boundary values a.e. on that circumference, one can, as in the preceding discussion, form the sets B, $\mathcal{O}$ and $\Omega(\rho)$ and do *step 2'*. One may then form the function $g(e^{i\vartheta})$ as in article 2; *here* it is bounded and measurable at least. The work of *step 2'* shows that $g(e^{i\vartheta})$ is *not* a.e. zero, while properties (ii)–(v) of article 2 *hold* for it (for the last one, see again the end of that article).

This is all we need.

4. Lemma about harmonic functions

Suppose we have a domain Ω regular for Dirichlet's problem, lying in the (open) unit disk Δ and having part of $\partial\Omega$ on the unit circumference. As in the last article, we write

$$\Gamma = \partial\Omega \cap \partial\Delta \quad \text{and} \quad \gamma = \partial\Omega \cap \Delta.$$

For the following discussion, let us agree to call ζ, $|\zeta| = 1$, a *radial accumulation point of* Ω if, for a *sequence* $\{r_n\}$ tending to 1, we have $r_n\zeta \in \Omega$ for each n. We then denote by Γ' the set of such radial accumulation points, noting that $\Gamma' \subseteq \Gamma$ with the inclusion frequently *proper*.

Lemma. (Jöricke and Volberg) Let $V(z)$, harmonic and bounded in Ω, be continuous up to γ, and suppose that

$$\lim_{\substack{r\to 1\\ r\zeta\in\Omega}} V(\zeta)$$

exists for almost all $\zeta\in\Gamma'$. Put $v(\zeta)$ *equal* to that limit for such ζ, and to *zero* for the *remaining* $\zeta\in\Gamma$. On γ, take $v(\zeta)$ equal to $V(\zeta)$. Then, for $z\in\Omega$,

$$V(z) \;=\; \int_{\partial\Omega} v(\zeta)\,\mathrm{d}\omega_\Omega(\zeta, z).$$

Proof. It suffices to establish the result for *real* harmonic functions $V(z)$, and, for those, to show that

$$V(z) \;\leqslant\; \int_{\partial\Omega} v(\zeta)\,\mathrm{d}\omega_\Omega(\zeta, z), \qquad z\in\Omega,$$

since the reverse inequality then follows on changing the signs of V and v.

By modifying $v(\zeta)$ on a subset of Γ having zero Lebesgue measure, we get a bounded Borel function defined on $\partial\Omega$. But on Γ, we have $\mathrm{d}\omega_\Omega(\zeta, z) \leqslant C_z|d\zeta|$ (see articles 2 and 3), so such modification cannot alter the value of $\int_{\partial\Omega} v(\zeta)\,\mathrm{d}\omega_\Omega(\zeta, z)$. *We may hence just as well take $v(\zeta)$ as a bounded Borel function* (on $\partial\Omega$) *to begin with.*

That granted, we desire to show that the integral just written is $\geqslant V(z)$. For this it seems necessary to hark back to the very foundations of integration theory. Call the *limit* of any *increasing sequence of functions continuous on $\partial\Omega$* an *upper function* (on $\partial\Omega$). There is then a *decreasing sequence* of *upper functions* $w_n(\zeta) \geqslant v(\zeta)$ such that

$$\int_{\partial\Omega} w_n(\zeta)\,\mathrm{d}\omega_\Omega(\zeta, z) \;\underset{n}{\longrightarrow}\; \int_{\partial\Omega} v(\zeta)\,\mathrm{d}\omega_\Omega(\zeta, z), \qquad z\in\Omega.$$

Indeed, corresponding to any *given* $z\in\Omega$, such a sequence is furnished by a basic construction of the Lebesgue–Stieltjes integral, $\omega_\Omega(\;\;, z)$ being a Radon measure on $\partial\Omega$. But then that sequence works also for *any other* $z\in\Omega$, since $\mathrm{d}\omega_\Omega(\zeta, z') \leqslant C(z, z')\mathrm{d}\omega_\Omega(\zeta, z)$ (Harnack).

Our inequality involving v and V will *thus be established*, provided that we can verify

$$V(z) \;\leqslant\; \int_{\partial\Omega} w_n(\zeta)\,\mathrm{d}\omega_\Omega(\zeta, z), \qquad z\in\Omega,$$

for each n. *Fixing*, then, any n, we write simply $w(\zeta)$ for $w_n(\zeta)$ and put

$$W(z) = \int_{\partial\Omega} w(\zeta)\, d\omega_\Omega(\zeta, z)$$

for $z \in \Omega$, making $W(z)$ *harmonic* there. Our task is to prove that

$$V(z) \leqslant W(z), \qquad z \in \Omega.$$

It is convenient to define $W(z)$ on *all* of $\bar{\Omega}$ by putting

$$W(\zeta) = w(\zeta), \qquad \zeta \in \partial\Omega.$$

At each $\zeta \in \partial\Omega$ we then have

$$\liminf_{\substack{z \to \zeta \\ z \in \bar{\Omega}}} W(z) \geqslant W(\zeta)$$

by the elementary approximate identity property of harmonic measure, since $w(\zeta)$, as limit of an *increasing* sequence of continuous functions, satisfies

$$\liminf_{\substack{\zeta \to \zeta_0 \\ \zeta \in \partial\Omega}} w(\zeta) \geqslant w(\zeta_0) \qquad \text{for } \zeta_0 \in \partial\Omega.$$

The function $W(z)$ enjoys a certain *reproducing property* in $\bar{\Omega}$. Namely, if the domain $\mathscr{D} \subseteq \Omega$ is also regular for Dirichlet's problem, with perhaps (and especially!) part of $\partial\mathscr{D}$ on $\partial\Omega$, we have

$$W(z) = \int_{\partial\mathscr{D}} W(\zeta)\, d\omega_{\mathscr{D}}(\zeta, z) \qquad \text{for } z \in \mathscr{D}.$$

To see this, take an increasing sequence of functions $f_k(\zeta)$ continuous on $\partial\Omega$ and tending to $w(\zeta)$ thereon, and let

$$F_k(z) = \int_{\partial\Omega} f_k(\zeta)\, d\omega_\Omega(\zeta, z), \qquad z \in \Omega.$$

Then the $F_k(z)$ tend monotonically to $W(z)$ in Ω by the monotone convergence theorem. That convergence actually holds *on* $\bar{\Omega}$ if we put $F_k(\zeta) = f_k(\zeta)$ on $\partial\Omega$; this, however, makes each function $F_k(z)$ *continuous on* $\bar{\Omega}$ besides being *harmonic* in Ω. *In the domain* $\mathscr{D}$, we therefore have

$$F_k(z) = \int_{\partial\mathscr{D}} F_k(\zeta)\, d\omega_{\mathscr{D}}(\zeta, z)$$

for each k. Another appeal to monotone convergence now establishes the corresponding property for W.

Fix any $z_0 \in \Omega$; we wish to show that $V(z_0) \leqslant W(z_0)$. For this purpose, we use the formula just proved with $\mathscr{D}$ equal to *the component* Ω_r *of* $\Omega \cap \{|z|<r\}$ *containing* z_0, where $|z_0|<r<1$. Because Ω is regular for Dirichlet's problem, so is each Ω_r; that follows immediately from the characterization of such regularity in terms of *barriers*, and, in the circumstances of the last article, can also be checked directly (cf. p. 360). We write

$$\Gamma_r = \partial\Omega_r \cap \Omega,$$

making Γ_r the union of some *open arcs* on $\{|\zeta|=r\}$, and then take

$$\gamma_r = \partial\Omega_r \sim \Gamma_r;$$

γ_r is a subset (perhaps proper) of $\gamma \cap \{|\zeta| \leqslant r\}$.

The function $V(z)$, given as harmonic in Ω and continuous up to γ, is certainly continuous up to $\partial\Omega_r$. Therefore, since $V(\zeta) = v(\zeta)$ on $\gamma \supseteq \gamma_r$, we have, for $z \in \Omega_r$,

$$V(z) = \int_{\gamma_r} v(\zeta)\,d\omega_{\Omega_r}(\zeta, z) + \int_{\Gamma_r} V(\zeta)\,d\omega_{\Omega_r}(\zeta, z).$$

At the same time, by the reproducing property of W,

$$W(z) = \int_{\gamma_r} W(\zeta)\,d\omega_{\Omega_r}(\zeta, z) + \int_{\Gamma_r} W(\zeta)\,d\omega_{\Omega_r}(\zeta, z), \qquad z \in \Omega_r.$$

We henceforth write $\omega_r(\ ,\ )$ *for* $\omega_{\Omega_r}(\ ,\ )$. Then, since on $\gamma_r \subseteq \partial\Omega$, $W(\zeta) = w(\zeta)$ is $\geqslant v(\zeta)$, the two last relations yield

$$W(z) - V(z) \geqslant \int_{\Gamma_r} (W(\zeta) - V(\zeta))\,d\omega_r(\zeta, z)$$

for $z \in \Omega_r$. Our idea is to now make $r \longrightarrow 1$ in this inequality.

For $|\zeta| = 1$, define

$$\Delta_r(\zeta) = \begin{cases} W(r\zeta) - V(r\zeta) & \text{if } r\zeta \in \Gamma_r, \\ 0 \quad \text{otherwise.} \end{cases}$$

Since $V(z)$ is given as *bounded*, the functions $\Delta_r(\zeta)$ are *bounded below*. *Moreover* (and this is the clincher),

$$\liminf_{r \to 1} \Delta_r(\zeta) \geqslant 0 \text{ a.e.}, |\zeta| = 1.$$

That is indeed *clear* for the ζ on the unit circumference *outside* Γ' (the set of radial accumulation points of Ω); since for such a ζ, $r\zeta$ cannot even belong to Ω (let alone to Γ_r) when r is near 1. Consider therefore a $\zeta \in \Gamma'$, and take any sequence of $r_n < 1$ tending to 1 with, wlog, *all the* $r_n\zeta$ *in* Ω

and even in their corresponding Γ_{r_n}. Then our hypothesis and the specification of v tell us that

$$V(r_n\zeta) \underset{n}{\longrightarrow} v(\zeta),$$

except when ζ belongs to a certain set of measure zero, independent of $\{r_n\}$. For such a sequence $\{r_n\}$, however,

$$\liminf_{n\to\infty} W(r_n\zeta) \geqslant W(\zeta) = w(\zeta)$$

as seen earlier, yielding, with the preceding,

$$\liminf_{n\to\infty} \Delta_{r_n}(\zeta) \geqslant w(\zeta) - v(\zeta) \geqslant 0.$$

The asserted relation thus holds *on* Γ' *as well*, save perhaps in a set of measure zero.

Returning to our fixed $z_0 \in \Omega$, we note that for $(1 + |z_0|)/2 < r < 1$ (say), we have, on Γ_r,

$$d\omega_r(\zeta, z_0) \leqslant K|d\zeta|$$

with K independent of r (just compare $\omega_r(\ \ ,\ \ )$ with harmonic measure for $\{|z| < r\}$). There are hence measurable functions $\mu_r(\zeta)$ defined on $\{|\zeta| = 1\}$ for these values of r, with $0 \leqslant \mu_r(\zeta) \leqslant K$ (and $\mu_r(\zeta) = 0$ for $r\zeta \notin \Gamma_r$), such that

$$\int_{\Gamma_r} (W(\zeta) - V(\zeta))\, d\omega_r(\zeta, z_0) = \int_{|\zeta|=1} \Delta_r(\zeta) r\mu_r(\zeta)|d\zeta|.$$

Here the products $\Delta_r(\zeta)r\mu_r(\zeta)$ are *uniformly bounded below* since the $\Delta_r(\zeta)$ are. And, by what has just been shown,

$$\liminf_{r\to 1} \Delta_r(\zeta) r\mu_r(\zeta) \geqslant 0 \quad \text{a.e., } |\zeta| = 1.$$

Thence, by *Fatou's lemma* (!),

$$\liminf_{r\to 1} \int_{|\zeta|=1} \Delta_r(\zeta) r\mu_r(\zeta)|d\zeta| \geqslant 0.$$

We have seen, however, that when $r > |z_0|$, $W(z_0) - V(z_0)$ is $\geqslant$ the left-hand integral in the previous relation. It follows therefore that

$$W(z_0) - V(z_0) \geqslant 0,$$

as was to be proven.

We are done.

Remark 1. When $V(z)$ is only assumed to be *subharmonic* in Ω but satisfies otherwise the hypothesis of the lemma, the argument just made shows that

$$V(z) \leqslant \int_{\partial\Omega} v(\zeta)\,\mathrm{d}\omega_\Omega(\zeta, z) \qquad \text{for } z \in \Omega.$$

Remark 2. In the applications made in article 3, the function $V(z)$ actually has a continuous extension to the open unit disk Δ with modulus *bounded, in* $\Delta \sim \Omega$, by a function of $|z|$ *tending to zero for* $|z| \longrightarrow 1$. That extension also has non-tangential boundary values a.e. on $\partial\Delta$. In these circumstances the lemma's *ad hoc* specification of $v(\zeta)$ on $\Gamma \sim \Gamma'$ is *superfluous*, for the non-tangential limit of $V(z)$ must *automatically be zero* at any $\zeta \in \Gamma \sim \Gamma'$ where it exists.

Remark 3. To arrive at the version of Jensen's inequality used in article 3, apply the relation from *remark 1* to the subharmonic functions $V_M(z) = \log^+|M\Phi(z)|$, referring to *remark 2*. That gives us

$$\max\left(\log|\Phi(z)|,\, \log\frac{1}{M}\right) \leqslant \int_{\partial\Omega} \max\left(\log|\Phi(\zeta)|,\, \log\frac{1}{M}\right)\mathrm{d}\omega_\Omega(\zeta, z)$$

for $z \in \Omega$. Then, since $|\Phi(z)|$ is bounded above, one may obtain the desired result by making $M \longrightarrow \infty$.

Addendum completed June 8, 1987.

Bibliography for volume I

Akhiezer, N.I. (also spelled Achieser). *Klassicheskaia problema momentov.* Fizmatgiz, Moscow, 1961. *The Classical Moment Problem.* Oliver & Boyd, Edinburgh, 1965.

Akhiezer, N.I. O vzveshonnom priblizhenii nepreryvnykh funktsiĭ na vsei chislovoĭ osi. *Uspekhi Mat. Nauk* **11** (1956), 3–43. On the weighted approximation of continuous functions by polynomials on the entire real axis. *AMS Translations* **22** Ser. 2 (1962), 95–137.

Akhiezer, N.I. *Theory of Approximation* (first edition). Ungar, New York, 1956. *Lektsii po teorii approksimatsii* (second edition). Nauka, Moscow, 1965. *Vorlesungen über Approximationstheorie* (second edition). Akademie Verlag, Berlin, 1967.

Benedicks, M. Positive harmonic functions vanishing on the boundary of certain domains in $\mathbb{R}^n$. *Arkiv för Mat.* **18** (1980), 53–72.

Benedicks, M. Weighted polynomial approximation on subsets of the real line. Preprint, Uppsala Univ. Math. Dept., 1981, 12pp.

Bernstein, S. *Sobranie sochineniĭ.* Akademia Nauk, USSR. Volume I, 1952; volume II, 1954.

Bernstein, V. *Leçons sur les progrès récents de la théorie des séries de Dirichlet.* Gauthier–Villars, Paris, 1933.

Bers, L. An outline of the theory of pseudo-analytic functions. *Bull. AMS* **62** (1956), 291–331.

Bers, L. *Theory of Pseudo-Analytic Functions.* Mimeographed lecture notes, New York University, 1953.

Beurling, A. Analyse spectrale des pseudomesures. *C.R. Acad. Sci. Paris* **258** (1964), 406–9.

Beurling, A. Analytic continuation across a linear boundary. *Acta Math.* **128** (1972), 153–82.

Beurling, A. *On Quasianalyticity and General Distributions.* Mimeographed lecture notes, Stanford University, summer of 1961.

Beurling, A. Sur les fonctions limites quasi analytiques des fractions rationnelles. *Huitième Congrès des Mathématiciens Scandinaves, 1934.* Lund, 1935, pp. 199–210.

Beurling, A. and Malliavin, P. On Fourier transforms of measures with compact support. *Acta Math.* **107** (1962), 291–309.

Boas, R. *Entire Functions.* Academic Press, New York, 1954.

Borichev, A. and Volberg, A. Uniqueness theorems for almost analytic functions. Preprint, Leningrad branch of Steklov Math. Institute, 1987, 39pp.

Brennan, J. Functions with rapidly decreasing negative Fourier coefficients. Preprint, University of Kentucky Math. Dept., 1986, 14pp.

Carleson L. Estimates of harmonic measures. *Annales Acad. Sci. Fennicae*, Series A.I. Mathematica **7** (1982), 25–32.

Cartan, H. *Sur les classes de fonctions définies par des inégalités portant sur leurs dérivées successives*. Hermann, Paris, 1940.

Cartan, H. and Mandelbrojt. S. Solution du problème d'équivalence des classes de fonctions indéfiniment dérivables. *Acta Math.* **72** (1940), 31–49.

Cartwright, M. *Integral Functions*. Cambridge Univ. Press, 1956.

Choquet, G. *Lectures on Analysis*. 3 vols. Benjamin, New York, 1969.

De Branges, L. *Hilbert Spaces of Entire Functions*. Prentice-Hall, Englewood Cliffs, NJ, 1968.

Domar, Y. On the existence of a largest subharmonic minorant of a given function. *Arkiv för Mat.* **3** (1958), 429–40.

Duren, P. *Theory of H^p Spaces*. Academic Press, New York, 1970.

Dym, H. and McKean, H. *Gaussian Processes, Function Theory and the Inverse Spectral Problem*. Academic Press, New York, 1976.

Dynkin, E. Funktsii s zadannoĭ otsenkoĭ $\partial f/\partial \bar{z}$ i teoremy N. Levinsona. *Mat. Sbornik* **89** (1972), 182–90. Functions with given estimate for $\partial f/\partial \bar{z}$ and N. Levinson's theorem. *Math. USSR Sbornik* **18** (1972), 181–9.

Gamelin, T. *Uniform Algebras*. Prentice-Hall, Englewood Cliffs, NJ, 1969.

Garnett, J. *Bounded Analytic Functions*. Academic Press, New York, 1981.

Garsia, A. *Topics in Almost Everywhere Convergence*. Markham, Chicago, 1970 (copies available from author).

Gorny, A. Contribution à l'étude des fonctions dérivables d'une variable réelle. *Acta Math.* **71** (1939), 317–58.

Green, George, Mathematical Papers of. Chelsea, New York, 1970.

Helson, H. *Lectures on Invariant Subspaces*. Academic Press, New York, 1964.

Helson, H. and Lowdenslager, D. Prediction theory and Fourier Series in several variables. Part I, *Acta Math.* **99** (1958), 165–202; Part II, *Acta Math.* 106 (1961), 175–213.

Hoffman, K. *Banach Spaces of Analytic Functions*. Prentice-Hall, Englewood Cliffs, NJ, 1962.

Jöricke, B. and Volberg, A. Summiruemost' logarifma pochti analiticheskoĭ funktsii i obobshchenie teoremy Levinsona–Kartraĭt. *Mat. Sbornik* **130** (1986), 335–48.

Kahane, J. Sur quelques problèmes d'unicité et de prolongement, relatifs aux fonctions approchables par des sommes d'exponentielles. *Annales Inst. Fourier* **5** (1953–54), 39–130.

Kargaev, P. Nelokalnye pochti differentsialnye operatory i interpoliatsii funktsiami s redkim spektrom. *Mat. Sbornik* **128** (1985), 133–42. Nonlocal almost differential operators and interpolation by functions with sparse spectrum. *Math. USSR Sbornik* **56** (1987), 131–40.

Katznelson, Y. *An Introduction to Harmonic Analysis*. Wiley, New York, 1968 (Dover reprint available).

Kellog, O. *Foundations of Potential Theory*. Dover, New York, 1953.

Khachatrian, I.O. O vzveshonnom priblizhenii tselykh funktsiĭ nulevoĭ stepeni mnogochlenami na deĭstvitelnoĭ osi. *Doklady A.N.* **145** (1962), 744–7. Weighted approximation of entire functions of degree zero by polynomials on the real axis. *Soviet Math (Doklady)* **3** (1962), 1106–10.

Khachatrian, I.O. O vzveshonnom priblizhenii tselykh funktsiĭ nulevoĭ stepeni mnogochlenami na deĭstvitelnoĭ osi. *Kharkovskiĭ Universitet, Uchonye Zapiski* **29**, Ser. 4 (1963), 129–42.

Koosis, P. Harmonic estimation in certain slit regions and a theorem of Beurling and Malliavin. *Acta Math.* **142** (1979), 275–304.

Koosis, P. *Introduction to H_p Spaces.* Cambridge University Press, 1980.

Koosis, P. Solution du problème de Bernstein sur les entiers. *C.R. Acad. Sci. Paris* **262** (1966), 1100–2.

Koosis, P. Sur l'approximation pondérée par des polynomes et par des sommes d'exponentielles imaginaires. *Annales Ecole Norm. Sup.* **81** (1964), 387–408.

Koosis, P. Weighted polynomial approximation on arithmetic progressions of intervals or points. *Acta Math.* **116** (1966), 223–77.

Levin, B. *Raspredelenie korneĭ tselykh funktsiĭ.* Gostekhizdat, Moscow, 1956. *Distribution of Zeros of Entire Functions* (second edition). Amer. Math. Soc., Providence, RI, 1980.

Levinson, N. *Gap and Density Theorems.* Amer. Math. Soc., New York, 1940, reprinted 1968.

Levinson, N. and McKean, H. Weighted trigonometrical approximation on the line with application to the germ field of a stationary Gaussian noise. *Acta Math.* **112** (1964), 99–143.

Lindelöf, E. Sur la représentation conforme d'une aire simplement connexe sur l'aire d'un cercle. *Quatrième Congrès des Mathématiciens Scandinaves, 1916.* Uppsala, 1920, pp. 59–90. [*Note:* The principal result of this paper is also established in the books by Tsuji and Zygmund (second edition), as well as in my own (on H_p spaces).]

Mandelbrojt, S. *Analytic Functions and Classes of Infinitely Differentiable Functions.* Rice Institute Pamphlet XXIX, Houston, 1942.

Mandelbrojt, S. *Séries adhérentes, régularisation des suites, applications.* Gauthier–Villars, Paris, 1952.

Mandelbrojt, S. *Séries de Fourier et classes quasi-analytiques de fonctions.* Gauthier–Villars, Paris, 1935.

McGehee, O., Pigno, L. and Smith, B. Hardy's inequality and the L^1 norm of exponential sums. *Annals of Math.* **113** (1981), 613–18.

Mergelian, S. Vesovye priblizhenie mnogochlenami. *Uspekhi Mat. Nauk* **11** (1956), 107–52. Weighted approximation by polynomials. *AMS Translations* **10** Ser 2 (1958), 59–106.

Nachbin, L. *Elements of Approximation Theory.* Van Nostrand, Princeton, 1967.

Naimark, M. *Normirovannye koltsa.*
First edition: Gostekhizdat, Moscow, 1956.
First edition: *Normed Rings,* Noordhoff, Groningen, 1959.
Second edition: Nauka, Moscow, 1968.
Second edition: *Normed Algebras.* Wolters-Noordhoff, Groningen, 1972.

Nehari, Z. *Conformal Mapping.* McGraw–Hill, New York, 1952.

Nevanlinna, R. *Eindeutige analytische Funktionen* (second edition). Springer, Berlin, 1953. *Analytic Functions.* Springer, New York, 1970.

Paley, R. and Wiener, N. *Fourier Transforms in the Complex Domain.* Amer. Math. Soc., New York, 1934.

Phelps, R. *Lectures on Choquet's Theorem.* Van Nostrand, Princeton, 1966.

Pollard, H. Solution of Bernstein's approximation problem. *Proc. AMS* **4** (1953), 869–75.

Riesz, F. and M. Uber die Randwerte einer analytischen Funktion. *Quatrième Congrès des Mathématiciens Scandinaves, 1916.* Uppsala, 1920, pp. 27–44. [*Note:* The material of this paper can be found in the books by Duren, Garnett, Tsuji, Zygmund (second edition) and myself (on H_p spaces).]

Riesz, F. and Sz-Nagy, B. *Leçons d'analyse fonctionnelle* (second edition). Akadémiai Kiadó, Budapest, 1953. *Functional Analysis.* Ungar, New York, 1965.

Riesz, M. Sur le problème des moments.
First note: *Arkiv för Mat., Astr. och Fysik* **16** (12) (1921), 23pp.
Second note: *Arkiv för Mat., Astr. och Fysik* **16** (19) (1922), 21pp.
Third note: *Arkiv för Mat., Astr. och Fysik* **17** (16) (1923), 52pp.

Rudin, W. *Real and Complex Analysis* (second edition). McGraw Hill, New York, 1974.

Shohat, J. and Tamarkin, J. *The Problem of Moments*. Math. Surveys No. 1, Amer. Math. Soc., Providence, RI, 1963.

Szegő, G. *Orthogonal Polynomials*. Amer. Math. Soc., Providence, RI, 1939; revised edition published 1958.

Titchmarsh, E. *Introduction to the Theory of Fourier Integrals* (second edition). Oxford Univ. Press, 1948.

Titchmarsh, E. *The Theory of Functions* (second edition). Oxford Univ. Press, 1939; corrected reimpression, 1952.

Tsuji, M. *Potential Theory in Modern Function Theory*. Maruzen, Tokyo, 1959; reprinted by Chelsea, New York, 1975.

Vekua, I. *Obobshchonnye analiticheskie funktsii*. Fizmatgiz, Moscow, 1959. *Generalized Analytic Functions*. Pergamon, London, 1962.

Volberg, A. Logarifm pochti-analiticheskoĭ funktsii summiruem. *Doklady A.N.* **265** (1982), 1297–302. The logarithm of an almost analytic function is summable. *Soviet Math* (Doklady) **26** (1982), 238–43.

Volberg, A. and Erikke, B., *see* Jöricke, B. and Volberg, A.

Widom, H. Norm inequalities for entire functions of exponential type. *Orthogonal Expansions and their Continuous Analogues*. Southern Illinois Univ. Press, Carbondale, 1968, pp. 143–65.

Yosida, K. *Functional Analysis*. Springer, Berlin, 1965.

Zygmund, A. *Trigonometric Series* (second edition of following item). 2 vols. Cambridge Univ. Press, 1959; now reprinted in a single volume. *Trigonometrical Series* (first edition of preceding). Monografje matematyczne, Warsaw, 1935; reprinted by Chelsea, New York, in 1952, and by Dover, New York, in 1955.

Index

Contents of volume II

Printed in the United States
By Bookmasters